SYSTEMS AND CIRCUITS FOR
ELECTRICAL ENGINEERING
TECHNOLOGY

McGRAW-HILL
BOOK COMPANY
New York
St. Louis
San Francisco
Auckland
Düsseldorf
Johannesburg
Kuala Lumpur
London
Mexico
Montreal
New Delhi
Panama
Paris
São Paulo
Singapore
Sydney
Tokyo
Toronto

CHARLES BELOVE
MELVYN M. DROSSMAN

Department of Electrical Engineering
Technology and Computer Science
New York Institute of Technology

Systems and Circuits for Electrical Engineering Technology

This book was set in Times Roman by Maryland Composition Incorporated.
The editors were Kenneth J. Bowman and J. W. Maisel;
the cover was designed by Ben Kann;
the production supervisor was Charles Hess.
The drawings were done by J & R Services, Inc.

Library of Congress Cataloging in Publication Data

Belove, Charles.
Systems and circuits for electrical engineering technology.

Includes index.
1. Electric circuits. 2. Electric networks.
I. Drossman, Melvyn, joint author. II. Title.
TK454.B3994 621.319′2 75-34190
ISBN 0-07-004430-9

SYSTEMS AND CIRCUITS FOR ELECTRICAL ENGINEERING TECHNOLOGY

1234567890DODO7832109876

To our wives
GOLDA and NORMA

CONTENTS

PREFACE

The technologies have undergone major changes during recent years, especially electrical technology in which the introduction of solid-state and then integrated-circuit devices has brought about a revolution in methods and practices. This book is motivated in large part by these changes and by recognition of the need for corresponding changes in the education of students who will be working in the field.

The book is oriented to the electrical engineering technology student; the mathematical level has been held to the minimum required for a correct presentation of the topics covered. Mathematical topics beyond high school algebra and trigonometry are introduced as needed so that no changes in curriculum are required.

In addition to its use in electrical technology programs, the book's scope makes it most suitable for a service course in electrical concepts for students who are not electrical engineering technology majors.

The major departure in content from that of other texts is a strong systems orientation reflecting the changing skills demanded by the introduction of integrated circuit technology and the emphasis on digital systems. The philosophy is to present a unified treatment of basic electric network theory. Thus, the analysis of nonlinear circuits, which is generally introduced in an electronics course, is covered in this text because it is a circuit analysis problem; similarly, controlled sources are introduced at an early stage.

A very comprehensive treatment of digital circuits, both combinational and sequential, is provided so that the student will be aware of their importance and not think of them as secondary to analog circuits. This material may be omitted entirely if the course is to follow the classical line covering only analog circuits.

Classical circuit theory is presented using the time-proven approach of introducing resistive circuits first for simplicity. Diode-resistor circuits are then discussed followed by circuits with energy storage elements. Methods for analyzing the frequency response and time response are presented next. Numerous examples are included in order to emphasize the importance of understanding the principles rather than simply the mechanical procedures. The use of the digital computer to analyze complicated circuits is covered so that the student will be equipped to deal effectively and realistically with such problems.

Feedback principles are approached from the point of view of operational amplifier circuits to prepare the student to use this new circuit element as well as to provide him or her with a basic background in feedback concepts. In addition, operational amplifiers are included in a section on hybrid systems which combine analog and digital techniques. The systems orientation is further emphasized by presenting magnetic circuits in terms of their applications to electric machines, relays, and measuring instruments.

The authors would like to acknowledge the contributions of a few of the many people who have helped bring this book to fruition. We thank Professors Edward Nelson and Milton Rosenstein of the New York Institute of Technology for their reading of the manuscript and suggestions, Professor David Tyrell of Middlesex Community College for his suggestions concerning large segments of the manuscript and Professor Tian Lih Teng of the New York Institute of Technology for preparing the solutions manual and checking the problems. Finally, we want to express our debt to the many students who have provided the stimulus for this work.

CHARLES BELOVE

MELVYN M. DROSSMAN

SYSTEMS AND CIRCUITS FOR ELECTRICAL ENGINEERING TECHNOLOGY

1

Essential Arithmetic, Units, and Notation

OBJECTIVES

Upon completion of this chapter you should be able to

1 Describe the mksa system of units and state the four basic quantities.
2 Perform calculations as required for problems involving linear equations and graph the equations.
3 Utilize scientific notation to effectively organize and complete numerical calculations with arbitrary numbers using a slide rule or calculator.
4 State the most commonly used decimal prefixes.
5 Define accuracy, precision, and tolerance and describe the preferred value system used by component manufacturers.
6 Perform calculations and conversions involving one system of units or several systems of units.

INTRODUCTION

In order to establish a firm foundation for our study of electrical technology, we must first consider several aspects of the mathematical language which is used by workers in the field. Experience has indicated that there are a few topics in basic mathematics which consistently require review and drill before more technical topics are considered. In this chapter, we shall consider these topics and provide examples and drill problems so that the student can acquire facility in these areas. Additional topics in mathematics will be taken up in later chapters as they are required.

1.1 UNITS

The importance of a thorough understanding of units cannot be stressed enough. Most of the problems with which technologists are concerned result in an answer which is a *number*. In almost every case the number has associated with it a *unit*. In all cases the information concerning the unit is as important as the number which gives the size of the quantity involved. For example, if you were told that an athlete ran 1 in 4, you would not have the slightest idea what he had accomplished. On the other hand if you were told that he ran 1 mile (mi) in 4 minutes (min) his accomplishment would be clear. A moment's thought will indicate that the frequent use of units in our daily lives is accomplished almost automatically. For example, if we purchase something at the store for 40 cents and give the storekeeper a $1 bill, we make use of our knowledge of the units of our money system in order to check the change. We know that the $1 bill is equivalent to 100 cents; thus our change should be 60 cents. Further, the change will undoubtedly be given to us in different coins. We must know the value of each of these in cents in order to check that we have received the correct amount and we do all this with very little effort. Our aim is to develop the same facility with technical systems of units.

1.1-1 Systems of Units

In scientific work, we must be able to describe whatever we do in such a way that anyone working in the same field will have no difficulty understanding exactly what we mean or in duplicating our results. In order to do this, we need a system of units which can be reproduced anywhere.

Today there are two principal systems of units in use. One of these is the English system used in our country. This is slowly being replaced by the *International System of Units* (abbreviated SI units) which includes the metric, or mksa, system used in electrical technology. The basic units of the mksa system are the meter (length), kilogram (mass), second (time), and ampere (current). The standards for these quantities have been established internationally and are the same all over the world.

TABLE 1.1-1 Basic Quantities

Quantity	Symbol	Unit	Abbreviation
Length	l	meter	m
Mass	m	kilogram	kg
Time	t	second	s
Current	i	ampere	A

The basic quantities along with their mksa units, usual mathematical symbols, and abbreviations are shown in Table 1.1-1. Other physical quantities such as power, energy, etc., are derived from the basic set. These derived quantities will be introduced as needed throughout the text.

Before concluding this section let us consider the relation between the magnitudes of the quantities involved in the English and mksa systems of units. Since most of us are familiar with the magnitude of the various quantities in the English system, this may provide a means for obtaining some feeling for the relative sizes of the quantities in the mksa system.

A comparison of some often used quantities is made in Table 1.1-2.

Later in this chapter we will consider calculations with units and conversion between systems.

1.2 ALGEBRA REVIEW

Most of the mathematics used in this text involves only algebra and trigonometry. A few topics from the calculus will be required and they will be introduced when needed. In this section we will review linear equations, a topic in algebra which is extremely important in what is to follow.

TABLE 1.1-2 Comparison of Some English and mksa Units

	English	mksa
Length	1 inch = 0.0254 meter	1 meter = 39.37 inch
Mass	1 slug = 14.6 kilogram	1 kilogram = 0.0684 slug
Force and weight	1 pound = 4.45 newton	1 newton = 0.224 pound
Energy	1 foot·pound = 1.356 joules	1 joule = 0.738 foot·pound

1.2-1 Linear Equations

A large percentage of the mathematics with which we will be concerned involves linear equations which graph as straight lines. Equations which graph as curves are called nonlinear equations. The slope-intercept form of the straight line equation is

$$y = mx + b \tag{1.2-1}$$

where y = dependent variable
x = independent variable
m = slope of the straight line
b = value of the y intercept

In order to illustrate the significance of the various terms, we consider a numerical example in which $m = 2$ and $b = 3$

$$y = 2x + 3 \tag{1.2-2}$$

The graph of this equation shown in Fig. 1.2-1 is easily constructed from the slope-intercept form by following four simple steps.

1 Set $x = 0$. Then $y = b = 3$. This gives the coordinate on the y axis.
2 Set $y = 0$. Then $x = -b/m = -3/2$. This gives the coordinate on the x axis.
3 Draw a straight line through the two points.
4 As a check, choose a convenient value for x (say $x = 1$), find y ($y = 5$), and see that this third point lies on the same straight line (it does).

Let us first note that in the graph the point where the line cuts the y axis is at $x = 0$, $y = b$. This point is determined by inspection when the equation is in slope-intercept form.

Second, note the slope of the line. In order to get at the significance of the slope, consider a change (increment) in x between points P_1 and P_2. There is a corresponding increment in y in the positive y direction. The x increment is called Δx where Δ (delta) means "the change in." From the diagram

$$\Delta x = x_2 - x_1 \tag{1.2-3}$$

The y increment corresponding to this Δx is

$$\Delta y = y_2 - y_1 \tag{1.2-4}$$

By definition, the slope of the line is the ratio of the change in y between points P_1 and P_2 to the corresponding change in x between these same two points. Thus

$$\text{Slope} = \frac{\Delta y}{\Delta x} = \frac{y_2 - y_1}{x_2 - x_1} = m \tag{1.2-5}$$

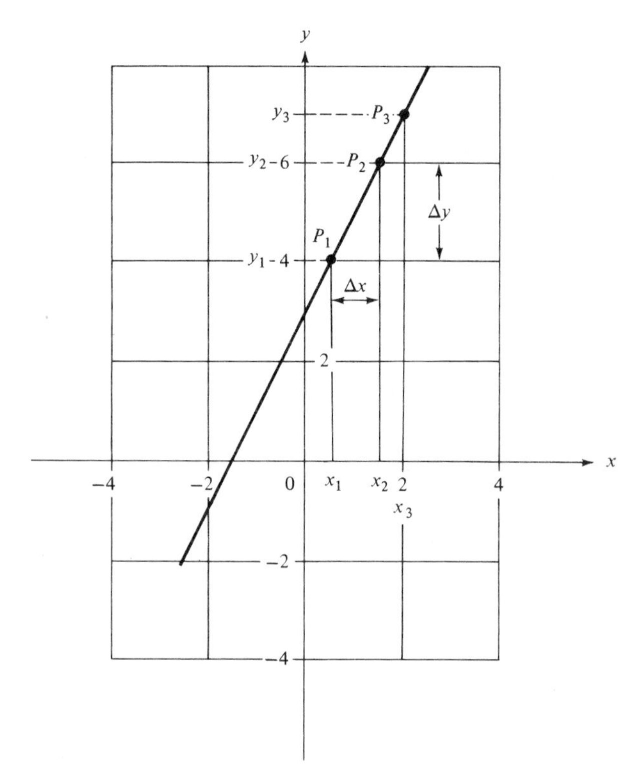

Fig. 1.2-1 Graph of $y = 2x + 3$.

For our example

$$m = \frac{y_2 - y_1}{x_2 - x_1} = \frac{6 - 4}{1.5 - 0.5} = 2 \tag{1.2-6}$$

This, of course, checks the value of the coefficient of x in Eq. (1.2-2). Since we are considering a straight line, the slope will be the same anywhere along the line. The slope will also be the same for any size Δx, as long as we remember to take the corresponding value of Δy. Thus, for a straight line, the slope is constant. Also note that the magnitude of the slope is a measure of the *steepness* of the line. For example, a line with a slope of 4 units would be twice as steep.

It is important to recognize the physical significance of the slope. It represents the *rate of change* of y with respect to x, that is, *the number of units change in y for a one unit change in x*. If we know the slope of a straight line, we can predict, for any point on the line, the change in y for *any* change in x.

Let us return to our example given in Eq. (1.2-2). Consider that we are starting at point P_1 in Fig. 1.2-1 and wish to determine the change in y if x changes by 1.5 units. Since the *rate of change* of y with respect to x is 2, all we need do is multiply the *change* in x (1.5 units) by the *rate of change* (2) to find the corresponding *change* in y (3 units). Note that we are not finding y, but we are finding the *change* in y, Δy. Thus the actual value of y at our new point P_3 is the value at P_1 ($y_1 = 4$) plus the change ($\Delta y = 3$) which gives the final value $y_3 = 7$.

In order to express this mathematically, we proceed as follows:

At point P_3:

$$y_3 = mx_3 + b \tag{1.2-7}$$

At point P_1:

$$y_1 = mx_1 + b \tag{1.2-8}$$

Subtracting

$$y_3 - y_1 = m(x_3 - x_1)$$

Add y_1 to both sides:

$$y_3 = y_1 + m(x_3 - x_1)$$

$$= y_1 + m\,\Delta x \tag{1.2-9}$$

Substituting the values $y_1 = 4$, $m = 2$, $\Delta x = 1.5$, we get

$$y_3 = 4 + (2)(1.5) = 7$$

The concept of slope as rate of change is an extremely important one, and we will be making considerable use of it as we proceed. In the preceding example, the slope was positive and the calculations very simple. In the next example, we consider a negative slope, which sometimes causes difficulty.

Example 1.2-1 Consider the equation

$$16x + 5y = 39$$

a Plot the equation.
b Use the slope concept to find the value of y when $x = -2$.

Solution

a The first step is to write the equation in slope-intercept form.

$$5y = -16x + 39$$

$$y = -\frac{16}{5}x + \frac{39}{5}$$

$$= -3.2x + 7.8$$

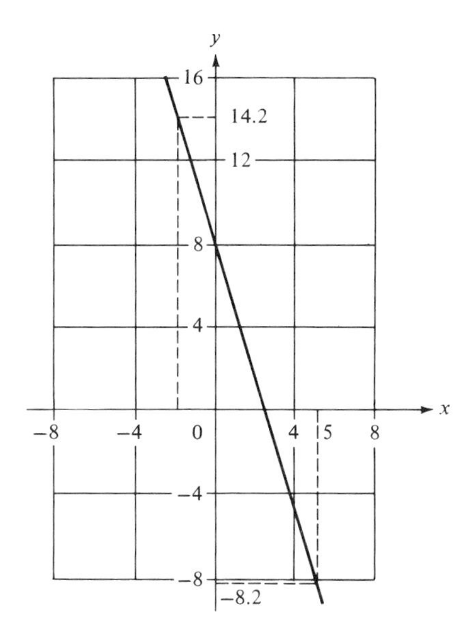

Fig. 1.2-2 Graph for Example 1.2-1.

Following the steps set down previously we have

1 With $x = 0$, $y = 7.8$.
2 With $y = 0$, $x = 2.44$.
3 The graph is shown in Fig. 1.2-2.
4 As a check take $x = 5$, then $y = (-3.2)(5) + 7.8 = -8.2$.

b Since the slope is -3.2, y *decreases* 3.2 units for each unit *increase* in x. If we use $x = 0$, $y = 7.8$ as our starting point, then x *decreases* by 2 units to the specified point. Thus the change in y is an *increase* of $2 \times 3.2 = 6.4$ units and the new value of y is $7.8 + 6.4 = 14.2$. To check this mathematically we use Eq. (1.2-9).

$$
\begin{aligned}
y_2 &= y_1 + m\,\Delta x \\
&= 7.8 + (-3.2)(-2) \\
&= 7.8 + 6.4 \\
&= 14.2
\end{aligned}
$$

////

1.3 NUMERICAL CALCULATIONS

One area that sometimes causes difficulty is in performing straightforward numerical calculations. Since it is so important that the student be able to carry out numerical calculations correctly, we will spend some time in this section providing hints on how to insure correct calculations and other appropriate topics. Use of the slide rule for multiplying and dividing is discussed in Appendix A for those readers who may require instruction.

1.3-1 Powers of Ten

In scientific work, the magnitudes of the quantities involved vary over wide ranges. In order to facilitate calculation when very small and very large numbers are involved, a scientific notation has been developed making use of powers of 10 so that numbers can all be expressed as a number of convenient magnitude (generally between 1 and 10) multiplied by a power of 10. The number which multiplies the power of 10 is called the *mantissa*[1] and the power to which ten is raised is called the *exponent.*

For numbers which are multiples of 10, the power is the number of 10s which must be multiplied together to obtain the number. The zero power is defined such that anything to the zero power is 1. In addition $10^{-n} = 1/10^n$. Some examples are

$$1{,}000 = 10^3$$

$$100 = 10^2$$

$$10 = 10^1$$

$$1 = 10^0$$

$$0.1 = 10^{-1}$$

$$0.01 = 10^{-2}$$

$$0.001 = 10^{-3}$$

RULES FOR FINDING THE POWER OF 10

In order to determine the proper power of 10, the following rules may be applied:

1 Decimals. Move the decimal point to the right as desired and count the number of places to the original point. The count is the proper negative power of 10.

Example

0.00001

$0.00001. = 10^{-5}$

since count = 5

[1] This name is also used to refer to the fractional part of the logarithm of a number which differs from this usage.

2 Large numbers. Move the decimal point to the left and count the number of places to the original point. The count is the proper positive power of 10.

Example

10,000

$1.0000. = 10^4$

since count = 4

For numbers which are not simply multiples of 10 the same rules apply.

Example

$$0.00432 = 4.32 \times 10^{-3}$$

$$0.137 = 1.37 \times 10^{-1}$$

$$1{,}260 = 1.26 \times 10^{3}$$

$$336{,}000 = 3.36 \times 10^{5}$$

LAWS OF EXPONENTS

The laws of exponents as applied to powers of 10 are particularly simple:

Multiplication:

$$10^m \times 10^n = 10^{m+n} \tag{1.3-1a}$$

Division:

$$\frac{10^m}{10^n} = 10^{m-n} \tag{1.3-1b}$$

Raising to a power:

$$(10^m)^n = 10^{m \times n} \tag{1.3-1c}$$

In all of these m and n may be any number. For cases of interest to us at this point, m and n will be integers.

Examples

$$(100)(10000) = (10^2)(10^4) = 10^{(2+4)} = 10^6$$

$$(100)(0.00001) = (10^2)(10^{-5}) = 10^{(2-5)} = 10^{-3}$$

$$\frac{100}{10000} = \frac{10^2}{10^4} = 10^{(2-4)} = 10^{-2}$$

$$\frac{100}{0.00001} = \frac{10^2}{10^{-5}} = 10^{[2-(-5)]} = 10^7$$

$$(100)^3 = (10^2)^3 = 10^{(2 \times 3)} = 10^6$$

$$(0.001)^4 = (10^{-3})^4 = 10^{(-3 \times 4)} = 10^{-12}$$

1.3-2 Arithmetic Operations with Arbitrary Numbers

Now we are ready to consider arithmetic operations with very large and very small numbers. Using scientific notation, the operations with the mantissas can be done separately from those involving the powers of 10. This makes the whole process easy to check and provides a means for adequate "bookkeeping."

Example 1.3-1 Do the indicated operations in the following problems. Express all answers in scientific notation.

1 Addition:

$$\begin{aligned} 720 + 1{,}340 &= 7.2 \times 10^2 + 13.4 \times 10^2 \\ &= (7.2 + 13.4) \times 10^2 \\ &= 20.6 \times 10^2 \\ &= 2.06 \times 10^3 \end{aligned}$$

Note that both numbers were converted so as to yield the same power of 10. This *must* be done for addition and subtraction since it is the only way we can factor out the power of 10.

$$\begin{aligned} 0.0036 + 0.00072 &= 36 \times 10^{-4} + 7.2 \times 10^{-4} \\ &= (36 + 7.2) \times 10^{-4} \\ &= 43.2 \times 10^{-4} \\ &= 4.32 \times 10^{-3} \end{aligned}$$

2 Subtraction:

$$\begin{aligned} 6{,}700 - 72{,}000 &= 6.7 \times 10^3 - 72 \times 10^3 \\ &= -65.3 \times 10^3 \\ &= -6.53 \times 10^4 \end{aligned}$$

3 Multiplication:

$$\begin{aligned} 6{,}700 \times 800 &= (6.7 \times 10^3)(8 \times 10^2) \\ &= (6.7 \times 8)(10^3 \times 10^2) \\ &= 53.6 \times 10^5 \\ &= 5.36 \times 10^6 \end{aligned}$$

Note that we do not have to use the same power of 10 for both numbers for multiplication. The power is chosen so that the mantissa will be convenient.

$$\begin{aligned} (0.0013)(0.04) &= (1.3 \times 10^{-3})(4 \times 10^{-2}) \\ &= (1.3 \times 4)(10^{-3} \times 10^{-2}) \\ &= 5.2 \times 10^{-5} \end{aligned}$$

4 Division:

$$\frac{6{,}400}{80{,}000} = \frac{6.4 \times 10^3}{8 \times 10^4}$$

$$= \frac{6.4}{8} \times \frac{10^3}{10^4}$$

$$= 0.8 \times 10^{-1}$$

$$= 8 \times 10^{-2}$$

$$\frac{0.00036}{0.03} = \frac{3.6 \times 10^{-4}}{3 \times 10^{-2}}$$

$$= \frac{3.6}{3} \times \frac{10^{-4}}{10^{-2}}$$

$$= 1.2 \times 10^{-2}$$

5 Powers:

$$(.0002)^3 = (2 \times 10^{-4})^3$$

Since $(ab)^n = a^n b^n$, this becomes

$$(2 \times 10^{-4})^3 = (2)^3 \times (10^{-4})^3$$

$$= 8 \times 10^{-12}$$

////

1.3-3 Estimating the Answer

In the preceding example, the arithmetic was quite simple and no special means were required to arrive at the correct answer. In more complicated cases, it is a good idea to estimate the answer before carrying through the detailed calculation. For example, consider multiplying 45 by 24. This is almost the same as multiplying 40 by 25 which by inspection is 1,000 ($40 = 4 \times 10$, so $40 \times 25 = 4 \times 25 \times 10 = 100 \times 10$). Thus the answer should be very close to 1,000. The actual product is 1,080. If the answer arrived at had been 10.8, we would have immediately suspected that a factor of 100 had been lost in the process and would have gone back to check.

In a more complicated case involving both multiplication and division, a good procedure to follow is

1 Convert all numbers to scientific notation.
2 Estimate the answer.
3 Use a slide rule or pocket or desk calculator to find the exact answer and check against the estimate.

This procedure is illustrated in the example which follows.

Example 1.3-5 An investigation leads to the formula

$$x = \frac{(0.0012)(370)}{(534)(0.07)}$$

Find x.

Solution

1 Scientific notation:

$$x = \frac{(1.2)(10^{-3})(3.7)(10^{2})}{(5.34)(10^{2})(7)(10^{-2})}$$

$$= \frac{(1.2)(3.7)}{(5.34)(7)} \times 10^{[(-3+2)-(2-2)]}$$

2 Estimate answer:

$$x \approx \frac{(1)(4)}{(40)} \times 10^{-1} = 10^{-2} = 0.01$$

3 Exact answer (slide rule accuracy):

$$x = 0.119 \times 10^{-1} = 0.0119$$

////

1.3-4 Electronic Pocket and Desk Calculators

As noted previously, portable electronic calculators are available at the time of this writing in a wide range of prices, styles, and capabilities. Photographs of some of these are shown in Fig. 1.3-1. The less expensive models perform the operations of addition, subtraction, multiplication, and division. Most of them automatically keep track of the decimal point while a few handle only integers so that the user must keep track of the decimal point. The number of significant figures varies, in some models running as high as 12. In almost all cases the number of significant figures is adequate for our purposes and the user will round off the final answer as desired.

More expensive models provide readouts in scientific notation, sometimes called floating decimal-point representation. Some models offer the user a choice of fixed or floating-point output. In addition to the four basic operations, some models perform other types of scientific calculations, such as square roots, squares, reciprocals, trigonometric, exponential, and logarithmic functions and can store a single number for later recall and use.

The most expensive models contain memory banks for retaining several intermediate values in a calculation and have other advanced features.

(*a*)

(*b*)

(*c*)

Fig. 1.3-1 Portable electronic calculators. (*a, Hewlett-Packard Co. Inc.; b, Texas Instruments Incorporated; c, Bowmar Instruments, Inc.*)

An appropriate type for the technologist would be a model utilizing floating decimal point notation and providing trigonometric, exponential, and logarithmic functions in addition to the basic arithmetic operations.

1.3-5 Decimal Prefixes

Certain powers of 10 arise so often that a system of prefixes has been devised to identify them so that they need not be written each time they are required. These which we use most in electrical technology are listed in Table 1.3-1.

Examples

$$1{,}500{,}000 \text{ ohms} = 1.5 \text{ megohm} = 1.5 \text{ M}\Omega$$

$$68{,}000 \text{ ohms} = 68 \text{ kilohms} = 68 \text{ k}\Omega$$

$$0.000033 \text{ farads} = 33 \text{ microfarads} = 33\ \mu\text{F}$$

$$0.002 \text{ henries} = 2 \text{ millihenries} = 2 \text{ mH}$$

$$4 \times 10^{-9} \text{ seconds} = 4 \text{ nanoseconds} = 4 \text{ ns}$$

$$6 \times 10^{-12} \text{ seconds} = 6 \text{ picoseconds} = 6 \text{ ps}$$

1.3-6 Precision and Error Considerations

In this section, we will consider certain aspects of the representation of data. Accuracy, tolerance, and the accepted conventions for conveying significant figures will be discussed along with techniques for the proper rounding of numbers.

Then we will show how the preferred values in the standard series used by resistor and capacitor manufacturers are obtained.

ACCURACY

When dealing with measurements, *accuracy* is defined as closeness to the truth. We are concerned with accuracy whenever we make a measurement. For

TABLE 1.3-1 Decimal Prefixes

Power of 10	Prefix	Abbreviation
10^{-12}	pico	p
10^{-9}	nano	n
10^{-6}	micro	μ
10^{-3}	milli	m
10^{3}	kilo	k
10^{6}	mega	M
10^{9}	giga	G
10^{12}	tera	T

example, suppose we order an item of furniture which appears as though it may just fit through our front door. If the fit is close enough we will have to be concerned with the accuracy of the ruler we use to measure the width of our door and the accuracy of the dimensions of the furniture as supplied by the manufacturer.

It is easy to envision a situation which would result in disaster. For example, if the quoted furniture dimension of interest is 36 in and our ruler indicates that the door measures 36.5 in we may feel that there is sufficient clearance and that the delivery will proceed smoothly. However, if our ruler is in error by 1 in so that the true door measurement is 35.5 in while the furniture dimension of 36 in is precise, the delivery will not proceed smoothly, and in fact the furniture will not go through the door. From the foregoing it is clear that we must know the accuracy of our measuring equipment.

Another important factor is the mathematical accuracy of the calculations we make using the results of measurements. In such cases we often make computations in which the mathematical operations are carried to several unnecessary decimal places. This extra work gives a false impression of the precision or degree of fineness of the measurement. For most of our work instruments are generally not correct to more than three significant figures, and the values marked on components are frequently correct to only two or even one significant figure.

SIGNIFICANT FIGURES

In technical work the precision of a number representing a measurement is expressed by the number of *significant figures* used. The number of significant figures is the number of digits in the number excluding leading zeros. Leading zeros are required in order to show the proper location of the decimal point. Some examples are

378	is correct to *three* significant figures
0.02	is correct to *one* significant figure
2,500	is correct to *four* significant figures
0.00250	is correct to *three* significant figures

The last digit in each of these numbers is understood to have a total uncertainty of 1 unit. Thus in the first example the actual number is between 377.5 and 378.5. In the second example the actual number is between 0.015 and 0.025.

ROUNDING

Since the precision of a calculated result cannot be greater than the precision of any of the numbers entering into the calculation the result of all computations should be *rounded off* accordingly. The following rules should be applied:

1 If the number *dropped* is greater than 5, the last figure *kept* is increased by 1. Thus 43.52 and 43.7 rounded to two significant figures are both 44.

2 If the last figure dropped is less than 5, the last figure kept is not changed. Thus 43.42 and 43.2 rounded to two significant figures are both 43.
3 If the number dropped is 5, add 1 to the last figure kept if it will make it *even*, otherwise do not change the last figure kept. Thus 43.5 rounded to two significant figures is 44 while 42.5 becomes 42.

In most applications, powers of 10 are used so that no zeros will appear beyond the uncertain digit. For example, if a resistor has been measured as 375,000 Ω correct to three significant figures, it can be written several ways, that is,

375 kΩ

0.375 MΩ

or 375×10^3 Ω

When the resistance is specified this way, there is no ambiguity in interpreting its value and the precision with which it is known. In this case it is understood to lie between 374.5 and 375.5 kΩ. Excluding values of precision components and precision instrument readings, most of the numbers with which we deal are not known to this much accuracy.

TOLERANCE AND PREFERRED VALUES

It is impossible for component manufacturers to produce every possible value which circuit designers might require. Instead they produce sets of *preferred values* with standard tolerances. A typical tolerance for carbon resistors as used in electronic circuits is ± 10 percent. This means that if we purchase a resistor which is labeled 10 kΩ, 10 percent, its resistance must lie between 9 kΩ and 11 kΩ, that is, within 10 percent of the nominal value, which is 10 kΩ. Other standard tolerances are ± 1, ± 5, and ± 20 percent. The 1 percent components are the most expensive of the four while the 20 percent group is the cheapest. For economic reasons, the 20 percent type will be found in many circuits, because it often turns out that the component values are not critical. We will have more to say about this later in the text when circuit design is discussed.

The derivation of the series of preferred values is interesting. We illustrate this using the R6 series which has six values that yield a maximum tolerance of ± 20 percent. Each of the values is obtained from a logarithmic progression based on an $n/6$ power of 10, that is, $10^{n/6}$, where n runs from 1 to 6. The development of the series is shown in Table 1.3-2.

The values of $10^{n/6}$ are rounded off to numbers each containing two significant figures which are easy to remember. Each of the preferred values may be multiplied by any power of 10 so that only the six numbers 10, 15, 22, 33, 47, and 68 are required to cover the entire range of 20 percent values. In a typical design, we may arrive at a certain value for a particular resistor. The error is then defined as the difference between the preferred value which is

TABLE 1.3-2 R6 Preferred Values

n	$10^{n/6}$	preferred value	maximum % error
0	1.000	1.0	
			±20
1	1.468	1.5	
			18.9
2	2.155	2.2	
			20
3	3.162	3.3	
			17.5
4	4.642	4.7	
			18.3
5	6.813	6.8	
			19.1
6	10.000	10	
			20
		15	
			18.9
		22	
		⋮	⋮

available and the actual desired value. The maximum percent error is found for each pair of numbers by choosing a value midway between the two preferred values and determining the error. For example, between 1.0 and 1.5 we choose the value 1.25. Then, for the lower value

$$\text{Percent error} = \frac{1.0 - 1.25}{1.25} \times 100 = \frac{-0.25}{1.25} \times 100 = -20 \text{ percent}$$

clearly the result is + 20 percent for the upper value because we choose our desired value midway between the two.

The series for which the tolerance is ± 10 percent is based on $n/12$ with n running from 1 to 12 while the ± 5 percent series is based on $n/24$ with n = 1 to 24. The 10 percent series includes all of the values in the 20 percent series in addition to intermediate values and the 5 percent series includes all values in both 10 percent and 20 percent series as well as intermediate values. Development of these series is left for the homework problems.

1.4 CALCULATIONS WITH UNITS

In addition to being an important part of every number we deal with, units are also very useful for a type of checking which often catches simple mistakes in algebra or conceptual errors. In this section we consider units in more detail, covering relations between systems of units, conversion of units within one system, and the important use of units for dimensional checking of equations.

1.4-1 Conversion of Units

Because of the fact that different systems are in use we often have to convert from one system to another. This can often be a confusing process prone to many errors unless careful "bookkeeping" methods are used. We illustrate the process in the following examples.

Example 1.4-1

a Convert 1 foot to centimeters.
b Convert 9 ounces to newtons.
c Convert 6 joules to inch-pounds
d Convert 60 miles per hour to kilometers per hour.

Solution

a The trick here is to recognize that we can multiply any number by 1 without changing its value. For this operation we use a ratio of units that reduces to 1, that is, for this example,

$$\frac{12 \text{ in}}{1 \text{ ft}} = 1 \qquad \text{and} \qquad \frac{2.54 \text{ cm}}{1 \text{ in}} = 1$$

The actual conversion is then written out as follows:

$$(1 \cancel{\text{ft}}) \left(\frac{12 \cancel{\text{in}}}{1 \cancel{\text{ft}}}\right)\left(\frac{2.54 \text{ cm}}{1 \cancel{\text{in}}}\right) = 30.5 \text{ cm}$$

Note how the dimensions ft and in cancel out as indicated by the lines through them. Also note that any of our ratios of units can be inverted without changing its value. This is illustrated in the next example.

b For this conversion, we use

$$\frac{16 \text{ oz}}{1 \text{ lb}} = 1 \qquad \text{and} \qquad \frac{1 \text{ lb}}{4.45 \text{ N}} = 1$$

Then

$$(9 \cancel{\text{oz}})\left(\frac{1 \text{ lb}}{16 \cancel{\text{oz}}}\right)\left(\frac{4.45 \text{ N}}{1 \cancel{\text{lb}}}\right) = 2.4 \text{ N}$$

Again, observe the manner in which this is written out so that all factors are accounted for and all unit cancellations are clearly indicated.

This kind of careful bookkeeping will help to avoid all kinds of unnecessary errors.

c $$(6\text{ J})\left(\frac{1\text{ ft}\cdot\text{lb}}{1.356\text{ J}}\right)\left(\frac{12\text{ in}}{1\text{ ft}}\right) = 53.1\text{ in}\cdot\text{lb}$$

d $$\left(60\,\frac{\text{mi}}{\text{h}}\right)\left(\frac{5{,}280\text{ ft}}{1\text{ mi}}\right)\left(\frac{12\text{ in}}{1\text{ ft}}\right)\left(\frac{0.0254\text{ m}}{1\text{ in}}\right)\left(\frac{1\text{ km}}{1{,}000\text{ m}}\right) = 96.7\,\frac{\text{km}}{\text{h}}$$

The student should practice this type of conversion until he is quite familar with it. ////

The next example illustrates conversions within one system of units.

Example 1.4-2

a Convert 1,200 grams to kilograms.
b Convert 62×10^8 centimeters to meters.
c Convert 0.076 hours to microseconds.
d Convert 0.016 microfarads to picofarads.

Solutions
As in the previous example, we multiply by 1 in the form of a ratio of units.

a $$(1{,}200\text{ g})\left(\frac{1\text{ kg}}{1{,}000\text{ g}}\right) = 1.2\text{ kg}$$

b $$(62 \times 10^8\text{ cm})\left(\frac{1\text{ m}}{100\text{ cm}}\right) = 62 \times 10^6\text{ m}$$

c $$(0.076\text{ h})\left(\frac{60\text{ min}}{1\text{ h}}\right)\left(\frac{60\text{ s}}{1\text{ min}}\right)\left(\frac{10^6\,\mu\text{s}}{\text{s}}\right) =$$

$$(7.6)(10^{-2})(6)(10^1)(6)(10^1)(10^6) = 274 \times 10^6$$

Note the use of scientific notation in this example to avoid decimal point errors.

d $$(0.016\,\mu\text{F})\left(\frac{10^6\text{ pF}}{1\,\mu\text{F}}\right) = 0.016 \times 10^6\text{ pF} = 16{,}000\text{ pF}$$

For conversions within the metric system, powers of 10 often have to be juggled around. The type of bookkeeping illustrated in this example will insure that they are handled properly. ////

1.4-2 Checking With Units

Every equation we write must have the property of dimensional homogeneity. In simpler language, this means that all terms which are added in the equation must have exactly the same units. This fact is used in two ways. First, to check calculations, and, second, to find the units of newly defined quantities. The student should get into the habit of checking units every time an equation is written.

For example, suppose an equation works out to be

$$R_T = \frac{R_1}{R_1 + R_2}$$

where R represents electrical resistance in ohms (Ω). Upon checking the dimensions of each side we immediately realize that an error has been made since the left side has dimensions of ohms while the right is dimensionless (Ω/Ω). What probably occurred is that a resistance factor multiplying the right side was lost somewhere along the line. A dimensionally correct formula would be

$$R_T = \frac{R_1 R_2}{R_1 + R_2}$$

Here the right side has dimensions of $(\Omega)^2/\Omega = \Omega$ as required.

It is important to note that correct dimensions *do not* guarantee that the equation is correct! In mathematical terms, correct dimensionality is a necessary, but not sufficient, condition that an equation be correct.

Dimensionality can often be used to jog the memory in order to recall an equation. For example, consider remembering the formula for velocity in terms of time and distance. We might initially remember it as $v = dt$. In terms of units this says that velocity has the units distance $\times$ time which is very obviously incorrect. Clearly, the only way that distance and time can be combined to yield the correct units is as the ratio d/t (for example, miles/hour) so that $v = d/t$ is the correct formula.

A further point to note again is that both sides must have the same units. If v is desired in m/s then d must be in meters (or if in another unit it must be converted to meters) and t must be in seconds. Although this may appear obvious and extremely simplistic, experience has indicated that students often make such mistakes. Correct application of the dimensionality rule will almost always detect such errors.

SUMMARY

1 The equation for a straight line is

$$y = mx + b$$

where m = slope and b = y intercept.

2 The *increment* Δx is used to specify a *change* in x, that is,

$$\Delta x = x_2 - x_1$$

3 The laws of exponents are

$$10^m \times 10^n = 10^{m+n}$$

$$\frac{10^m}{10^n} = 10^{m-n}$$

$$(10^m)^n = 10^{m \times n}$$

4 The most important decimal prefixes are: mega $= 10^6$, kilo $= 10^3$, milli $= 10^{-3}$, micro $= 10^{-6}$, nano $= 10^{-9}$, pico $= 10^{-12}$

5 When dealing with equations describing any physical phenomena, the dimensions of each term must be the same.

PROBLEMS[1]

1.2-1 Graph the following equations

a $y = 6x + 5$
b $y = -20x + 11$
c $i = 100v + 10^3$
d $v + 2i = 7$
e $y - 6x = 24$

1.2-2 For each of the equations of Prob 1.2-1, find m and b.

1.2-3 A straight line passes through (3, 4) and y increases 3 units for each 6-unit increase in x. Find the equation of the line.

1.2-4 A straight line passes through (10, 20) and x decreases 5 units for each 10-unit increase in y. Find the equation of the line.

1.3-1 Express the following as powers of 10.

a 1,000,000
b 0.00001
c 10,000
d 0.0000001

1.3-2 Express the following as a number between 1 and 10 times a power of 10.

a 1,365,000
b 0.0000127
c 14,000
d 0.03

[1] Problems are numbered to correspond with the sections to which they apply.

1.3-3 Reduce all of the following to a single power of 10.

a (100) (1000)
b (0.01) (0.0001)
c (0.001) (10,000)
d (0.001) (10^6)

e $\dfrac{100}{10{,}000}$

f $\dfrac{100}{0.0001}$

g $\dfrac{0.01}{10^6}$

h $\dfrac{100}{10^8}$

1.3-4 Do the following calculations. All results are to be in terms of powers of 10.

a $(10{,}000)^3$
b $(0.0001)^{1/2}$
c $(1{,}000{,}000)^{1/3}$

d $\dfrac{(0.01)^2(1000)^3}{0.001}$

e $\dfrac{[(100)(0.001)]^4}{(10{,}000)^{-2}}$

f $\dfrac{(0.0001)^{-3}}{10^6}$

1.3-5 Do the following calculations. All results are to be a number between 1 and 10 times a power of 10. Estimate all answers in advance.

a $\dfrac{(62)(3.47)}{(0.012)(23{,}000)}$

b $\dfrac{(1.6)^2(24)^{1/2}}{12}$

1.3-6 Repeat Prob. 1.3-5 for the following.

a $\dfrac{(3{,}700)(0.046)}{(82)(6{,}720{,}000)}$

b $\dfrac{(7.92)(0.73)^2}{(42)^3}$

c $\dfrac{(7.3)(630)}{(42)}$

d $\dfrac{(1.3)(.062)}{(70)(8.3)}$

1.3-7 Express the following using the standard decimal prefixes in Table 1.3-1.

a 0.0001 F
b 0.01 H
c 10,000 Ω
d 637,000 g
e 0.00427 N
f 1,362,000 in

1.3-8 Round off the following to the indicated number of significant figures.

a 376,000 to two figures
b .00454 to two figures
c 1235 to three figures
d 1245 to three figures
e 0.00765 to one figure
f 0.17 to one figure

1.3-9 Derive the R12 preferred value series for which the tolerance is ± 10 percent.

1.4-1 Perform the following conversions.

a 3 s to μs
b 0.01 F to μF
c 1.7 min to s
d 3.2 h to s
e 0.0000063 s to ns
f 0.0000063 s to μs

1.4-2 Convert

a 732,000 Ω to MΩ
b 0.016 μF to nF
c 78 mH to H
d 1 m^2 to cm^2
e 1.320 cm^2 to m^2
f 7.3 lb to N

1.4-3 Check each of the terms below to determine if the final dimension is length.

a 6 m/20 mi
b (3 mi/h) (20 s)
c (67 cm) (20 mi/h)
d (80 m^2) (6 cm^{-1})
e (7 cm/s)/20 s
f (20 mi) (2 μs)

1.4-4 Check each of the equations below for dimensional homogeneity. Units are t in seconds, v in cm/s, x in cm.

a $t_4 = \dfrac{t_1 t_2}{t_1 + t_2}$

b $x_4 = v_3 t_3 + x_0$

c $v_3 = \dfrac{v_1}{v_0 - v_5}$

d $t_7 = \dfrac{v_7}{x_7}$

2
Signals

OBJECTIVES

Upon completion of this chapter, you should be able to

1 Define the unit step function.
2 Combine step functions to generate pulse waveforms.
3 Define pulse amplitude and duration.
4 For a pulse train, define period and prf.
5 Indicate how groups of pulses can generate a code.
6 Write the equation for the general form of the decaying exponential and the rising exponential and define the amplitude and time constant.
7 Sketch the graph of the exponentials in objective 6, indicating significant values.
8 Write the equation for the general form of a sinewave and define the amplitude, period, frequency, and phase angle.
9 Sketch the graph of the sinewave in objective 8.

INTRODUCTION

In order to keep pace with modern engineering practice, a text such as this must present material which is applicable to modern systems and devices, both digital and analog. Because of the heavy emphasis on systems today we must approach as many topics as possible from a systems points of view wherein we consider a complete system as a single entity without regard for the internal components which go to make up the system. For now, we will consider systems

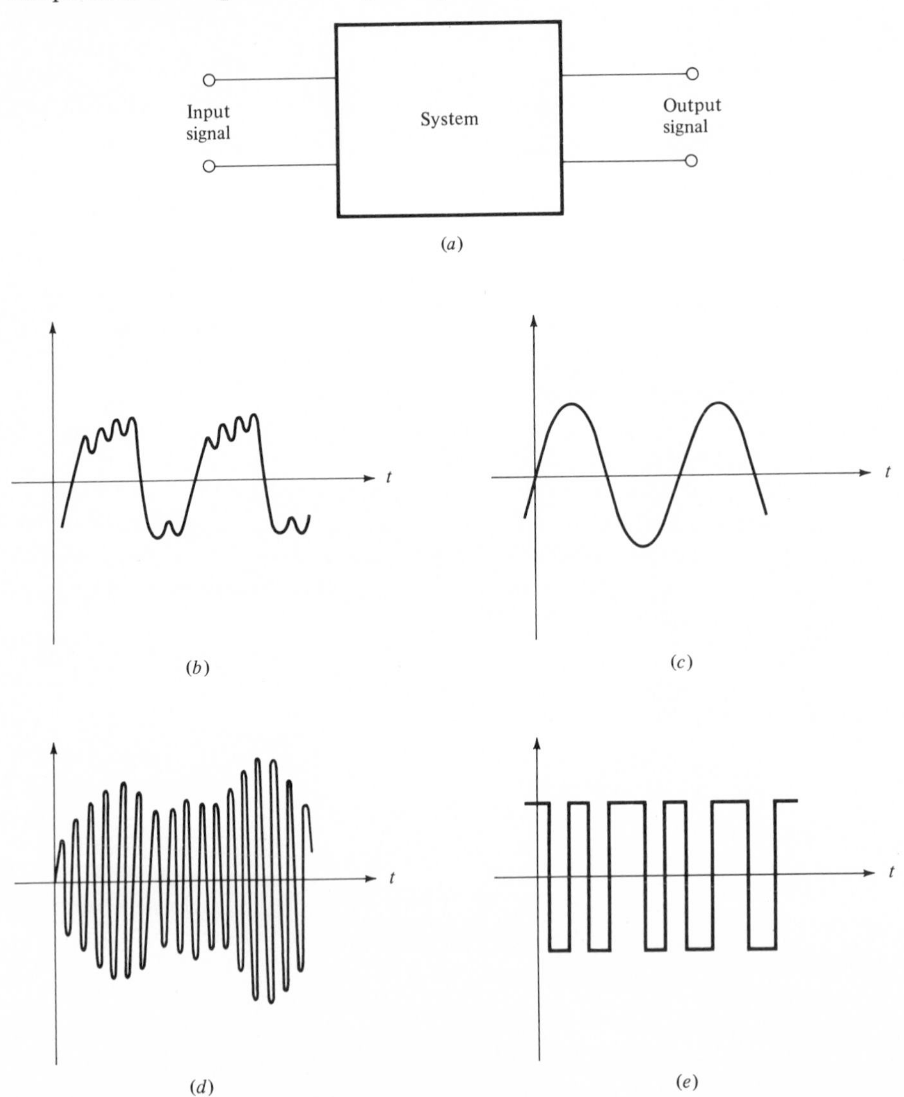

Fig. 2.1 Systems and signals. (*a*) System block diagram, (*b*) speech signal, (*c*) sine wave, (*d*) AM wave, (*e*) digital signal.

which have a single input and a single output as shown in Fig. 2-1*a*. A little reflection will show that many of the electrical and electronic devices which we take for granted in our daily lives are complete systems which can be categorized in this way. For example, the radio receiver which we all use can be thought of as a system in which the input signal is the electromagnetic wave picked up by the antenna and the output signal is the sound wave generated by the loudspeaker. The most important characteristic of these signals is that they are carriers of *information.* The system can profitably be thought of as *processing* the input signal so as to perform any required modifications of the information which it contains.

Signals come in a variety of shapes and sizes. In this chapter we will introduce some of the more important types of signals used in modern digital and analog systems and consider some of the ways in which information is carried by these signals.

Some typical signals are shown in Figs. 2-1*b*, *c*, *d*, and *e*. Part *b* of the figure illustrates what a speech wave might look like when displayed on an oscilloscope after being converted to a voltage by means of a microphone. This is classified as an *analog* signal; the information is contained in the time variation of the signal amplitude. Analog signals can be represented by a series of sine waves (sinusoids) one of which is shown in part *c* of the figure. The sine function is extremely important in electric circuit theory (in most power generation systems, the supply voltage is generated in the form of a sine wave) and we will discuss it in more detail in a later section.

Figure 2-1*d* shows an amplitude-modulated signal as used in conventional AM radio. This is the actual waveform transmitted as an electromagnetic wave.

Finally, in part *e* of the figure, we have shown a pulse signal. In digital systems, such as the digital computer, information is transmitted throughout the system by series of pulses such as those shown.

The signals introduced in this chapter will be used throughout the remainder of the text as inputs to circuits being studied; our problem will be to compute the resulting output signal.

2.1 DC AND THE STEP FUNCTION

DC SIGNALS

One of the most important of the signals from the point of view of learning circuit theory is that called *dc.* The letters actually stand for *direct current,* but are often used to indicate any signal which is constant, that is, does not vary with time. We will begin our study of circuits in Chap. 4 using dc currents and voltages as the basic signals because at that point we do not want to have the additional complication of time variation to worry about. When considered as a carrier of information, however, dc is about the least useful signal that one can imagine. Recall that a dc current or voltage is constant. For example, a

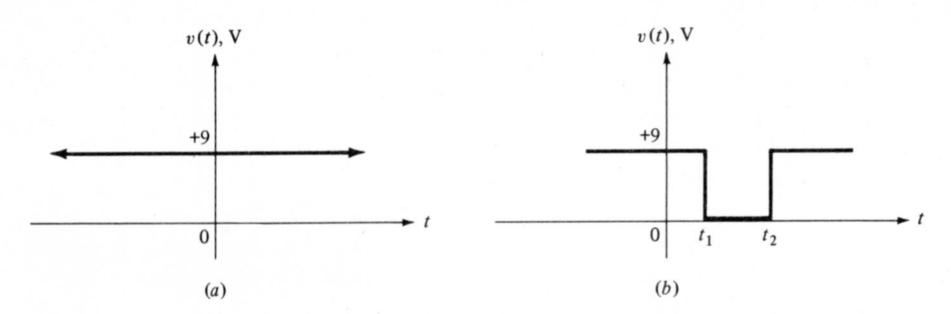

Fig. 2.1-1 Dc signals. (*a*) Battery voltage; (*b*) switched battery voltage.

battery is a source of dc voltage. A typical battery, such as one used in a transistor radio, has a voltage of 9 V. Thus the only information available in this dc voltage is the value +9 V. This signal is extremely uninformative; the battery, in reality, is used as a source of power rather than a carrier of information.

We can graphically illustrate the lack of information in this voltage by plotting a graph of the battery voltage vs. time as shown in Fig. 2.1-1*a* where the graph clearly shows that the value +9 V is all that we obtain from the battery from the moment of its manufacture until it runs down.

The amount of information carried by dc currents and voltages can be increased by switching them *on* and *off* resulting in a signal which is no longer dc. The times at which the switching occurs then become additional pieces of information carried by the signal. This process eventualy leads to *pulse* signals which will be discussed in the next section. A switched signal is shown in Fig. 2.1-1*b*. This particular one contains three pieces of information which are the magnitude, +9, and times t_1 and t_2 at which it is switched off and on.

THE STEP FUNCTION

Switched dc signals are often described mathematically using the *step function* as the basic signal. Fig. 2.1-2*a* shows a circuit which applies 1 V to the system by means of a switch. The arrow indicates that the switch contact moves from terminal *a* to terminal *b* at a time designated as $t = 0$. The waveform of the voltage $v_i(t)$ will then be as shown in Fig. 2.1-2*b*. This is the *unit step function* which is given the symbol $u(t)$. Mathematically, $u(t)$ can be expressed as

$$u(t) = \begin{cases} 0 & t < 0 \\ 1 & t > 0 \end{cases} \tag{2.1-1}$$

The step function is left undefined at the precise instant $t = 0$, when the jump takes place.

Often one has a shifted step function starting at a time other than zero as shown in Fig. 2.1-2*c*. For this case a slightly different notation is used; it is,

$$v_i(t) = u(t - t_0) = \begin{cases} 0 & t < t_0 \\ 1 & t > t_0 \end{cases} \tag{2.1-2}$$

The time at which the jump takes place is found by setting the *argument* $(t - t_0)$ to zero. Thus the jump occurs when

$$t - t_0 = 0$$

or at

$$t = t_0$$

If the actual step of interest has a magnitude other than 1 we simply multiply the unit step by the actual voltage or current. For example, the voltage output of a 9-V transistor battery switched on at $t = 0$ would be written

$$v_B(t) = 9u(t)$$

By combining positive and negative shifted steps it is possible to generate various waveforms as shown in the following example.

Example 2.1-1 Express the pulse waveform shown in Fig. 2.1-3*a* in terms of step functions.

Solution The idea here is to express $v(t)$ as a *sum* of step functions. Since we have two jumps, one at $t = 3$ s and the second at $t = 6$ s, we expect that two shifted steps should be sufficient to describe $v(t)$. The jump at

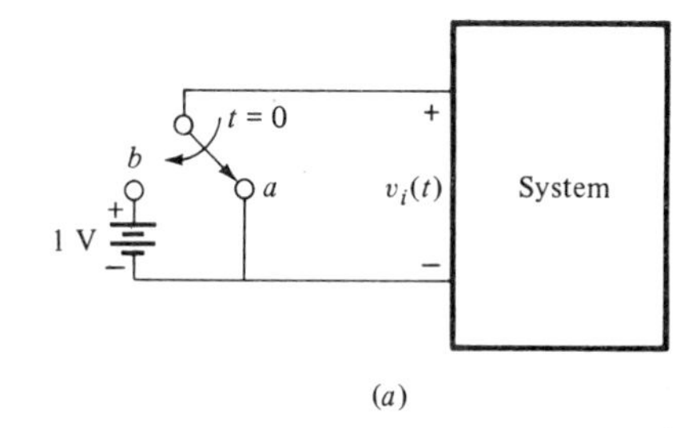

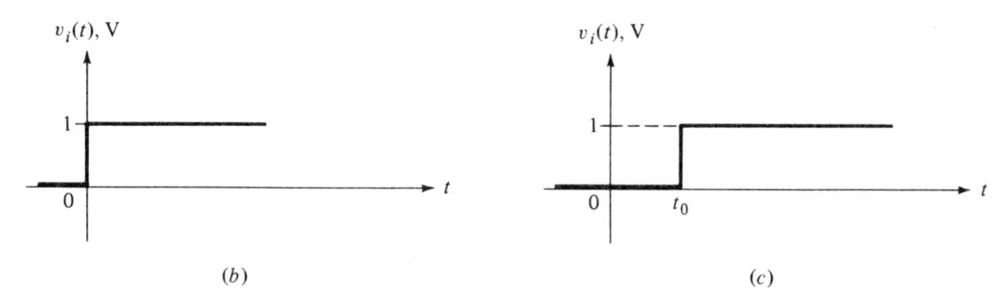

Fig. 2.1-2 The step function. (*a*) Circuit for generating a step function. (*b*) Unit step function beginning at $t = 0$. (*c*) Unit step beginning at $t = t_0$.

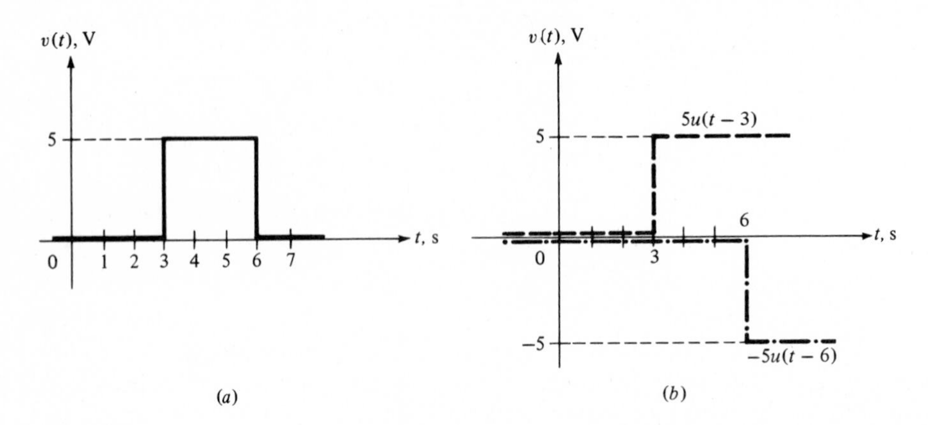

Fig. 2.1-3 Waveforms for Example 2.1-2. (*a*) Pulse waveform; (*b*) individual step functions which add up to the original waveform.

$t = 3$ s can be described by $5u(t-3)$. At $t = 6$ s, the waveform returns to zero. This can be accounted for by adding a *negative* step that starts at $t = 6$ s, that is, $-5u(t-6)$. Then

$$v(t) = 5u(t-3) - 5u(t-6)$$

The individual terms are sketched in Fig. 2.1-3*b*. The student should ascertain that the two terms do indeed add up to the original pulse waveform. ////

2.2 PULSE SIGNALS

SINGLE PULSES

The basic component of any digital signal is the pulse shown in Fig. 2.2-1*a*. For convenience, we represent it as a voltage and we note that it is characterized by its amplitude (or height) V_0, starting time t_1, and duration, or width, $t_2 - t_1$.

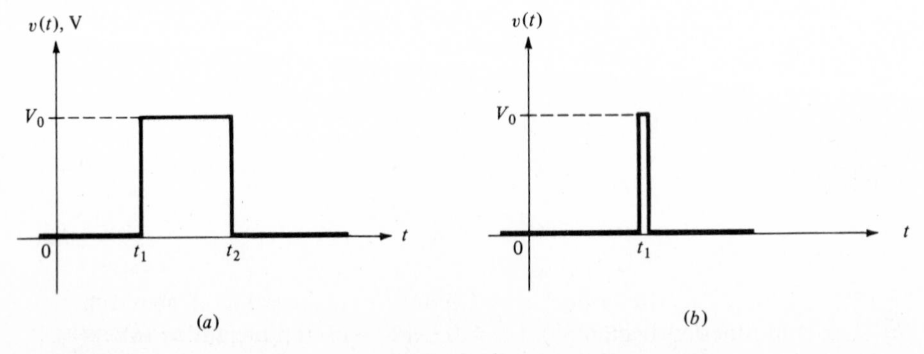

Fig. 2.2-1 Digital signals. (*a*) Single pulse; (*b*) narrow pulse.

Any (or more than one) of these properties can be varied from pulse to pulse in order to convey information.

In many systems the pulse width is a constant, and much less than the time interval between pulses. When this is the case, only the amplitude and starting time of the pulse are important, and the pulse is represented as shown in Fig. 2.2-1*b*. In digital communications systems only one amplitude is used, and it corresponds to the binary number 1. The binary number 0 is indicated by the absence of a pulse at a particular time.

PULSE TRAINS

A constant-amplitude pulse train consists of a series of identical pulses generated at regular intervals of Δt s as shown in Fig. 2.2-2*a*. The time interval Δt from the beginning of one pulse to the beginning of the next is called the *period* of the pulse train. The duration, or width, of the pulse is symbolized by the

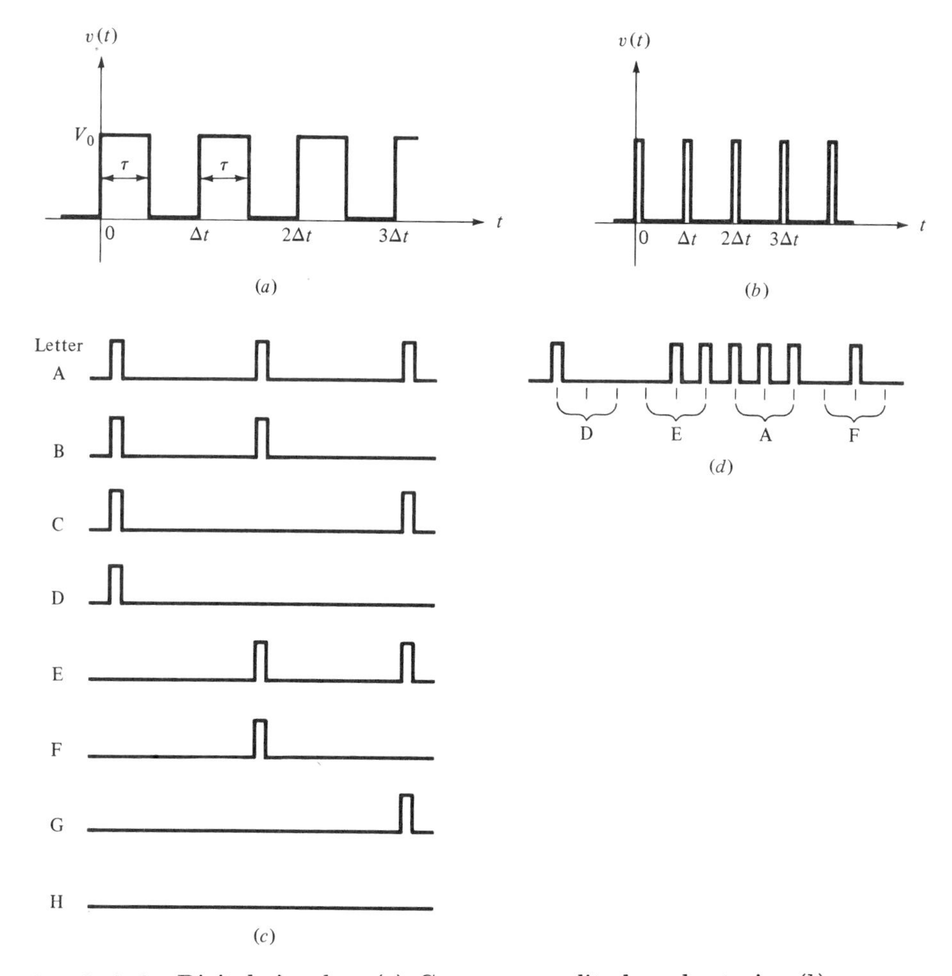

Fig. 2.2-2 Digital signals. (*a*) Constant-amplitude pulse train; (*b*) narrow-width pulse train; (*c*) code; (*d*) message *DEAF*.

greek letter tau, τ. If the pulse width is much less than the period, the pulse train appears as in Fig. 2.2-3*b*

The *pulse repetition frequency* (prf) sometimes called the *pulse repetition rate* (prr) is simply the reciprocal of the period, i.e., since the period is Δt s, we have

$$\text{prf} = \frac{1}{\Delta t} \text{ pulses/second} \tag{2.2-1}$$

For such pulse trains, the information content is practically zero since no characteristic of either signal is varied. To convey information we need only agree on a "code" which is implemented by omitting certain pulses. As an example, consider using each set of three pulses as a group, with each pulse in the group allowed to be either zero or full amplitude. With this scheme we can transmit $2^3 = 8$ different messages. Each message can be anything we wish. As long as we agree on it in advance, all users of the code will understand what we are sending. For example, if we represent the first eight letters of the alphabet as shown in Fig. 2.2-2*c*, then the message DEAF would be transmitted as shown in Fig. 2.2-2*d*. This type of signal is used in digital computers, and many ingenious codes have been developed for different applications.

In another type of pulse system, the desired information is contained in the amplitude of the pulses. Usually the pulse amplitudes vary directly with the signal. Such signals are called *pulse-amplitude modulated* (PAM) signals and are used in some types of communication systems. An example is shown in Fig. 2.2-3 in which the pulse width is much less than the period as is the case in most practical systems.

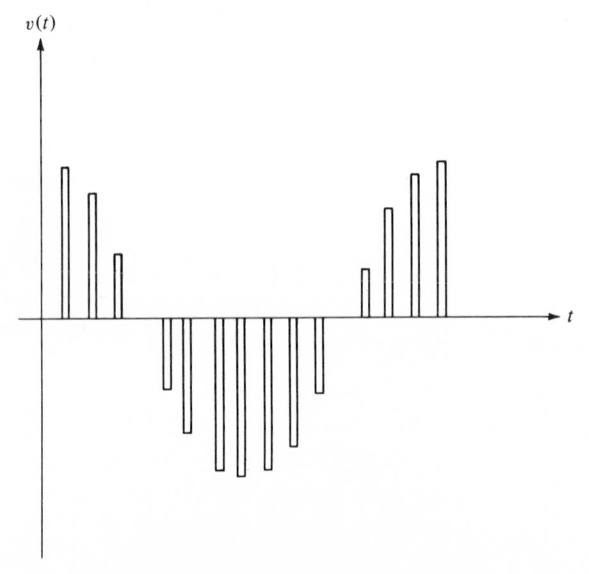

Fig. 2.2-3 Pulse-amplitude-modulated signal.

2.3 REAL EXPONENTIALS

The exponential function is extremely important in engineering. Most beginning technology students, however, will not have encountered it in previous math courses. In order to introduce it we begin by considering a slight variation of the scientific notation studied in Chap. 1, that is,

$$y = 10^t \tag{2.3-1}$$

Here y is the dependent variable, and the independent variable t, which is the exponent, represents time. Only positive values of time are considered in this section. This function is tabulated and plotted in Fig. 2.3-1*a*. On the linear scale used, the function is almost coincident with the t axis up to about $t = 4$ s and rises very steeply thereafter.

The exponential function of most interest to us uses the base of the natural (naperian) system of logarithms instead of 10. This magic number is $e = 2.71828\ldots$, which we can round off to 2.72. Its importance stems from the fact that it turns up in the solutions of many problems involving natural events. A few of these are population growth, growth and decay of current and voltage in inductive and capacitive circuits, and motion of masses and springs restrained by viscous friction.

We next consider

$$y = e^t = 2.72^t \tag{2.3-2}$$

This is tabulated and plotted in Fig. 2.3-1*b*. One phenomenon described by this curve is the growth of population under the simplifying assumption that the growth at any time is proportional to the population existing at that time. The steep rise predicted by the curve is what causes such grave concern to statesmen and demographers.

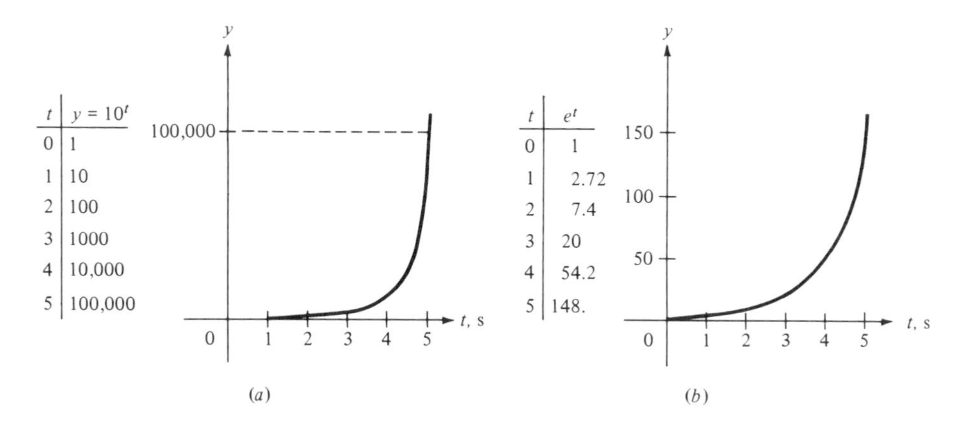

t	$y = 10^t$
0	1
1	10
2	100
3	1000
4	10,000
5	100,000

t	e^t
0	1
1	2.72
2	7.4
3	20
4	54.2
5	148.

Fig. 2.3-1 Exponential growth functions: (*a*) 10^t; (*b*) e^t.

The two functions just considered were growth functions. In electric circuit theory, we are more concerned with decay functions in which the exponent is negative. Consider

$$y = 10^{-t} \tag{2.3-3}$$

This is plotted in Fig. 2.3-2*a*. Note that the function decays very rapidly from its initial value of 1. In fact, it is down to 1 percent of its initial value when $t = 2$. Next consider

$$y = e^{-t} \tag{2.3-4}$$

This decay function is plotted in Fig. 2.3-2*b*. The waveform describes the decay process that takes place when a capacitor is discharged through a resistor. We shall have more to say about this in later chapters. For now, we will point out those features of the curve that are used to characterize it. To do this, we consider a more general form of the exponential function, expressed as a voltage for convenience

$$v(t) = V_0 e^{-t/\tau} \tag{2.3-5}$$

Here $v(t)$ is the voltage, V_0 is a constant called the amplitude, t is the independent variable, time, and τ is the *time constant*, to be explained shortly. Equation (2.3-5) is plotted in Fig. 2.3-3 which may be considered as a universal exponential decay curve. From the curve we can explain the significance of the time constant. Consider the value of $v(t)$ when $t = \tau$. This value is

$$v(\tau) = V_0 e^{-1} \approx \frac{V_0}{2.72} \approx 0.37 V_0 \tag{2.3-6}$$

Since the decay curve starts off at $v(t) = V_0$ when $t = 0$, Eq. (2.3-6) indicates that it will decay to 37 percent of its initial value at a time equal to the time constant. Thus the time constant can be used as a measure of the speed of the decay. For brevity, an exponential decay waveform is often specified by simply

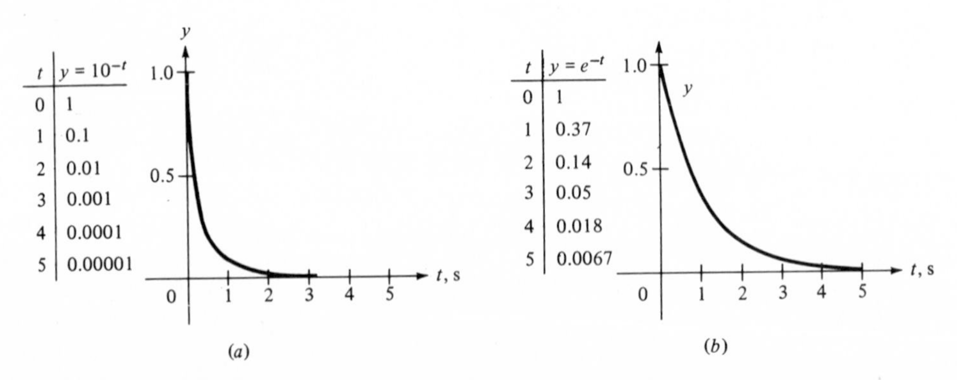

t	$y = 10^{-t}$
0	1
1	0.1
2	0.01
3	0.001
4	0.0001
5	0.00001

t	$y = e^{-t}$
0	1
1	0.37
2	0.14
3	0.05
4	0.018
5	0.0067

Fig. 2.3-2 Exponential decay functions. (*a*) 10^{-t}; (*b*) e^{-t}.

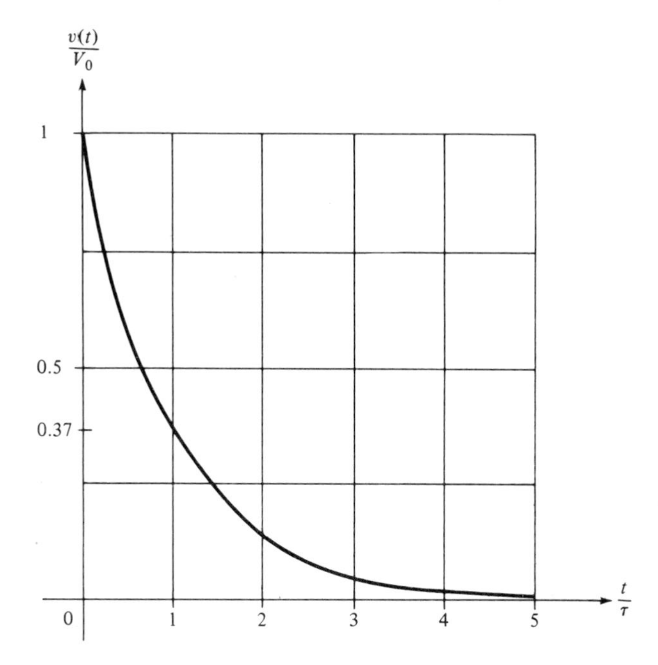

Fig. 2.3-3 Plot of decaying exponential.

stating its amplitude and time constant. This will be illustrated in the next example.

Another point on the exponential decay which is worth remembering is the point at which $t = 5\tau$:

$$v(5\tau) = V_0 e^{-5} \approx 0.007 V_0 \tag{2.3-7}$$

In words, this says that when $t = 5\tau$, the waveform has decayed to less than 1 percent of its initial value, and for all practical purposes, the decay is complete. Values of e^{-t} for intermediate values of t can be found in Appendix C.

In addition to the exponential decay, we often encounter an exponential rise (which is different from the exponential growth considered previously). The rise is characterized by the equation

$$v(t) = V_0(1 - e^{-t/\tau}) \tag{2.3-8}$$

This is plotted in Fig. 2.3-4. For this curve, we have, after one time constant,

$$v(t) = V_0(1 - e^{-1}) \approx 0.63 V_0 \tag{2.3-9}$$

Thus the significance of the time constant for the exponential rise is that it represents the time at which the waveform has risen to 63 percent of its final value.

For the point where $t = 5\tau$, we find

$$v(t) = V_0(1 - e^{-5}) \approx 0.99 V_0 \tag{2.3-10}$$

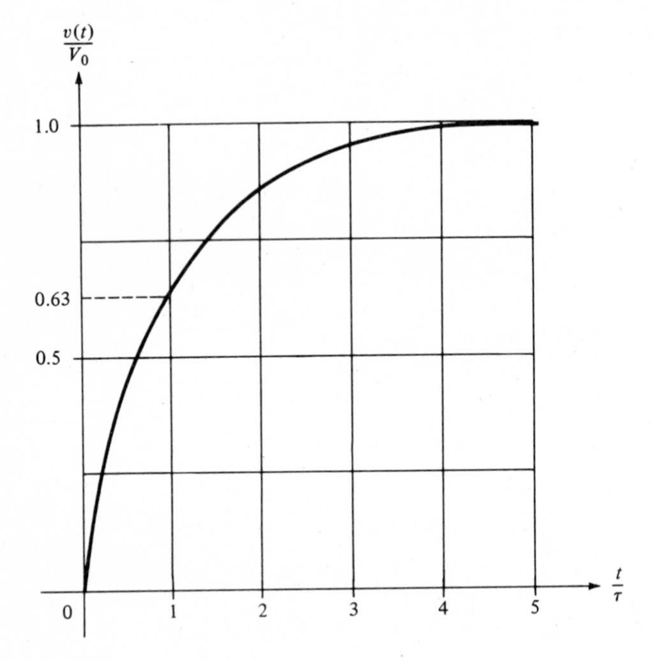

Fig. 2.3-4 Plot of rising exponential.

So after five time constants the exponential rise is up to 99 percent of its final value. In the examples which follow, we consider both types of exponentials with actual numerical values.

Example 2.3-1 An exponentially decaying voltage waveform is characterized by an amplitude of 20 V and a time constant of 3 ms. Write the equation for this voltage and sketch it to suitable scales.

Solution Using Eq. (2.3-5) we note that $V_0 = 20$ V and $\tau = 3 \times 10^{-3}$ s. Thus

$$v(t) = 20e^{-t/3\times10^{-3}} \text{ V} \tag{2.3-11}$$

In order to plot Eq. (2.3-11) we tabulate below some significant values.

t, ms	$v(t)$, V
0	20
3	7.4
6	2.7
15	0.2

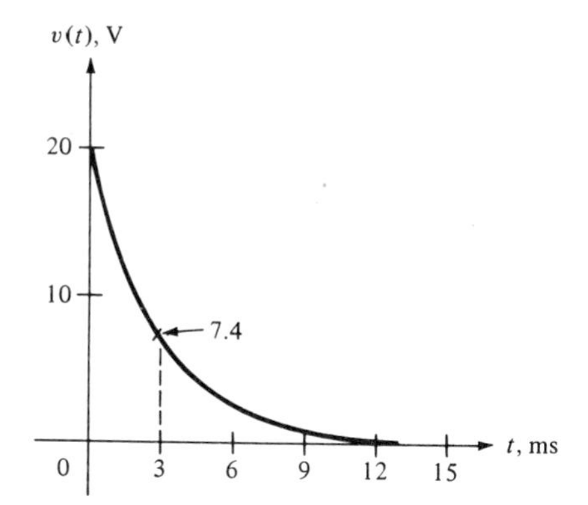

Fig. 2.3-5 Waveform for Example 2.3-1.

The student should verify the computed values using the table of exponentials in Appendix C. The waveform is shown in Fig. 2.3-5. ////

Example 2.3-2 An exponentially rising current has an amplitude of 10 mA. Write the equation for the waveform with a time constant of 10 μs and plot it for time constants of 10, 20, 30 μs.

Solution Referring to Eq. (2.3-8) with $I_0 = 10$ mA in place of V_0 and $\tau = 10\ \mu$s, we have

$$i(t) = 10(1 - e^{-t/10\times 10^{-6}}) \text{ mA} \tag{2.3-12}$$

For the other time constants the exponent is changed accordingly. Pertinent values of $i(t)$ are tabulated below and plotted in Fig. 2.3-6.

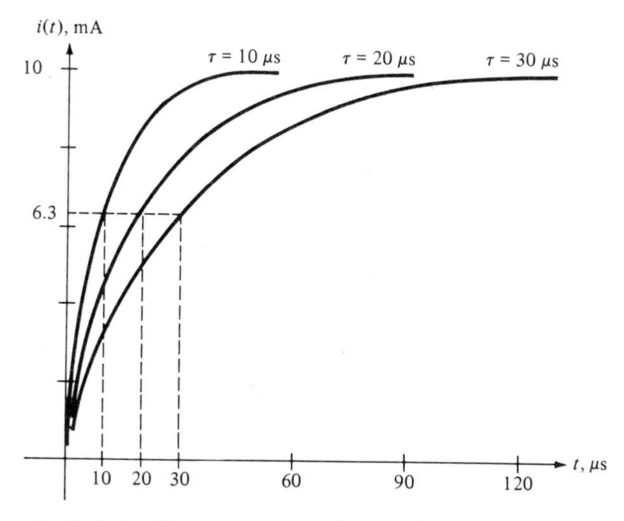

Fig. 2.3-6 Waveforms for Example 2.3-2.

	$i(t)$		
t, μs	τ = 10 μs	τ = 20 μs	τ = 30 μs
0	0	0	0
10	6.3		
20		6.3	
30			6.3
50	9.9		
100		9.9	
150			9.9

If desired, intermediate values can be filled into the table by making use of tables of the function e^{-t} in Appendix C. The student should complete the table for practice. ////

2.4 SINE-WAVE SIGNALS

In this section we consider the sine wave. As noted previously, this is the waveform used in power systems and thus is the waveform of the ac (alternating current) voltage which is present at the electric outlets in our homes.

The sine wave is the most basic of a class of signals which are called *periodic.* The property that all periodic waveforms share is that they repeat themselves every T s, where T is called the period. In mathematical terms, this is expressed simply as

$$v(t) = v(t + T) \tag{2.4-1}$$

In words this equation states that the value of the time-varying voltage v is exactly the same at time $t + T$ as it was at time t. The largest nonrepeating portion of a periodic waveform is called a *cycle*. Clearly if one cycle occurs every T s then Eq. (2.4-1) is true.

The alternating voltage shown in Fig. 2.4-1a can be written in terms of sines or cosines. Since the cosine is the base for a shorthand method we are going to introduce later, we use it to write

$$v(t) = V_m \cos (\omega t + \varphi) \tag{2.4-2}$$

The three parameters which completely define $v(t)$ are:

1 The *maximum,* or *peak,* value V_m. Since the cosine function has a maximum value of 1, V_m is the maximum value that $v(t)$ can attain.
2 The *angular frequency* $\omega = 2\pi/T$ radians per second (rad/s). More often we use the cyclic frequency f. This is the reciprocal of the period,

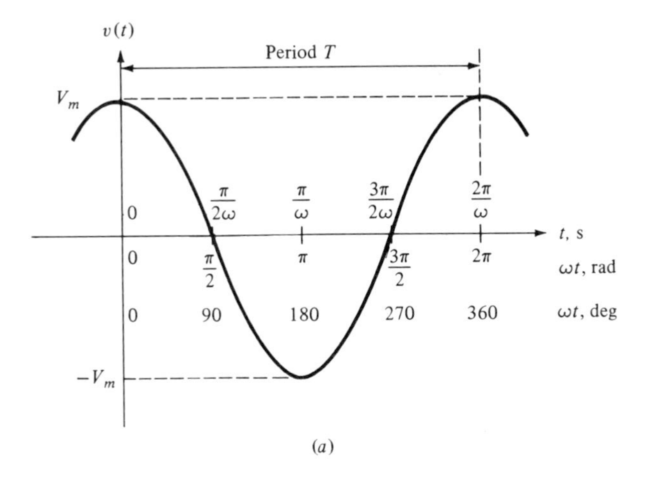

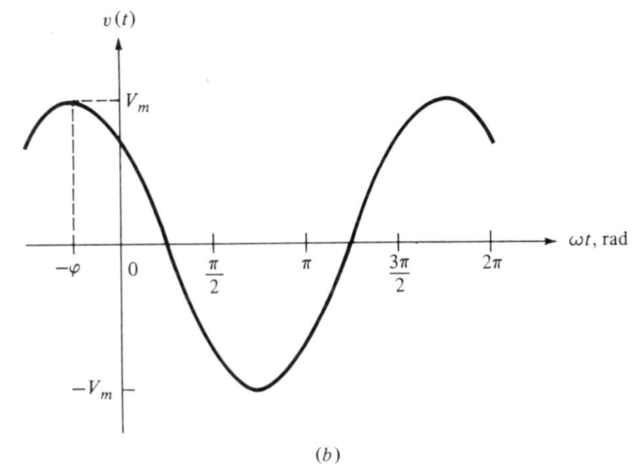

Fig. 2.4-1 Sinusoidal signals: (*a*) cosine wave; (*b*) cosine wave at arbitrary phase angle φ.

that is, $f = 1/T$, and, since the units of T are seconds per cycle, the units of f are cycles per second. Further, the unit for cycles per second is the hertz (Hz). The angular and cyclic frequencies are related by the formula

$$\omega = 2\pi f \tag{2.4-3}$$

For those who may not be familar with radian measure, we note that 2π rad $= 360°$. Since a sine or cosine wave goes through a complete cycle when its argument ωt goes through $360°$ or 2π rad, we have (since one cycle takes T s)

$$\omega T = 2\pi$$

solving for ω,

$$\omega = \frac{2\pi}{T} \tag{2.4-4}$$

Noting that $T = 1/f$ we then get Eq. (2.4-3).

3 The *phase angle* φ. For the cosine wave shown in Fig. 2.4-1*a*, the phase angle $\varphi = 0$. Not all cosine waves have the value 1 at $t = 0$. As shown in Fig. 2.4-1*b*, the *phase angle* is the angle between the positive maximum closest to $t = 0$ and the point $t = 0$. When this maximum occurs to the left of $t = 0$, the angle takes the plus sign in the expression, otherwise it takes a minus sign. For the cosine wave in Fig. 2.4-1*b*, the expression would be $V_m \cos(\omega t + 45°)$.

The examples which follow illustrate some typical calculations involving sinusoids. ////

Example 2.4-1 For the voltage

$$v(t) = 150 \cos(377t - 72°) \text{ V}$$

Find

a The amplitude
b The angular frequency
c The cyclic frequency
d The period
e The phase angle

Solution

a $V_m = 150$ V
b $\omega = 377$ rad/s

c $f = \dfrac{\omega}{2\pi} = \dfrac{377}{6.28} = 60$ Hz

d $T = \dfrac{1}{f} = \dfrac{1}{60} = 0.0167 \text{ s} = 16.7 \text{ ms}$

e $\varphi = -72°$ ////

Example 2.4-2 A voltage is described by the equation

$$v(t) = 60 \cos 377t \text{ V}$$

Find $v(t)$ when $t = 0.001$ s.

Solution

$$\omega t = (377)(0.001) = 0.377 \text{ rad}$$

Converting radians to degrees,

$$(0.377 \text{ rad}) \frac{360°}{2\pi \text{ rad}} = 21.6°$$

Finally, then

$$v(0.001) = 60 \cos 21.6° = 60(0.93) = 55.8 \text{ V} \qquad ////$$

Example 2.4-3 Find $v(t)$ at $t = 0.001$ s for the voltage of Example 2.4-2 if it has a phase angle of $+60°$.

Solution With this phase angle, the equation of the voltage is

$$v(t) = 60 \cos (377t + 60°) \text{ V}$$

Note that the argument of the cosine has two terms, $377t$ having units of radians and 60° having units of degrees. It is very bad practice to mix units this way, but it is usually done in this case. Degrees are always written with the degree symbol (°). Before any calculations are made both terms must be converted to the same units.

Since we have, from Example 2.4-1, that $\omega t = 21.6°$, the argument of the cosine is $21.6° + 60° = 81.6°$. Then

$$v(0.001) = 60 \cos 81.6° = 60(0.146) = 8.76 \text{ V} \qquad ////$$

SUMMARY

1 The unit step function occurring at time $t = t_0$ is defined by the equation

$$u(t - t_0) = \begin{cases} 0 & t < t_0 \\ 1 & t > t_0 \end{cases}$$

2 In a pulse train consisting of pulses spaced Δt s apart, the pulse repetition frequency (prf) is

$$prf = \frac{1}{\Delta t} \text{ pulses/s}$$

3 The important exponential signals are the decaying exponential

$$v(t) = V_0 e^{-t/\tau}$$

where V_0 is the amplitude and τ is the time constant. At a time equal to one time constant $v(t) = 0.37V_0$ and at five time constants $v(t) \approx 0.01V_0$.

The rising exponential

$$v(t) = V_0(1 - e^{-t/\tau})$$

At a time equal to one time constant, $v(t) = 0.63V_0$; and at five time constants, $v(t) \approx 0.99V_0$.

4 The basic sinusoidal signal for our purposes is

$$v(t) = V_m \cos(\omega t + \varphi)$$

where V_m is the peak value, ω is the angular frequency in radians per second, and φ is the phase angle.

PROBLEMS

2.1-1 Sketch the following functions.

a $10\,u(t)$
b $5\,u(t-7)$
c $3\,u(-t)$
d $4\,u(5-t)$
e $tu(t)$
f $(t-1)u(t-1)$
g $25[u(t-3) - u(t-7)]$
h $25[u(t-3) + u(t-7)]$
i $[u(t) - u(t-\pi/2)] \cos t$

2.1-2 Write equations for the waveforms in Fig. P2.1-2 using step functions and any other algebraic functions you feel are appropriate.

2.2-1 In a certain pulse train, the pulse width τ is 10 μs. The time interval between pulses is 4τ. Find the prf and sketch the pulse train.

2.2-2 Using the scheme of Fig. 2.2-2 we wish to devise a code to transmit the ten digits 0, 1, 2, . . . , 8, 9. How many pulses are required in each set?

2.2-3 Repeat Prob. 2.2-2 for the English alphabet.

2.3-1 A certain voltage is described by the equation $v(t) = 30e^{-t/(20\times10^{-6})}$ V. What are the amplitude and time constant? Neatly sketch the voltage using graph paper.

2.3-2 An exponentially decaying current has an initial value (amplitude at $t = 0$) of 8 mA and a time constant of 10 μs. Using graph paper, neatly sketch the waveform.

2.3-3 An exponentially decaying voltage has a time constant of 5 ms. When $t = 5$ ms, $v(t) = 8$ V. Find the amplitude of the voltage at $t = 0$ and write the equation.

2.3-4 An exponentially decaying current has an initial value of 10 mA. At $t = 3$ ms, $i(t) = 1$ mA. Find the time constant and write the equation for $i(t)$.

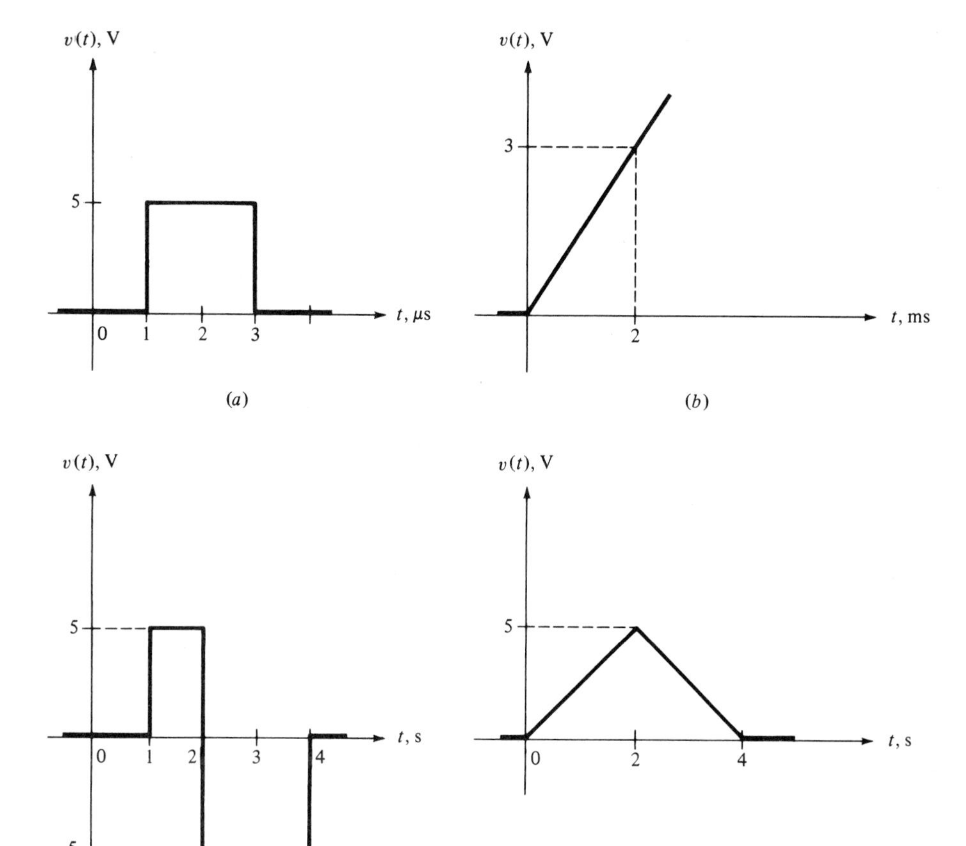

Fig. P2.1-2.

2.3-5 An exponentially rising voltage is described by

$$v(t) = 10(1 - e^{-t/3\times10^{-3}}) \text{ V}$$

What are its initial value, final value, and time constant? Sketch the waveform on graph paper.

2.3-6 A certain current is described by

$$i(t) = 3 - 7e^{-t/2} \text{ A}$$

Is this a rise or a decay? Sketch the waveform on graph paper.

2.3-7 The equation of a voltage is

$$v(t) = 10 + 15e^{-t/3} \text{ V}$$

Is this a rise or a decay? Sketch the waveform on graph paper.

2.3-8 An exponentially rising current has an initial value of zero, a final value of 10 mA and a value of 2 mA after 2 ms. Find the time constant and write the equation for $i(t)$.

2.4-1 A sinusoidal voltage is described by the equation

$$v(t) = 50 \cos (6{,}280t - 32°) \text{ mV}.$$

Find its amplitude, angular frequency, cyclic frequency, period, and phase angle. Sketch the wave on graph paper being sure to scale the axes properly.

2.4-2 Sketch the following functions on the same set of axes.

a $v(t) = 10 \cos 377t$
b $v(t) = 10 \cos (377t + \pi/6)$
c $v(t) = 10 \cos (377t + \pi/3)$

2.4-3 A cosine voltage waveform has a maximum value of 440 V and zero phase angle. Find $v(t)$ when

a $\omega t = 30°$
b $\omega t = 0.2$ rad
c $\omega t = 220°$

2.4-4 A current is given by $i(t) = 200 \sin (628t + 80°)$ mA. Find its instantaneous value when

a $t = 5$ ms
b $t = 50$ ms

2.4-5 A cosine voltage waveform has a phase angle of 30° and an instantaneous value of 130 V when $\omega t = \frac{\pi}{4}$. What is its peak value?

2.4-6 Write the equation for the sinusoids shown in Fig. P2.4-6 using the cosine function.

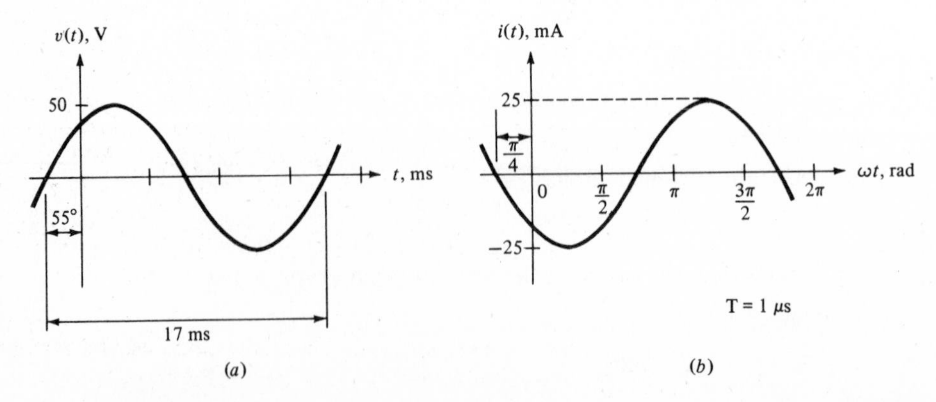

Fig. P2.4-6.

3
Electrical Quantities

OBJECTIVES

Upon completion of this chapter, you should be able to

1 List the most important electric quantities.
2 State the units of measurement and symbols used for the electric quantities.
3 Express the relationships among the electric quantities.
4 Relate the electric quantities to basic physical quantities.
5 Describe the source of electric charge.
6 Indicate the correct values, both magnitude and sign, of electric quantities whose reference directions are shown in diagrams.

INTRODUCTION

In this chapter we shall try to provide an understanding of what electricity is. In order to facilitate this task we shall make frequent use of analogs, or systems that are similar in certain respects to the one we are discussing. It is important to remember that an analog may be very different from the original system in all but a single or limited number of characteristics. It is used to explain new concepts by relating them to others with which we are already familiar.

3.1 ATOMIC THEORY

We have all had some experience with electricity. We know it is invisible and odorless and that it can result in a shock which may vary from a mild tingling sensation to a fatal experience. This information, however, does not enable us to use electricity in a productive way. In order to achieve this goal we must be able to apply our knowledge of electricity in such a way as to provide quantitative information. In order to accomplish this, we begin by considering the atomic theory of matter.

The atomic theory of matter is a description of the structure of materials based on numerous experiments concerning the physical and chemical properties of matter. One historical group of experiments involves observed forces of attraction and repulsion between various materials after they have been rubbed together. The atomic theory postulates the existence of two kinds of electric *charges* which account for these forces; the forces are said to be due to static electricity.

According to atomic theory, all matter is constructed of atoms of which there are somewhat over one hundred kinds. An *element* is a material all of whose atoms are identical; all materials are either elements or compounds consisting of different kinds of atoms that have joined in specific ways to form *molecules*. The chemical and physical properties of materials can be explained in terms of their atomic structure.

Every atom is composed of a dense central region, the nucleus, containing two kinds of heavy particles called protons and neutrons, and various kinds of lighter particles with which we shall not be concerned. Orbiting around the nucleus, at certain relatively fixed distances from it, are a number of relatively light particles called electrons. A group of electrons which are at approximately the same distance from the nucleus constitute a "shell." The entire atom is submicroscopic in size but may be envisioned somewhat like a miniature solar system with the nucleus in place of the sun and the electrons in place of the planets as shown in Fig. 3.1-1. The protons and neutrons are each of the same mass and together account for almost all of the mass of the atom. The protons each have one type of electric charge, called *positive*, while the electrons each have a second type of electric charge, called *negative*. Opposite kinds of charge

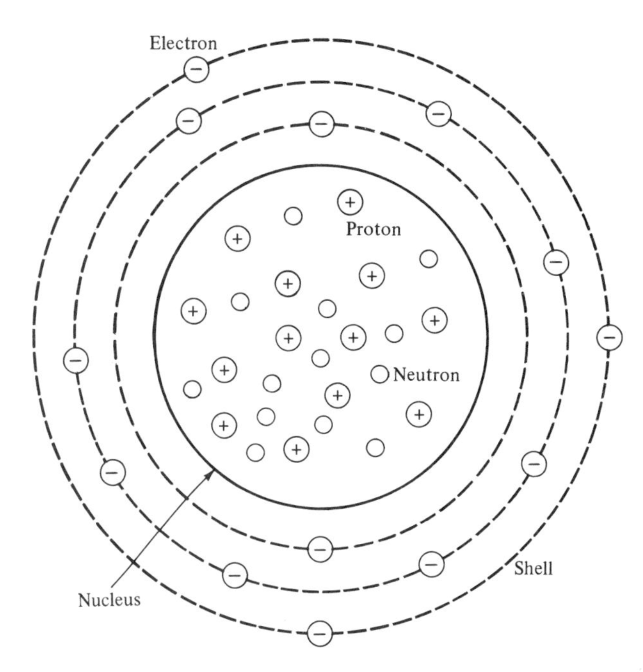

Fig. 3.1-1 Atomic structure.

attract each other while two charges of the same type repel each other. Every neutral atom has as many electrons as protons. The amount of charge on each electron is equal in magnitude to the amount of charge on each proton although they are of opposite sign. The neutrons have no electric charge and the atom is therefore electrically neutral. The *forces* due to static electricity result when electrons are caused to move from one material to another by rubbing the two materials against each other. This leaves one material with a net negative charge because it has excess electrons and the other with a net positive charge because it has excess protons. Because the two materials have net charges of opposite sign they attract each other. Since the charges are not moving this is called static electricity.

3.2 ELECTRIC CHARGE

We have seen that electric charge is an inherent part of all matter. Normally, there are equal amounts of positive and negative charge in every atom and this balanced situation results in the cancellation of their effects. It is, however, possible to isolate charges so that we can study their effects.

There are a number of ways that charges can be isolated. Rubbing different materials together can dislodge electrons from one material and cause them to become attached to another material. This technique is used in experiments with static electricity.

Electrons are frequently transferred from one atom to another in chemical reactions thereby forming ions which are atoms having a deficit or excess of electrons and a concomitant positive or negative charge. When in solution, these ions are free to move about as charge carriers. It is also possible to cause electrons to become detached from atoms and move about between the atoms in certain materials, notably in metals which are generally good conductors of electricity.

The most important characteristic of charges is the force that exists between them—a force of attraction between opposite charges and a force of repulsion between like charges. These forces are similar to those between magnets in which unlike poles attract each other while like poles repel each other.

It is found that the force on one charge due to another is proportional to the product of the two charges and inversely proportional to the square of the distance between them. A positive force is taken as a repulsive force because two positive charges repel each other as shown in Fig. 3.2-1a.

The letter q is generally used to represent a quantity of charge. If we let r represent the distance between two charges q_1 and q_2, the force on one charge due to the other is

$$f = k\frac{q_1q_2}{r^2} \tag{3.2-1}$$

where k is a constant of proportionality which depends on the system of units that are used.

Notice that if one of the charges, q_1 or q_2, is negative the force becomes negative. Since we have taken a repulsive force as positive, a negative force implies an attractive force, i.e., a force in the opposite direction. This agrees

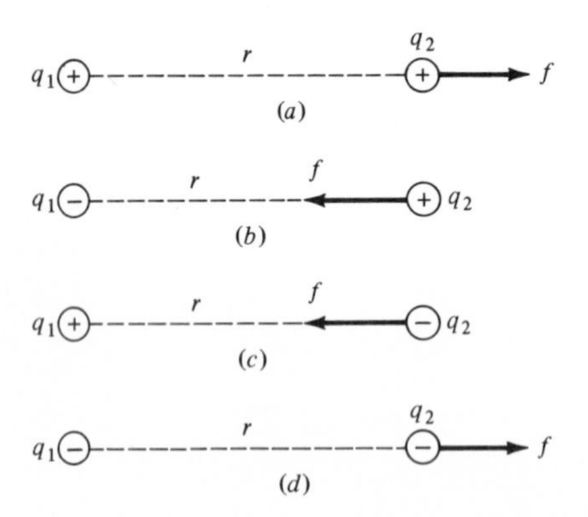

Fig. 3.2-1 Force on one charge due to a second charge. (*a*) Two positive charges; (*b*) force on positive charge due to negative charge; (*c*) force on negative charge due to positive charge; (*d*) two negative charges.

with the result obtained in each case of Fig. 3.2-1 and illustrates that Eq. (3.2-1) gives the correct result in all cases providing we interpret a negative force in one direction as a positive force of the same magnitude in the opposite direction. In the case of the force between two charges, the directions are relative, i.e., attractive or repulsive, rather than absolute such as left, right, up, or down. This must be the case since moving both charges so that their separation is unchanged does not change the force although it may change its absolute direction.

It would be natural to select the amount of charge on a proton as the basic unit of charge since it is a positive quantity and the smallest unit that normally exists independently. The problem in doing this is that the proton charge (which is equal to the magnitude of the electron charge) is a very small quantity. The basic unit of charge in the mksa system is taken as the coulomb (C) which is that amount of charge which when placed 1 meter (m) from an equal charge experiences a repulsive force of 1 newton (N). Thus, in the mksa system, the formula for the force on a charge q_1 due to a second charge q_2 located a distance r from q_1 is

$$f = k\frac{q_1q_2}{r^2} \tag{3.2-2}$$

where q_1 and q_2 are in coulombs and r is in meters and

$$k = 9.0 \times 10^9 \ \frac{\text{N}\cdot\text{m}^2}{\text{C}^2} \tag{3.2-3}$$

To illustrate the size of 1 C of charge, we note that if e is the charge on one electron then

$$1 \text{ C} = -6.25 \times 10^{18}e \tag{3.2-4}$$

that is, it takes 6.25×10^{18} electrons to make up 1 C of charge. This indicates how small the magnitude of the electron charge is. From Eq. (3.2-4) we find that the charge on one electron is

$$e = -\frac{1}{6.25 \times 10^{18}} = -1.6 \times 10^{-19} \text{ C} \tag{3.2-5}$$

Example 3.2-1 Two charges of +3 mC (millicoulombs) and −6 mC are separated by a distance of 3 cm. Find the force on each of the charges.

Solution The repulsive force on each charge due to the other is given by Eq. (3.2-2):

$$f = k\frac{q_1q_2}{r^2}$$

Therefore

$$f = 9.0 \times 10^9 \frac{\text{N}\cdot\text{m}^2}{\text{C}^2} \frac{(3\text{mC})(-6\text{ mC})}{(3\text{ cm})^2}$$

$$= \frac{9.0 \times 10^9\text{ N}\cdot\text{m}^2}{\text{C}^2} \frac{(3 \times 10^{-3}\text{ C})(-6 \times 10^{-3}\text{ C})}{(3 \times 10^{-2}\text{ m})^2}$$

$$= -18.0 \times 10^7\text{ N}$$

Hence, there is an attractive force of 18.0×10^7 N (which is the same as a repulsive force of -18.0×10^7 N) on each charge. ////

3.3 DIRECTED QUANTITIES

Certain quantities have a direction associated with them as well as a magnitude, or numerical measure. A force, for example, has a direction as well as a magnitude (strength) whereas mass has only a magnitude. Some quantities, such as charge, are measured as algebraic quantities; that is, they may be positive or negative although they do not have a direction. A quantity which has no direction associated with it is called a *scalar*; one which has a direction associated with it is called a *vector*. The *magnitude* of a vector or an algebraic quantity is the numerical value associated with it; i.e., without regard for direction or algebraic sign.

If two vector quantities have the same magnitude but opposite directions, then one is said to equal the negative of the other. In our study of electric circuits we shall often have to determine the values of vector quantities whose directions will be limited to one of two possible directions which are opposite each other. The force on one charge due to another is an example of such a situation: the force can be attractive or repulsive. In such a situation we can often *arbitrarily* assign a direction to the quantity and then compute its value based on this direction; the resulting numerical value will be positive or negative. Had we arbitrarily chosen the opposite direction the numerical value would be equal in magnitude but opposite in sign. We call this arbitrarily chosen direction a *reference direction* because it permits us to refer to the vector quantity with no ambiguity as to its direction. For example, if we know that a road runs north-south and we want to know how far a car has gone along it we might ask, "How far north have you gone?" If the car has actually traveled 10 miles (mi) south the scientific reply would be "−10 mi." The reference direction in this case is north. Had we simply asked, "How far have you gone?" and received the reply "10 mi," we would have no idea of the direction. For this reason, we shall be using reference directions in much of our work.

3.4 ELECTRIC CURRENT

In materials which are categorized as good electrical conductors, charges are free to move about with great ease. In general, charges are in constant random motion due to thermal energy. Each charge will move in a particular direction, collide with another charge, and move off in a new direction as shown in Fig. 3.4-1*a*. As a result of this random motion there will be no net change in charge distribution.

Under certain conditions forces can be applied to the charges so that there is a greater tendency to move in a specified direction. Then there will be a net *drift* in that direction superimposed on the random motions of the charge carriers as shown in Fig. 3.4-1*b*. *Electric current* occurs when there is such a *net* movement of charge from one place to another.

We can visualize current flow as being similar to the flow of water in a pipe; instead of measuring the number of gallons of water passing a given cross section of the pipe per second we are concerned with the number of coulombs of charge that pass a given cross section of the conductor per second.

The letter I is generally used to symbolize average current. In the mksa system, current is measured in amperes (A), where

$$1 \text{ A} = 1 \frac{\text{coulomb}}{\text{second}} = 1 \frac{\text{C}}{\text{s}} \tag{3.4-1}$$

It is important to recognize that current has a direction just as the flow of water has a direction. At this point a problem arises in using this analogy because there are two kinds of charge–positive and negative—but only one kind of water. We will explain the effects of the two types of charge by assuming a hypothetical situation which illustrates the differences in their behavior.

Suppose we have a solution containing positive ions (atoms with a net positive charge) which we pump through a tube as shown in Fig. 3.4-2*a*. The ions passing a cross section of the tube (shown by the dotted line in the figure) constitute a current flow, denoted I, in the same direction as the charges are moving.

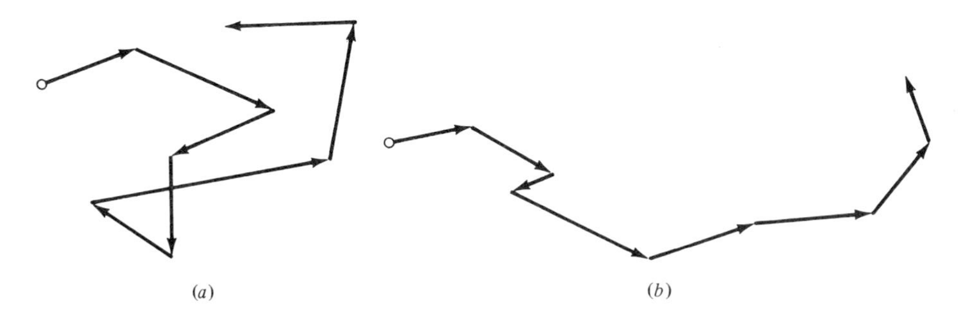

Fig. 3.4-1 Motion of charge carrier: (*a*) random; (*b*) with applied force.

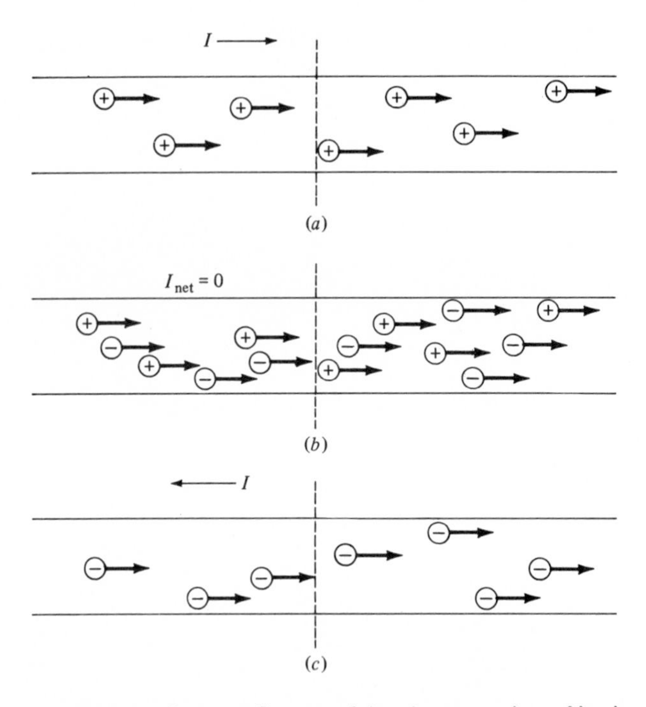

Fig. 3.4-2 Current flow resulting from motion of ionic fluid. (*a*) Positive ion solution; (*b*) solution with positive and negative ions; (*c*) negative ion solution.

Now assume that we add negative ions to the solution so that it has as much negative charge as positive charge. When this solution is pumped through the tube, as illustrated in Fig. 3.4-2*b*, there will be *zero* current flow since as many negative charges will cross any given cross section as positive charges resulting in a net charge flow of zero. Since the positive charges are still flowing we must conclude that the negative charges are causing a current equal in magnitude to I but in the *opposite* direction as shown in Fig. 3.4-2*c*. The two opposing currents cancel each other resulting in a net current $I_{\text{net}} = 0$.

We conclude that positive charges moving in a given direction constitute a positive current flow in that direction; negative charges moving in a given direction constitute a positive current flow in the direction opposite to that of the flow of negative charges.

Electric currents in circuits generally flow through metallic conductors. The atoms in a metal are in fixed locations but free electrons are easily caused to move in the space between the atoms. Thus, electric currents in metals are the result of electron flow. Since electrons have negative charge, the direction of the current flow is opposite the direction of the electron flow as illustrated in Fig. 3.4-2*c*.

The average current in a given direction is computed as the amount of charge crossing a given cross-sectional divided by the time interval during which the charge flow is measured. Mathematically we have

$$I = \pm \frac{q(t_2) - q(t_1)}{t_2 - t_1} = \frac{\Delta q}{\Delta t} \tag{3.4-2}$$

where the plus sign applies if the directions of I and charge flow are the same, and the minus sign applies if they are opposite. $\Delta q = q(t_2) - q(t_1)$ is the amount of charge crossing the area in the time interval Δt (from t_1 to t_2). Figure 3.4-3 shows how this equation automatically takes care of directions for all charge polarities and charge flow directions. In this figure, it is assumed that the magnitude of charge that has passed a given cross section at $t_1 = 1$ s is 2 C and at $t_2 = 3$ s it is 8 C. Note that the sign of the charge is either positive or negative and the formula has a positive or negative sign depending on whether the charge flow is in the same direction or opposite to the reference current flow direction.

Example 3.4-1 The current flowing through a wire is measured and found to have a constant value of 5 A. Such a current whose value does not vary with time is called direct current (dc). Find the amount of charge transported along the wire in 3 min by this current.

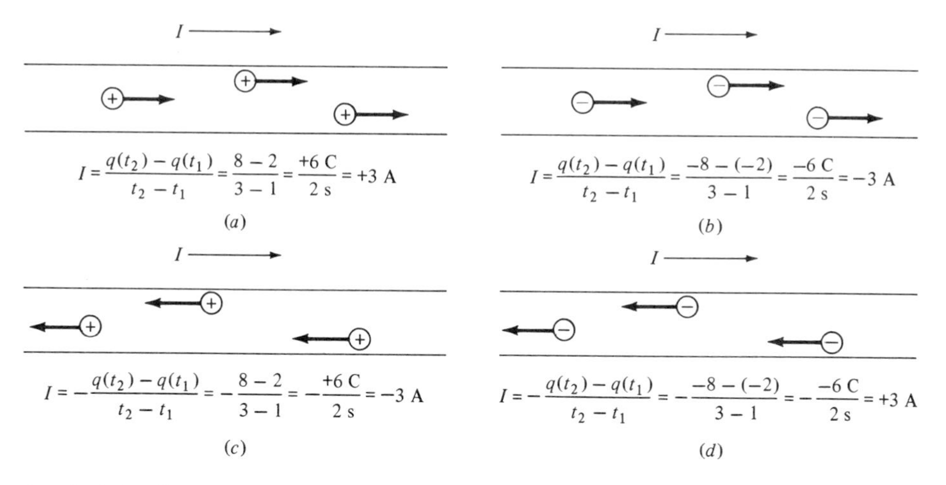

Fig. 3.4-3 Current computation illustrating effects of charge flow direction and charge polarity. (*a*) Current reference direction same as charge flow direction, positive charge. (*b*) Current reference direction same as charge flow direction, negative charge. (*c*) Current reference direction opposite charge flow direction, positive charge. (*d*) Current reference direction opposite charge flow direction, negative charge.

Solution Equation (3.4-2) states that

$$I = \frac{\Delta q}{\Delta t}$$

Thus

$$\Delta q = I \, \Delta t$$

where

$$\Delta t = 3 \text{ min} \times 60 \frac{\text{s}}{\text{min}} = 180 \text{ s}$$

Therefore

$$\Delta q = 5 \text{ A} \times 180 \text{ s} = 900 \text{ C}$$

This is the amount of charge transported along the wire in 3 min. ////

3.5 INSTANTANEOUS VS. DC QUANTITIES

Currents and voltages generally change with time. The usual notation for a current, for example, emphasizes this fact by indicating explicitly that the current is a function of time by writing it as $i(t)$. Frequently the symbol i alone is used as shorthand for $i(t)$.

For many purposes we shall be interested in the average, or mean, value of a current over some given interval of time. The symbol I is generally used to represent the average value of a current whose instantaneous value is represented by i or $i(t)$.

In many applications an electrical quantity has a constant value which is independent of time. Such a quantity is called a direct-current (dc) quantity. The current in Example 3.4-1 was dc. The average value of a dc quantity is equal to the instantaneous value so uppercase letters are generally used to represent dc quantities as well as average quantities.

We now turn our attention to the relationship between current and charge. This was done for dc current in Sec 3.4; in this section the relationship will be extended to time-varying currents.

The graph of a dc current is shown in Fig. 3.5-1*a*. Equation (3.4-2)

$$I = \pm \frac{q(t_2) - q(t_1)}{t_2 - t_1} = \frac{\Delta q}{\Delta t} \tag{3.4-2}$$

leads to the formula for the incremental change in charge during an increment of time:

$$\Delta q = I \, \Delta t \tag{3.5-1}$$

From Eq. (3.5-1) it is clear that the increment in charge transported through a

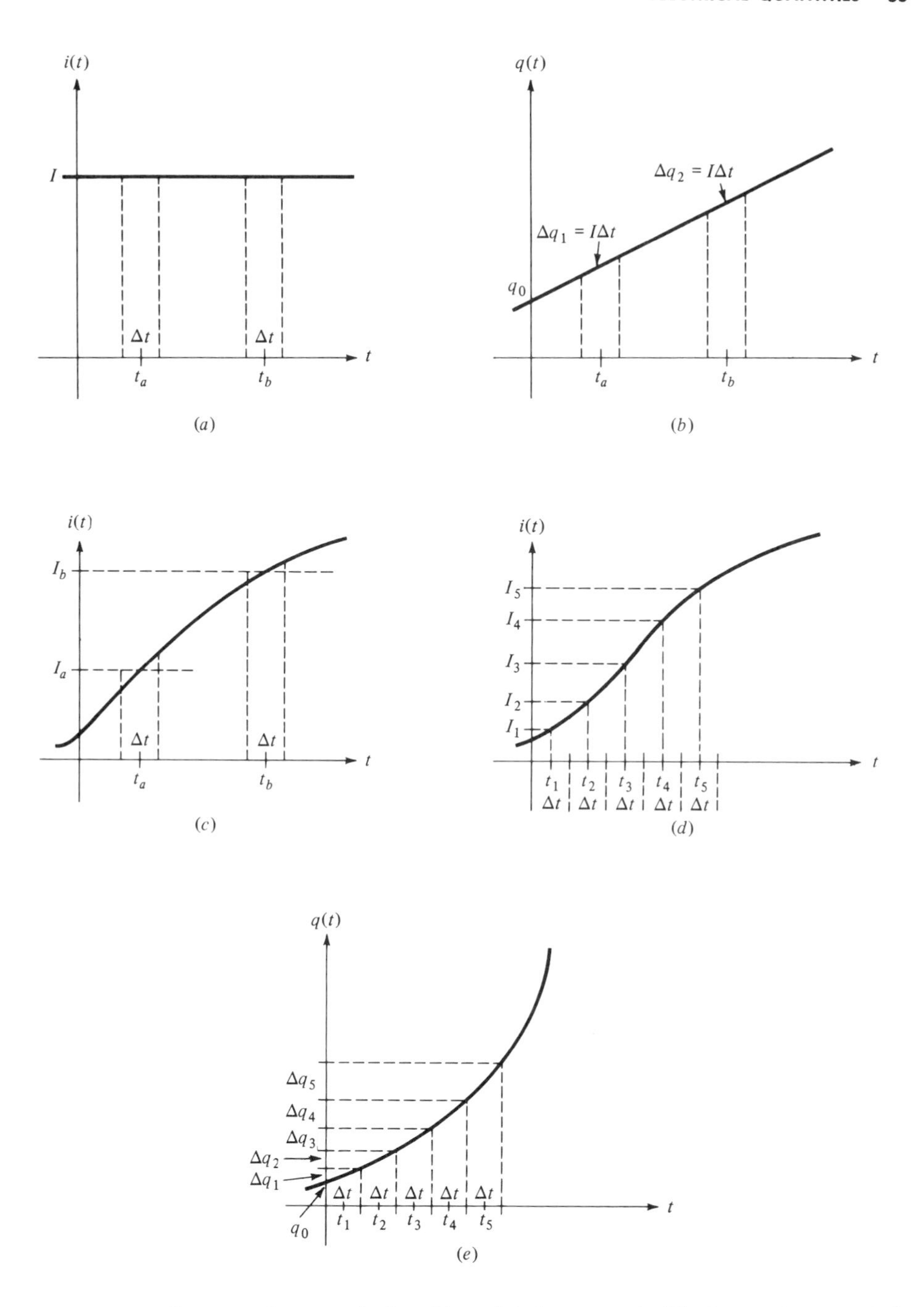

Fig. 3.5-1 Current-charge relationships: (*a*) dc current; (*b*) charge transported by dc current; (*c*) time-varying current; (*d*) time-varying current showing increments of time; (*e*) charge transported by time-varying current.

given cross section of the wire is proportional to the increment of time and is independent of the particular time at which the increment occurs. The graph of $q(t)$ implied by Eq. (3.5-1) for the current shown in Fig. 3.5-1*a* is shown in Fig. 3.5-1*b*. q_0 is the initial value of charge at $t = 0$; its value cannot be determined from Eq. (3.5-1) but must be given separately. Equation (3.5-1) indicates that the change in charge, Δq, during an interval of time, Δt is equal to the area under the current curve during this interval (see Fig. 3.5-1*a*). Equation (3.4-2) indicates that the value of the current is equal to the slope of the charge curve.

The same principles that were discussed above for the dc current of Fig. 3.5-1*a* can be applied to the time-varying current of Fig. 3.5-1*c* provided that Δt is limited to a value which is small enough so that $i(t)$ can be treated as if it were constant over the interval Δt. This is shown for two different times t_a and t_b in Fig. 3.5-1*c*. The value of $i(t)$ during the interval surrounding t_a can be approximated by $i(t_a) = I_a$, and during the interval surrounding t_b it can be approximated by $i(t_b) = I_b$. The important distinction between the time-varying and dc case is that Δt must be very small for the time-varying case so that the current does not change by much over the interval; for the dc case the current is constant and the interval may be of any size. With Δt adequately small, $\Delta q_a = I_a \Delta t$ and $\Delta q_b = I_b \Delta t$ as shown in Fig. 3.5-1*c*. From Fig. 3.5-1*c* it is seen that $\Delta q_a < \Delta q_b$ because $I_a = i(t_a)$ is less than $I_b = i(t_b)$. This is a second important difference between the time-varying and dc cases: the incremental changes in charge for equal increments in times are the same for the dc case (Fig. 3.5-1*a*) but differ for the time-varying case (Fig. 3.5-1*c*). By dividing the time axis into many small intervals, as shown in Fig. 3.5-1*d*, it is possible to compute and plot successive values of $q(t)$ if the initial value q_0 is given; this is illustrated in Fig. 3.5-1*e* where

$$\Delta q_n = i(t_n)\,\Delta t \tag{3.5-2}$$

Δq_n is the increment of charge that is transported through a given cross section of wire by the current $i(t)$ during an interval Δt surrounding t_n; it is therefore appropriate to denote Δq_n as $\Delta q(t_n)$. When this is substituted into Eq. (3.5-2) the current becomes

$$i(t_n) = \frac{\Delta q(t_n)}{\Delta t}, \tag{3.5-3}$$

which is a mathematical statement that the current value at any time is equal to the slope of the charge curve at that time. Graphically, the slope of the charge curve is not constant for the time-varying current as it is for the dc current; it is found by taking the slope of the tangent line to the charge curve. The concept of slope will be treated again in more detail in later chapters.

Example 3.5-1 The curve of charge passing a given cross-sectional area of a wire as a function of time is shown in Fig. 3.5-2. During what time interval, if any, is there a constant current? What is its value (if it exists)?

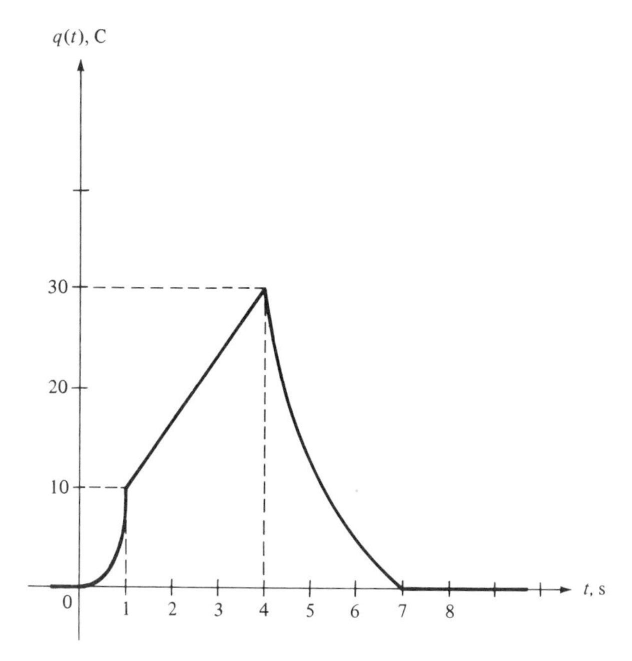

Fig. 3.5-2 Charge passing cross-sectional area of wire.

Solution We see that the curve of charge as a function of time is a straight line only during the interval from $t = 1$ to $t = 4$. Therefore, there is a constant current other than zero only during this interval. The value of the current is given by the slope of the charge curve:

$$I = \frac{\Delta q}{\Delta t} = \frac{q(t_2) - q(t_1)}{t_2 - t_1} = \frac{30\text{ C} - 10\text{ C}}{4\text{ s} - 1\text{ s}} = \frac{20\text{ C}}{3\text{ s}} = 6.67\text{ A} \qquad ////$$

3.6 WORK AND ENERGY

We all have an intuitive idea of what it is to do work and there would be general agreement that energy is expended in doing work. We must, however, provide a more precise definition of these concepts.

Energy can exist in many forms: electric, mechanical, acoustic (sound), thermal (heat), radiant (light), and numerous others. Furthermore energy can be converted from one form to another; a light bulb converts electric energy into light and heat energy while a generator can be used to convert mechanical energy, in the form of rotational motion, into electric energy.

A device used to convert energy from one form to another is called an *energy conversion device*. In general, energy conversion and transmission applications fall into two categories. The first involves large amounts of energy used for its ability to do work and to provide power for lighting, etc. The second category is concerned with relatively small amounts of energy used to transmit information by varying the amount of energy as a function of time. We call this time-varying energy a *signal* and the energy conversion devices used for these applications are generally called *transducers.* A microphone, for example, is a transducer that converts acoustic energy into electric energy. It is important that a transducer provide an output signal that varies in the same way as the input signal. We refer to this ability as the *fidelity* of the transducer; the fidelity is a measure of how well the output signal matches the input signal. A system with poor fidelity *distorts* or changes the signal thereby changing the information.

We have spoken, so far, about energy but we have not defined it. Because of the basic nature of energy we take it as an undefined term and instead rely on the various comments already made and point out that energy is necessary for something to happen; we can only perceive something by being affected by energy, whether it be light or sound or any other form of energy to which we are sensitive.

Work is associated with the expenditure of mechanical energy. If we carry a heavy object up a flight of stairs we have worked and used up energy. We can quantify our concept of mechanical energy by defining it in terms of other quantities. Work is done when a body is moved over a distance by applying a force to it in the direction of motion. The symbol w is used to denote energy; it is computed as

$$w = fd \tag{3.6-1}$$

where f is the applied force and d is the distance moved in the direction of f.

Energy is measured in joules in the mksa system and a joule is defined as 1 newton-meter (N·m). Thus a force of 1 N applied over a distance of 1 m corresponds to an energy expenditure of 1 J. If one were to push against a stationary wall with a force of 1,000 N for 1 h, no energy would have been used since there was no motion.

If a weight is lifted from a table, energy is expended in lifting it the given distance by applying a force opposite that of gravity. If the weight is returned to the table, the net energy used must be zero since the net distance the weight has been moved is zero. Consequently, we must conclude that whatever energy was used in lifting the weight up has been returned in the process of replacing it. This can be understood more easily if we first learn a very important and basic law of physics called the *law of conservation of energy.* This law states that the total amount of energy in an isolated system is constant (provided there are no atomic reactions which do permit a conversion between matter and energy—which results in great amounts of nuclear energy). When we lift the weight from the table we have expended energy. This energy has not disappeared, rather it has increased the energy associated with the weight. The weight, when it is above the table has a greater energy than when it is on the table. We call this form of energy *potential energy* since it is not being used but is available for use. Potential energy is determined by the position of an object. When we release the weight it falls back to the table landing with a loud noise and possibly breaking the table. The noise and the breaking of the table are a result of the potential energy that was in the weight by virtue of its position above the table. This energy must be gone when the weight is back on the table. It has been partially converted to acoustic energy; the rest has been transmitted to the table.

Example 3.6-1 A force of 50 N is required to lift the weight of a pile driver. How high must it be lifted so as to transmit 1 kJ of energy to a pile on which it drops?

Solution The energy is equal to the force multiplied by the distance. Therefore, the distance is

$$d = \frac{\text{energy}}{\text{force}} = \frac{1 \text{ kJ}}{50 \text{ N}} = \frac{1{,}000 \text{ J}}{50 \text{ N}} = 20 \text{ m} \qquad ////$$

3.7 POTENTIAL DIFFERENCE

We have seen that a charge in the vicinity of another charge experiences a force due to the second charge. (The second charge also experiences an equal but opposite force due to the first charge.) If we place one or more charges at certain fixed positions a test charge brought into their vicinity and held in a fixed position will experience a net force which is the sum of the forces due to each of the other charges. This force will, in general, vary as a function of the position of the test charge. Furthermore, energy will be required to move the test charge since a force will be required to oppose the force due to the other fixed charges.

If the test charge is to be moved in a direction opposing the force due to the fixed charges a force will have to be applied in the direction of motion; this requires the expenditure of energy by whatever causes the motion of the test charge. This energy must go into the charge that has been moved; in other words, the test charge is at a higher energy level after it has been moved just as a weight, when it is lifted, is at a higher energy level. The energy of the test charge, like that of the weight, is dependent on the position of the test charge and is called potential energy. Furthermore, just as the weight will fall to a position of lower potential energy when it is released, so the test charge will move in the direction of the force due to the fixed charges to a position of lower potential energy.

The potential energy of an object that has been lifted from a table is computed as the force required to lift it, which is the weight of the object, multiplied by the vertical distance it is lifted. Had we lifted the object from the floor the potential energy that we calculate would be increased because the vertical distance it is lifted would be increased by the height of the table. In both cases the object is at the same position even though we calculate two different potential energies. This strange situation is resolved by recognizing that potential energy is not an absolute, but a relative, quantity. It is true that the object does have more energy relative to the floor than to the table. You can verify this by dropping a weight on your toe from a distance of 2 in and again from a distance of 5 ft—you will directly experience the difference in energy.

The same relative characteristic holds for electric potential energy as for the mechanical potential energy which is its analog. For this reason we always consider potential energy difference *between two points*; unless we specify what our reference is it doesn't make sense to speak of potential energy at a single point. The difference in the potential energy of a test charge moved from one point to another will be proportional to the charge because the force on the test charge due to the other charges is proportional to the test charge. We define the electric potential difference between two points as the difference in energy level that a unit test charge would undergo while being moved from one point to the other. Note that potential difference is a function of the position of the two points. Thus, if a test charge of 2 C were used the energy difference would be doubled. The potential difference would be the same, however, since it is computed as the difference in energy divided by the charge. To find the difference in energy level of a charge moved between the two points we must multiply the charge by the potential difference. Thus, the potential difference between two points, generally symbolized by the letter v, is defined as

$$v = \frac{\Delta w}{\Delta q} \tag{3.7-1}$$

where Δw is the difference in energy level of a charge Δq when it is moved from one point to the other. A potential difference has a polarity associated with it which indicates which of the two points is the point at which a unit positive charge would be at a higher energy level; this point is indicated by a positive sign while the point of lower energy has a negative sign associated with it as shown in Fig. 3.7-1 where point a is at the higher reference energy level.

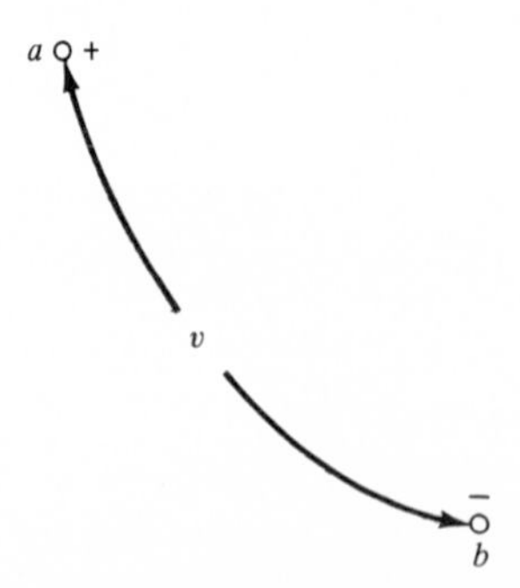

Fig. 3.7-1 Notation for potential difference.

Potential difference, in the mksa system, is measured in volts (V). The volt is defined by

$$1 \text{ volt (V)} = 1 \frac{\text{joule}}{\text{coulomb}} = 1 \frac{\text{J}}{\text{C}} \tag{3.7-2}$$

Usually, potential difference is called *voltage* and this is the word we will use from this point on.

Example 3.7-1 A charge of 5 C gains 4 J of energy in moving from point a to point b. Find the voltage difference V between points a and b.

Solution

$$V = \frac{W}{Q} = \frac{4 \text{ J}}{5 \text{ C}} = 0.8 \text{ V}$$

where point b is 0.8 V higher in voltage than point a since a positive charge gains energy in moving from point a to point b. ////

Example 3.7-2 Two J of energy are delivered to the box shown in Fig. 3.7-2 over a period of 10 s by the 5 mA current. Find the potential difference V shown in the figure.

Solution The total amount of charge moved from the left side of the box to the right is

$$\begin{aligned} q &= 5 \text{ mA} \times 10 \text{ s} \\ &= 5 \frac{\text{mC}}{\text{s}} \times 10 \text{ s} = 50 \times 10^{-3} \text{ C} \end{aligned}$$

Since 2 J of energy are delivered to the box, this charge must lose 2 J of energy in going through the box. The potential difference, V, must therefore be

$$V = -\frac{2 \text{ J}}{50 \times 10^{-3} \text{ C}} = -40 \text{ V}$$

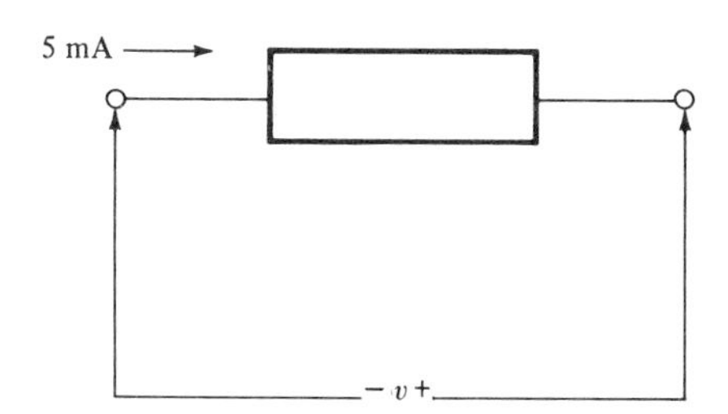

Fig. 3.7-2 Circuit segment for Example 3.7-1.

The negative sign is required because the reference polarity for V indicates a lower energy level on the left whereas the energy level is actually higher on the left. ////

Different notations are used for voltage polarity. One common notation is of the form "the voltage drop v_{xy}"; this refers to the potential difference between two points x and y with the reference plus sign at x and the reference minus sign at y. Similarly the term "voltage rise v_{xy}" refers to a potential difference between two points x and y with the reference minus sign at x and the reference plus sign at y. In what follows, the term voltage without any modifier will *always* refer to a drop in potential, usually called a *voltage drop*. Thus v_{xy} will *always* mean a *reference voltage drop* from x to y. This will be considered in more detail in Chap. 4 when we discuss the basic circuit laws.

3.8 POWER

Power is the rate at which energy is used or supplied. If the total amount of energy that has been supplied to a system as a function of time is $w(t)$, then the instantaneous power delivered to the machine at time t is

$$p(t) = \frac{\Delta w}{\Delta t} \qquad (3.8\text{-}1)$$

where Δw is the change in the amount of energy supplied during a short interval Δt around time t. Power is measured in watts (W) in the mksa system where

$$1 \text{ watt (W)} = 1 \frac{\text{joule}}{\text{second}} = 1 \frac{\text{J}}{\text{s}} \qquad (3.8\text{-}2)$$

We shall now derive a formula for the power delivered to an element in an electric circuit. Suppose a current $i(t)$ flows through an element with a voltage drop $v(t)$ across it in the direction of current flow as shown in Fig. 3.8-1. The charge transferred through the element in a short period Δt around time t is

$$\Delta q(t) = i(t)\,\Delta t \qquad (3.8\text{-}3)$$

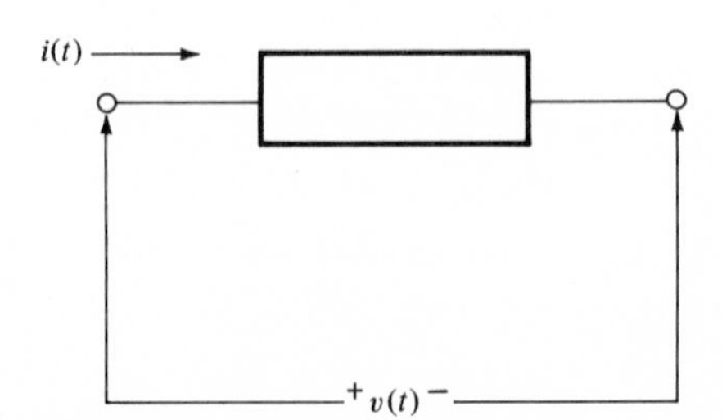

Fig. 3.8-1 Circuit element for power computation.

while the energy delivered to the element is, from Eq. (3.7-1),

$$\Delta w(t) = v(t)\,\Delta q(t) \tag{3.8-4}$$

Note that this energy is *delivered* to the element since the charge is going from a point of higher potential energy to one of lower potential energy—thus the difference must be delivered to the element. The power is found by using Eqs. (3.8-3) and (3.8-4) in Eq. (3.8-1) as follows:

$$p(t) = \frac{\Delta w(t)}{\Delta t} = \frac{v(t)\,\Delta q(t)}{\Delta t} = \frac{v(t)\,i(t)\,\Delta t}{\Delta t}$$

or

$$p(t) = v(t)\,i(t) \tag{3.8-5}$$

Note that in the formula $p(t)$ is power delivered to (or absorbed by) the element, $v(t)$ is the potential difference across the element, and $i(t)$ is the current through the element in the direction from higher to lower potential. A negative value of $p(t)$ indicates that power is delivered by the element rather than to it.

Example 3.8-1 Find the power delivered to the element of Fig. 3.8-2 if $V = 8$ V and $I = 40$ mA. How much energy is delivered to the element in 2 min?

Solution

$$P = VI$$

$$P = (8 \text{ V})(40 \text{ mA}) = 320 \text{ mW}$$

The energy delivered to the element is

$$W = Pt$$

where t is the time interval. Hence

$$W = 320 \times 10^{-3} \text{ W} \times 2 \text{ min} \times 60 \frac{\text{s}}{\text{min}} = 38.4 \text{ J}$$ ////

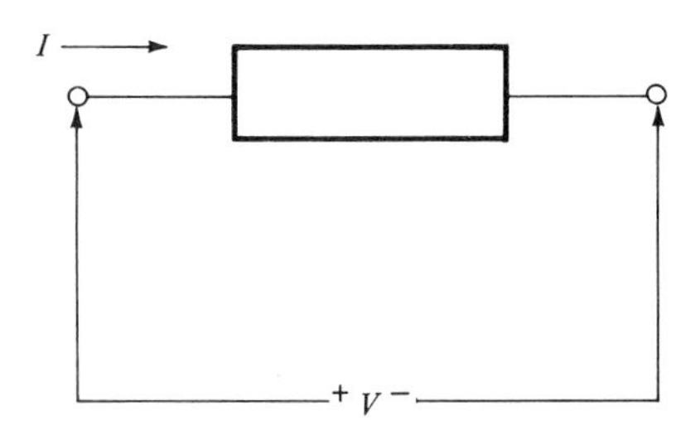

Fig. 3.8-2 Circuit element for Example 3.8-1.

SUMMARY

Electric Quantities

Quantity	Symbol	Unit and its abbreviation
Charge	q	coulomb (C)
Current	i	ampere (A)
Energy	w	joule (J)
Voltage (potential difference)	v	volt (V)
Power	p	watt (W)
Force†	f	newton (N)
Distance†	r, x	meter (m)
Time†	t	second (s)

† These quantities are not electrical but are involved in formulas for other electrical quantities.

FORMULAS

1 Force on charge due to second charge

$$f = k \frac{q_1 q_2}{r^2}$$

2 Electron charge

$$e = -1.6 \times 10^{-19} \text{ C}$$

3 Current

$$i = \frac{\Delta q}{\Delta t}$$

4 Potential difference

$$v = \frac{\Delta w}{\Delta q}$$

5 Power

$$p = \frac{\Delta w}{\Delta t} = vi$$

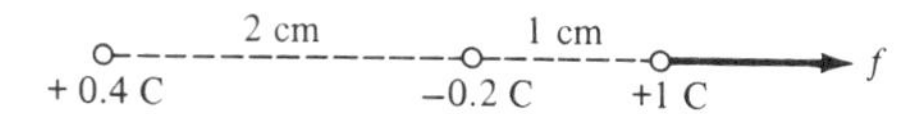

Fig. P3.2–5.

PROBLEMS

3.2-1 How many electrons are required to provide 0.02 C of charge? What will the polarity of this charge be?

3.2-2 Compute the magnitude and direction of the force between electrons which are 10^{-10} m apart. What will the force be if the distance is doubled?

3.2-3 Find the distance between two charges of 5 C and −4 C if the force acting on one due to the other is 15 N.

3.2-4 A charge of 4 C attracts a second charge which is 3 m from the first charge with a force of 8 N. Find the magnitude of the second charge.

3.2-5 Three charges are located as shown in Fig. P3.2-5. Find the force f shown in the figure. *Hint:* f is the sum of the forces acting on the + 1 C charge due to each of the other two charges acting separately.

3.4-1 Electrons travel along a wire at the rate of 9×10^{20} electrons/min from left to right. What is the current and what is its direction?

3.4-2 A current of 25 mA flows through a wire. How long does it take for 50 C to pass a given cross-sectional area of the wire?

3.4-3 A charge of −5 C travels around a wire loop whose circumference is 30 cm at a speed of 1,200 m/min as shown in Fig. P3.4-3. Find the current I.

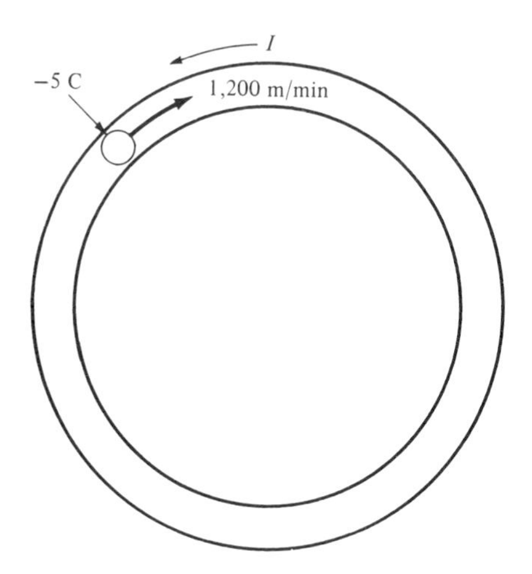

Fig. P3.4–3.

3.4-4 The density of free electrons in a wire is 10^{15} electrons/mm^3. Compute the current flowing in a wire whose diameter is 5 mm if the electrons are moving at a speed of 3×10^8 m/s.

3.4-5 A solution of positive ions flows through a tube whose cross-sectional area is 5 mm. Each ion has a charge equal to twice the magnitude of an electron charge. If the solution is pumped at a rate of 20 cm^3/s, what must be the density of ions in the solution if the current is to be 3 mA?

3.5-1 A dc current of 15 mA flows for 5 s. How much charge is transported as a result?

3.5-2 A graph of the charge that has flowed across a cross-sectional area of a wire is shown in Fig. P3.5-2. Draw a graph of the current flow in the same direction.

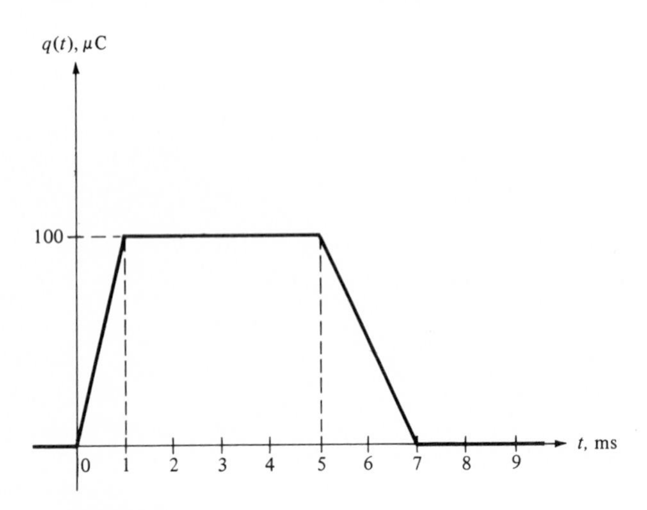

Fig. P3.5-2.

3.5-3 The amount of charge transferred across a given cross-sectional area of wire is given by

$$q(t) = t\,u(t) - (t - 4)\,u(t - 4)$$

Draw a graph of the current.

3.5-4 Find the charge that has flowed across a given cross section of wire if the current flow in the same direction is given by

$$i(t) = (1 + t)[u(t) - u(t - 2)]$$

3.6-1 Compute the potential energy of a mass 5 m above the ground which required a force of 4.3 N to be lifted to its position relative to the ground.

3.6-2 What is the difference in potential energy of the mass of Prob. 3.6-1 when it is on a 1.4-m-high table as compared with a position on the ground.

3.7-1 Find the difference in the energy level of a 3 C positive charge when it is moved from one point to a second point with a potential difference of 2 V between them, the second point being at a higher potential.

3.7-2 Determine the change in potential energy of a −8 C charge when it is moved from point a to point b in Fig. P3.7-2.

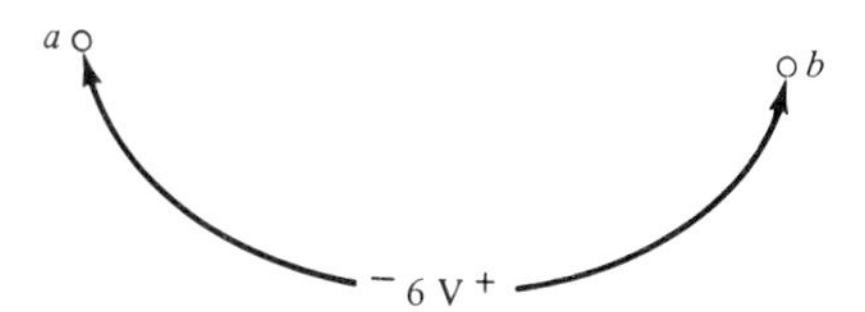

Fig. P3.7–2.

3.7-3 Find the potential difference v_{ab} if a −5 C charge gains 18 J of energy in being moved from point a to point b in Fig. P3.7-3.

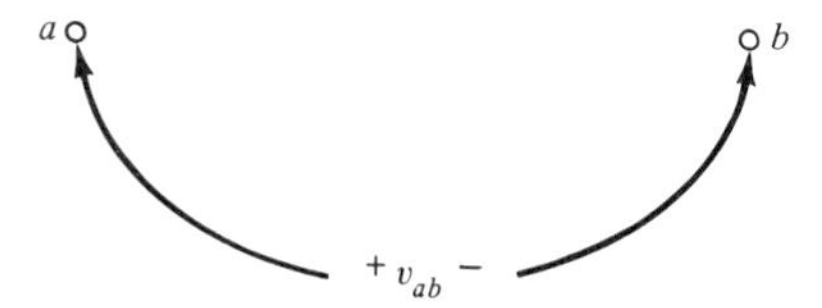

Fig. P3.7–3.

3.7-4 A charge gains 40 J of energy when moving through a potential drop of 18 V. What is the value of the charge?

3.7-5 A current of 4 mA flows through an element having a potential drop of 10 V in the direction of current flow. How much energy is transferred in a 1 s interval? Is the energy delivered to or from the element?

3.8-1 Find the power in the circuit element of Fig. P3.8-1. Is this power delivered to or from the element?

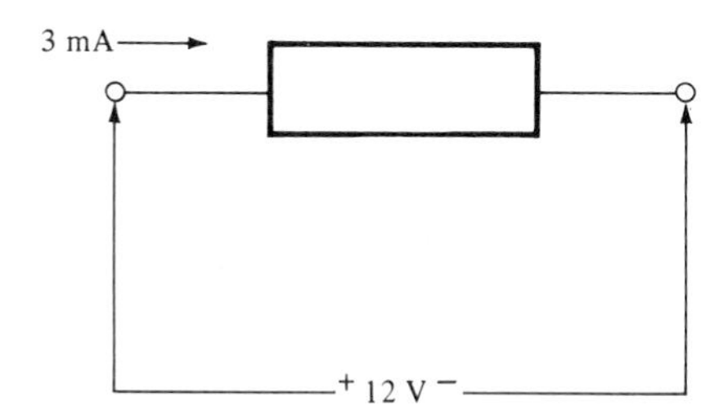

Fig. P3.8–1.

3.8-2 An electric generator delivers 5 kW of electric power at a voltage of 100 V. What is the current flowing through the generator? How much energy is generated in an hour?

3.8-3 A circuit element dissipates 10 W and has a voltage of 30 V across it. Find the current through the element and its direction.

3.8-4 25 mA flows through a device which absorbs 2 W of power. Find the voltage across the device and its polarity (relative to the current flow direction).

3.8-5 A 9-V battery stores 3,600 J of energy. How long will the battery last if 10 mA of current are drawn? If the current drain is 5 mA? What is the maximum average current drain if the battery is to last for at least 50 h?

4
Resistance

OBJECTIVES

Upon completion of this chapter, you should be able to

1 Determine the resistance of a conductor when given its dimensions and resistivity.
2 Determine the conductance of a resistor.
3 Use wire tables to determine the resistance of a given length of wire of given gauge.
4 Determine the change in resistance of a conductor with temperature variations.
5 Compute the resistance of a sheet resistance.
6 Identify the value and tolerance of a resistor using the color code.

4.1 RESISTIVITY

Every material has *resistance.* The resistance of a material is, in fact, one of its basic physical characteristics. Materials having a very high resistance are called insulators, those with a very low resistance are called conductors. The term resistance, as well as its circuit symbol shown in Fig. 4.1-1a, is motivated by the intuitive concept that a resistance resists or impedes the flow of current.

A *resistor* is a two-terminal circuit element which is characterized by its resistance. The voltage across it is related to the current through it by Ohm's law:

$$v = Ri \tag{4.1-1}$$

where R is the resistance measured in ohms (Ω). From this equation, we see that 1 Ω is the equivalent of 1 V/A.

The *conductance* of a resistor is the reciprocal of its resistance and is measured in mhos (℧). It is usually symbolized by the letter G. Thus, for any resistance R, we have

$$G = \frac{1}{R} \tag{4.1-2}$$

The hydraulic analog for current flow is useful in developing a physical feeling for resistance. Recall that fluid flow is the analog of current while pressure is the analog of voltage. The amount of fluid flow through a pipe is proportional to the pressure just as current flow through a resistance is proportional to voltage. The smaller the pipe diameter, the greater the pressure will have to be to achieve the same flow. Furthermore, the longer the pipe, the greater the pressure drop will be from one end to the other. Work must be done to pump water through a pipe because of the friction between the fluid and the walls of the pipe. This friction generates heat which is dissipated; the electrical energy absorbed by a resistance is converted to heat and dissipated in the same way.

Just as the hydraulic resistance of a pipe depends on its length and its cross sectional area, so the electrical resistance of a wire depends on its length and cross sectional area. In the case of the pipe there is one additional factor which affects the resistance to flow; that is the coefficient of friction between the pipe wall and the fluid.

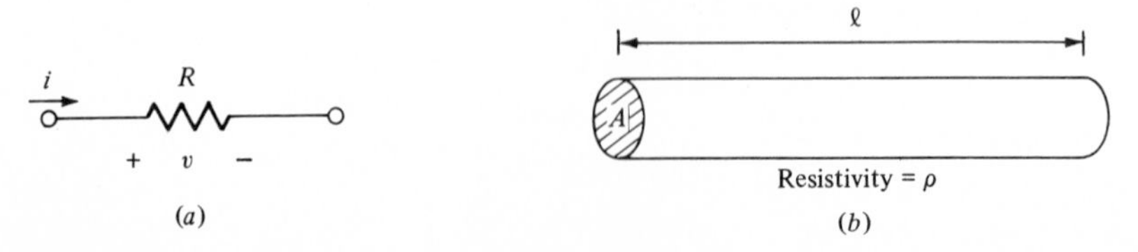

Fig. 4.1-1 Resistance. (a) Circuit symbol; (b) parameters for computation of resistance.

In the electrical case we find a completely analogous situation. The resistance is proportional to the length of the wire, and inversely proportional to the cross sectional area of the wire. There is a constant of proportionality, analogous to the coefficient of friction for the pipe, which depends on the material. This constant of proportionality is called the *resistivity* of the material; it is symbolized by the lowercase Greek letter rho (ρ) and has a very high value for insulators and a very low value for good conductors. A table of resistivity values for various materials is included in Appendix C. The resistance of a conductor may be computed from the following formula which summarizes the previous discussion:

$$R = \rho \frac{\ell}{A} \tag{4.1-3}$$

where ℓ and A are the length and cross-sectional area of the conductor as shown in Fig. 4.1-1*b*. The resistance R would be measured between connections made to the two ends of the conductor. If the length and area are measured in meters and square meters respectively, then ρ must be given in units of ohm-meters

Example 4.1-1 Compute the resistance of a conductor which is 1 m long, which has a 4×4 mm square cross section and a resistivity of 10 $\Omega \cdot m$. What will the resistance be if each cross-sectional dimension is doubled?

Solution

$$R = \rho \frac{\ell}{A} = \frac{(10\ \Omega\cdot\text{m})\ (1\ \text{m})}{(4\ \text{mm})\ (4\ \text{mm})} \times \frac{1{,}000\ \text{mm}}{\text{m}} \times \frac{1{,}000\ \text{mm}}{\text{m}}$$

$$= 625\ \text{k}\Omega$$

If each cross-sectional dimension is doubled, the area is quadrupled, and the resistance becomes one-quarter of its original value. Thus

$$R = 156.25\ \text{k}\Omega$$

////

4.2 USE OF WIRE TABLES

It is frequently necessary to determine the resistance of wires used to connect electric devices. Because the problem is so common, the resistance per unit length of standard wire types is tabulated in wire tables along with other characteristics of the wire.

The American Wire Gauge (AWG) is the most common method for specifying the size of a wire in the United States. In addition to the wire gauge, the diameter and/or cross-sectional area of the wire is also frequently specified. The diameter is often given in mils where 1 mil is equal to one one-thousandth (0.001) in. The area is generally specified in circular mils; this value is found by squaring the diameter given in mils. Thus the cross-sectional area of a wire whose diameter is 3 mil is 9 circular mil (cmil).

The resistance is generally specified in ohms per 1,000 ft at a specified temperature, generally 20 or 25°C (68 or 77°F). A wire table is included in Appendix C of this book.

Example 4.2-1 Find the resistance at 20°C of 1 mi of copper wire whose diameter is the smallest standard diameter equal to or greater than 0.01 in.

Solution The diameter of the wire in mils is

$$0.01 \text{ in} \times \frac{1 \text{ mil}}{0.001 \text{ in}} = 10 \text{ mil}$$

and its area is $10^2 = 100$ cmil. From the table in Appendix C we find the smallest standard wire satisfying this requirement is AWG 30 with an area of 100 cmil and a resistance of 105.0 Ω/1,000 ft at 20°C. Then

$$1 \text{ mi} \times 5{,}280 \frac{\text{ft}}{\text{mi}} \times \frac{105.0\ \Omega}{1{,}000 \text{ ft}} = 554.4\ \Omega$$ ////

4.3 RESISTANCE CHANGES WITH TEMPERATURE

The resistance of most materials varies with temperature. In general, the resistance increases with increasing temperature. Although the variation of resistance with temperature is not linear, it can be approximated fairly well by a linear relationship over a relatively large range around normal room temperature. The slope of this straight line approximation is symbolized by the Greek letter α. It is called the temperature coefficient of resistance and has the units of $°\text{C}^{-1}$ or $°\text{F}^{-1}$ depending on whether centigrade or fahrenheit units are being used. If the resistance of a conductor at 20°C, which is normal room temperature, is R_{20} Ω, then its resistance at temperature T °C is

$$R_T = R_{20}[1 + \alpha(T - 20)]\ \Omega \tag{4.3-1}$$

A table of temperature coefficients of resistance for various materials is included in Appendix C.

Example 4.3-1 Find the resistance of the wire of Example 4.2-1 at 180°F.

Solution Since the temperature coefficient of resistance is given in units of $°\text{C}^{-1}$ we convert to centigrade. The conversion formula is

$$T_C = (T_F - 32)\frac{1}{1.8}$$

where T_C is the temperature in °C and T_F is the temperature in °F. The validity of this formula may be checked using the freezing point of water,

32°F = 0°C, and the boiling point of water, 212°F = 100°C. The result for T_F = 180°F is

$$T_C = (180 - 32)\frac{1}{1.8} = 82.2°\text{C}$$

Substitute this, the value $R_{20} = 554.4\ \Omega$ found in Example 4.2-1, and the temperature coefficient of resistance for copper, $\alpha = 0.0039/°\text{C}$ from the table, in Eq. (4.3-1):

$$R_{180} = 554.4[1 + 0.0039\ (82.2 - 20)] = 689\ \Omega \qquad ////$$

4.4 SHEET RESISTANCE

In the construction of thin film or thick film integrated circuits, thin layers of resistive material are often formed on an insulating substrate as shown in Fig. 4.4-1.

Using Eq. (4.1-3) with $A = wt$, the resistance between the two conductive connectors is

$$R = \rho \frac{\ell}{wt} \qquad (4.4\text{-}1)$$

Since the thickness of such layers is generally constant in a given integrated circuit, it is common to rewrite this equation in the form

$$R = \frac{\rho}{t}\frac{\ell}{w} \qquad (4.4\text{-}2)$$

The term ρ/t is called the sheet resistance R_s; its unit is Ω.

Example 4.4-1 Find the resistance between the connectors of the two thin-film resistors shown in Fig. 4.4-2 if the sheet resistance is $R_s = 100\ \Omega$.

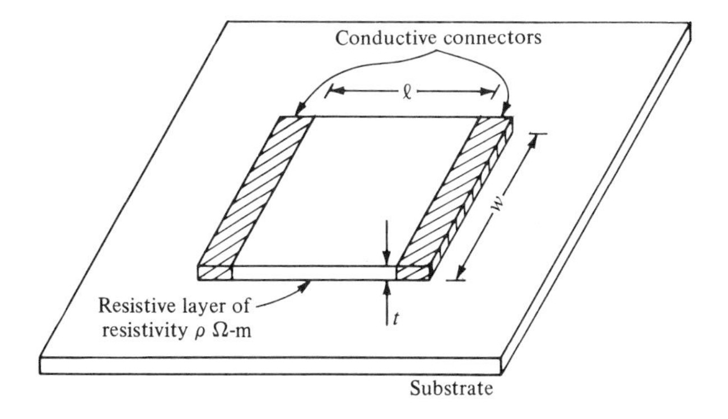

Fig. 4.4-1 Sheet resistance.

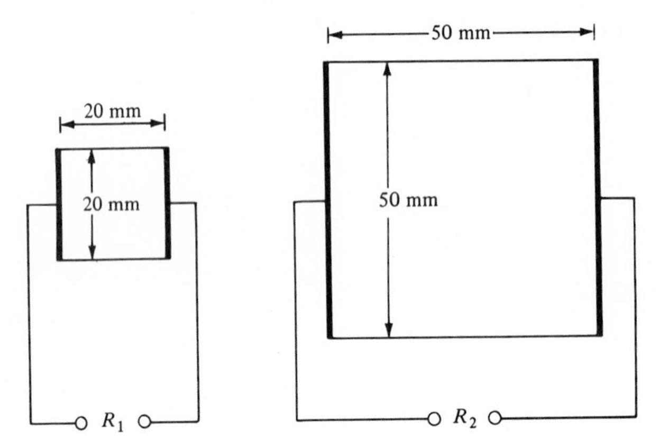

Fig. 4.4-2 Thin-film resistors for Example 4.4-1.

Solution For the first square we have

$$R_1 = 100\ \Omega \frac{0.02\ \text{m}}{0.02\ \text{m}} = 100\ \Omega$$

For the second square we have

$$R_2 = 100\ \Omega \frac{0.05\ \text{m}}{0.05\ \text{m}} = 100\ \Omega$$

In the preceding example we found that the resistance of a square is independent of its size. For this reason, the sheet resistance is often expressed in units of ohms per square (or $\Omega/\square$). This is equivalent to Ω but it serves to remind one that it is the resistance per square of a layer of resistive material of constant thickness. ////

4.5 RESISTOR COLOR CODE

One type of commercially available resistor is shown in Fig. 4.5-1. The size of a resistor is correlated more with its power dissipating capacity than its resistance value.

Most resistors are fabricated from a molded carbon composition or from high resistance wire wound on a core and then covered with a vitreous enamel glaze. The composition resistors are cheaper while the wire wound resistors are capable of being manufactured to tighter tolerances. Wire wound resistors are frequently held to a tolerance of 1 percent and are used in precision applications such as analog computers.

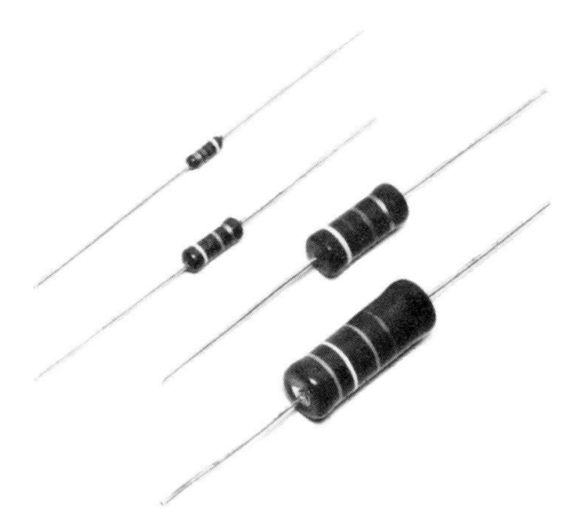

Fig. 4.5-1 Resistors. (*Sprague Electric Company.*)

Composition resistors are the most widely used type in general electronics applications. Very small resistors are available whose power dissipation is limited to $\frac{1}{8}$ watt (W), while larger ones are available which can dissipate $\frac{1}{4}$, $\frac{1}{2}$, 1, and 2 W. Higher wattage values are also available but are not quite as common in electronic applications.

The resistance value and the tolerance of composition resistors are generally indicated by a color code on the body of the resistor. The value is indicated by three or four rings or bands of color around the resistors as shown in Fig. 4.5-2*a*. The resistance value and tolerance is determined as indicated in this figure using the table shown in Fig. 4.5-2*b*.

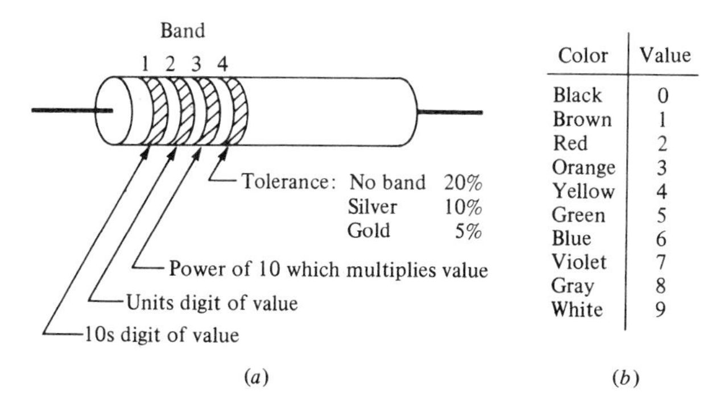

Color	Value
Black	0
Brown	1
Red	2
Orange	3
Yellow	4
Green	5
Blue	6
Violet	7
Gray	8
White	9

Fig. 4.5-2 Composition resistor color code. (*a*) Location of bands; (*b*) table of color values.

Example 4.5-1 Find the resistance and tolerance of a resistor having the following four colored bands starting with the one nearest the end of the resistor: red, violet, orange, gold. Find the range of values in which this resistance will lie.

Solution From Fig. 4.5-2 we determine

red = 2, violet = 7, orange = 3, gold = 5%

Thus the nominal resistance value is $27 \times 10^3\ \Omega = 27\ \text{k}\Omega$ and the actual value lies in the range $27\ \text{k}\Omega \pm 1.35\ \text{k}\Omega$. Therefore $25.65\ \text{k}\Omega \leq R \leq 28.35\ \text{k}\Omega$.

////

SUMMARY

1 Resistivity is symbolized by ρ and its unit is $\Omega \cdot m$.

2 The formula for resistance is $R = \rho \dfrac{\ell}{A}\ \Omega$

3 The unit of resistance is the ohm (Ω):

$$1\ \Omega = 1\ \frac{V}{A}$$

4 The formula for the sheet resistance of a square of material of thickness t and resistivity ρ is

$$R_s = \frac{\rho}{t}\ \frac{\Omega}{\square}$$

5 The formula for the resistance of wire at temperature T °C in terms of its resistance at 20°C is

$$R_T = R_{20}\left[1 + \alpha(T - 20)\right]$$

where α is the temperature coefficient of resistance.

PROBLEMS

4.1-1 Determine the resistance between the two ends of the bar shown in Fig. P4.1-1 if the resistivity of the material from which it is made is 5 $\Omega \cdot$cm.

4.1-2 Find the conductance of the bar of Prob. 4.1-1.

4.1-3 A bar is identical to the one in Prob. 4.1-1 except for its length which is unknown. If the resistance between the two ends of this bar is found to be 100 Ω, what is its length?

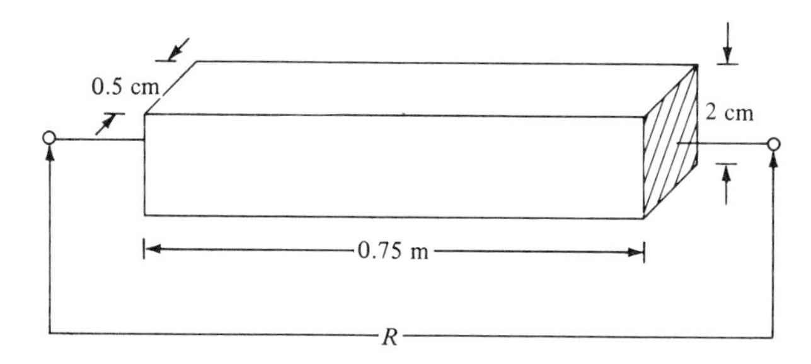

Fig. P4.1-1.

4.1-4 A strain gage is used to measure very small changes in length by stretching a wire and measuring the change in its resistance. Consider the wire shown in Fig. P4.1-4. The solid lines show the wire before it is stretched; the dotted lines after it is stretched.

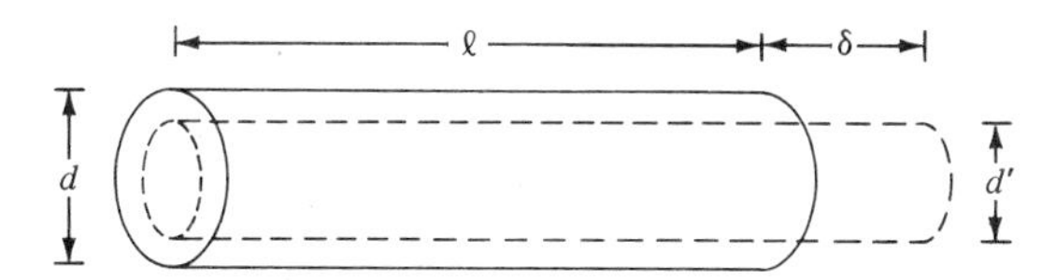

Fig. P4.1-4.

a Write a formula for R_0, the resistance of the wire before it is stretched

b Show that the value of d'^2, the square of the diameter of the wire after it is stretched an amount δ, is

$$d'^2 = \frac{d^2}{1 + \delta/l}$$

Hint: The volume of the wire does not change.

c Write a formula for the new resistance R after the wire is stretched.

d Show that

$$\frac{\Delta R}{R_0} = 2\frac{\delta}{l} + \frac{\delta^2}{l^2} \qquad \text{where } \Delta R = R - R_0$$

Is the change in resistance proportional to the change in length? If not, is there any condition for which it is approximately proportional? If so, what is this condition?

4.2-1 A copper wire has a diameter of 0.008 in.

a Find its area in circular mils.

b What is the closest standard gage?

c Find the resistance of 400 yd of this gage wire at 20°C.

4.2-2 Determine the gage of the thinnest copper wire which will have a resistance no greater than 20 Ω for 50 ft at 20°C.

4.3-1 Find the resistance of the wire of Prob. 4.2-1 at 100°C.

4.3-2 Repeat Prob. 4.2-2 for a temperature of 86°F rather than 20°C.

4.4-1 Find the sheet resistance of a 1-mm-thick layer of material whose resistivity is 5 kΩ·m.

4.4-2 A rectangular layer of resistive material has a sheet resistance of 100 Ω/□. Its dimensions are 2 × 5 cm. Find the resistance between conductive connectors at each end of the 5-cm dimension.

4.4-3 Find the dimensions of a rectangular layer of resistive material whose area is to be 1 cm^2 and whose resistance, from one end to the other, is to be 1 k Ω if R_s = 50 Ω/□.

4.5-1 Starting at the band nearest the edge of a composition resistor, what colors will the bands be for a 15 MΩ, 10 percent resistance value.

4.5-2 Starting at the end band of a resistor the colors are brown, black, and blue. What is the value of the resistance? What is its tolerance?

5
Basic Resistive Network Laws

OBJECTIVES

Upon completion of this chapter you should be able to

1 Define the terms electric circuit, electric network, terminal, branch, node, and loop.
2 Describe the analytic and graphical representation of the terminal characteristic of a two-terminal circuit element.
3 Determine the power supplied to a two-terminal element in terms of the voltage across it and current through it.
4 Describe the terminal behavior of an independent current source.
5 State Kirchhoff's current law.
6 Describe the terminal behavior of an independent voltage source.
7 State Kirchhoff's voltage law.
8 Describe the terminal behavior of a resistance and state Ohm's law.

INTRODUCTION

In this textbook we will be concerned with electric circuits or networks. An *electric circuit* is a system consisting of electric devices connected together with wires. The term *electric network* is essentially synonymous with the term *electric circuit,* although some authors reserve the word network for more complex interconnections of elements and refer to simpler interconnections as circuits. We shall use the two terms interchangeably.

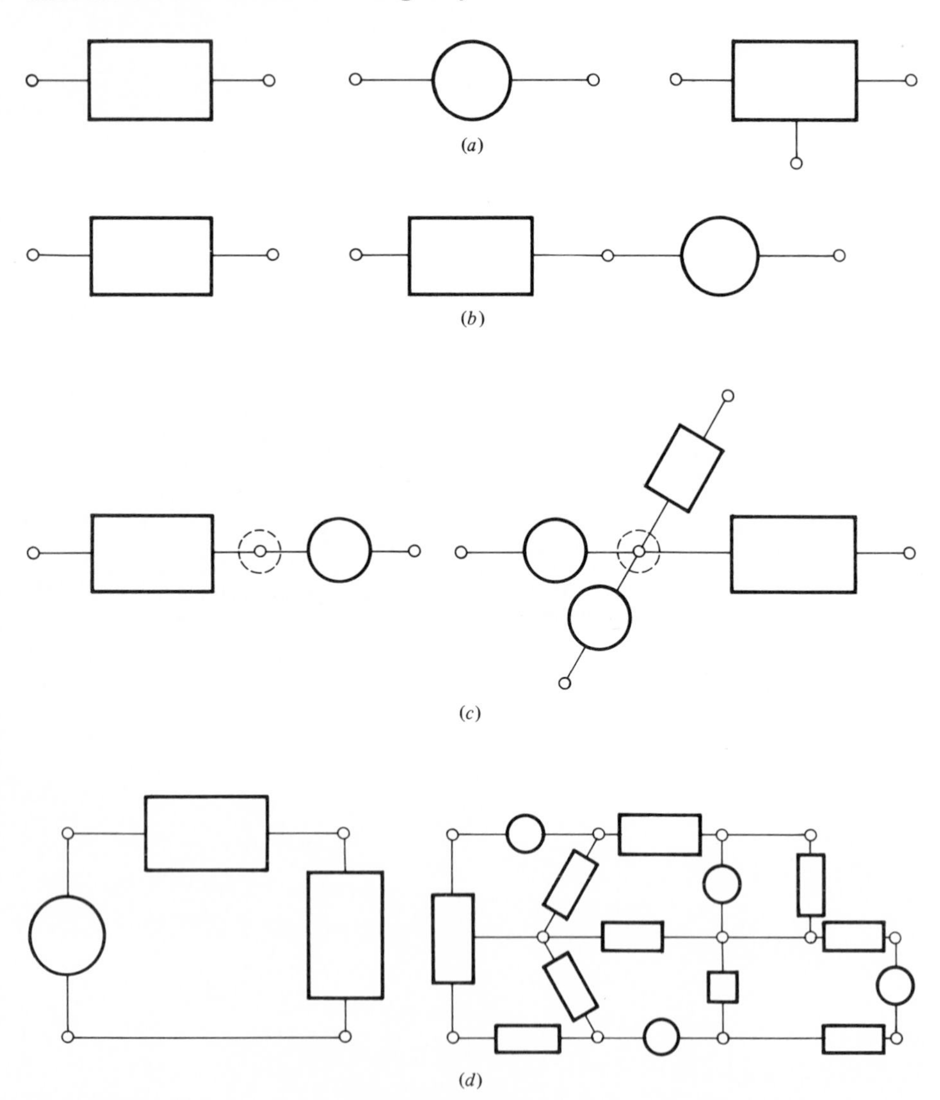

Fig. 5.1-1 Network concepts. (*a*) Electric elements; (*b*) branches; (*c*) nodes; (*d*) circuits.

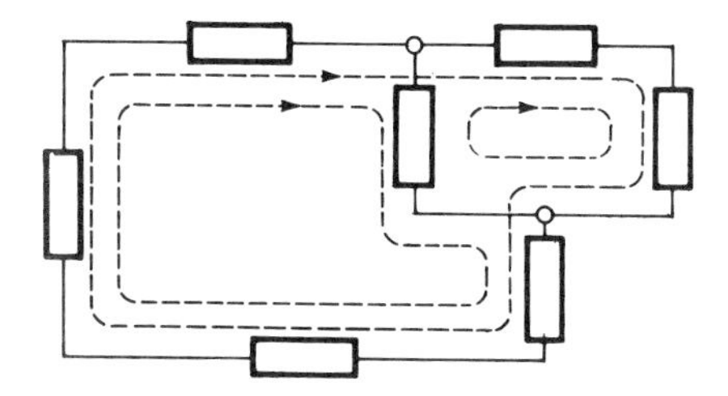

Fig. 5.1-2 Network with loops indicated.

5.1 BRANCHES, NODES, AND LOOPS

Electric circuits are composed of elements which have *terminals.* Elements are interconnected by attaching their terminals together in various configurations using wires. Any portion of a network which has two, and only two, external terminals is called a *branch.* A point at which two or more branches are connected together is called a *node*; Fig. 5.1-1 illustrates these concepts. Note that the interconnection of two elements may be considered as a node between two branches or the two elements may be considered as a single branch, in which case the interconnection of the two elements is not considered to be a node.

Any closed path that can be traced around a set of network branches without crossing or retracing itself is called a *loop*; Fig. 5.1-2 illustrates a number of loops in a circuit in which the branches are indicated by boxes and connecting wires.

In drawing electric circuits it is necessary to distinguish between wires which cross each other with no connection and those which do connect. Figure 5.1-3 illustrates the conventions used.

Example 5.1-1 Indicate all the possible nodes and loops for the circuit of Fig. 5.1-4*a*.

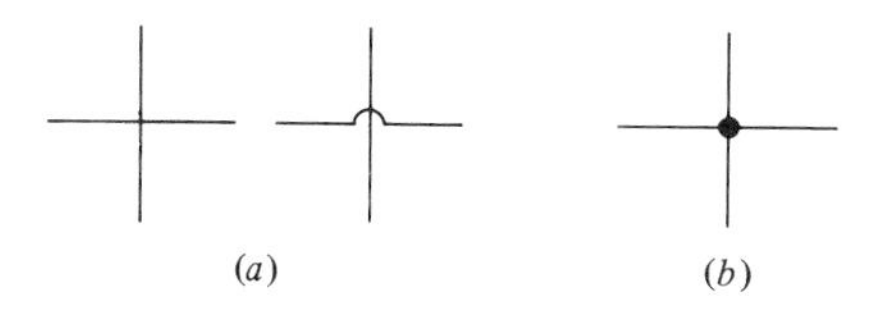

Fig. 5.1-3 Conventions used to indicate unconnected crossing wires and connected wires. (*a*) Wire crossing—no connection; (*b*) wires connected.

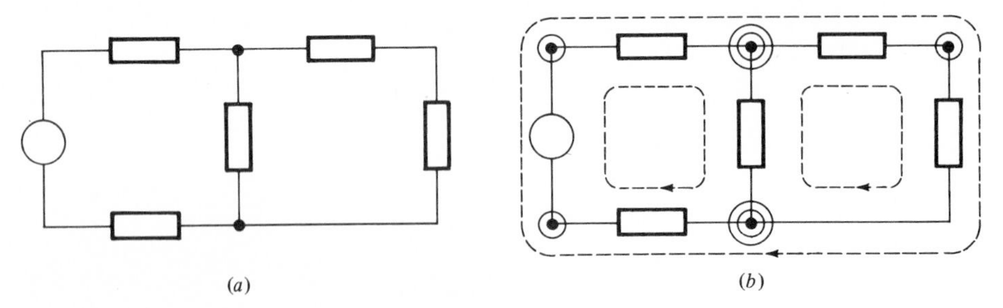

Fig. 5.1-4 Example 5.1-1. (*a*) Circuit; (*b*) nodes and loops.

Solution See Fig. 5.1-4*b*. The nodes are all indicated by circles. Two of them are indicated by double circles. The other three nodes, indicated by single circles would normally be optional; if the elements on either side of one of these nodes are considered parts of a single branch, then the connection is an internal part of the branch rather than a node. It is, in any case, a possible node and must be considered for this example. The three possible loops are indicated by the dotted loops in the figure. ////

5.2 TWO-TERMINAL ELEMENTS

All electric circuit elements have at least two terminals. The behavior of an element is defined by the relationships between the various terminal currents and voltages. The simplest and most often used type of element has two terminals.

The behavior of a two-terminal element may be defined by the relationships between the potential difference, or voltage, across the element and the current through it. This relationship is called the *terminal characteristic* of the device. It is assumed that whatever charge *enters* one terminal of any two-terminal element *leaves* through the other terminal; this means that whatever current flows *in* one terminal of a two-terminal device flows *out* the other terminal. Thus we can refer to the current *through* the device rather than indicating whether it is the current entering or leaving. Figure 5.2-1*a* illustrates a two-terminal device and the current and voltage conventions used to describe its behavior.

The terminal characteristic of a two-terminal device may be presented graphically; the voltage and current are plotted with either one as the independent variable and the other one as the dependent variable. Figure 5.2-1*b* illustrates the terminal characteristic of the element shown in Fig. 5.2-1*a*. The current is shown as the dependent variable in this case. Notice that Fig. 5.2-1*b* is incomplete without Fig. 5-2.1*a*, which indicates the reference directions for the terminal voltage and current.

A second method for specifying the terminal characteristic of a two-terminal device is the analytic technique in which an equation which relates the terminal

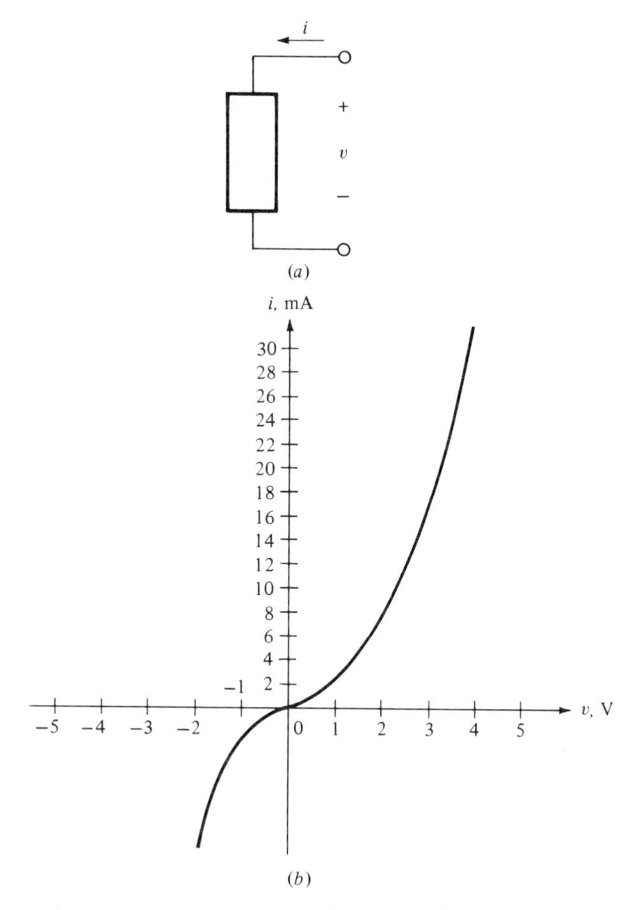

Fig. 5.2-1 Two-terminal elements. (*a*) Two-terminal device with terminal voltage and current indicated; (*b*) terminal characteristic.

current and voltage is given. For the element described in Fig. 5.2-1 this equation is:

$$i = \begin{cases} 2v^2 & v > 0 \\ -2v^2 & v < 0 \end{cases} \tag{5.2-1}$$

where i is in milliamperes (mA) and v is in volts (V). This equation is incomplete without a diagram, such as that of Fig. 5.2-1*a* to indicate the current and voltage reference directions.

These two methods of describing the terminal characteristic of a two-terminal element—the graphical and analytic—are the most generally useful methods. Other methods are available and are useful for certain applications;

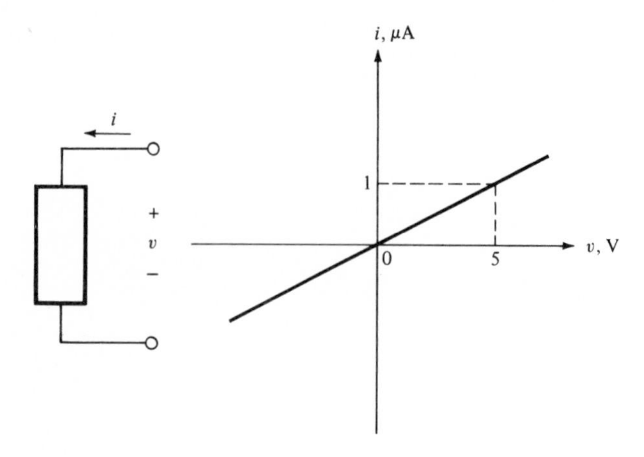

Fig. 5.2-2 Terminal characteristic for Example 5.2-1.

a set of tabulated values, for example, might be very useful for computer analysis of a circuit containing the element. Since our principle concern will be analysis of networks without the use of computers, we shall rely most heavily on the analytic and graphical representations of the terminal characteristic.

Example 5.2-1 Find the analytic representation for the terminal characteristic whose graphical representation is given in Fig. 5.2-2.

Solution Refer to Sec. 1.2-1. Since the graph is that of a straight line its equation is of the form

$$i = mv + b \tag{5.2-2}$$

When $v = 0$, $i = 0$, therefore $b = 0$. When $v = 5$ V, $i = 1\ \mu$A, therefore $m = 0.2\ \mu$A/V. Hence

$$i = 0.2v \tag{5.2-3}$$

where v is in volts and i is in microamperes. ////

5.3 POWER

The formula for the power delivered to a two-terminal device was derived in Chap. 3. Using the reference directions given in Fig. 5.2-1*a*, the power is

$$p = vi \tag{5.3-1}$$

A negative value for p indicates that the element is supplying power to the remainder of the circuit; a positive value indicates that the remainder of the circuit is delivering power to the element. In any circuit some elements will supply power while others will absorb power. The total power supplied must equal the total power absorbed at every instant.

Certain devices, such as batteries, can convert chemical energy into electric energy and thereby supply electric power. Other devices, such as electric heaters, convert electric power into heat which is dissipated into the atmosphere; these devices absorb electric power. Other kinds of devices can absorb and store energy which is later delivered back to the remainder of the circuit.

Example 5.3-1 For the circuit of Fig. 5.3-1 find (*a*) the power, P_x, delivered to element X and (*b*) the voltage, V_x, across element X.

Solution

a Since the current leaving each element is the same as the current entering it, the current through each element is the same, 2 mA flowing around the circuit in a clockwise direction. The power delivered to element A is

$$P_A = (-8\text{ V}) \times (2\text{ mA}) = -16 \times 10^{-3}\text{ W} \tag{5.3-2}$$

where the minus sign is due to the reversal of the voltage polarity relative to the current direction when compared with the reference directions specified for Eq. (5.3-1). The power delivered to element B is

$$P_B = +(5\text{ V}) \times (2\text{ mA}) = +10 \times 10^{-3}\text{ W} \tag{5.3-3}$$

The net power delivered to all the elements in the circuit must be zero, i.e., power supplied must equal power absorbed; therefore

$$P_A + P_B + P_X = 0 \tag{5.3-4}$$

$$-16 \times 10^{-3}\text{ W} + 10 \times 10^{-3}\text{ W} + P_X = 0 \tag{5.3-5}$$

$$P_X = 6 \times 10^{-3}\text{ W} = 6\text{ mW} \tag{5.3-6}$$

b The power delivered to element X is given by the equation

$$P_X = (2\text{ mA}) \times V_x \tag{5.3-7}$$

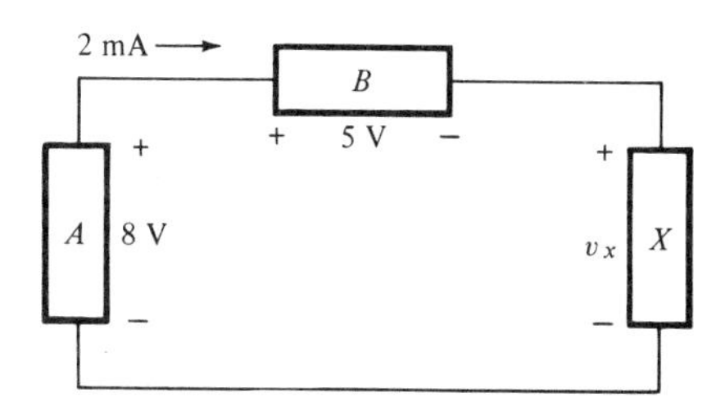

Fig. 5.3-1 Circuit for Example 5.3-1.

Substituting $P_X = 6$ mW we find that

$$V_x = 3 \text{ V} \tag{5.3-8}$$

////

5.4 INDEPENDENT SOURCES

Thus far, in this chapter, we have discussed general properties of two-terminal elements. In this section we will describe two particular kinds of elements: the independent current source and the independent voltage source.

An electrical source, either current or voltage, is classified as an *active* device because it can supply electric energy to the circuit to which it is connected; it is a *source* of electric energy. On the other hand, a *passive* device is one which cannot supply electric energy other than that which has previously been delivered to it; that is, it may be able to return electric energy which it has previously absorbed but it cannot supply any *additional* electric energy as an active element can.

The term "source" in this context can lead to confusion. These sources are not sources of energy; they simply convert energy from some nonelectric form to the electric form. A battery, for example, converts chemical energy into electric energy, a generator converts mechanical energy into electric energy, and solar cells convert light energy into electric energy. Thus the electric sources are only sources when we consider *only* the electric behavior. The overall behavior of the device does satisfy the principle of conservation of energy, i.e., no new energy is created. Despite the fact that these devices are actually energy converters, they are nevertheless called sources because they are capable of supplying electric energy, and it is only this function that is of concern in electric network analysis and design.

CURRENT SOURCE

The *independent current source*[1] is a two-terminal device whose terminal current is some specified function of time which is independent of the circuitry connected to it. The function is determined by the source: a direct current (dc) source provides a constant current; an alternating current (ac) source provides a sinusoidally varying current; various other functions are available as well. The terminal voltage for the current source can have any value; it is determined by the circuitry connected to the terminals of the current source.

The circuit symbol for the current source is shown in Fig. 5.4-1*a* and the graphical representation of its terminal characteristic is shown in Fig. 5.4-1*b*. Note that the reference directions for $v(t)$ and $i(t)$ have been chosen the same as those for the resistance element. Note, also, that the characteristic is shown for a typical instant of time and that it moves up and down as $i_s(t)$ changes.

[1] The term "current source" is synonymous with "independent current source" unless otherwise indicated.

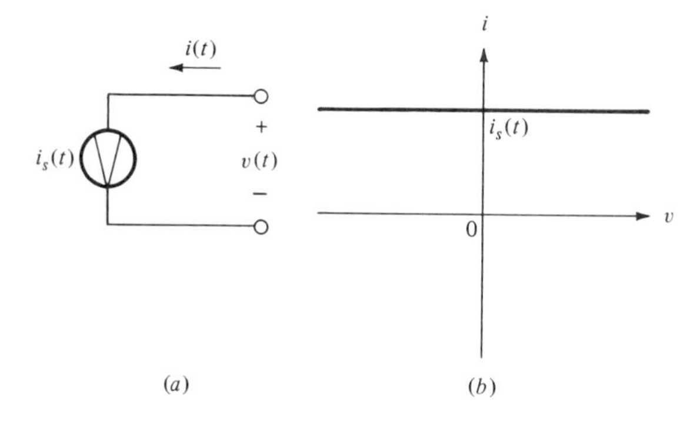

Fig. 5.4-1 Independent current source. (*a*) Circuit symbol; (*b*) terminal characteristic.

VOLTAGE SOURCE

The *independent voltage source* is a two-terminal device whose terminal voltage is some specified function of time which is independent of the circuitry connected to it. There are dc and ac voltage sources and a variety of other waveforms including square waves, pulses, ramps, sawtooths, and triangular waves are available. The terminal current of the voltage source[1] can have any value; it is determined by the circuitry connected to the terminals of the voltage source.

The circuit symbol for the voltage source is shown in Fig. 5.4-2*a*; the graphical representation of its terminal characteristic is shown in Fig. 5.4-2*b*. Note that the characteristic is shown for a typical instant of time and that it moves back and forth as $v_s(t)$ changes.

The dc current source is analogous to a constant volume pump. It maintains a constant current independent of the voltage across it just as the constant

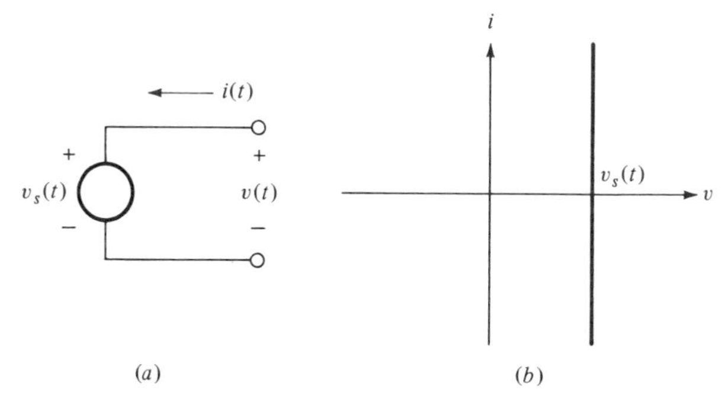

Fig. 5.4-2 Independent voltage source. (*a*) Circuit symbol; (*b*) terminal characteristic.

[1] The shortened term "voltage source" is synonymous with "independent voltage source" unless otherwise indicated.

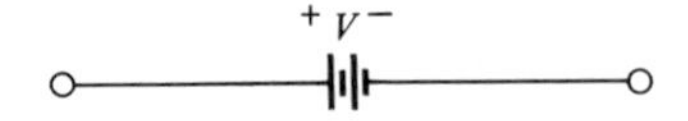

Fig. 5.4-3 Dc voltage source.

volume pump keeps a constant flow of fluid (constant volume per unit time) independent of the pressure against which it is pumping. A dc voltage source, on the other hand, is analogous to a constant pressure pump which maintains a fixed pressure head. A battery is a physical device which approximates a dc voltage source. Because of its frequent use there is a special circuit symbol which may be used for a dc source; it is shown in Fig. 5.4-3.

5.5 IDEAL ELEMENTS

The current and voltage sources just described are both ideal elements. An *ideal element* is one which represents a mathematical relationship rather than a physical device. The ideal sources we have defined approximate the electric behavior of these physical devices. A physical device, if it is to be described with great accuracy, generally requires very complex mathematical equations for its description. The analysis of a network consisting of such devices would become extremely difficult and it would be almost impossible to gain insight into the behavior of the network.

Ideal elements, on the other hand, are described by rather simple mathematical relationships which are easily grasped. When we look at a network diagram consisting of these elements we can, after learning the concepts of network analysis, comprehend the way in which the network functions with relative ease.

The obvious question that arises is, "what good is it to analyze a network of ideal elements when we are building networks of real physical elements?" To answer this question we must introduce the concept of a model. The ideal elements can be used to model or approximate the behavior of physical devices. Very often an ideal element is defined in such a way that it approximates the function of a given physical device. The ideal element represents the major features of the operation of the physical device; for this reason it may provide only a gross approximation to the behavior of the physical element. This approximation will be quite adequate for many applications. More accurate models can be used if greater accuracy is needed but they will generally involve a number of ideal elements to model a given physical device. The greater complexity of the more accurate model will make the analysis more difficult; in general the analyst must reach a balance between the accuracy of his analysis and the difficulty of performing the analysis, all the while taking into account the required accuracy and cost.

In our discussion of models, we should point out that any representation of a device is a model. An equation, or set of equations, which describes a system is a model of the system. Likewise, the graph of a device's terminal characteristic is also a model of the device.

5.6 THE RESISTANCE ELEMENT

The resistance element discussed in Chap. 4 is probably the most commonly found element in electric and electronic networks. It is a passive element; it absorbs electrical energy which it converts into heat which in turn is dissipated into the surroundings. Figure 4.5-1 shows a common type of resistor. The behavior of these resistors is approximated, generally quite well, by the *ideal resistance element.*

A *resistance* is a two-terminal device whose terminal voltage is directly proportional to its terminal current. The constant of proportionality is the resistance of the element; it is generally symbolized by the letter R and is measured in ohms. The unit of measurement, ohms, is abbreviated using the uppercase Greek letter omega (Ω). The resistance circuit symbol is shown in Fig. 5.6-1*a* along with the reference directions for the terminal voltage and current.

The relationship between the terminal voltage and terminal current, Ohm's law (named after the French scientist who first stated it), is repeated here:

$$v(t) = R\,i(t) \tag{5.6-1}$$

By simple algebraic manipulation, this equation can be written in either of the following two forms as well:

$$i(t) = \frac{v(t)}{R} \tag{5.6-2}$$

or

$$R = \frac{v(t)}{i(t)} \tag{5.6-3}$$

The *ohm* is defined as that resistance across which a current of one ampere produces a potential difference of one volt. Symbolically,

$$1\ \Omega = 1\ \frac{\text{V}}{\text{A}} \tag{5.6-4}$$

LINEAR NETWORKS

The terminal characteristic of the resistance element is illustrated in Fig. 5.6-1*b*. Each of the circuit elements we have studied thus far, the current source, the voltage source, and the resistance is said to be a *linear* circuit element because it is described by a linear volt-ampere (*vi*) relationship; i.e., the graph

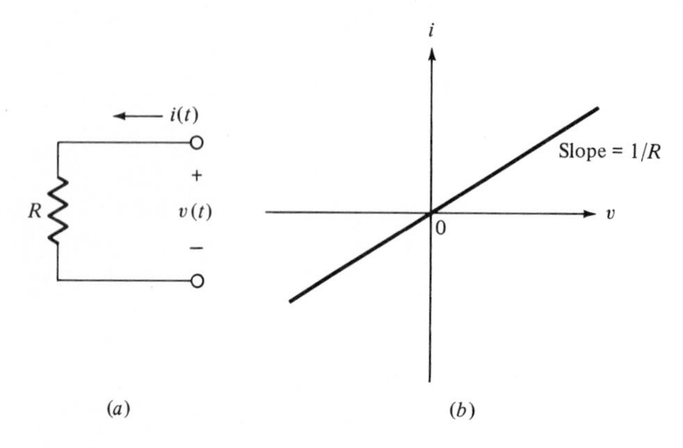

Fig. 5.6-1 The resistance element. (*a*) Circuit symbol; (*b*) terminal characteristic.

of the terminal characteristic is a straight line (see Figs. 5.4-1*b*, 5.4-2*b*, 5.6-1*b*). A circuit which consists of only linear circuit elements is called a *linear circuit* or network. The analysis of a linear network is generally simpler than the analysis of a nonlinear network (one which contains at least one element which is not linear) because all the voltages and/or currents in a linear network are related by linear equations.

Example 5.6-1 The voltage, $v(t)$, across the resistance of Fig. 5.6-1*a* is found to be 7.0 V while the current, $i(t)$, through it, is measured and found to be 2.0 mA.

a Find its resistance.
b What will the current be if the voltage is changed to 5,000 V?

Solution

a Use Ohm's law:

$$R = \frac{v}{i} = \frac{7.0 \text{ V}}{2.0 \text{ mA}} = \frac{7.0 \text{ V}}{2.0 \times 10^{-3} \text{ A}} = 3.5 \times 10^3 \ \Omega$$

$$= 3.5 \text{ k}\Omega$$

b Using the other form of Ohm's law,

$$i = \frac{v}{R} = \frac{5000 \text{ V}}{3.5 \text{ k}\Omega} = \frac{5 \times 10^3 \text{ V}}{3.5 \times 10^3 \ \Omega} = 1.43 \text{ A}$$

////

Example 5.6-2 A current of 8.5 A flows through a 15-kΩ resistance.

a Find the voltage across the resistance.
b Describe the polarity of this voltage.
c What is the power dissipated by the resistance?

Solution

a Use Ohm's law:

$$V = RI = 15\ \text{k}\Omega \times 8.5\ \text{A} = 127.5\ \text{kV} \approx 128\ \text{kV}$$

b It is a voltage drop in the direction of the current.

c Use the equation for power:

$$P = VI = 127.5\ \text{kV} \times 8.5\ \text{A} \approx 1080\ \text{kW}$$ ////

BRANCH EQUATIONS

Ohm's law defines the terminal characteristic of a resistance. The terminal characteristics of the sources and resistances in a network provide a set of equations which are used in the analysis of the network. If the network contains b branches there are $2b$ unknown variables: a branch voltage and a branch current for each branch. Therefore $2b$ equations are required to completely analyze the network. The terminal characteristic equations provide b of these, one for each branch. These b equations are called the *branch equations*. The remaining b equations are obtained using two general network laws called Kirchhoff's laws which are considered in the next two sections.

5.7 KIRCHHOFF'S CURRENT LAW

Kirchhoff's current law (KCL) relates all of the currents entering or leaving a node. Since charge cannot be stored in a node and since current is a flow of charge, it is necessary that the sum of the currents flowing into a node must equal the sum of the currents leaving (flowing out of) the node. Alternatively, recognizing that a current flowing out of a node can be represented by a negative current of the same magnitude flowing into the node, we may phrase the KCL as follows:

The algebraic sum of the currents entering a node is zero.

This is symbolically stated as

$$\sum_{node} i = 0 \tag{5.7-1}$$

where the uppercase Greek letter sigma ($\sum$) stands for "summation of." The reference directions for the currents in this formula may all be chosen entering the node or they may all be chosen leaving the node, so long as all are chosen in the same relative direction.

Example 5.7-1 Find the current i leaving the node shown in Fig. 5.7-1.
Solution Choose (arbitrarily) currents leaving the node as the positive reference direction. Then we have

$$-5\ \text{A} + 3\ \text{A} + i = 0$$

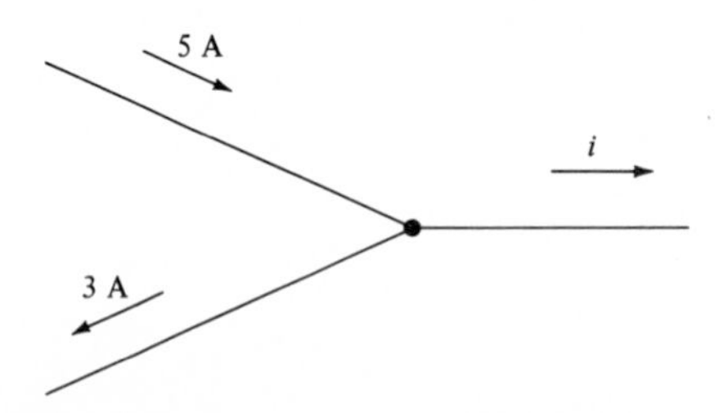

Fig. 5.7-1 Node for Example 5.7-1.

which results in

$i = 2\text{ A}$

Had we chosen the positive reference direction to be currents entering the node we would have

$5\text{ A} - 3\text{ A} - i = 0$

which results in the same solution

$i = 2\text{ A}$ ////

A very useful generalization of Kirchhoff's current law results from the fact that no electric circuit element stores a net charge. As a result all the charge that enters any device, or group of devices, must leave the device or group of devices. Since current is a flow of charge, the current entering any device or group of devices must equal the current leaving the device or devices. Once again making use of the relationship between current direction and algebraic sign we can state the generalized Kirchhoff current law as follows:

The algebraic sum of the currents entering any closed surface is equal to zero.

Applying this to the system shown in Fig. 5.7-2, for example, results in

$i_1 + i_2 - i_3 = 0$

This statement may also be rephrased in terms of the currents leaving the closed surface; either way it is symbolically represented by

$$\sum_{\text{closed surface}} i = 0 \qquad (5.7\text{-}2)$$

This generalized form of Kirchhoff's current law includes the original form as a special case obtained by considering a closed surface which encloses the node.

Example 5.7-2 A transistor is a three-terminal electronic device used in amplifier circuits. The circuit diagram of a transistor shown in Fig. 5.7-3 indicates the base (B), collector (C), and emitter (E) leads. The collector

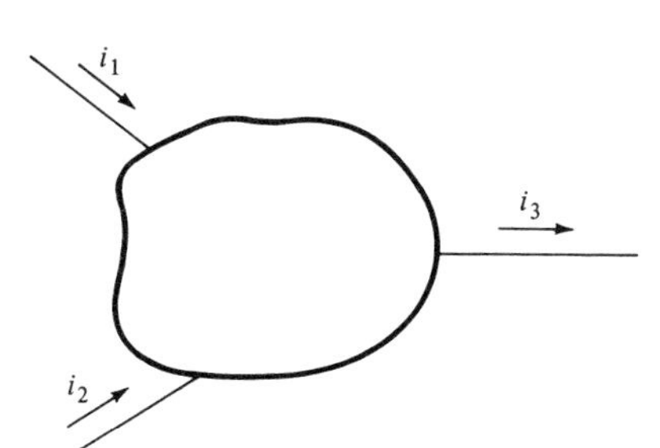

Fig. 5.7-2 Currents entering and leaving a closed surface.

current, i_C, is slightly less than, but proportional to, the emitter current, i_E. The ratio of the two is

$$\alpha = \frac{i_C}{i_E} \tag{5.7-3}$$

Normally α is in the range from 0.95 to 0.99. The ratio of collector to base current is defined as

$$\beta = \frac{i_C}{i_B} \tag{5.7-4}$$

a Find a formula for β in terms of α.
b What is β if α = 0.95?, 0.99?

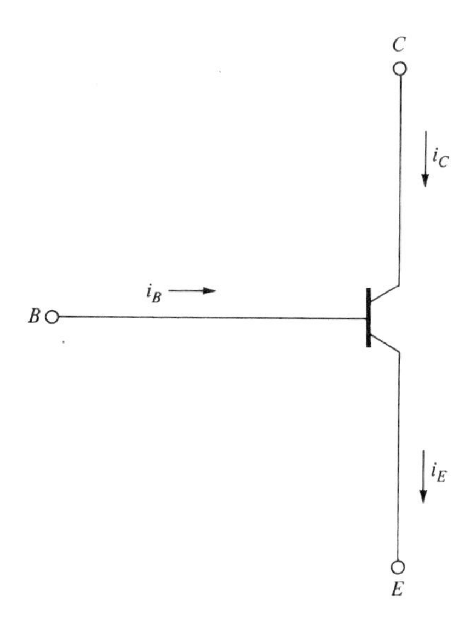

Fig. 5.7-3 Transistor schematic symbol.

Solution

a Consider the currents entering the closed surface enclosing the transistor: By Kirchhoff's current law (generalized form),

$$i_C + i_B - i_E = 0 \tag{5.7-5}$$

From Eq. (5.7-3)

$$i_E = \frac{i_C}{\alpha} \tag{5.7-6}$$

Substituting Eq. (5.7-6) into Eq. (5.7-5) we have

$$i_C + i_B - \frac{i_C}{\alpha} = 0$$

$$i_B = i_C\left(\frac{1}{\alpha} - 1\right) = i_C\left(\frac{1 - \alpha}{\alpha}\right)$$

$$\beta = \frac{i_C}{i_B} = \frac{\alpha}{1 - \alpha} \tag{5.7-7}$$

b For $\alpha = 0.95$

$$\beta = \frac{0.95}{1 - 0.95} = \frac{0.95}{0.05} = \frac{95}{5} = 19$$

and for $\alpha = 0.99$

$$\beta = \frac{0.99}{1 - 0.99} = \frac{0.99}{0.01} = 99$$

////

NUMBER OF INDEPENDENT CURRENT LAW EQUATIONS

In analyzing an electric network, it is possible to write a Kirchhoff's current law equation for each node. It will always turn out that these equations are not all independent; in fact, if there are n nodes there will be $n - 1$ independent nodal equations. These equations may be written for any $n - 1$ nodes. The equation for the remaining node will express the same information as the other $n - 1$ equations do and should not be used in the solution of a network problem.

5.8 KIRCHHOFF'S VOLTAGE LAW

Thus far we have found that there are, in a network with b branches and n nodes, $2b$ unknown branch voltages and currents (b of each). In order to find these unknowns, we have available b terminal characteristic relations for the branches, and $n - 1$ independent Kirchhoff current law equations. The remaining $b - (n - 1)$ equations are obtained using Kirchhoff's voltage law.

Kirchhoff's voltage law (KVL) is based on the fact that the potential energy associated with a fixed charge depends only on its position. Thus, if a charge is moved from a given point through any path back to the same point its potential energy will be unchanged. Kirchhoff's voltage law may be stated as follows:

The sum of the voltage rises around any closed loop equals the sum of the voltage drops around the closed loop.

Alternatively, since a voltage rise is a negative voltage drop, Kirchhoff's voltage law may be stated as:

The algebraic sum of the voltage drops around any closed loop is zero.

Symbolically, this is represented by

$$\sum_{\text{closed loop}} v = 0 \qquad (5.8\text{-}1)$$

Example 5.8-1 Find the voltage V_x in the circuit of Fig. 5.8-1.

Solution Equate the sum of the voltage drops around the loop, in a clockwise direction, to zero.

$$-40 \text{ V} + 25 \text{ V} + V_x = 0$$

This results in

$$V_x = 15 \text{ V}$$

////

Example 5.8-2 The circuit of Fig. 5.8-2 contains a current source as well as a voltage source.

a Find V_C.
b Find R.

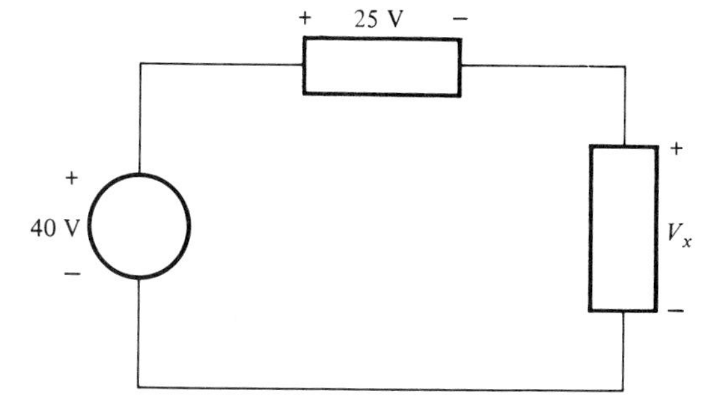

Fig. 5.8-1 Circuit for Example 5.8-1.

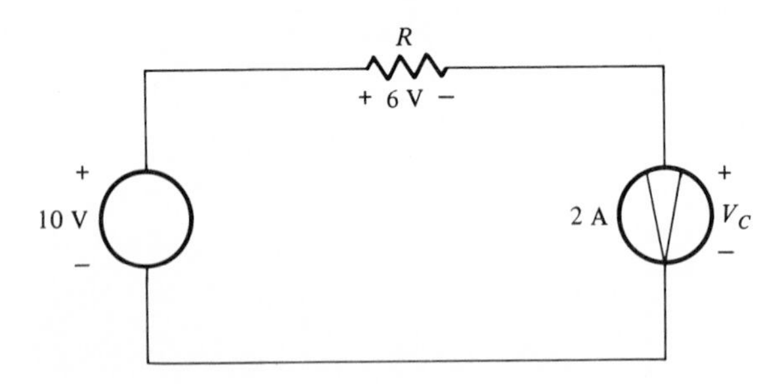

Fig. 5.8-2 Circuit for Example 5.8-2.

Solution

a Using Kirchhoff's voltage law, equate the sum of the voltage drops going clockwise around the loop to zero:

$$-10\text{ V} + 6\text{ V} + V_C = 0$$

Solving,

$$V_C = 4\text{ V}$$

b The current through all the elements is the same and must be equal to the current source value of 2 A. Use Ohm's law to find R:

$$R = \frac{V}{I} = \frac{6\text{ V}}{2\text{ A}} = 3\ \Omega$$

////

Example 5.8-3 For the circuit of Fig. 5.8-3 find I, V_1, and V_2.

Solution By Ohm's law,

$$V_1 = 20 \times 10^3\, I \tag{5.8-2}$$

and

$$V_2 = 30 \times 10^3\, I \tag{5.8-3}$$

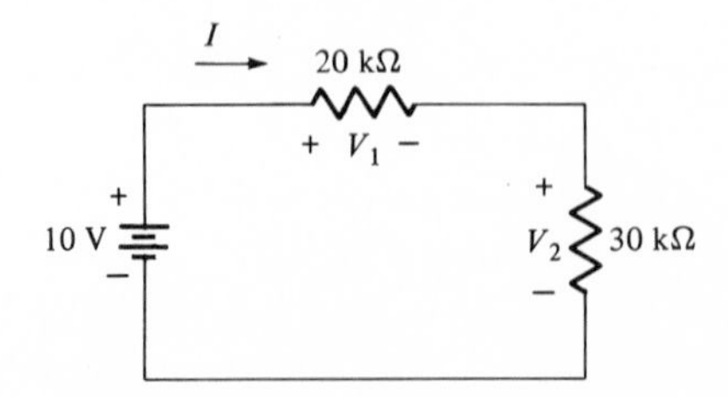

Fig. 5.8-3 Circuit for Example 5.8-3.

By Kirchhoff's voltage law,

$$-10 + V_1 + V_2 = 0 \tag{5.8-4}$$

Substitute Eqs. (5.8-2) and (5.8-3) into Eq. (5.8-4):

$$-10 + 20 \times 10^3 I + 30 \times 10^3 I = 0$$

which results in

$$I = \frac{10}{50 \times 10^3} = 0.2 \text{ mA}$$

Substitute this result back in Eqs. (5.8-2) and (5.8-3):

$$V_1 = 20 \times 10^3 \,\Omega \times 0.2 \text{ mA} = 4 \text{ V}$$

$$V_2 = 30 \times 10^3 \,\Omega \times 0.2 \text{ mA} = 6 \text{ V}$$

Check using KVL by substituting these values into Eq. (5.8-4)

$$-10 \text{ V} + 4 \text{ V} + 6 \text{ V} = 0$$ ////

SELECTION OF INDEPENDENT LOOPS

In writing the KCL equations it is necessary to select the loops such that each loop contains at least one branch that is not part of any other loop; otherwise the resulting equations will not be an independent set.

Example 5.8-4 Find a set of nodes and loops which may be used to write a set of independent equations to analyze the circuit of Fig. 5.8-4.

Solution There are five branches and thus ten equations are required to solve for each branch current and voltage. Five equations are obtained from the terminal characteristics of the branches. There are three nodes: a, b, and

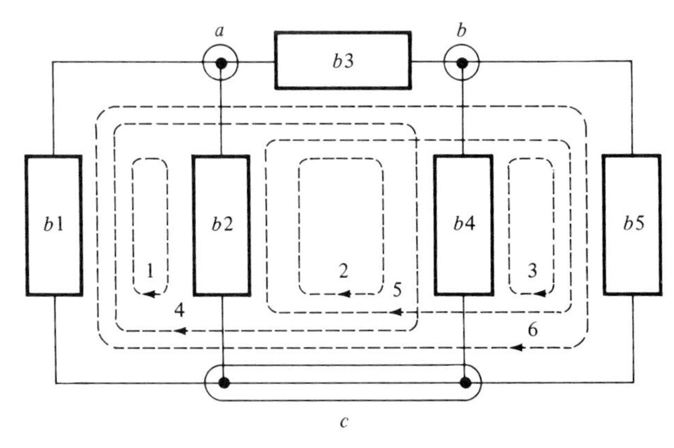

Fig. 5.8-4 Circuit for Example 5.8-4.

c. Independent Kirchhoff's current law equations can be written for any two of these. The remaining three equations must be written using the Kirchhoff voltage law around three of the six loops shown in the figure. The loops used must, however, be an independent set which means that every loop must contain at least one branch which is not part of any other loop. Thus loops 1, 2, and 3 constitute an independent set while loops 1, 2, and 4 are not an independent set. Some other independent sets are 1, 4, and 6 or 2, 5, and 6. ////

Example 5.8-5 Find the branch currents I_1, I_2, and I_3 in the circuit of Fig. 5.8-5.

Solution Use Kirchhoff's current law at node a:

$$I_1 - I_2 - I_3 = 0 \tag{5.8-5}$$

Write Kirchhoff voltage law equations for loops 1 and 2 using Ohm's law to find the branch voltages as shown in Fig. 5.8-5 (note that volts, milliamperes, and kilohms are the units used in this example):

$$-120 + 2I_1 + 6I_2 = 0 \tag{5.8-6}$$

$$-6I_2 + 4I_3 + 8I_3 = 0 \tag{5.8-7}$$

Use Eq. (5.8-5) to solve for I_1 and substitute into Eqs. (5.8-6) and (5.8-7):

$$2(I_2 + I_3) + 6I_2 = 120 \tag{5.8-8}$$

$$12\,I_3 = 6I_2 \tag{5.8-9}$$

Solve for I_2 from Eq. (5.8-9), substitute into Eq. (5.8-8) and collect terms:

$$18\,I_3 = 120 \tag{5.8-10}$$

$$I_3 = 6.67 \text{ mA.}$$

Substitute this value into Eq. (5.8-9) and solve:

$$I_2 = 13.33 \text{ mA}$$

These values can be substituted into Eq. (5.8-5) resulting in:

$$I_1 = 20 \text{ mA}$$

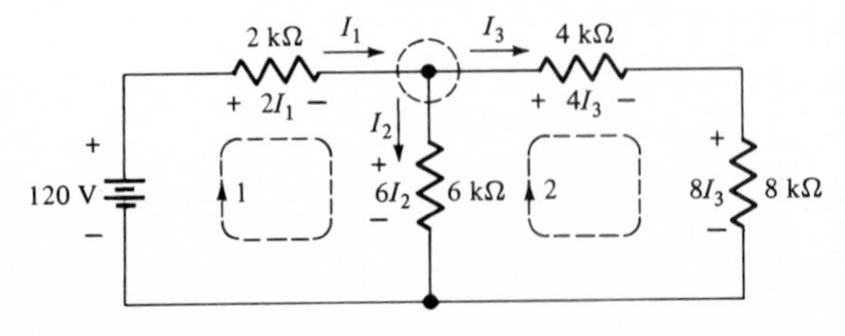

Fig. 5.8-5 Circuit for Example 5.8-5.

These results may be checked by substitution into the KVL equations (5.8-6) and (5.8-7):

$$-120 + 2 \times 20 + 6 \times 13.33 = 0$$

$$-120 + 40 + 80 = 0$$

and

$$-6 \times 13.33 + 4 \times 6.67 + 8 \times 6.67 = 0$$

$$-80 + 26.67 + 53.33 = 0$$

////

SUMMARY

1 *vi characteristics*

Current source:

$$i(t) = i_s(t)$$

Voltage source:

$$v(t) = v_s(t)$$

Resistance:

$$v(t) = Ri(t)$$

2 In a network of b branches and n nodes there are $2b$ unknown branch voltages and currents. There are

b branch equations obtained from the terminal characteristics

$n - 1$ node equations obtained using Kirchhoff's current law

$b - (n - 1)$ loop equations obtained using Kirchhoff's voltage law

3 Kirchhoff's current law:

$$\sum_{\text{Closed surface}} i = 0$$

4 Kirchhoff's voltage law:

$$\sum_{\text{Closed loop}} v = 0$$

PROBLEMS

5.1-1 In the circuit of Fig. P5.1-1:

a Encircle all the nodes. How many are there?
b How many branches are there?
c Trace out all the possible loops. How many are there?

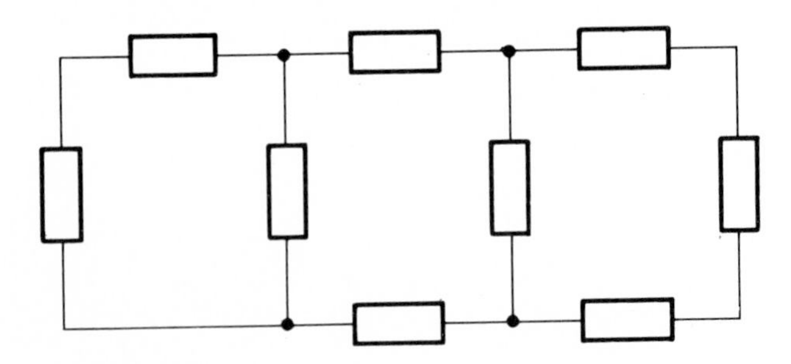

Fig. P5.1-1.

5.1-2 Repeat Prob. 5.1-1 for the circuits of Fig. P5.1-2.

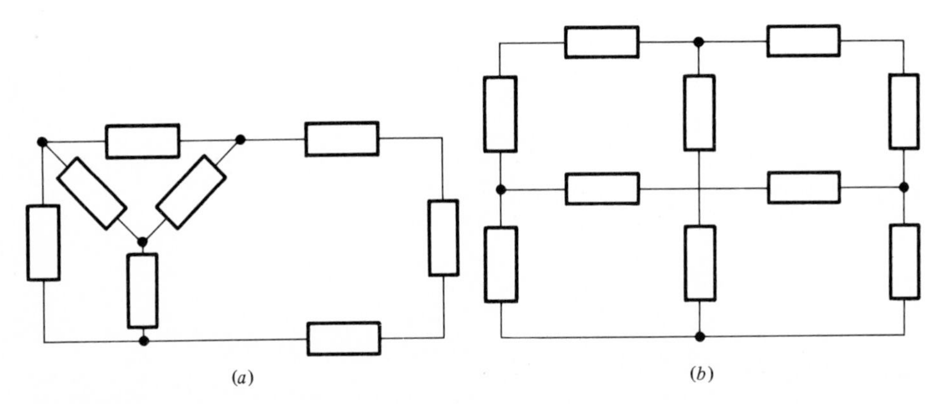

Fig. P5.1-2.

5.2-1 Plot the graphical representation of the terminal characteristics of the two-terminal circuit elements whose analytic representations are given below. All voltages are in volts and currents are in milliamperes.

a $i = 0.1\,v - 2$
b $i = 10\,v^2 u(v)$

5.2-2 Find analytic representations of the terminal characteristics whose graphical representations are shown in Fig. P5.2-2.

5.2-3 The values of terminal voltage and current for a two terminal circuit element are tabulated below.

v, V	−10	−9	−8	−7	−6	−5	−4	−3	−2	−1	0
i, μA	−1000	−810	−640	−490	−360	−250	−160	−90	−40	−10	0

v, V	1	2	3	4	5	6	7	8	9	10
i, μA	10	40	90	160	250	360	490	640	810	1000

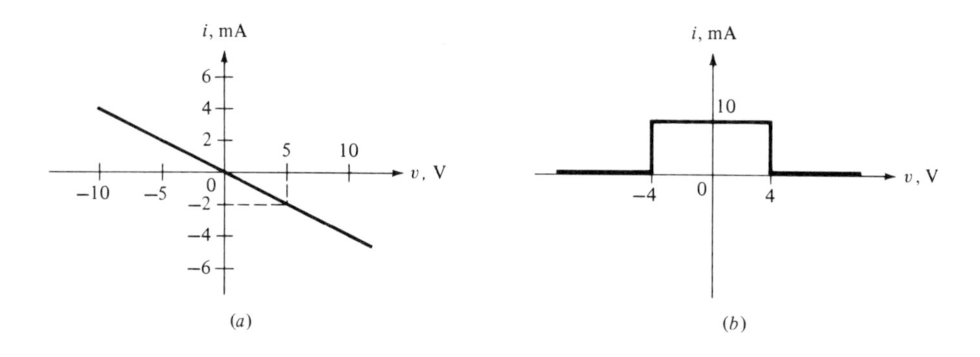

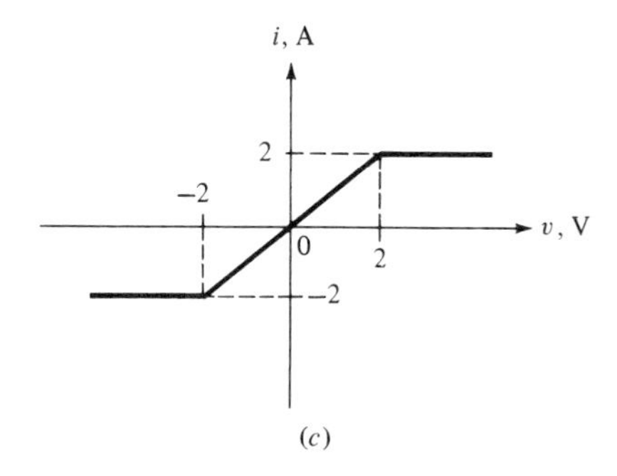

Fig. P5.2-2.

a Draw the graph of the terminal characteristic.
b Find an analytic representation for this terminal characteristic.
c Find the current that would flow through this element if the voltage were 4.5 V.
d What would the voltage be if the current were $-750\ \mu$A?

5.3-1 For the circuit element shown in Fig. P5.3-1

a Does it absorb or deliver power?
b How much?

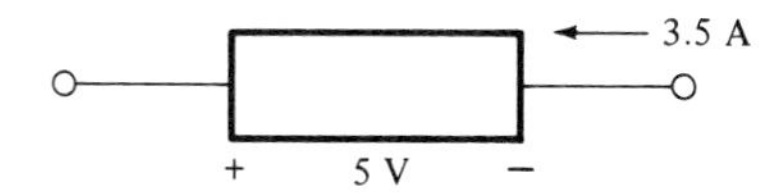

Fig. P5.3-1.

5.3-2 For the circuit of Fig. P5.3-2 the source, S, has a voltage of 20 V. A current of 5 A flows through all the elements.

a How much power does the source supply?
b If element A absorbs 20 W of power what is the voltage V_A across it?
c How much power does element B absorb?
d What is the voltage V_B?

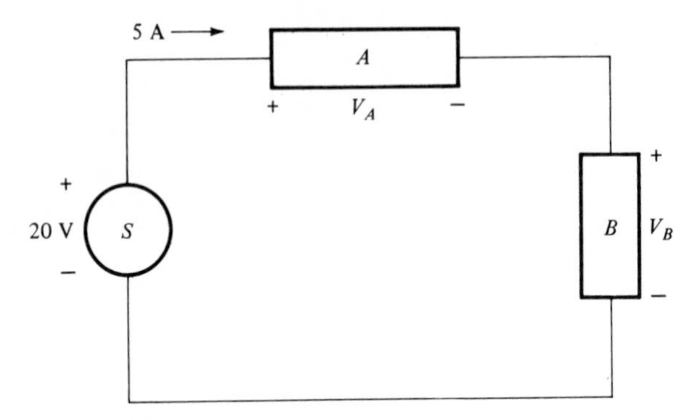

Fig. P5.3-2.

5.3-3 A solar cell is connected to supply electric power to a device as shown in Fig. P5.3-3. The light is of constant brightness and supplies 100 J of energy every hour.

a If $V = 4$ V, what is the maximum value that I can have?
b The value of I is measured and found to be 5 mA. How might this be explained?

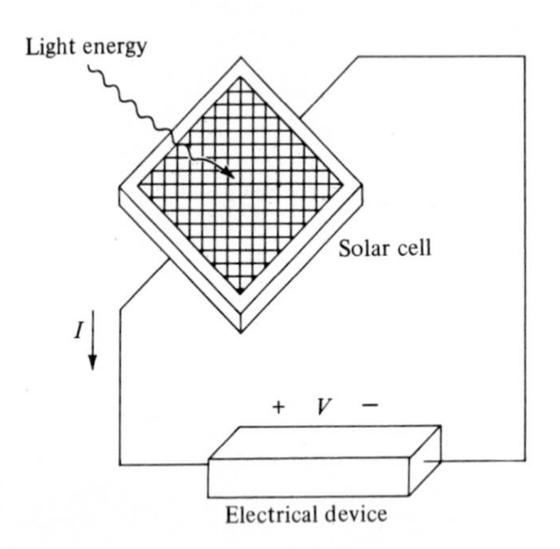

Fig. P5.3-3.

5.3-4 The density of water is 1 g/cm³ and there are 29.57 cm³ per fluid ounce. One calorie of heat is required to raise the temperature of 1 g of water 1°C and 1 cal is equal to 4.186 J. What current would an immersion heater which is to boil an 8-oz cup of water in 2 min have to draw if it is plugged into a 120 V supply?

5.5-1 The characteristic of the device shown in Fig. P5.5-1*a* is shown in Fig. P5.5-1*b*.

a Find a circuit model which approximates the behavior of this device in the vicinity of the origin.
b Over what range of voltages and/or currents is the model a good one?

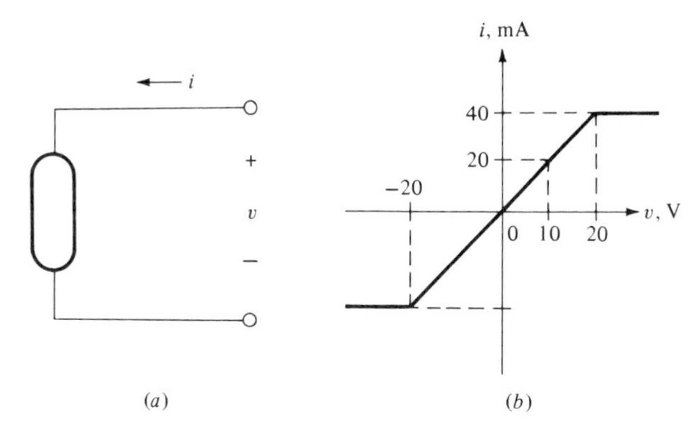

Fig. P5.5-1 (*a*) Device; (*b*) terminal characteristic.

5.6-1 If the voltage across a 5-kΩ resistance is 24 sin (1,000 t) what is the current through it?

5.6-2 A dc voltage of 10 V applied across a resistance results in a current of 400 μA. What is the value of the resistance?

5.6-3 A unit step of voltage at $t = 0$ applied to an element results in an exponentially decaying current

$$i = 5\,e^{-4t}\text{ mA}.$$

Can this element be a resistance? If yes, what is its value?

5.6-4 A current of 20 mA is applied to a 27-kΩ resistance as shown in Fig. P5.6-4.

a Find the voltage V.
b Find the power delivered to the resistance.

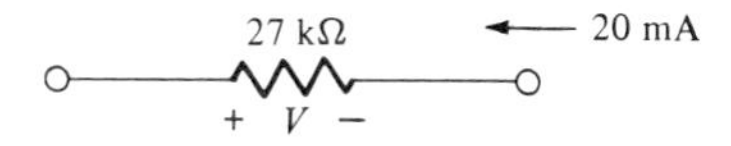

Fig. P5.6-4.

5.6-5 The power delivered to a 100-Ω resistance is 4 W. Find the current flowing through the resistance.

5.6-6 What is the voltage across a 20-Ω resistance which is dissipating 8 mW of power?

5.7-1 For the circuit of Fig. P5.7-1 find the current I.

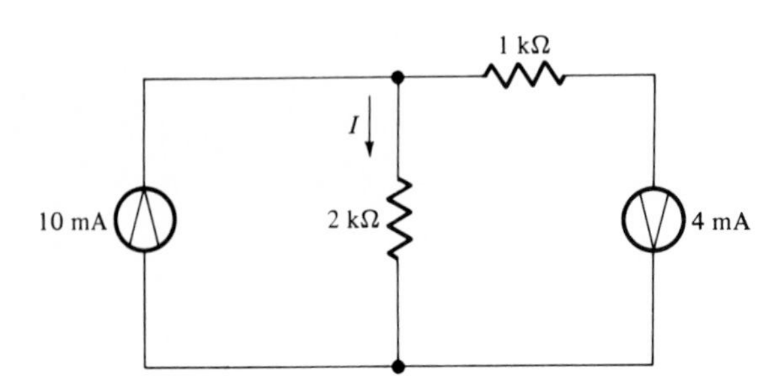

Fig. P5.7-1.

5.7-2 For the circuit of Fig. P5.7-2 find

a I_R using Ohm's law
b I_B using Kirchhoff's current law
c The power supplied by the dc voltage source

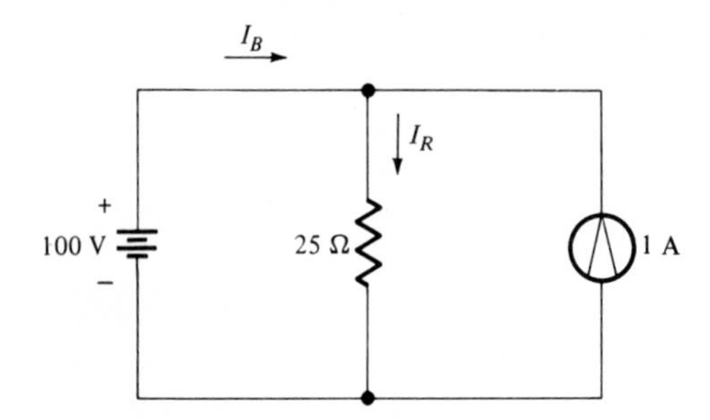

Fig. P5.7-2.

5.7-3 For the circuit of Fig. P5.7-3 find

a I_1 using Ohm's law
b I_2 by applying Kirchhoff's current law to a closed surface enclosing the right end of the circuit
c I_3 using Kirchhoff's current law
d V_3 using Ohm's law.

5.7-4 For the circuit of Fig. P5.7-4 find the

a Current I
b Voltage V

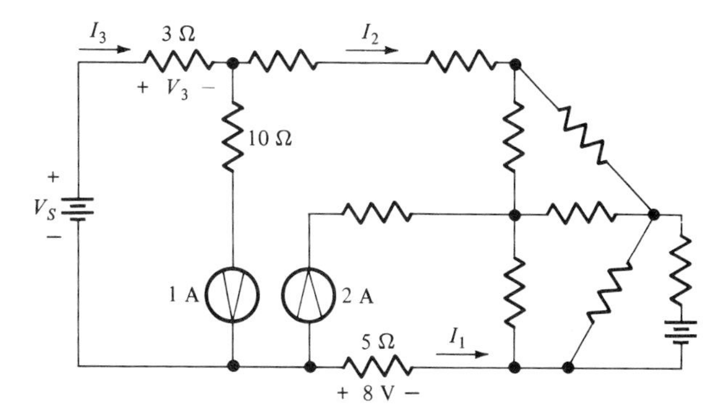

Fig. P5.7-3.

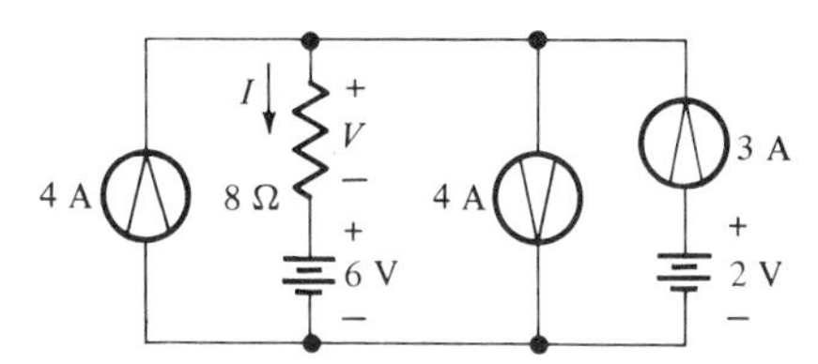

Fig. P5.7-4.

5.7-5 How many independent KCL equations can be written for the circuit of Fig. P5.7-5?

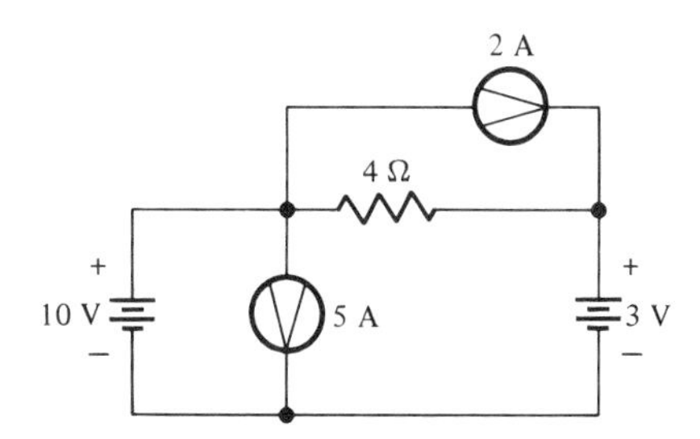

Fig. P5.7-5.

5.8-1 Find V_x for the circuit of Fig. P5.8-1.

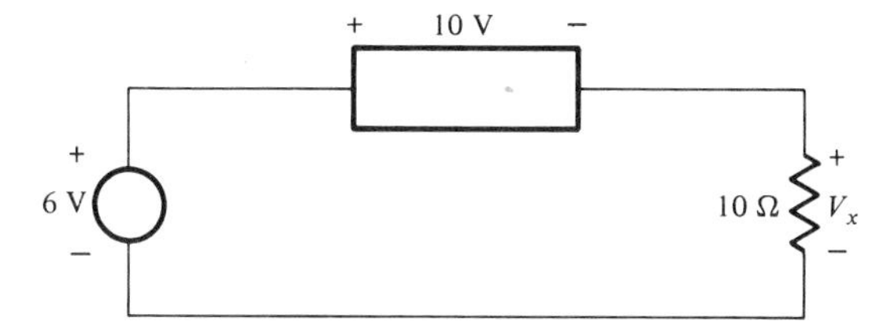

Fig. P5.8-1.

5.8-2 For the circuit of Fig. P5.8-2

a Find V_R
b Find V_C
c Find the power supplied by the voltage source
d Find the power supplied by the current source
e Find the kind of circuit this is, i.e.; what its function is

Hint: your automobile battery may require servicing with this circuit.

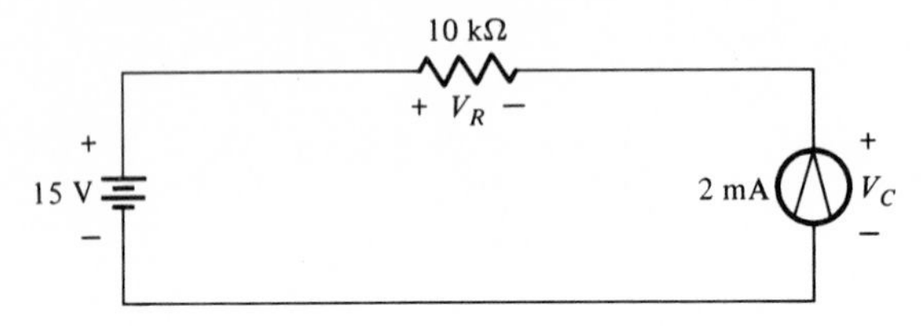

Fig. P5.8-2.

5.8-3 For the circuit of Fig. P5.8-3 find

a I
b V_1
c V_2

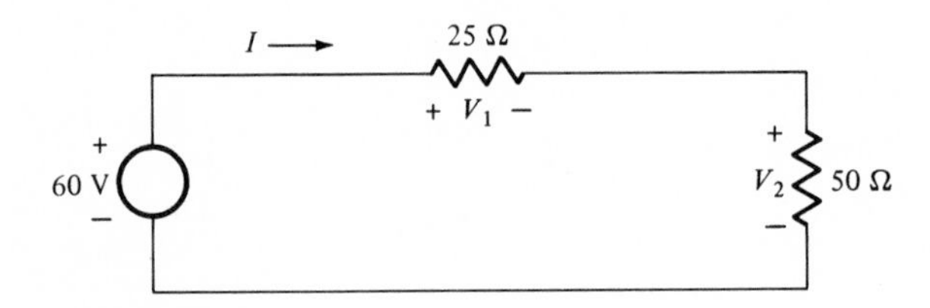

Fig. P5.8-3.

5.8-4 For the circuit of Fig. P5.8-4 find

a I_1, I_2, and I_3
b V

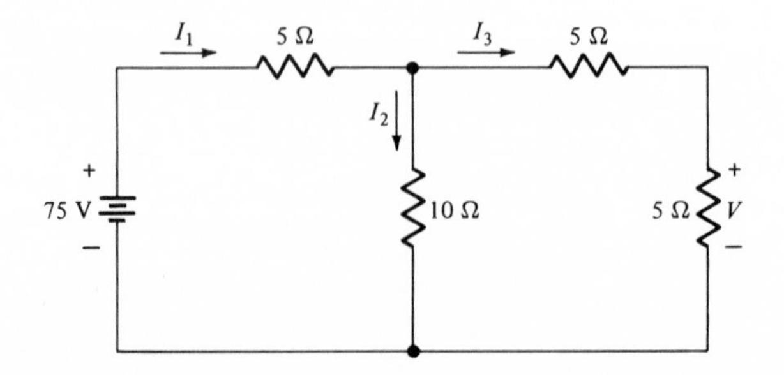

Fig. P5.8-4.

5.8-5 For the circuits of Fig. P5.8-5 find the indicated voltages and/or currents.

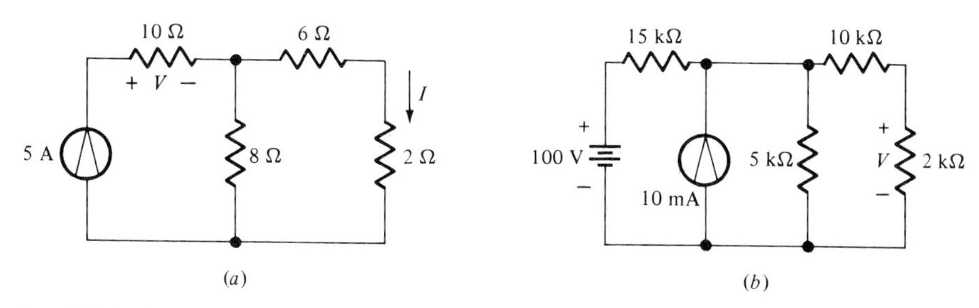

Fig. P5.8-5.

6

Resistive Networks: Applications of the Basic Laws

OBJECTIVES

Upon completion of this chapter, you should be able to

1 Analyze a single node-pair network to determine all voltages and currents.
2 Find the equivalent resistance of two or more parallel resistances.
3 Analyze a single loop network to determine all voltages and currents.
4 Find the equivalent resistance of two or more series resistances.
5 Apply the voltage divider relation to an appropriate network in order to determine unknown voltages.
6 Apply the current divider relation to an appropriate network in order to determine unknown currents.
7 Use ECAP to solve simple networks.

INTRODUCTION

In the last chapter, we considered the voltage-current behavior of sources and resistances. In this chapter, we will extend these ideas to include circuits consisting of various combinations of sources and resistances. Our aim is to develop facility in making routine circuit calculations and to provide a basis from which the student can begin to obtain an intuitive understanding of the behavior of resistive circuits.

At the time that this is being written, many technologists have access to computer programs which analyze extremely complicated networks and print out in seconds answers to network problems which formerly took days or longer to solve. However, the computer solution seldom provides "insight" into network behavior and it is this insight that we wish to foster.

6.1 ANALYSIS OF SINGLE NODE-PAIR NETWORKS

When networks consist of more than a few elements, their analysis requires methods which are carefully organized. One such method is called the *node-voltage* method, and we begin to consider it in this section by applying it to a single node-pair network. Such a network is shown in Fig. 6.1-1. In the network, the four unspecified branches are said to be in *parallel* because they share a pair of common nodes, labelled a and b.

The circuit which we are going to consider is shown in Fig. 6.1-2. The parallel branches are specified as an ideal current source and two resistors. The object of our analysis is to find all of the unknown currents and voltages in the circuit. The known quantities are the strength of the current source I_i, and the

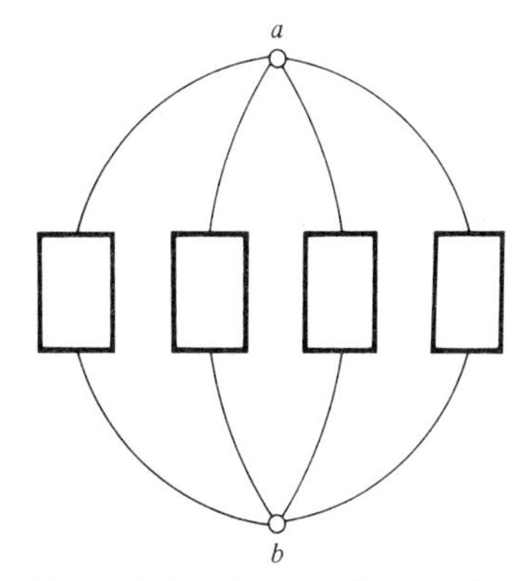

Fig. 6.1-1 A single node-pair network consisting of four branches.

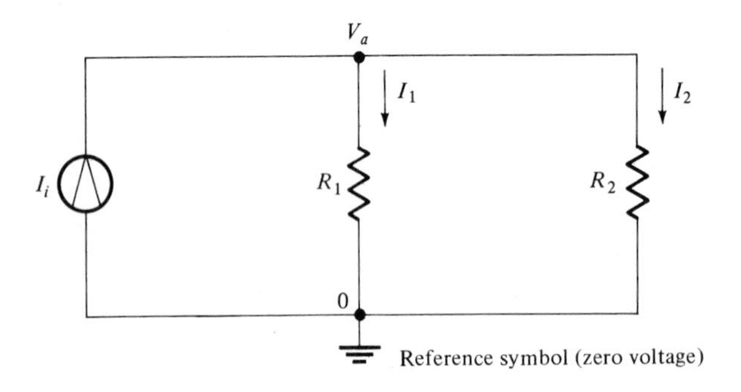

Fig. 6.1-2 Single node-pair network.

values of the resistances R_1 and R_2. The unknowns are the currents I_1 and I_2 and the voltage at node a, designated V_a. This voltage is measured with respect to the other node, designated 0, which is considered to be the zero reference (sometimes called *ground*) node at zero voltage. Thus, the *reference* polarity of V_a is positive with respect to ground and we note that V_a is really shorthand for V_{a0}, the *voltage drop* from node a to node 0.

Since the reference polarity for V_a is positive, the currents I_1 and I_2 are shown flowing from node a to ground, in accordance with Ohm's law (Sec. 5.6). This is important, since it insures correct polarities in our result. To apply the node-voltage method, we write KCL (Kirchhoff's current law) at each node (except for the ground node) in terms of the previously assigned reference node voltages and the element *vi* relations (Ohm's law for resistances). The node voltages are then the unknowns for which the KCL equations are solved. We illustrate using the circuit of Fig. 6.1-2.

For the circuit we write KCL at node a as follows:

$$I_i = I_1 + I_2 \tag{6.1-1}$$

Now, the voltage across R_1 is V_a, and the voltage across R_2 is also V_a (in a *parallel* circuit, the voltage is the same across *all* elements). Thus, we use Ohm's law to write

$$I_1 = \frac{V_a}{R_1} \qquad I_2 = \frac{V_a}{R_2} \tag{6.1-2}$$

When dealing with parallel circuits, it is convenient to work with the reciprocal of resistance. This is called conductance; its unit is the *mho* (ohm spelled backward) which is symbolized by ℧ (inverted omega). Thus, we have

$$G_1 = \frac{1}{R_1} \qquad G_2 = \frac{1}{R_2} \tag{6.1-3}$$

and Eq. (6.1-2) can be written

$$I_1 = G_1V_a \qquad I_2 = G_2V_a \tag{6.1-4}$$

Now we substitute Eq. (6.1-4) into Eq. (6.1-1) to obtain

$$I_i = (G_1 + G_2)V_a \tag{6.1-5}$$

Since I_i, G_1, and G_2 are known, this equation can be solved for V_a. Then Eq. (6.1-2) or Eq. (6.1-4) can be solved for I_1 and I_2, completing the solution. At this point, we note that if only the voltage V_a is required, inspection of Eq. (6.1-5) indicates that we can replace the two parallel conductances by a single conductance of value

$$G = G_1 + G_2 \tag{6.1-6}$$

G is called the *equivalent* conductance. A little thought will show that Eq. (6.1-6) can be extended to any number of conductances in *parallel* so that, for example, for four conductances in parallel:

$$G = G_1 + G_2 + G_3 + G_4 \tag{6.1-7}$$

Example 6.1-1 In the circuit of Fig. 6.1-2

$I_i = 5$ A $\qquad R_1 = 10\ \Omega \qquad R_2 = 5\ \Omega$

Find (*a*) V_a, (*b*) I_1, and I_2, (*c*) G

Solution Solving Eq. (6.1-5) for V_a

$$V_a = \frac{I_i}{G_1 + G_2}$$

a Since $G_1 = 1/R_1 = 1/10 = 0.1$ ℧ and $G_2 = 1/R_2 = 1/5 = 0.2$ ℧

$$V_a = \frac{5}{0.1 + 0.2} = 16.7 \text{ V}$$

b To find I_1 and I_2, we use Ohm's law, Eq. (6.1-2)

$$I_1 = \frac{V_a}{R_1} = \frac{16.7}{10} = 1.67 \text{ A}$$

$$I_2 = \frac{V_a}{R_2} = \frac{16.7}{5} = 3.34 \text{ A}$$

Checking, using KCL

$$I_1 + I_2 = 1.67 + 3.34 = 5.01 \text{ A}$$

This checks the given value of 5 A to sufficient accuracy for our purposes.

c We have, from *a*

$$G = G_1 + G_2 = 0.1 + 0.2 = 0.3 \text{ ℧}$$ ////

6.1-1 Parallel resistance

The calculation of the equivalent resistance of a group of parallel resistors arises so often in practice that we will devote this entire section to this topic. Refer to Eq. (6.1-6). If we call the *equivalent resistance*, R, then

$$G = \frac{1}{R} = \frac{1}{R_1} + \frac{1}{R_2} = \frac{R_2 + R_1}{R_1R_2} \tag{6.1-8}$$

Solving for R, we have

$$R = \frac{R_1R_2}{R_1 + R_2} \tag{6.1-9}$$

This result is used so often that it should be memorized. From this point on, we will abbreviate this result as $R = R_1 \parallel R_2$ which should be read as "R_1 in parallel with R_2" and which implies the calculation shown in Eq. (6.1-9).

If more than two resistors are involved, we use Eq. (6.1-7) to find

$$\frac{1}{R} = \frac{1}{R_1} + \frac{1}{R_2} + \frac{1}{R_3} + \frac{1}{R_4} \tag{6.1-10}$$

This can be generalized to include any number of parallel resistors. Very often, when more than two resistors are involved, it is expedient to combine them two at a time. This will be illustrated in the examples which follow.

Example 6.1-2 Find the equivalent resistance of the circuits of Fig. 6.1-3*a* and *b*.

Solution

a The best procedure here is to use Eq. (6.1-9).

$$R_{T1} = R_1 \parallel R_2 = \frac{R_1R_2}{R_1 + R_2} = \frac{(6.8)(3.3)}{6.8 + 3.3} = 2.22\ \Omega$$

This can be checked using Eq. (6.1-8)

$$\frac{1}{R_{T1}} = \frac{1}{R_1} + \frac{1}{R_2} = \frac{1}{6.8} + \frac{1}{3.3} = 0.147 + 0.303 = 0.45$$

$$R_{T1} = \frac{1}{0.45} = 2.22\ \Omega$$

b There are several ways to find R_{T2}. One way is to combine the 1.5 kΩ and 4.7 kΩ resistors to find equivalent resistance R_{T3}. This is then combined with the 10 kΩ resistance to find R_{T2} as shown in Fig. 6.1-3*c*. Proceeding as outlined, we have (note that all values are in kΩ, so that

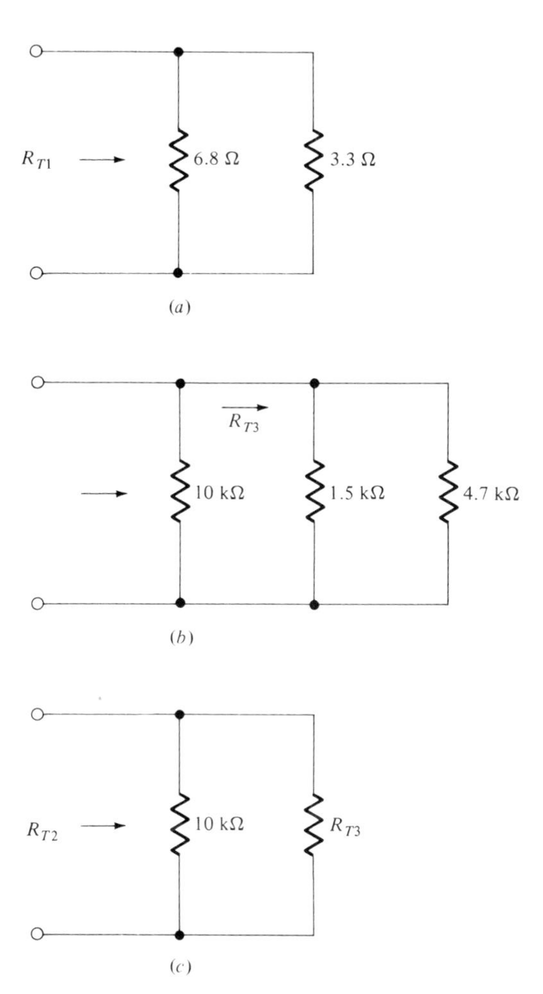

Fig. 6.1-3 Circuits for Example 6.1-2.

we are dimensionally correct and do not have to carry the 10^3 factors through the calculation)

$$R_{T3} = 1.5 \,||\, 4.7 = \frac{(1.5)\,(4.7)}{1.5 + 4.7} \approx 1.14 \text{ k}\Omega$$

$$R_{T2} = 10 \,||\, R_{T3} = 10 \,||\, 1.14 = \frac{(10)\,(1.14)}{10 + 1.14} \approx 1.02 \text{ k}\Omega$$

The student should check this example by first combining the 10 kΩ and 4.7 kΩ resistances, then their equivalent with the 1.5 kΩ resistance.

////

Example 6.1-3 Use the technique of parallel resistance combination to find V_a in the circuit of Example 6.1-1.

Solution Returning to Example 6.1-1, we proceed by first finding the equivalent resistance.

$$R = R_1 \,||\, R_2 = 10 \,||\, 5 = \frac{(10)(5)}{10+5} = \frac{50}{15} = 3.33\ \Omega$$

Next, we use Ohm's law to find V_a:

$$V_a = RI_i = (3.33)(5) = 16.7\ \text{V}$$

////

Before proceeding to our final example, there are a few points to be made concerning parallel resistance calculations. The first point to note is that *the equivalent resistance is always less than any of the individual resistances!* Whenever a parallel resistance combination is made, this point should be checked immediately.

The second point concerns *equal* resistances in parallel. Consider the case when two equal resistances R_1 are connected in parallel. The equivalent resistance is

$$R = \frac{R_1R_1}{R_1 + R_1} = \frac{R_1^2}{2R_1} = \frac{R_1}{2} \tag{6.1-11}$$

For the general case where n resistances R_1 are in parallel, we use the generalized form

$$\frac{1}{R} = \frac{1}{R_1} + \frac{1}{R_2} + \frac{1}{R_3} + \cdots + \frac{1}{R_n}$$

Since $R_1 = R_2 = R_3 = \cdots = R_n$

$$\frac{1}{R} = \frac{1}{R_1} + \frac{1}{R_1} + \frac{1}{R_1} + \cdots + \frac{1}{R_1} = \frac{n}{R_1}$$

Thus $R = \dfrac{R_1}{n}$ (6.1-12)

From the above, we can make the general statement that the equivalent resistance of any number of *equal* parallel resistances is the value of one of the resistances divided by their number. This idea will be utilized in the following example.

Example 6.1-4 In the circuit of Fig. 6.1-4*a*, find all currents.

Solution We proceed by first finding the equivalent resistance R_T as "seen by" the battery. To combine R_3 and R_4, we note that they are equal

I_{34}
I_T
I_1
I_2
I_3
I_4
$V_i = 100$ V
$R_1 = 2\ \Omega$
$R_2 = 2\ \Omega$
$R_3 = 4\ \Omega$
$R_4 = 4\ \Omega$
R_T

(*a*)

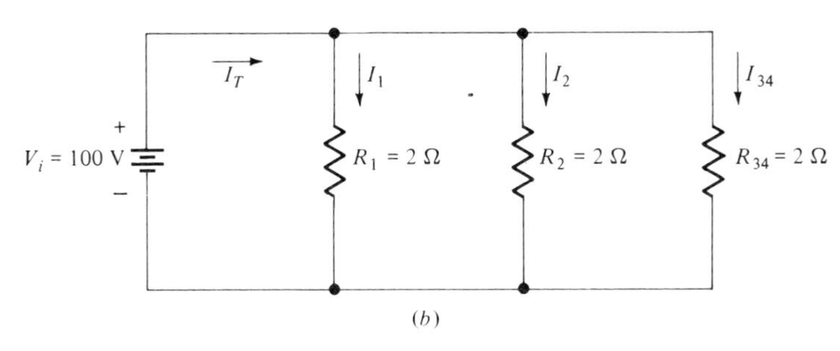

(*b*)

Fig. 6.1-4 Circuits for Example 6.1-4.

so their equivalent resistance R_{34} is simply one half of either one, that is, $2\,\Omega$. The equivalent circuit is then as shown in Fig. 6.1-4*b*. This is particularly simple since it consists of three 2-Ω resistances in parallel so that, by inspection, we see that $R_T = 2/3\ \Omega$.

Next we use Ohm's law to find I_T:

$$I_T = \frac{V_i}{R_T} = \frac{100}{0.67} = 150 \text{ A}$$

Now refer to Fig. 6.1-4b. Since the voltage across all three resistances is the same (100 V), we may conclude that the currents are the same. Thus,

$$I_1 = I_2 = I_{34} = \frac{100 \text{ V}}{2\ \Omega} = 50 \text{ A}$$

Since I_3 and I_4 flow in identical 4 Ω resistances which each have 100 V across them, we must have

$$I_3 = I_4 = \frac{100 \text{ V}}{4\ \Omega} = 25 \text{ A}$$

Checking KCL for I_T we have

$$I_T = I_1 + I_2 + I_3 + I_4 = 50 + 50 + 25 + 25 = 150 \text{ A} \qquad ////$$

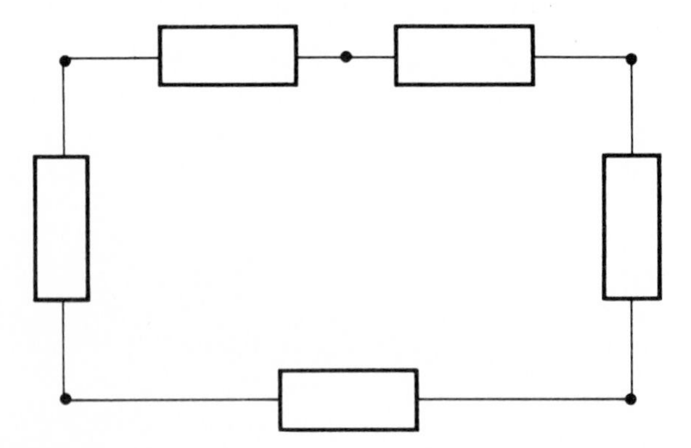

Fig. 6.2-1 Series circuit.

6.2 ANALYSIS OF SINGLE-LOOP NETWORKS

Thus far in this chapter, we have considered parallel circuits. The other common variety of basic circuit is the *series* circuit and the arrangement is shown in Fig. 6.2-1. In a series circuit, the *current* is the same *through* each branch of the circuit. Recall that in a parallel circuit, the voltage is the same *across* each branch of the circuit. The method of analysis usually used for the analysis of series circuits is called the *loop-current* method and is based on KVL (Kirchhoff's voltage law). We will illustrate the method by applying it to the circuit of Fig. 6.2-2.

The circuit consists of a dc voltage source in series with two resistors. The reference direction for the unknown current I is taken arbitrarily to be flowing clockwise. In accordance with this current direction and Ohm's law, the voltage drops across the two resistors will then have the reference polarities shown for V_1 and V_2. A negative value for I will indicate that positive current flows counterclockwise, i.e., opposite the arbitrarily chosen clockwise direction.

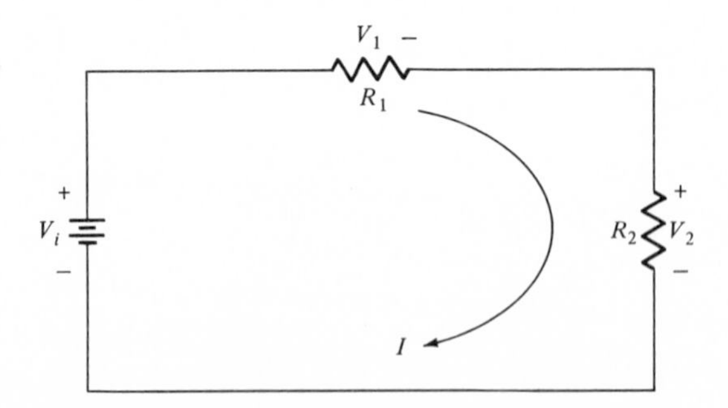

Fig. 6.2-2 Series circuit for loop-current analysis.

To apply the loop-current method we write KVL around the loop in terms of the previously assigned current and the element *vi* relations. The current is then the unknown for which the KVL equation is solved. For the circuit of Fig. 6.2-2 KVL yields

$$V_i = V_1 + V_2 \tag{6.2-1}$$

Since the current is the same throughout the series circuit, we use Ohm's law to write

$$V_1 = IR_1 \qquad V_2 = IR_2 \tag{6.2-2}$$

Next we substitute Eq. (6.2-2) into Eq. (6.2-1) to obtain

$$\begin{aligned} V_i &= IR_1 + IR_2 \\ &= I(R_1 + R_2) \end{aligned} \tag{6.2-3}$$

Since V_i, R_1, and R_2 are known, this equation can be solved for I. Then Eq. (6.2-2) is used to find V_1 and V_2, completing the solution.

Notice that if the current I is the only answer required, we can replace the two series resistances by a single equivalent resistance of value

$$R = R_1 + R_2 \tag{6.2-4}$$

It can be shown (Prob. 6.2-1) that Eq. (6.2-4) can be extended to any number of resistances in series, so that, for example, if we have four resistances *in series*, the equivalent resistance is

$$R = R_1 + R_2 + R_3 + R_4 \tag{6.2-5}$$

From this equation we see that if all of the resistances are the same, say R_1, then $R = 4R_1$ for the case of four resistances or, for n equal resistances in series,

$$R = nR_1 \tag{6.2-6}$$

Example 6.2-1 In the circuit of Fig. 6.2-2

$$V_i = 9 \text{ V}$$

$$R_1 = 47 \ \Omega$$

$$R_2 = 68 \ \Omega$$

Find *a* I

b V_1 and V_2

Solution

a Using Eq. (6.2-3)

$$I = \frac{V_i}{R_1 + R_2} = \frac{9}{47 + 68} = \frac{9}{115} = 0.0783 \text{ A}$$

b Using Eq. (6.2-2)

$$V_1 = IR_1 = (.0783)(47) = 3.68 \text{ V}$$

$$V_2 = IR_2 = (.0783)(68) = 5.32 \text{ V}$$

Checking KVL we must have $V_i = V_1 + V_2 = 3.68 + 5.32 = 9.0$ V ////

Example 6.2-2 In the circuit of Fig. 6.2-3 find

a I

b V_3, V_4, V_5, V_6

Solution

a Applying KVL to the loop, we add voltage drops clockwise.

$$-V_1 + V_3 + V_2 + V_4 - V_5 = 0 \tag{6.2-7}$$

The circuit diagram indicates that $V_1 = 22$ V, $V_2 = 9$ V. Applying Ohm's law to the three resistances, noting carefully the reference directions for the voltages and the assumed direction for I, we have

$$V_3 = R_3I \qquad V_4 = R_4I \qquad V_5 = -R_5I \tag{6.2-8}$$

Substituting into Eq. (6.2-7)

$$-V_1 + R_3I + R_4I + V_4 - (-R_5I) = 0 \tag{6.2-9}$$

Solving

$$(R_3 + R_4 + R_5)I = V_1 - V_4$$

$$I = \frac{V_1 - V_4}{R_3 + R_4 + R_5} = \frac{22 - 9}{3.3 + 6.8 + 10} = \frac{13}{20.1} = 0.647 \text{ mA}$$

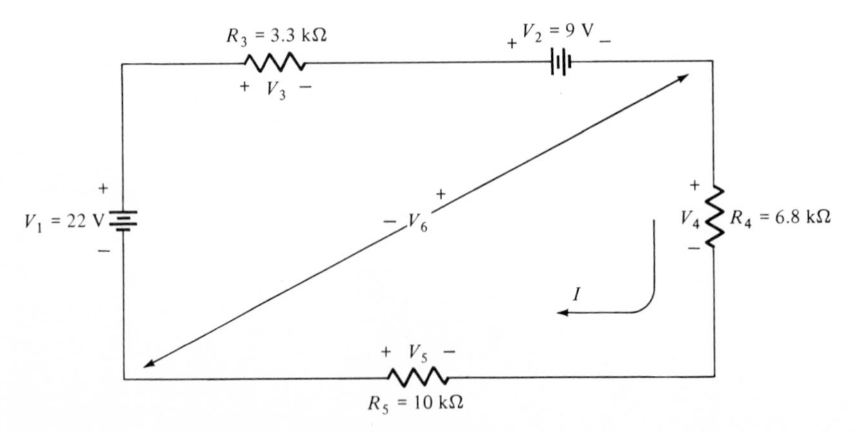

Fig. 6.2-3 Circuit for Example 6.2-2.

Note that we have used units of V, kΩ, and mA in Ohm's law. This is often a very convenient set of units when dealing with electronic circuits.

b From Eq. (6.2-8), with units of V, kΩ, and mA

$$V_3 = R_3 I = (3.3)(0.647) = 2.13 \text{ V}$$

$$V_4 = R_4 I = (6.8)(0.647) = 4.4 \text{ V}$$

$$V_5 = -R_5 I = (-10)(0.647) = -6.47 \text{ V}$$

Checking KVL we have $V_3 + V_4 - V_5 = 2.13 + 4.4 - (-6.47) = 13$ V. In order to find V_6 we write KVL around the loop including V_6 as follows:

$$-V_1 + V_3 + V_2 + V_6 = 0$$

$$V_6 = V_1 - V_3 - V_2$$

$$= 22 - 2.13 - 9$$

$$= 10.87 \text{ V}$$

The student should check this by tracing around the loop including V_4 and V_5. Also note carefully the way in which the reversed reference polarity of V_5 is handled. ////

Circuits that are not simple series or parallel can sometimes be reduced to that form by using series and parallel resistor combination. Consider the circuit of Fig. 6.2-4*a*. If parallel resistors R_2 and R_3 are replaced by their equivalent resistance R_T the circuit is reduced to the series circuit shown. This is easily analyzed using the methods presented in this chapter.

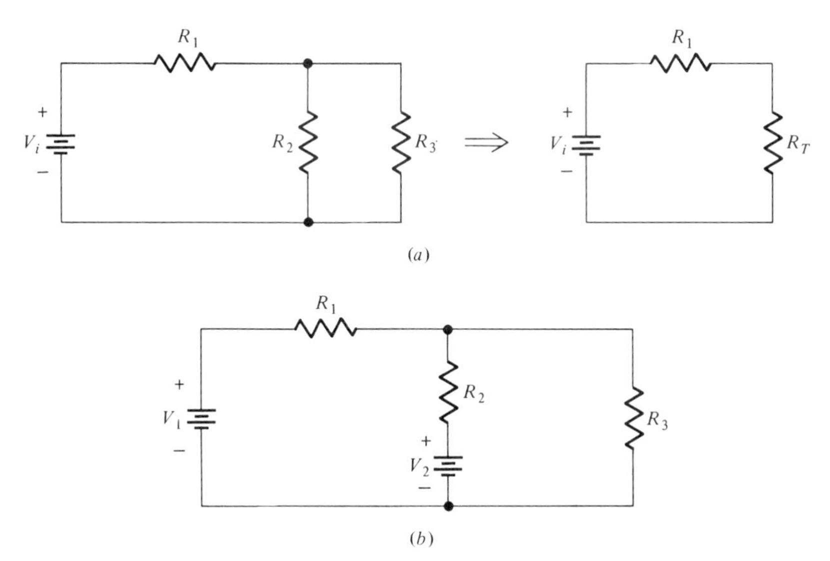

Fig. 6.2-4 Circuits which are not series or parallel.

On the other hand, the circuit of Fig. 6.2-4*b* can not be reduced to simple series or parallel form. This type of circuit will be considered in Chap 8.

6.3 VOLTAGE DIVIDER NETWORKS

The voltage divider shown in Fig. 6.3-1 is a circuit which appears very often in electronic systems. In the circuit, V_i represents the signal source and the voltage V_2 across R_2 is the desired output. Our analysis will seek a relation between V_2 and V_i, that is, output and input. When we find this relation, we will be able to use it to find the output in response to any input.

There are several ways to perform the analysis. Perhaps the most instructive is to apply the node-voltage method of Sec. 6.1. Note that the circuit has two nodes in addition to the reference; however, only the voltage at node *b* is an independent variable, because the voltage at *a* is known. We proceed by writing KCL at the independent node *b*. Since we have no advance knowledge of the relative values of V_i and V_2, we assume I_1 in the direction shown. The reference polarity of V_{R1} and the direction of I_2 are chosen to produce positive signs in the Ohm's law equations.

Writing KCL at node *b*,

$$I_1 + I_2 = 0 \tag{6.3-1}$$

Now we must find I_1 and I_2 in terms of the voltages. For I_2 we apply Ohm's law to get $I_2 = V_2/R_2$. For I_1 we must be careful in applying Ohm's law to use the voltage *across* R_1, with the indicated reference polarity. Thus, applying Ohm's law and KVL,

$$I_1 = \frac{V_{R1}}{R_1} = \frac{V_2 - V_i}{R_1} \tag{6.3-2}$$

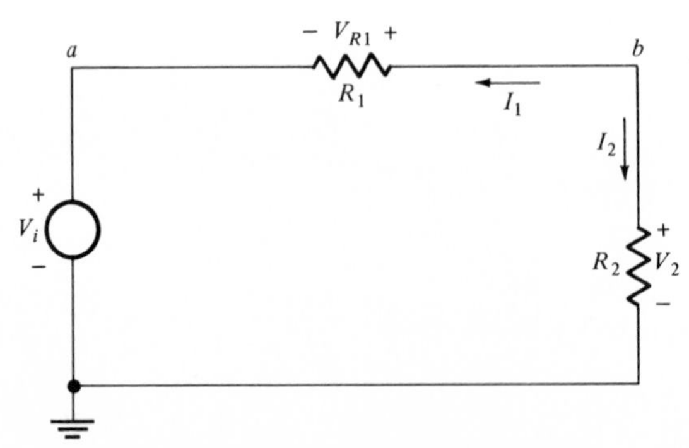

Fig. 6.3-1 The voltage divider.

Substituting this into Eq. (6.3-1), we find

$$\frac{V_2 - V_i}{R_1} + \frac{V_2}{R_2} = 0 \tag{6.3-3}$$

Collecting terms

$$V_2\left(\frac{1}{R_1} + \frac{1}{R_2}\right) - V_i\left(\frac{1}{R_1}\right) = 0$$

Solving

$$V_2 = \frac{R_2}{R_1 + R_2} V_i \tag{6.3-4}$$

This is the desired relation. It is often convenient to consider the *voltage ratio* $A_v = V_2/V_i$, called the *voltage gain*, or *transfer function*. From Eq. (6.3-4)

$$A_v = \frac{V_2}{V_i} = \frac{R_2}{R_1 + R_2} \tag{6.3-5}$$

In words, this formula states that the voltage gain equals the ratio of the output resistance to the total resistance. Before we proceed to examples, there are a few points to note about this formula. First, we see that as long as R_1 and R_2 are positive and not infinite, A_v will be less than 1. Thus, the voltage gain of the circuit has a *maximum* value of 1 and the output voltage is *always* less than the input. Secondly, resistances R_1 and R_2 can represent the equivalent resistance of a combination of other resistances. This point will be explored further in the examples which follow.

Example 6.3-1 In the circuit of Fig. 6.3-1 $R_2 = 6.8$ kΩ and $R_1 = 2.2$ kΩ. Find the voltage gain and use it to find V_2 if $V_i = 27$ V.

Solution To find A_v we use Eq. (6.3-5) with all resistances in kΩ

$$A_v = \frac{R_2}{R_1 + R_2} = \frac{6.8}{2.2 + 6.8} = \frac{6.8}{9} \approx 0.76$$

Then, from the definition of voltage gain,

$$V_2 = A_v V_i = (0.76)(27) = 20.5 \text{ V}$$

////

Example 6.3-2 In the circuit of Fig. 6.3-2 find V_1, V_2, and V_3.

Solution For this problem, we restate the voltage divider rule in a form appropriate to the circuit (see Prob. 6.3-2):

$$V_n = \frac{R_n V_i}{R_T}$$

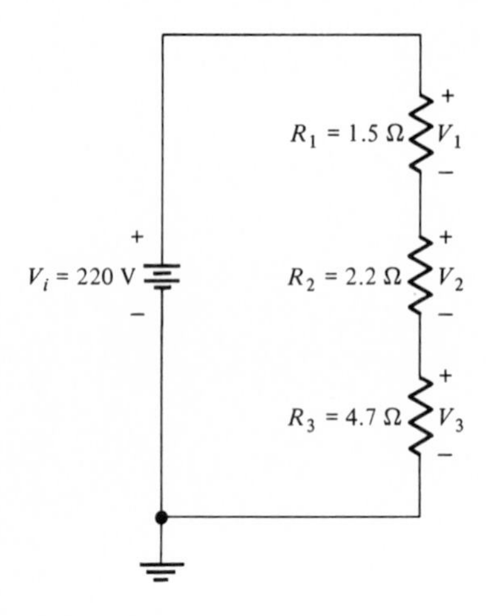

Fig. 6.3-2 Circuit for Example 6.3-2.

where $n = 1, 2, \text{ or } 3$

$$\begin{aligned} \text{and } R_T &= R_1 + R_2 + R_3 \\ &= 1.5 + 2.2 + 4.7 \\ &= 8.4\ \Omega \end{aligned}$$

$n = 1$:

$$V_1 = \frac{R_1}{R_T} V_i = \frac{(1.5)(220)}{8.4} = 39.3 \text{ V}$$

$n = 2$:

$$V_2 = \frac{R_2}{R_T} V_i = \frac{(2.2)(220)}{8.4} = 57.6 \text{ V}$$

$n = 3$:

$$V_3 = \frac{R_3}{R_T} V_i = \frac{(4.7)(220)}{8.4} = 123.1 \text{ V}$$

Checking KVL:

$$V_i = V_1 + V_2 + V_3$$

$$220 = 39.3 + 57.6 + 123.1$$

////

Example 6.3-3 Find V_2 in the circuit of Fig. 6.3-3.

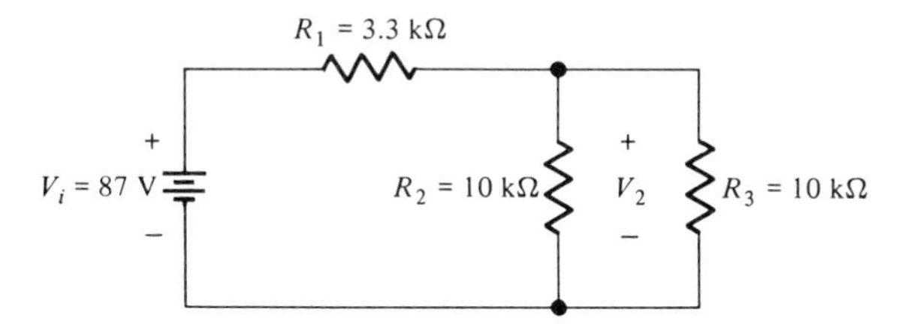

Fig. 6.3-3 Circuit for Example 6.3-3.

Solution At first glance, we might attack this problem by writing

$$V_2 = \frac{R_2}{R_1 + R_2} V_i = \frac{10\,(87)}{10 + 3.3} = 65.4 \text{ V}$$

This method is, of course, wrong. The mistake arises because R_3 has been omitted from consideration. Comparison of Fig. 6.3-1, from which the formula was derived, with Fig. 6.3-3 indicates that $R_2 \parallel R_3$ in the latter takes the place of R_2 in the original. Then the correct relation is

$$V_2 = \frac{(R_2 \parallel R_3)\,V_i}{R_1 + (R_2 \parallel R_3)} = \frac{(10 \parallel 10)\,(87)}{3.3 + (10 \parallel 10)} = \frac{(5)\,(87)}{3.3 + 5} = 52.4 \text{ V}$$

This example points up the fact that any resistance connected in parallel with either R_1 or R_2 in Fig. 6.3-1 must be taken into account. ////

6.4 CURRENT DIVIDER CIRCUITS

A circuit which performs the same function for currents as the voltage divider does for voltages is called a *current divider* and the basic arrangement is shown in Fig. 6.4-1. In the circuit, the input current I_i divides between R_1 and R_2. We analyze the circuit to find the divider ratio.

From Ohm's Law,

$$V = I_i\left(R_1 \parallel R_2\right) = I_i \frac{R_1 R_2}{R_1 + R_2} \tag{6.4-1}$$

Also, noting that each resistance has V volts across it,

$$V = I_1 R_1 = I_2 R_2 \tag{6.4-2}$$

Comparing these equations

$$I_1 R_1 = I_2 R_2 = I_i \frac{R_1 R_2}{R_1 + R_2}$$

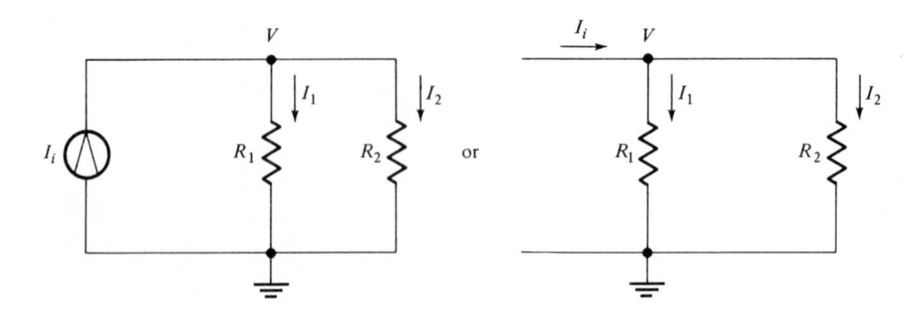

Fig. 6.4-1 Current divider circuit.

from which

$$I_1 = \frac{R_2}{R_1 + R_2} I_i \qquad I_2 = \frac{R_1}{R_1 + R_2} I_i \tag{6.4-3}$$

Or in terms of current gain A_i,

$$A_{i1} = \frac{I_1}{I_i} = \frac{R_2}{R_1 + R_2} \qquad A_{i2} = \frac{I_2}{I_i} = \frac{R_1}{R_1 + R_2} \tag{6.4-4}$$

Note the similarity to the voltage divider gain formula. The difference lies in the fact that the numerator of the current gain formula contains the resistance through which the desired current *does not* flow.

As long as R_1 and R_2 are positive, the current gain cannot exceed 1 and neither I_1 nor I_2 can be larger than I_i. Also, as in the case of the voltage divider, R_1 and R_2 may each represent the equivalent resistance of a number of other resistances.

Example 6.4-1 In Fig. 6.4-1, $R_1 = 6.8\ \text{k}\Omega$, $R_2 = 2.2\ \text{k}\Omega$, and $I_i = 10$ mA. Find I_1 and I_2.

Solution Using Eq. (6.4-4)

$$A_{i1} = \frac{R_2}{R_1 + R_2} = \frac{2.2}{6.8 + 2.2} = 0.245$$

$$A_{i2} = \frac{R_1}{R_1 + R_2} = \frac{6.8}{6.8 + 2.2} = 0.755$$

$$I_1 = A_{i1} I_i = (0.245)(10) = 2.45 \text{ mA}$$

$$I_2 = A_{i2} I_i = (0.755)(10) = 7.55 \text{ mA}$$

Checking KCL, $I_1 + I_2 = I_i$; $2.45 + 7.55 = 10$ mA. ////

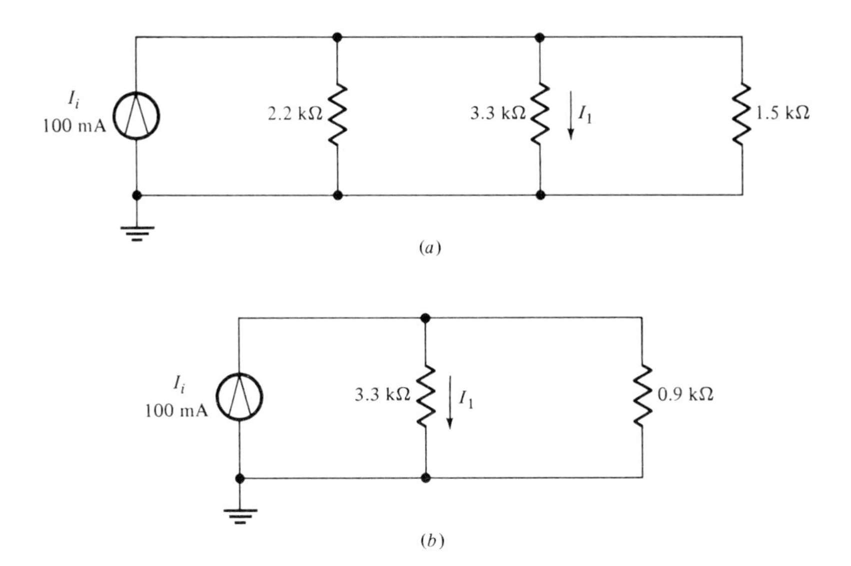

Fig. 6.4-2 Circuits for Example 6.4-2: (*a*) original circuit; (*b*) reduced circuit.

Example 6.4-2 Find I_1 in the circuit of Fig. 6.4-2*a*.

Solution The circuit must first be reduced to the form of Fig. 6.4-1. In order to do this we note that the currents through the 2.2 kΩ and 1.5 kΩ resistors are not required. Since they are in parallel they can be replaced by a single resistance of 2.2 || 1.5 = 0.9 kΩ as shown in Fig. 6.4-2*b*. From this circuit

$$A_{i1} = \frac{0.89}{3.3 + 0.89} = 0.212$$

Thus,

$$I_1 = A_{i1} I_i = (0.212)(100) = 21.2 \text{ mA} \qquad ////$$

Example 6.4-3 Find I_1 in the circuit of Fig. 6.4-3*a*.

Solution This circuit requires two applications of the current divider principle. We first reduce the circuit by combining the R_2-R_3 parallel branch as shown in Fig. 6.4-3*b*. Then we use the current divider to find I_2. Once I_2 is known the current divider is used again to find I_1. The steps are as follows.

From Fig. 6.4-3*b*

$$I_2 = \frac{R_4 I_i}{(R_{23} + R_1) + R_4} = \frac{2(10)}{8 + 2} = 2 \text{ A}$$

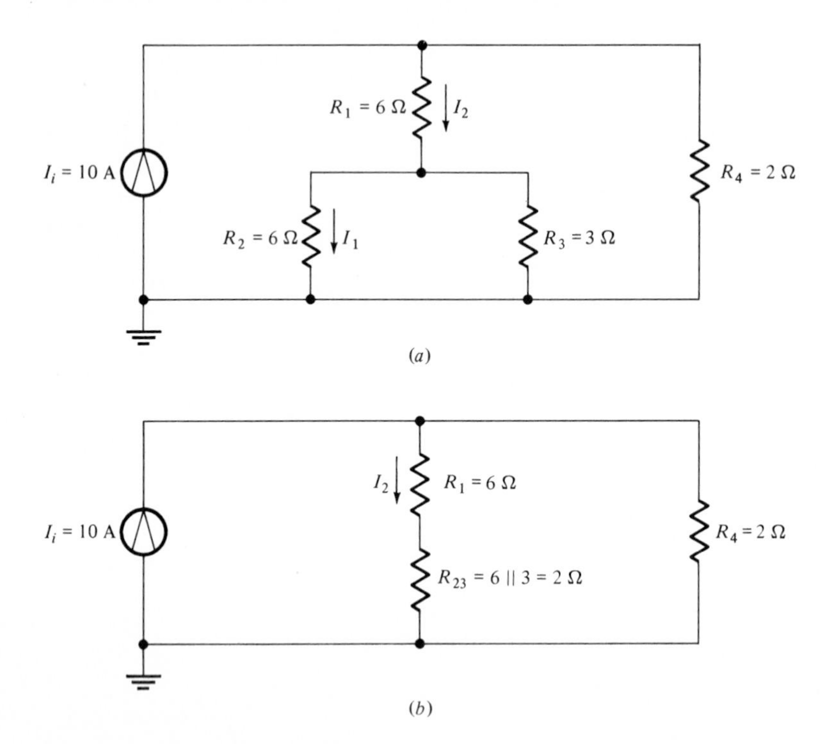

Fig. 6.4-3 Circuits for Example 6.4-3: (*a*) original circuit; (*b*) simplified circuit.

From Fig. 6.4-3*a*

$$I_1 = \frac{R_3}{R_3 + R_2} I_2 = \frac{3(2)}{3+6} = 0.67 \text{ A} \qquad ////$$

6.5 COMPUTER ANALYSIS OF NETWORKS—THE ECAP PROGRAM†

As we mentioned in the introduction to this chapter, computer programs designed to perform the analysis of electric networks are available at this time. In this section we will describe one of the most popular of these programs, IBM's ECAP (Electronic Circuit Analysis Program). In practice, ECAP is used to solve relatively complicated circuits. In the discussion which follows, it is applied to simple problems in order to teach the use of ECAP without the additional complication of a difficult circuit problem.

Briefly, when using ECAP, the technologist takes the schematic diagram of the circuit to be analyzed, and converts it to a standard ECAP equivalent

† This section may be omitted if appropriate computer facilities are not available.

circuit form. The information from the ECAP circuit is transferred to a coding sheet, using the ECAP language, then to punched cards, or directly to the computer via a typewriter terminal if the program is being used in a time-sharing system. When the computer receives the information from either the card deck or the typewriter, the ECAP program automatically formulates the network node equations, solves them, and prints or otherwise displays the results. The technologist does not have to go through the labor of setting up and solving equations; he can spend his time more profitably interpreting and utilizing the results.

From the practical point of view, ECAP is even more useful because it can be used to determine changes in currents and voltages when circuit elements or signals vary. With this capability, the program can be used to find element values which will lead to a specified voltage or current and thus becomes an extremely powerful design tool.

In this section we will present a brief introduction to the language of ECAP and examples of its use in the analysis of the types of networks considered in this chapter. It is important that the user learn the ECAP language thoroughly because communication with the computer must be exactly according to the ECAP rules or the computer will not recognize the commands.

THE SIMPLIFIED ECAP STANDARD BRANCH

The circuit *branch* is the basic unit used by ECAP. Each branch begins and ends on a node. The programmer specifies all of the circuit branches, and the program proceeds from this information to the solution.

A simplified form of the ECAP standard branch for dc analysis is shown in Fig. 6.5-1. This branch *must* contain a resistance and may, in addition, contain

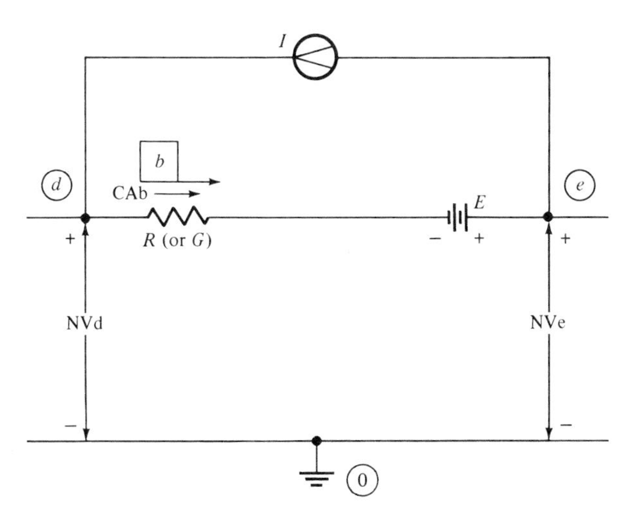

Fig. 6.5-1 The simplified ECAP standard branch.

independent current and/or voltage sources as explained in detail below. The various features of the diagram are:

Branches. Each branch must contain a resistance R (or conductance G), which must be nonzero. If independent sources are present, they are designated by E and/or I as shown. Reference directions for these sources are discussed later in Example 6.5-1. The circuit branches may be numbered in any order, but consecutive numbers beginning with 1 must be used. The branch number b is placed in a box, with an arrow on the box indicating the arbitrary reference direction from the initial node d to the final node e.

Nodes and Node Voltages. The network nodes (d and e in the figure) are numbered in any order using consecutive numbers beginning with 0. ECAP always assumes that node 0 is the reference (or ground) node. To distinguish node numbers from branch numbers, node numbers should be encircled. Node voltages in the figure are specified as NVd and NVe with respect to ground.

Element Currents. The element current in branch b (called CAb by ECAP) is the current flowing through R from the initial node d to the final node e. Note that our original choice of direction determines which is the initial and which the final node.

To prepare an ECAP circuit, branches and nodes are numbered and current directions are assumed. This information is transferred to a coding sheet using the ECAP language and then either to punched cards or a typewriter terminal. The procedure is illustrated in the following example.

Example 6.5-1 Find all currents and voltages in the circuit of Fig. 6.5-2a using ECAP.

Solution The ECAP circuit is shown in Fig. 6.5-2b. The circuit contains three nodes and four branches, one of which contains a voltage source. Note that all nodes and branches have been carefully numbered. The coding sheet is shown in Fig. 6.5-3, and an explanation of the numbered lines on the sheet follows†.

Line 1 Data required by the computer center to identify the user, the program, (ECAP in this case), etc. More than one line may be required.

Line 2 A command statement which tells the ECAP program that a dc analysis is desired.

Line 3 The letter C specifies a comment. Any text following the C is printed

† The column numbers on the coding sheet must be observed carefully if punched cards are used. Terminal input will be illustrated in the next example.

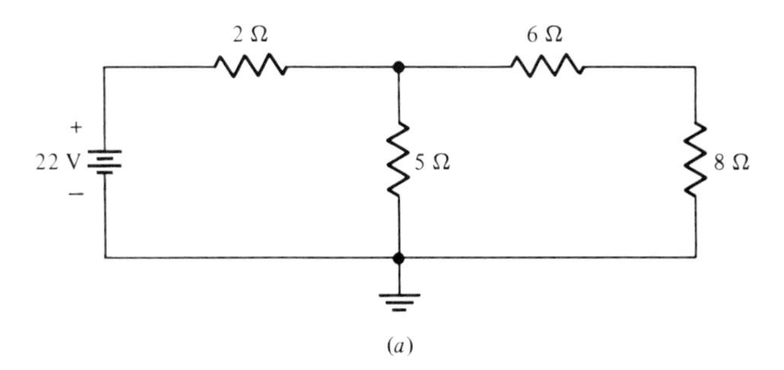

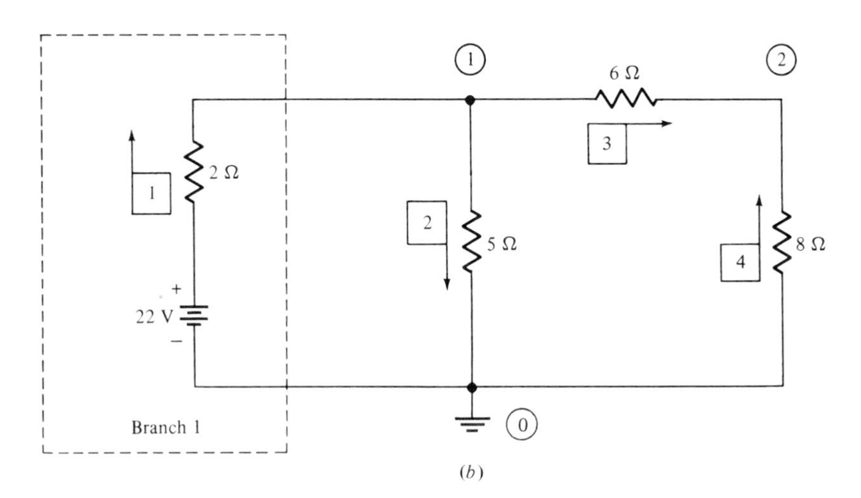

Fig. 6.5-2 (*a*) Circuit for Example 6.5-1. (*b*) Circuit prepared for ECAP.

as part of the output but has no other effect. Comments may be used anywhere in the program. A blank comment card or line is inserted after line five to provide a space in the printout.

Line 4 This line describes branch 1 to the program: Its initial node is 0, and its final node is 1. It contains a 2-Ω resistance and a 22-V voltage source. If the voltage source were reversed, we would write $E = -22$.

To determine the proper sign for a branch voltage source, pretend to *short-circuit* the branch. If the resulting current flow through the resistance is in the same direction as the preassigned reference arrow, the sign is positive; otherwise it is negative. For branch current sources (see Fig. 6.5-1) we pretend to disconnect the branch from the rest of the circuit. If the current from the source flows through the resistance in the same direction as the preassigned reference arrow, it is positive; otherwise, it is negative.

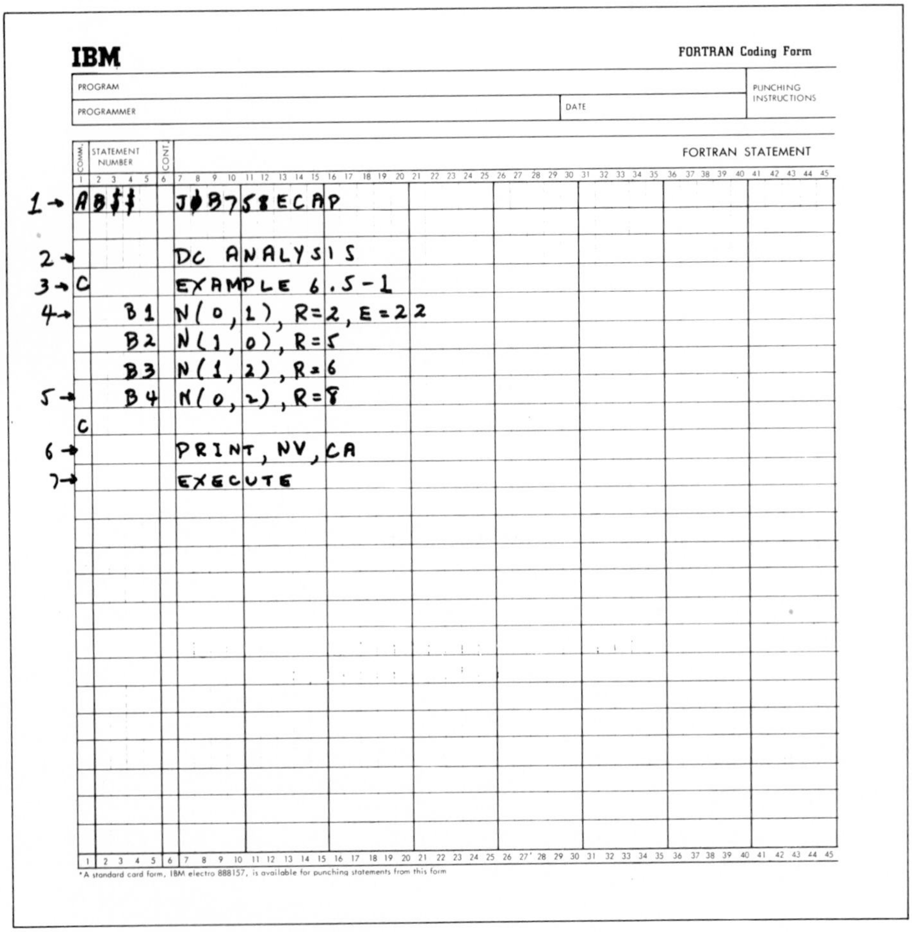

IBM FORTRAN Coding Form

PROGRAM | PUNCHING INSTRUCTIONS

PROGRAMMER | DATE

```
COMM. | STATEMENT NUMBER | CONT. | FORTRAN STATEMENT
1→  AB$$        JØB758ECAP

2→              DC ANALYSIS
3→  C           EXAMPLE 6.5-1
4→        B1    N(0,1),R=2,E=22
          B2    N(1,0),R=5
          B3    N(1,2),R=6
5→        B4    N(0,2),R=8
    C
6→              PRINT,NV,CA
7→              EXECUTE
```

*A standard card form, IBM electro 888157, is available for punching statements from this form

Fig. 6.5-3 Coding sheet for Example 6.5-1. Each line represents one punched card.

Line 5 Specifies branch 4: initial node 0, final node 2, and resistance of 8 Ω.

Line 6 Tells ECAP that all node voltages (NV) and element currents (EC) or (CA) are to be printed as output.

Line 7 Command statement indicating that the program is to be executed.

Figure 6.5-4 shows the computer printout. The node voltages and element currents are listed in rows according to the numbering scheme used on the ECAP circuit. All numerical answers are in standard computer "floating point" form. The number following the E (sometimes D is used) indicates the power of

```
ECAP          10:32    12/07/73  FRIDAY       I02



NV  1         NV  2        EC  1         EC  2        EC  3         EC  4
1.426E 01     8.148E 00    3.870E 00     2.852E 00    1.019E 00    -1.019E 00
```

Fig. 6.5-4 Computer printout for Example 6.5-1.

10 by which the number preceding the E is to be multiplied. Thus $I_4 = EC\ 4 = -1.019$ A indicating that $+1.019$ A flows from node 2 to node 0. ////

Example 6.5-2 Find all currents, voltages, and power dissipated in each branch in the circuit of Fig. 6.5-5*a*.

Solution The ECAP circuit is shown in Fig. 6.5-5*b* and the typewriter terminal input in Fig. 6.5-6. Note that line numbers must be used when the program is run using a typewriter terminal for input.

In the program, the current in branch 1 is given with a negative sign in

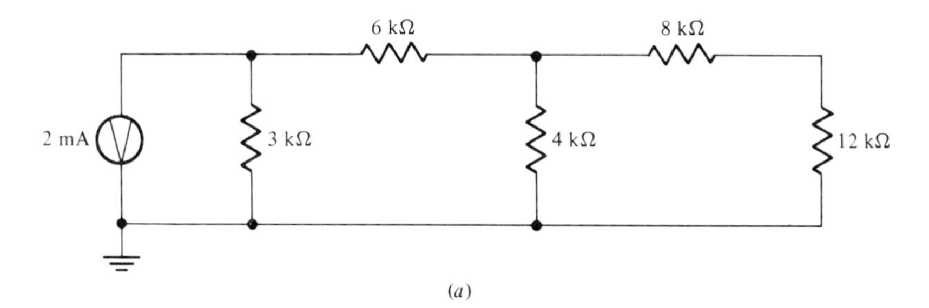

(*a*)

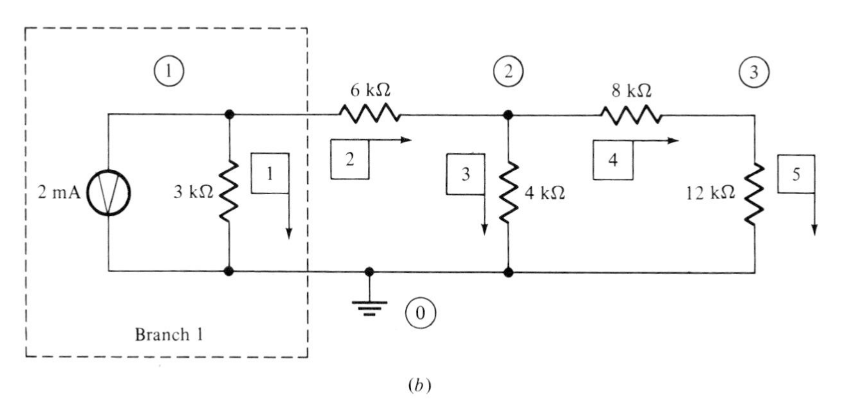

(*b*)

Fig. 6.5-5 (*a*) Circuit for Example 6.5-2. (*b*) Circuit prepared for ECAP.

```
100   DC ANALYSIS
200C  EXAMPLE 6.5-2
300   B1 N(1,0),R=3E03,I=-2E-03
400   B2 N(1,2),R=6E03
500   B3 N(2,0),R=4E03
600   B4 N(2,3),R=8E03
700   B5 N(3,0),R=12E03
800C
900   PRINT,NV,CA,BP
1000  EXECUTE
1100  END

RUN **ECAP
```

Fig. 6.5-6 Typewriter terminal input for Example 6.5-2.

accordance with the procedure outlined in Example 6.5-1, and the PRINT statement calls for branch power (BP) in addition to voltages and currents. The final statement, RUN **ECAP, typed without a line number calls the ECAP program. This may be somewhat different at other computer installations.

The printout of results is shown in Fig. 6.5-7. Verification of the results is left for the homework problems. ////

Additional capabilities of the ECAP program will be explored in later chapters after the pertinent theory has been developed.

```
ECAP          10:34   12/07/73   FRIDAY      I02



 NODE VOLTAGES
  1- 3   -4.541E 00   -1.622E 00   -9.730E-01
 ELEMENT CURRENTS
  1- 5   -1.514E-03   -4.865E-04   -4.054E-04   -8.108E-05   -8.108E-05
 ELEMENT POWER LOSSES
  1- 5    6.872E-03    1.420E-03    6.574E-04    5.259E-05    7.889E-05


END OF RUN

PROCESSING      1 UNITS
```

Fig. 6.5-7 Computer printout for Example 6.5-2.

SUMMARY

1 To solve a network using the node-voltage method, apply KCL at the appropriate nodes, and solve for the unknown node voltages. The unknown currents are then found using Ohm's law.
2 Conductances in parallel have an equivalent conductance

$$G = G_1 + G_2 + G_3 + \cdots$$

3 Two resistances in parallel have an equivalent resistance

$$R = R_1 \| R_2 = \frac{R_1 R_2}{R_1 + R_2}$$

4 The equivalent resistance of any number of parallel resistances is *always less than* any of the individual resistances.
5 To solve a network using the loop-current method, apply KVL around appropriate loops, and solve for the unknown loop currents. The unknown voltages are then found using Ohm's law.
6 Resistances in series have an equivalent resistance

$$R = R_1 + R_2 + R_3 + \cdots$$

7 The voltage ratio of a voltage divider network is

$$A_v = \frac{V_2}{V_1} = \frac{R_2}{R_1 + R_2}$$

where R_2 is the resistance across which V_2 appears.
8 The current ratio of a current divider network is

$$A_i = \frac{I_2}{I_1} = \frac{R_1}{R_1 + R_2}$$

where R_1 is the resistance through which I_2 *does not* flow.
9 In the ECAP program, the network being analyzed is described to the computer in terms of its branches. The program then does all of the work.

PROBLEMS

6.1-1 In Fig. P6.1-1, indicate which resistors are in parallel.

6.1-2 In the circuit of Fig. 6.1-2 $I_i = 20$ mA, $R_1 = 1.8$ kΩ, and $R_2 = 3.3$ kΩ. Find V_a, I_1, I_2 and G. Be sure to check KCL.

6.1-3 In the circuit of Fig. 6.1-2 $I_i = 50$ mA, $G_1 = 5$ m℧, and $G_2 = 2$ m℧. Find V_a, I_1, I_2, and R.

6.1-4 In the circuit of Fig. 6.1-2 $I_i = 30$ mA, $V_a = 20$ V, and $R_1 = 1$ kΩ. Find R_2, I_1, and I_2, and check your result as many ways as possible.

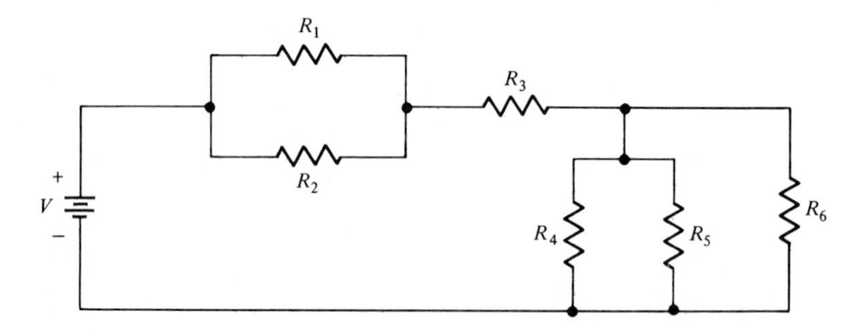

Fig. P6.1-1.

6.1-5 Find R_T for each of the circuits of Fig. P6.1-5.

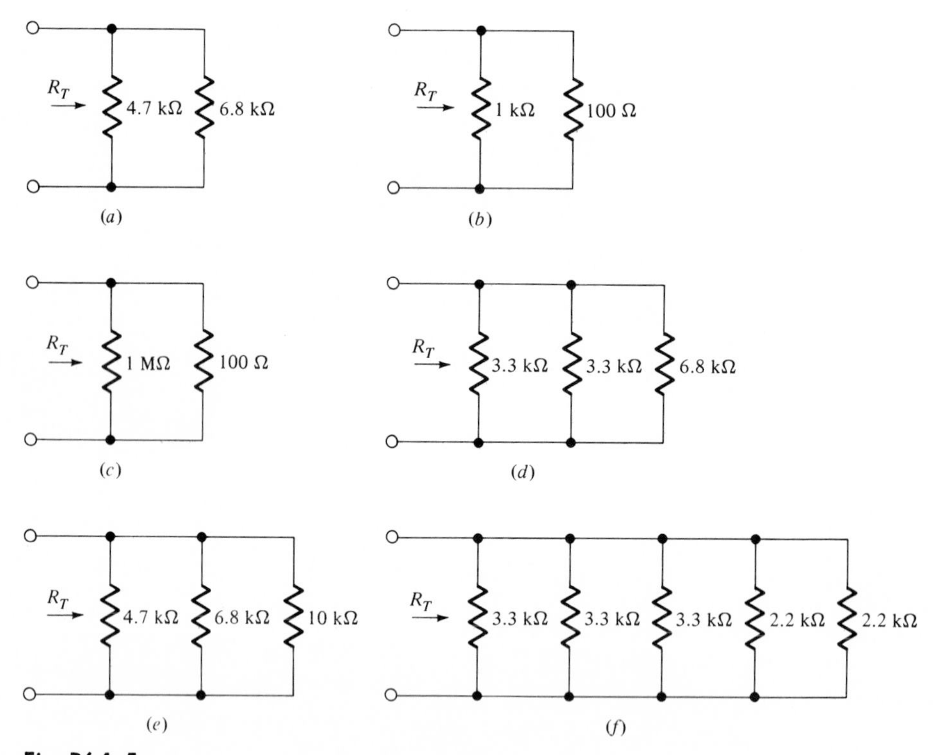

Fig. P6.1-5.

6.1-6 In the circuit of Fig. P6.1-6, $R_T = 20$ kΩ and $R_1 = 25$ kΩ. Find R_2.

6.1-7 In the circuit of Fig. P6.1-7, find all currents and find the power dissipated in the 4.7 kΩ resistor.

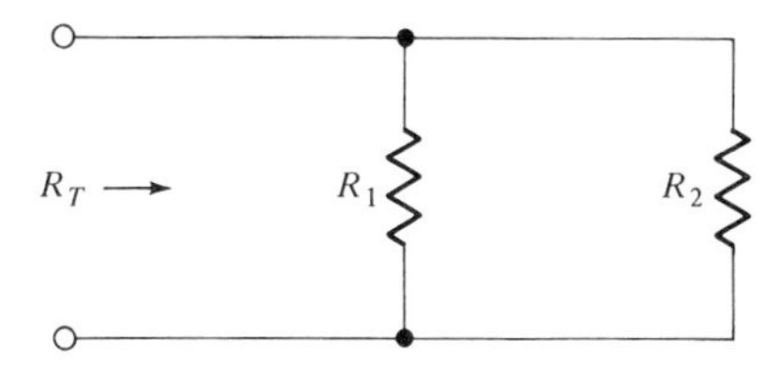

Fig. P6.1-6.

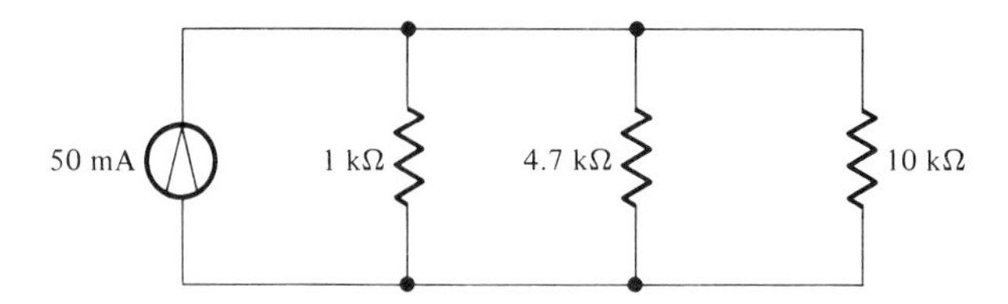

Fig. P6.1-7.

6.1-8 In the circuit of Fig. P6.1-8, find V_1, V_2, I_1, I_2, and I_3.

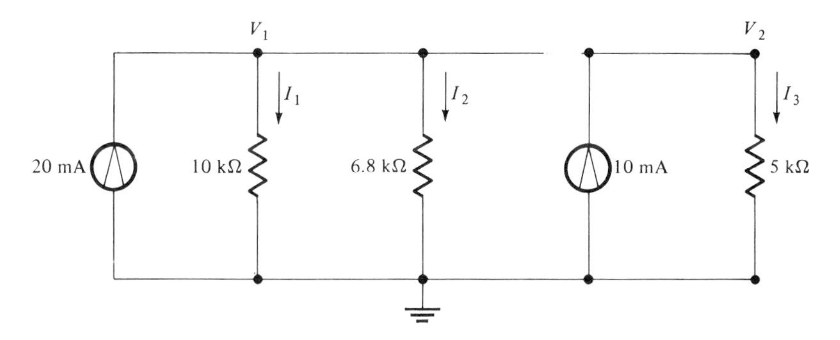

Fig. P6.1-8.

6.1-9 Repeat Prob. 6.1-8 for the circuit of Fig. P6.1-9.

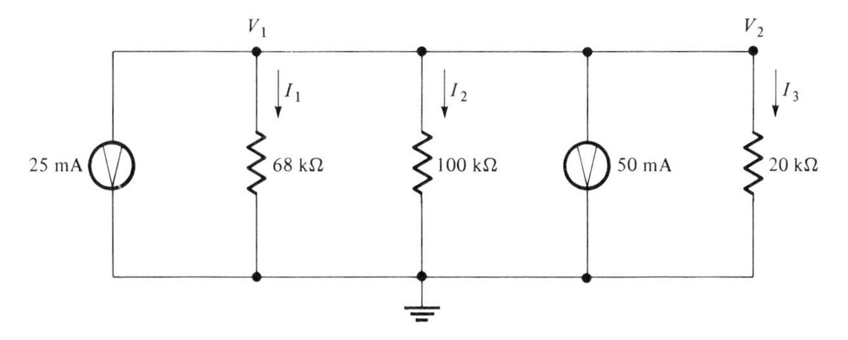

Fig. P6.1-9.

6.2-1 Prove, using KVL, that the equivalent resistance of four resistances in series is

$$R = R_1 + R_2 + R_3 + R_4$$

6.2-2 In Fig. 6.2-2, $V_i = 75$ V, $R_1 = 2.2$ kΩ, $R_2 = 6.8$ kΩ. Find I, V_1, V_2, the power drawn from the battery, and the power dissipated in each resistance. Check your answers wherever possible.

6.2-3 In Fig. 6.2-2, $V_i = 3$ V, $R_1 = 10$ kΩ, and $R_2 = 68$ kΩ. Find I, V_1, and V_2.

6.2-4 In Fig. 6.2-2 $V_i = 6$ V, $I = 20$ mA, $R_1 = 175$ Ω. Find R_2, V_1, and V_2.

6.2-5 Fig. P6.2-5 shows a circuit in which both series and parallel resistor combinations appear. Find all currents and voltages.

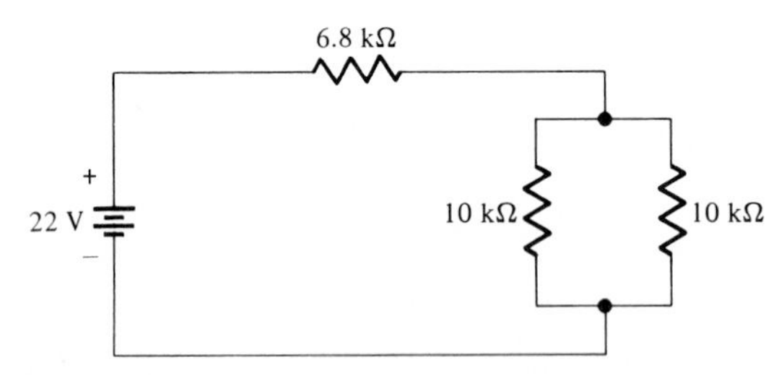

Fig. P6.2-5.

6.2-6 In the circuit of Fig. P6.2-6, find I, V_1, V_2, V_3, and V_4.

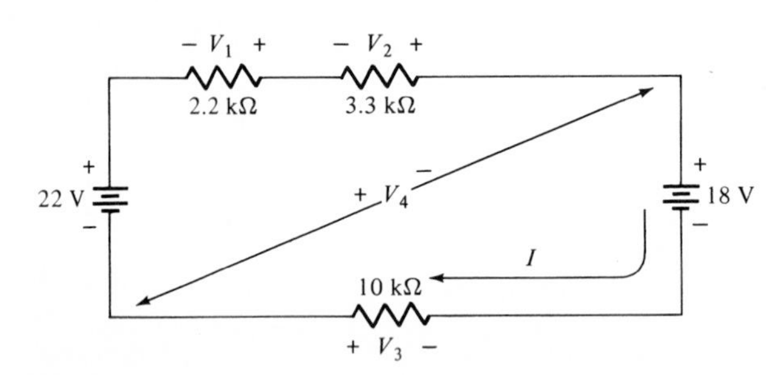

Fig. P6.2-6.

6.2-7 Repeat Prob. 6.2-6 for the circuit of Fig. P6.2-7.

6.2-8 Find all indicated unknowns in the circuit of Fig. P6.2-8.

6.3-1 In the circuit of Fig. 6.3-1, $V_i = 10$ V, $R_1 = 10$ kΩ and $R_2 = 6.8$ kΩ. Find V_2 and V_{R1} and check that KVL is satisfied.

6.3-2 Prove the modified voltage divider rule given in Example 6.3-2 and use it to find V_1, V_2, V_3, and V_4 in the circuit of Fig. P6.3-2.

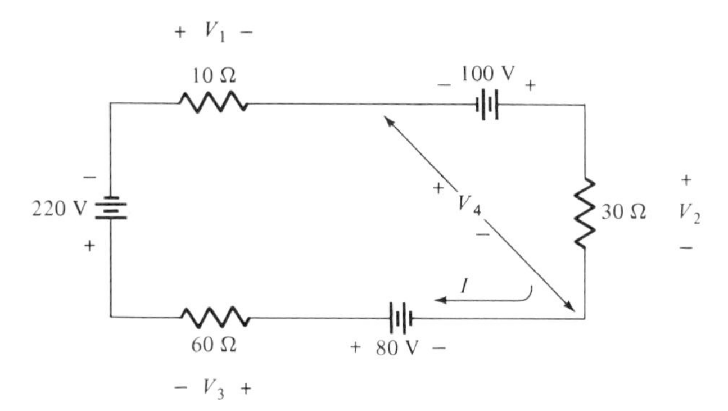

Fig. P6.2-7.

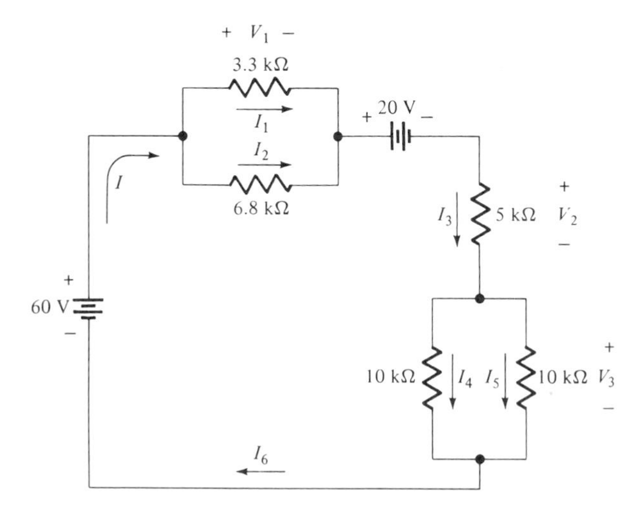

Fig. P6.2-8.

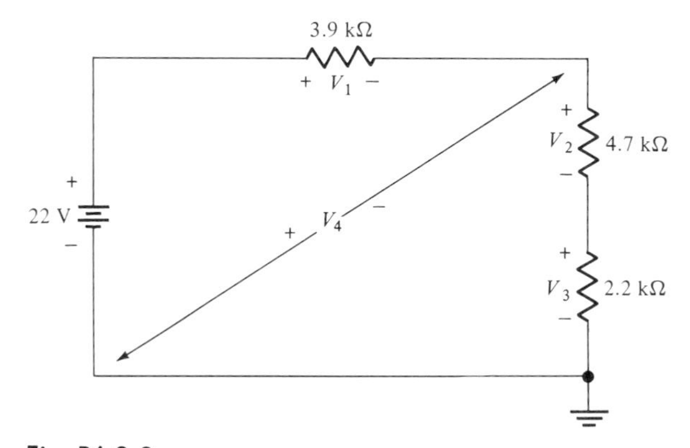

Fig. P6.3-2.

6.3-3 In the circuits of Fig. P6.3-3 find V_2.

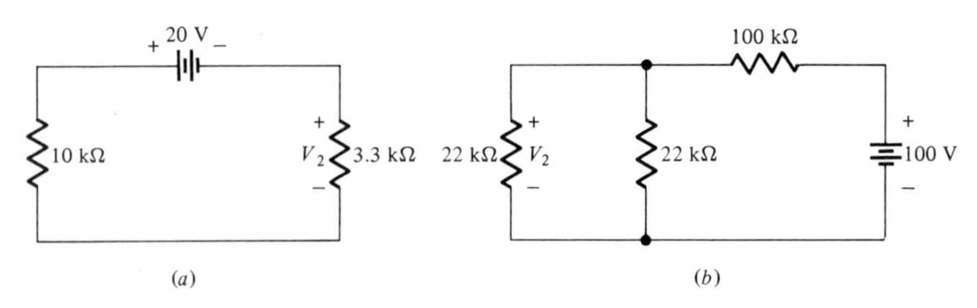

Fig. P6.3-3.

6.3-4 Find V_2 in the circuit of Fig. P6.3-4.

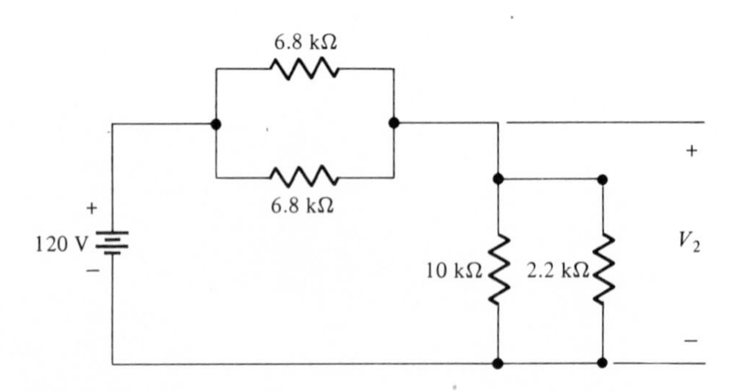

Fig. P6.3-4.

6.3-5 Find V_2 in the circuit of Fig. P6.3-5. Use the voltage divider relation twice. (Be careful!)

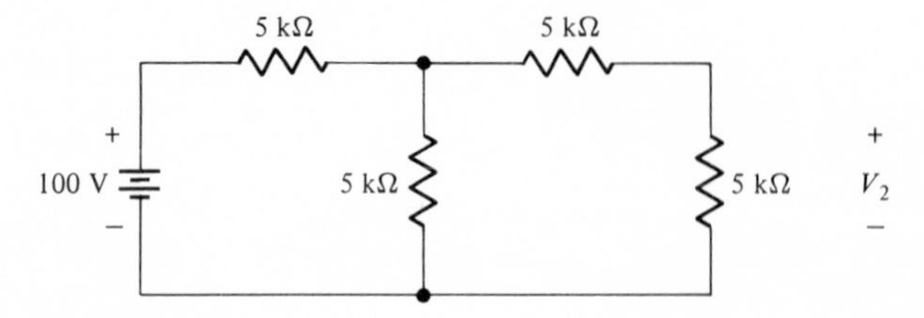

Fig. P6.3-5.

6.3-6 Design a voltage divider as in Fig. 6.3-1 so that the input resistance $R_1 + R_2$ is 10 kΩ and $A_v = 0.6$.

6.3-7 Design a voltage divider as in Fig. 6.3-1 so that $A_v = 0.3$ and $R_1 = 600\ \Omega$.

6.3-8 Design a loaded voltage divider as in Fig. 6.3-3 so that the input resistance is 10 kΩ and $A_v = 0.6$. The load resistance is $R_3 = 12$ kΩ. [*Hint*: For the circuit the input resistance is $R_1 + (R_2 \| R_3)$.]

6.4-1 In the circuit of Fig. 6.4-1, $R_1 = 10$ kΩ, $R_2 = 68$ kΩ, and $I_1 = 150$ mA. Find I_1, I_2, and V.

6.4-2 In the circuit of Fig. P6.4-2 find I_2.

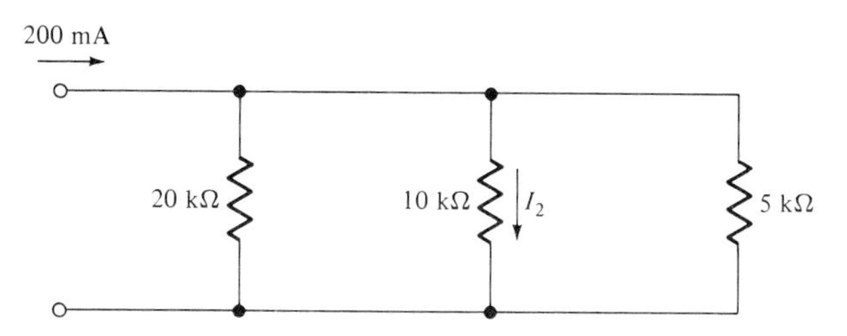

Fig. P6.4-2.

6.4-3 In the circuit of Fig. P6.4-3 find I_2 and V.

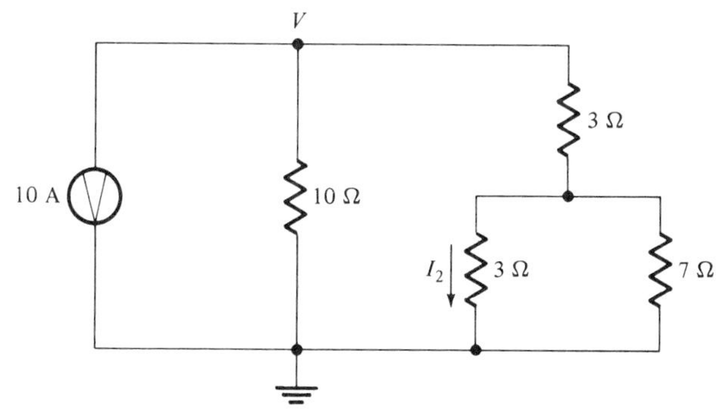

Fig. P6.4-3.

6.4-4 In the circuit of Fig. P6.4-4 find I_1, I_2, I_3, and I_4.

6.4-5 In the circuit of Fig. P6.4-5 find V_i.

6.4-6 In the circuit of Fig. P6.4-6 find I_1, I_2, I_3, and I_4.

6.4-7 Design a current divider as in Fig. 6.4-1 so that the input resistance $(R_1 \| R_2)$ is 100 kΩ and $A_i = I_2/I_i = 0.5$.

6.4-8 Design a current divider as in Fig. 6.4-1 so that $A_i = I_2/I_i = 0.2$ and $R_1 = 68$ kΩ.

6.5-1 Verify all of the computer results in Example 6.5-1.

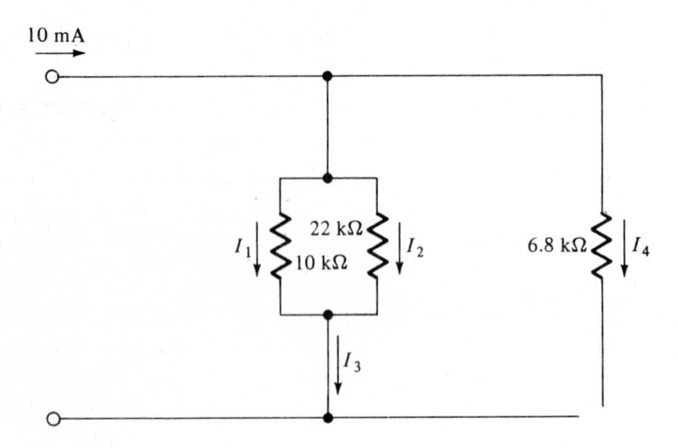

Fig. P6.4-4.

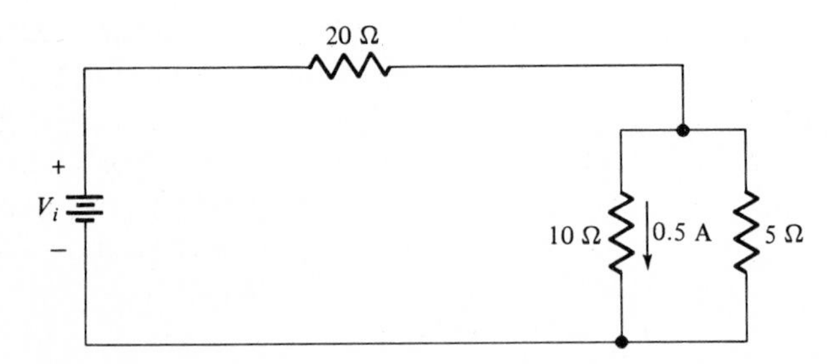

Fig. P6.4-5.

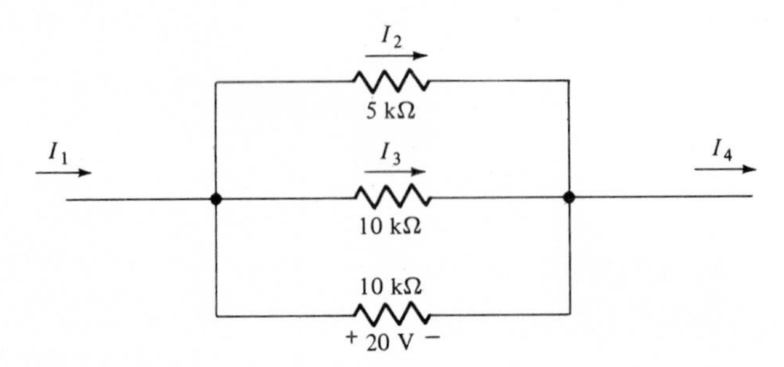

Fig. P6.4-6.

6.5-2 Repeat Prob. 6.5-1 for Example 6.5-2.

6.5-3 Find all currents and voltages in the circuit of Fig. P6.3-4 using any available computer circuit analysis program.

6.5-4 Repeat Prob. 6.5-3 for the circuit of Fig. P6.4-2.

6.5-5 Repeat Prob. 6.5-3 for the circuit of Fig. P6.4-3.

7
Dc Instruments

OBJECTIVES

Upon completion of this chapter, you should be able to

1 Describe the operation of the D'Arsonval movement.
2 Calculate the error introduced when a meter of known characteristics is used to measure current.
3 Calculate meter shunt resistance for a given movement.
4 Calculate the multiplier resistance required to convert a given movement to a voltmeter.
5 Define the current sensitivity of a D'Arsonval movement.
6 Calculate the error introduced when a meter of known characteristics is used to measure voltage.
7 Describe the operation of series and shunt ohmmeters.
8 Describe the operation of the Wheatstone bridge.
9 Describe qualitatively the iron-vane and electrodynamometer movements.

INTRODUCTION

When working with electric circuits, we usually use indicating instruments to tell us what the voltages and currents are throughout the circuit. For dc measurements, such instruments come in two types. The first is classified as *analog* where the output indication is continuously variable. Often the position of a pointer gives the desired output indication. Sometimes a pen or stylus gives a permanent record of the variable being measured.

The second type of instrument is *digital* and is characterized by an output display that gives the measured value in decimal numeric values. At the time of this writing, digital instruments are becoming more and more popular due to the availability and economy of integrated circuits. In general, for routine dc measurements, digital meters are much more accurate and reliable than their analog counterparts.

In this chapter we will consider the basic analog meter and its digital counterpart. Wherever possible we will develop the theory necessary to assess the effect of the connection of the instrument on the circuit variable being measured. This effect is usually called *loading*. Sufficient theory will be provided so that the student will be able to perform laboratory experiments involving these instruments.

7.1 THE D'ARSONVAL METER MOVEMENT

This movement is basically an electromagnetic device which is dc current-sensitive, and is used in both ammeters and voltmeters. A sketch of its principal parts is shown in Fig. 7.1-1. Current enters through the upper spring, goes through the moving coil, and exits through the lower spring (not shown in the diagram). This current creates a magnetic field which interacts with the field produced by the permanent magnet. The result is that a torque is produced which tends to rotate the moving coil. This torque is opposed by a counter torque provided by the springs. The instrument is designed so that when $I = 0$ the pointer will point to zero on the scale. The springs are adjusted so that the net torque produced when the current is at its maximum desired value will move the pointer so that it points to the desired full-scale value on the scale. A torque which moves the pointer upscale results only when current flows in the indicated direction. It can be shown that the pointer deflection is directly proportional to the current, so that the scale is linear.

A typical D'Arsonval movement without any external resistances is manufactured to deflect full scale for a current of 50 μA. Full-scale currents from 50 μA to 1 mA are available. For these meters the movement resistance usually lies between 50 Ω and 1,000 Ω. The current required for full-scale deflection is called the *current sensitivity*, which along with the resistance, is usually used to specify the characteristics of a movement.

It is important that the technologist understand how to determine the

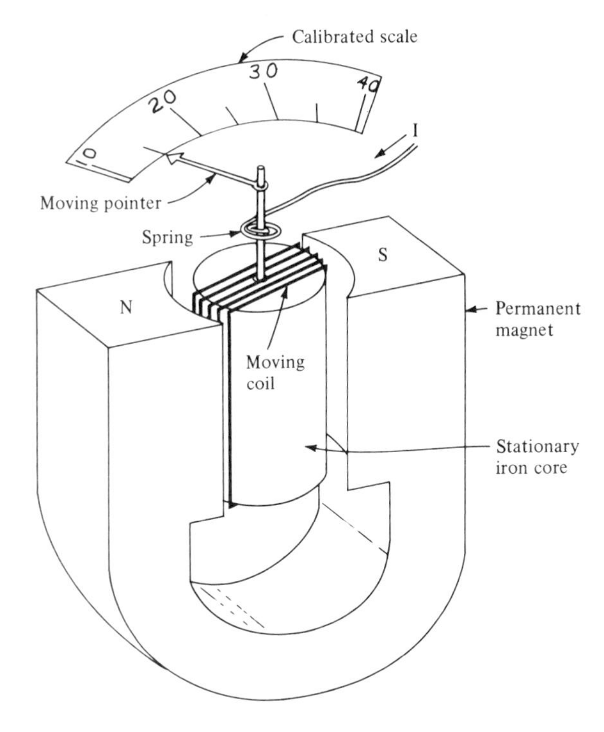

Fig. 7.1-1 The D'Arsonval meter movement.

effect of connecting such a meter into a circuit. This will be considered in the next section.

7.2 AMMETERS

To use the D'Arsonval movement for dc current measurement, the meter is connected directly into the wire through which the desired current is flowing, i.e., in *series*. Most movements have a + sign on the terminal at which the current must enter. Connecting the meter the wrong way may result in physical damage because the pointer will be forced against the stop which is usually provided to prevent excess motion of the coil.

It is convenient to model a real ammeter by a circuit consisting of an *ideal* ammeter having zero resistance in series with a resistance R_m which is equal to the resistance of the real meter. If we know the characteristics of the meter, we can predict the loading effect of the ammeter in a specific circuit.

Figure 7.2-1 shows a circuit in which an ammeter has been connected to measure the current in load resistor R_L. Note that the meter is connected in *series* with R_L, with the + sign where the current enters.

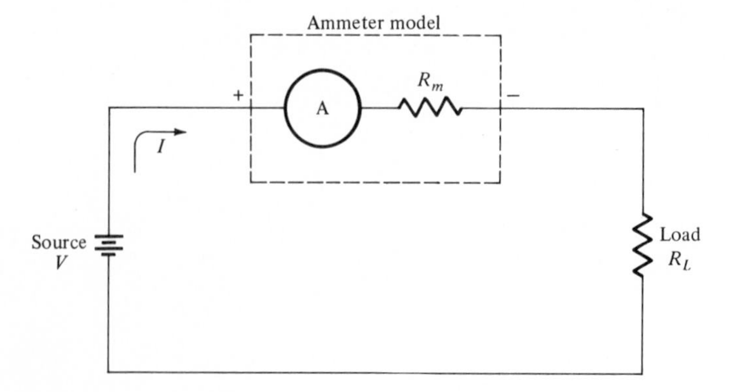

Fig. 7.2-1 Ammeter used to measure current.

When full-scale current flows there will be a voltage drop across the meter

$$V_m = I_m R_m \tag{7.2-1}$$

where V_m = voltage drop at full-scale, V
I_m = full scale current (current sensitivity), mA
R_m = meter resistance, kΩ

For most ammeters in most circuits, this voltage is small, and causes little error. However, there are situations where significant error is introduced, as in the example which follows.

Example 7.2-1 A 1-mA 200-Ω movement is to be used to measure the current in the circuit of Fig. 7.2-1 where we have a 1-V source feeding a 1 kΩ load.

a Find the error introduced due to the insertion of the meter.
b Find the error introduced if the source is increased to 10 V and the load to 10 kΩ.

Solution The percent error is defined by the equation

$$\text{Percent error} = \frac{I_{\text{meas}} - I_{\text{true}}}{I_{\text{true}}} \times 100$$

where I_{meas} = meter reading (measured value), mA
I_{true} = actual value (with meter not connected), mA

The numerator, $I_{\text{meas}} - I_{\text{true}}$, represents the difference between the measured value and the true value, and is called the *absolute error* or *absolute deviation*. If the presence of the meter did not alter the current being measured, this quantity would be zero.

Using Ohm's law, we find

$$I_{\text{meas}} = \frac{V}{R_m + R_L} \tag{7.2-2}$$

$$I_{\text{true}} = \frac{V}{R_L} \tag{7.2-3}$$

Substituting these equations into Eq. (7.2-1) we obtain an expression for the error in terms of the circuit parameters.

$$\text{Percent error} = \frac{I_{\text{meas}} - I_{\text{true}}}{I_{\text{true}}} \times 100$$

$$= \frac{V/(R_m + R_L) - V/R_L}{V/R_L} \times 100$$

$$\text{Percent error} = \frac{-R_m}{R_m + R_L} \times 100 \tag{7.2-4}$$

Before we substitute values into Eq. (7.2-4) we note that the percent error will always be negative, since $R_m/(R_m + R_L)$ will always be between 0 and +1. A little thought will show that this is quite reasonable, since a negative error means that the presence of the meter has increased the resistance and thus caused a *decrease* in the current in the series circuit.

Finally, substituting into Eq. (7.2-4) with all resistance values in kΩ

a

$$\text{Percent error} = \frac{-0.2}{0.2 + 1} \times 100 = -16.7 \text{ percent}$$

b

$$\text{Percent error} = \frac{-0.2}{10 + 0.2} \times 100 = -1.96 \text{ percent}$$

For the higher resistance circuit the error is almost negligible. Thus, the criterion for small errors due to ammeter insertion is that the circuit resistance should be much larger than the meter resistance. ////

METER SHUNTS

The maximum current which a D'Arsonval movement can carry is equal to its current sensitivity. In order to measure higher currents, we can connect a shunt (*parallel* resistance) across the meter so that only a part of the current will flow through the meter while the additional current flows through the shunt.

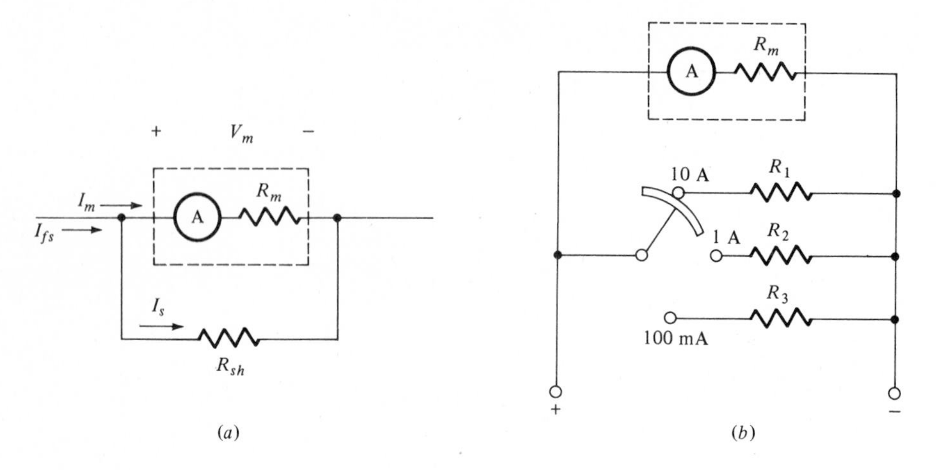

Fig. 7.2-2 Ammeter. (*a*) Shunt arrangement to increase range; (*b*) multirange ammeter.

This additional resistance is shown in Fig. 7.2-2*a*. The value of R_{sh} is calculated using full-scale values as follows:

The current through the shunt I_s will be, using KCL

$$I_s = I_{fs} - I_m \tag{7.2-5}$$

where I_{fs} = desired full-scale current

I_m = current sensitivity of meter movement

Since R_{sh} and R_m are in parallel, the voltage across each must be the same, that is,

$$I_m R_m = I_s R_{sh} \tag{7.2-6}$$

Usually, I_{fs}, I_m, and R_m are known and we would like to solve Eqs. (7.2-5) and (7.2-6) so that R_{sh} is given in terms of them. To do this we solve Eq. (7.2-6) for R_{sh}

$$R_{sh} = \frac{I_m R_m}{I_s}$$

Next substitute for I_s from Eq. (7.2-5) to get

$$R_{sh} = \frac{I_m R_m}{I_{fs} - I_m} \tag{7.2-7}$$

which is the desired relation. Typical values are given in the following example.

Example 7.2-2 It is desired to use a 1-mA 50-Ω D'Arsonval movement to measure 1 A full scale. Find the required shunt resistance in the circuit of Fig. 7.2-2*a*.

Solution The given values are substituted into Eq. (7.2-7) being sure to keep all quantities in proper units. Thus

$$R_{sh} = \frac{(0.001)(50)}{1 - 0.001} = \frac{0.05}{0.999} \approx 0.05\ \Omega \qquad ////$$

In practice, it is desirable to have a multirange ammeter, so that one can select from several different ranges. Various shunt arrangements are possible; one with separate shunts selected by a switch is shown in Fig. 7.2-2*b*. The switch must be of the *make-before-break** type, in order to prevent the possibility of excessive current through the meter movement during range switching. Each range might have its own scale, or one scale might serve with separate multiplying factors for each range. Each of the shunt resistances is determined exactly as R_2 for the 1-A range in Example 7.2-2. (Values for R_1 and R_3 are calculated in Prob. 7.2-3.) In other types of multirange ammeters a series connection of shunts is used. As the meter is switched to higher current ranges, more of the shunt resistance is shorted by the switch. This will be explored further in the problems.

When using multirange ammeters, they should be set initially on their highest current range in order to avoid damage due to excessive current. A lower range can then be selected to give a suitable deflection.

7.3 VOLTMETERS

Since the D'Arsonval movement is current sensitive, it is not suitable for the measurement of voltage without modification. The required modification consists of adding a series resistance R_s (called a *multiplier* resistance) as shown in Fig. 7.3-1*a*. This resistance is chosen so that full-scale current flows through the movement when the desired full-scale voltage V_{fs} is applied across the terminals. We can find the value of this resistance in terms of known parameters by writing KVL around the loop formed by the voltmeter and the voltage being measured as follows:

$$V_{fs} = I_m(R_m + R_s) \qquad (7.3\text{-}1)$$

where V_{fs} = desired full scale voltage
I_m = current sensitivity of meter movement
R_m = resistance of movement
R_s = added series resistance

* The name arises from the fact that, for example, the 1-A terminal is connected to the moving contact *before* the 10-A terminal is disconnected. This avoids the possibility of the meter being without a shunt.

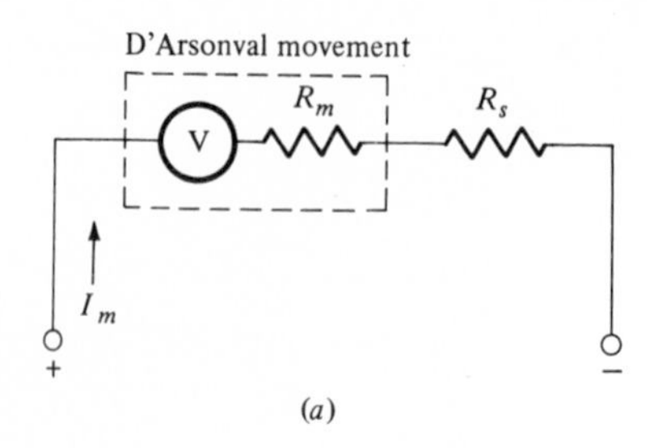

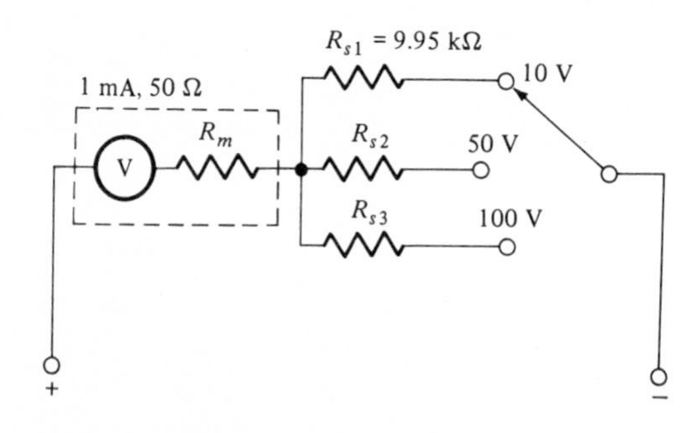

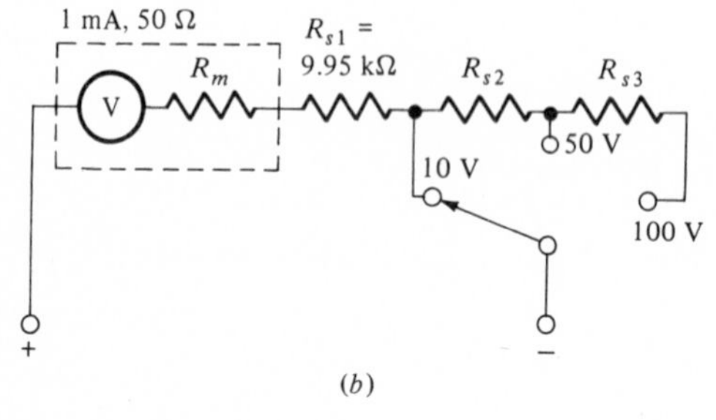

Fig. 7.3-1 Voltmeters. (*a*) Multiplier; (*b*) multirange switching.

In a typical case, V_{fs}, R_m, and I_m are known and R_s is to be determined. Thus, solving for R_s

$$R_s = \frac{V_{fs} - I_m R_m}{I_m} \tag{7.3-2}$$

Numerical values are included in the following example.

Example 7.3-1 The 1-mA 50-Ω movement of Example 7.2-2 is to be used to measure 10 V full scale. Find the required multiplier resistance.

Solution Equation (7.3-2) is used, being sure to keep all given quantities in proper units of V, Ω, and A. Thus

$$R_s = \frac{10 - (0.001)(50)}{0.001} = \frac{10 - 0.05}{0.001}$$

$$= 9.95 \times 10^3 \ \Omega$$

$$= 9.95 \ \text{k}\Omega$$

////

At this point we note that in contrast with the ammeter, the voltmeter is connected *across*, or *in parallel with*, the element or circuit for which the voltage is to be measured. The + terminal of the meter must be connected to the positive side of the voltage being measured in order to obtain an upscale reading.

In order to construct a multi-range voltmeter, various switching arrangements can be used. Two possibilities are shown in Fig. 7.3-1*b* using the 1-mA 50-Ω movement of Example 7.3-1. Calculation of the multiplier resistors is left for the problems.

As with multirange ammeters, precautions should be observed when connecting a multi-range voltmeter across a circuit. The range switch should *always* be turned to the highest voltage range *before* connecting the meter. The range can then be lowered to obtain an adequate deflection.

VOLTMETER SENSITIVITY—THE OHMS-PER-VOLT RATING

Most commercial voltmeters using the D'Arsonval movement will include an *ohms-per-volt* rating which is usually printed on the face of the meter below the scales. This rating is extremely important, as it allows us to immediately determine the current sensitivity of the movement used in the voltmeter, and the total resistance of the meter for any of its ranges. This, in turn, allows us to readily calculate the *loading* effect of the meter on the original circuit.

The sensitivity of the meter, in ohms per volt, is simply the reciprocal of the current sensitivity I_m,

$$\text{Ohms per volt} = \frac{1}{I_m} \tag{7.3-3}$$

In order for a voltmeter to have little effect on the circuit being measured, it should draw as little current as possible. Thus, higher ohms-per-volt ratings are usually more desirable.

The total meter resistance is the ohms-per-volt rating times the full scale voltage for the range being used. This resistance is constant regardless of the magnitude of the voltage being measured. Use of the sensitivity is illustrated in the following example.

Example 7.3-2

a Find the sensitivity of the 1-mA 50-Ω movement of Example 7.3-1.
b Find the total meter resistance for each of the ranges shown in Fig. 7.3-1*b*.

Solution

a

$$\text{Sensitivity} = \frac{1}{I_m} = \frac{1}{10^{-3}} = 1{,}000\ \frac{\Omega}{\text{V}}$$

b

$$R_T = \frac{\Omega}{\text{V}} \times V_{fs}$$

10-V range:

$$R_T = (1{,}000)(10) = 10\ \text{k}\Omega$$

50-V range:

$$R_T = (1{,}000)(50) = 50\ \text{k}\Omega$$

100-V range:

$$R_T = (1{,}000)(100) = 100\ \text{k}\Omega$$

////

METER LOADING AND OTHER ERRORS

The measurement error which results because of the voltmeters effect on the circuit depends on the ratio of the circuit resistance to the voltmeter resistance. Consider Fig. 7.3-2. Clearly, if R_2 and R_T are of the same order of magnitude, the error introduced will be high, because the original circuit without the voltmeter is considerably different from the circuit when the voltmeter is connected. On the other hand, if R_T is much greater than R_2, the disturbance will be small. The next example illustrates these points.

Example 7.3-3 In the circuit of Fig. 7.3-2, $R_1 = R_2 = 330\ \text{k}\Omega$, $V_1 = 100\ \text{V}$, and the voltmeter uses a 50-μA movement and has a full-scale range of 50 V.

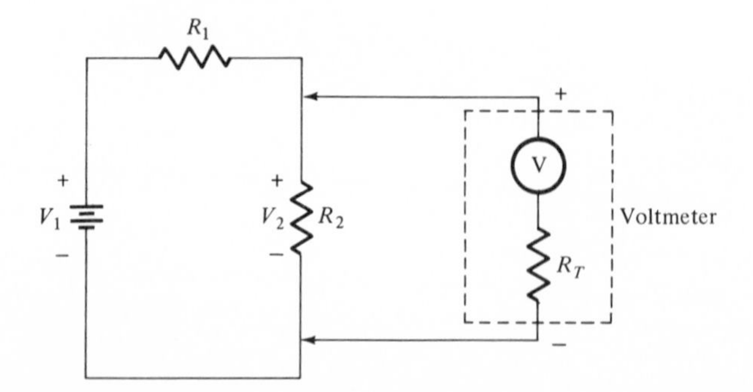

Fig. 7.3-2 Circuit for determining voltmeter loading.

a Find the sensitivity and total resistance of the meter.
b Find the voltage across R_2 without the meter.
c Find the voltage indicated by the meter when it is connected.
d Find the percent error due to voltmeter loading.
e Repeat *b*, *c*, and *d* if $R_1 = R_2 = 33\ \text{k}\Omega$.

Solution

a

$$\text{Sensitivity} = \frac{1}{I_m} = \frac{1}{50 \times 10^{-6}} = \frac{20\ \text{k}\Omega}{\text{V}}$$

$$R_T = \left(20\,\frac{\text{k}\Omega}{\text{V}}\right)(50\ \text{V}) = 1\ \text{M}\Omega$$

b Using the voltage-divider formula,

$$V_2 = V_1 \frac{R_2}{R_1 + R_2} = 100\left(\frac{330}{330 + 330}\right) = 50\ \text{V}$$

c The resistance of the parallel connection of R_2 and the meter is (keeping all resistances in units of MΩ)

$$R_{eq} = \frac{R_T R_2}{R_T + R_2} = \frac{(1)\,(0.33)}{(1 + 0.33)} = 0.248\ \text{M}\Omega$$

Since the meter measures the voltage across R_{eq}, (this time resistances are in kΩ for convenience).

$$V_{2m} = V_1 \frac{R_{eq}}{R_1 + R_{eq}} = 100\left(\frac{248}{330 + 248}\right) = 43\ \text{V}$$

d The error is $43 - 50 = -7$ V. Thus

$$\text{Percent error} = \frac{-7}{50} \times 100 = -14\ \text{percent}$$

e

$$V_2 = 100\left(\frac{33}{33 + 33}\right) = 50\ \text{V}$$

$$R_{eq} = \frac{(1)\,(.033)}{1 + .033} = .032\ \text{M}\Omega$$

$$V_{2m} = 100\left(\frac{32}{33 + 32}\right) = 49\ \text{V}$$

$$\text{Percent error} = \frac{49 - 50}{50} \times 100 = -2\ \text{percent}$$

Clearly, the loading effect decreases as the ratio of meter resistance to circuit resistance increases. The student can further convince himself of this by repeating the calculations above for a 1-mA movement. ////

In the previous example, the error was due totally to loading of the circuit by the meter. Readings may also depart from the true value because of inaccuracy of the movement itself. This is usually specified by the manufacturer and typical values are ± 1, 2, 3, and 5 percent *of full scale*. This means that for a 50-V scale and ± 2 percent accuracy the possible error is $(\pm.02)(50) = \pm 1$ V. Thus the true value of a reading of 40 V on this scale will actually lie between 39 V and 41 V. For a 5-V reading, the actual value will lie between 4 V and 6 V. These numbers indicate clearly that meter ranges should be chosen so that readings are as close to full-scale as possible. Readings below 10 percent of full scale are avoided whenever possible because of the large errors involved.

Other causes of error in measurements are called *systematic* errors which may be attributed to miscalibration, nonlinearity, undetected shift of the zero point, or bias on the part of the experimenter. This type of error usually tends to be in one direction and to have a relatively constant magnitude. The previously discussed errors were *random* in nature, as indicated by the ± specification.

These *instrument errors* should be carefully distinguished from *instrument loading* which was illustrated in Example 7.3-2. Not too much can be done about instrument errors unless one has a very large budget with which to buy better instruments. However, instrument loading can be corrected for or minimized by careful design of the measuring method.

7.4 RESISTANCE MEASUREMENT

THE SERIES OHMMETER

For routine measurement of resistance the *series* ohmmeter circuit of Fig. 7.4-1*a* is used. The accuracy is not impressive since it lies in the 10 to 20 percent range for most meters. However, for many applications this is adequate.

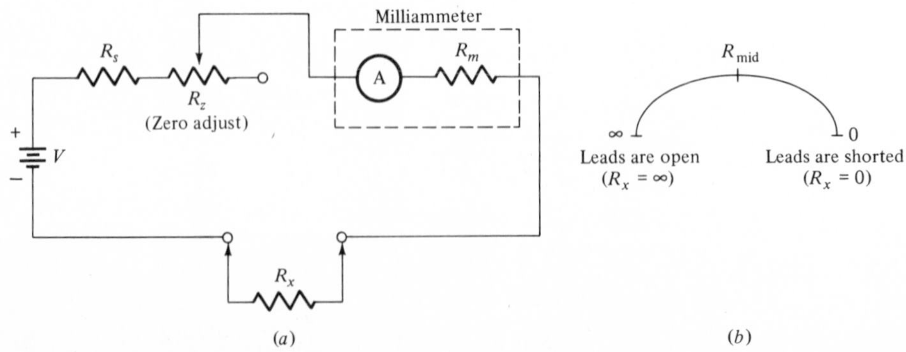

Fig. 7.4-1 Series ohmmeter. (*a*) Circuit; (*b*) scale.

If the terminals are left open so that $R_x = \infty$, no current will flow, and the meter will not register. Thus the left end of the scale corresponds to $R_x = \infty$ as shown in Fig. 7.4-1*b*. When the terminals are shorted (connected directly together), $R_x = 0$ and the remaining resistance $R_s + R_z + R_m$ is chosen so that full-scale current flows. Thus the right end of the scale (full-scale) corresponds to $R_x = 0$. The required resistance can be found by noting that the meter current is

$$I = \frac{V}{R_s + R_z + R_m + R_x} \tag{7.4-1}$$

where R_z is a small variable resistance used as a *zero adjust* control to calibrate the ohmmeter and compensate for battery aging by bringing the pointer to zero when the input test leads are shorted. Thus, the full-scale current occurs when $R_x = 0$:

$$I_m = \frac{V}{R_s + R_z + R_m} \tag{7.4-2}$$

Exactly one-half of this value will flow when $R_x = R_s + R_z + R_m$ and this is the midscale resistance R_{mid}. Thus R_s is chosen a small amount less than the desired midscale value to allow for R_z and R_m and then V is chosen to give the desired full-scale current.

One disadvantage of the series ohmmeter can be seen by considering the scale shown in Fig. 7.4-1*b*. The left half of the scale covers all values from R_{mid} to ∞, and the right half all values from 0 to R_{mid}. Thus the scale is very nonlinear and accuracy is poor, especially at the high resistance end of the scale where the values crowd together. Usually, ohmmeters are provided with multiple ranges so that different midscale values are available. In the example which follows, typical values are calculated.

Example 7.4-1 A 50-μA 200-Ω meter movement is used in the series ohmmeter circuit of Fig. 7.4-1*a* along with a 3-V battery. Find suitable values for R_s and R_z so that the midscale resistance is 10 kΩ.

Solution In order to have $R_{\mathrm{mid}} = 10$ kΩ we must have

$$R_s + R_z + R_m = 10 \text{ k}\Omega$$

Since $R_m = 200\ \Omega = 0.2$ kΩ

$$R_s + R_z = 9.8 \text{ k}\Omega$$

For the zero adjust, we assume that an adjustment of ± 25 percent of full scale will be adequate. Thus if we use a 5 kΩ variable resistance for R_z and adjust R_s so that exactly full scale current flows when the variable resistance R_z is at one-half of its maximum value, we will achieve the desired result.

Then,

$$R_s + \frac{5 \text{ k}\Omega}{2} = 9.8 \text{ k}\Omega$$

$$R_s = 7.3 \text{ k}\Omega$$

The final design values are then

$R_s = 7.3 \text{ k}\Omega$ fixed resistance

$R_z = 5 \text{ k}\Omega$ variable resistance ////

At this point we must warn the student that the ohmmeter (of *any* type) must *never be connected to a circuit in which sources are active.* The active circuit may try to force a current through the movement which is much larger than its full-scale value and thus cause permanent damage.

THE SHUNT OHMMETER

For relatively low values of resistance, a somewhat different arrangement, called a *shunt* ohmmeter is used. The circuit is shown in Fig. 7.4-2*a* where we see that it differs from the series ohmmeter in that the meter movement is directly in shunt (parallel) with the unknown resistance R_x. The switch is included so that power will not be drawn from the battery when the ohmmeter is not in use.

For this ohmmeter, the condition $R_x = 0$ places a direct short across the meter movement so that $R_x = 0$ corresponds to zero meter indication as shown in Fig. 7.4-2*b*. When $R_x = \infty$ maximum current flows in the meter movement and R_s and R_z are adjusted so that this equals the current sensitivity of the meter movement.

Equation (7.4-2) holds for the shunt ohmmeter when $R_x = \infty$ with R_z now called an *infinity adjust* resistor. As contrasted to the series instrument, a midscale reading occurs when $R_x = R_m \,||\, (R_s + R_z)$ so that the shunt ohmmeter tends to have a relatively low midscale reading.

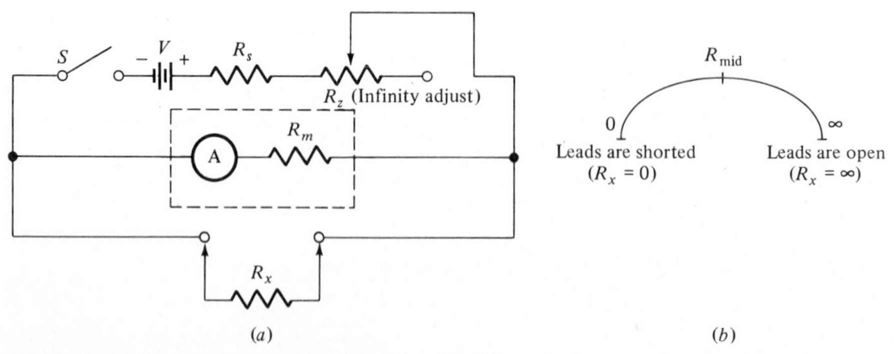

Fig. 7.4-2 Shunt ohmmeter. (*a*) Circuit; (*b*) scale.

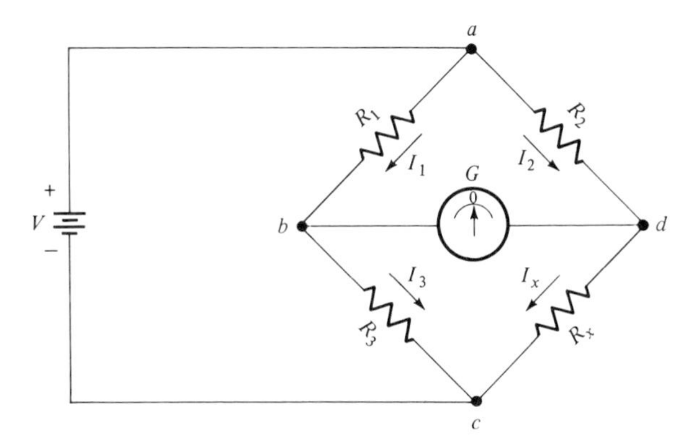

Fig. 7.4-3 Wheatstone bridge circuit.

BRIDGE MEASUREMENT OF RESISTANCE

When ohmmeter measurements do not provide the required accuracy we must turn to some sort of *bridge* measuring device, the *Wheatstone Bridge* being one of the most popular types. The basic circuit is shown in Fig. 7.4-3, where R_1, R_2 and R_3 are standard resistors which are known very accurately, often to 0.01 percent, R_x is the unknown resistance, and G is a sensitive zero-center galvanometer (microammeter).

In operation, one of the standard resistances is varied until the galvanometer current is zero. This is called the *null*, or *balanced*, condition for the bridge. Since the meter current is zero at balance, the voltage across it is also zero. Then, using Ohm's law and KVL we obtain two equations, as follows:

$$I_1R_1 = I_2R_2$$

$$I_3R_3 = I_xR_x \tag{7.4-3}$$

Noting that the meter current is zero, application of KCL at nodes b and d yields the current equations

$$I_1 = I_3$$

$$I_2 = I_x \tag{7.4-4}$$

Now we divide the two equations in Eq. (7.4-3) to get

$$\frac{I_1R_1}{I_3R_3} = \frac{I_2R_2}{I_xR_x}$$

Noting Eq. (7.4-4) the currents cancel and the balance condition becomes

$$\frac{R_1}{R_3} = \frac{R_2}{R_x} \tag{7.4-5}$$

Fig. 7.4-4 Commercial Wheatstone bridge. (*Leeds & Northrup Corp.*)

Finally, solving for R_x

$$R_x = \frac{R_3R_2}{R_1} \tag{7.4-6}$$

The value of the unknown can be calculated from Eq. (7.4-6) and is independent of the battery voltage V. In commercial versions of the Wheatstone bridge as shown in Fig. 7.4-4 switches are used to select suitable values of the standard resistances and the value of the unknown resistance is read directly from the dials when the bridge is balanced.

7.5 MULTIMETERS

We have seen that the basic D'Arsonval meter movement can be used to measure current, voltage, and resistance by properly choosing the associated circuitry. With the use of gangs of switches, all three types of measurements, with several ranges for each type, can be performed with one instrument. Such instruments are called volt-ohm-milliammeters, usually abbreviated VOM. They usually have provisions for both ac and dc measurements and are battery operated for portability. A photograph of one popular VOM is shown in Fig. 7.5-1 and a schematic diagram is shown in Fig. 7.5-2. In the schematic, the Range switch is shown in the 500/1,000 V position and the Function switch is at +dc. Tracing the schematic leads to the voltmeter circuit shown in Fig. 7.5-3. The circuits corresponding to other switch positions will be considered in the problems. Note that this VOM provides 8 dc voltage, 6 dc current, and 3 resistance ranges in addition to various ac ranges. The rated accuracy for dc measurements is 2 percent of full scale and the meter sensitivity is 20,000 Ω/V.

One drawback of the previously described VOM is its relatively low Ω/V rating. Other, more expensive, VOMs utilize solid state devices called field-effect transistors (FETs) to increase the effective meter resistance to about 10 MΩ on *all* ranges. Thus the FET VOM will have negligible loading effect on almost any circuit to which it is connected.

The most sophisticated multimeters available at the time of this writing are digital multimeters, an example of which is shown in Fig. 7.5-4. These

Fig. 7.5-1 Simpson 260 series VOM. (*Simpson Electric Co.*)

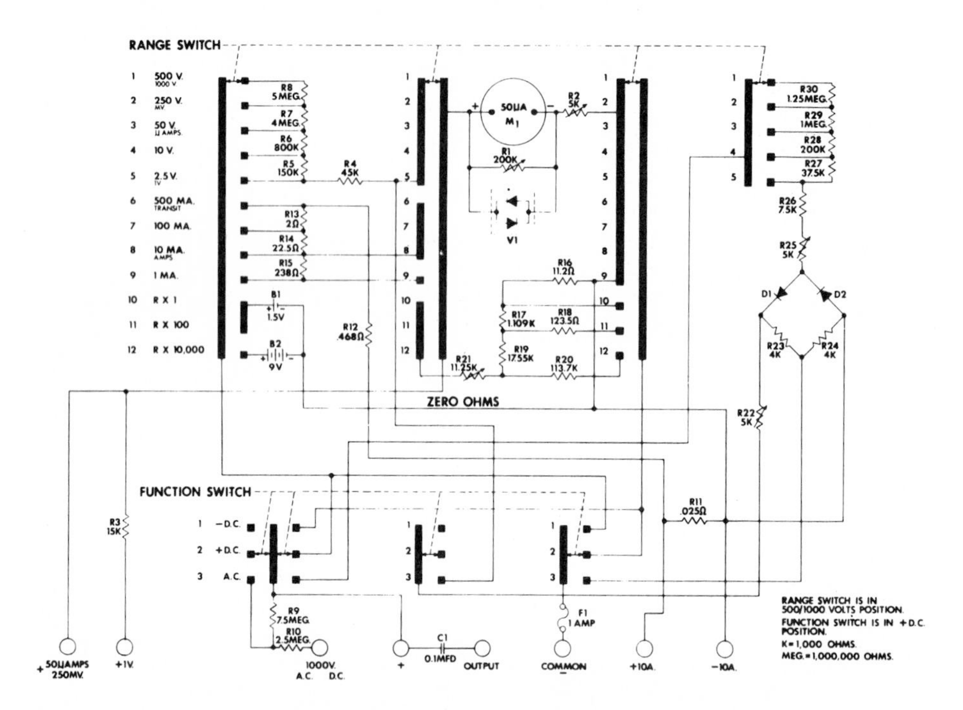

Fig. 7.5-2 Schematic of Simpson 260 VOM. (*Simpson Electric Co.*)

instruments operate on completely different principles than the previously described VOMs. They utilize complex digital circuits to provide digital displays and feature input resistances of 1,000 MΩ or more and extremely high accuracy, typically 0.05 percent of full scale. As might be expected, they are more expensive than the other types of VOM.

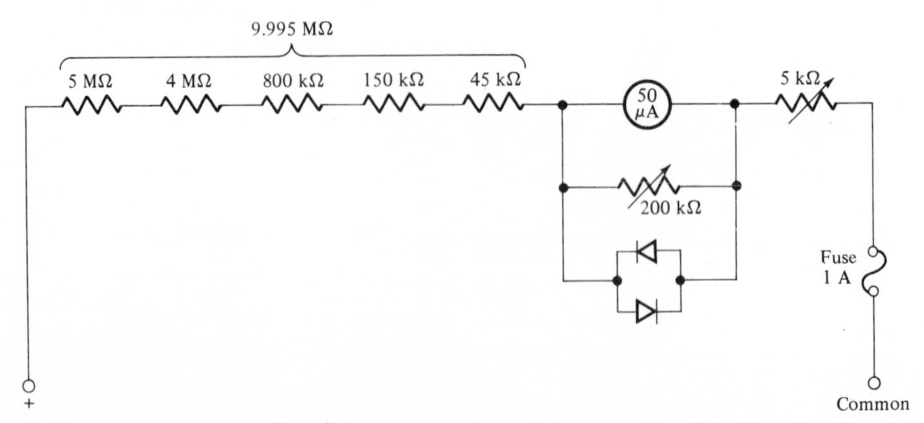

Fig. 7.5-3 Voltmeter circuit of Simpson 260 on 500-V range. Note: At 20,000 Ω/V, the total resistance on this range should be 20,000 Ω/V × 500 V = 10 MΩ.

Fig. 7.5-4 Digital multimeter. (*Weston Instruments, Inc.*)

7.6 OTHER TYPES OF ANALOG METER

7.6-1 The Iron-Vane Meter Movement

A second type of meter movement utilizing magnetic effects is called the *iron-vane* mechanism. When two strips of magnetic material are placed in a magnetic field they are magnetized alike so that a force of repulsion exists between them. Fig. 7.6-1 shows such a movement with concentric circular vanes. When current flows through the coil a magnetic field results and the vanes are magnetized as shown. The repulsive force causes the movable vane to move the pointer in an upscale direction. Increased current causes an increase in the force and moves the pointer farther upscale. If the current reverses, the N and S poles in both vanes are interchanged but the repulsive force remains in the same direction so that the pointer again moves upscale. Thus this type of movement reads the same regardless of the current direction and can be used to measure both ac and dc. When used on dc corrections must be made to account for the residual magnetism in the iron vanes. A photograph showing the construction of a commercial-vane movement is shown in Fig. 7.6-2.

7.6-2 The Electrodynamometer Movement

This type of movement is more expensive than the D'Arsonval or iron-vane types but can be used as a wattmeter for both dc and ac. The basic arrangement is shown in Fig. 7.6-3. The magnetic field in which the moving coil rotates is

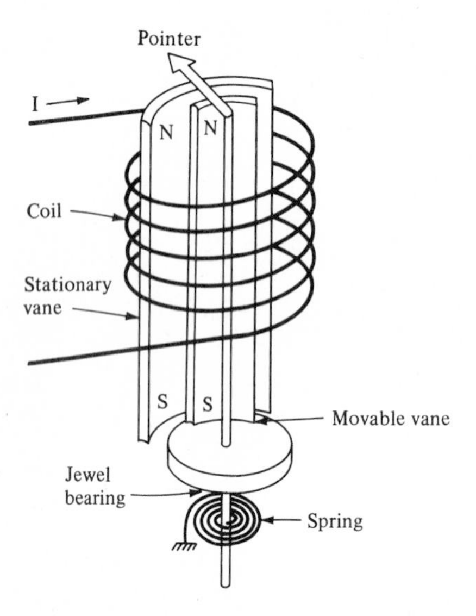

Fig. 7.6-1 Iron-vane movement.

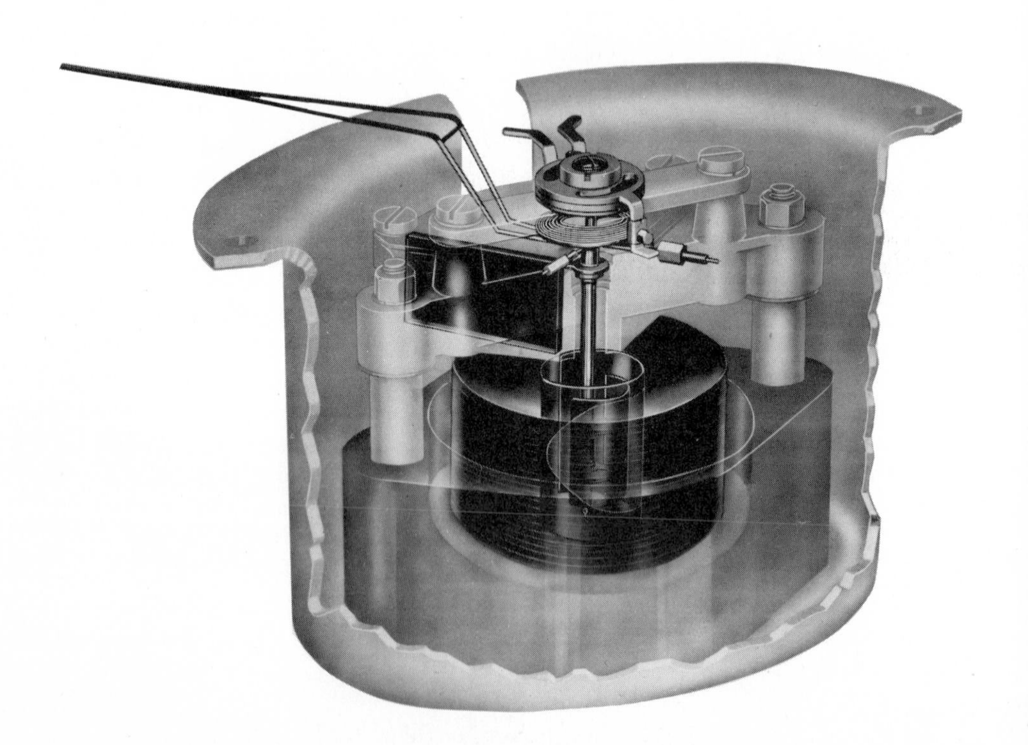

Fig. 7.6-2 Iron-vane movement. (*Weston Instruments, Inc.*)

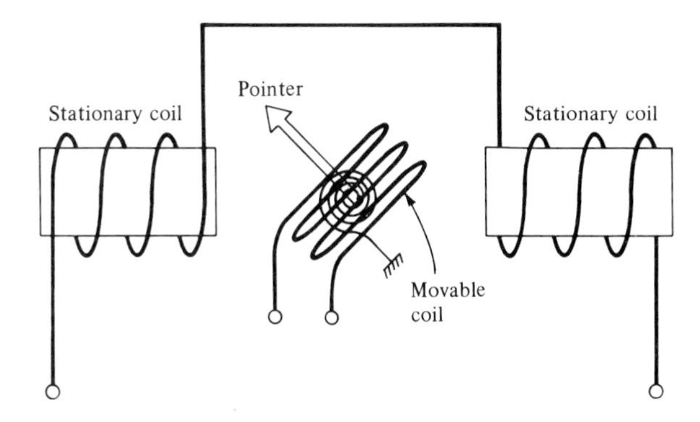

Fig. 7.6-3 The electrodynamometer mechanism.

provided by the current flowing in the stationary coils. If the movable and stationary coils are in series, the same current flows through both and since the torque is proportional to the product of the currents the movement of the pointer will also be proportional to this product which represents the square of the current. The resulting scale, called a *square-law* scale, is nonlinear. Also, if the current reverses, it does so in both coils simultaneously so the direction of torque does not change. Thus the movement can be used for both dc and ac.

When this movement is used as a wattmeter, the line current is passed through the stationary coil (called the current coil) and the line voltage is applied to the movable coil (the potential coil) through a multiplier resistance (see Fig. 7.6-4a). Thus the current in the potential coil is proportional to the line voltage and as noted in the previous paragraph the movement of the pointer is proportional to the product of the two currents. The meter reading is then pro-

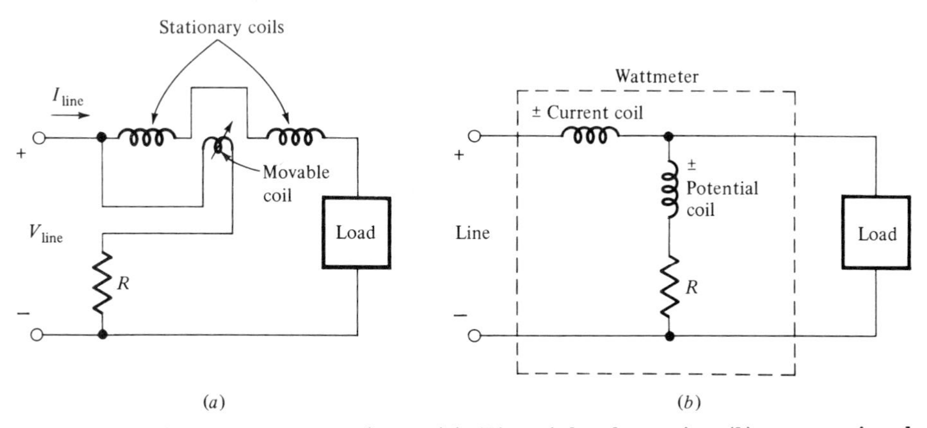

Fig. 7.6-4 Wattmeter connections. (*a*) Pictorial schematic; (*b*) conventional schematic.

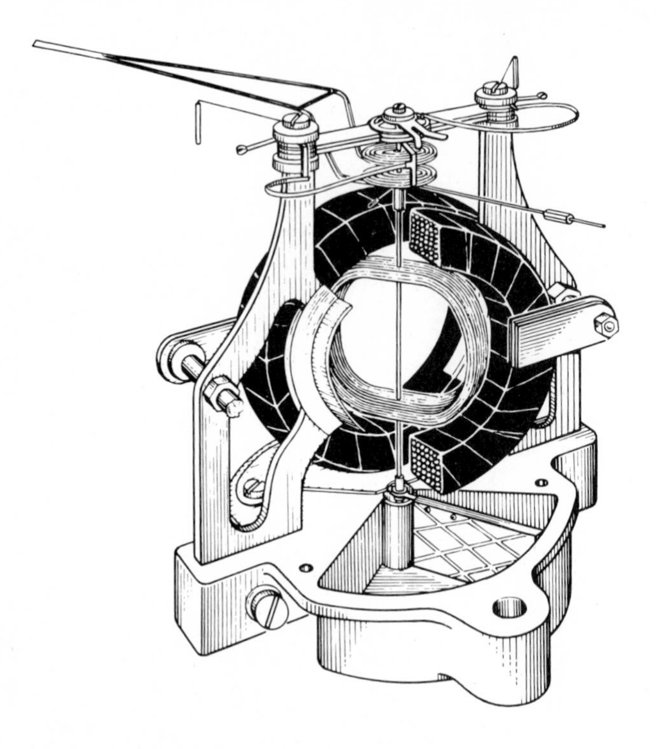

Fig. 7.6-5 Electrodynamometer movement. (*Weston Instruments, Inc.*)

portional to the power. A cutaway drawing of a commercial electrodynamometer movement is shown in Fig. 7.6-5.

SUMMARY

1 The D'Arsonval movement is a current-sensitive device which is used with shunts to measure current and with multipliers to measure voltage.
2 The sensitivity of a D'Arsonval movement is the current required for full-scale deflection.
3 An ammeter is always connected in *series*, and its resistance should be low compared to the circuit resistance.
4 A voltmeter is always connected in *parallel*, and its resistance should be high compared to the circuit resistance.
5 The resistance of a given voltmeter is found by multiplying its sensitivity (ohms-per-volt) rating by the full-scale voltage.
6 Ohmmeters must *never* be connected to an energized circuit.
7 The Wheatstone bridge is used for precision measurement of resistance.

PROBLEMS

7.2-1 A 1-mA 200-Ω movement reads 0.92 mA when exactly 0.9 mA is flowing. What is the percent error?

7.2-2 A 50-μA 1,000-Ω D'Arsonval movement is to be used to measure current in the circuit of Fig. 7.2-1. In the circuit $V = 4$ V and $R = 100$ kΩ. Find the error introduced due to the insertion of the meter.

7.2-3 Find values for R_1 and R_3 in Fig. 7.2-2*b*. (Refer to Example 7.2-2 in which the value of R_2 was found.)

7.2-4 A 50-μA 1,000-Ω movement is to be used in a milliammeter having ranges of 10 mA, 25 mA, 50 mA, and 100 mA. Draw a circuit and show all component values. What voltage is across the meter when full scale current flows for each range?

7.2-5 A 1-mA 100-Ω movement is to be used in a multirange ammeter to measure full-scale currents of 1 A, 5 A, 10 A, and 50 A. Draw a circuit and show all component values.

7.2-6 A dc milliammeter has a full scale reading of 5 mA. If the resistance on this scale is 2 Ω, what shunt resistance is required to extend the range to 5 A?

7.3-1 Find suitable values for R_{s2} and R_{s3} in both circuits of Fig. 7.3-1*b*.

7.3-2 A voltmeter with a full scale value of 30 V is to be made up using a 5-mA 20-Ω movement. Find the required multiplier resistance and the ohms-per-volt rating.

7.3-3 Repeat Prob. 7.3-2 for a 50-μA 1,000-Ω movement.

7.3-4 A certain movement has a voltage of 100 mV across it at full scale. If the meter resistance is 10 Ω what value of multiplier resistance will convert it to a 100-V meter?

7.3-5 Repeat Example 7.3-3 for a 1-mA movement and compare the results.

7.3-6 A 50-V voltmeter has an accuracy of ± 2 percent of full scale. When the meter reads 29 V, what is the range of the true value?

7.3-7 A multirange voltmeter has a sensitivity of 20 kΩ/V. Find the full scale current and the resistance of each range if the ranges are 2.5, 10, 50, and 250 V.

7.3-8 A circuit consists of two 50-kΩ resistors in series connected to a 100-V source. A 100-V voltmeter with a resistance of 2 MΩ is used to measure the voltage across one of the resistors. Find the percent error in the measured value.

7.4-1 A series ohmmeter is to use the following:

1 100-μA 1,000-Ω movement

2 Zero-adjust variable resistor of value 2 kΩ
3 3-V battery
4 Series resistor to be determined

a Find $R_s + R_z$ and the resistance R_x required for full-scale, 3/4 scale, 1/2 scale, and 1/4 scale deflection.

b Draw the scale to be used with the meter.

7.4-2 For the shunt ohmmeter, prove that the midscale reading occurs when $R_x = R_m \,||\, (R_s + R_z)$.

7.4-3 Design a shunt ohmmeter using a 1-mA, 50-Ω movement, variable zero-adjust resistor of 200-Ω 2-V battery, and shunt resistor to be determined. Find $R_s + R_z$ and the unknown resistance R_x for full, 3/4, 1/2, and 1/4 scale deflection and draw the scale.

7.4-4 The Wheatstone Bridge shown in Fig. P7.4-4 is balanced. Find R_x.

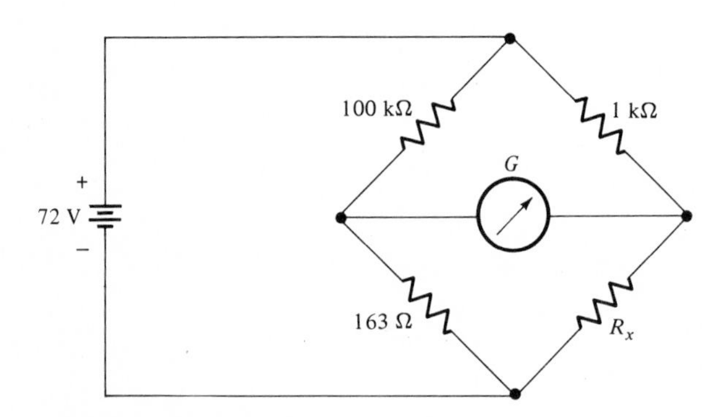

Fig. P7.4-4.

7.5-1 In the multimeter circuit of Fig. 7.5-2, trace the 10-V dc voltmeter range and draw the resulting circuit.

7.5-2 Repeat Prob. 7.5-1 for the 1-mA dc range.

8
Linear Network Theorems

OBJECTIVES

Upon completion of this chapter you should be able to

1 State the linearity principle.
2 Apply the superposition principle to the analysis of multisource networks.
3 Use loop and node equations for the analysis of two-loop and two-node networks.
4 Find the Thevenin and/or Norton equivalent of a given network.
5 Write an ECAP program for a given network.

INTRODUCTION

All of the circuits we have analyzed up to this point consisted of resistors and sources connected in such a way that the circuits could be reduced to a single source and a single resistance. These were relatively simple circuits and the basic tools required for their analysis were Ohm's law and KCL and KVL. When the circuits to be analyzed are more complex, and cannot be reduced to simple form the tools are the same, but we must resort to more advanced network analysis techniques.

In this chapter we will consider several of the most often-used of these techniques.

8.1 LINEARITY AND SUPERPOSITION

The circuits which we have been considering were all *linear* circuits consisting of linear elements. A linear element is an element that has a straight-line vi characteristic. The only passive element we have considered thus far is the resistance, which has a vi characteristic given by Ohm's law:

$$v(t) = Ri(t)$$

This is obviously linear.

In addition, we must specify that if our circuits contain dependent sources, they must be linear. Such sources have output currents or voltages proportional only to the first power of another current or voltage in the network. Dependent sources such as $i_1 = 3v_x^2$ are *nonlinear*, and a network containing such a source cannot be analyzed using the techniques to be described.

Consider that we are analyzing a linear network consisting only of resistors and two independent current sources I_1 and I_2. There are no voltage sources, and no dependent sources. Further, let us assume that the response of interest is the voltage V_1. A specific circuit answering this description is shown in Fig. 8.1-1a. The analysis will yield an equation

$$V_1 = K_1I_1 + K_2I_2 \tag{8.1-1}$$

The coefficients K_1 and K_2 depend only on the resistances in the network. Since these are linear, K_1 and K_2 will be constant and V_1 is then a linear function of I_1 and I_2, as we would expect from a linear network.

One consequence of linearity is that the response V_1 is proportional to the excitations I_1 and I_2. Thus, if I_1 and I_2 are both doubled, the response is doubled. This is true not only of V_1, but of all other currents and voltages in the network, as well as the outputs of any dependent sources. This principle can be extended to a linear network of any complexity. In the general case, the *linearity principle* can be stated as follows:

If all independent sources are multiplied by a constant, then all of the re-

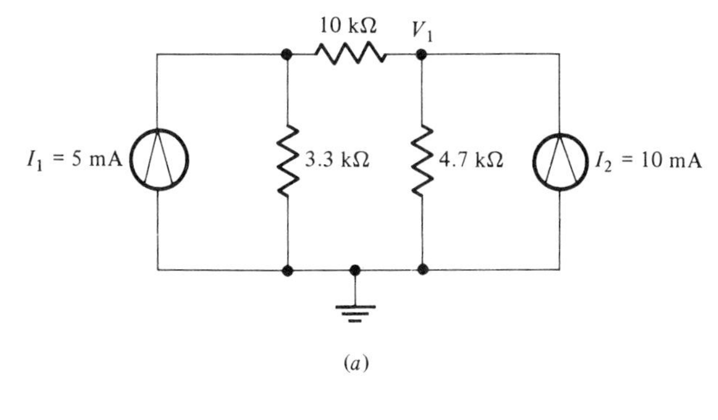

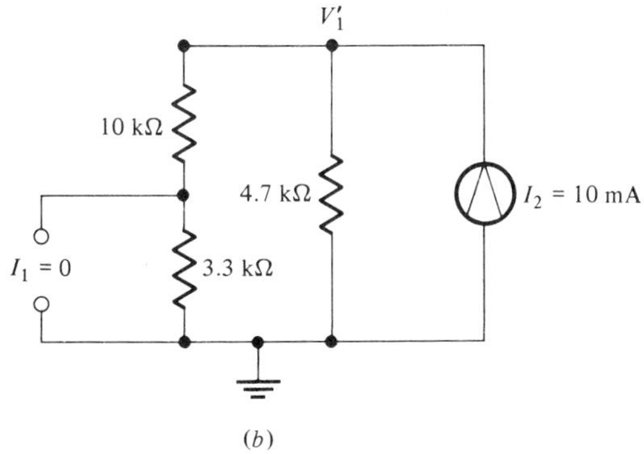

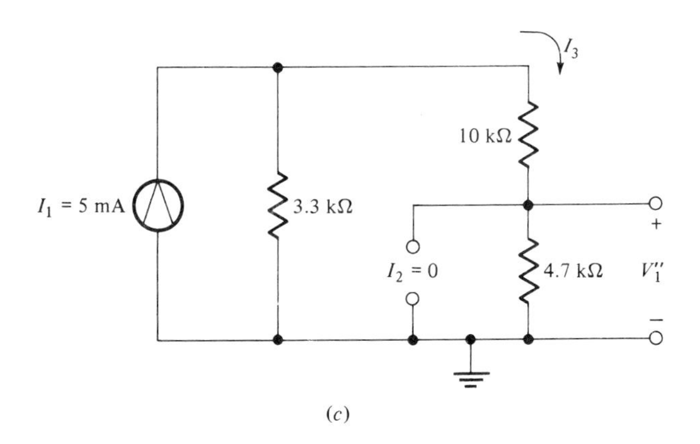

Fig. 8.1-1 Circuit for Example 8.1-1. (*a*) Circuit; (*b*) circuit for $I_1 = 0$; (*c*) circuit for $I_2 = 0$.

sponses (currents and voltages) in the network will be multiplied by the same constant.

Probably the most important consequence of linearity is the superposition theorem. In order to obtain it we return to the response Eq. (8.1-1) and consider each source separately. If we pretend to replace the independent source I_1 by an open circuit and call the resulting response V_1', then

$$V_1' = V_1 \bigg|_{I_1=0} = K_2 I_2 \tag{8.1-2}$$

We next pretend to open-circuit source I_2 and call the resulting response V_1'' so that

$$V_1'' = V_1 \bigg|_{I_2=0} = K_1 I_1 \tag{8.1-3}$$

Now we add Eqs. (8.1-2) and (8.1-3) to get

$$\begin{aligned} V_1' + V_1'' &= K_1 I_1 + K_2 I_2 \\ &= V_1 \end{aligned} \tag{8.1-4}$$

Thus the response V_1 can be found as the *sum* of two components, each due to one source *acting alone*, while the other is set to zero. This is the principle of superposition applied to the network which led to Eq. (8.1-1). It is extremely useful because it reduces a complicated problem to a series of much simpler problems. A general statement of the *superposition principle* is as follows:

In a linear network containing more than one source (current or voltage), any response can be found by adding the response due to each source acting alone, all others being set to zero (short circuits for voltage sources, open circuits for current sources).

Thus, if there are five sources, we analyze five circuits, each containing only one source. Often the circuits containing only one source can be solved by inspection while the original circuit containing all of the sources would require the use of more complicated techniques and the solution of simultaneous equations. The utility of this theorem is demonstrated in the following examples.

Example 8.1-1 In the circuit of Fig. 8.1-1*a* find the response V_1 using superposition.

Solution Using superposition we analyze two circuits, each with only one source, and then simply add the results. The first circuit, with $I_1 = 0$ is shown in Fig. 8.1-1*b*. This is particularly easy to solve, since all we need to do is find the total resistance and then apply Ohm's law. We do this in one step, using kΩ and mA.

$$\begin{aligned} V_1' &= R_{eq} I_2 = [(10 + 3.3) \parallel 4.7](10) \\ &= \frac{(13.3)(4.7)(10)}{(13.3 + 4.7)} = 34.7 \text{ V} \end{aligned}$$

The second circuit to be analyzed, with $I_2 = 0$, is shown in Fig. 8.1-1c. Here we use the current divider to find I_3 and then Ohm's law to find V_1''.

$$I_3 = \frac{3.3}{10 + 4.7 + 3.3}(5 \text{ mA}) = 0.92 \text{ mA}$$

$$V_1'' = I_3(4.7) = (0.92 \text{ mA})(4.7 \text{ k}\Omega) = 4.31 \text{ V}$$

Finally, the desired voltage is the *sum*,

$$\begin{aligned} V_1 &= V_1' + V_1'' \\ &= 34.7 + 4.31 \\ &\approx 39 \text{ V} \end{aligned}$$

////

Example 8.1-2 The circuit of Fig. 8.1-2a has both a voltage source and a current source. Find V_2 using superposition.

Solution Let V_2' be the response to V_1 with $I_1 = 0$. The corresponding

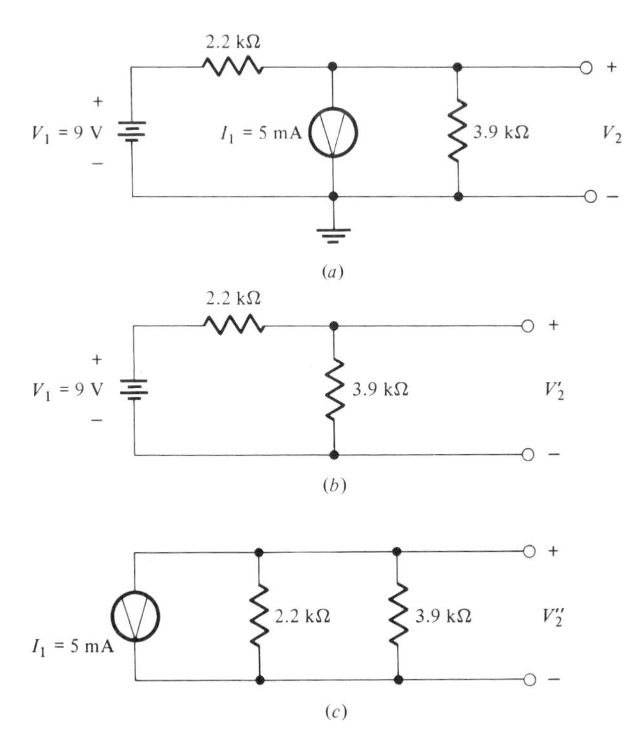

Fig. 8.1-2 Circuit for Example 8.1-2. (*a*) Circuit; (*b*) circuit for $I_1 = 0$; (*c*) circuit for $V_1 = 0$.

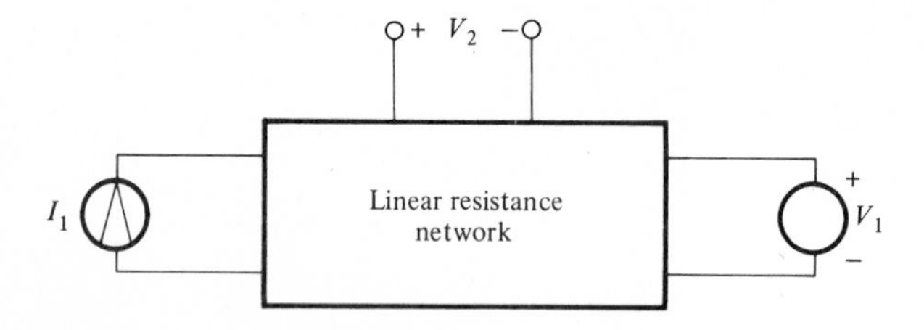

Fig. 8.1-3 Circuit for Example 8.1-3.

circuit is shown in Fig. 8.1-2*b*. Using the voltage-divider formula

$$V_2' = \left(\frac{3.9}{3.9 + 2.2}\right)9 = 5.75 \text{ V}$$

V_2'' is the response to I_1 with $V_1 = 0$ and the corresponding circuit rearranged for clarity is shown in Fig. 8.1-2*c*. Using Ohm's law

$$V_2'' = -\left(2.2 \,||\, 3.9\right)5 = -(1.41)\,5 = -7.03 \text{ V}$$

Finally, $V_2 = V_2' + V_2''$
$= 5.75 - 7.03$
$= -1.28$ V ////

Example 8.1-3 In the circuit of Fig. 8.1-3, V_1 and I_1 are independent sources, and V_2 is the response. A voltmeter is connected to read V_2 and the following data are observed:

a When $V_1 = 20$ V and $I_1 = 3$ A; $V_2 = 6$ V.
b When $V_1 = 0$ V and $I_1 = 2$ A; $V_2 = 1$ V.

Find V_2 when $V_1 = 8$ V and $I_1 = 5$ A.

Solution Since the network is linear, we can write an equation similar to Eq. (8.1-1).

$$V_2 = AV_1 + BI_1$$

The constants A and B are found from the experimental data. Substituting the measured values

a $6 = 20\,A + 3\,B$
b $1 = 0\,A + 2\,B$

Solving these two equations leads to the values

$$B = 0.5 \text{ and } A = 0.225$$

Thus, the response equation is

$$V_2 = 0.225\ V_1 + 0.5\ I_1$$

For the desired conditions, $V_1 = 8$ V and $I_1 = 5$ A, we obtain

$$V_2 = 0.225(8) + 0.5(5) = 4.3\ \text{V}$$ ////

8.2 TWO-LOOP AND TWO NODE-PAIR NETWORKS

In Chap. 6 we solved single node-pair networks using the node-voltage method and single-loop networks using the loop-current method. In this section both of these methods will be extended to more complicated circuits.

TWO NODE-PAIR NETWORKS

Consider the network of Fig. 8.2-1 in which there are two independent nodes labeled a and b, and two independent current sources. The reference node is, in order to simplify the analysis, taken as the node which has the largest number of branches connected to it. Our problem is to find all voltages and currents in the network and the decision we have to make concerns the method to be used. For this circuit we use the node-voltage method in order to illustrate it. At the end of this section we will discuss reasons why one method may be preferable to another in specific cases.

Recall from Chap. 6 that the node-voltage method involves application of KCL at each independent node. The currents in the KCL equation are expressed in terms of the node voltages and element vi relations. The KCL equations are then solved for the node voltages which are, in turn, used to find all currents.

There is a definite sequence of steps to be followed in the application of the method which we illustrate using the circuit of Fig. 8.2-1.

Step 1 Choose a reference node and identify and label the voltage at each independent node. For our example, there are two independent nodes, a and b, having node voltages V_a and V_b; both with respect to ground, which is assumed at zero voltage.

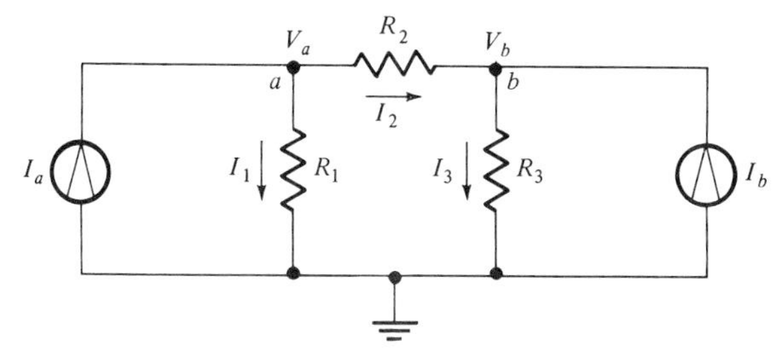

Fig. 8.2-1 Two node-pair network.

Step 2 In order to write KCL equations we assume branch currents through all of the resistances as shown in the figure. Currents through any resistors connected to ground are always assumed to flow in that direction, i.e., from the node to ground. All other currents can be assumed to flow in any direction; for example, I_2 could have been chosen as flowing from node b to node a.

Step 3 Express the branch currents in terms of the node voltages. Here we have to be careful to apply Ohm's law correctly. Thus

$$I_1 = \frac{V_a}{R_1} \tag{8.2-1}$$

$$I_2 = \frac{V_a - V_b}{R_2} \tag{8.2-2}$$

$$I_3 = \frac{V_b}{R_3} \tag{8.2-3}$$

Step 4 Write a KCL equation for each independent node

Node a:

$$I_1 + I_2 = I_a \tag{8.2-4}$$

Node b:

$$I_3 - I_2 = I_b \tag{8.2-5}$$

Step 5 Substitute the Ohm's law equation into the KCL equations to obtain a set of simultaneous equations in the unknown node voltages. Using Eqs. (8.2-1), (8.2-2), and (8.2-3) in Eqs. (8.2-4) and (8.2-5) we have

$$\frac{V_a}{R_1} + \frac{V_a - V_b}{R_2} = I_a$$

$$\frac{V_b}{R_3} - \frac{V_a - V_b}{R_2} = I_b$$

Combining terms:

$$V_a\left(\frac{1}{R_1} + \frac{1}{R_2}\right) - V_b\left(\frac{1}{R_2}\right) = I_a \tag{8.2-6}$$

$$-V_a\left(\frac{1}{R_2}\right) + V_b\left(\frac{1}{R_2} + \frac{1}{R_3}\right) = I_b \tag{8.2-7}$$

This is the set of two simultaneous equations which describe the circuit. They must be solved for V_a and V_b using determinants or some other method.

The coefficients of the voltages in these equations are related to the physical structures of the network. The coefficient of V_a in Eq. (8.2-6), the equation for node a, represents the total conductance *connected to* node a, and the coefficient of $-V_b$ represents the conductance connected *between* nodes a and b. In Eq. (8.2-7), the equation for node b, the coefficient of V_b represents the total conductance connected to node b while the coefficient of $-V_a$ represents the conductance between b and a (which is obviously the same as that between a and b). These interpretations of the coefficients can be used as a check when equations are written for any network.

Step 6 Solve for the branch currents using Eqs. (8.2-1) through (8.2-3). This completes the solution. In a practical case it would usually not be necessary to solve for all of the unknown voltages and currents.

Step 7 Check all answers.

A numerical example follows.

Example 8.2-1 In the circuit of Fig. 8.2-1, $R_1 = \frac{1}{2}\,\Omega$, $R_2 = \frac{1}{3}\,\Omega$, $R_3 = \frac{1}{6}\,\Omega$, $I_a = 2$ A, and $I_b = 4$ A. Find all other currents and voltages.

Solution We follow the sequence of steps previously set down.

Steps 1 and *2* are as shown on the diagram.

Step 3 The Ohm's law equations are

$$I_1 = 2\,V_a$$

$$I_2 = 3(V_a - V_b)$$

$$I_3 = 6\,V_b$$

Step 4 Applying KCL at nodes a and b,

Node a:

$$I_1 + I_2 = 2$$

Node b:

$$I_3 - I_2 = 4$$

Step 5 Substituting from step 3 into step 4:

$$2\,V_a + 3(V_a - V_b) = 2$$

$$6\,V_b - 3(V_a - V_b) = 4$$

Combining terms,

$$5\,V_a - 3\,V_b = 2$$

$$-3\,V_a + 9\,V_b = 4$$

Solving with determinants:*

$$V_a = \frac{\begin{vmatrix} 2 & -3 \\ 4 & 9 \end{vmatrix}}{\begin{vmatrix} 5 & -3 \\ -3 & 9 \end{vmatrix}} = \frac{18 + 12}{45 - 9} = \frac{30}{36} = 0.833 \text{ V}$$

$$V_b = \frac{\begin{vmatrix} 5 & 2 \\ -3 & 4 \end{vmatrix}}{36} = \frac{20 + 6}{36} = 0.722 \text{ V}$$

Step 6

$$I_1 = 2(0.833) = 1.67 \text{ A}$$

$$I_2 = 3(0.833 - 0.722) = 0.333 \text{ A}$$

$$I_3 = 6(0.72) = 4.33 \text{ A}$$

Step 7 Checking KCL (step 4) we have

$$I_1 + I_2 = 1.67 + 0.333 = 2.00 \text{ A}$$

$$I_3 - I_2 = 4.33 - 0.333 = 4.00 \text{ A}$$

These results are seen to agree with the KCL equations in step 4. This type of checking should become a habit, since it often turns up numerical errors made along the way. ////

The next example illustrates the node-voltage method when a voltage source is present.

Example 8.2-2 The circuit of Fig. 8.2-2 contains a voltage source and a

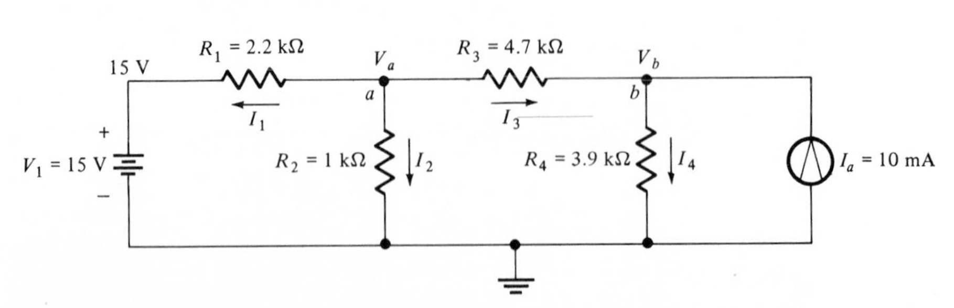

Fig. 8.2-2 Two node-pair network with voltage source.

* For students who have not studied determinants, a brief exposition is given in Appendix B.

current source. Find all currents and voltages using the node-voltage method.

Solution We again follow the sequence of steps given.

Step 1 The reference node is chosen as indicated on the diagram as that node with the largest number of branches connected to it and is marked with the ground symbol. The two independent nodes are marked a and b respectively, with node voltages V_a and V_b. The node between the battery and R_1 is *not* independent because its voltage is *known*.
Step 2 Branch currents are assumed as shown in the diagram.
Step 3 The branch currents are (units V, kΩ, mA)

$$I_1 = \frac{V_a - 15}{2.2}$$

$$I_2 = \frac{V_a}{1}$$

$$I_3 = \frac{V_a - V_b}{4.7}$$

$$I_4 = \frac{V_b}{3.9}$$

Step 4 The KCL equations are

Node a:

$$I_1 + I_2 + I_3 = 0$$

Node b:

$$I_4 - I_3 = 10$$

Step 5 The node-voltage equations are obtained by substituting the branch current equations into the KCL equations. This yields

$$\frac{V_a - 15}{2.2} + V_a + \frac{V_a - V_b}{4.7} = 0$$

$$\frac{V_b}{3.9} - \frac{V_a - V_b}{4.7} = 10$$

Combining terms and rearranging:

$$V_a\left(\frac{1}{2.2} + 1 + \frac{1}{4.7}\right) - \frac{V_b}{4.7} = \frac{15}{2.2}$$

$$-\frac{V_a}{4.7} + V_b\left(\frac{1}{3.9} + \frac{1}{4.7}\right) = 10$$

Simplifying the coefficients:

$$1.667\ V_a - 0.213\ V_b = 6.818$$

$$-0.213\ V_a + 0.469\ V_b = 10$$

Solving by the determinant method:

$$V_a = \frac{\begin{vmatrix} 6.818 & -0.213 \\ 10 & 0.469 \end{vmatrix}}{\begin{vmatrix} 1.667 & -0.213 \\ -0.213 & 0.469 \end{vmatrix}} = \frac{3.198 + 2.130}{0.782 - 0.045} = \frac{5.328}{0.737} = 7.229\text{ V} \approx 7.23\text{ V}$$

$$V_b = \frac{\begin{vmatrix} 1.667 & 6.818 \\ -0.213 & 10 \end{vmatrix}}{0.737} = \frac{16.670 + 1.452}{0.737} = 24.589\text{ V} \approx 24.6\text{ V}$$

These calculations were carried out using a pocket electronic calculator; hence the additional significant digits in the intermediate results.

Step 6 The individual branch currents are, from step 3:

$$I_1 = \frac{7.229 - 15}{2.2} = -\ 3.532\text{ mA}$$

$$I_2 = \frac{V_a}{1} = 7.229\text{ mA}$$

$$I_3 = \frac{7.229 - 24.589}{4.7} = -3.694\text{ mA}$$

$$I_4 = \frac{24.589}{3.9} = 6.305\text{ mA}$$

Step 7 Checking the KCL equations from step 4:

$$I_1 + I_2 + I_3 = -3.532 + 7.229 - 3.694 = 0.003 \approx 0\text{ mA}$$

$$I_4 - I_3 = 6.305 - (-3.694) = 9.999 \approx 10\text{ mA}$$

////

TWO-LOOP NETWORKS

In this section we will extend the loop-current method described in Chap. 6 to more complicated networks. Consider the circuit of Fig. 8.2-3. Although

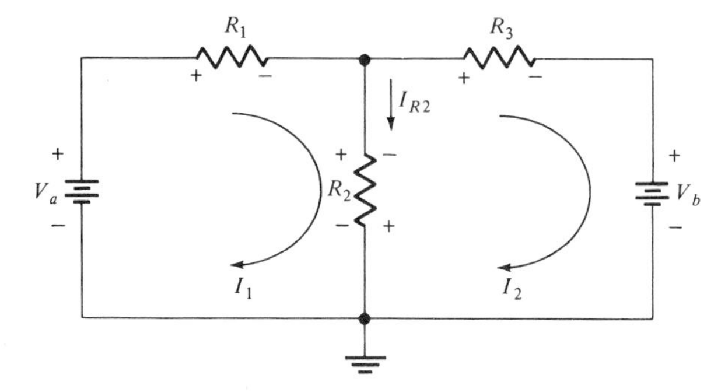

Fig. 8.2-3 Two-loop network.

this network can be solved by the node-voltage method using only *one* node equation we will use it to illustrate the loop-current method. The idea is to write a set of KVL equations for each "window" of the network in terms of fictitious "loop" currents and the element *vi* relations. The sequence of steps to be followed is applied to the circuit of Fig. 8.2-3 as follows:

Step 1 If possible, draw the network in such a way that no conductors cross each other then assign "loop" currents to each of the "windows" in the diagram. These currents should be given clockwise reference directions for uniformity.* In the circuit of Fig. 8.2-3 there are two windows and the clockwise loop-currents are designated I_1 and I_2.

Step 2 Assign polarities to the voltages resulting from the loop currents flowing in each element. If two different loop currents pass through one element, the voltage will be the sum of two components whose polarities must be carefully observed.

In the circuit of Fig. 8.2-3 the polarities are indicated on the diagram. In accordance with Ohm's law, they are positive at the terminal at which the loop current enters. Note that R_2 has two voltage components associated with it, one in loop 1 due to I_1, and one in loop 2 due to I_2. The polarities of these two voltages are opposite to each other and we will have to observe this carefully when writing the KVL equations.

Step 3 Write a KVL equation for each loop by starting at one point and following the loop current around the loop, adding all voltages algebraically until the starting point is reached. For loop 1 of the circuit of Fig. 8.2-3 we begin at the lower left hand corner and obtain the following equation

$$-V_a + R_1I_1 + R_2I_1 - R_2I_2 = 0 \tag{8.2-8}$$

Note carefully that when we trace through a voltage *rise*, as in the first and last terms, they take a negative sign while the signs of voltage drops are

* These instructions are not mandatory, but are the safest way for beginning students to learn the method without unnecessary complications.

positive. Also note that the last two terms in the equation represent the voltage across *one* resistor, R_2. Since the loop currents go through R_2 in different directions, the polarities of the voltages due to the two currents are opposite. For loop 2, starting at the ground end of R_2, we obtain

$$R_2I_2 - R_2I_1 + R_3I_2 + V_b = 0 \tag{8.2-9}$$

Again note that R_2 has two voltages across it, one due to I_1 flowing through it in one direction, the other due to I_2 flowing in the other direction. The term $-R_2I_1$ represents the effect on loop 2 of the current in loop 1 and is called the *coupling,* or *mutual* voltage.

Rearranging Eqs. (8.2-8) and (8.2-9), we have the desired KVL equations.

$$(R_1 + R_2)I_1 - R_2I_2 = V_a \tag{8.2-10}$$

$$-R_2I_1 + (R_2 + R_3)I_2 = -V_b \tag{8.2-11}$$

Note the symmetrical form of these equations. The coefficient of I_1 in Eq. (8.2-10) represents the total resistance connected in series around loop 1, that is, $R_1 + R_2$. The coefficient of I_2 represents the resistance common to loops 1 and 2, that is, R_2. When loop currents are all taken clockwise, the mutual terms always have a negative sign. Similar interpretations apply to the coefficients in Eq. (8.2-11).

Step 4 Solve the simultaneous KVL equations for the unknown loop currents. Mutual branch currents and all voltage drops are then found as required. The method is illustrated by the following examples.

Example 8.2-3 In the circuit of Fig. 8.2-3 $V_a = 9$ V, $V_b = 22$ V, $R_1 = 2.2$ kΩ, $R_2 = 15$ kΩ, and $R_3 = 4.7$ kΩ. Find all currents and voltages and check wherever possible.

Solution Following the steps set down above we have the following:

Steps 1 and 2 Loop currents and the resulting voltage polarities are shown on the circuit diagram.

Step 3 The KVL equations are (units: V, kΩ, mA)

Loop 1:

$$-9 + 2.2\,I_1 + 15\,I_1 - 15\,I_2 = 0$$

Loop 2:

$$15\,I_2 - 15\,I_1 + 4.7\,I_1 + 22 = 0$$

Rearranging and combining:

$$17.2\,I_1 - 15\,I_2 = 9$$

$$-15\,I_1 + 19.7\,I_2 = -22$$

Step 4 Solving with determinants:

$$I_1 = \frac{\begin{vmatrix} 9 & -15 \\ -22 & 19.7 \end{vmatrix}}{\begin{vmatrix} 17.2 & -15 \\ -15 & 19.7 \end{vmatrix}} = \frac{177.3 - 330}{338.8 - 225} = \frac{-152.7}{113.8} = -1.34 \text{ mA}$$

$$I_2 = \frac{\begin{vmatrix} 17.2 & 9 \\ -15 & -22 \end{vmatrix}}{113.8} = \frac{-378.4 + 135}{113.8} = \frac{-243.4}{113.8} = -2.14 \text{ mA}$$

The negative signs indicate that the actual currents flow in directions opposite to the assumed clockwise loop currents. Using the results,

$$I_{R2} = I_1 - I_2 = 0.8 \text{ mA}$$

$$V_{R2} = R_2 I_{R2} = (15)(0.8) = 12 \text{ V}$$

$$V_{R1} = R_1 I_1 = (2.2)(-1.34) = -2.95 \text{ V}$$

$$V_{R3} = R_3 I_2 = (4.7)(-2.14) = -10.06 \text{ V}$$

Checking with KVL we must have

$$V_a = V_{R1} + V_{R2} = 12 - 2.95 = 9.05 \text{ V} \approx 9 \text{ V}$$

also

$$V_b = -V_{R3} + V_{R2} = -(-10.06) + 12 = 22.06 \text{ V} \approx 22 \text{ V}$$

These check with sufficient accuracy for our purposes. ////

Example 8.2-4 The circuit of Fig. 8.2-4 contains a current source in addition to a voltage source. Use the loop current method to find all currents and voltages.

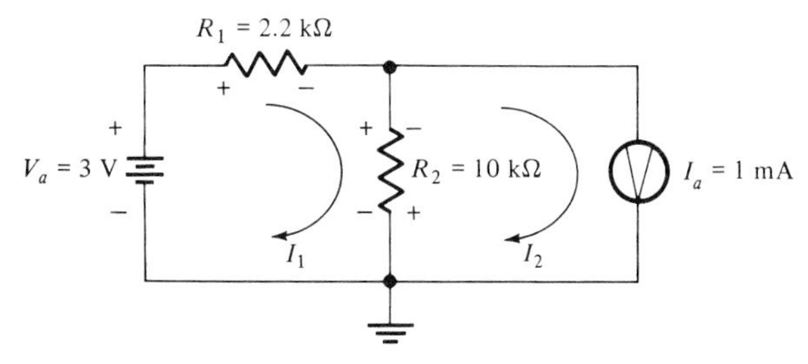

Fig. 8.2-4 Circuit for Example 8.2-4.

Solution Again we follow the prescribed sequence of steps, indicating differences from the previous example because of the presence of the current source as we proceed:

Step 1 The loop currents are assigned as shown on the diagram. Note that loop current I_2 *must* be the same as the source current I_a by virtue of the fact that the current source is *ideal* and I_2 is the only loop current flowing through the branch containing the source.
Step 2 The voltage polarities due to the loop currents are indicated on the diagram.
Step 3 The KVL equations are written for loops which do not contain current sources, for this example only loop 1. Thus the KVL equation for loop 1 is

$$-3 + 2.2I_1 + 10I_1 - 10I_2 = 0$$

$$12.2I_1 - 10I_2 = 3$$

The loop containing the current source is often called a "dummy" loop and from it we write

$$I_2 = I_a = 1 \text{ mA}$$

This is substituted into the KVL equation for loop 1 and we have

$$12.2\,I_1 - (10)(1) = 3$$

$$12.2\,I_1 = 13$$

$$I_1 = 1.066 \text{ mA}$$

Thus even though at first glance the network appears to have two loops so that two simultaneous KVL equations will result, we end up with only one KVL equation because the current in the dummy loop is known. The rest of the currents and voltages are easily found. Thus

$$V_{R1} = R_1I_1 = (2.2)(1.066) = 2.34 \text{ V}$$

$$V_{R2} = R_2(I_1 - I_2) = 10(1.066 - 1) = 0.66 \text{ V}$$

Checking

$$V_a = V_{R1} + V_{R2} = 2.34 + 0.66 = 3.00 \text{ V}$$ ////

CHOICE BETWEEN NODE AND LOOP METHODS

It is not possible to state a rule which can be used to determine the method to be used for a specific circuit. The usual procedure is to use that method which leads to the smallest number of equations that must be solved simultaneously. As an example, consider the network of Fig. 8.2-5. If node analysis is considered, we see that the circuit has two independent nodes, labeled 1 and 2 on the diagram. The node between the battery and R_1 and R_5 is *not* independent. Thus two node

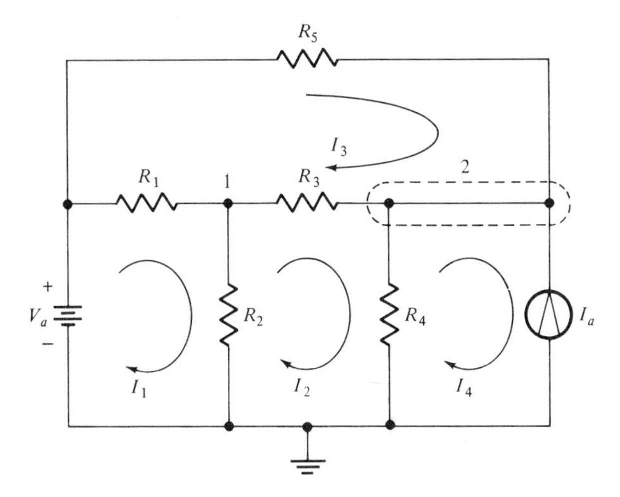

Fig. 8.2-5 Circuit for loop or node analysis.

equations will suffice. On the other hand, there are four windows, one of which is a dummy, so that three loop equations are required. Thus, on the basis of least number of simultaneous equations, the node-voltage method would be used for this circuit. For networks involving three independent node-pairs or three loops the reader is referred to Appendix B.

8.3 TERMINAL CHARACTERISTICS AND EQUIVALENT CIRCUITS

Suppose we have a dc electric system which is completely enclosed except for two terminals at which we can make any electrical measurements we wish. We should like to determine as much as possible about the system inside the box from these measurements. In general, it is not possible to determine exactly what the unknown system is. However, if the system is *linear,* we can find a circuit which is exactly *equivalent,* as far as behavior at the terminals is concerned, from two voltage-current measurements. In what follows, we assume that the system is an electric circuit. However, all conclusions hold for any electric system or device.

MEASUREMENT OF TERMINAL CHARACTERISTICS

Since the circuit is linear, its terminal characteristic will be a straight line. Therefore, two different volt-ampere measurements will provide two points which are sufficient to determine the straight line characteristic. These measurements can be made by connecting an adjustable voltage source in series with an ammeter across the terminals of the network as shown in Fig. 8.3-1*a*. The voltage source is adjusted to two different values and the current is measured for each of these voltages. (It is assumed that the voltage across the ammeter is negligible

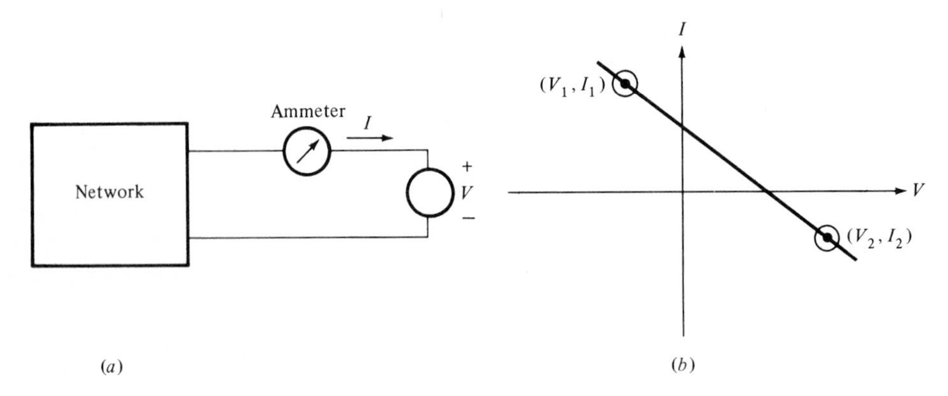

Fig. 8.3-1 Measurement of network terminal characteristic. (*a*) Volt-ampere measurement; (*b*) terminal characteristic.

in this procedure.) Figure 8.3-1*b* shows two typical points along with the terminal characteristic drawn through them.

LINEARITY AND EQUIVALENCE

We have described a way of finding the terminal characteristic of a linear network. It is important to note the relevance of the linearity of the circuit; as a consequence of this property it is possible to determine the complete terminal characteristic from two simple measurements. If a network is not known to be linear, we cannot conclude anything about its behavior except at those points where we have made measurements.

The original question we asked is how much can be determined about the network in the box. The following argument indicates that we can, in the case of a linear network, go only so far as determining an equivalent circuit. Let us assume that the terminal characteristic is that of a 4-Ω resistance. We cannot conclude that the network in the box is a 4-Ω resistance; it might be two 2-Ω resistances in series, a 3-Ω and a 1-Ω resistance in series, two 8-Ω resistances in parallel, or any one of an infinite number of other combinations which result in an *equivalent* resistance of 4-Ω.

Thus we conclude that

1 For a linear circuit only, we can find an equivalent circuit for the one in the box from two measured points.
2 The equivalent circuit will not necessarily be identical to the one in the box.
3 The circuits are equivalent only to the extent that they have the same volt-ampere characteristic at the terminal pair at which the measurements are made.

In succeeding sections we will show how to find the equivalent circuit.

8.3-1 The Thevenin Equivalent Circuit

Consider the following question: Given a linear network which may contain resistances and sources, is there any network structure which can always be used as an equivalent circuit with respect to a given terminal pair? We have just illustrated a method for finding the terminal characteristic of such a network and pointed out that it will always be a straight line. Therefore, we can change the question as follows: Is it possible to find a network structure whose *vi* characteristic can be made to fit any given straight line? We will prove that the answer to this question is yes by presenting such a circuit.

The equation of the straight line *vi* characteristic shown in Fig. 8.3-2*a* is of the form

$$i = mv + b \tag{8.3-1}$$

This equation, which relates the two variables i and v, has two parameters: the slope m, and the i intercept b (the value of i when $v = 0$). The equivalent circuit will have to contain at least two independent elements whose values can be adjusted so that these parameters for the equivalent circuit characteristic are equal to the corresponding parameters for the given network characteristic. One very simple circuit, consisting of a voltage source in series with a resistance, which satisfies this requirement is shown in Fig. 8.3-2*b*. This circuit is known as the Thevenin equivalent after the man who first described its properties.

We know that b is the i intercept, the value of i when $v = 0$. Physically, it corresponds to the current that would flow through a short circuit placed across the terminals of the given network. (The short circuit across the terminals forces the condition $v = 0$.) Thus $b = I_s$ where I_s represents the short-circuit current.

The slope m is given by the equation

$$m = -\frac{I_s}{V_0} \tag{8.3-2}$$

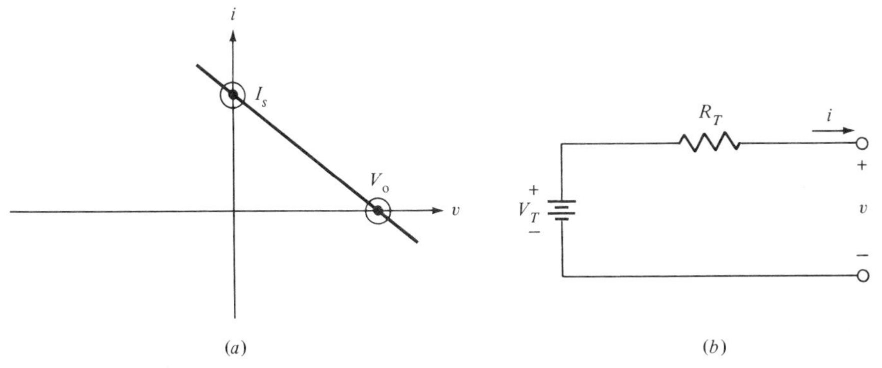

Fig. 8.3-2 Thevenin equivalent. (*a*) Terminal characteristic; (*b*) Thevenin equivalent circuit.

as can be seen from Fig. 8.3-2*a*. V_0 is the voltage intercept, the value of v when $i = 0$. Physically, it is the voltage that exists across the terminals of the given network when they are open-circuited. (The open circuit forces the condition $i = 0$.)

These values of b and m when substituted into Eq. (8.3-1) result in

$$i = -\frac{I_s}{V_0} v + I_s \qquad (8.3\text{-}3)$$

If we multiply both sides of Eq. (8.3-3) by $-V_0/I_s$ we find

$$-\frac{V_0}{I_s} i = v - V_0 \qquad (8.3\text{-}4)$$

which can be rearranged to provide

$$v = V_0 - \frac{V_0}{I_s} i \qquad (8.3\text{-}5)$$

We shall now write the KVL equation for the Thevenin equivalent circuit shown in Fig. 8.3-2*b*. This results in

$$v = V_T - R_T i \qquad (8.3\text{-}6)$$

If we compare this with Eq. (8.3-5) we conclude that the Thevenin circuit is equivalent to the one whose characteristic is shown in Fig. 8.3-2*a* if

$$V_T = V_0 \qquad (8.3\text{-}7)$$

and

$$R_T = \frac{V_0}{I_s} \qquad (8.3\text{-}8)$$

This result is quite general, since we have only specified that the unknown network contain resistances and sources, with no restrictions as to their number. Thus, as far as the pair of terminals of interest is concerned, we can replace such a network, no matter how complicated, by its Thevenin equivalent. We will often find it convenient to do this in order to simplify problems in network analysis.

In the remainder of this section we will consider methods for finding the parameters of the Thevenin equivalent circuit.

THEVENIN EQUIVALENT OF A GIVEN NETWORK

The Thevenin equivalent circuit for a given network may be determined from the circuit diagram of the given network. The replacement of certain portions of a complicated network by their Thevenin equivalents is frequently useful in simplifying the analysis of the overall network.

One method for finding the Thevenin equivalent involves computing the open-circuit voltage and short-circuit current and then proceeding as we did when given the network terminal characteristic. This method will be illustrated in Example 8.3-2.

We will now present a simple method for finding the Thevenin resistance, R_T, directly.

If all of the independent sources in the given circuit are replaced by zero values, the resulting network is a purely resistive circuit. The equivalent resistance, i.e., the resistance that would be measured across the terminal pair, is called the input resistance, R_{in}. The Thevenin resistance, R_T, is equal to R_{in}. This is so because the Thevenin equivalent circuit reduces to the Thevenin resistance when the Thevenin voltage source, V_T, is set to zero (see Fig. 8.3-2*b*). Remember, when using this method, that a voltage source is set to zero by replacing it by a short circuit while a current source is set to zero by replacing it by an open circuit.

We now have three circuit parameters which can be computed in order to find the Thevenin equivalent. They are the open circuit voltage V_0, the short circuit current I_s, and the input resistance R_{in}. The Thevenin parameters can be computed from any two of these three quantities using the following equations:

$$V_T = V_0 = R_{\text{in}} I_s \tag{8.3-9}$$

$$R_T = R_{\text{in}} = \frac{V_0}{I_s} \tag{8.3-10}$$

Since the determination of the Thevenin equivalent from a circuit diagram is such a useful technique, we will illustrate it with several examples.

Example 8.3-1 Find the Thevenin equivalent of the circuit of Fig. 8.3-3*a* across terminals *a–b*.

Solution The Thevenin voltage is the open-circuit voltage drop from terminal *a* to *b*. It is computed using the voltage divider formula:

$$V_T = \frac{2\text{ k}\Omega}{2\text{ k}\Omega + 8\text{ k}\Omega} \times 24\text{ V} = 4.8\text{ V} \tag{8.3-11}$$

The Thevenin resistance is found as the equivalent resistance across terminals *a-b* with the voltage source replaced by a short circuit as shown in Fig. 8.3-3*b*. This resistance is computed using the formula for parallel resistances.

$$R_T = \frac{(8\text{ k}\Omega)(2\text{ k}\Omega)}{8\text{ k}\Omega + 2\text{ k}\Omega} = 1.6\text{ k}\Omega \tag{8.3-12}$$

The resulting Thevenin equivalent is shown in Fig. 8.3-3*c*.

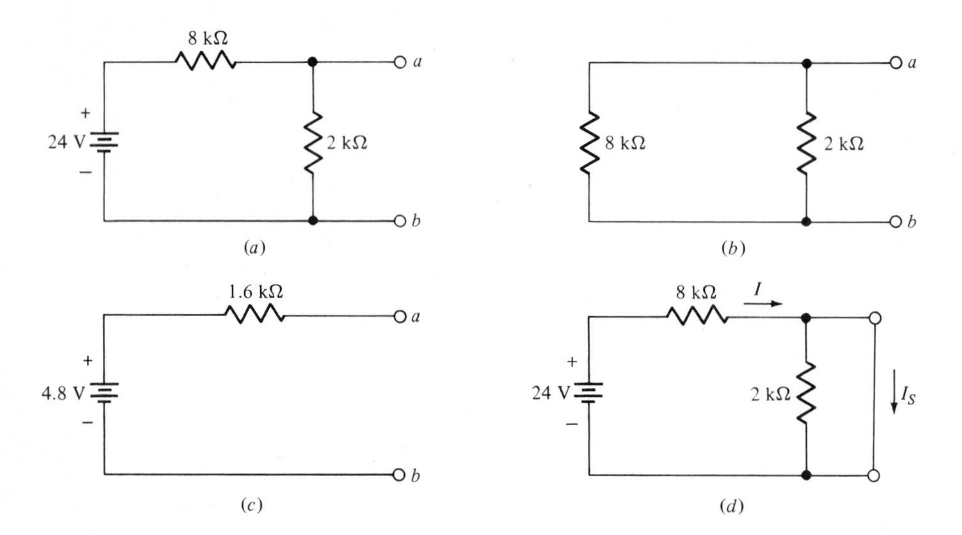

Fig. 8.3-3 Circuits for Example 8.3-1. (*a*) Given circuit; (*b*) circuit with voltage source set to zero; (*c*) Thevenin equivalent; (*d*) circuit for calculating short-circuit current.

As a check, we can compute the short-circuit current from terminal *a* to terminal *b* for the original circuit and the Thevenin equivalent.

In the circuit of Fig. 8.3-3*d*, which is the original circuit with the output shorted, the current I will all flow through the short circuit so that no current flows through the 2 kΩ resistance (why?). Thus $I = I_s$ and Ohm's law yields

$$I_s = \frac{24 \text{ V}}{8 \text{ k}\Omega} = 3 \text{ mA} \qquad (8.3\text{-}13)$$

Applying Ohm's law to the Thevenin equivalent of Fig. 8.3-3*c* with a short circuit between points *a* and *b* we obtain

$$I_s = \frac{4.8 \text{ V}}{1.6 \text{ k}\Omega} = 3 \text{ mA} \qquad (8.3\text{-}14)$$

in agreement with the short-circuit current for the original circuit. ////

Example 8.3-2 Find the Thevenin equivalent at terminals *a*–*b* of the circuit shown in Fig. 8.3-4*a*.

Solution We will use superposition (Sec. 8.1) to compute the open-circuit voltage and short-circuit current due to each source acting alone, i.e., with the other source replaced by a zero value. The total open-circuit voltage and short-circuit current are then obtained by taking the algebraic sum of the components due to each source.

Figure 8.3-4*b* illustrates the equivalent circuit with the current source replaced by an open circuit (recall that a zero current source is equivalent to an open circuit). The open-circuit voltage, V_{ab}, is easily computed using the voltage-divider relationship:

$$V_{ab1} = \frac{6\text{ k}\Omega}{6\text{ k}\Omega + 4\text{ k}\Omega} \times 24\text{ V} = 14.4\text{ V} \tag{8.3-15}$$

The short-circuit current is computed using Ohm's law applied to the circuit of Fig. 8.3-4*b* but with a short circuit connected from *a* to *b*. Note that the 6 kΩ resistance has no effect (why?). Using Ohm's law we find

$$I_{ab1} = \frac{24\text{ V}}{4\text{ k}\Omega} = 6\text{ mA} \tag{8.3-16}$$

The next step is to compute the open-circuit voltage and short-circuit current due to the current source. The equivalent circuit with the voltage source replaced by a short circuit (recall that a zero voltage source is equivalent to a short-circuit) is shown in Fig. 8.3-4*c*.

The parallel combination of the 4 kΩ and 6 kΩ resistances can be replaced by a single equivalent resistance having a value of

$$\frac{(4\text{ k}\Omega)(6\text{ k}\Omega)}{4\text{ k}\Omega + 6\text{ k}\Omega} = 2.4\text{ k}\Omega \tag{8.3-17}$$

The open-circuit voltage is computed using Ohm's law:

$$V_{ab2} = -(2.4\text{ k}\Omega)(10\text{ mA}) = -24\text{ V} \tag{8.3-18}$$

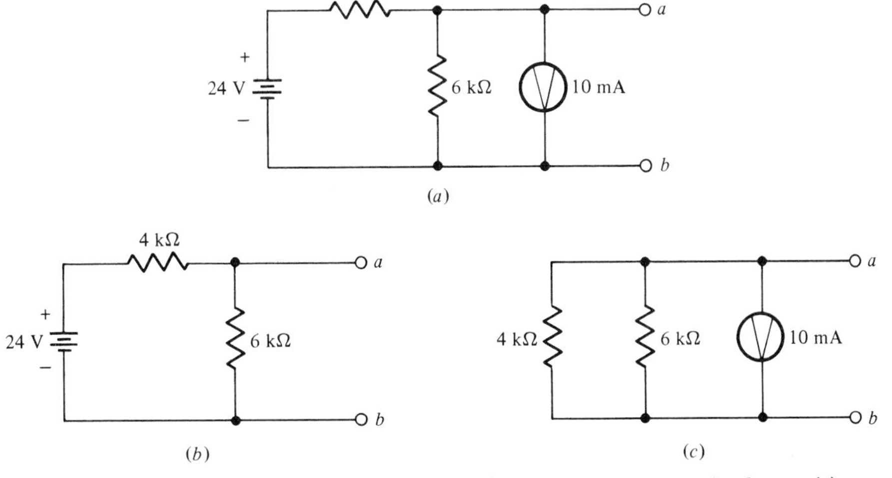

Fig. 8.3-4 Example 8.3-2. (*a*) Circuit; (*b*) zero current equivalent; (*c*) zero voltage equivalent.

Note that the current is flowing from the bottom to the top through the resistance. Thus the polarity of the *actual* output voltage is opposite to the *reference* polarity (drop from a to b); hence the minus sign.

To find the short-circuit current note that in the circuit of Fig. 8.3-4c all of the current from the 10 mA source *must* flow through the short-circuit connected between terminals a and b (why?). Thus

$$I_{ab2} = -10 \text{ mA}$$

Note the minus sign which is a result of the difference between the *reference* direction for I_{ab2} (from a to b) and the *actual* direction of current due to the 10 mA source through the short circuit.

Finally, we compute the total open-circuit voltage and short-circuit current as

$$V_{ab} = V_{ab1} + V_{ab2} = 14.4 - 24 = -9.6 \text{ V} \tag{8.3-19}$$

$$I_{ab} = I_{ab1} + I_{ab2} = 6 - 10 = -4 \text{ mA} \tag{8.3-20}$$

Therefore

$$V_T = V_{ab} = -9.6 \text{ V} \tag{8.3-21}$$

$$R_T = \frac{V_0}{I_s} = \frac{-9.6 \text{ V}}{-4 \text{ mA}} = 2.4 \text{ k}\Omega \tag{8.3-22}$$

The resulting circuit is shown in Fig. 8.3-5. ////

Example 8.3-3 Verify that the Thevenin resistance calculated in Example 8.3-2 is the input resistance of the circuit of Fig. 8.3-4a.

Solution Figure 8.3-6 shows the circuit of Fig. 8.3-4a with both sources replaced by zero values. From this circuit it is clear that the input resistance, R_{ab}, is the equivalent resistance of the parallel combination of the 4 kΩ and 6 kΩ resistances. This value was computed in Example 8.3-2 as 2.4 kΩ which is the same as the value calculated for R_T. ////

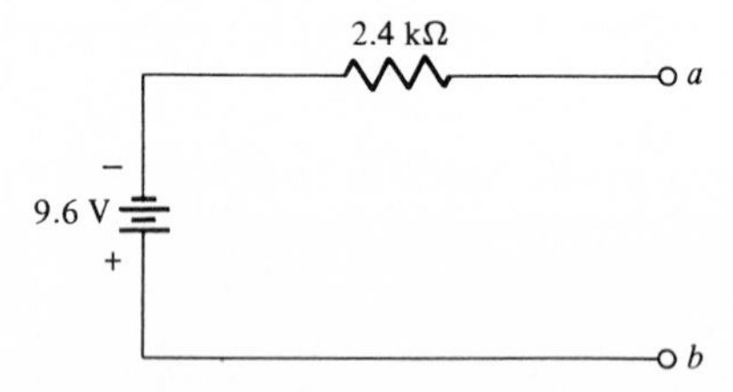

Fig. 8.3-5 Thevenin equivalent circuit for Example 8.3-2.

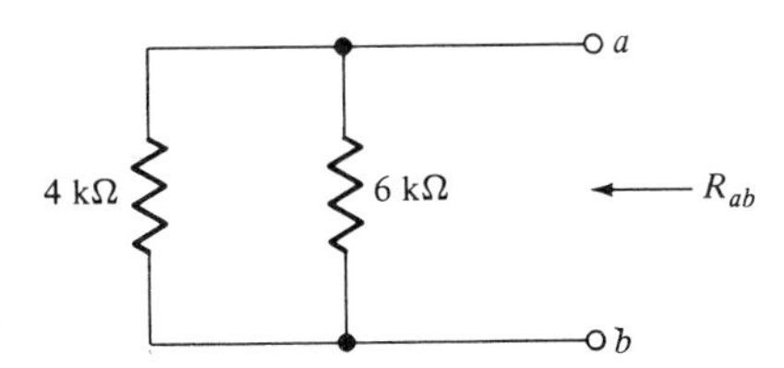

Fig. 8.3-6 Equivalent circuit for Example 8.3-3.

Example 8.3-4 Find the Thevenin equivalent across terminals a–b for the circuit of Fig. 8.3-7a.

Solution The first step in this solution is to determine which two of the three quantities (open-circuit voltage, short-circuit current, or equivalent resistance) should be computed. The short-circuit current is a good choice since the relatively complicated portion of the circuit connected across the output terminals is shorted out and therefore has no effect.

$$I_{ab} = \frac{20 \text{ V}}{8 \text{ k}\Omega} = 2.5 \text{ mA} \tag{8.3-23}$$

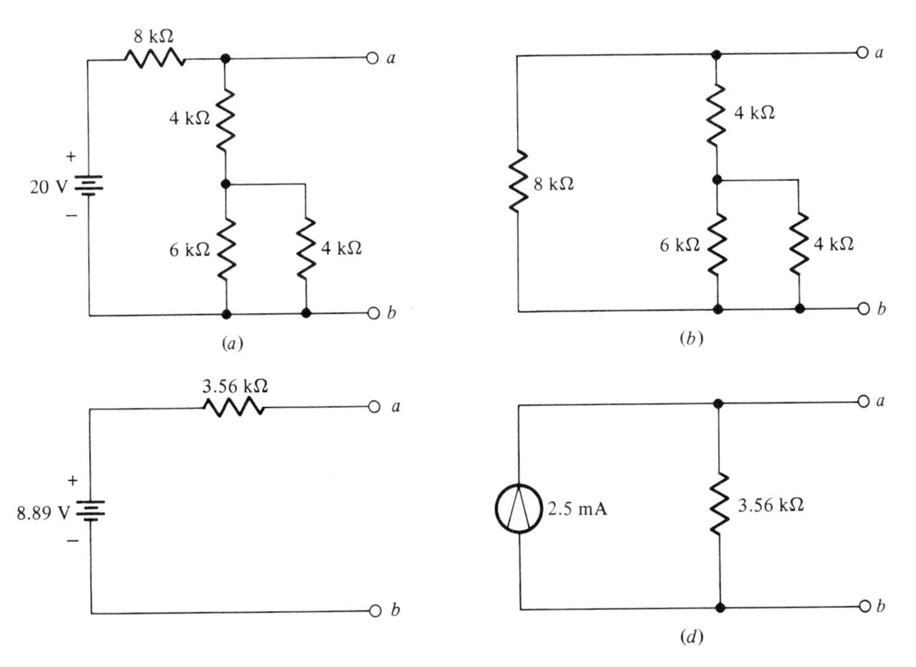

Fig. 8.3-7 Circuit for Example 8.3-4. (a) Circuit; (b) zero source equivalent; (c) Thevenin equivalent; (d) Norton equivalent.

We will compute the input resistance as the second quantity. The resistive circuit with the source set to zero is shown in Fig. 8.3-7*b*. The parallel combination of the 6 kΩ and 4 kΩ resistances can be replaced by a single 2.4 kΩ equivalent resistance. The series combination of this resistance and the 4 kΩ resistance is a 6.4 kΩ resistance which is in parallel with the 8 kΩ resistance. The resulting input resistance is

$$R_{in} = \frac{(8\text{ k}\Omega)(6.4\text{ k}\Omega)}{8\text{ k}\Omega + 6.4\text{ k}\Omega} = 3.56\text{ k}\Omega \tag{8.3-24}$$

The Thevenin voltage is computed as

$$V_T = R_{in}I_s = (3.56\text{ k}\Omega)(2.5\text{ mA}) = 8.9\text{ V} \tag{8.3-25}$$

which together with the Thevenin resistance

$$R_T = R_{in} = 3.56\text{ k}\Omega \tag{8.3-26}$$

define the Thevenin equivalent shown in Fig. 8.3-7*c*. ////

8.3-2 Norton Equivalent Circuits

There is another form which the equivalent circuit may take called the Norton equivalent after the engineer who first described it. The circuit is shown in Fig. 8.3-8, and it consists of a current source in parallel with a resistance.

The *vi* relation for this circuit is found by writing KCL at the upper node. This yields

$$i + \frac{v}{R_N} - I_N = 0$$

Solving for v we obtain

$$v = I_N R_N - iR_N \tag{8.3-27}$$

This is identical in form to Eq. (8.3-8), the *vi* characteristic of the Thevenin circuit. For the two circuits to be identical, the slopes and i intercepts must be the same that is,

$$V_T = I_N R_N \tag{8.3-28}$$

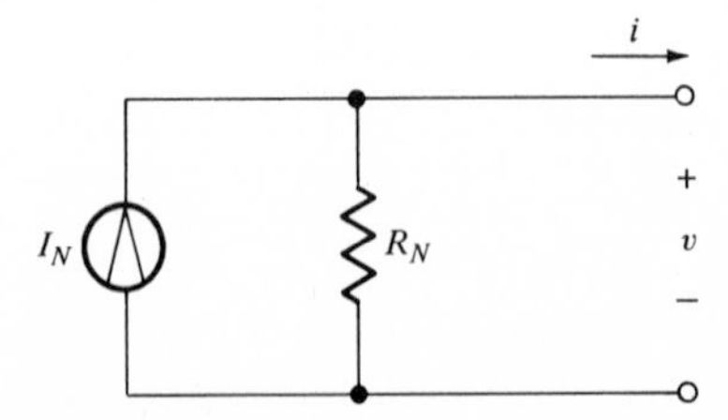

Fig. 8.3-8 Norton equivalent circuit.

and

$$R_T = R_N \tag{8.3-29}$$

When these two equations are satisfied the Thevenin and Norton circuits are equivalent at their external terminals and thus both are equivalent to the original linear circuit.

The physical significance of the parameters of the Norton circuit is similar to that of the Thevenin circuit. For the Thevenin circuit, the voltage source represents the open-circuit voltage at the terminals of interest. For the Norton circuit the current source represents the short-circuit current at these terminals. For both circuits the resistance represents the resistance of the original circuit with all sources dead (voltage sources replaced by short circuits and current sources by open circuits).

The Norton equivalent of a given circuit can be found using the same techniques as we used for the Thevenin circuit. An example follows.

Example 8.3-5 Find the Norton equivalent of the circuit of Fig. 8.3-7*a* (the Thevenin equivalent was found in Example 8.3-4).

Solution To find the Norton current, we connect a short circuit from a to b and calculate the resulting short-circuit current. Since no current will flow through the resistance network connected between a and b because of the short, the calculation simply involves Ohm's law. Thus

$$I_N = \frac{20}{8} = 2.5 \text{ mA}.$$

The Norton resistance R_N is the same as the Thevenin resistance found in Example 8.3-4, that is, $R_N = 3.56$ kΩ. The Norton circuit is shown in Fig. 8.3-7*d*.

As a check, we find the open-circuit voltage of the Norton circuit to be (2.5 mA) (3.56 kΩ) = 8.9 V. This agrees with the value found in Example 8.3-4. ////

8.4 ECAP SOLUTION OF COMPLICATED CIRCUITS

In Sec. 8.2 we considered the solution of two-loop and two node-pair networks using the loop current and node voltage methods of analysis. When a network has more loops or nodes the solution becomes very tedious and the arithmetic involved becomes quite lengthy and difficult to carry out on paper. For such networks the computer solution is fast and accurate and allows the technologist to concentrate on the interpretation of the solution rather than the mechanics involved. In this section we will extend the discussion of the ECAP program begun in Sec. 6.5 and illustrate how it is used to obtain repeated solutions for different parameter values.

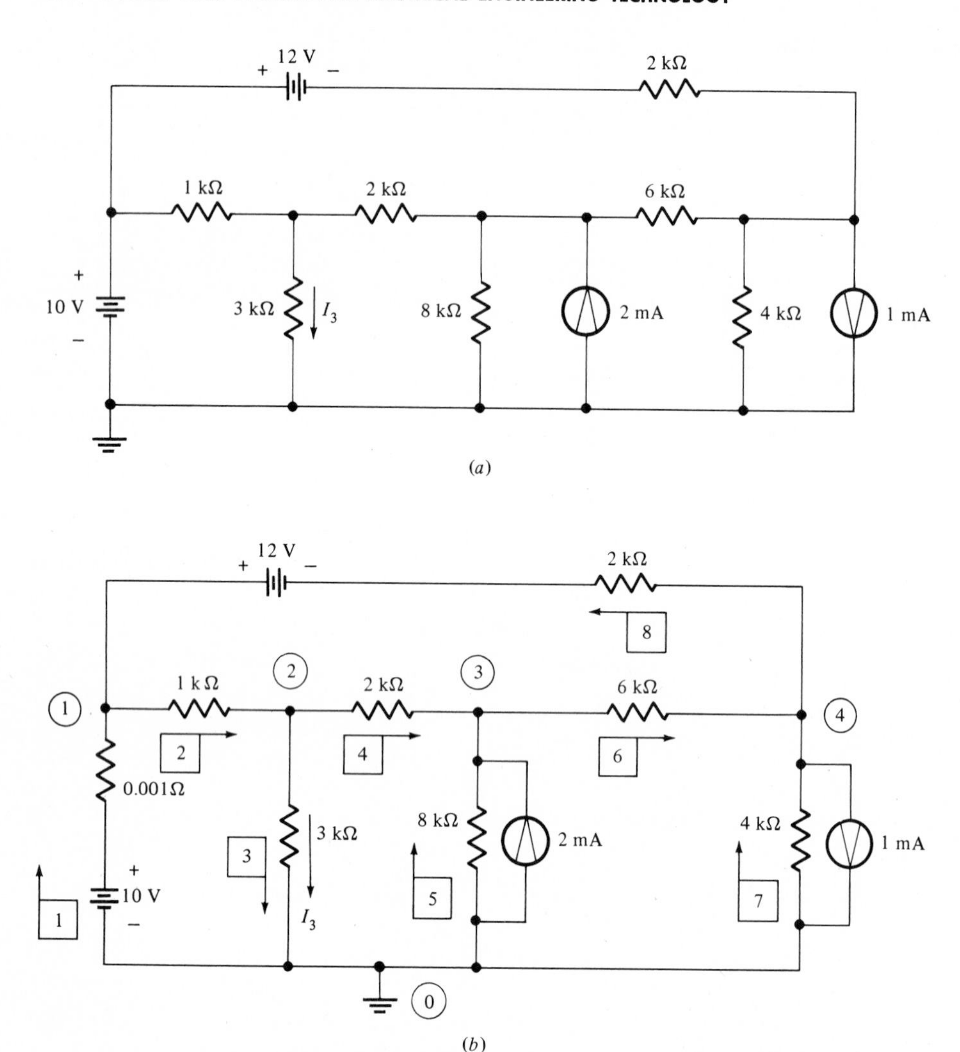

Fig. 8.4-1 Example 8.4-1. (*a*) Original circuit; (*b*) ECAP circuit.

Example 8.4-1 Use the ECAP program to find I_3 in the circuit of Fig. 8.4-1*a*.

Solution The ECAP circuit is shown in Fig. 8.4-1*b*. Note that the 10-V battery in the original circuit does not have a resistance in series with it and thus does not constitute a proper ECAP branch. In order to accommodate the battery we add a 0.001-Ω resistance in series with it. This is small enough so that it will not affect the solution but will satisfy the ECAP requirement that each branch have a nonzero resistance. The rest of the ECAP circuit follows the rules set down in Sec. 6.5.

```
BKEX          13:00    09/20/74   FRIDAY      I05

100   DC ANALYSIS
200   C   EXAMPLE 8.4-1
300   B1  N(0,1), R=.001, E=10
400   B2  N(1,2), R=1E3
500   B3  N(2,0), R=3E3
600   B4  N(2,3), R=2E3
700   B5  N(0,3), R=8E3, I=-2E-3
800   B6  N(3,4), R=6E3
900   B7  N(0,4), R=4E3, I=1E-3
1000  B8  N(4,1), R=2E3, E=12
1100  C
1200  PRINT, CA3
1300  EXECUTE

RUN **ECAP
```

(*a*)

```
ECAP          13:01    09/20/74   FRIDAY      I05


 EC  3
 2.453E-03


END OF RUN
```

(*b*)

Fig. 8.4-2 Example 8.4-1. (*a*) Terminal input; (*b*) output.

The terminal input is shown in Fig. 8.4-2*a*. The student should convince himself that the signs of the voltage and current sources are correct. The statement in line 1,200 differs from previous examples in that it asks ECAP to print out only the value of element current 3, which is the desired answer. Otherwise, the program follows the examples in Sec. 6.5.

Note that the original network has three independent nodes so that three node equations must be solved simultaneously for a solution. Additional branches would increase this number. With ECAP, each additional branch simply adds one line to the program.

The printout of the solution is shown in Fig. 8.4-2*b*. The output is indicated as EC3 (element current 3) and is seen to be 2.453 mA. ////

8.4-1 The MODIFY routine

The MODIFY command allows us to repeat an immediately preceding DC ANALYSIS with modified parameter values. As we will see in the next example, it is necessary to add only the new parameter values. Values that remain the same need not be repeated.

The MODIFY routine can be used in a number of ways in both analysis and design. In analysis, it can be used to determine the effects of component tolerances on a particular current or voltage in the network. If a circuit has been designed and all component values are available, the MODIFY routine can be used to find the effect of changes in one or more circuit parameters on any or all of the currents and voltages in the network.

In design, one way that the MODIFY routine might be used occurs when all but a few components can readily be determined on paper. The undetermined components can then be found by using the MODIFY routine to scan appropriate ranges of these components until the desired response is obtained.

In using this routine a parameter is varied over a range by specifying the smallest value of the range, the largest value, and the total number of values required in the range. For example, the entry

```
100    B6    R=1E3(9)10E3
```

tells the computer that the resistor in branch 6 starts at 1,000 Ω and ends at 10,000 Ω. There are nine intervals in between so that the resistance change per step is

$$\frac{10{,}000 - 1{,}000}{9} = 1{,}000\ \Omega$$

Thus, the resistor starts at 1,000 Ω and progresses in steps of 1,000 Ω up to and including 10,000 Ω. When this is used in a MODIFY routine, a DC solution is produced for each value of the range.

The MODIFY routine is illustrated in the following example.

Example 8.4-2 Use the MODIFY routine to find the variation in I_3 of the circuit of Fig. 8.4-1*a* if the 3 kΩ resistor in branch 3 varies from 1 kΩ to 9 kΩ in 2 kΩ steps.

Solution Since the nominal solution of 3 kΩ was obtained in Example 8.4-1 we use the program in Fig. 8.4-2*a* for our nominal solution. The only addition required to take account of the parameter variation is the addition of the three statements shown in Fig. 8.4-3*a*. These statements are inserted between the EXECUTE and END statements of the original program by using appropriate line numbers as shown. When they are added to the original program, a value of I_3 is calculated for each of the values of the modified resistor. The output is shown in Fig. 8.4-3*b*. Note that the nominal solution is given first and then repeated during the course of the parameter iteration because the 3 kΩ parameter value is included in the iteration range. ////

```
1310 MODIFY
1320 B3, R=1E3(4)9E3
1330 EXECUTE
```

(*a*)

```
RUN **ECAP

ECAP        13:11    09/20/74   FRIDAY      105



 EC   3
 2.453E-03

R    = 1.000E 03
 5.100E-03
R    = 3.000E 03
 2.453E-03
R    = 5.000E 03
 1.615E-03
R    = 7.000E 03
 1.204E-03
R    = 9.000E 03
 9.593E-04
```

(*b*)

Fig. 8.4-3 Example 8.4-2. (*a*) MODIFY routine; (*b*) output.

SUMMARY

1 The linearity principle states that in a linear circuit, multiplication of all independent sources by a constant will result in all other voltages and currents being multiplied by the same constant.

2 The superposition theorem states that in a linear circuit containing more than one source, any response can be found by adding the responses due to each source acting alone, all others being set to zero (short circuits for voltage sources, open circuits for current sources).

3 The steps to be followed in the node-voltage method are:

a Choose a reference node and label all independent nodes.

b Assume branch currents through all resistances.

c Express the branch currents in terms of the node voltages.

d Write KCL equations for each independent node in terms of the branch currents.

e Substitute the Ohm's law equations into the KCL equations and solve the simultaneous equations for the node voltages.

f Solve for the branch currents using Ohm's law.

g Check all answers.

4 The steps to be followed in the loop-current method are:

a Assign clockwise loop currents to each window in the network.
b Assign polarities to the voltages caused by the loop currents flowing in each element.
c Write a KVL equation clockwise around each loop.
d Solve the simultaneous KVL equations for the unknown loop currents.
e Find element voltages using Ohm's law.
f Check all answers.

5 The Thevenin equivalent of a circuit consists of a voltage source which has a voltage equal to the open-circuit voltage at the terminals of interest in series with a resistance equal to the resistance of the "dead" network as measured at the terminals.
6 The Norton equivalent of a circuit consists of a current source which has a current equal to the current through a short-circuit across the terminals of interest in parallel with a resistance equal to the resistance of the "dead" network as measured at the terminals.

PROBLEMS

8.1-1 In the circuit of Fig. P8.1-1, find I_1 using superposition.

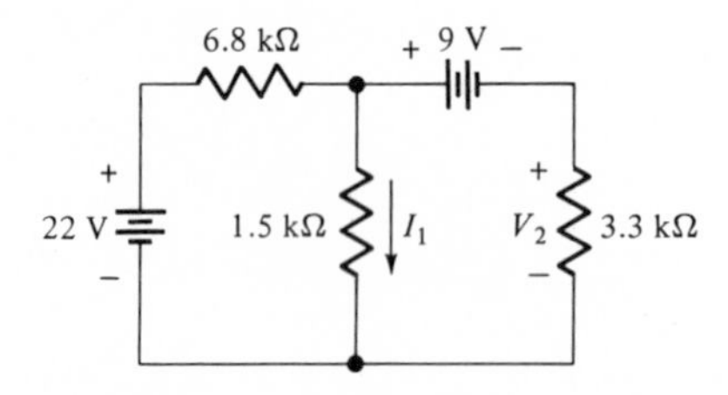

Fig. P8.1-1.

8.1-2 In the circuit of Fig. P8.1-1, find V_2 using superposition.

8.1-3 In the circuit of Fig. P8.1-3, find I_2 using superposition.

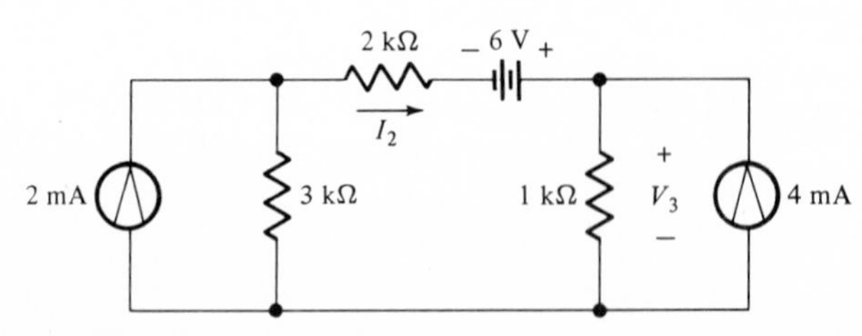

Fig. P8.1-3.

8.1-4 Use superposition to find V_3 in the circuit of Fig. P8.1-3.

8.1-5 In the circuit of Fig. P8.1-5 the following data are measured:

When $I_1 = 3$ mA and $V_2 = 6$ V, $I_4 = 2$ mA.

When $I_1 = 2$ mA and $V_2 = 8$ V, $I_4 = 7$ mA.

Find the value of I_4 when $I_1 = 4$ mA and $V_2 = 2$ V.

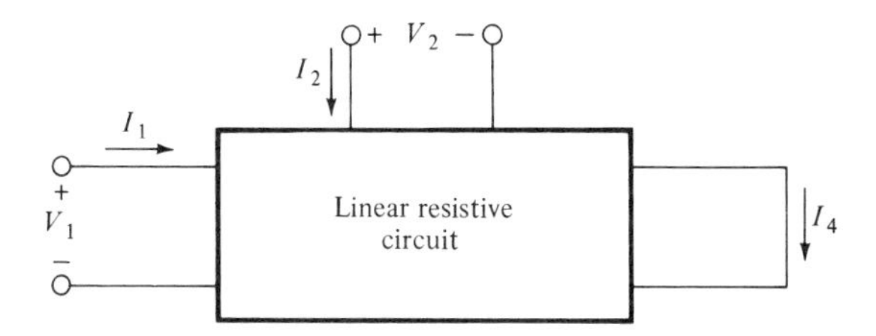

Fig. P8.1-5.

8.2-1 Use the node voltage method to find all currents and voltages in the circuit of Fig. P8.2-1.

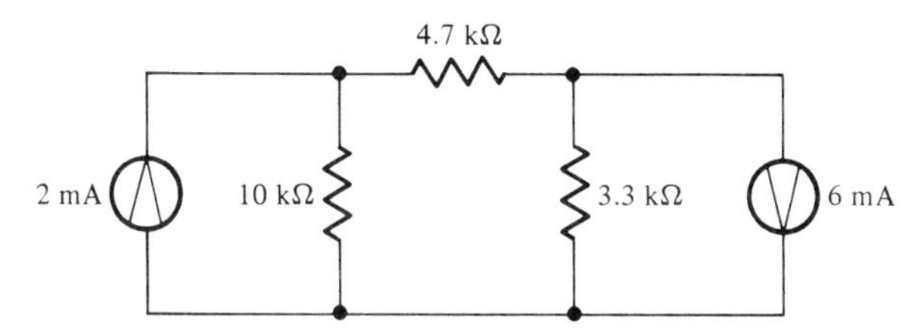

Fig. P8.2-1.

8.2-2 Repeat Prob. 8.2-1 for the circuit of Fig. P8.2-2.

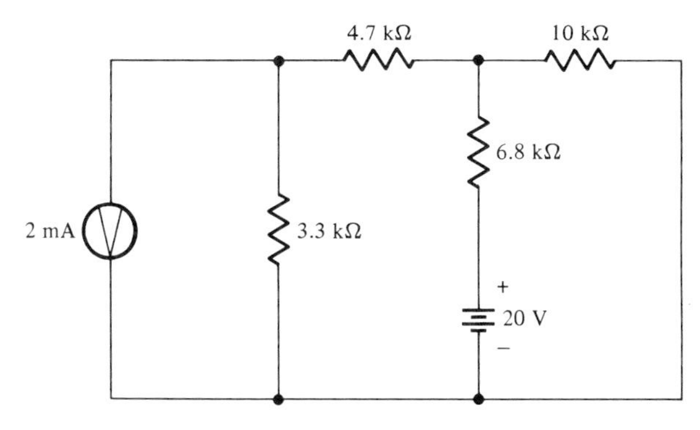

Fig. P8.2-2.

8.2-3 Repeat Prob. 8.2-1 for the bridge network of Fig. P8.2-3.

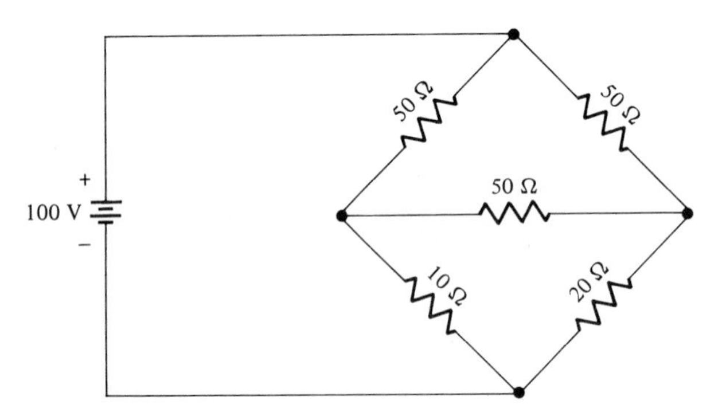

Fig. P8.2-3.

8.2-4 Use the loop-current method to find all currents and voltages in the circuit of Fig. P8.2-4.

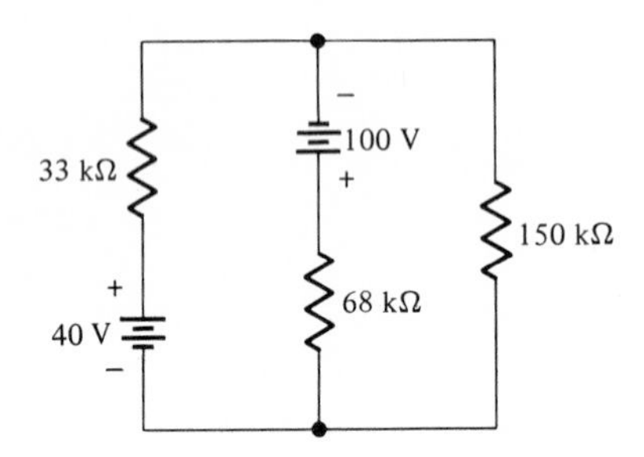

Fig. P8.2-4.

8.2-5 Write, but do not solve, the three loop equations for the bridge circuit of Fig. P8.2-3.

8.3-1 The terminal characteristic of the network shown in Fig. 8.3-1*a* is a straight line that intercepts the voltage axis at 5 V and the current axis at 3 mA. Find the Thevenin equivalent for the circuit.

8.3-2 Three volt-ampere measurements are made at the terminals of the network shown in Fig. 8.3-1*a*. The measured values are 5 V, 1 mA; 11 V, 3 mA; 20 V, 5 mA. Is it possible to find an equivalent linear circuit for the given network? If yes, find it; otherwise explain why you can't.

8.3-3 Find the Thevenin equivalent for the network whose terminal characteristic is shown in Fig. P8.3-3.

8.3-4 Find the Thevenin equivalent for the network shown in Fig. P8.3-4.

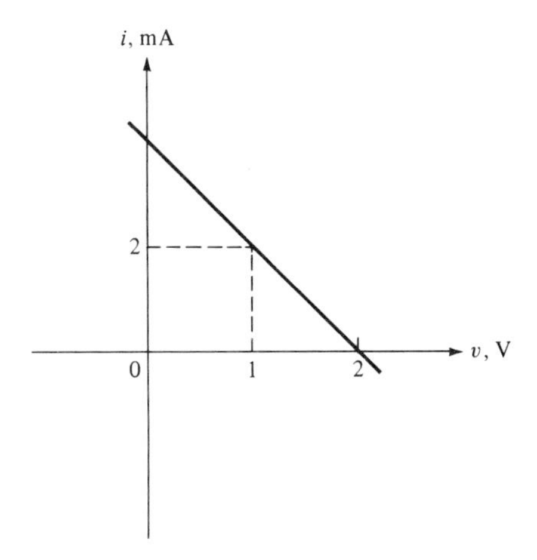

Fig. P8.3-3.

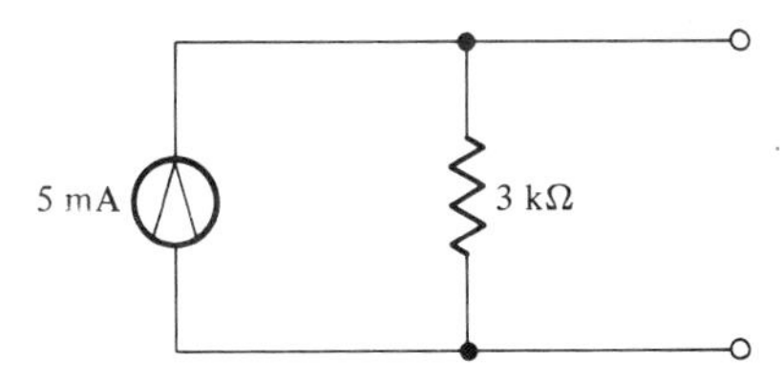

Fig. P8.3-4.

8.3-5 Find the Thevenin equivalent for the network shown in Fig. P8.3-5.

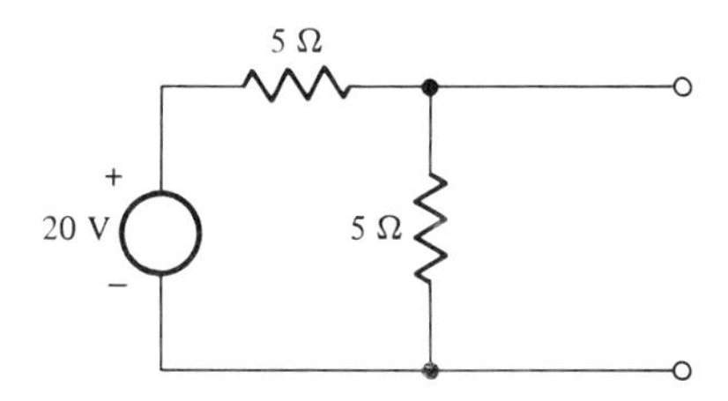

Fig. P8.3-5.

8.3-6 Find the Thevenin equivalent for the circuit shown in Fig. P8.3-6.

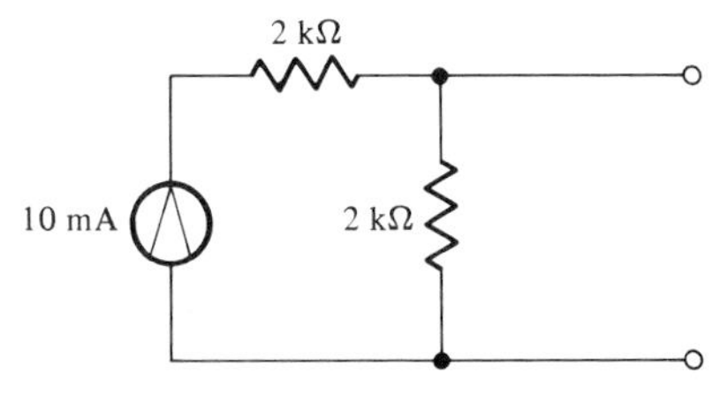

Fig. P8.3-6.

8.3-7 Find the Thevenin and Norton equivalents for the circuit shown in Fig. P8.3-7.

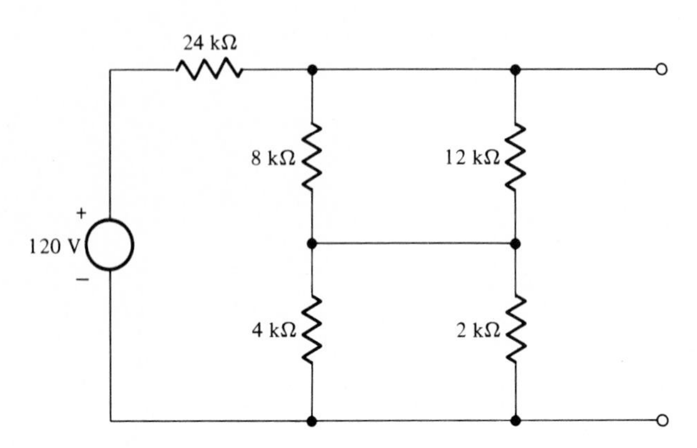

Fig. P8.3-7.

8.3-8 Find the Thevenin and Norton equivalents for the circuit shown in Fig. P8.3-8.

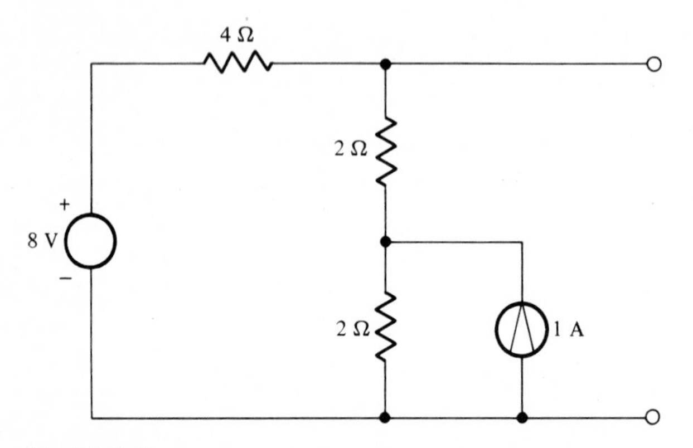

Fig. P8.3-8.

8.3-9 Find the power delivered to the 5 Ω resistance in the circuit shown in Fig. P8.3-9. *Hint*: Replace everything but the 5 Ω resistor by a Thevenin circuit.

8.3-10 Find the voltage V_0 for the circuit shown in Fig. P8.3-10. Make use of Thevenin's theorem.

8.3-11 Find the Thevenin and Norton equivalents for the circuit of Fig. P8.3-11.

8.3-12 Find the Thevenin equivalent for the network shown in Fig. P8.3-12. *Hint*: Replace the series combination of sources by a single equivalent source. This may require some careful thought.

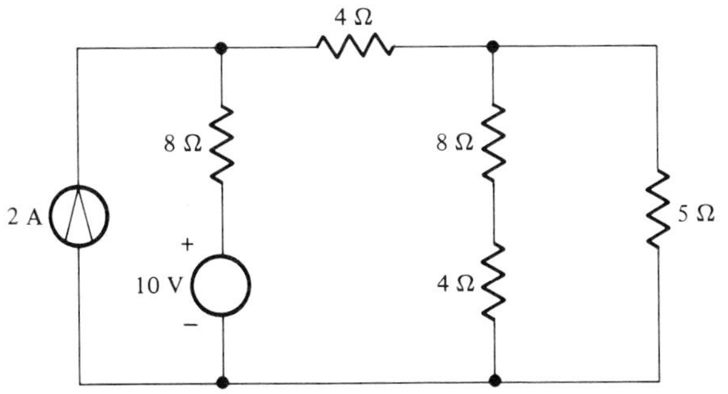

4 Ω
8 Ω
8 Ω
2 A
5 Ω
+
10 V
−
4 Ω

Fig. P8.3-9.

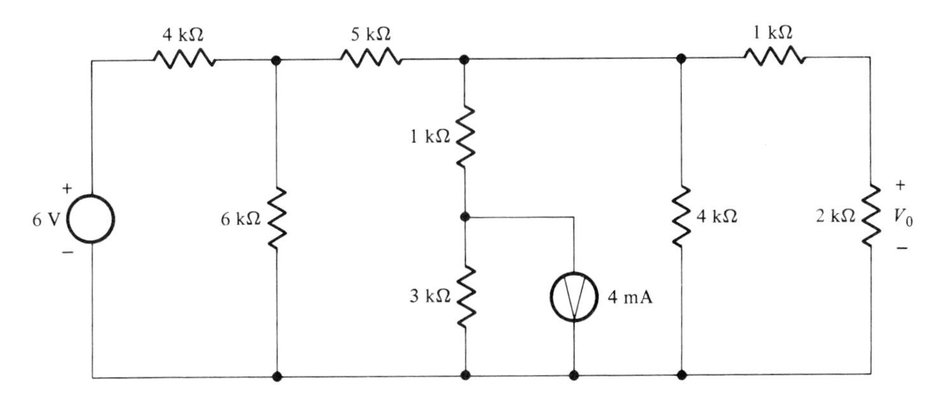

Fig. P8.3-10.

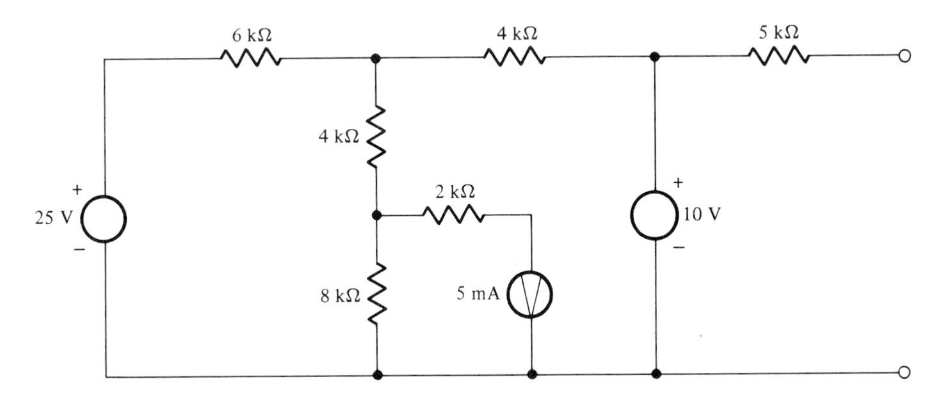

Fig. P8.3-11.

8.4-1 Write an ECAP program to solve the circuit of Fig. P8.4-1. If you have a computer available, run the program.

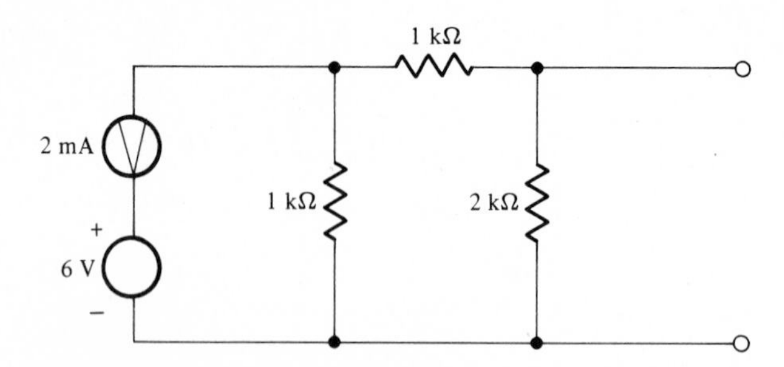

Fig. P8.3-12.

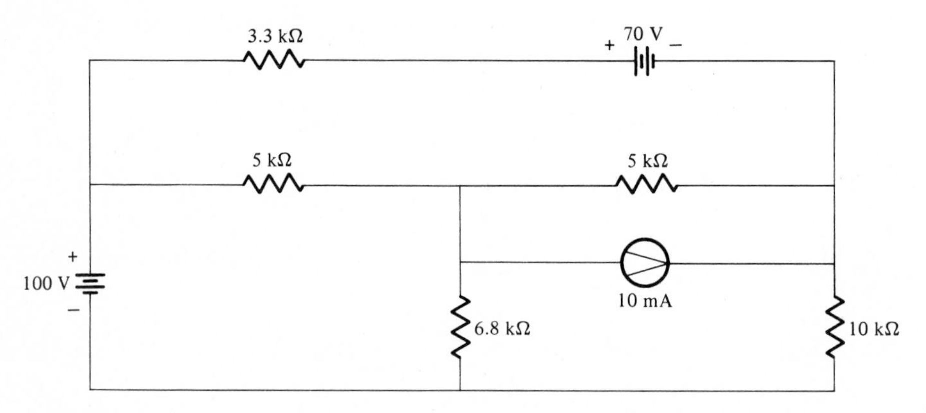

Fig. P8.4-1.

9
Diodes

OBJECTIVES

Upon completion of this chapter, the student should be able to

1 Draw the circuit symbol and a typical *vi* characteristic for a semiconductor diode.
2 Draw the circuit symbol and *vi* characteristic of the ideal diode.
3 Use graphical methods to analyze a resistive circuit containing a single nonlinear two-terminal element.
4 Find a piecewise linear model for a nonlinear two-terminal element.
5 Analyze a resistive circuit containing one or more ideal diodes.
6 Draw the circuit symbol and *vi* characteristic for a typical zener diode.
7 Describe the operation of a zener diode voltage regulator circuit.

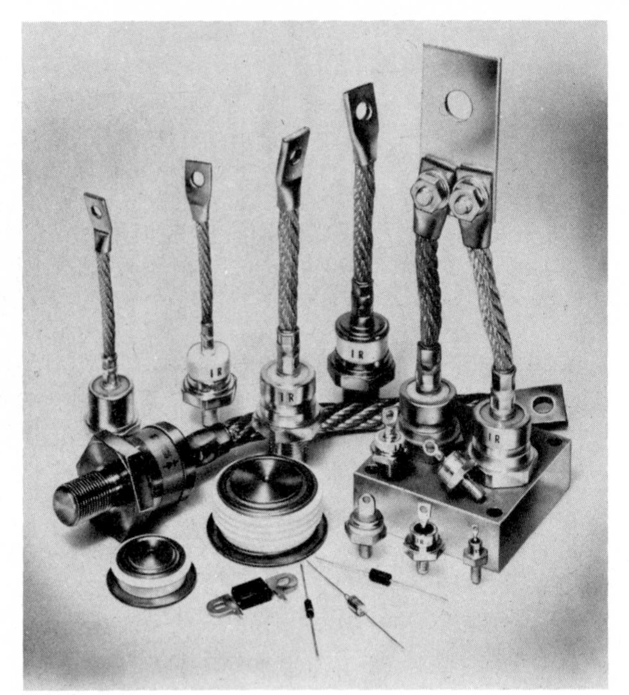

Fig. 9-1 Semiconductor diodes. (*International Rectifier Corporation.*)

INTRODUCTION

The semiconductor diode is an electronic device which is widely used in many electric and electronic circuits and systems. It is fabricated from a material whose electrical conductivity is between that of a conductor and an insulator, generally germanium or silicon. For this reason it is called a semiconductor diode. Figure 9-1 is a photograph of some semiconductor diodes. It is the first device we shall study which is useful because of its nonlinear properties. We will find that new analysis methods must be used in dealing with this nonlinear element.

The semiconductor diode also serves as motivation for defining a new abstract circuit element: the ideal diode. The ideal diode is defined as an idealization of the physical diode which it approximates.

9.1 MEASUREMENT OF SEMICONDUCTOR DIODE VI CHARACTERISTIC

We have learned that the behavior of a two-terminal element can be represented by its terminal, or *vi*, characteristic. In order to find the *vi* characteristic for an element it is necessary to measure different combinations of element voltage and

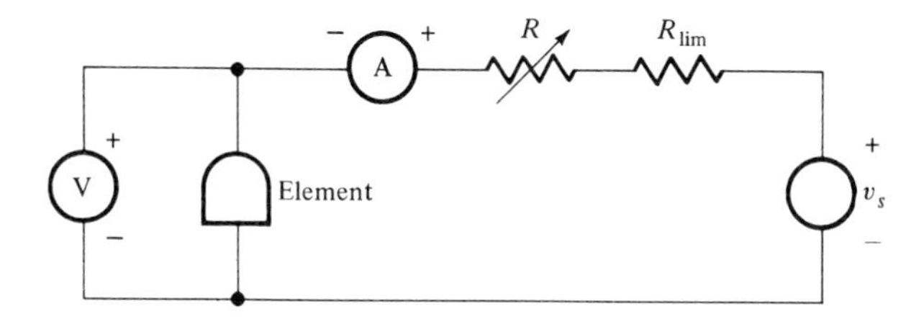

Fig. 9.1-1 Circuit for measurement of *vi* characteristic of element.

current. The graph obtained from these measurements is the terminal characteristic.

A circuit which facilitates this measurement is shown in Fig. 9.1-1. The voltmeter, V, measures the voltage across the element while the ammeter, A, measures the current through the element plus the current through the voltmeter. Normally, the current through the voltmeter is much smaller than that through the element; this permits us to neglect the voltmeter current. We must check that this is, indeed, the case; this is done by disconnecting the voltmeter temporarily and determining the change in ammeter reading. If it is more than a negligible amount, then the loading effect of the voltmeter is not negligible (see Chap. 7).

An alternative circuit for measuring the *vi* characteristic when the voltmeter loading is not negligible is shown in Fig. 9.1-2 where the ammeter measures the current through the element while the voltmeter measures the sum of the voltages across the element and across the ammeter. Normally, the voltage across the ammeter can be neglected because it is much smaller than the voltage across the element. This must be tested by temporarily shorting out the ammeter (by placing a short circuit across its terminals) and noting the change in the voltmeter reading. If the change is not negligible, it is necessary to change either the voltmeter or the ammeter. A higher resistance voltmeter may be tried in the circuit of Fig. 9.1-1 or a lower resistance ammeter may be tried in the circuit of

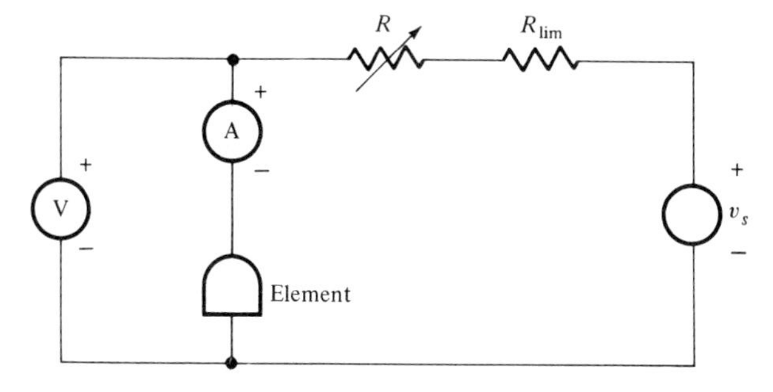

Fig. 9.1-2 Alternate circuit for measuring *vi* characteristic of element.

Fig. 9.1-2; it would be most unusual, however, to have this problem with both circuits.

If the element whose characteristic is being measured is a fixed resistor, this check need be made only once since the element and meter resistances all remain fixed (assuming meter scales are not changed). For other types of elements, and for the semiconductor diode in particular, it is necessary to check this loading effect for every pair of *vi* readings since the element does not behave like a constant resistance. It may be necessary to make some readings using the circuit of Fig. 9.1-1 and others with the circuit of Fig. 9.1-2.

The terminal or *vi* characteristic is obtained by plotting the graph of a series of different *vi* measurements made by varying v_s and/or R. One additional factor that must be considered in this procedure is the limitations of the element whose characteristic is being measured. All elements have some maximum current carrying capability, a maximum terminal voltage, and a maximum power dissipating capability; exceeding any of these ratings will generally damage the element. Therefore, it is necessary to determine these maximum ratings and to insure that they are not exceeded. The resistance R_{lim} in Figs. 9.1-1 and 9.1-2 is used as a current limiting resistance; its value is chosen such that the maximum current capability of the element cannot be exceeded. The maximum value of v_s is chosen so as not to exceed the maximum voltage rating of the element and R is varied being careful that the maximum power rating of the element is not exceeded.

Example 9.1-1 We wish to measure the *vi* characteristic of a 25-W resistor R_x whose nominal value is 10 kΩ. The characteristic is to be measured for terminal voltages from −500 to +500 V. We have a 500-V supply, a 10-MΩ variable resistor, a voltmeter with full-scale readings of 1,000 V, 300 V, 100 V, 30 V, 10 V, 3 V, and 1 V with a sensitivity of 1,000 Ω/V, and a milliameter with full scale readings of 100, 20, 10, 2, 1, and 0.2 mA which uses a 200-Ω 20-μA meter movement. Specify a circuit and procedure for finding the terminal characteristic.

Solution The internal resistance of the voltmeter on the 1-V scale is 1 kΩ which will severely load the resistance R_x whose nominal value is 10 kΩ. The ammeter resistance is 20 Ω or less depending on the range setting (check this by determining the resistance of the shunt for the 0.2-mA range). The voltage across the ammeter, if it is connected in series with the resistance R_x, will be much less than the voltage across R_x. Therefore we use the circuit of Fig. 9.1-3 with $V_s = +500$ V or $V_s = -500$ V; R is the 10-MΩ variable resistor.

By varying R it is possible to vary the voltage across R_x according to the formula

$$V = \frac{R_x \parallel R_m}{(R_x \parallel R_m) + R} V_s$$

where R_m is the voltmeter resistance for the scale being used.

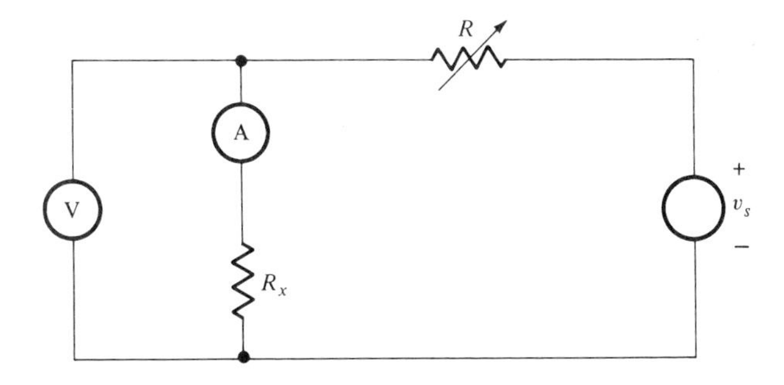

Fig. 9.1-3 Circuit for Example 9.1-1.

When R is set at its maximum value, the lowest voltmeter range would be used. In this case $R_m = 1\ \mathrm{k\Omega}$ and $R_x \parallel R_m \approx 1\ \mathrm{k\Omega}$. Then

$$V = \frac{R_x \parallel R_m}{(R_x \parallel R_m) + 10 \times 10^6} 500 \approx \frac{10^3}{10^3 + 10^7} 500 = 0.05\ \mathrm{V}$$

When R is set to zero, the 1,000-V voltmeter range must be used. For this case

$$V = \frac{R_x \parallel R_m}{(R_x \parallel R_m) + 0} 500 = 500\ \mathrm{V}$$

In addition we have the corresponding negative values; the value zero is obtained by replacing V_s by a short circuit. ////

SEMICONDUCTOR DIODE

Before we measure the terminal characteristic of the semiconductor diode, also called the *pn* junction diode, we should have some idea of its behavior and its maximum ratings.

We will characterize the diode's behavior by its terminal characteristic; for the moment we will find it adequate to point out that it acts like a very low resistance for current in one direction, called the forward direction, and acts like a very large resistance for current flow in the opposite direction, called the reverse direction. The circuit symbol used to represent the diode is shown in Fig. 9.1-4*a*; the arrowhead points in the direction of forward current flow. One side of the diode is called the *p* side, the other the *n* side, these are labeled in the figure. The usual reference directions for current and voltage are also shown in the figure. The *vi* characteristic for a typical silicon semiconductor diode is shown in Figs. 9.1-4*b* and 9.1-4*c* which show the complete characteristic and an enlarged view in the vicinity of the origin.

The diode is said to be forward-biased when the values of the reference voltage and current are positive; it is reverse biased when these values are negative. From the slope of the diode characteristic we see that the diode acts like a very large resistance for small forward bias as well as reverse bias while it

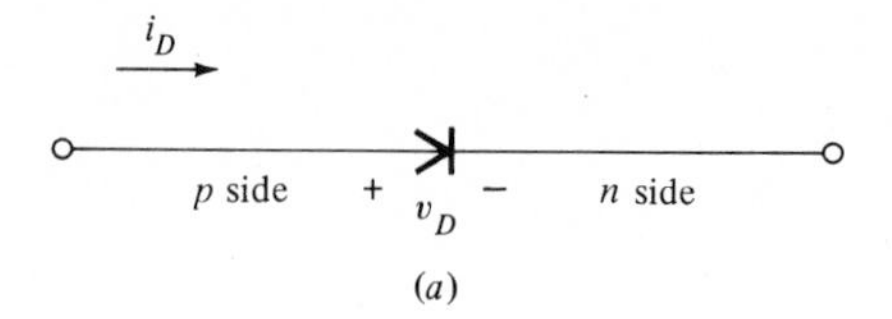

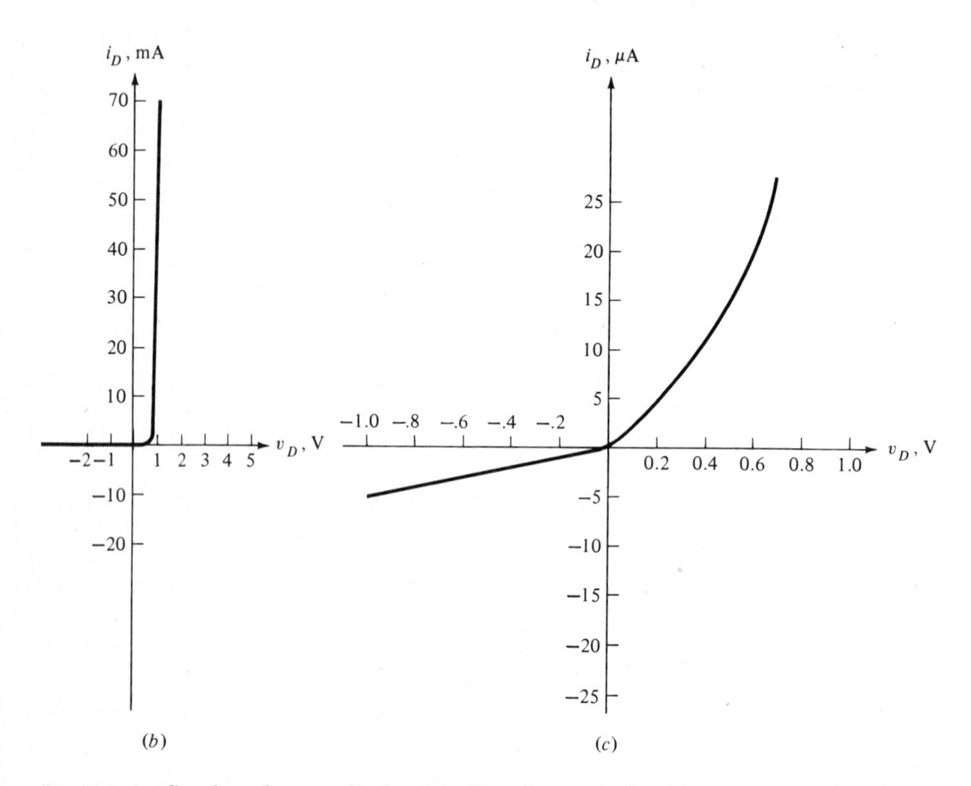

Fig. 9.1-4 Semiconductor diode. (*a*) Circuit symbol with current and voltage reference direction shown; (*b*) *vi* characteristic; (*c*) enlarged view of *vi* characteristic in vicinity of origin.

acts like a small resistance for larger forward bias. In between these two regions there is a small region in which a smooth transition occurs. Because of this difference in behavior two different circuits must be used to measure the diode *vi* characteristic: that of Fig. 9.1-5*a* for reverse bias or very small forward bias and that of Fig. 9.1-5*b* for large positive bias. In the middle of the transition region either circuit will operate satisfactorily but the loading should be checked to ensure that the correct circuit is being used because the transition region covers such a small range.

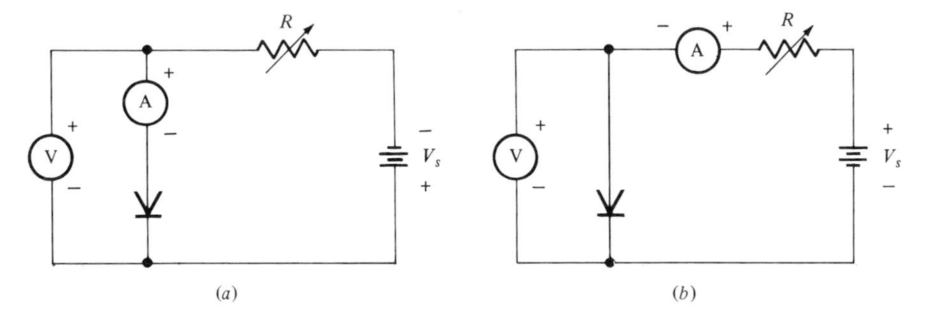

Fig. 9.1-5 Circuits for measurement of diode *vi* characteristic. (*a*) Reverse bias (V and A negative) and small forward bias; (*b*) forward bias (V and A positive).

9.2 GRAPHICAL ANALYSIS AND THE LOAD LINE

We are now able to describe any two-terminal element by its *vi* characteristic which we have learned to measure. When we proceed to analyze a circuit containing a diode, or any other nonlinear element (an element whose *vi* characteristic is not a straight line), we find ourselves with a problem. We cannot use Ohm's law to describe the device nor can we describe it by a simple formula similar to those used for voltage or current sources.

Consider the circuit of Fig. 9.2-1*a* which contains a diode whose terminal characteristic is shown in Fig. 9.2-1*b*. One approach we might take in analyzing this circuit is to find an analytic representation for the diode terminal characteristic whose graphical representation is shown in Fig. 9.2-1*b*; it will be of the

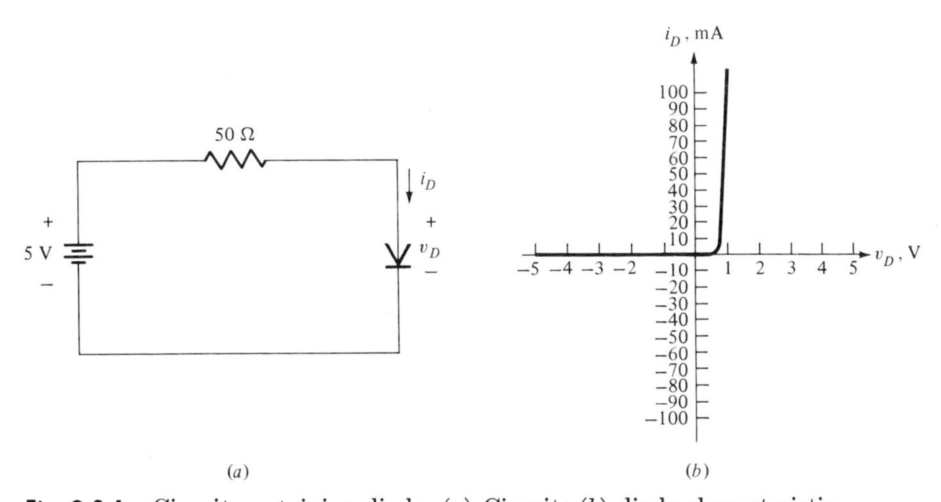

Fig. 9.2-1 Circuit containing diode. (*a*) Circuit; (*b*) diode characteristic.

form (see Sec. 5.2):

$$i_D = f(v_D) \tag{9.2-1}$$

The remainder of the circuit of Fig. 9.2-1a (the voltage source and resistance connected in series) imposes a second constraint on the diode voltage and current (see Sec. 8.3):

$$v_D = 5 - 50i_D \tag{9.2-2}$$

The particular values of v_D and i_D, denoted by V_D and I_D, respectively, which exist in the circuit of Fig. 9.2-1a must satisfy both Eqs. (9.2-1) and (9.2-2) simultaneously. Substituting Eqs. (9.2-1) into (9.2-2) results in

$$V_D = 5 - 50f(V_D) \tag{9.2-3}$$

The solution of this equation is the diode voltage V_D which actually exists in the circuit of Fig. 9.2-1a. The corresponding current is found by solving Eq. (9.2-2) which results in

$$I_D = \frac{5 - V_D}{50} \tag{9.2-4}$$

In order to find numerical values for I_D and V_D it is necessary that f be specified as a given function; in Eq. (9.2-1) we have specified only the form of the terminal characteristic equation. In general, it is very difficult to find the actual function f. Therefore we shall solve the simultaneous Eqs. (9.2-1) and

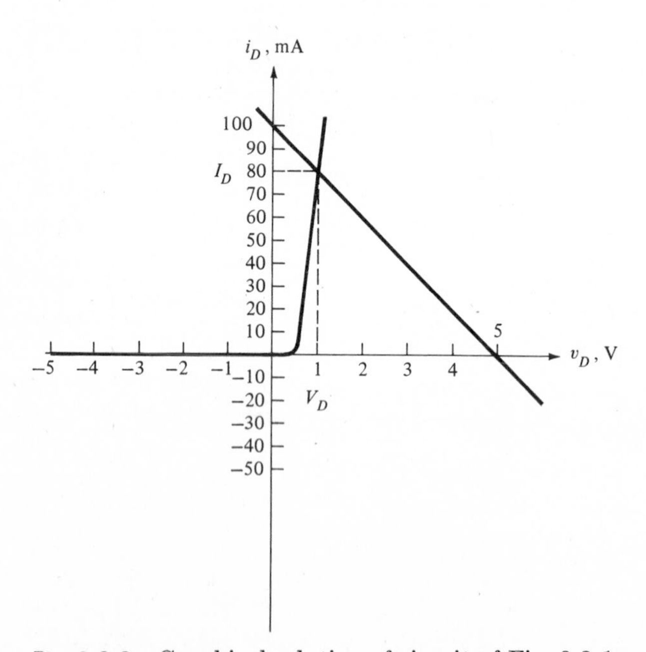

Fig. 9.2-2 Graphical solution of circuit of Fig. 9.2-1.

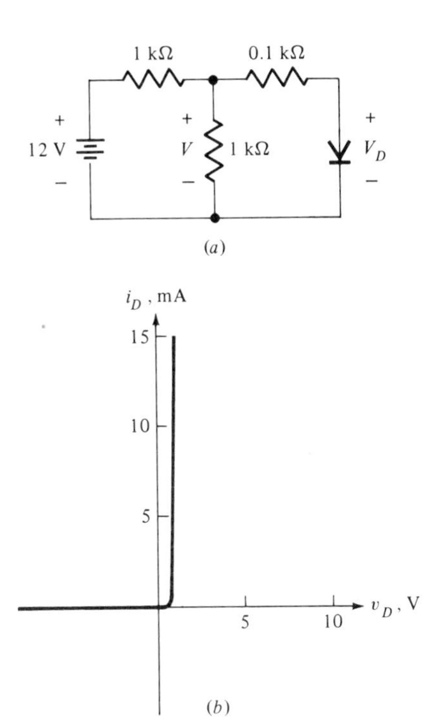

Fig. 9.2-3 Circuit and characteristic for Example 9.2-1. (*a*) Circuit; (*b*) diode *vi* characteristic.

(9.2-2) graphically. Recall that two simultaneous equations in two unknowns can be solved by finding the intersection of their graphs. An alternative way of stating the same thing is: The diode current and voltage must lie on the diode characteristic which is the graph of Eq. (9.2-1); the diode current and voltage must also lie on the graph of Eq. (9.2-2) which is shown plotted on the same axes as the diode characteristic in Fig. 9.2-2. The graph of Eq. (9.2-2) is a straight line since Eq. (9.2-2) is the terminal characteristic of a linear circuit. Methods for drawing this graph were covered in Sec. 8.3.

The only point which lies on both graphs is at the intersection of the two graphs. Thus this point represents the operating point of the circuit. The approximate values of V_D and I_D at this point are $V_D = 1.0$ V and $I_D = 80$ mA.

Example 9.2-1 The diode in the circuit of Fig. 9.2-3*a* has the characteristic shown in Fig. 9.2-3*b*. Find the diode voltage, V_D, and the voltage V.

Solution The key to solving this problem is to replace the circuitry connected to the diode by its Thevenin equivalent as shown in Fig. 9.2-4*a* (check that the circuit shown is indeed the proper Thevenin equivalent). This makes the problem identical, except for numerical values, to the one

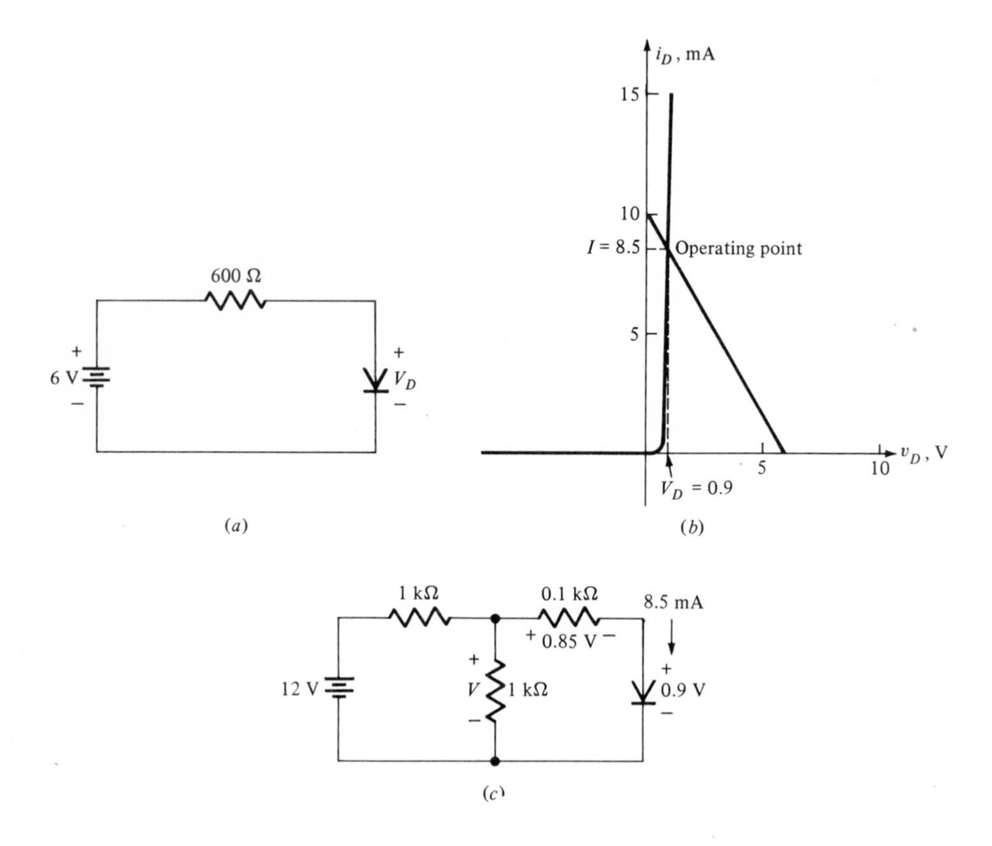

Fig. 9.2-4 Solution of Example 9.2-1. (*a*) Simplified circuit using Thevenin equivalent; (*b*) graphical analysis; (*c*) original circuit with diode current and voltage shown.

we have just finished. The terminal characteristic of this Thevenin equivalent is shown on the same set of axes as the diode characteristic in Fig. 9.2-4*b*. The diode voltage, V_D, corresponding to the intersection of these two graphs is found to be approximately 0.9 V on this diagram.

From Fig. 9.2-4*b* the diode current is found to be about 8.5 mA. The diode current and voltage are applied to the original circuit from which, using Ohm's law and KVL the voltage V is computed:

$$V = (0.1\ \text{k}\Omega)\,(8.5\ \text{mA}) + 0.9\ \text{V} = 1.75\ \text{V}$$

As a check we compute the current through the resistance across which V is measured using Ohm's law:

$$I = \frac{1.75\ \text{V}}{1\ \text{k}\Omega} = 1.75\ \text{mA}$$

By KCL the current through the voltage source and resistance in series with it is

$$I_s = 1.75 \text{ mA} + 8.5 \text{ mA} = 10.25 \text{ mA}$$

and the voltage drop across that resistance is

$$(1 \text{ k}\Omega)(10.25 \text{ mA}) = 10.25 \text{ V}$$

When this is added to the 1.75 V value of V the sum equals the 12 V source voltage thus satisfying KVL around the left-hand loop. ////

LOAD LINE

From the example and the material preceding it we see that the key to analyzing a circuit with a nonlinear element is to find the intersection of the nonlinear element's *vi* characteristic with the linear *vi* characteristic of the remainder of the circuit.

The linear *vi* characteristic of the remainder (linear part) of the network is called the *load line*. The load line is most easily found by obtaining the Thevenin equivalent of the linear portion of the circuit as discussed in Sec. 8.3.

9.2-1 Graphical Analysis With Time-varying Signals

Consider the circuit of Fig. 9.2-5*a*. The terminal characteristic for the diode is shown in Fig. 9.2-5*b*. The sinusoidal voltage source is generally considered to be the signal component of the input voltage because it is time-varying while the 2-V dc-voltage source is assumed to be a *bias* source. By this we mean that it determines on which portion of the nonlinear elements characteristic we are operating but is not itself part of the signal.

The analysis of the circuit consists, in this case, of finding the diode voltage $v_D(t)$. This is accomplished, in theory, by drawing a separate load line for each instantaneous value of $v_1(t)$, finding the corresponding $v_D(t)$, and plotting the resultant curve. In practice we do this for a series of points which we connect by a smooth curve.

The first value of $v_1(t)$ we might consider is the value zero. The load line for this value is shown in Fig. 9.2-5*b*; it is the one which is labeled $v_1 = 0$. The intersection of this line with the diode characteristic is called the quiescent (Q) point because it corresponds to the zero signal condition; it is labeled Q in the diagram. The corresponding value of $v_D(t)$ is found to be about 1 V. This quiescent condition occurs at $100t = 0$, π, and 2π rad.

Load lines are also shown for $100t = \pi/2$ and $3\pi/2$. At $100t = \pi/2$, $v_1(t) = 3$ V and the total applied voltage is 5 V; the corresponding diode voltage at which this load line intersects the diode characteristic is $v_D = 1.9$ V. For $t = 3\pi/2$, $v_1(t) = -3$ V and the total applied voltage is -1 V. The corresponding load line lies in the third quadrant and intersects the diode characteristic right on the voltage axis so $v_D = -1.1$ V.

By drawing additional load lines for intermediate points (which are not

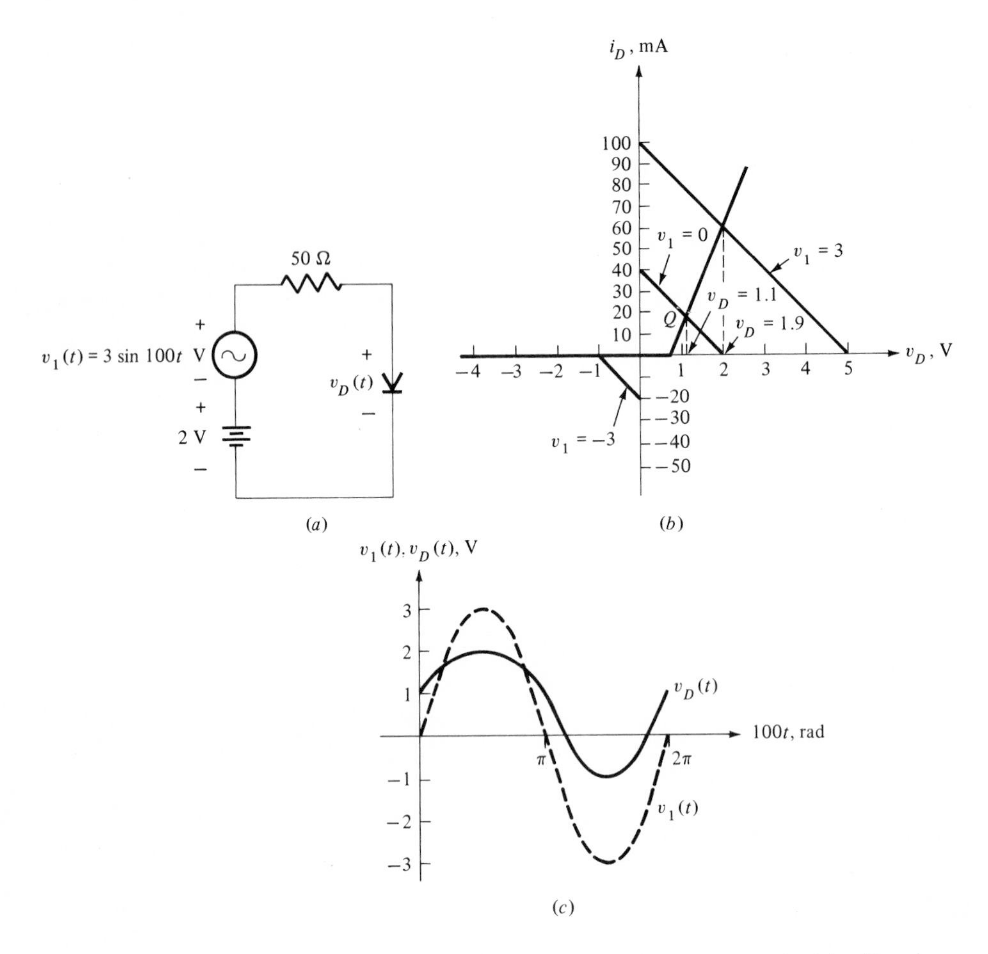

Fig. 9.2-5 Analysis of nonlinear circuit with time-varying signal. (*a*) Circuit; (*b*) graphical analysis; (*c*) voltage waveforms.

shown in the figure) enough points are obtained to plot the curve of output voltage $v_D(t)$ shown, together with the input voltage, in Fig. 9.2-5*c*.

Note that the output voltage is not symmetrical like the input voltage but is a distorted sine wave. This is due to the nonlinearity of the diode characteristic.

9.3 PIECEWISE-LINEAR CHARACTERISTICS

The graphical analysis method is applicable to linear, as well as nonlinear, circuits. It is almost never used for linear circuits because of the tediousness of the procedure—the necessity of drawing accurate graphs to scale, etc.

The disadvantages of the graphical method have motivated the development of other methods which do not require graphs. One important technique is based on the use of a piecewise-linear characteristic which approximates the given terminal characteristic.

The piecewise-linear approximation is obtained by fitting a set of straight-line segments to the curve. There is no unique approximation for a given nonlinear characteristic. A number of possible approximations for a given characteristic are shown in Fig. 9.3-1. From this figure it is evident that a more accurate approximation can be obtained by using a greater number of straight-line segments. Subsequent discussion will indicate that the analysis procedure becomes more complex as we increase the number of segments in the approximating piecewise-linear characteristic. Accordingly, the minimum number of segments consistent with the required accuracy is used.

Two segments provide an adequate approximation to the semiconductor diode characteristic for most applications. A number of commonly used piecewise-linear characteristics are shown in Fig. 9.3-2. The actual characteristic is shown in part *a* and by the dashed line in the remainder of the figure. The slope of the reverse-biased portion is exaggerated for clarity. The piecewise-linear

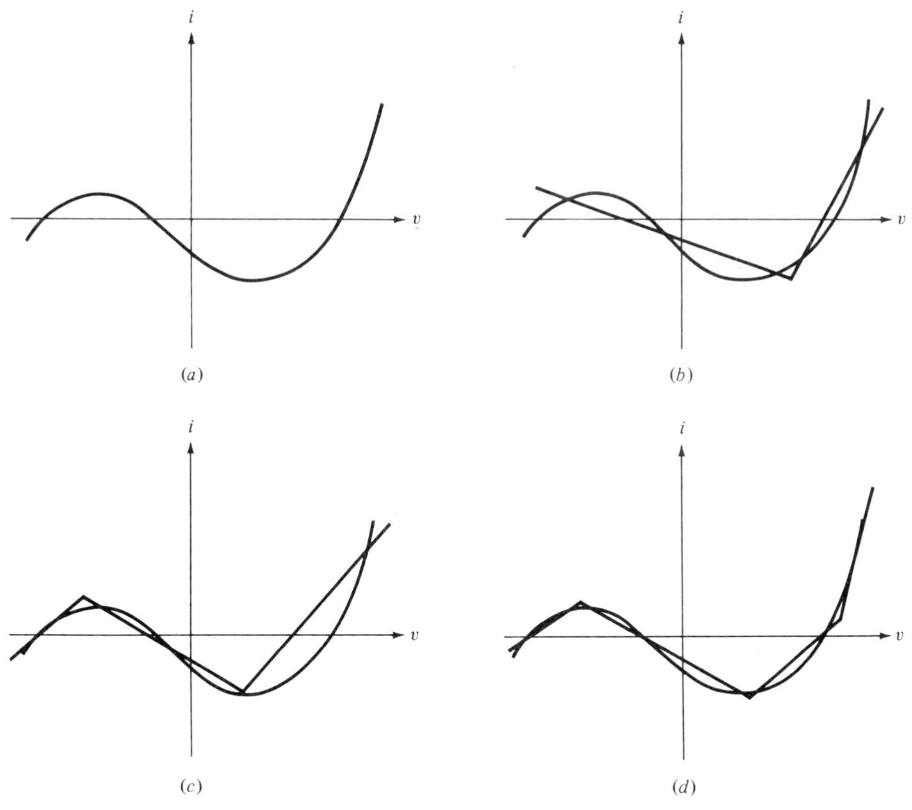

Fig. 9.3-1 Piecewise-linear approximations. (*a*) Two segment; (*b*) an alternate two segment; (*c*) three segment; (*d*) four segment.

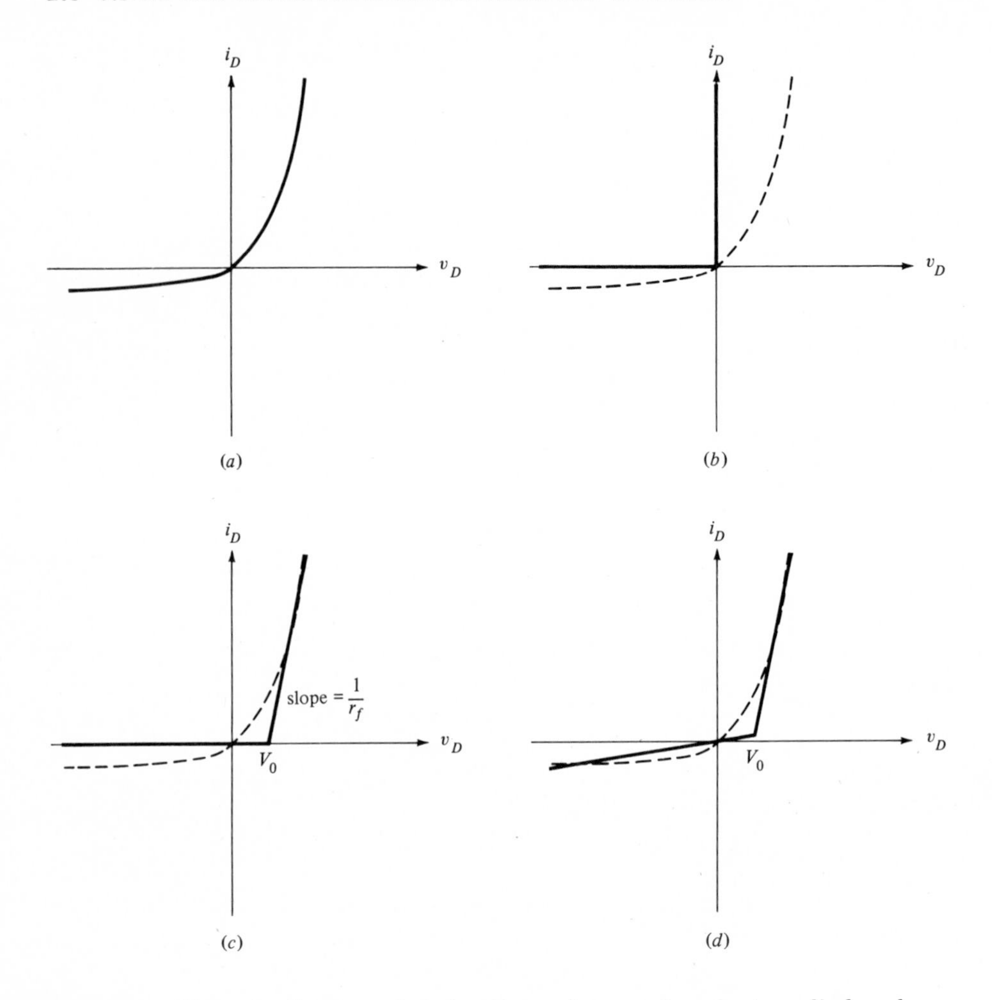

Fig. 9.3-2 Piecewise-linear approximations for semiconductor diode characteristic. (*a*) Actual characteristic; (*b*) simplest approximation; (*c*) better approximation; (*d*) best approximation.

approximations in parts *b* to *d* of the figure are successively better approximations. We shall see, in Sec. 9.5, that the analysis becomes more complex as the approximation becomes better. As a result the simplest approximation which provides the required degree of accuracy is used.

9.4 THE IDEAL DIODE

In Sec. 5.5 we discussed the use of ideal elements to model physical circuit elements. The use of these ideal elements simplifies the analysis of circuits because secondary effects of minor importance are omitted from consideration.

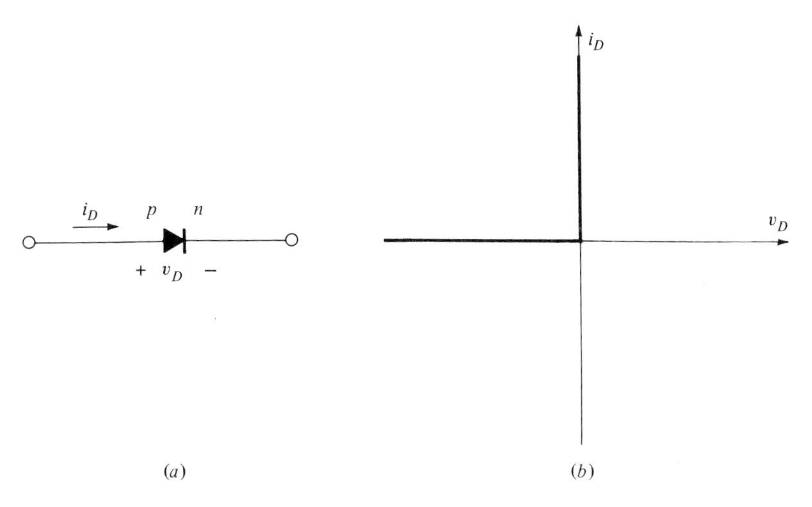

Fig. 9.4-1 Ideal diode. (*a*) Circuit symbol; (*b*) *vi* characteristic.

In this section we are going to introduce a new ideal element—the ideal diode. Just as the resistance element approximates the behavior of a physical resistor, so the ideal diode approximates the behavior of a semiconductor diode. The major feature of the semiconductor diode's behavior is its nonlinearity; the ideal diode represents this most important aspect of the diode behavior while eliminating the other effects. As a result the ideal diode behavior is a rather gross approximation of the behavior of the physical diode; nevertheless, it is adequate for many purposes.

The circuit symbol and usual reference current direction and voltage polarity are shown in Fig. 9.4-1*a* while the ideal diode characteristic is shown in Fig. 9.4-1*b*. Notice that we use a closed arrowhead for the ideal diode and an open arrowhead for the physical diode so as to avoid confusion. As can be seen from the figure, the diode acts like a short circuit or zero resistance in the forward direction and an open circuit or infinite resistance in the reverse direction. We say that the diode is forward-biased when it is conducting (current flowing from p side to n side) and reverse-biased when it is cut off or not conducting (n side at positive voltage with respect to p side).

The continuous nonlinear nature of the physical diode's *vi* characteristic makes it necessary to use graphical analysis. The nonlinearity of the ideal diode characteristic is concentrated at a single point, the origin. As a consequence, a circuit containing an ideal diode can be analyzed using linear circuit analysis techniques applied to two equivalent circuits; one in which the diode is replaced by an open circuit and the other in which the diode is replaced by a short circuit. This is illustrated in the following example.

Example 9.4-1 Determine the output voltage, $v_0(t)$, for the circuit of

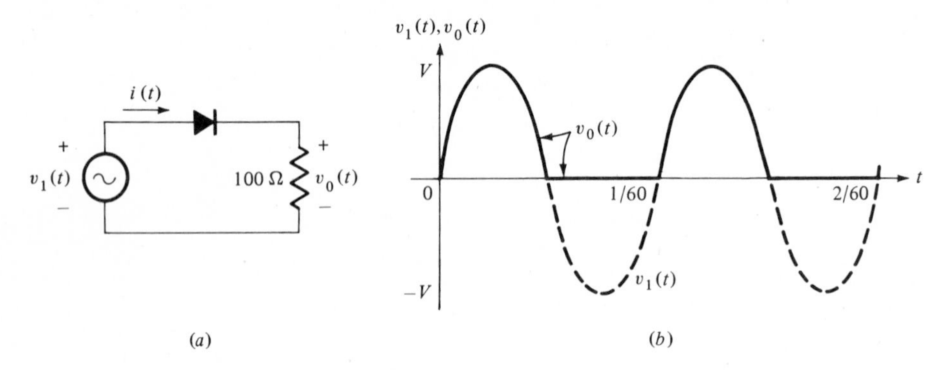

Fig. 9.4-2 Circuit and voltages for Example 9.4-1. (*a*) Circuit; (*b*) voltages.

Fig. 9.4-2*a* if the input voltage is given by

$$v_1(t) = V_m \sin 120\pi t$$

Solution When $v_1(t)$ is positive, $i(t)$ is positive and the diode is forward biased and thus acts like a short circuit. When $v_1(t)$ is negative, $i(t)$ would tend to be in a direction opposite to the reference direction. Since current cannot flow in the reverse direction, the diode acts like an open circuit resulting in a zero current. For positive values of $v_1(t)$, $v_0(t) = v_1(t)$ since the diode voltage is zero. (Why?) When $v_1(t)$ is negative, $v_0(t) = 0$ since $i(t) = 0$.

A plot of $v_1(t)$ (dotted curve) and $v_0(t)$ (solid curve) is shown in Fig. 9.4-2*b* What is the diode voltage when $v_1(t)$ is negative? This quantity is of importance because diode's have a peak inverse voltage (PIV) rating which must not be exceeded if the diode is not to be damaged. ////

MULTIPLE DIODE CIRCUITS

The graphical analysis of circuits with multiple nonlinear elements is extremely tedious and, as a result, is very rarely done. In the case of ideal diodes, the analysis of multidiode circuits can be performed using the same method as was used for a circuit with a single diode; the only difference is that all combinations of open and short circuit equivalents for the diodes must be considered.

Example 9.4-2 Determine the output voltage, $v_0(t)$, for the circuit of Fig. 9.4-3*a* if the input voltages $v_1(t)$ and $v_2(t)$ are shown in Fig. 9.4-3*b*.

Solution At time t_1 diode D_1 is forward-biased due to $v_1(t)$ while diode D_2 is reverse-biased since its *n* side is at +1 V [due to $v_0(t) = v_1(t)$] and its *p* side is at $v_2(t) = 0$. At t_2 the roles of the diodes are reversed since the values of $v_1(t)$ and $v_2(t)$ are interchanged. At t_3 both diodes are forward-biased since the *p* sides are both at +1 V and current flows through them and the 1 kΩ resistance causing $v_0(t) = 1$ V. At t_4 diode D_1 is forward-biased by $v_1(t) = +2$ V resulting in $v_0(t) = +2$ V. Diode D_2 is reverse-biased since its *n* side

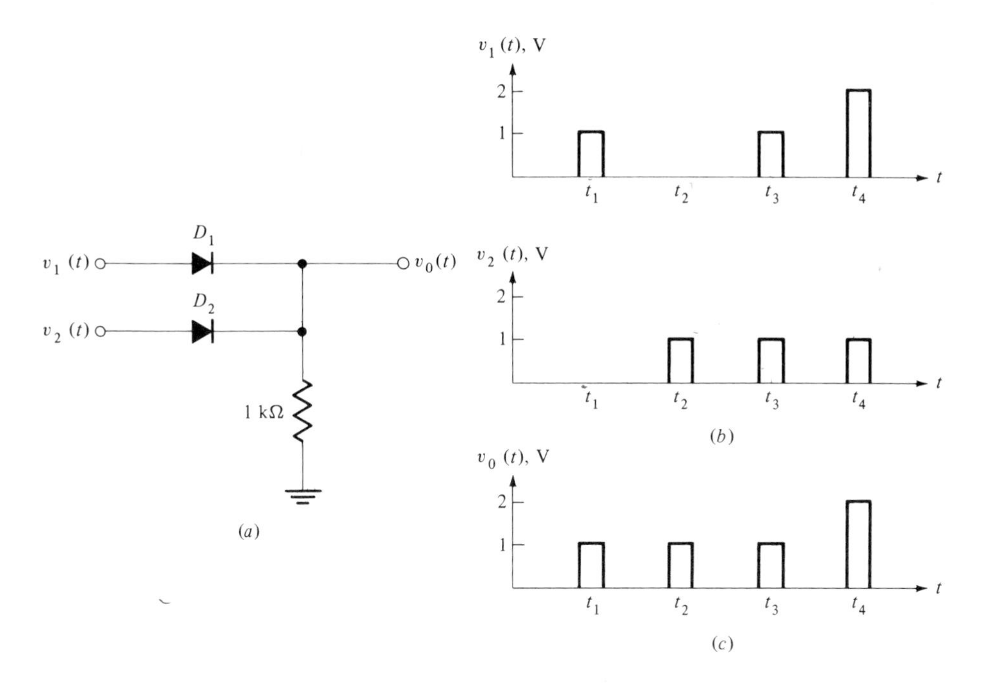

Fig. 9.4-3 Circuit and voltages for Example 9.4-2. (*a*) Circuit; (*b*) input voltages; (*c*) output voltages.

is at $v_0(t) = +2$ V while its *p* side is at $v_2(t) = +1$ V. In between pulses both diodes are at their transition points (diode voltage and current are both zero). A plot of the output voltage is shown in Fig. 9.4-3*c*. ////

9.5 PIECEWISE-LINEAR MODELS

The piecewise-linear approximation for a nonlinear terminal characteristic was discussed in Sec. 3 of this chapter. At that point we indicated that piecewise-linear approximations would be used as the basis for a nongraphical method of analyzing circuits containing nonlinear elements.

In this section we shall use the ideal diode element to generate a circuit model which is described by the piecewise linear characteristic which in turn approximates the given *vi* characteristic. This circuit model, which we call the *piecewise-linear model*, can be analyzed using the techniques we have just discussed for the analysis of circuits containing ideal diodes.

We will describe the techniques for finding a circuit model for a given piecewise-linear characteristic by using the piecewise-linear approximations for the semiconductor diode shown in Fig. 9.3-2. We will start with the simplest model and work toward the most complex one.

The piecewise-linear approximation of Fig. 9.3-2*b* is identical to the characteristic of the ideal diode. Therefore the ideal diode is the simplest model for

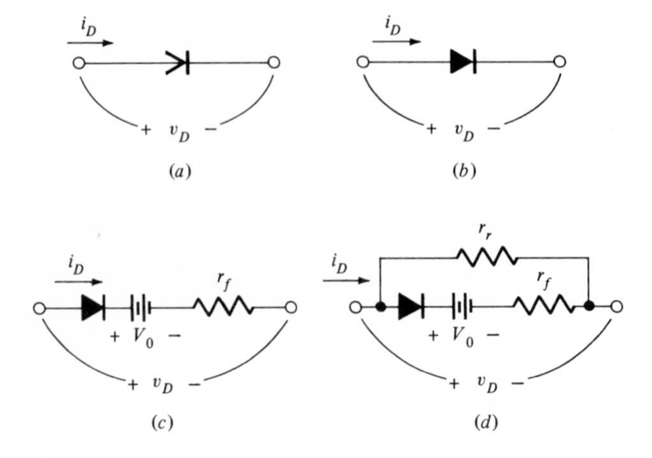

Fig. 9.5-1 Piecewise-linear models for semiconductor diode. (*a*) Physical diode symbol; (*b*) ideal diode model; (*c*) more complex model; (*d*) most complex model.

the semiconductor diode. It is shown in Fig. 9.5-1*b*. The physical diode symbol is shown in Fig. 9.5-1*a*.

A somewhat better approximation for the semiconductor diode characteristic is shown in Fig. 9.3-2*c*. The differences between it and the one in Fig. 9.3-2*b* are the slope of the forward-biased portion of the characteristic and the location of the transition point. The slope corresponds to a resistance, denoted r_f, whose value is given by the reciprocal of the slope of the piecewise-linear approximation for $v_D > V_0$:

$$r_f = \frac{1}{\left.\dfrac{\Delta i_D}{\Delta v_D}\right|_{v_D > V_0}} \tag{9.5-1}$$

(The reciprocal of the slope is used because the graph is a plot of i_D as a function of v_D.) This resistance is called the forward resistance of the diode. The model for the semiconductor diode is the series combination of an ideal diode, the resistance r_f, and a voltage source V_0, called the forward conducting drop, which displaces the transition point. This model is shown in Fig. 9.5-1*c*.

The added feature in the piecewise-linear characteristic in Fig. 9.3-2*d* is a nonzero slope for the reverse-biased portion of the characteristic. The resistance, denoted r_r, for this portion of the piecewise-linear approximation is given by the reciprocal of the slope for $v_D < V_0$:

$$r_r = \frac{1}{\left.\dfrac{\Delta i_D}{\Delta v_D}\right|_{v_D < V_0}} \tag{9.5-2}$$

This resistance is called the reverse resistance of the diode; it is much larger than the forward resistance as can be seen by comparing the slopes of the forward and reverse portion of the characteristic. A resistance of value r_r in parallel with the model of Fig. 9.5-1*c* provides the required piecewise-linear characteristic. The complete model is shown in Fig. 9.5-1*d*.

It is instructive to check that this model, shown in Fig. 9.5-1*d*, does indeed have the piecewise-linear characteristic shown in Fig. 9.3-2*d*. First of all, we note that the transition point occurs when the ideal diode has zero current through it and zero voltage across it. Consequently the voltage across r_r is V_o and the current through it, by Ohm's law, is

$$i_D = \frac{V_0}{r_r} \tag{9.5-3}$$

When $i_D = 0$, the ideal diode is reverse-biased due to V_0 and therefore acts like an open circuit. The current through r_r and r_f is zero and therefore $v_D = 0$. Therefore, the characteristic passes through the origin as it should.

The equivalent resistance when the ideal diode is open-circuited, which will occur when $v_D < V_0$, is r_r. This agrees with the piecewise-linear characteristic.

Finally, we must check the equivalent resistance when the diode is conducting, i.e. when $v_D > V_0$. The equivalent resistance is found by replacing the ideal diode and V_0 by a short circuit resulting in r_r and r_f in parallel. The equivalent resistance is

$$r_r \parallel r_f = \frac{r_r r_f}{r_r + r_f} \approx \frac{r_r r_f}{r_r} = r_f \tag{9.5-4}$$

since $r_r \gg r_f$. This value is also in agreement with the characteristic.

Having checked that the model provides the proper resistance values, that its characteristic passes through the origin, and the transition occurs at the proper voltage we may conclude that it does satisfy the characteristic of Fig. 9.3-2*d*.

A circuit containing semiconductor diodes can be analyzed by replacing each semiconductor diode by a piecewise-linear model. The resulting circuit is then analyzed using the analytic techniques discussed in the previous section. The model used for the diode should be the simplest that provides sufficient accuracy for the given application.

Frequently the manufacturer of semiconductor diodes will provide typical values for the piecewise-linear model parameters r_f, also called the forward resistance, and r_r, also called the reverse resistance. The voltage V_0, called the forward conducting voltage or "break" voltage, at which the transition from reverse to forward behavior occurs depends on the material from which the transistor is fabricated. The most common materials are silicon for which $V_0 \approx 0.7$ V and germanium for which $V_0 \approx 0.2$ V. Typical values for the other parameters are $r_f = 20\ \Omega - 100\ \Omega$ and $r_r = 100\ \text{k}\Omega - 10\ \text{M}\Omega$.

9.6 RECTIFIER CIRCUIT ANALYSIS

A rectifier circuit is used to convert ac current, which has a zero average value, to current which has a nonzero average value. We will consider two types of rectifier circuits: half-wave rectifiers and full-wave rectifiers.

9.6-1 Half-Wave Rectifier

The circuit for a half-wave rectifier is shown in Fig. 9.6-1*a* in which $v_1(t) = V_m \sin 120\pi t$, $V_m = 155$ V, $R_L = 100\ \Omega$, and the diode is a silicon diode with $r_f = 50\ \Omega$ and $r_r = 2$ MΩ. The voltage $v_1(t)$ is shown in Fig. 9.6-1*b*.

The simplest model for the diode (an ideal diode) results in the circuit already analyzed in Example 9.4-1.

The next more complicated model results in the equivalent circuit of Fig.

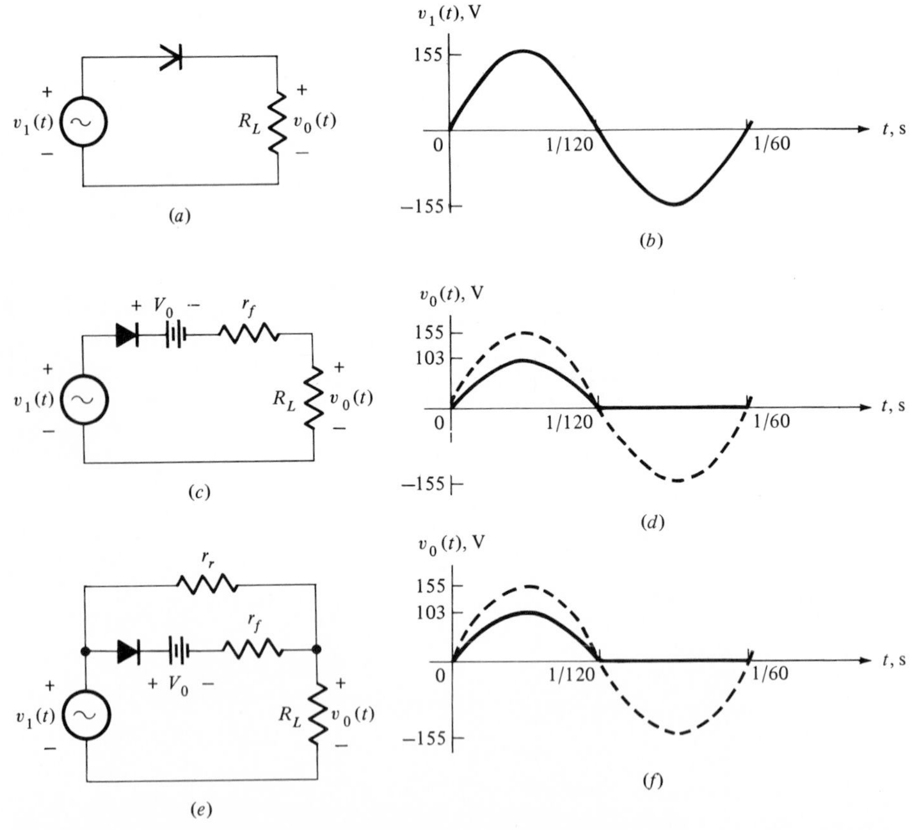

Fig. 9.6-1 Half-wave rectifier analysis. (*a*) Half-wave rectifier circuit; (*b*) input voltage; (*c*) equivalent circuit incorporating diode forward resistance and forward conducting voltage; (*d*) output voltage for circuit of *c*; (*e*) equivalent circuit incorporating all diode parameters; (*f*) output voltage for circuit of *e*.

9.6-1*c*. The break voltage V_0 is about 0.7 V for silicon. This value is so small that it is negligible on the scale required for the input voltage. Consequently, we neglect it. When the diode is conducting the circuit is a voltage divider; the output voltage amplitude is given by:

$$V_0 = \frac{R_L}{R_L + r_f} V_m = \frac{100}{100 + 50} 155 = 103 \text{ V} \tag{9.6-1}$$

When the diode is cut off the output voltage is zero. A graph of the output voltage is shown in Fig. 9.6-1*d*. We see that the forward resistance of the diode has a significant effect on the output.

Fig. 9.6-1*e* shows an equivalent circuit in which the diode is replaced by a model which incorporates the reverse resistance. When the ideal diode in the model is conducting the output is the same as it is for the circuit of Fig. 9.6-1*c*. When the ideal diode is open circuited the output is determined by the voltage divider relationship:

$$\begin{aligned} v_0(t) &= \frac{R_L}{R_L + r_r} v_1(t) \\ &= \frac{100}{100 + 2 \times 10^6} v_1(t) \\ &= 0.5 \times 10^{-4} v_1(t) \end{aligned} \tag{9.6-2}$$

which is negligible. The output is shown on the graph of Fig. 9.6-1*f*.

From the foregoing analysis we see that the only diode parameter which has a significant effect on the output of this circuit is the diode's forward resistance. This result is not universally true for all circuits; in general r_r can be neglected (replaced by an open circuit) if it is much greater than the other resistances in the circuit, r_f can be neglected (replaced by a short circuit) if it is much smaller than the other resistances in the circuit, and V_0 can be neglected (replaced by a short circuit) if it is much less than the other voltages in the circuit. Even this is only a guideline and a more complete analysis of the situation must be made if there is any doubt as to the validity of neglecting any parameters. On the other hand, a great deal of effort can be wasted by retaining parameters which can, with a little thought, be seen to have a negligible effect on the output.

The function of the rectifier circuit is to produce a signal with a nonzero average value. This is done in the half-wave rectifier by eliminating the negative portion of the wave. The average value of a half-wave rectified sinusoidal signal of peak value V_m is V_m/π. The average value of $v_0(t)$ in Fig. 9.6-1*f* is $V_{\text{avg}} = 103/\pi = 32.8$ V.

9.6-2 Full-Wave Bridge Rectifier

The full-wave bridge rectifier provides an output voltage with double the average value of a half-wave rectifier if the same input voltage is applied to both circuits.

A rectifier combined with a filter forms a dc power supply whose output voltage is equal to the average value of the output of the rectifier circuit. (The dc power supply will be discussed later after we have covered the information required to discuss filters.) Since the average value of the full-wave bridge rectifier output is double that of the half-wave rectifier, that is, $V_{avg} = 2V_m/\pi$ where V_m is the peak value and V_{avg} is the average value, we may say that it is twice as efficient; it is also more expensive since it uses four diodes instead of the one used in the half-wave rectifier. The circuit of the full-wave bridge rectifier is shown in Fig. 9.6-2*a*; v_1, R_L, and the diodes are assumed to be identical to those in the half-wave rectifier. Our problem is to find $v_0(t)$.

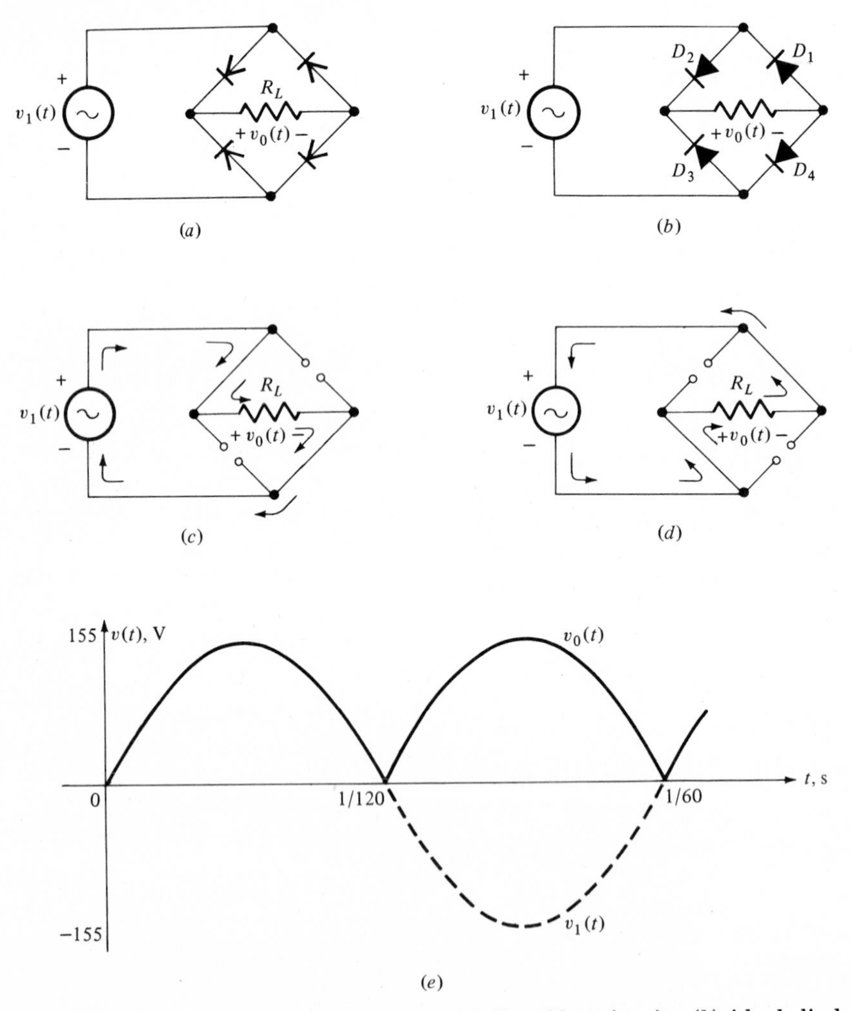

Fig. 9.6-2 Full-wave bridge rectifier. (*a*) Rectifier circuit; (*b*) ideal diode model; (*c*) equivalent circuit for $v_1(t) > 0$; (*d*) equivalent circuit for $v_1(t) < 0$; (*e*) voltage waveforms based on ideal diode model.

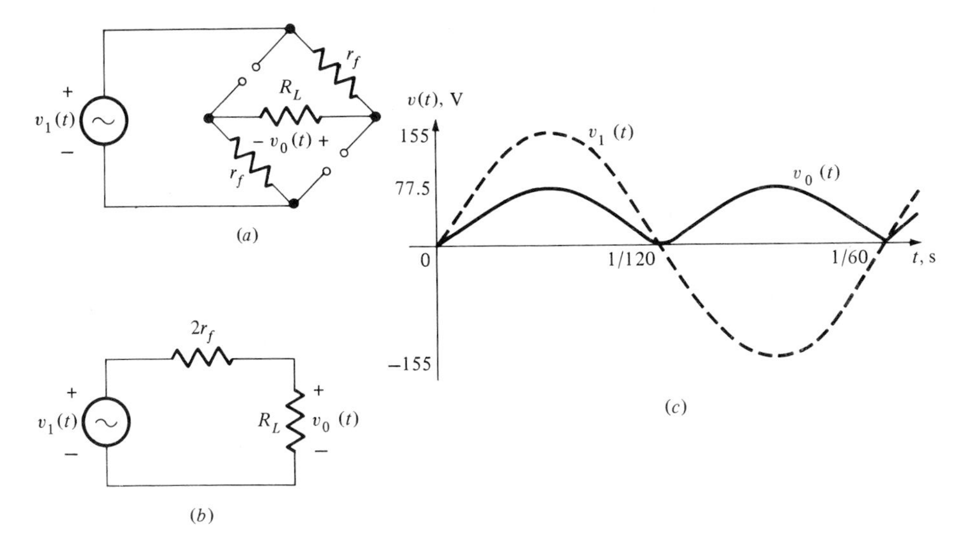

Fig. 9.6-3 Accurate analysis of full-wave rectifier. (*a*) Circuit model incorporating diode forward resistance; (*b*) simplified equivalent circuit; (*c*) voltage waveforms.

We can gain insight into the basic functioning of the circuit by using the simplest diode model, the ideal diode, for our first approximation. The resulting circuit is shown in Fig. 9.6-2*b*. Because of the larger number of diodes in this circuit we should first determine the states of the various diodes. If $v_1(t)$ is positive two of the diodes, D_2 and D_4, will conduct while the other two, D_1 and D_3, will be reverse biased as shown in Fig. 9.6-2*c* which also shows the flow of positive current in the circuit. By Kirchhoff's voltage law around the resulting loop it is easily seen that $v_0(t) = v_1(t)$.

When $v_1(t)$ becomes negative the states of the diodes are reversed as shown in Fig. 9.6-2*d* which also shows the flow of positive current. From this figure we can see, using KVL, that $v_0(t) = -v_1(t)$. The resulting output voltage, based on the ideal diode approximation for the semiconductor diodes, is shown in Fig. 9.6-2*e*.

The next step is to consider the effects of a more accurate model for the diode. Based on our experience with the half-wave rectifier it would seem reasonable to consider the diode's forward resistance and to neglect its reverse resistance and forward conducting voltage drop (V_0). The equivalent circuit corresponding to Fig. 9.6-2*c* is shown in Fig. 9.6-3*a*. This figure may be redrawn and simplified by combining the two resistances of value r_f which are in series with R_L to provide the circuit of Fig. 9.6-3*b* from which, using the voltage divider relationship, we find the peak value of output voltage to be:

$$V_{om} = \frac{R_L}{R_L + 2r_f} V_1 = \frac{100\ \Omega}{100\ \Omega + 2(50\ \Omega)} (155\ \text{V}) = 77.5\ \text{V} \tag{9.6-3}$$

By symmetry we see that the other half of the waveform follows the same equation except for the change in sign. The resulting waveforms are shown in Fig. 9.6-3*c*.

It is left as an exercise for the student to show that inclusion of the diodes' reverse resistance and/or forward conducting voltage drop would not significantly alter this result.

9.7 ZENER-DIODE TERMINAL CHARACTERISTIC

If the semiconductor diode is reverse biased and the negative bias is made larger and larger in magnitude, a point will be reached at which there is a sudden increase in current. Operation in this region causes permanent damage to the usual type of semiconductor diode and is therefore avoided. Certain diodes, called *zener diodes,* are specifically manufactured for operation in this region. The reverse-bias voltage at which this sudden increase in current occurs is called the *zener voltage* V_z. The zener voltage is determined by certain factors in the manufacturing process and zener diodes are available which cover a large range of zener voltages from about one to a few hundred volts.

The circuit symbol, reference current and voltage directions, and typical *vi* characteristic for a zener diode are shown in Fig. 9.7-1.

The important characteristics of the zener-diode terminal characteristic are as follows: for voltages greater than $-V_z$ the behavior of the zener diode is similar to that of an ordinary semiconductor diode; at $v = -V_z$ there is a very sharp increase in current. This transition is much sharper than the transition to the forward conduction state and the slope of the curve for $v < -V_z$ is a large positive value substantially greater than the slope of the curve in the forward conduction region. The slope of the curve in the zener breakdown region

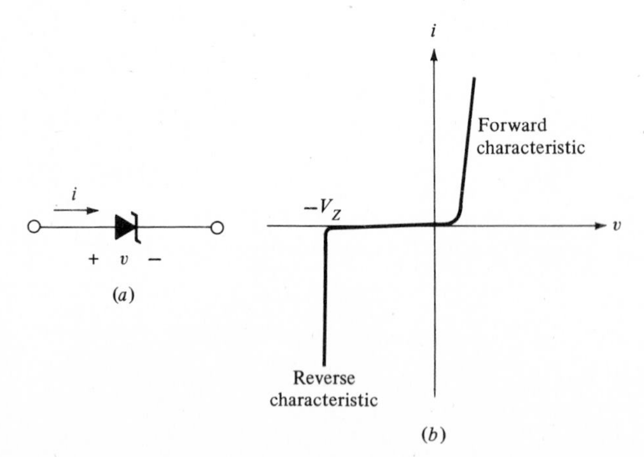

Fig. 9.7-1 Zener diode. (*a*) Circuit symbol; (*b*) *vi* characteristic.

($v < -V_z$) typically corresponds to a resistance of a few ohms as compared to the forward diode resistance which is in the order of tens of ohms.

9.7-1 Zener-Diode Voltage Regulator Circuit

As a consequence of the fact that in the reverse region the terminal voltage V_z is almost independent of the current, the zener diode is very useful as a voltage reference element and in circuits used to maintain constant voltages.

A voltage regulator is a circuit used to provide a constant output voltage despite fluctuations in input voltage and load. The term *load* refers to the device to which this constant voltage is being supplied. A variation in load refers to a change in the current being drawn by the device. In this section we will study the effectiveness of a zener diode voltage regulator in maintaining a constant output voltage with varying load.

If our voltage supply were a constant ideal voltage source there would be no problem. Figure 9.7-2*a* shows such a source feeding a load R_L which varies between 5 and 15 kΩ with a nominal value of 10 kΩ. Using KVL we see that $V_0 = 10$ V no matter what the value of R_L is.

Problems arise when we introduce a more realistic model for the physical voltage source used in actual practice; such a model incorporates a small resistance in series with the ideal voltage source which represents its internal resistance. We will assume a 1 kΩ internal resistance, and that a nominal output voltage (for $R_L = 10$ kΩ) of 10 V is required. Using the voltage divider relation-

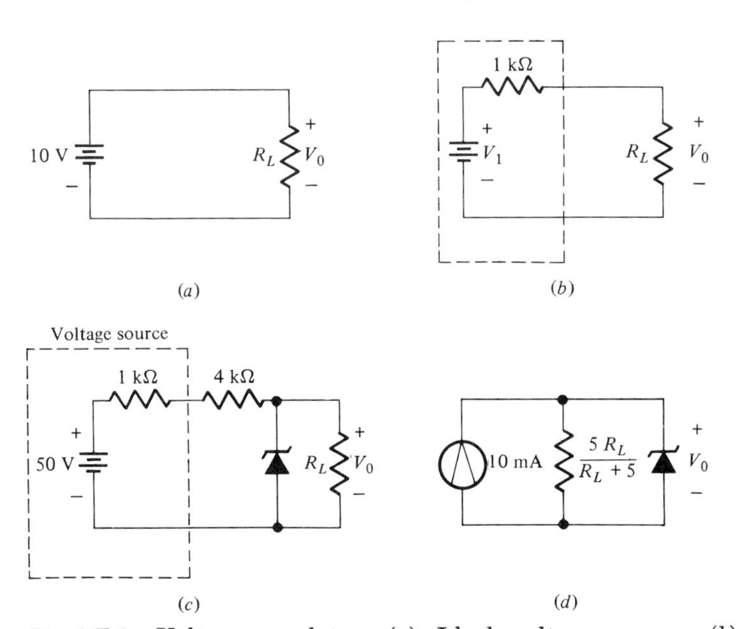

Fig. 9.7-2 Voltage regulator. (*a*) Ideal voltage source; (*b*) practical voltage source; (*c*) zener-diode voltage regulator; (*d*) equivalent circuit for voltage regulator.

ship for the circuit of Fig. 9.7-2*b* results in

$$V_0 = \frac{R_L}{R_L + 1} V_1 \tag{9.7-1}$$

In order to maintain V_0 at 10 V with R_L at its nominal value of 10 kΩ this formula leads to the value $V_1 = 11$ V. Next we use the same formula with V_1 set at 11 V and R_L changed to 5 kΩ and we find $V_0 = 9.17$ V. With V_1 set at 11 V and R_L changed to 15 kΩ we find $V_0 = 10.3$ V. Thus the overall voltage change is $10.3 - 9.17 = 1.13$ V as the load varies between its minimum and maximum values.

The circuit for a zener-diode regulated voltage supply is shown in Fig. 9.7-2*c* where the diode has a rated zener voltage of 10 V. In order to analyze this circuit graphically we replace the original circuit by an equivalent circuit in which the linear part is replaced by its Norton equivalent as shown in Fig. 9.7-2*d*. (Refer back to Sec. 8.4 and check the Norton parameters.) The open circuit, or Thevenin, voltage across the Norton equivalent circuit is given by

$$V_T = \frac{50\, R_L}{R_L + 5} \tag{9.7-2}$$

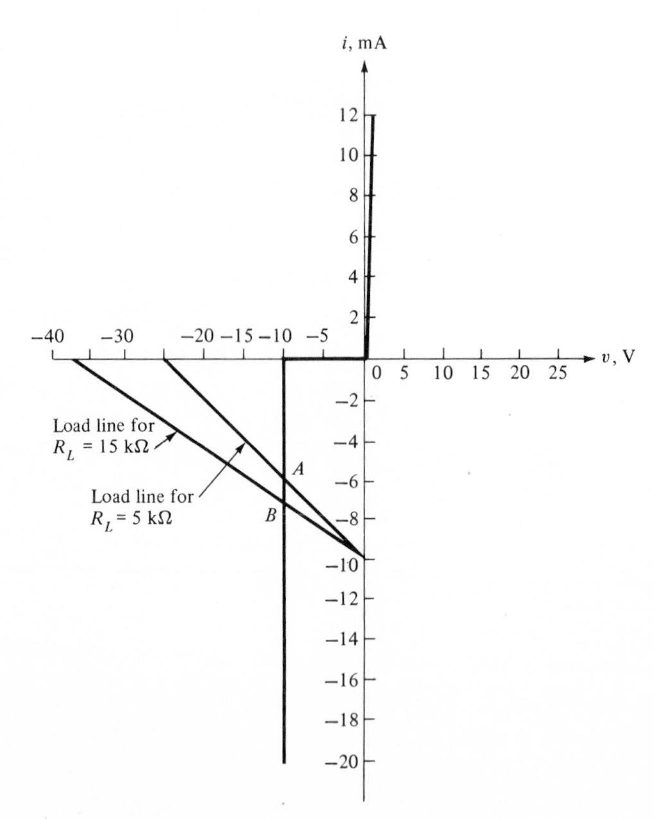

Fig. 9.7-3 Voltage regulator analysis.

which yields 25 V for $R_L = 5$ kΩ and 37.5 V for $R_L = 15$ kΩ. Each of these values leads to a different load line; the two load lines are shown plotted on the zener diode characteristic in Fig. 9.7-3. With the given variation in load, the operating point of the diode will always be between points A and B on the reverse characteristic. The voltage difference between the intersections of the two load lines with the zener diode characteristic is not measurable on the characteristic of Fig. 9.7-3. Thus the change in the regulated output voltage is negligible for a change in load resistance from 5 to 15 kΩ while the change in the voltage output before regulation is 3.75 V.

Quantitative values for this problem can be computed using a piecewise linear model for the zener diode. The piecewise linear approximation for a zener-diode characteristic is shown in Fig. 9.7-4*c*. For voltages more negative than the zener voltage, $-V_z$, the slope becomes very large—normally on the order of ten times the slope in the forward biased region. We denote the reciprocal of this slope, which has the dimension of ohms, by r_z.

The corresponding piecewise-linear model is shown in Fig. 9.7-4*d*. It is like that of the ordinary diode but has an additional branch similar to that for the

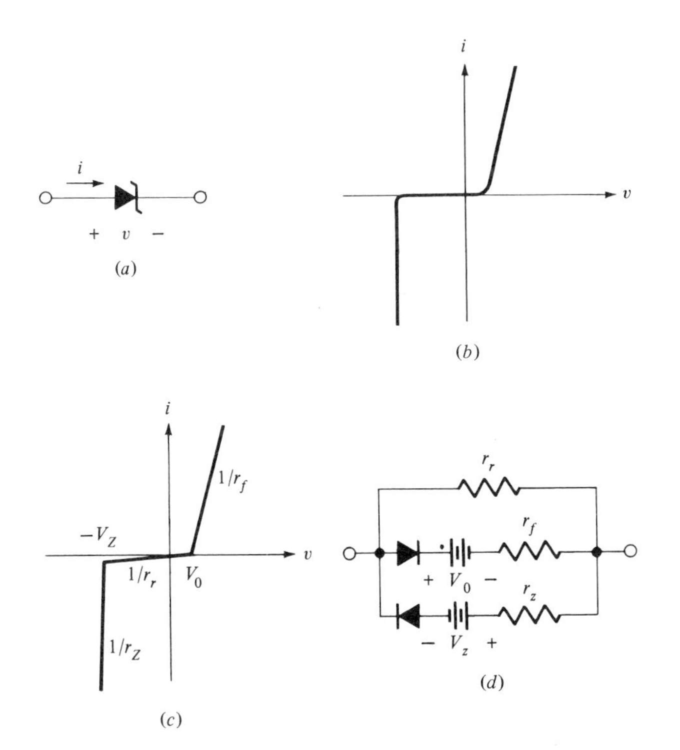

Fig. 9.7-4 Zener diode piecewise-linear model. (*a*) Symbol and reference directions; (*b*) terminal characteristic; (*c*) piecewise-linear approximation; (*d*) piecewise-linear model.

forward biased region except that the parameters are different and the ideal diode and voltage source are in the opposite direction. Examination of the piecewise-linear approximation indicates the validity of this circuit; the zener region is similar to the forward region except it is of opposite polarity and numerical values are different.

A simplifying approximation for this circuit which is very frequently valid is that r_r is large enough to be considered infinite. Then r_r is replaced by an open circuit in Fig. 9.7-4*d*.

The remainder of the analysis of the zener-diode voltage regulator using the piecewise-linear model is left as an exercise for the student (see Prob. 9.7-4).

SUMMARY

1 A *piecewise-linear characteristic* approximates a nonlinear *vi* characteristic with a sequence of straight line segments.
2 A *piecewise-linear model* is a circuit consisting of ideal diodes and linear elements which models a nonlinear element.
3 The ideal diode acts like a short circuit in the forward direction and an open circuit in the reverse direction.
4 A rectifier converts an ac voltage to a voltage with an average, or dc, value. The full-wave rectifier is more efficient than the half-wave rectifier. For a sinusoid of peak value V_m the average value of a half-wave rectified signal is V_m/π; for the full-wave rectified signal it is $2V_m/\pi$.
5 The zener diode is used as a voltage reference and in voltage regulator circuits. A voltage regulator is used to maintain a constant output voltage despite varying input voltages and loads.

PROBLEMS

9.1-1 A 10-V (full-scale) voltmeter with a sensitivity of 1,000 Ω/V and a 50-mA (full-scale) milliammeter which uses a 0.5-mA 1-kΩ meter movement are to be used with a 100-V dc voltage source to measure the terminal characteristic of a nominal 100-Ω resistor. Both meters are accurate to 1 percent of full scale.

a Determine the overall resistance of each meter.

b Draw the circuit diagram to be used. Use any additional fixed or variable resistors that may be required.

c Compute the values, or ranges of values, of all additional resistors so that no meter readings will exceed the full scale values.

d Determine the maximum power the 100-Ω resistor will have to dissipate for these measurements.

9.2-1 Use graphical analysis to find V_1 and I_1 for each of the circuits of Fig. P9.2-1*c*, *d*, and *e* if the diode characteristic is that shown in Fig. P9.2-1*a* and expanded around the origin in Fig. P9.2-1*b*.

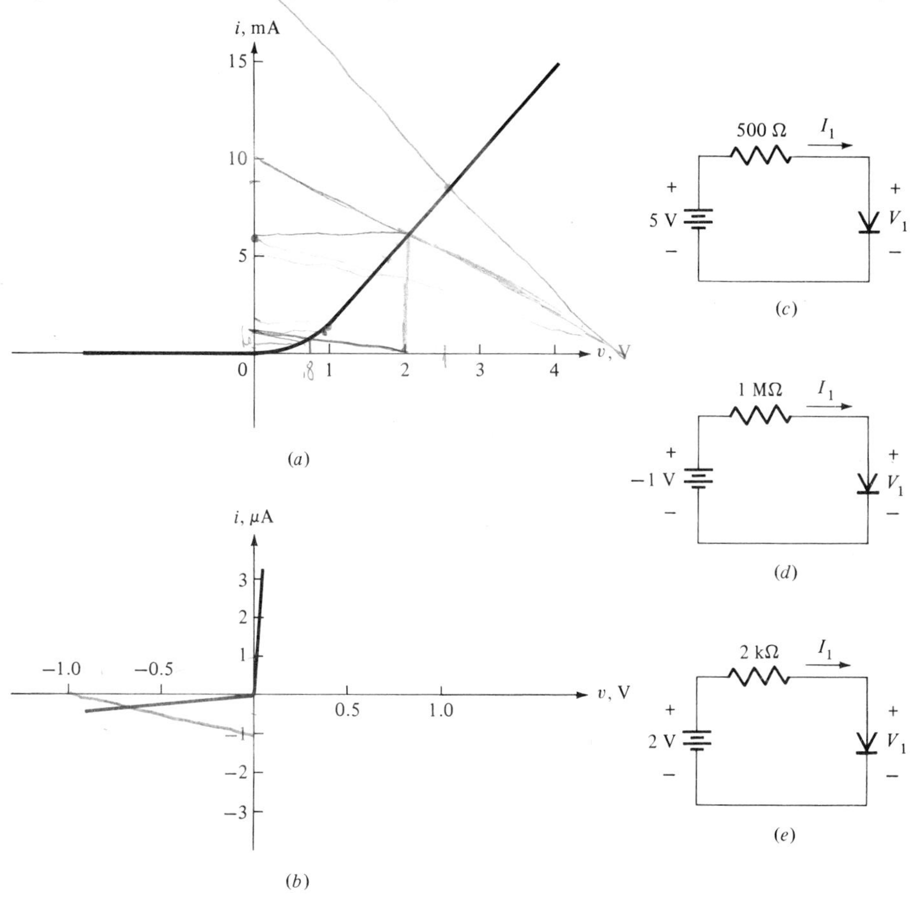

Fig. P9.2-1.

9.2-2 Use graphical analysis to find V_1 and I_1 in the circuit of Fig. P9.2-2 if the diode characteristic is that shown in Fig. P9.2-1.

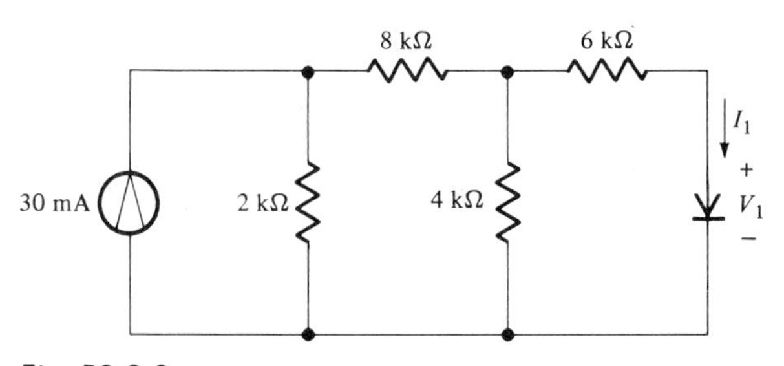

Fig. P9.2-2.

9.2-3 The terminal characteristic for the diode in the circuit of Fig. P9.2-3 is shown in Fig. P9.2-1. Use graphical analysis to draw a sketch of $v_D(t)$ with important values labeled for

a $V = 6$ V and $R = 300\ \Omega$
b $V = 1$ V and $R = 1\ \text{k}\Omega$

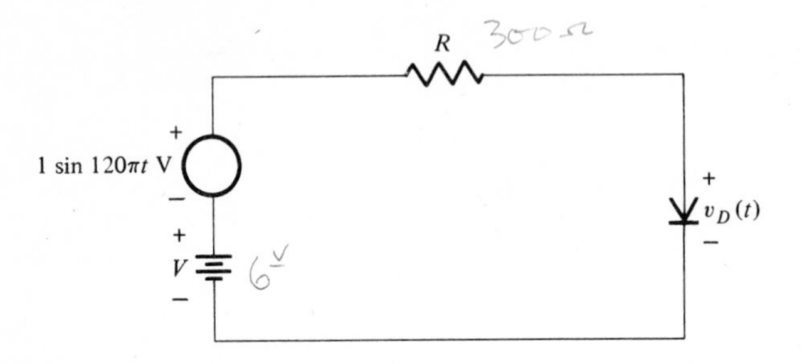

Fig. P9.2-3.

9.3-1 Draw a two-segment piecewise-linear characteristic that passes through the origin and approximates the diode characteristic of Fig. P9.2-1.

9.4-1 Find the indicated voltages and/or currents in the circuits of Fig. P9.4-1. Note that all of the diodes are ideal.

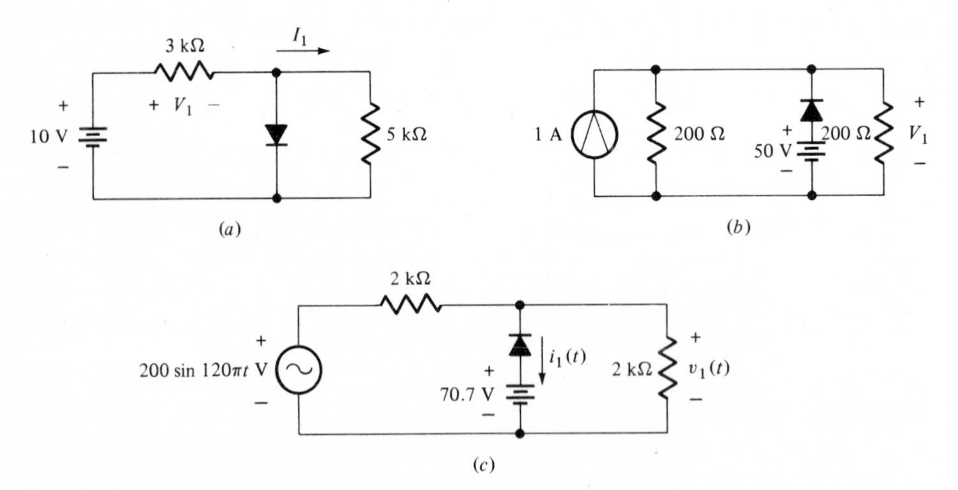

Fig. P9.4-1.

9.5-1 Find the piecewise-linear diode model whose terminal characteristic is the piecewise-linear characteristic found as the solution to Prob. 9.3-1.

9.5-2 For the nonlinear element shown in Fig. P9.5-2*a* whose *vi* characteristic is shown in Fig. P9.5-2*b* find

a A piecewise-linear approximation for its *vi* characteristic.
b A piecewise-linear model based on this approximation.

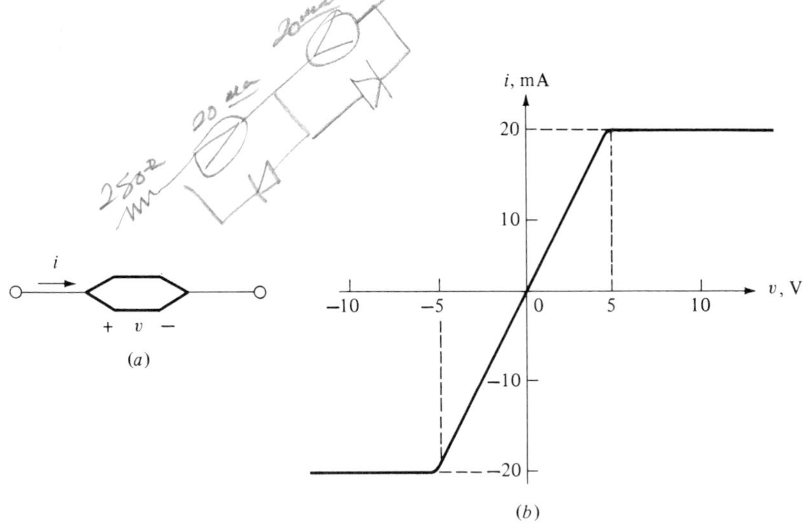

Fig. P9.5-2.

9.5-3 Repeat Prob. 9.5-2 for Fig. P9.5-3.

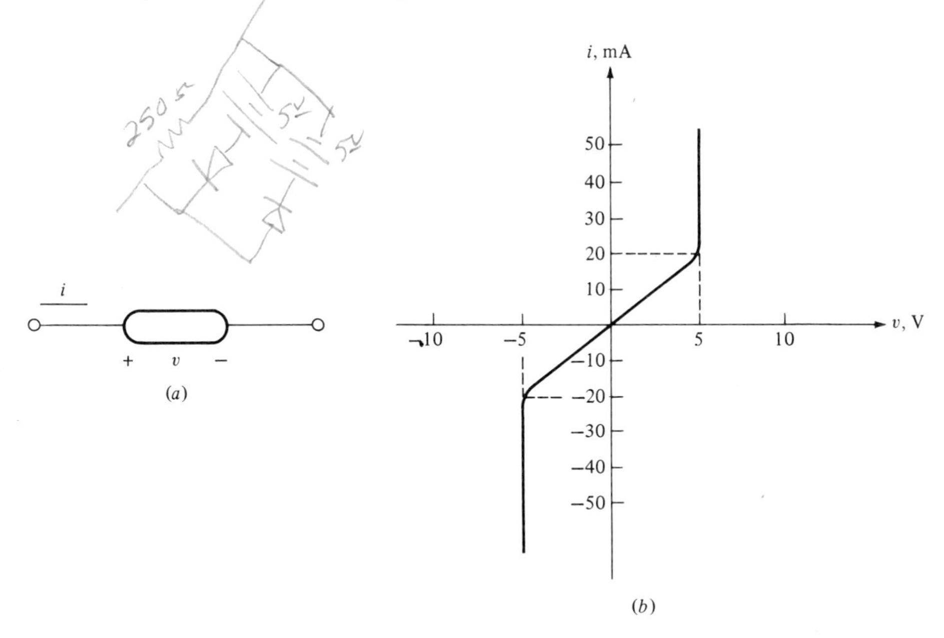

Fig. P9.5-3.

9.5-4 Repeat Prob. 9.5-2 for Fig. P9.5-4.

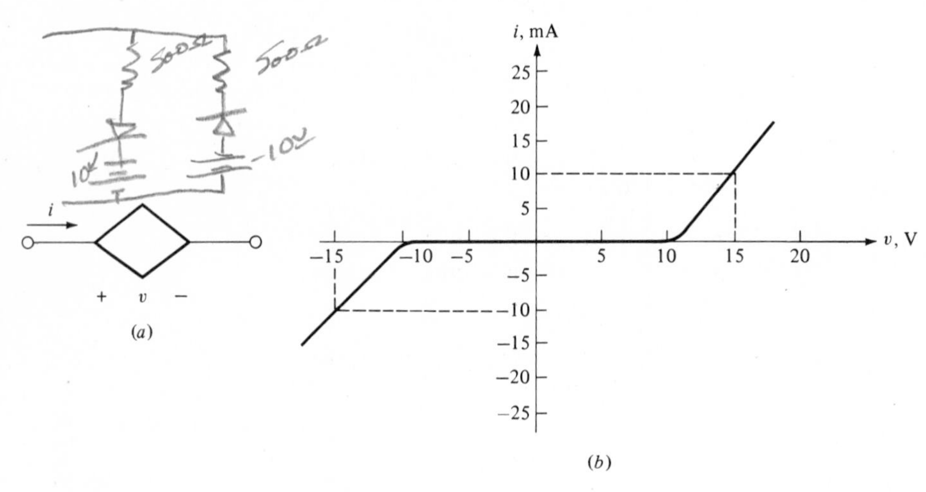

Fig. P9.5-4.

9.5-5 Analyze the circuits of Figs. P9.2-1*c*, *d*, and *e* using the piecewise-linear model found in Prob. P9.5-1.

9.5-6 Analyze the circuits shown in Fig. P9.5-6 which incorporate the nonlinear elements of Prob. 9.5-2, 9.5-3, and 9.5-4. Find the indicated currents and/or voltages. Draw a sketch of time varying signals and label important points.

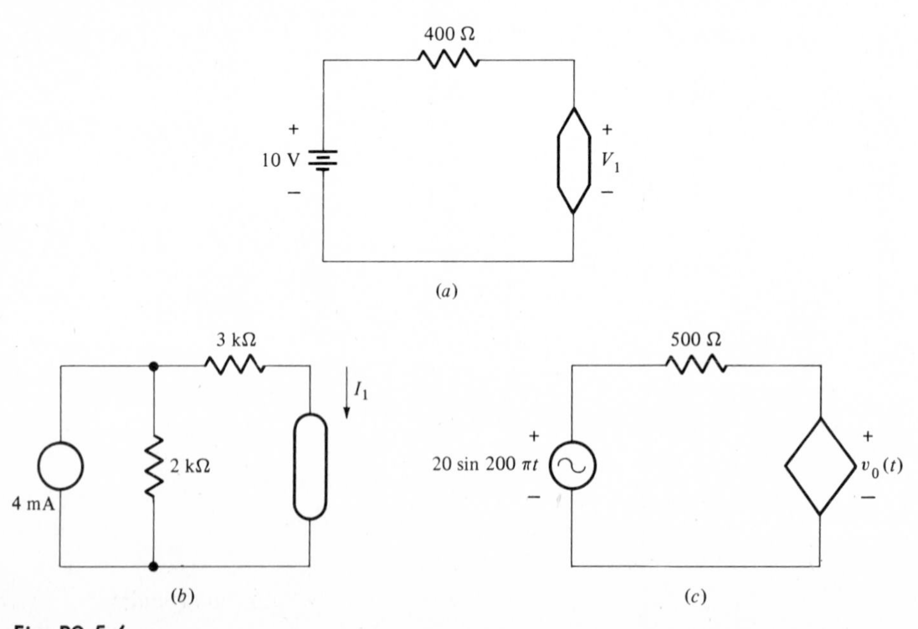

Fig. P9.5-6.

9.6-1 Determine whether the effects of the diodes' reverse-resistance and/or forward-conducting voltage are significant in the operation of the full-wave bridge rectifier of Fig. 9.6-2*a*. *Hint*: Although this problem nay be solved by analyzing the circuit with all these factors included the question may also be answered by determining the Thevenin resistance across the terminals to which R_L is connected and by using superposition to find the voltage output due to the forward conducting voltages with $v_1(t) = 0$ after appropriate simplifications have been made for r_r and r_f.

9.6-2 For each of the circuits of Fig. P9.6-2 sketch $v_o(t)$ and label important values. Do this using a piecewise-linear model which

a Is simply an ideal diode

b Incorporates those factors (forward resistance = 50 Ω, reverse resistance = 1 MΩ, forward-conducting voltage = 0.7 V) which are significant in the circuit operation. Explain why you have omitted any of these factors when you do so.

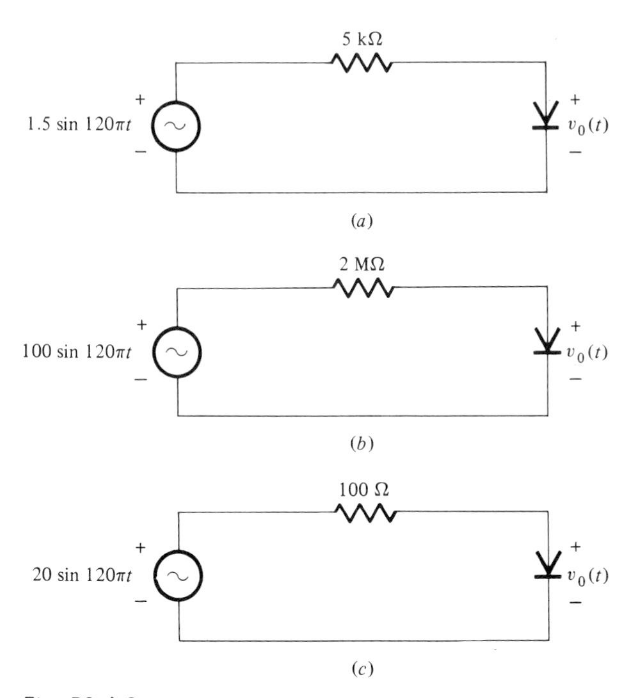

Fig. P9.6-2.

9.7-1 Use graphical analysis to find the range of values of V_L for the circuit of Fig. P9.7-1*a* if R_L varies between 12 Ω and 36 Ω. The terminal characteristic for the zener diode is shown in Fig. P9.7-1*b*.

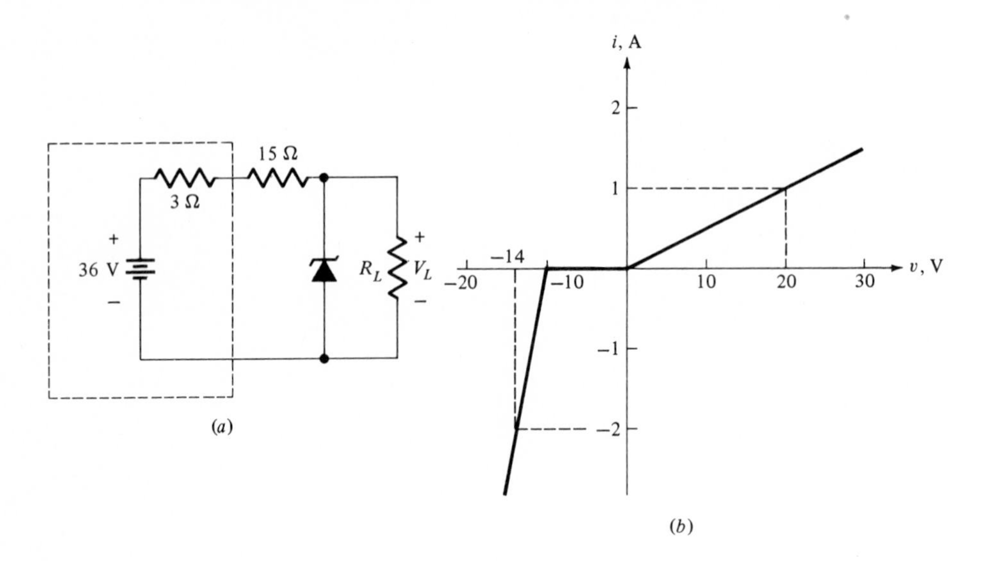

Fig. P9.7-1.

9.7-2 For the typical zener diode characteristic shown in Fig. 9.7-1*b*

a Find a three-segment piecewise-linear characteristic which approximates the given characteristic.

b Find a piecewise-linear model for the zener diode based on this approximation. Assume the slope of the leftmost portion of the characteristic is $1/r_z$.

9.7-3 Determine the indicated voltages for the circuits of Fig. P9.7-3 assuming the zener-diode parameters are $r_f = 20\ \Omega$, $r_r = 1\ \text{M}\Omega$, $r_z = 1\Omega$, $V_z = 10$ V and the diode is made of silicon. If the voltage varies with time sketch it and label important values. Make appropriate approximations.

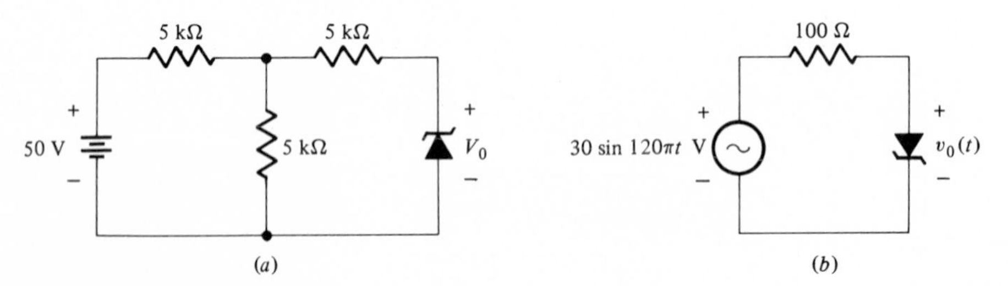

Fig. P9.7-3.

9.7-4 Assume the zener diode in the voltage regulator of Fig. 9.7-2 has the characteristics $r_f = 20\ \Omega$, $r_r = 2\ \text{M}\Omega$, $r_z = 2\ \Omega$, $V_z = 10$ V, and is made of silicon. Find V_0 for $R_L = 5$ kΩ, 10 kΩ, and 15 kΩ.

9.7-5 The voltage regulator reduces variations in output voltage due to input voltage variations as well as variations in the load. Assume the 50-V dc voltage source in the circuit of Fig. 9.7-2*c* has an ac source of value $5 \sin 120\pi t$ V connected in series with it. Find the output voltage across the load $R_L = 10\ \text{k}\Omega$ assuming the same zener diode parameters as were given in Prob. 9.7-3.

10
Linear Two-Port Networks

OBJECTIVES

Upon completion of this chapter, you should be able to

1 Find the transfer function of a given network.
2 Define the power amplification of an amplifier.
3 Define power, voltage, and current gain in decibels.
4 Calculate gain in decibels for different power, voltage, or current ratios.
5 Define and give examples of application of the controlled source.
6 Calculate gain and output voltage and current for ideal voltage and current amplifiers.
7 Describe input and output resistance as applied to ideal voltage and current amplifiers.
8 Calculate gain and output voltage and current of amplifiers with input and output resistance taken into account.
9 Define and give examples of the application of the hybrid parameters.
10 Calculate gain for cascaded two-ports.

INTRODUCTION

In previous chapters we have discussed the laws which govern current and voltage distribution in networks containing resistances and independent sources. Such networks cannot *amplify* voltages or currents. In order to provide amplification, transistors or other active devices are required. These devices are modeled by circuits containing *controlled* sources, the new element being introduced in this chapter.

Most of the networks are classified as *two-ports*. The general representation for a two-port is shown in Fig. 10.1-1 where the two ports are the input and output terminal pairs. In this chapter, we consider methods used for the description and analysis of such networks.

10.1 Transfer Functions

In the two-port network shown in Fig. 10.1-1, the network in the box, if it is linear, can be described mathematically by a transfer function which represents the ratio of the output signal to the input signal. This is a natural way to characterize such a system, since we can easily find the output in response to any input, given the transfer function.

We first introduced the idea of the transfer function in Sec. 6.3 in connection with the voltage divider network of Fig. 6.3-1. At that point, our interest centered on the transfer voltage ratio

$$A_v = \frac{v_2}{v_1} \tag{10.1-1}$$

where A_v is the transfer function for the voltage divider network. If we solve for the output voltage v_2, the result is

$$v_2 = A_v v_1 \tag{10.1-2}$$

and we see that the output is simply the product of the transfer function and the input. In this case, as shown in Sec. 6.3, the transfer function is a real number that lies between 0 and 1 so that the output has the same waveshape as the input, except that it may be reduced in amplitude.

> ***Example 10.1-1*** Consider the voltage divider network shown in Fig. 10.1-2*a*. Assume that $R_1 = R_2 = 10\ \text{k}\Omega$. Find $v_2(t)$ if $v_1(t) = 10e^{-2t}$ V.

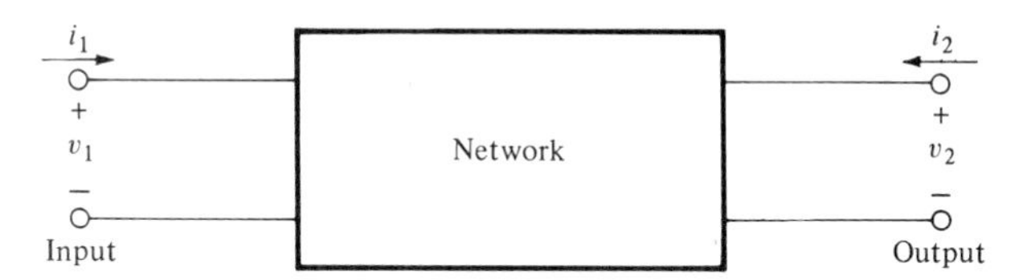

Fig. 10.1-1 Two-port network.

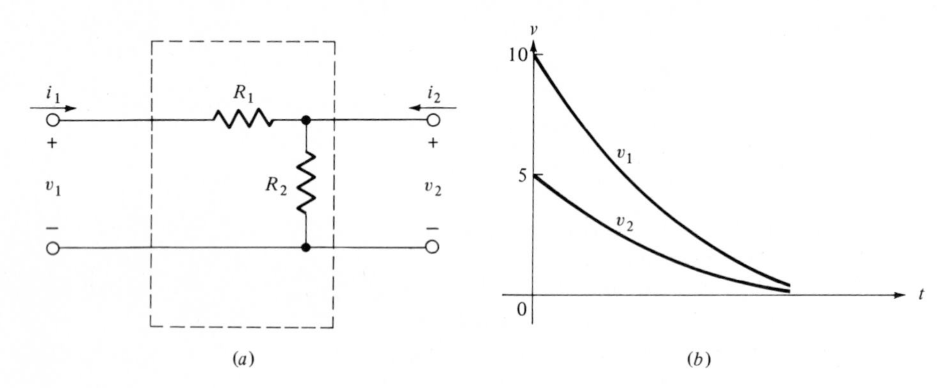

Fig. 10.1-2 (*a*) Network; (*b*) waveforms.

Solution If $i_2 = 0$, we can use the voltage divider formula

$$A_v = \frac{v_2}{v_1} = \frac{R_2}{R_1 + R_2} = \frac{1}{2} \tag{10.1-3}$$

Then

$$v_2(t) = A_v v_1(t) = \tfrac{1}{2} \times 10e^{-2t} = 5e^{-2t} \text{ V} \tag{10.1-4}$$

and we see that the output has the same waveshape as the input, but one-half the amplitude as shown in Fig. 10.1-2*b*. ////

This example is deceptively simple because we assumed in the course of the solution that the output current i_2 was zero. In fact, this assumption was made in the original derivation of the voltage-divider relation. In general, when dealing with a two-port, the output current will not be zero and the specification of the transfer function becomes much more complicated. In a later section we will find that it takes four transfer functions to completely specify a two-port. However, in many cases, a single transfer function will suffice as shown in the following example.

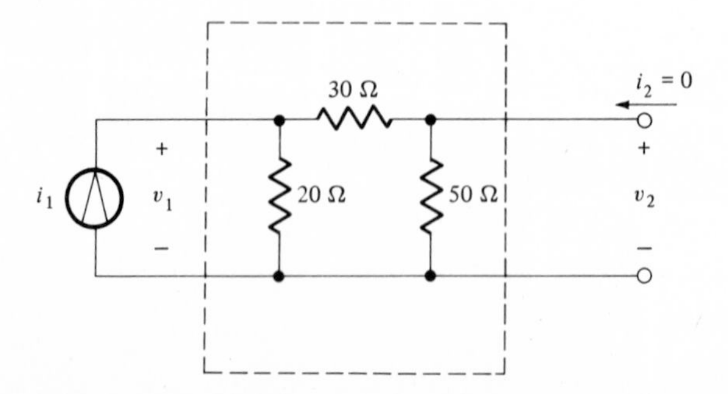

Fig. 10.1-3 Two-port for Example 10.1-2.

Example 10.1-2 In the circuit of Fig. 10.1-3, find

a The transfer resistance $R_T = v_2/i_1$, with $i_2 = 0$.

b The output voltage $v_2(t)$ when the input current is $i_1(t) = 5 \sin \omega t$ mA.

Solution

a Up to now we have considered transfer functions which were voltage ratios or current ratios. Here we are interested in a transfer function which relates an output voltage to an input current and thus has dimensions of volts per ampere, or ohms.

We can find this transfer function, denoted R_T, by writing it in the following way:

$$R_T = \left(\frac{v_1}{i_1}\right)\left(\frac{v_2}{v_1}\right) = \frac{v_2}{i_1} \tag{10.1-5}$$

This "chain" type of calculation reduces a relatively complicated problem to two simpler problems since each of the terms in the chain is calculated separately as follows:

Using Ohm's law

$$\frac{v_1}{i_1} = 20 \parallel (30 + 50) = \frac{(20)(80)}{(100)}$$

Using the voltage-divider relation

$$\frac{v_2}{v_1} = \frac{50}{30 + 50} = \frac{(50)}{(80)}$$

Finally, carrying out the indicated multiplication

$$R_T = \frac{v_2}{i_1} = \frac{(20)(80)}{(100)} \frac{(50)}{(80)} = 10 \text{ V/A}$$

In this case, it is best to think of R_T as 10 V/A rather than 10 Ω. The units immediately suggest that R_T is 10 V at the output terminals per ampere into the input terminals which is the proper way to view a *transfer* resistance.

b
$$\begin{aligned} v_2(t) &= R_T i_1(t) \\ &= (10)(5 \times 10^{-3} \sin \omega t) \\ &= 0.05 \sin \omega t \text{ V} \end{aligned}$$

Once again, the output voltage has exactly the same waveshape as the input current. ////

When we deal with networks which contain only resistances, the transfer functions are always real numbers and the output waveshape is always a scaled replica of the input waveshape. In later chapters, when we add inductors and capacitors to the network, we will find more complicated transfer functions which will modify the input waveshape so that the output waveshape is often different from the input waveshape.

10.2 AMPLIFIERS AND APPLICATIONS

In most systems involving signal transmission, amplification is required either because input signals are too small to actuate the load directly or because losses in signal strength occur due to long transmission distances.

A familiar example is the radio receiver. In this system, the signal picked up by the antenna has a power level measured in microwatts and a voltage level measured in microvolts. In order to actuate the load which consists of one or more loudspeakers requiring perhaps 0.05 to 50 or more watts for proper operation, the power level must be amplified by large factors, often 1×10^6 or more. This amplification, or gain, is provided by electronic amplifiers.

Another example is the long-distance telephone circuit. The power level at the output of the telephone instrument is only a few milliwatts. As this signal is transmitted along the telephone cable it is subject to power loss because of the resistance of the cable. In order to maintain sufficient signal strength for satisfactory operation it is necessary to place amplifiers at suitable points along the cable.

Numerous other examples exists in all fields of technology. Some of these will be considered in later chapters.

AMPLIFIER GAIN

The basic function of an amplifier is to provide amplification of power, voltage, or current. A block diagram of such an amplifier is shown in Fig. 10.2-1. In a power amplifier the output power from the amplifier is greater than the input power. The power amplification is expressed as

$$A_p = \frac{\text{output signal power}}{\text{input signal power}} \tag{10.2-1}$$

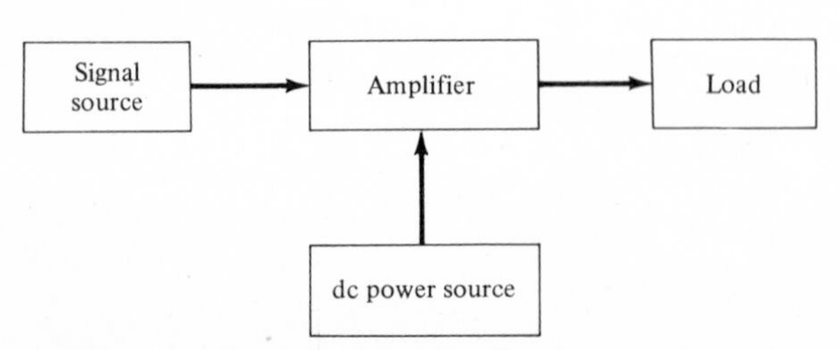

Fig. 10.2-1 Amplifier block diagram.

Since the power out is greater than the power in it might appear that the principle of conservation of energy is being violated. This is not so because the additional power is supplied by the dc power source shown in the figure. The power from this source is converted to signal power in the amplifier.

In the sections which follow, we will begin to look at electronic amplifiers from the two-port point of view.

10.3 GAIN AND THE DECIBEL

In this section we introduce the decibel (dB), a logarithmic unit often used in the specification and measurement of amplifier performance. This unit was originally introduced because it was found that the ear perceives, as approximately equal, changes in sound intensity which are in the same *ratio*, so that a change from 5 mW to 10 mW would seem the same as a change from 50 mW to 100 mW (because 5/10 = 50/100). Since these changes in the ear's response are approximately equally spaced on a logarithmic scale, it is appropriate that a logarithmic unit be used for expressing power levels and ratios in communication circuits. A change of 0.5 to 1 dB is about the smallest discernible change for the average person.

A further reason for the use of the decibel is that the use of a logarithmic unit converts the operation of multiplication into simple addition. This arises frequently in amplifier work. In addition, the logarithmic unit compresses a very large range of numbers into a much smaller range, a decided convenience in many instances.

The *power gain* of an amplifier measured in decibels is†

$$G_p = 10 \log \frac{P_2}{P_1} \qquad \text{dB} \tag{10.3-1}$$

where log means $\log_{10}$ and P_2 and P_1 are output and input power, respectively.

This definition is used in two ways: first, to express a power *ratio* in logarithmic units, and second, to express a power *level* with respect to a fixed reference power level. One often-used reference level is 1 milliwatt, for which the units are called dBm. Thus for a power level in decibels with respect to 1 mW, we have, with P_2 in watts,

$$\begin{aligned} G_p &= 10 \log \frac{P_2}{10^{-3}} \\ &= 10 \log (P_2 \times 10^3) \qquad \text{dBm} \end{aligned} \tag{10.3-2}$$

DECIBEL VOLTAGE GAIN

The decibel was originally defined in terms of power gain. Because of its usefulness as a logarithmic measure, it is generally applied directly to voltage

† Henceforth we will use the symbol G for decibel gain and A for amplification ratio.

gain and current gain based on the following: If P_2 and P_1 in Eq. (10.3-1) are dissipated in identical resistances R, then

$$P_2 = \frac{V_2^2}{R} = I_2^2R \qquad \text{and} \qquad P_1 = \frac{V_1^2}{R} = I_1^2R \tag{10.3-3}$$

Substituting Eq. (10.3-3) into Eq. (10.3-1),

$$G_v = 10 \log \frac{V_2^2}{V_1^2} = 20 \log \frac{V_2}{V_1} \qquad \text{dB} \tag{10.3-4}$$

and

$$G_i = 10 \log \frac{I_2^2}{I_1^2} = 20 \log \frac{I_2}{I_1} \qquad \text{dB} \tag{10.3-5}$$

where we have used $\log X^2 = 2 \log X$.

From the foregoing, we see that voltage and current gains in decibels are equal to power gain only if the resistances in which the powers are dissipated are equal. However, due to long usage, Eqs. (10.3-4) and (10.3-5) are usually used for voltage and current gain regardless of the resistances.

Example 10.3-1

a Find G_v in decibels for $V_2/V_1 = 0.1, 1, 2, 5, 10, 20, 100, 1000$.
b Find G_p for $P_2/P_1 = 1, 2, 10$.
c Find G_p in dBm for $P_2 = 10^{-6}$ W and 10 W.

Solution Some of the calculations are shown below. All of the results are tabulated at the end.

a

$V_2/V_1 = 0.1$

$$G_v = 20 \log \frac{V_2}{V_1} = 20 \log 0.1 = 20 \log 10^{-1} = -20 \log 10 = -20 \text{ dB}$$

$V_2/V_1 = 1$

$$G_v = 20 \log 1 = 20(0) = 0 \text{ dB}$$

$V_2/V_1 = 2$

$$G_v = 20 \log 2 = 20(0.3) = 6 \text{ dB}$$

$V_2/V_1 = 1{,}000$

$$G_v = 20 \log 1{,}000 = 20 \log 10^3 = 60 \log 10 = 60 \text{ dB}$$

Note that when the output voltage is greater than the input voltage the gain in decibels is a positive number; when v_2 is less than v_1, it is a negative number; and when $v_2 = v_1$ the gain in decibels is zero.

V_2/V_1	0.1	1*	2*	5*	10*	20	100	1000
G_v, dB	−20	0	6	14	20	26	40	60

Those entries marked with an asterisk (*) are worth memorizing. Decibel values for many round numbers can easily be found from these in the following way: consider $V_2/V_1 = 4{,}000$. Then

$$\begin{aligned} G_v &= 20 \log 4{,}000 = 20 \log [(2)\ (2)\ (1000)] \\ &= 20 \log 2 + 20 \log 2 + 20 \log 1000 \\ &= 6 + 6 + 60 \\ &= 72 \text{ dB} \end{aligned}$$

b

$$\frac{P_2}{P_1} = 1 \qquad G_p = 10 \log 1 = 0 \text{ dB}$$

$$\frac{P_2}{P_1} = 2 \qquad G_p = 10 \log 2 = 3 \text{ dB}$$

$$\frac{P_2}{P_1} = 10 \qquad G_p = 10 \log 10 = 10 \text{ dB}$$

Note that these values are one-half the values for voltage-ratios which are numerically equal.

c

$$P_2 = 10^{-6} \text{ W}$$

$$\begin{aligned} G_p &= 10 \log (P_2 \times 10^3) = 10 \log (10^{-6} \times 10^3) = 10 \log 10^{-3} \\ &= -30 \log 10 \\ &= -30 \text{ dBm} \end{aligned}$$

$$P_2 = 10 \text{ W}$$

$$\begin{aligned} G_p &= 10 \log (10 \times 10^3) = 10 \log 10^4 = 40 \log 10 \\ &= 40 \text{ dBm} \end{aligned}$$

////

10.4 CONTROLLED SOURCES

In order to model a real amplifier, we must add the *controlled source* to our list of elements. The sources introduced in Chap. 4 were *independent* sources, whose current or voltage was independent of any of the currents or voltages in the network, i.e., the current or voltage was determined by the source itself. The voltage or current from a controlled source, on the other hand, *depends on* a voltage or current at some point in the network. In this sense it is dependent on the form of the network and also the independent sources; it is sometimes called a *dependent* source.

We have occasion to use both current and voltage sources which can in turn be controlled by a current or voltage. Thus there are four possibilities:

1 *Voltage-controlled voltage source.* This is shown in Fig. 10.4-1*a*. In the diagram v_x is the *controlling voltage* and K_1 is the factor by which the controlled source multiplies v_x. The current i will be determined by the circuit connected to the controlled source.

2 *Current-controlled voltage source* This is shown in Fig. 10.4-1b. Here i_x is the controlling current and K_2 is the factor by which this current must be multiplied in order to obtain the voltage of the source. Note that K_2 must have the units of resistance (Ω), so that when multiplied by a current, the result will be volts.

3 *Voltage-controlled current source* This is shown in Fig. 10.4-1*c*. Here v_x is the controlling voltage and K_3 is the factor by which v_x is multiplied in order to obtain the source current. Note that K_3 must have units of conductance and that the voltage v across the source will depend on the external circuit.

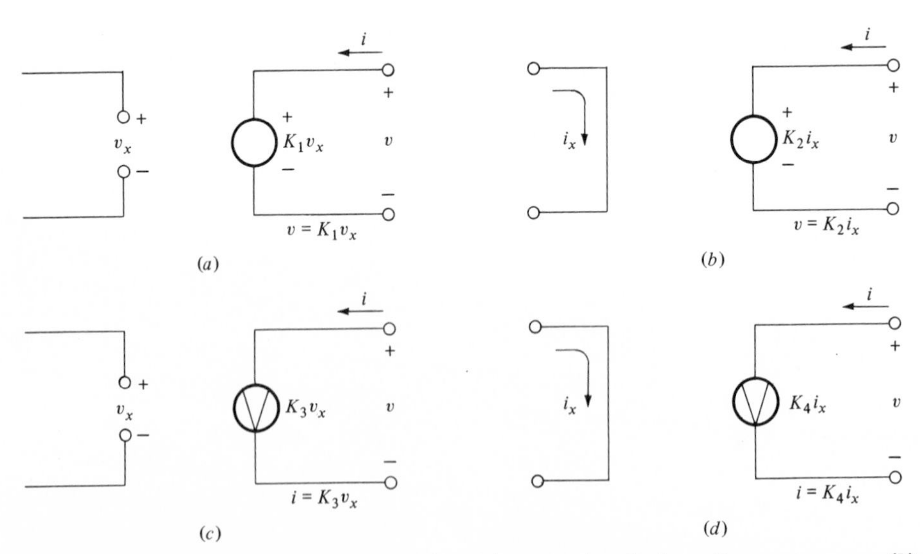

Fig. 10.4-1 Controlled sources. (*a*) Voltage-controlled voltage source; (*b*) current-controlled voltage source; (*c*) voltage-controlled current source; (*d*) current-controlled current source.

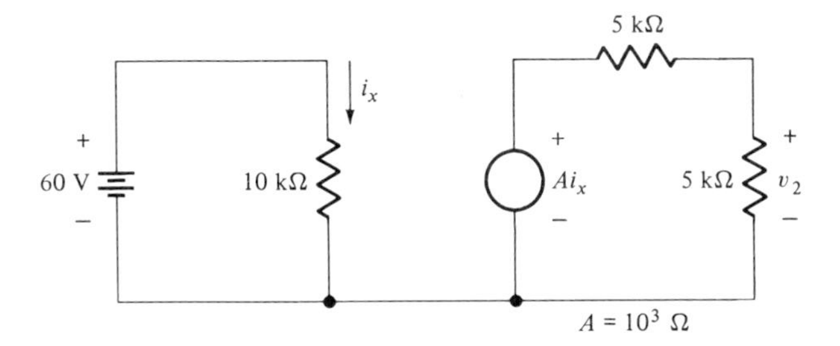

Fig. 10.4-2 Circuit for Example 10.4-1.

4 *Current-controlled current source.* This is shown in Fig. 10.4-1*d*. Note that K_4 must be dimensionless.

In the following sections we will show how these sources are used to model actual amplifiers.

Example 10.4-1 The circuit of Fig. 10.4-2 contains an independent voltage source of 60 V and a current-controlled voltage source. Find the output voltage v_2.

Solution We must first find the controlling current i_x. Using Ohm's law

$$i_x = \frac{60}{10} = 6 \text{ mA}$$

Thus the voltage output of the controlled source is $10^3 \times 6 \times 10^{-3} = 6$ V. Next we use the voltage divider:

$$v_2 = \frac{5}{5+5} \times 6 = 3 \text{ V}$$

Note that it was necessary to find the numerical value of the controlling current before the solution could be completed. ////

10.5 THE IDEAL VOLTAGE AMPLIFIER

The ideal voltage amplifier takes the form shown in Fig. 10.5-1*a*. The voltage-controlled voltage source provides an output voltage v_2 which is directly proportional to v_1, that is,

$$v_2 = A_v v_1 \tag{10.5-1}$$

where the constant A_v is the *voltage* amplification factor. In most cases of interest A_v will be a real number greater than unity. The *vi* characteristic of the output

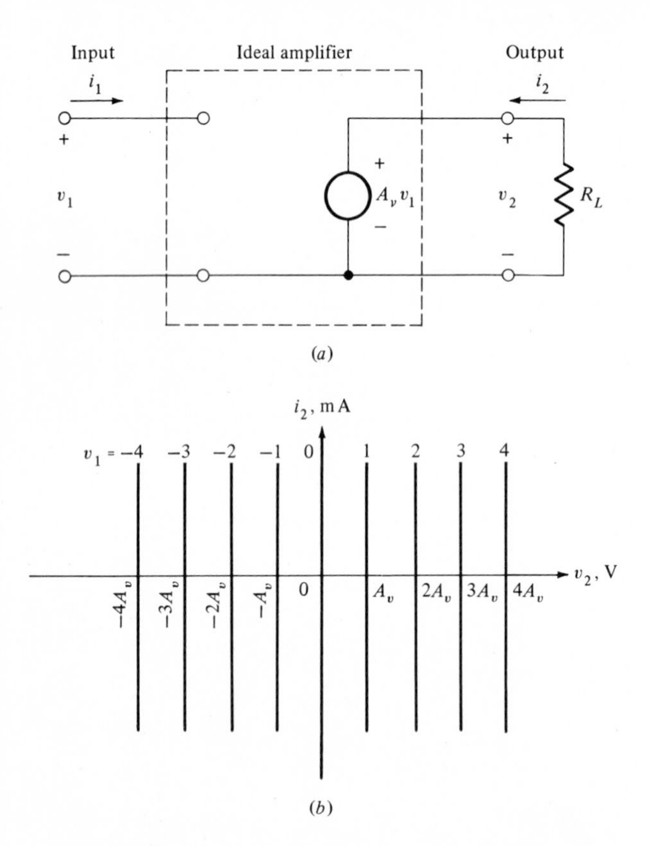

Fig. 10.5-1 Ideal voltage amplifier. (*a*) Circuit; (*b*) *vi* characteristic.

circuit is shown in Fig. 10.5-1*b*. This differs from previous *vi* characteristics in that it consists of a family of vertical straight lines with the input voltage v_1 as a parameter. Note that the output voltage is independent of the output current and the output current is determined by the external circuit. For the amplifier in the figure, taking note of the indicated reference directions

$$i_2 = \frac{-v_2}{R_L} = \frac{-A_v v_1}{R_L} \tag{10.5-2}$$

Example 10.5-1 In a certain ideal amplifier, the amplification factor is $A_v = 150$, the load is $R_L = 600\ \Omega$, and the input signal is 10 mV. Find

a The output voltage v_2, current i_2, and power P_2
b Power gain

Solution

a From Eq. (10.5-1)

$$v_2 = A_v v_1 = (150)(0.01) = 1.5 \text{ V}$$

From Eq. (10.5-2)

$$i_2 = -\frac{v_2}{R_L} = -\frac{1.5}{1{,}000} = -1.5 \text{ mA}$$

The output power is

$$P_2 = \frac{v_2^2}{R_L} = \frac{(1.5)^2}{1{,}000} = 2.25 \text{ mW}$$

b The input power is zero since $i_1 = 0$. Thus the power gain is infinite for the ideal voltage amplifier. ////

10.5-1 Input Resistance

An important characteristic of any voltage amplifier is its input resistance, i.e., the resistance seen by the signal source. For the ideal amplifier the input resistance is infinite, i.e., an open circuit. Practical voltage amplifiers always have finite input resistance, typically ranging from kilohms to several tens of megohms. Figure 10.5-2 shows the circuit diagram of a voltage amplifier with finite input resistance R_i. Included in the diagram is a practical signal source v_i with Thevenin resistance R_s driving the amplifier. Note carefully that the controlled source amplifies the input voltage v_1. This will always be less than the signal source voltage v_i because of the voltage divider network formed by R_s and R_i. This effect is considered in the next example.

Example 10.5-2 The amplifier of Example 10.5-1 has $A_v = 150$ and $R_L = 600\ \Omega$. Assume that its input resistance is 2 kΩ and that the 10 mV

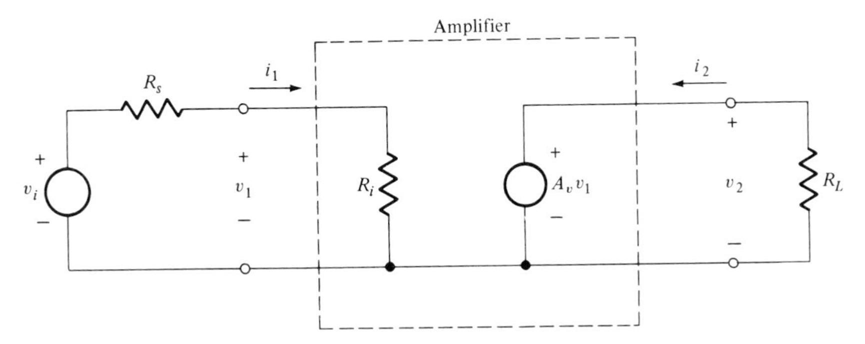

Fig. 10.5-2 Voltage amplifier with input resistance.

signal source has 600 Ω internal resistance. Find

a The overall gain† $G_{vT} = v_2/v_i$ in dB
b v_2 (compare with Example 10.5-1.)

Solution

a From the circuit of Fig. 10.5-2

$$\frac{v_2}{v_1} = A_v \qquad \text{and} \qquad \frac{v_1}{v_i} = \frac{R_i}{R_s + R_i}$$

Thus

$$A_{vT} = \frac{v_2}{v_i} = \left(\frac{v_2}{v_1}\right)\left(\frac{v_1}{v_i}\right) = A_v\left(\frac{R_i}{R_s + R_i}\right) = \frac{(150)(2{,}000)}{(600 + 2{,}000)} = 115$$

in dB

$$G_{vT} = 20 \log (115) = 20 \log (1.15 \times 10^2)$$
$$= 20 \log 1.15 + 20 \log 10^2$$
$$= 1.2 + 40 = 41.2 \text{ dB}$$

b $v_2 = A_{vT}v_i = (115)(0.01) = 1.15$ V

This is smaller than the 1.5-V output voltage in Example 10.5-1 where the input resistance was taken as infinite. Otherwise, the circuits were the same. ////

10.5-2 Output Resistance

Another important characteristic of real voltage amplifiers is that they do not have zero output resistance. Figure 10.5-3 shows the circuit diagram of a voltage

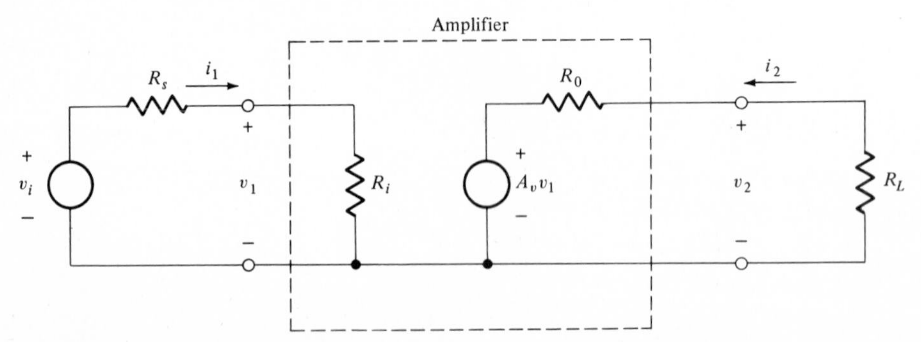

Fig. 10.5-3 Voltage amplifier with input and output resistance.

† Henceforth, we will use the subscript T to indicate total, or overall, gain.

amplifier with both input and output resistance. Both of these elements tend to reduce the gain from the theoretical maximum value A_v which is the gain when both are not present.

The amplifier model in Fig. 10.5-3 is adequate for many practical calculations of amplifier performance.

Example 10.5-3 In the amplifier of Example 10.5-2 the output resistance is $R_0 = 200\ \Omega$. Find

a $G_{vT} = v_2/v_i$ in decibels
b v_2 (compare with Examples 10.5-1 and 10.5-2)

Solution

a From Fig. 10.5-3 we have for the output circuit

$$\frac{v_2}{v_1} = \frac{A_v R_L}{R_0 + R_L} \tag{10.5-3}$$

For the input circuit

$$\frac{v_1}{v_i} = \frac{R_i}{R_s + R_i} \tag{10.5-4}$$

The overall gain A_{vT} is

$$A_v = \frac{v_2}{v_i} = \left(\frac{v_2}{v_1}\right)\left(\frac{v_1}{v_i}\right) = A_v\left(\frac{R_L}{R_0 + R_L}\right)\left(\frac{R_i}{R_s + R_i}\right)$$

$$= 150\left(\frac{600}{200 + 600}\right)\left(\frac{2{,}000}{600 + 2{,}000}\right) = 86.5 \tag{10.5-5}$$

In decibels

$$G_{vT} = 20 \log (86.5) = 20 \log (8.65 \times 10)$$

$$= 20 \log 8.65 + 20 \log 10$$

$$= 18.8 + 20 = 38.8 \text{ dB}$$

b The output voltage is

$$v_2 = A_{vT} v_1 = (86.5)(0.01) = 0.865 \text{ V} \tag{10.5-6}$$

As expected, the ouput voltage is further reduced compared to the previous examples due to the output resistance. ////

Note that the "chain" type of calculation done in Eq. (10.5-5) serves several purposes. First, it breaks the complicated transfer function v_2/v_i into two separate simpler transfer functions v_2/v_1 and v_1/v_i which are multiplied together. These simple transfer functions usually result when the input and output circuits are isolated as they are in this model. Second, it places clearly in evidence those factors which exert most influence on the overall gain. This is important in design, where tradeoffs often have to be made in order to meet specifications.

10.5-3 Circuit Model for a Phototube

In this section we consider an example which illustrates the utility of the controlled source when it is desired to find a circuit model for a device which is not completely electrical. We are going to find a model for a *phototube*, a light-sensitive electronic vacuum tube in which current flow depends not only on the voltage, but also on the intensity of the light which is incident on the cathode, or light-gathering surface of the internal structure of the tube. This light causes electrons to be emitted from the cathode which are attracted to the anode and thus cause a current to flow. A schematic diagram is shown in Fig. 10.5-4*a* of a phototube connected in series with a battery and load resistor.

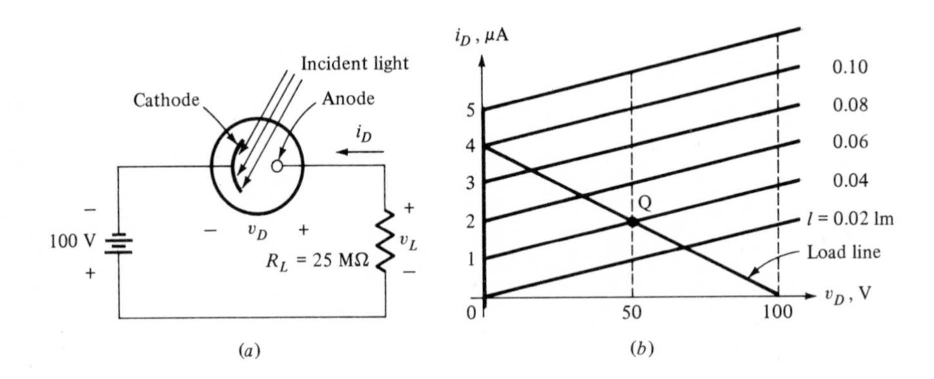

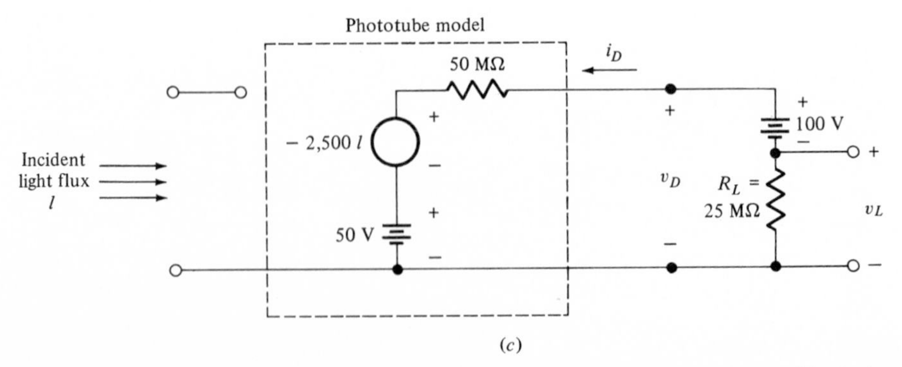

Fig. 10.5-4 Modeling a phototube. (*a*) Circuit; (*b*) *vi* characteristic; (*c*) derived circuit model.

A set of linearized characteristics for a typical phototube is shown in Fig. 10.5-4*b* and we see that they are very similar to the *vi* characteristics of the ideal amplifier shown in Fig. 10.5-1 except that the phototube characteristics have a slope. This slope can be accounted for by including an output resistance in series with the controlled source. The parameter in the curve family is the light flux, measured in lumens (abbreviated lm).

Our aim is to find a model for the phototube, and also to predict the effects of changes in light intensity on the load voltage v_L. We begin by noting that in the first quadrant, the phototube voltage is a linear function of both the current and light intensity. Thus we can write

$$v_D = Al + Bi_D + C \tag{10.5-7}$$

where l is the incident light intensity in lumens. The coefficients A, B, and C are found from the *vi* characteristics by noting that if we keep i_D constant at a value of 2 μA, then

$$A = \frac{\Delta v_D}{\Delta l} = \frac{100\text{ V} - 0\text{ V}}{0.02\text{ lm} - 0.06\text{ lm}} = \frac{100\text{ V}}{-0.04\text{ lm}} = -2{,}500\text{ V/lm} \tag{10.5-8}$$

If we keep l constant at 0.04 lm, then

$$B = \frac{\Delta v_D}{\Delta i_D} = \frac{100\text{ V} - 0\text{ V}}{3\ \mu\text{A} - 1\ \mu\text{A}} = \frac{100\text{ V}}{2\ \mu\text{A}} = 50\text{ M}\Omega \tag{10.5-9}$$

Now we have

$$v_D = -2{,}500l + 50 \times 10^6 i_D + C$$

In order to find the value of C, we can evaluate this equation at any point on the *vi* characteristic. For convenience, we use the point $i_D = 1\ \mu\text{A}$, $v_D = 50$ V, $l = 0.02$ lm. This yields

$$50 = (-2{,}500)(0.02) + (50 \times 10^6)(1 \times 10^{-6}) + C$$

$$C = 50\text{ V}$$

Finally, then, Eq. (10.5-7) becomes

$$v_D = -2{,}500l + 50 \times 10^6 i_D + 50 \tag{10.5-10}$$

The student should check several points on the *vi* characteristic to convince himself that this is correct.

Equation (10.5-10) suggests that the phototube voltage is made up of the *sum* of three component voltages and can thus be modeled by a three-element series circuit. The first voltage, $-2{,}500l$ is represented by a controlled source whose controlling variable is the light intensity l. The second, $50 \times 10^6 i_D$, represents the voltage drop across a 50-MΩ resistor, and the third, a 50-V battery. The series circuit which satisfies Eq. (10.5-10) and is our desired circuit model for the phototube is shown in Fig. 10.5-4*c*, along with the load resistance R_L and 100-V battery of the original circuit.

In order to predict the effect of changes in light level on the load voltage, we construct the load line (see Chap. 9). This represents the Thevenin characteristic of the circuit connected to the phototube, i.e., the load resistance R_L and the 100-V battery. The load line intersects the v_D axis at 100 V and the i_D axis at 100 V/25 MΩ = 4 μA. This is shown in Fig. 10.5-4*b*. If the ambient light level is 0.04 lm then the Q point (Sec. 9.2-1) is at $i_D = 2\ \mu A$, $v_D = 50$ V. To find the load voltage under ambient light conditions we note that when $i_D = 2\ \mu A$, $v_L = -i_D R_L = -(2\ \mu A)(25\ M\Omega) = -50$ V.

In order to find how v_L changes when the light level changes from its ambient value, we make use of the circuit model of Fig. 10.5-4*c*. Using the voltage divider formula, and being very careful with reference polarities, we find

$$v_L = \frac{25\ M\Omega}{50\ M\Omega + 25\ M\Omega}(-50 - 2{,}500l) = -\frac{1}{3}(50 + 2{,}500l) \qquad (10.5\text{-}11)$$

The change in v_L in terms of the change in l is found from this equation as

$$\frac{\Delta v_L}{\Delta l} = -\frac{2500}{3} = -833\ V/lm$$

This gives the *change* in v_L from its Q point value per lumen of incident light. For example, if $\Delta l = 0.01$ lm, that is, if l increases from the ambient value of 0.04 lm to 0.05 lm then $\Delta v_L = -8.33$ V.

The technique presented in this section can be used to model and predict behavior for any device whose characteristics can be linearized without departing too much from the actual characteristics.

10.6 CURRENT AMPLIFIERS

Amplifiers which utilize junction transistors as amplifying elements are best modeled by current amplifiers. The ideal current amplifier and its *vi* char-

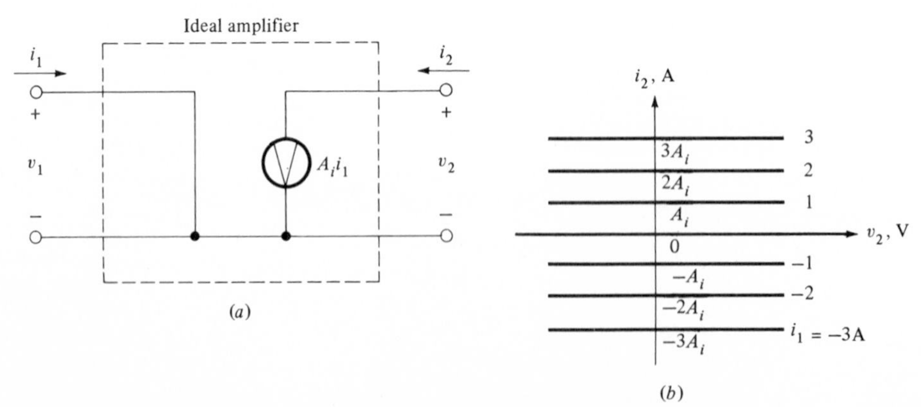

Fig. 10.6-1 Ideal current amplifier. (*a*) Circuit; (*b*) *vi* characteristic.

acteristic are shown in Fig. 10.6-1. Note that the ideal current amplifier has a short circuit ($R_i = 0$) across its input terminals, in contrast to the open-circuited input ($R_i = \infty$) of the ideal voltage amplifier.

From the *vi* characteristic we observe that the current from the controlled source is A_i amperes for each ampere into the input terminals so that

$$A_i = \frac{i_2}{i_1} \tag{10.6-1}$$

where A_i is the current-amplification factor. Note that the output current is independent of the output voltage; it depends only on input current according to Eq. (10.6-1). The output voltage will depend on the circuit connected to the output terminals.

Example 10.6-1 An ideal current amplifier is characterized by a current-amplification factor $A_i = 350$ and a load $R_L = 50\ \Omega$. The input current $i_1 = 1\ \mu\text{A}$. Find

a Output current i_2, voltage v_2, and power P_2
b Power gain

Solution

a From (10.6-1)

$$i_2 = A_i i_1 = (350)(10^{-6}) = 0.35 \text{ mA}$$

$$v_2 = -i_2 R_L = -(0.35)(0.05) = -0.0175 \text{ V}$$

$$P_2 = i_2^2 R_L = (0.35)^2(50) = 6.13\ \mu\text{W}$$

b Since the input power is zero because $R_i = 0$, the power gain is infinite. ////

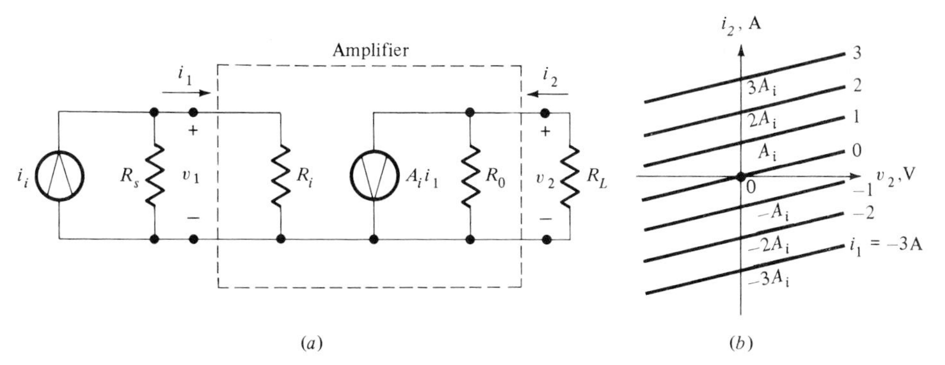

Fig. 10.6-2 Current amplifier with input and output resistance. (*a*) Circuit; (*b*) Output characteristic.

Practical current amplifiers can often be modeled with sufficient accuracy by an ideal current amplifier with input and output resistance as shown in Fig. 10.6-2*a*. The graphical characteristics of this circuit are shown in Fig. 10.6-2*b*. These should be compared with Fig. 10.6-1*b*. The effect of the output resistance is to cause each of the family of straight lines to have a slope equal to the reciprocal of the output resistance. Typical gain calculations are illustrated in the following example.

Example 10.6-2 In the circuit of Fig. 10.6-2, $R_s = 50\ \Omega$, $R_i = 10\ \Omega$, $A_i = 350$, $R_0 = 10\ \text{k}\Omega$, $R_L = 1\ \text{k}\Omega$, and $i_i = 5\ \mu\text{A}$. Find

a Overall current gain $G_{iT} = i_2/i_i$ in decibels
b i_2
c Power gain in decibels

Solution

a For overall gain A_{iT} we use a "chain" equation as follows:

$$A_{iT} = \frac{i_2}{i_i} = \left(\frac{i_2}{i_1}\right)\left(\frac{i_1}{i_i}\right)$$

where

$$\frac{i_2}{i_1} = A_i\left(\frac{R_0}{R_0 + R_L}\right)$$

and

$$\frac{i_1}{i_i} = \left(\frac{R_s}{R_s + R_i}\right)$$

Thus

$$A_{iT} = A_i\left(\frac{R_0}{R_0 + R_L}\right)\left(\frac{R_s}{R_s + R_i}\right)$$

$$= 350\left(\frac{10}{10 + 1}\right)\left(\frac{50}{50 + 10}\right)$$

$$= 265$$

In decibels,

$$G_{iT} = 20 \log 265 = 20 \log (2.65 \times 10^2) = 20 \log 2.65 + 20 \log 10^2$$

$$= 8.5 + 40 = 48.5\ \text{dB}$$

b $i_2 = A_{iT} i_i = (265)(5) = 1325\ \mu\text{A} \approx 1.32\ \text{mA}.$

c The power dissipated in the load is

$$P_2 = i_2^2 R_L = (1.33 \times 10^{-3})^2 (10^3) = 1.77 \times 10^{-3}\ \text{W}$$

The power delivered by the signal source to the amplifier is

$$P_1 = i_1^2 R_i$$

where

$$i_1 = i_i \frac{R_s}{R_s + R_i} = (5)\left(\frac{50}{50 + 10}\right) = 4.17\ \mu\text{A}$$

Thus

$$P_1 = (4.17 \times 10^{-6})^2 (10) = 1.74 \times 10^{-10}\ \text{W}$$

and

$$A_p = \frac{P_2}{P_1} = \frac{1.77 \times 10^{-3}}{1.74 \times 10^{-10}} \approx 10^7$$

$$G_p = 10 \log 10^7 = 70\ \text{dB}$$

////

10.7 HYBRID PARAMETERS

When we considered two terminal networks, we found that their terminal behavior could be completely characterized by a single linear equation involving the voltage and current at the terminals as variables. When we consider a two-port network, we have four variables to consider; voltage and current at both input and output terminals. Therefore, we would expect that the mathematical description of the behavior of the two-port would involve more than one simple equation. In fact, two equations are required to completely specify the terminal behavior of the two-port. These two equations can be written in a number of different ways. For example, we can write

$$i_1 = f_1(v_1, v_2)$$

$$i_2 = f_2(v_1, v_2)$$

or

$$v_1 = f_3(i_1, i_2)$$

$$v_2 = f_4(i_1, i_2)$$

or the variables can be mixed. One popular version of the two equations uses output current i_2 and input voltage v_1 as the dependent variables with input current and output voltage the independent variables. The resulting set of equations, called "hybrid" equations, turns out to be particularly appropriate for

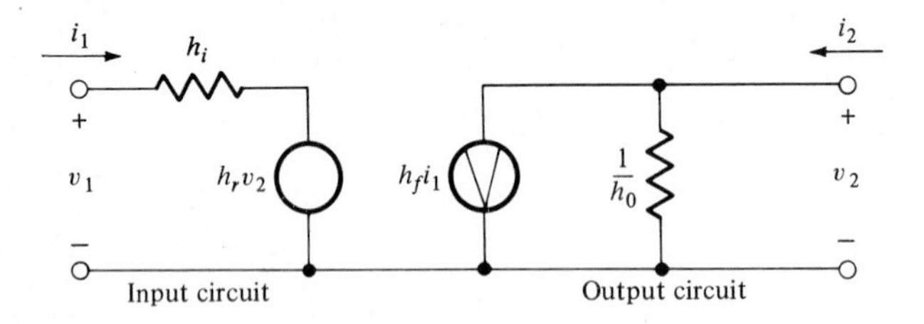

Fig. 10.7-1 Hybrid equivalent circuit.

transistor models. These equations for a linear two-port with no independent sources are

$$v_1 = h_i i_1 + h_r v_2 \tag{10.7-1a}$$

$$i_2 = h_f i_1 + h_o v_2 \tag{10.7-1b}$$

The subscripts identify the physical nature of the parameter, and can best be explained by reference to the hybrid equivalent circuit shown in Fig. 10.7-1. This circuit is derived from the hybrid equations in the following way: Consider Eq. (10.7-1*a*) which leads to the input circuit. Each term is a voltage and the equation is an expression of KVL. The first term on the right-hand side indicates that h_i is dimensionally a resistance and the term represents the voltage drop across the resistance due to current i_1 flowing through it. The second term represents a voltage-controlled voltage source, with h_r the factor by which v_2 is multiplied. Each of these terms corresponds to one of the elements in the input circuit.

The output circuit is derived from Eq. (10.7-1*b*) which is an expression of KCL. The first term on the right represents a current-controlled current source where h_f is the factor by which i_1 is multiplied. In the second term h_o must have the dimensions of conductance since the term represents a current.

PHYSICAL INTERPRETATION OF PARAMETERS

The physical meaning of the h parameters can be explained by considering both the equations and the equivalent circuit. For example, consider h_i. From Eq. (10.7-1*a*) we can write

$$h_i = \left.\frac{v_1}{i_1}\right|_{v_2=0} \tag{10.7-2}$$

This indicates that h_i is the input resistance with the output short-circuited ($v_2 = 0$). The subscript i stands for *input*. Similarly, h_r is dimensionless and represents the *reverse* voltage ratio with the input open-circuited ($i_1 = 0$).

Definitions for all four parameters are as follows:

$$h_i = \left.\frac{v_1}{i_1}\right|_{v_2=0} = \text{short-circuit } \textit{input} \text{ resistance} \tag{10.7-3}$$

$$h_r = \left.\frac{v_1}{v_2}\right|_{i_1=0} = \text{open-circuit } \textit{reverse} \text{ voltage gain} \tag{10.7-4}$$

$$h_f = \left.\frac{i_2}{i_1}\right|_{v_2=0} = \text{short-circuit } \textit{forward} \text{ current gain} \tag{10.7-5}$$

$$h_o = \left.\frac{i_2}{v_2}\right|_{i_1=0} = \text{open-circuit } \textit{output} \text{ conductance} \tag{10.7-6}$$

These definitions, when applied to the current amplifier of Fig. 10.6-2 yield the following equivalences:

$$h_i = R_i$$

$$h_r = 0$$

$$h_f = A_i$$

$$\frac{1}{h_o} = R_o$$

Thus the hybrid model fits the current amplifier perfectly

The equivalent circuit of Fig. 10.7-1 is useful for several reasons: First, it isolates the input and output circuits, their interaction being accounted for by the two controlled sources; second, the two parts of the circuit are in a form which makes it easy to take into account source and load circuits.

The definitions also suggest methods of measurement for the parameters. For example, Eq. (10.7-5) indicates that h_f can be measured by placing a short circuit[1] across the output (so that $v_2 = 0$), applying a small known input current i_1 and measuring the resulting output current i_2. The parameter h_f is then the ratio of these currents.

The hybrid equivalent circuit is used in a number of ways. It can be used like the Thevenin and Norton equivalent circuits, i.e., to reduce a complicated circuit to an equivalent form which is easier to handle. For the hybrid circuit the equivalence is at both input and output terminals. Another use is for modeling of new or unknown devices. For example, if we are given an amplifier or other two-port in a "black box" where we have no idea of the circuitry inside the black box, we can perform the measurements indicated previously in order to determine the four h parameters, and thus have a circuit model for the black box. This model can then be used to determine the effects of different sources

[1] This cannot always be done in practice since it may cause physical damage.

and loads or for any other purpose which involves the behavior of the black box at the two pairs of terminals. Clearly, we cannot use our hybrid model to predict the behavior of the individual circuit elements inside the box.

Example 10.7-1.

a Find the hybrid parameters for the voltage divider circuit of Fig. 10.7-2*a*.

b Verify the voltage divider formula for the case when $i_2 = 0$.

Solution

a The simplest way to find the parameters is to apply the definitions in Eqs. (10.7-3) through (10.7-6).

In order to find h_i we short circuit the output as shown in Fig. 10.7-2*b*. This short circuits resistor R_2 so that

$$h_i = \left.\frac{v_1}{i_1}\right|_{v_2=0} = R_1 \tag{10.7-7}$$

To find h_r we turn the circuit around as shown in Fig. 10.7-2*c*. Since $i_1 = 0$ there is no voltage drop across R_1 and we must have $v_1 = v_2$. Then

$$h_r = \left.\frac{v_1}{v_2}\right|_{i_1=0} = 1 \tag{10.7-8}$$

For h_f we again consider Fig. 10.7-2*b*. Since no current flows through

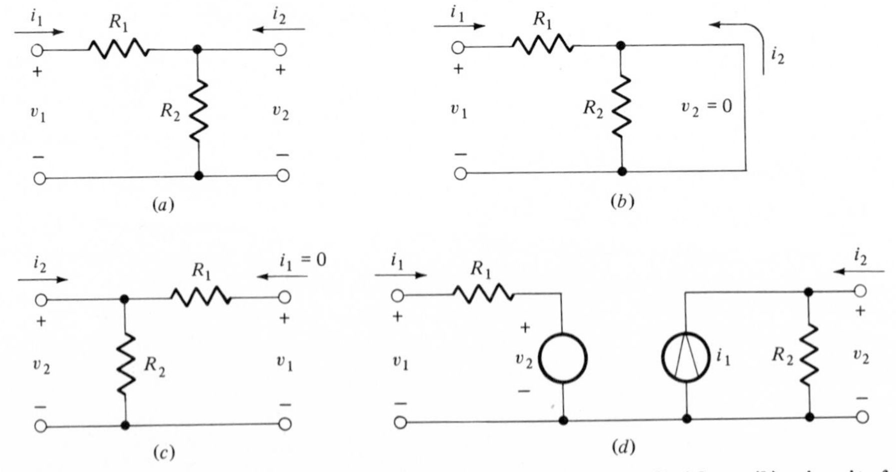

Fig. 10.7-2 Circuits for Example 10.7-1. (*a*) Voltage divider; (*b*) circuit for $v_2 = 0$; (*c*) circuit for $i_1 = 0$; (*d*) hybrid circuit.

R_2 we must have $i_2 = -i_1$ so that

$$h_f = \frac{i_2}{i_1}\bigg|_{v_2=0} = -1 \tag{10.7-9}$$

Finally, for h_0 we again consider Fig. 10.7-2*c*. With the input terminals open we have simply

$$h_o = \frac{i_2}{v_2}\bigg|_{i_1=0} = \frac{1}{R_2} \tag{10.7-10}$$

The hybrid equations (10.7-1) become

$$v_1 = R_1 i_1 + v_2 \tag{10.7-11}$$

$$i_2 = -i_1 + \frac{v_2}{R_2} \tag{10.7-12}$$

The equivalent circuit is shown in Fig. 10.7-2*d*. The student will note that it is more complicated than the original circuit but has the advantage that input and output are isolated.

b The voltage divider relation can be verified by setting $i_2 = 0$ in Eq. (10.7-12). Then

$$i_1 = \frac{v_2}{R_2}$$

This relation is substituted into Eq. (10.7-11):

$$v_1 = \frac{R_1 v_2}{R_2} + v_2 = v_2\left(\frac{R_1 + R_2}{R_2}\right)$$

Finally

$$\frac{v_2}{v_1} = \frac{R_2}{R_1 + R_2} \tag{10.7-13}$$

////

10.7-1 Cascaded Two-Ports

It often becomes necessary to connect the output of one two-port to the input of another as shown in Fig. 10.7-3*a*. They are then said to be connected in *cascade*. If each individual two-port is the model for a practical current amplifier as discussed in Sec. 10.6 then the cascade takes the form shown in Fig. 10.7-3*b* where a source and load have been added to the cascade. Gain calculations are carried out using a chain equation as illustrated in the following example.

Example 10.7-2 In the circuit of Fig. 10.7-3*b*, both amplifiers have $h_i = 1$ kΩ, $1/h_o = 10$ kΩ, and $h_f = 150$. The source has $v_i = 10\ \mu$V and $R_s = 500\ \Omega$. The load is $R_L = 500\ \Omega$. Find the output voltage.

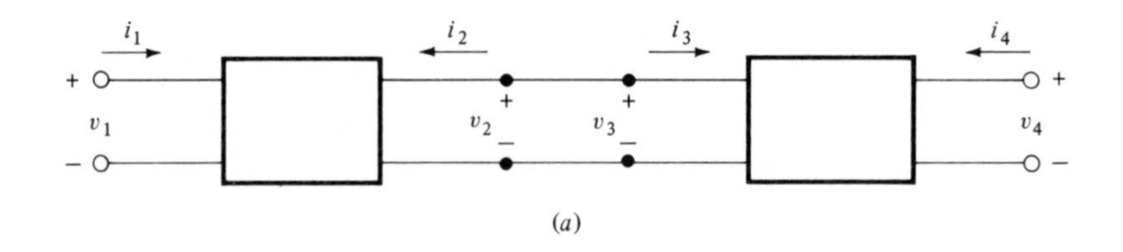

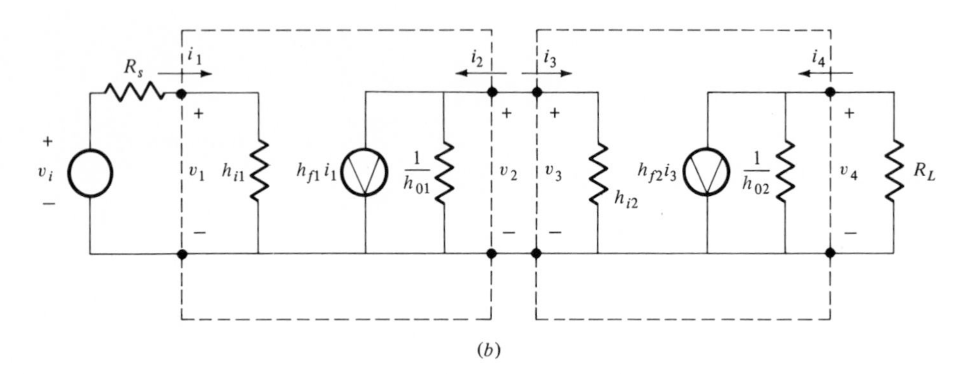

Fig. 10.7-3 Two-port networks. (*a*) Cascade connection; (*b*) cascade of current amplifiers.

Solution We first find the voltage gain $A_{vT} = v_4/v_i$. This is done using a chain equation:

$$A_{vT} = \frac{v_4}{v_i} = \left(\frac{v_4}{i_3}\right)\left(\frac{i_3}{i_2}\right)\left(\frac{i_2}{i_1}\right)\left(\frac{i_1}{v_i}\right) \tag{10.7-14}$$

The four individual terms in the chain are:

1 Using Ohm's law with all resistors in kΩ,

$$\frac{v_4}{i_3} = -h_{f2}\left(\frac{1}{h_{o2}} \,\|\, R_L\right) = -150\left(10 \,\|\, 0.5\right) = \frac{-(150)(5)}{(10.5)}$$

2 By inspection, $i_3 = -i_2$ so that $i_3/i_2 = -1$

3 Using the current divider formula, and taking careful note of current directions,

$$\frac{i_2}{i_1} = h_{f1}\left(\frac{1/h_{o1}}{h_{i2} + 1/h_{o1}}\right) = \frac{(150)(10)}{(11)}$$

4 Again using Ohm's law,

$$\frac{i_1}{v_i} = \frac{1}{R_s + h_{i1}} = \frac{1}{(1.5)}$$

Substituting into Eq. (10.7-14) we obtain

$$A_{vT} = \frac{v_4}{v_i} = \frac{(-150)(5)(-1)(150)(10)}{(10.5)(11)(1.5)} = 6{,}500$$

Finally, since $v_i = 10\ \mu\text{V}$,

$$v_4 = A_{vT}v_i$$
$$= (6.5)(10^3)(10)(10^{-6}) = 65 \times 10^{-3} = 65\ \text{mV} \qquad ////$$

GAIN OF CASCADED AMPLIFIERS

When amplifiers are connected in cascade, the overall gain can often be found simply by adding the individual gains expressed in decibels. To illustrate this, consider the system shown in Fig. 10.7-4 where two voltage amplifiers are connected in cascade. If $R_{i2} \gg R_{01}$, we can write

$$A_{vT} = \frac{v_0}{v_i} \approx \left(\frac{v_0}{v_1}\right)\left(\frac{v_1}{v_i}\right) = A_{v1}A_{v2} \qquad (10.7\text{-}15)$$

where A_{vi} and A_{v2} are the open-circuit individual gains.

Then, converting to decibels

$$G_{vT} = 20 \log A_{vT} = 20 \log (A_{v1}A_{v2}) = 20 \log A_{v1} + 20 \log A_{v2}$$
$$= G_{v1} + G_{v2} \qquad (10.7\text{-}16)$$

If the input resistance of the second amplifier were not much greater than the output resistance of the first, there would be a loss in gain due to the loading effect and the overall gain calculation would be more complicated.

Example 10.7-3 Three identical voltage amplifiers are cascaded. Each has an open-circuit gain of 10, an input resistance of 10 kΩ, and an output resistance of 100 Ω. Find the overall gain.

Solution Since $R_i \gg R_o$ between stages, we can use Eq. (10.7-16). Thus

$$G_{v1} \approx 20 \log 10 = 20\ \text{dB}$$

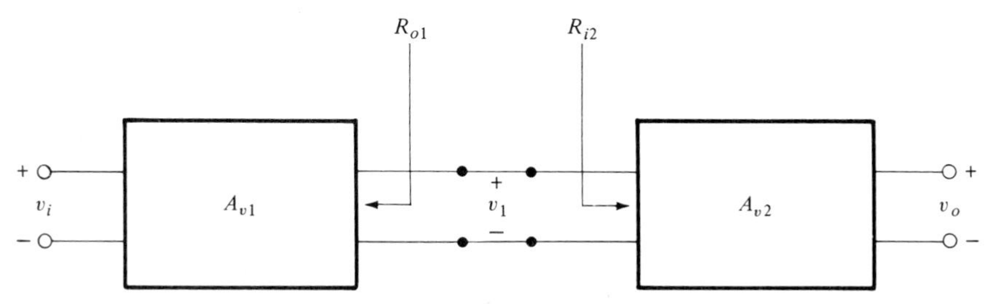

Fig. 10.7-4 Cascade voltage amplifiers.

and

$$G_{vT} = G_{v1} + G_{v2} + G_{v3} = 3 \times 20 = 60 \text{ dB} \qquad ////$$

SUMMARY

1 The transfer function represents the ratio of an output quantity to an input quantity.
2 Amplifiers are used to provide gain, i.e., an increase in the voltage, current, or power level of a signal.
3 The decibel is a logarithmic unit. For voltage gain it is defined as $G_v = 20 \log_{10} (V_2/V_1)$.
4 The output voltage or current of a controlled source depends on a voltage or current at some other point in the network.
5 The ideal voltage amplifier has infinite input resistance and zero output resistance. In a practical voltage amplifier, both of these resistances are finite and nonzero.
6 The ideal current amplifier has zero input resistance and infinite output resistance. In a practical current amplifier both of these resistances are finite and nonzero.
7 The hybrid parameters and equivalent circuit are used to reduce complicated circuits to simpler form and to model devices.
8 When amplifiers are cascaded, the overall gain in decibels is equal to the sum of the individual gains in decibels, assuming that loading can be neglected.

PROBLEMS

10.1-1 Find the voltage ratio transfer function for the circuit of Fig. P10.1-1.

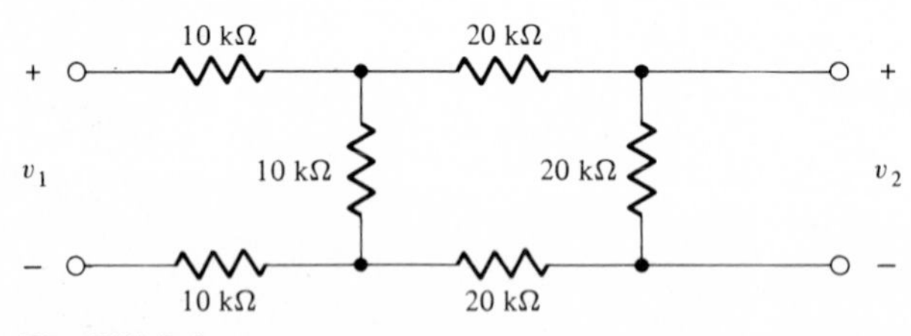

Fig. P10.1-1.

10.1-2 In the circuit of Fig. P10.1-1 the output voltage v_2 is to be $10 \cos 377t$ V. What input voltage is required to produce this output?

10.1-3 Find the short-circuit transfer conductance $G_T = i_2/v_1 \big|_{v_2=0}$ for the circuit of Fig. P10.1-3.

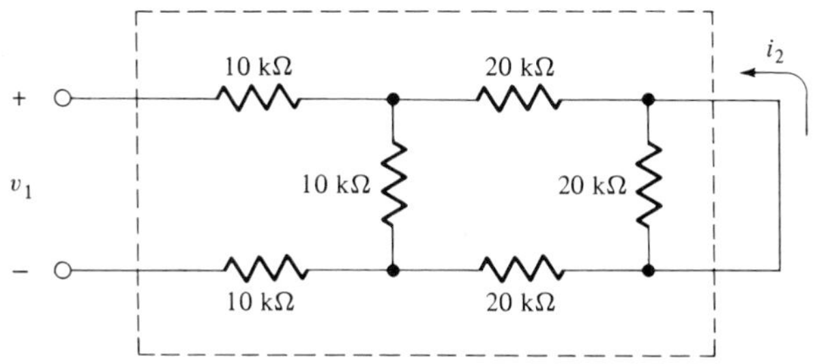

Fig. P10.1-3.

10.1-4 In the circuit of Fig. 10.1-3, the input voltage is 10 $[u(t) - u(t-2)]$ V. Find and sketch the short-circuit current.

10.2-1 In a certain system, the output power is 2 hp and the input power is 1 W. Find the power gain.

10.3-1 The output voltage of an amplifier is 30 V when the input is 2 mV. The input resistance is 10 kΩ and the load resistance is 600 Ω.
Find

a Power gain in decibels
b Input and output power in dBm
c Voltage gain in decibels

10.3-2 A current amplifier has an input resistance of 10 Ω and an output resistance of 100 kΩ. When the input current is 2 μA and the load resistance is 5 kΩ, the output current is 10 mA. Find

a Power gain in decibels
b Input and output power in dBm
c Current gain in decibels

10.3-3 The input power to an audio transformer is 27 W and the output power is 25 W. What is the power loss in decibels?

10.3-4 What is the voltage across 600 Ω at zero dBm? 10 dBm?

10.3-5 An amplifier has a power gain of 120 dB. If the input power is 5 mW, what is the output power?

10.3-6 In a 100-W amplifier, the hum level is 80 decibels below full output. How much hum power does this represent?

10.3-7 An amplifier has a power output of 34 dBm and a power gain of 50 decibels. The input resistance of the amplifier is 25 kΩ. Find the input voltage.

10.3-8 An amplifier has an input resistance of 10 kΩ and output resistance of 600 Ω. The output power is 60 W when the input voltage is 2.3 V.

Find

a Power input
b Power gain in decibels
c Voltage gain in decibels

10.4-1 In the circuit of Fig. P10.4-1 find the output current i_2 and the current gain in decibels.

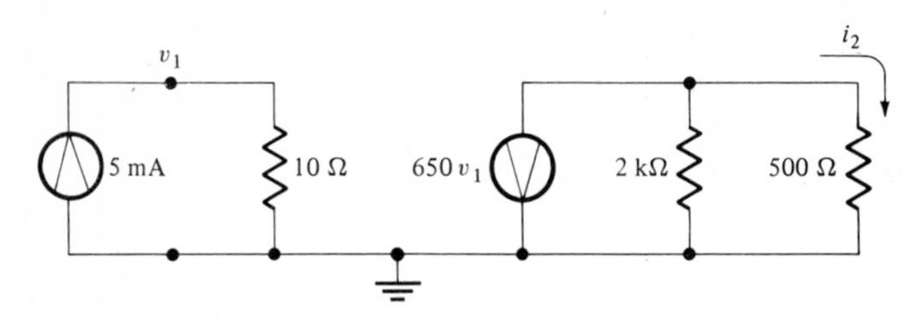

Fig. P10.4-1.

10.4-2 In the circuit of Fig. P10.4-2, $v(t) = 30 \cos 377t$ V. Find $v_2(t)$.

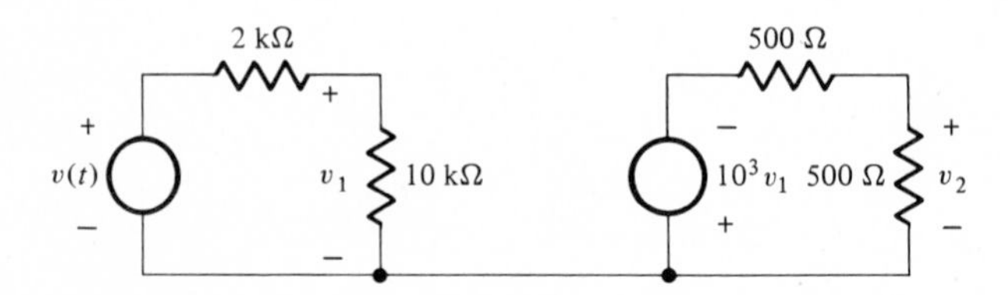

Fig. P10.4-2.

10.5-1 An ideal voltage amplifier is connected to a load of 2.2 kΩ. The input voltage is 10 mV and the output power is 200 mW. Find the output voltage and the voltage amplification factor.

10.5-2 A voltage amplifier has an input resistance of 20 kΩ, amplification factor of 500, and output resistance of 1000 Ω. The signal source has an internal resistance of 1 kΩ and the load resistance is 470 Ω. Find the overall voltage gain.

10.5-3 In the amplifier of Prob. 10.5-2 the output power is 500 mW. Find the input voltage and the power gain.

10.5-4 An amplifier as in Fig. 10.5-3 has $R_i = 1$ kΩ, and $R_0 = 500$ Ω. The source has $R_s = 50$ Ω and the load is a loudspeaker rated at 10 Ω. If the power to the loudspeaker is to be 10 W and the source voltage is 10 mV, find the required voltage amplification factor.

10.5-5 The circuit of a *difference* amplifier is shown in Fig. P10.5-5. The amplification factor operates on the *difference* between the two input signals. In the circuit $v_2 = 0$, $v_1 = 10 \sin \omega t$ mV, $A_v = 5{,}000$, $R_s = 100$ kΩ, $R_0 = 200$ Ω, and $R_L = 4.7$ kΩ. Find

a Voltage gain
b Current gain
c Power gain

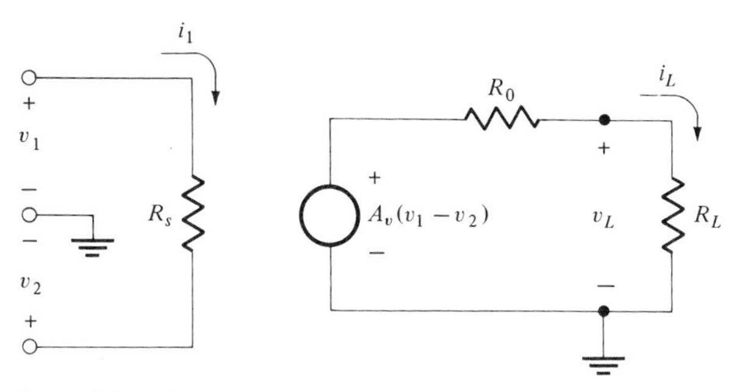

Fig. P10.5-5.

10.5-6 In the difference amplifier of Prob. 10.5-5 v_1 and v_2 are as shown in Fig. P10.5-6. Sketch the output waveform.

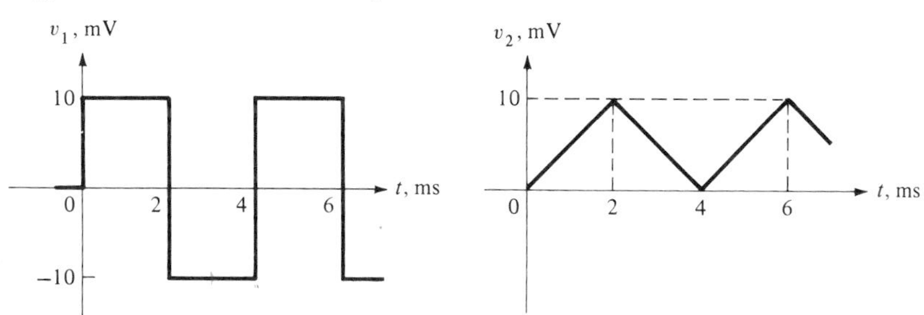

Fig. P10.5-6.

10.6-1 In the current amplifier of Fig. 10.6-2, $i_i = 10 \sin \omega t$ mA, $R_s = 10$ kΩ, $R_i = 100$ Ω, $A_i = 10^4$, $R_0 = 10$ kΩ, and $R_L = 10$ Ω. Find i_2 and the current gain.

10.6-2 In the amplifier of Prob. 10.6-1 the average power into R_L is to be 10 W. Find the required values of P_i and i_i.

10.6-3 A certain current amplifier has $R_s = 5$ kΩ, $R_i = 50$ Ω, $R_0 = 5$ kΩ, and $R_L = 10$ Ω. The output power is to be 5 W and the input current is 2 mA. Find the required value of A_i.

10.7-1 The equations

$$v_1 = R_{11}i_1 + R_{12}i_2$$

$$v_2 = R_{21}i_1 + R_{22}i_2$$

define four *open-circuit resistance* parameters similar to the hybrid parameters. Discuss the physical significance of these parameters. [*Hint*: see Eqs. (10.7-3) through (10.7-6).] Draw an equivalent circuit similar to Fig. 10.7-1.

10.7-2 A black box has two pairs of terminals with reference directions as shown in Fig. 10.1-1. When the output terminals are short-circuited and a 1-mA current is applied to the input terminals, the short-circuit current is 5 mA and the voltage across the input terminals is 25 mV. When the input terminals are left open and a 1-mV source is applied at the output, the voltage across the input terminals is 0.1 mV and the current drawn from the 1 mV source is 250 μA. Find the hybrid parameters for the black box.

10.7-3 Find the hybrid parameters for the circuit of Fig. P10.7-3.

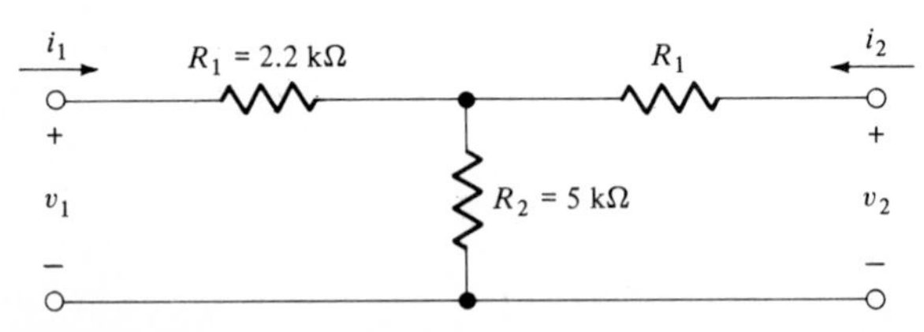

Fig. P10.7-3.

10.7-4 In the circuit of Fig. 10.7-3 the amplifiers are identical with $R_L = R_s = 50\ \Omega$, $h_i = 250\ \Omega$, $1/h_o = 100\ \text{k}\Omega$, and $h_f = 350$. Find the overall gain $A_v = v_4/v_i$.

10.7-5 A voltage amplifier circuit model is shown in Fig. P10.7-5. In the circuit $R_i = 1\ \text{M}\Omega$, $R_0 = 600\ \Omega$, and $A_v = 250$. Find the overall gain if two of these amplifiers are cascaded to drive a load resistance of 10 kΩ.

10.7-6 Repeat Prob. 10.7-5 for a load resistance of 600 Ω.

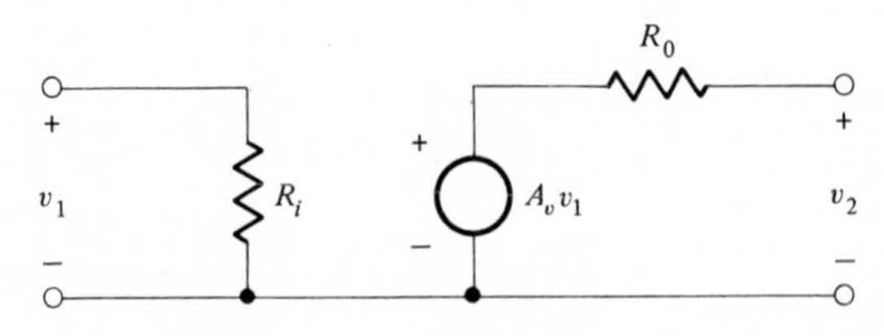

Fig. P10.7-5.

10.7-7 Repeat Prob. 10.7-6 with R_i changed to 600 Ω.

11

Amplifiers: Piecewise-Linear and Incremental Models

OBJECTIVES

Upon completion of this chapter, you should be able to

1 Linearize a set of nonlinear characteristics and derive a piecewise-linear model.
2 Construct the load line for a given amplifier on the device characteristics.
3 Find the operating point on the load line.
4 Describe saturation and cutoff.
5 Find the input signal voltage required to produce a specified output current swing.
6 Find the maximum signal that can be accommodated by a given amplifier without distortion.
7 Describe the linear incremental hybrid model for the transistor and find the model parameters from the characteristic curves.
8 Find the small signal gain and input impedance of a single-stage amplifier.

INTRODUCTION

In this chapter we bring together the theory and techniques studied previously and apply them to the analysis of actual amplifier circuits. The active devices which provide amplification are inherently nonlinear, and exact analysis of circuits containing these devices is often not practicable. Such circuits are usually analyzed by one of a number of methods, depending on the magnitude of the signal. If the signal amplitude is comparable to the maximum that the amplifier can accommodate, we call it a large signal, and for dc and large-signal analysis, graphical techniques are used. When accurate characteristics are available and accurate results are required, graphical methods offer a way of taking into account the inherent nonlinearity of the amplifying device. If only the gross nonlinearities are of interest and extreme accuracy is not required, a piecewise-linear model can be derived from a piecewise-linear approximation of the characteristic and all circuit variables can be found using standard circuit analysis techniques, without recourse to graphical methods, or a combination of graphical and analytical methods can be used.

If the signal is small compared to the maximum, we call it a small signal, and for such signals linear circuit models are derived from the characteristics and conventional circuit analysis techniques are applied.

We begin by considering the graphical characteristics of a typical amplifying element, the bipolar junction transistor, and finding a suitable piecewise-linear model. A combination of graphical and analytical methods will be applied to the analysis of a typical transistor amplifier and various aspects of amplifier operation will be considered.

Finally, we will consider the linear incremental model which is used for small-signal analysis. This model is derived from the characteristic curves of the transistor and uses the controlled source introduced in Chap. 10.

It is important to keep in mind that the purpose of this chapter is to present circuit aspects of electronic amplifiers. Such topics as the physics of semiconductor devices, biasing, and compensation will be covered in an electronics course which is normally taken concurrently with, or after, the basic circuits course.

11.1 PIECEWISE-LINEAR MODELS—THE TRANSISTOR

In Chap. 9 we introduced the *pn* junction diode, a nonlinear semiconductor device. When two *pn* junctions are formed simultaneously in a single piece of silicon the resulting device is a transistor* which exhibits amplifying properties. The two possible configurations (called *npn* and *pnp*) are shown in Fig. 11.1-1*a* and their circuit symbols are shown in Fig. 11.1-1*b*. Note that the transistor is a three-terminal device, the external terminals being called the emitter E, base B,

* The full name is *bipolar junction transistor*

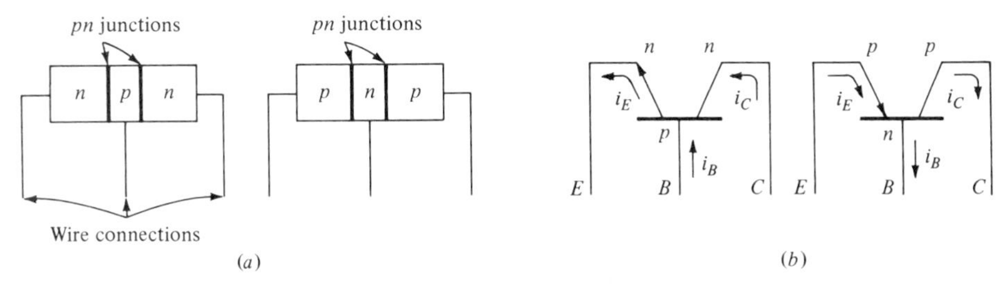

Fig. 11.1-1 The bipolar junction transistor. (*a*) Physical configuration of *npn* and *pnp*; (*b*) circuit symbols.

and collector C. The arrow on the emitter terminal indicates the direction of positive dc emitter current in normal operation: *into* the emitter for the *pnp* transistor and *out* of the emitter for the *npn* transistor. The corresponding base and collector currents in normal operation are as shown on the diagram. All currents in the *npn* transistor are opposite to those in the *pnp* transistor. Both types are used in practice but the *npn* variety is somewhat easier to manufacture in integrated circuit form so it is used more often.

CURRENT RELATIONS IN THE TRANSISTOR

The transistor is a device that acts primarily as a current amplifier. We therefore begin by applying Kirchhoff's current law. For either *pnp* or *npn* this yields

$$i_E = i_B + i_C \tag{11.1-1}$$

In normal operation, all three are positive currents.

The amplification arises because in linear operation, the major component of the collector current is directly proportional to the base current, and the constant of proportionality is usually much greater than one. Thus if the collector current is considered as the output, and the base current as the input, we have amplification. The constant of proportionality between the base and collector currents is called β; thus we have

$$i_C \approx \beta i_B \tag{11.1-2}$$

The current amplification factor β typically ranges between 25 and 350 and actually is not constant because of second-order effects which come into play. In addition, there is a very small collector current that flows when $i_B = 0$. In what follows, we will neglect these effects and assume that β is constant.

If we substitute i_C from Eqs. (11.1-2) into (11.1-1), we find the relation between emitter current i_E and base current i_B:

$$i_E \approx (\beta + 1)i_B \tag{11.1-3}$$

For values of β in the normal range (25 to 350) $\beta \approx \beta + 1$ so that $\beta_{iB} \approx (\beta + 1)i_B$. Thus $i_C \approx i_E$ and both of these are much larger than the base current i_B.

In the next section, we will consider actual measured characteristic curves for the transistor, along with a method for deriving a piecewise-linear model from these characteristics. As noted in the introduction this model will be used for circuit calculations in which extreme accuracy is not required and in which only gross nonlinearities are to be taken into account.

11.1-1 Obtaining Piecewise-Linear Model Parameters from Transistor Characteristics

THE COLLECTOR-EMITTER CIRCUIT

There are a number of ways in which the *vi* characteristics of a transistor can be plotted. The most often used is called the collector characteristic and is shown in Fig. 11.1-2*a*. In this plot of i_C vs. v_{CE} the base current i_B is the parameter and the value of β is approximately 100. Note that this family of curves is very similar to the *vi* characteristic of the ideal current amplifier with output resistance shown in Fig. 10.6-2. There are several points to be noted. The first is that the actual device characteristic of Fig. 11.1-2*a* lies only in the first quadrant, due to the physical nature of the *pn* junctions. Thus, in order to use the transistor as a linear amplifier, capable of amplifying ac signals, we are going to have to arrange the circuit to take account of this. We do this by providing *bias*, to be explained shortly. The second point is that the area adjacent to the i_C axis is highly nonlinear (this is called the *saturation* region) and will have to be avoided if we are to achieve linear amplification. The area in the region $v_{CE} > 20$ V also exhibits considerable nonlinearity (due to avalanche breakdown) and must be avoided. In addition, the collector current cannot become negative in normal operation, so the $i_C = 0$ axis defines the lower limit on collector current, called *cutoff*. A third point concerns the value of β. It is not uncommon to have the β's of individual transistors of one type vary by a factor of 5 to 1 or more. Because of this special precautions must be taken in design in order to make circuit performance somewhat independent of this variation. In view of these variations the collector curves provided by manufacturers can only indicate average behavior of a particular transistor type. In order to develop a piecewise-linear model the curves are *linearized* as shown in Fig. 11.1-2*b*. It is a relatively simple matter to determine the value of β from the characteristics by noting that for a fixed value of v_{CE},

$$\Delta i_C = \beta \, \Delta i_B$$

Thus

$$\beta = \left.\frac{\Delta i_C}{\Delta i_B}\right|_{\Delta v_{CE}=0} \tag{11.1-4}$$

This ratio can be found from the piecewise-linear curves of Fig. 11.1-2*b* by determining the change in i_C corresponding to a known change in i_B at one value of v_{CE}. For the particular transistor whose characteristics are given in the

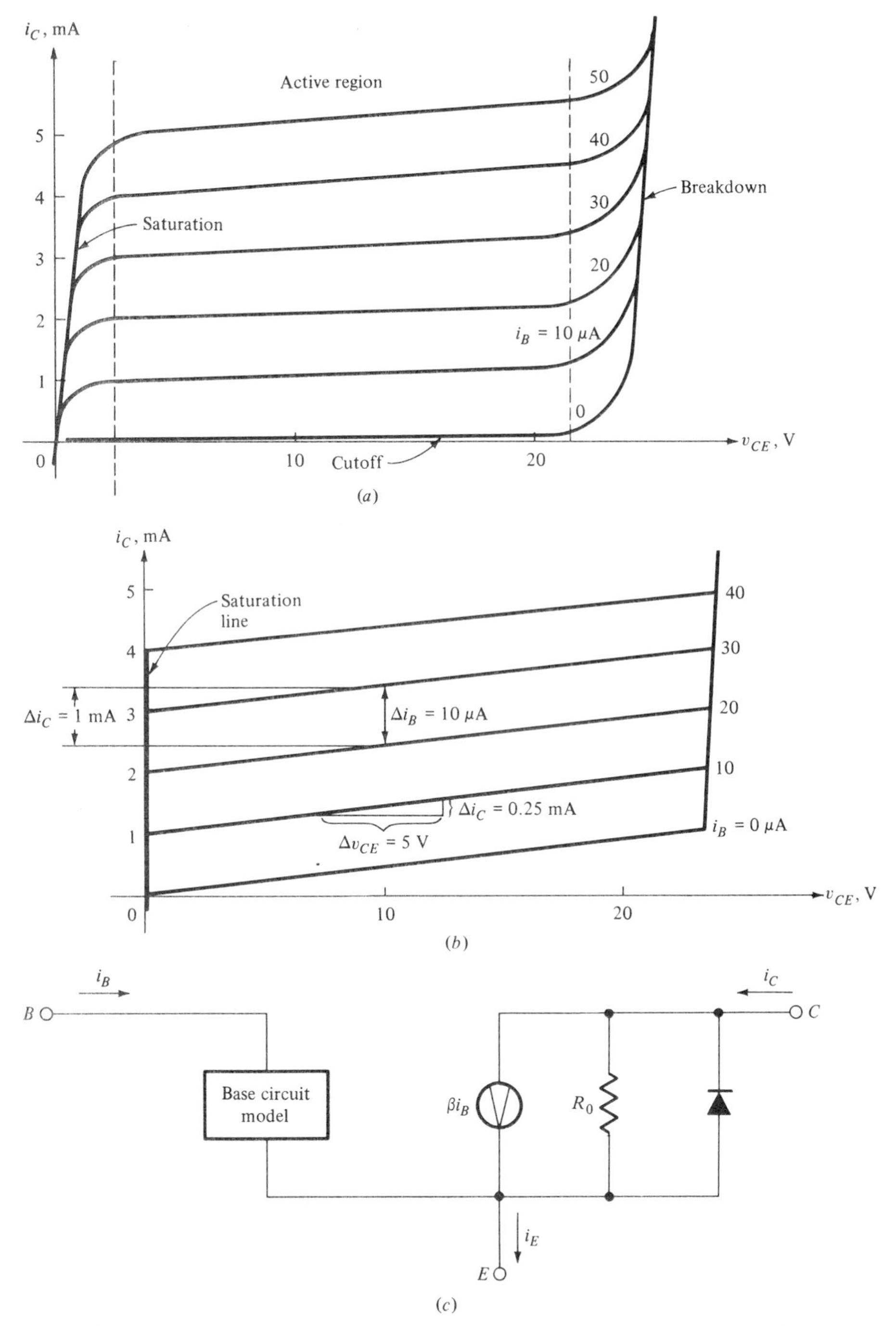

Fig. 11.1-2 Collector characteristics of *npn* transistor. (*a*) Actual characteristic; (*b*) piecewise-linear approximation; (*c*) piecewise-linear model.

figure, we see that a 10 μA change in i_B at $v_{CE} \approx 10$ V causes a 1 mA change in i_C. Thus, for this transistor,

$$\beta = \left.\frac{\Delta i_C}{\Delta i_B}\right|_{\Delta v_{CE}=0} = \frac{1 \text{ mA}}{10\ \mu\text{A}} = \frac{1 \text{ mA}}{0.01 \text{ mA}} = 100$$

The slope of the characteristics can be interpreted as the output resistance (see Sec. 10.6) and the value of this resistance is found by measuring the inverse slope of any one of the parallel characteristics. The construction is shown on the $i_B = 10\ \mu$A characteristic and the resistance is given by

$$R_0 = \left.\frac{\Delta v_{CE}}{\Delta i_C}\right|_{\Delta i_B=0} \approx \frac{5 \text{ V}}{0.25 \text{ mA}} = 20 \text{ k}\Omega \tag{11.1-5}$$

Except for the behavior in the saturation, breakdown, and cutoff regions, the collector-emitter circuit can be modeled by a current-controlled current source in parallel with R_0, as shown in Fig. 11.1-2*c*. This has the same form as the ideal current amplifier with output resistance which we studied in Sec. 10.6.

Since the breakdown region is always avoided in linear operation, we do not bother to include any elements in our model to account for this effect. The cutoff effect is taken care of by the base-emitter circuit model to be described shortly.

The saturation region in the curves of Fig. 11.1-2*a* is represented as a saturation line on the $v_{CE} = 0$ axis in the linearized curves of Fig. 11.1-2*b*. This approximation is reasonable for many applications and considerably simplifies the piecewise-linear model. The effect is accounted for by the ideal diode connected between collector and emitter. With this diode present the voltage v_{CE} cannot become negative because this would cause the diode to turn on which in turn causes the voltage across it to become zero.

THE BASE-EMITTER CIRCUIT

For the base-emitter circuit, the characteristic curves usually plotted are i_B vs. v_{BE} with v_{CE} as the parameter as shown in Fig. 11.1-3*a*. Note that only one curve is shown, rather than a family of curves, as for the collector circuit. The reason for this is that the characteristics for values of v_{CE} greater than a few tenths of a volt are identical for all practical purposes.

This characteristic is almost identical to the *pn* diode characteristic shown in Fig. 9.1-4. Thus we can use the piecewise-linear model developed for the diode in Sec. 9.5 as a model for the base-emitter circuit. Recall that this model consists of an ideal diode in series with a resistance r_f (often about 1 kΩ for silicon transistors) and a voltage source V_0 (0.7 V for silicon units). The complete piecewise-linear transistor model is shown in Fig. 11.1-3*b*. The ideal diode in the model of the base accounts for collector current cutoff because it only allows i_B to flow into the base; i_B cannot flow in the reverse direction. Thus the βi_B current can only flow in the direction indicated and i_C cannot become negative.

The states of the ideal diodes in the model depend on the particular operating

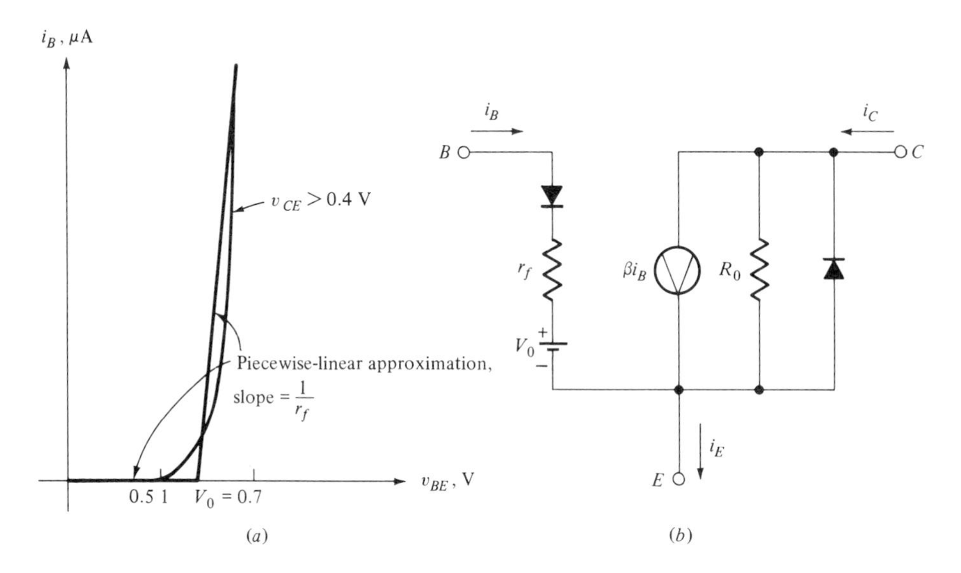

Fig. 11.1-3 Characteristics of an *npn* transistor. (*a*) i_B-versus-v_{BE} curve; (*b*) complete piecewise-linear model.

conditions. In saturation both diodes are short circuits, while under cutoff conditions, both diodes are open circuits. In the active region, the base-emitter diode is a short circuit while the collector-emitter diode is an open circuit. If it is known that the transistor is operating only in the active region, for example, the diodes can be replaced by a short circuit and an open circuit, thus simplifying the model. In the next section we will use this model to calculate amplifier performance.

11.1-2 Load Lines and Operating Points

The amplifier circuit to be analyzed is shown in Fig. 11.1-4*a*. The transistor is connected as a *common-emitter* amplifier. The name common-emitter is applied because the emitter is connected to both the base and collector (through resistors and sources) and is thus *common* to both the base and collector circuitry.

The V_{BB} battery and R_b are included in the base-emitter circuit to insure that the base-emitter junction is forward-biased and to provide an appropriate value of dc base bias current, I_{BQ}. The V_{CC} battery and R_c are included in the collector-emitter circuit to ensure that $v_{CE} \geq 0$ under all operating conditions.

Our analysis of the circuit will be based on the piecewise-linear model. Once the model parameters have been obtained from measured characteristics as in the previous section, all circuit calculations can be carried out using standard circuit analysis techniques. However, we are going to use a combination of analytical and graphical methods because of the greater insight furnished by the graphical approach.

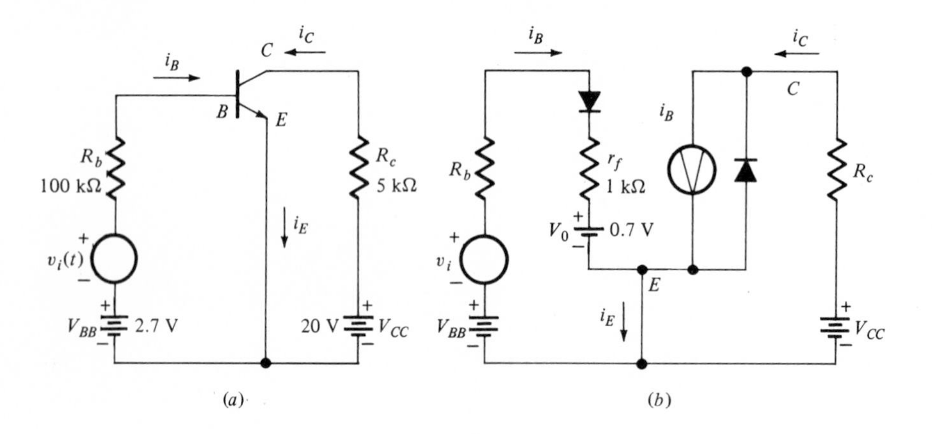

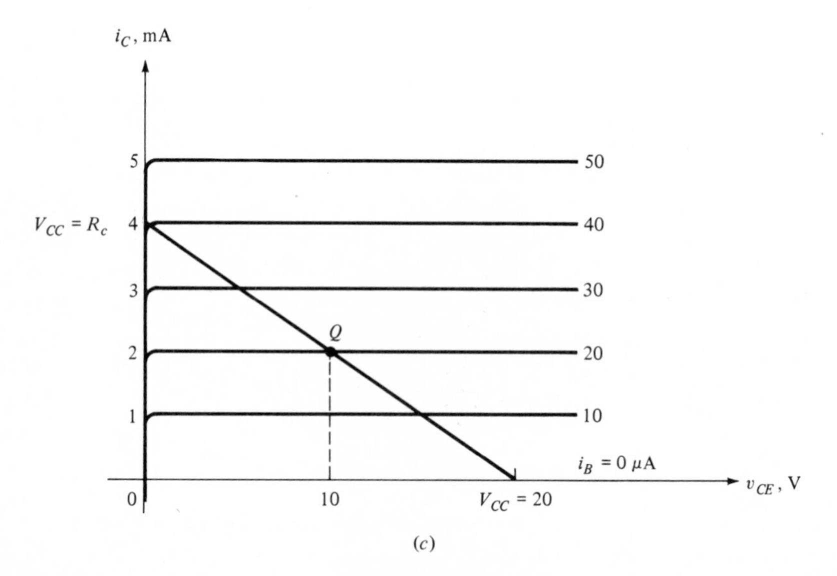

Fig. 11.1-4 Common-emitter amplifier. (*a*) Circuit; (*b*) piecewise-linear model; (*c*) collector characteristic for $\beta = 100$.

Our aim in the analysis is to first plot the *operating path* of the circuit on the collector characteristic. When this is done we will be able to tell if we are operating in an appropriate region of the characteristics and if we are meeting the given specifications on output signal voltage and current. The circuit with the transistor replaced by its piecewise linear model is shown in Fig. 11.1-4*b* and the piecewise-linear characteristics for the transistor are shown in Fig. 11.1-4*c*. Note that the characteristics are horizontal, which means that we are assuming R_0 to be so large as to be negligible. This is often the case in practice.

We proceed as in Sec. 9.2 to draw the load line, which is the *vi* characteristic of the circuit connected to the collector-emitter terminals. The equation for the load line can be obtained by writing KVL around the collector-emitter loop. Tracing counterclockwise from the collector, this yields

$$v_{CE} - V_{CC} + i_C R_c = 0$$

Rearranging,

$$v_{CE} = V_{CC} - i_C R_c \tag{11.1-6}$$

We can plot the load line most easily by finding its intersections with the axes of the collector characteristic. If we set $v_{CE} = 0$ we find the intersection with the i_C axis to be

$$i_C \Big|_{v_{CE}=0} = \frac{V_{CC}}{R_c} \tag{11.1-7}$$

Setting $i_C = 0$ we find the intersection with the v_{CE} axis

$$v_{CE} \Big|_{i_C=0} = V_{CC} \tag{11.1-8}$$

These points and the load line are shown on the linearized collector characteristic of Fig. 11.1-4*c*. For this circuit, the load line is the locus of all possible operating points. Note that i_C can vary from 0 to a maximum of 4 mA and v_{CE} can vary from 0 to 20 V. Admissible pairs of values of i_C and v_{CE} must, of course, lie on the load line.

When we apply a signal in the base circuit i_B will vary with the signal. These variations are amplified in the collector circuit. In order for the operating point of the collector circuit to go through its maximum possible range the base current i_B must lie in the range from 0 to 40 μA. A 40 μA change in i_B corresponds to a 4 mA change in i_C. This amplification agrees with the given value $\beta = 100$. If i_B increases above 40 μA, the collector current does not increase: it stays at 4 mA because of the convergence of all the curves in the family. When this happens, the transistor is said to be saturated.

In order to determine the output waveform in response to a given input waveform, we need to first determine the quiescent (Q) point, i.e., the operating point when no signal is applied. This point is determined by conditions in the base circuit, which we can find by applying KVL around the base-emitter loop. For this calculation we assume that the signal $v_i(t) = 0$ so that we can determine the quiescent base current I_{BQ}. Thus, tracing clockwise,

$$-V_{BB} + I_{BQ}R_b + V_{BEQ} = 0$$

$$I_{BQ} = \frac{V_{BB} - V_{BEQ}}{R_b} \tag{11.1-9}$$

For normal linear operation, the base-emitter junction is forward-biased and, as shown by the piecewise-linear circuit of Fig. 11.1-4*b*, has a conducting voltage drop of 0.7 V and a resistance of about 1,000 Ω which is negligible compared with $R_b = 100$ kΩ. Using these values in Eq. (11.1-9) we get

$$I_{BQ} = \frac{2.7 - 0.7}{10^5} = 2 \times 10^{-5}\,\text{A} = 20\,\mu\text{A}. \tag{11.1-10}$$

The value of collector current is, from Eq. (11.1-2)

$$I_{CQ} = \beta I_{BQ} = 2{,}000\,\mu\text{A} = 2\,\text{mA} \tag{11.1-11}$$

The collector-emitter voltage is from Eq. (11.1-6)

$$\begin{aligned} V_{CEQ} &= V_{CC} - I_{CQ}R_c \\ &= 20 - 2 \times 5 \\ &= 10\,\text{V} \end{aligned} \tag{11.1-12}$$

The Q point is plotted on the collector characteristic of Fig. 11.1-4*c*.

11.1-3 Signal Amplitudes and Distortion

COLLECTOR CURRENT VARIATION

The next step in our analysis of the common-emitter amplifier is to find the maximum possible collector current variation (often called the "swing") which can be provided without distortion and the base circuit signal required to produce this variation.

We consider first the collector circuit. Note that the Q point in Fig. 11.1-4*c* is placed directly in the center of the load line for this example. Thus the collector current can swing from its quiescent value of 2 mA up to a maximum value of 4 mA before saturation, and it can swing down to a minimum collector current of 0 mA which corresponds to collector current cutoff (as noted previously, because of the characteristics of the collector-base junction, the collector current cannot become negative in normal operation). Thus the maximum available ac swing in the collector current is 4 mA peak-to-peak or 2 mA peak. In practice the figure will be somewhat less because of the nonlinearities which we are neglecting.

Before we begin, we must be sure that we understand the notation used. Up to now, currents and voltages have been labeled as either *total* values, that is, $i_C(t)$ (lowercase i, uppercase subscript) or dc values, for instance, I_C. It is often convenient to separate the ac (time varying) component of the current or voltage from its dc value. Thus, the total current $i_C(t)$ is the sum of a dc component and an ac component:

$$i_C(t) = I_C + i_c(t) \tag{11.1-13}$$

For most practical purpose, the dc component I_C when a signal is present is equal to the zero signal quiescent value I_{CQ}. Here $i_c(t)$ (lowercase subscript)

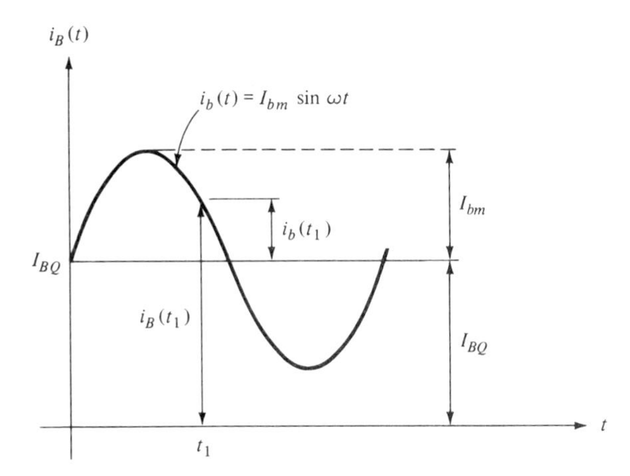

Fig. 11.1-5 Graphical interpretation of notation. $i_B(t)$ = total base current; $i_b(t)$ = ac component of $i_B(t)$; I_{BQ} = quiescent (dc) component of $i_B(t)$; I_{bm} = peak value of sine wave.

represents the ac component of the collector current, which by definition has zero average value. A graphical interpretation of this notation is shown in Fig. 11.1-5.

Now let us consider the ac signal required in the base circuit to achieve various values of collector current swing. Assume that $v_i(t)$ is such that the ac part of the base current is $i_b(t) = 10 \sin \omega t\ \mu$A. The total base current is then $i_B(t) = 20 + 10 \sin \omega t\ \mu$A. The resulting collector current can be found graphically, as in Sec. 9.2-1. The construction is shown in Fig. 11.1-6*a*. The waveform of i_b, for example, is plotted by constructing the $i_b(t)$ axis (ac component) parallel to the load line and locating the $i_b = 0$ axis directly in line with the Q point, where the ac component is zero. The i_c and v_{ce} axis are set up similarly and the waveforms are obtained by projecting a point on the i_b waveform for a particular time first to the load line and then to the i_c and v_{ce} waveforms at the same value of t. This is illustrated on the figure for point a on the i_b waveform at time t_1. We see that the ac collector current and voltage are both undistorted replicas of the ac base current. For the ac collector current we have, since there is no distortion,

$$i_c(t) = \beta i_b(t) = (100)(10 \sin \omega t) = 1000 \sin \omega t\ \mu\text{A}$$
$$= 1 \sin \omega t\ \text{mA} \tag{11.1-14}$$

The total collector current is

$$i_C(t) = I_{CQ} + i_c(t) = 2 + 1 \sin \omega t\ \text{mA} \tag{11.1-15}$$

The graphical construction indicates that the ac collector-emitter voltage has a

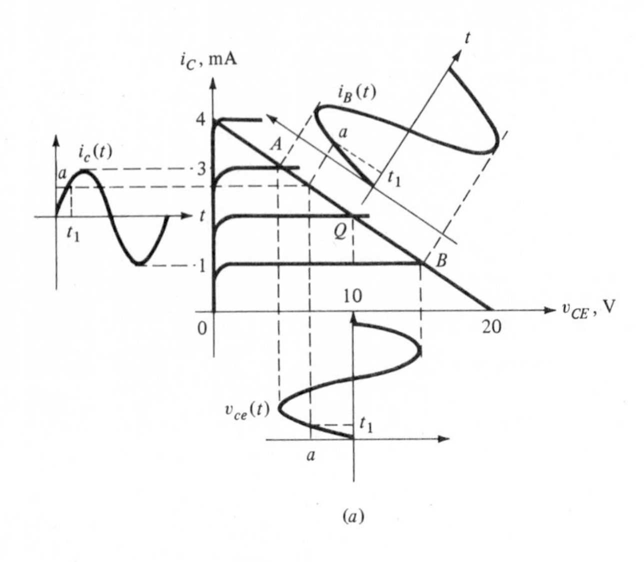

(a)

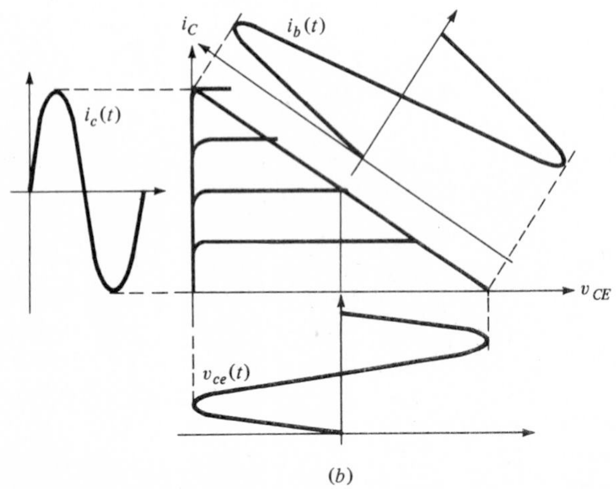

(b)

Fig. 11.1-6 Waveforms in the *CE* amplifier. (*a*) Less than maximum possible swing; (*b*) maximum symmetrical swing.

polarity opposite to that of $i_b(t)$. This is also brought out by applying Eq. (11.1-6)

$$
\begin{aligned}
v_{CE}(t) &= V_{CC} - i_c(t)R_c \\
&= V_{CC} - (I_{CQ} + i_c(t))R_c \\
&= 20 - (2 + 1 \sin \omega t)(5) \\
&= 20 - 10 - 5 \sin \omega t \\
&= 10 - 5 \sin \omega t \text{ V} \qquad (11.1\text{-}16)
\end{aligned}
$$

Note that the signal causes the operating point to move back and forth between points A and B on the load line.

Now, if we increase $v_i(t)$ so that $i_b(t)$ becomes 20 sin ωt μA we will cause the transistor operating point to traverse the entire load line, from one end to the other, as the base current goes through each cycle. This is called *maximum symmetrical swing*. The graphical construction is shown in Fig. 11.1-6*b* and again we have faithful reproduction of the base current waveform. In practice, there will be some distortion of the positive half-cycle because of the curvature of the actual characteristics in the region near the i_C axis.

In the previous two cases, there was no distortion of the waveform and in terms of the piecewise-linear model of Fig. 11.1-4*b* the base-emitter diode was a short circuit and the collector-emitter diode was an open circuit. If $v_i(t)$ is increased further (to produce 30 μA peak $i_b(t)$) we encounter both saturation and cutoff. Let us consider the various waveforms as the sinewave goes through its cycle. We will analyze this case in terms of the behavior of the piecewise-linear model of Fig. 11.1-4*b* and at the same time graphically in Fig. 11.1-7.

While $v_i(t)$ is going through the first half-cycle, the base-emitter diode is

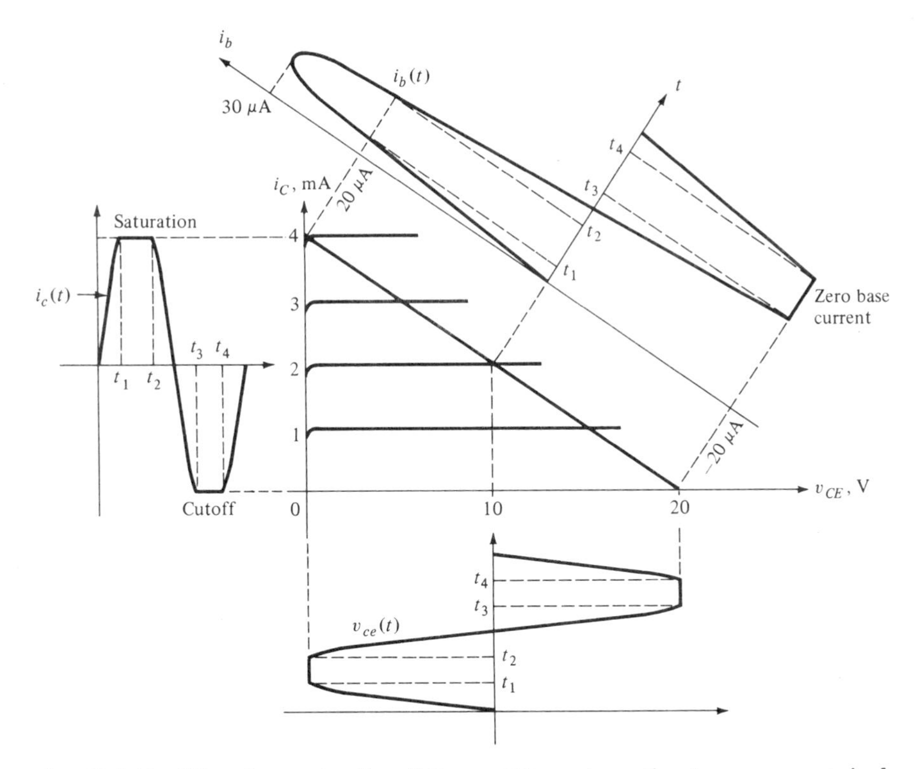

Fig. 11.1-7 Waveforms in the *CE* amplifier when the base current is large enough to produce significant distortion.

forward-biased so base current flows during the complete half-cycle. The collector current, when it reaches its saturation value of 4 mA at time t_1 will stay at 4 mA until time t_2. Thus part of the positive half of the sine wave is clipped off due to saturation as shown in the figure. During this time interval, the collector-emitter diode changes state and becomes a short circuit. The conditions which lead to this can be found by noting that the saturation current is

$$I_{C,\mathrm{sat}} = \frac{V_{CC}}{R_c} \tag{11.1-17}$$

The corresponding base current is

$$I_{B,\mathrm{sat}} = \frac{I_{C,\mathrm{sat}}}{\beta} = \frac{V_{CC}}{\beta R_c} \tag{11.1-18}$$

In saturation, we have

$$i_B \geq I_{B,\mathrm{sat}} \tag{11.1-19}$$

From the base-emitter loop in Fig. 11.1-4*b*, we have, using Ohm's law

$$i_B = \frac{v_i + V_{BB} - V_0}{R_b + r_f} \tag{11.1-20}$$

Using Eq. (11.1-20) in the inequality Eq. (11.1-19) we have

$$\frac{v_i + V_{BB} - V_0}{R_b + r_f} \geq \frac{V_{CC}}{\beta R_c}$$

Rearranging, the input voltage v_i which causes saturation is

$$v_i(t) \geq \frac{V_{CC}(R_b + r_f)}{\beta R_c} - V_{BB} + V_0 \tag{11.1-21}$$

The equals sign in this equation will hold at times t_1 and t_2.

While $v_i(t)$ is going through its negative half-cycle, the base current will tend to reverse when the total base current reaches zero at time t_3. However, at this point the base-emitter diode changes state and becomes an open circuit, the total base current i_B remaining at zero until time t_4. The condition for this to occur can be obtained from Eq. (11.1-20) by noting that we must have $i_B < 0$ for the diode to be an open circuit. Thus cutoff occurs when

$$\frac{v_i + V_{BB} - V_0}{R_b + r_f} \leq 0$$

This simplifies to the condition

$$v_i \leq V_0 - V_{BB} \tag{11.1-22}$$

The portion of the base current waveform between t_3 and t_4 will be clipped off because in this interval the inequality Eq. (11.1-22) is satisfied so that $i_B = 0$.

Since zero base current causes zero collector current, the collector is at *cutoff* and the part of its waveform between t_3 and t_4 is also clipped, as shown. The collector-emitter voltage waveform is also clipped correspondingly. Note that the time t_1 in all three waveform diagrams is the same as are t_2, t_3, and t_4.

It is important to recognize, in an example such as this, that we are considering an idealized situation and have approximated the actual transistor characteristics by a family of straight lines. In practice the waveform distortion in Fig. 11.1-7 would not be as sharply defined and would begin to be apparent at somewhat smaller values of $i_b(t)$ because of the curvature of the actual characteristics. The graphical construction when the actual device characteristics are used proceeds in exactly the same fashion as in Fig. 11.1-7. The base current waveform is plotted first, then appropriate time intervals are chosen, and the corresponding collector current and voltage at each time is determined by projecting from the base current at that time to the load line and then to the collector current and voltage waveforms. When this is done carefully we obtain a more accurate picture of the waveform distortion.

INPUT VOLTAGE REQUIREMENTS

In the previous discussion we used the base current as our input. It is of interest to find the amplitude of $v_i(t)$ required to produce a specific $i_b(t)$. To find $v_i(t)$ we write KVL around the base-emitter circuit:

$$-V_{BB} - v_i(t) + R_b i_B(t) + V_{BEQ} = 0 \tag{11.1-23}$$

We have assumed that $v_{be}(t)$, the time variation of $v_{BE}(t)$ is negligible so that $v_{BE}(t) \approx V_{BEQ} = 0.7$ V. The validity of this assumption will be checked later. The next step is to separate $i_B(t)$ into its dc and ac components, I_B and $i_b(t)$. Assuming linear operation $I_B = I_{BQ}$ resulting in

$$-V_{BB} - v_i(t) + R_b I_{BQ} + R_b i_b(t) + V_{BEQ} = 0 \tag{11.1-24}$$

This equation can be separated into two equivalent equations by applying superposition to the circuit. One equation relates the dc quantities and the other the ac quantities. They are

dc:

$$-V_{BB} + R_b I_{BQ} + V_{BEQ} = 0 \tag{11.1-25}$$

ac:

$$v_i(t) = R_b i_b(t) \tag{11.1-26}$$

A useful physical interpretation comes about if we draw a circuit for which each of these applies. The two circuits are shown in Fig. 11.1-8. Again we must emphasize that the ac circuit is valid *only* if $v_i(t) \gg v_{be}(t)$. Since, in the linear region, the total v_{BE} will only vary from about 0.6 to 0.8 V, we see that the *ac* component, $v_{be}(t)$, can be no more than about 100 mV peak. In our example,

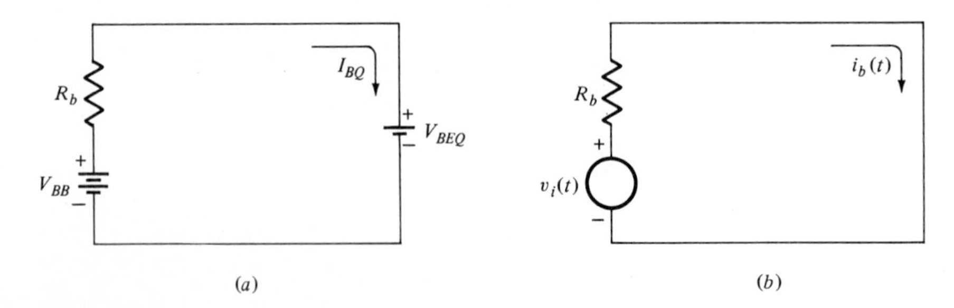

Fig. 11.1-8 Equivalent base-emitter circuits. (*a*) dc only; (*b*) ac only, valid for $v_i(t) \gg v_{be}(t)$.

the smallest base current was 10 μA peak and $R_b = 100$ kΩ. Then

$$v_i(t) = R_b i_b(t) = (10^5)(10)(10^{-6}) \sin \omega t = 1 \sin \omega t \text{ V}$$

Since this is 10 times the maximum possible peak $v_{be}(t)$ our initial assumption is justified and we can neglect the time variation of v_{BE}. However, for smaller signals, the assumption will not be justified and we must take the variation of v_{BE} into account. This is the "small signal" case which will be covered in Sec. 11.2.

In the example which follows, we consider a variation in the amplifier which we have been analyzing.

Example 11.1-1 In the circuit of Fig. 11.1-4*a*, R_b is changed to 150 kΩ. Find

a All quiescent currents and voltages.
b The maximum undistorted collector current swing.
c The base circuit conditions for the collector current found in part *b*.

Solution

a The quiescent base current is

$$I_{BQ} = \frac{V_{BB} - V_{BEQ}}{R_b} = \frac{2.7 - 0.7}{0.15} = 13.3\ \mu\text{A}.$$

The corresponding collector current and collector-emitter voltage are:

$$I_{CQ} = \beta I_{BQ} = (100)(13.3) = 1{,}330\ \mu\text{A} = 1.33 \text{ mA}$$

$$V_{CEQ} = V_{CC} - I_{CQ}R_c = 20 - (1.33)(5) = 13.35 \text{ V}$$

The load line and Q point are shown in Fig. 11.1-9.

b From the load line we see that the collector current can swing 1.33 mA down to cutoff and 2.67 mA up to saturation. Thus the maximum undistorted *symmetrical* collector current swing is 1.33 mA.

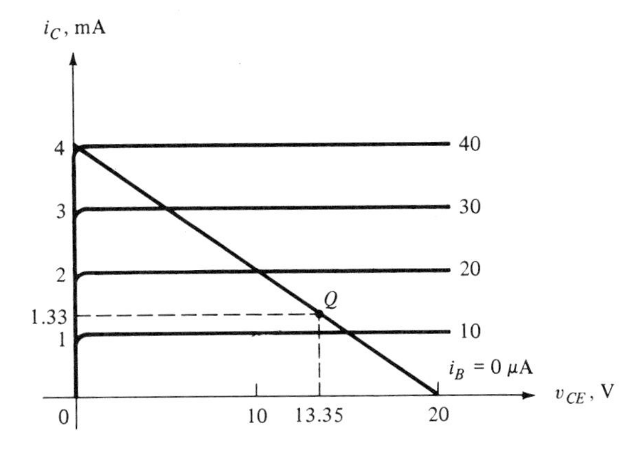

Fig. 11.1-9 Q point for Example 11.1-1.

c In order to have an ac collector current $i_c(t) = 1.33 \sin \omega t$ mA we must have a base current

$$i_b(t) = \frac{1.33}{100} \sin \omega t = 0.0133 \sin \omega t \ \text{mA}$$

The signal voltage required to produce this base current is, from Eq. (11.1-26)

$$v_i(t) = R_b i_b(t) = (150)(0.0133 \sin \omega t) = 2 \sin \omega t \ \text{V}$$

This is the maximum input signal that can be accommodated without distortion. As larger signals are applied, first cutoff and then saturation will be encountered. ////

11.2 SMALL-SIGNAL ANALYSIS: THE LINEAR INCREMENTAL MODEL

In many applications, transistor amplifiers are used to amplify very small signals, often well below a volt. Such signals exist, for example, at the input to most radio and TV receivers because the signal picked up by the antenna is extremely small. When the transistor is called upon to amplify such signals, the operating point for the signal only moves a very small amount on either side of the Q point. Under these conditions distortion is negligible and the amplifier is *linear* for all practical purposes. We then derive a *linear incremental model* which is based on the properties of the transistor measured at the Q point. As we shall see some of these properties actually depend on the Q point parameters.

Our first step in deriving the linear incremental model is to observe the large-signal piecewise-linear model which we derived in Sec. 11.1 and which is repeated for convenience in Fig. 11.2-1a. The base-emitter pn junction is modeled

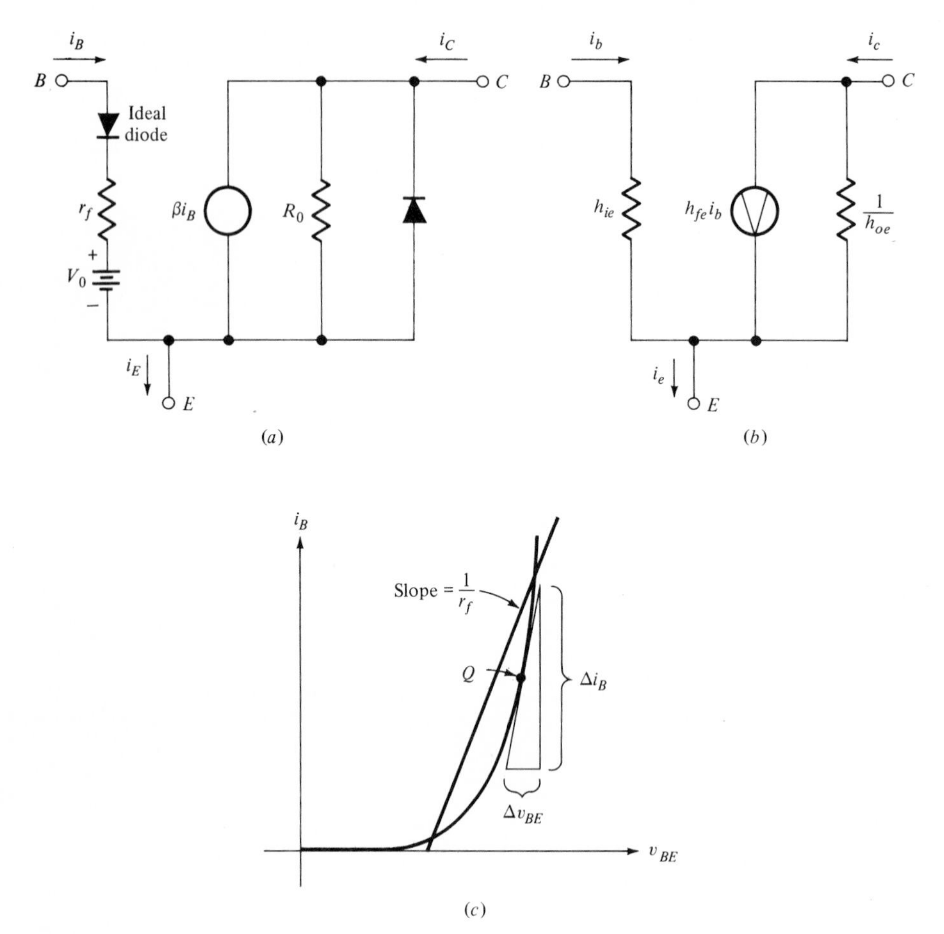

Fig. 11.2-1 Circuit models for the *npn* transistor. (*a*) Piecewise-linear large-signal model; (*b*) small-signal model; (*c*) input characteristic.

by an ideal diode in series with a resistor and a battery. When we were analyzing the response to large signals, we neglected the small forward resistance of the base-emitter junction and assumed that it was adequately modeled by the 0.7-V battery. But since we are concerned with amplification of very small signals, we must now take into account this small resistance because the ac voltage developed across it may be comparable to the small ac input signal. Thus for small signals the original model is replaced by the hybrid model (see Fig. 10.7-1) shown in Fig. 11.2-1*b*. In this model, we are interested *only* in *small* ac signals. Thus the battery and ideal diode in the base circuit of the piecewise-linear model do not appear since they are short circuits for ac. The collector circuit diode is an open circuit. The input resistance h_{ie}† is slightly

† The additional subscript *e* identifies the common-emitter connection.

different from the diode forward resistance r_f discussed in connection with piecewise-linear models in Sec. 9.5. In the present case, h_{ie} is the *incremental* resistance measured at the Q point. In terms of the base-emitter vi characteristic, it is the inverse slope of the curve evaluated at the Q point.

$$h_{ie} = \left.\frac{\Delta v_{BE}}{\Delta i_B}\right|_{Q\text{point}} \tag{11.2-1}$$

This is illustrated graphically in Fig. 11.2-1*c* along with the *average* inverse slope line $1/r_f$, which defines the resistance used in the piecewise-linear model. The student should note carefully the difference between r_f, which represents an *average* over the whole characteristic and h_{ie}, which represents the inverse slope at one point on the characteristic.

The hybrid parameter h_{re}, which represents the reverse transmission is zero. This is a consequence of the fact that one curve suffices to describe the complete input family, indicating that changes in v_{CE} have no effect on the input circuit.

In order to determine the output circuit parameters, h_{fe} and h_{oe} we use the output characteristic shown in Fig. 11.2-2*a*. Because the small-signal parameters are measured at the Q point, it is convenient to expand the curves in the vicinity of the Q point as shown by the dashed circle in Fig. 11.2-2*a*. The expanded curves which are straight lines for all practical purposes, with a new set of axes, shifted by the quiescent values, are shown in Fig. 11.2-2*b*. Since the Q point represents the zero ac signal point, the new axes are labeled to correspond to ac currents and voltages. We find the hybrid parameters by applying the definitions:

$$h_{fe} = \left.\frac{\Delta i_C}{\Delta i_B}\right|_{v_{CE}=V_{CEQ}} = \left.\frac{i_c}{i_b}\right|_{v_{ce}=0} = \frac{2\text{ mA}}{0.02\text{ mA}} = 100 \tag{11.2-2}$$

This can be evaluated from either set of curves in Fig. 11.2-2 and the graphical construction is shown on both.

For the output resistance, we have

$$\frac{1}{h_{oe}} = \left.\frac{\Delta v_{CE}}{\Delta i_c}\right|_{i_B=I_{BQ}} = \left.\frac{v_{ce}}{i_c}\right|_{i_b=0} \approx \frac{5\text{ V}}{0.25\text{ mA}} = 20\text{ k}\Omega \tag{11.2-3}$$

This represents the inverse slope of the curves at the Q point as shown in Fig. 11.2.-2*a* or the equivalent inverse slope of the linear characteristics in Fig. 11.2-2*b*. In each case the graphical construction is shown.

For silicon transistors, the input resistance depends on the quiescent current and has been found to be

$$h_{ie} \approx \frac{0.025\,\beta}{I_{EQ}} \tag{11.2-4}$$

Units are kilohms if I_{EQ} is in milliamperes. As noted before, $h_{re} \approx 0$.

In the collector circuit, we note that the short-circuit current gain is

$$h_{fe} \approx \beta \tag{11.2-5}$$

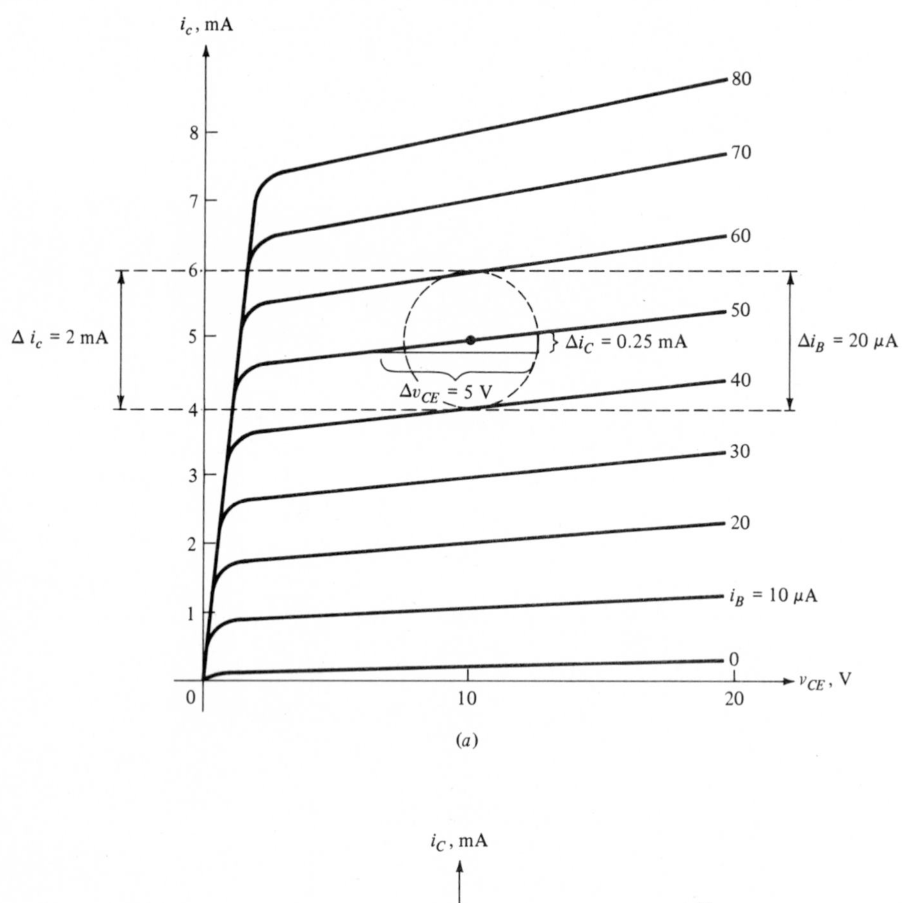

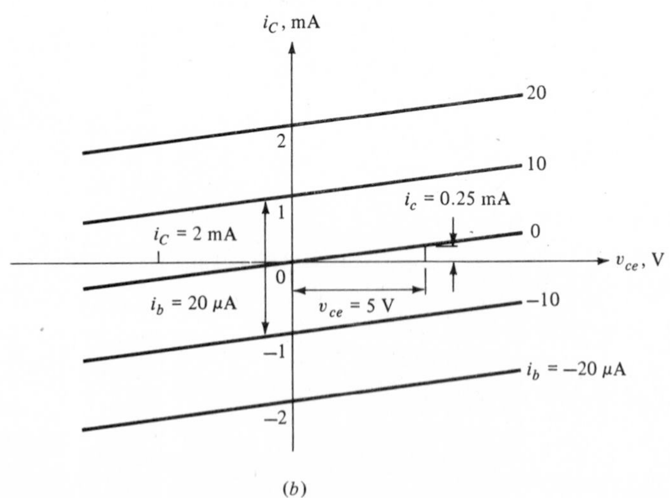

Fig. 11.2-2 Hybrid parameters from measured characteristics. (*a*) Complete characteristics; (*b*) expanded view near the *Q* point.

For most silicon transistors, the resistance $1/h_{oe}$ is typically 10 to 50 kΩ which is often large enough compared to the other resistances present to be considered an open circuit.

Typical amplifier calculations using the small-signal equivalent circuit are presented in the example which follows.

Example 11.2-1 In the amplifier of Fig. 11.2-3*a* the signal current is $2 \sin \omega t$ μA, $I_{EQ} = 2$ mA, and resistor R_c represents the load. Find

a The small-signal current gain $A_i = i_l/i_i$ and the ac output current i_l.
b The input resistance seen by the source.

Solution

a The small-signal equivalent circuit shown in Fig. 11.2-3*b* is obtained directly from the amplifier circuit by replacing the transistor by its

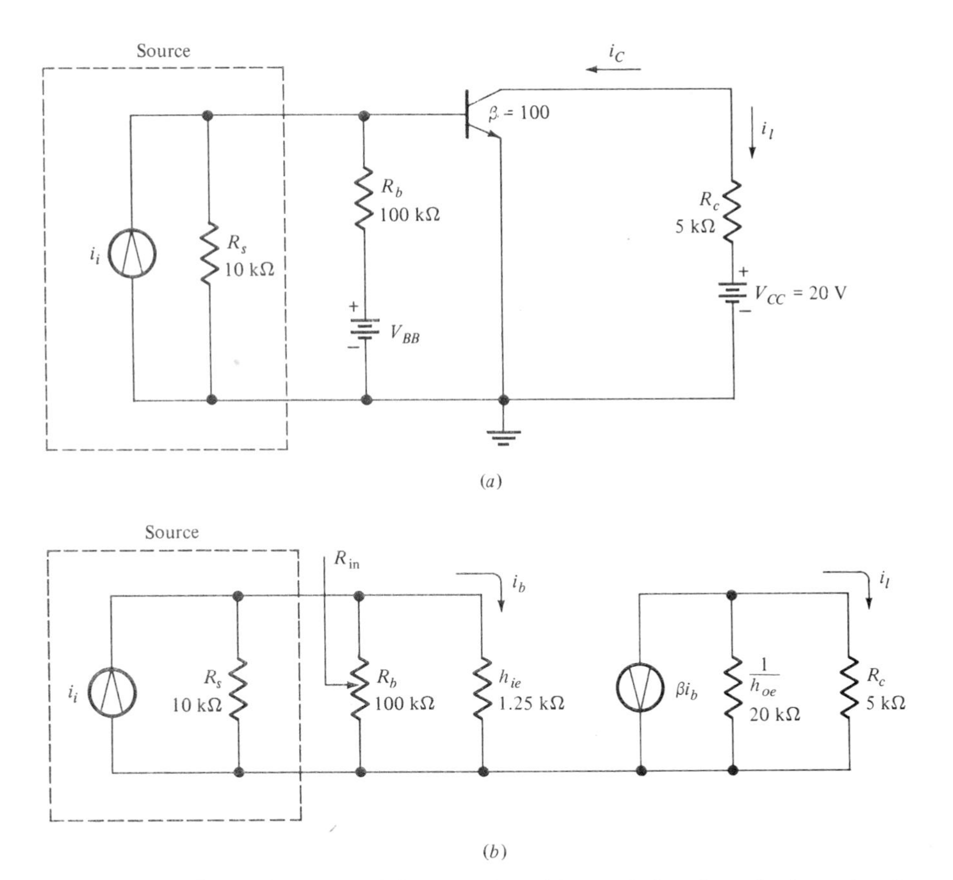

Fig. 11.2-3 Example 11.2-1. (*a*) Amplifier; (*b*) small-signal equivalent circuit.

small-signal hybrid equivalent and replacing all batteries by short circuits (why are batteries short circuits for ac?).

In order to find h_{ie}, we need to first find I_{EQ}. In this example it is given as $I_{EQ} \approx 2$ mA. Then

$$h_{ie} = \frac{\beta\,(0.025)}{I_{EQ}} = \frac{(100)\,(.025)}{(2)} = 1.25\ \text{k}\Omega$$

With the value for h_{ie} established, the circuit is complete and we can evaluate the required current gain. We use a "chain" calculation as in Sec. 10.1-1.

$$A_i = \frac{i_l}{i_i} = \left(\frac{i_l}{i_b}\right)\left(\frac{i_b}{i_i}\right) \tag{11.2-6}$$

From the output circuit, using the current divider formula:

$$\frac{i_l}{i_b} = -\beta\,\frac{1/h_{oe}}{R_c + 1/h_{oe}} = -100\left(\frac{20}{25}\right) = -80 \tag{11.2-7}$$

From the input circuit the ratio i_b/i_i is also found using the current divider formula:

$$\frac{i_b}{i_i} = \frac{R_b \,||\, R_s}{h_{ie} + (R_b \,||\, R_s)} = \frac{9.1}{1.25 + 9.1} = 0.88 \tag{11.2-8}$$

Thus the small-signal ac current gain A_i is

$$A_i = (-80)\,(0.88) = -70.4 \tag{11.2-9}$$

This means that the ac load current i_l is 70.4 times larger than the ac input current i_i,

$$i_l = (-70.4)\,(2 \sin \omega t) = -140.8 \sin \omega t \quad \mu\text{A}$$

In practice, this might not be sufficient load current and additional "stages" of amplification would have to be added. This will be considered in a later chapter. The negative sign indicates a phase reversal between input and output.

b The input resistance seen by the source is indicated as R_{in} on the circuit diagram. This is simply the parallel combination of R_b and h_{ie}.

$$R_{in} = R_b \,||\, h_{ie} = 100 \,||\, 1.25 = 1.23\ \text{k}\Omega$$ ////

SUMMARY

1 The piecewise-linear circuit model for an amplifying device is obtained from the linearized characteristic curves for the device. The model ele-

ments represent average behavior over the whole active range of the device.

2 The load line represents the *vi* characteristic of the linear circuit connected to the nonlinear amplifying device.

3 The quiescent (Q) point is that point on the load line at which the circuit variables appear when the signal is zero.

4 In most amplifying devices, saturation and cutoff characteristics limit the available signal excursion.

5 The small-signal circuit model (or linear incremental model) for an amplifying device is obtained from the linearized characteristic curves in the immediate vicinity of the Q point. The model elements are representative of the device behavior in that small region.

PROBLEMS

11.1-1 In a certain transistor $\beta = 150$ and $i_B = 20\ \mu\text{A}$. Find i_C and i_E.

11.1-2 In a power transistor $\beta = 20$ and $i_E = 150$ A. Find i_B and i_C.

11.1-3 The transistor amplifier shown in Fig. 11.1-4*a* has the following parameter values: $\beta = 50$, $R_b = 2\ \text{k}\Omega$, $R_c = 3\ \text{k}\Omega$, $V_{CC} = 15$ V.

a Sketch the linearized collector characteristics for the transistor.
b Plot the load line on the same graph as *a*.
c Select V_{BB} so that i_C undergoes maximum symmetrical swing.

Find the swing in i_C, i_B, and v_{CE} for this condition.

d Find the maximum input signal $v_i(t) = V_{im} \sin \omega t$ which will not produce a distorted output.

11.1-4 In the circuit of Fig. 11.1-4*a* $\beta = 100$, $R_b = 1\ \text{k}\Omega$, $V_{BB} = 1.4$ V, and $v_i(t) = V_{im} \sin \omega t$.

a If $V_{im} = 0.5$ V, sketch $i_B(t)$ and $i_C(t)$.
b If $V_{im} = 1.35$ V, sketch $i_B(t)$ and $i_C(t)$.
c Find the maximum V_{im} for undistorted i_C.

11.1-5 In the circuit of Prob. 11.1-4, $V_{CC} = 10$ V, $V_{im} = 1$ V, and $R_c = 100\ \Omega$. Sketch $i_C(t)$.

11.1-6 Fig. P11.1-6 shows a transistor connected in the common-base configuration. The emitter-base junction can be represented by a piecewise-linear equivalent circuit consisting of a 0.7-V battery in series with a 20-Ω resistor and an ideal diode. Find V_{EBQ} for $R_e = 1\ \text{k}\Omega$ and $10\ \text{k}\Omega$ if $V_{EE} = 6$ V.

11.1-7 In the circuit of Fig. P11.1-7 find the value of V_{BB} which will bring the transistor just to the edge of saturation.

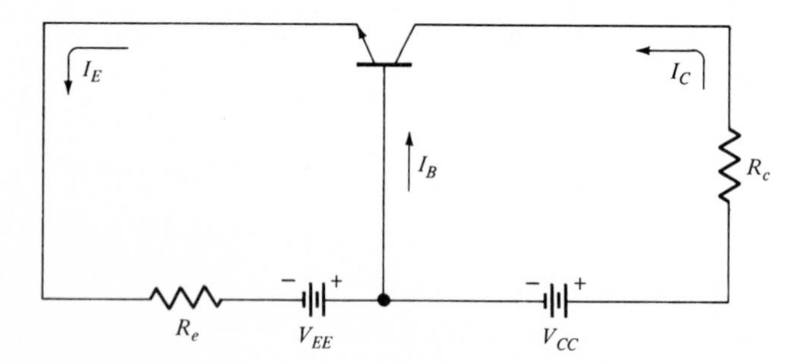

Fig. P11.1-6.

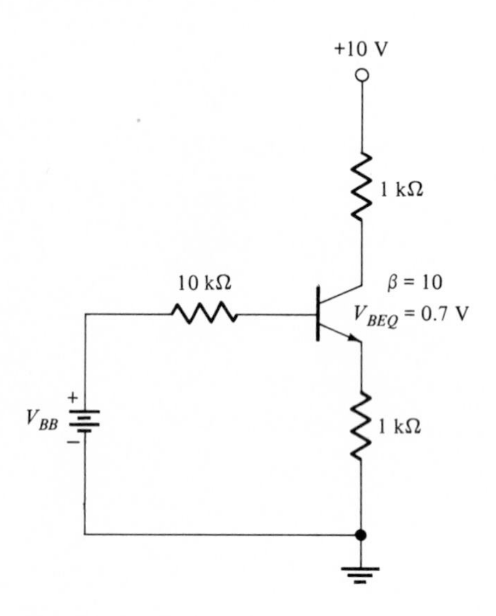

Fig. P11.1-7.

11.1-8 Find quiescent conditions in the circuit of Fig. P11.1-8 for $R_b = 1$ kΩ and 10 kΩ.

11.1-9 In the circuit of Fig. P11.1-8 $R_b = 1\,0$ kΩ. Find quiescent conditions for $\beta = 20$ and 100.

11.1-10 In the circuit of Fig. P11.1-10 find V_{BB} for maximum symmetrical collector current swing.

11.2-1 Fig. P11.2-1 shows a set of collector characteristics for a transistor. Estimate h_{fe} and $1/h_{oe}$ at $I_{CQ} = 10$ mA, $V_{CEQ} = 15$ V. Find h_{ie} at this operating point.

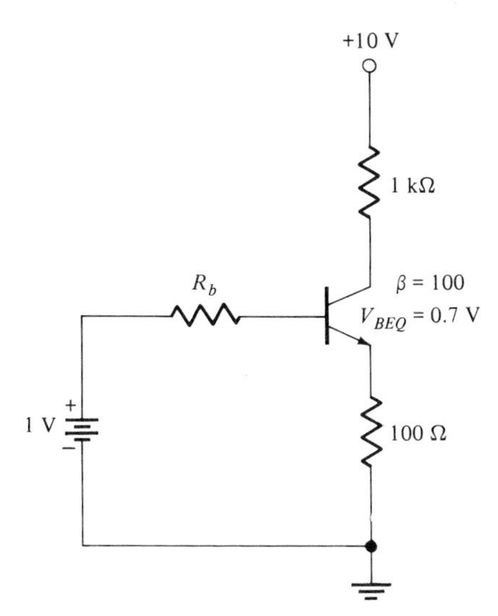

Fig. P11.1-8.

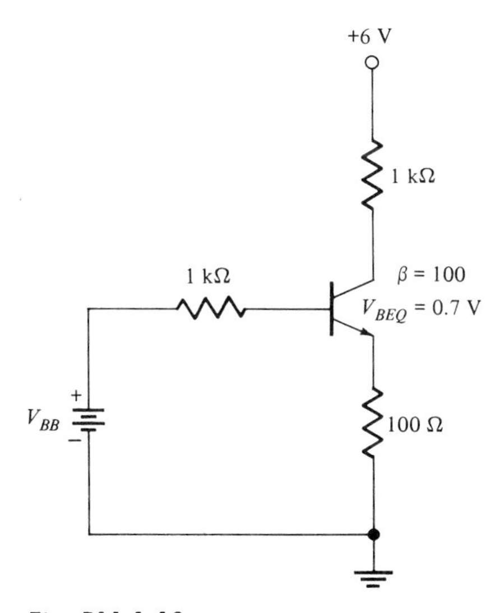

Fig. P11.1-10.

11.2-2 The transistor of Prob. 11.2-1 is used in the circuit of Fig. 11.2-3*a*. Find the current gain $A_i = i_l/i_i$ and the input resistance R_{in} seen by the current source.

11.2-3 In the circuit of Fig. 11.2-3*a* the parameters are changed to: R_s = 2.2 kΩ, R_b = 39 kΩ, V_{BB} = 4.7 V, β = 20, R_c = 1 kΩ, and V_{CC} = 6 V. Find the current gain $A_i = i_l/i_i$ and the input resistance R_{in} seen by the current source.

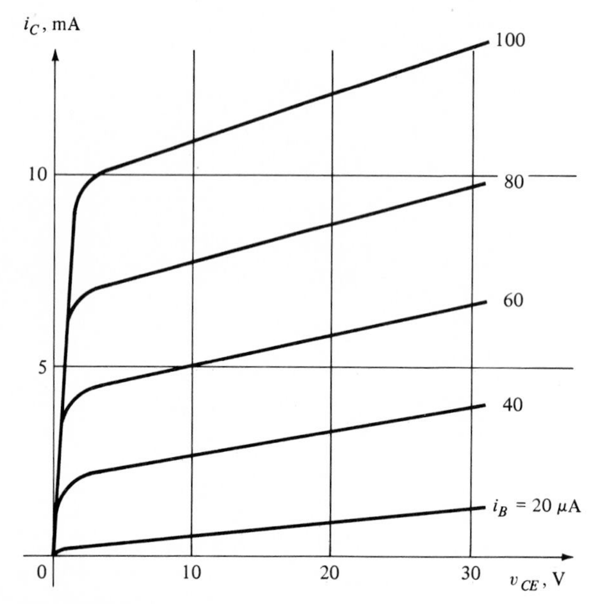

Fig. P11.2-1.

12
Switching Algebra

OBJECTIVES

Upon completion of this chapter, you should be able to

1 List and define the operations in the switching algebra.
2 Determine the switching function for a given switching circuit and draw the switching circuit for a given switching function.
3 Find the truth table for a given specification.
4 Find a switching function for a given truth table.
5 Minimize a switching function using the theorems of the switching algebra.

INTRODUCTION

The switching algebra is a symbolic logic which is very useful in the design of digital electronic circuits and switching circuits. (A switching circuit is an electric circuit composed of switches and/or relays, which are electrically controlled switches. These switches or relays control the flow of power to a load which may consist of lamps or telephones, etc. Such circuits have been widely used by the telephone industry.) The switching algebra was developed by Claude Shannon during the late 1930s and is based on a symbolic logic originally presented by George Boole in the 1850s.

A symbolic logic is a system of representing and manipulating the logical relationships between statements using precisely defined symbols. These statements which are called *propositions* and may be either true or false, are represented by variables (letters). The logical operations and relationships are represented by the words AND, OR, NOT, IMPLY, and EQUAL. The unique contribution of Boole's work, which is partially maintained in Shannon's adaptation, is the use of the operations of ordinary algebra in the symbolic logic. The interpretations are, of course, different.

12.1 BASIC CONCEPTS OF SWITCHING ALGEBRA

In this section, we will show how variables can be used to represent propositions; i.e., a statement which may be true or false. For example, we might define the variable S by the statement

S = it is snowing

when defined this way a variable may have one of two possible values: a value of zero means the proposition is false (it is *not* snowing) while a value of one means the proposition is true (it *is* snowing). At any moment, a variable must have one, and only one, of these two values.

12.1-1 Switching Circuit Concepts

It is possible to construct a switching circuit which corresponds to any given switching logic equation. Associated with each independent variable in the equation is a switch in the circuit; the dependent variable in the equation may be represented by a lamp. The independent variables are circuit inputs while the dependent variable is a circuit output.

A zero value corresponds to an open switch or an unlighted lamp while a one value corresponds to a closed switch or a lighted lamp.* These concepts are illustrated in Fig. 12.1-1.

* This is the so-called transmission concept. In the hindrance concept, which is not used in this text, a zero value corresponds to a closed switch or lit lamp while a one value corresponds to an open switch or unlighted lamp.

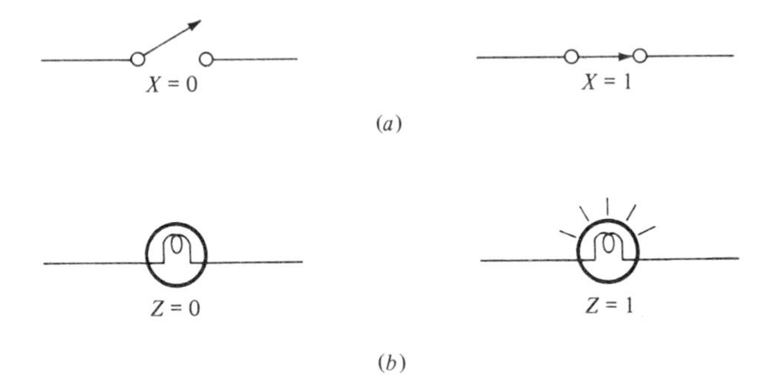

Fig. 12.1-1 Representation of values in switching circuit. (*a*) Independent variable—input; (*b*) dependent variable—output.

We shall frequently provide the switching circuit representation of expressions or equations in this chapter. This is done because the circuit diagram frequently provides physical insight lacking in the mathematical formulaton.

12.1-2 Switching Algebra Operations

In this section we will define and illustrate the basic operations used in the switching algebra. The first is called the AND operation.

AND

The AND operation is exemplified by the sentence "I will go swimming if it is hot AND I feel well." The word AND is represented by *multiplication.* The product of two variables is indicated by a dot between them or by simply writing them next to each other, that is,

$$Z = X \cdot Y = XY$$

The sentence in the example may be represented by the logic equation

$$S = HW \tag{12.1-1}$$

where S = I will go swimming.
H = It is hot.
W = I feel well.

As a consequence of the meaning of the word AND we conclude that it is true that I will go swimming ($S = 1$) if, and only if, it is true that it is hot ($H = 1$) AND it is true that I feel well ($W = 1$). This provides the definition for multiplication which is summarized in Table 12.1-1. This is called a truth table; it lists all combinations of values of the independent variables and the corresponding value of the dependent variable for each such combination. The

TABLE 12.1-1
Truth Table for
AND Operation

H	W	$S = HW$
0	0	0
0	1	0
1	0	0
1	1	1

last line of the table corresponds to $S = 1$ if $H = 1$ AND $W = 1$ in the statement above.

The second line of the table for example, interpreted in words, is; If it is NOT hot ($H = 0$), and I am well ($W = 1$), I will NOT go swimming ($S = 0$).

For the switching circuit we must find some way of connecting two switches and a lamp and power supply so that the lamp will light when, and only when, both switches are closed, i.e., equal to one. Clearly, this will be the case if the two switches are connected in series as shown in Fig. 12.1-2*a*.

OR

The OR operation is exemplified by the sentence

I will stay home if it snows OR if it is cold.

The word OR is represented by the *addition* operation. Thus the sentence in this example is represented by the logic equation

$$H = S + C \qquad (12.1\text{-}2)$$

where H = I will stay home.
S = It snows.
C = It is cold.

As a consequence of the meaning of the word OR† we determine from the example sentence that it is true that I will stay home ($H = 1$) if either it snows ($S = 1$) OR it is cold ($C = 1$) OR both are true. Therefore, the truth table for the OR operation is as shown in Table 12.1-2.

It is worthwhile to note that the definition of logical multiplication AND agrees with the definition for arithmetic multiplication while the definition for logical addition (OR) differs from that of arithmetic addition only when both variables equal one; the difference is due to the logical, rather than numerical, significance of the values and operations.

† OR refers to the *inclusive* OR, frequently written as X and/or Y rather than the *exclusive* OR which is often written as either X or Y but not both. The algebraic symbol for exclusive OR is $\oplus$.

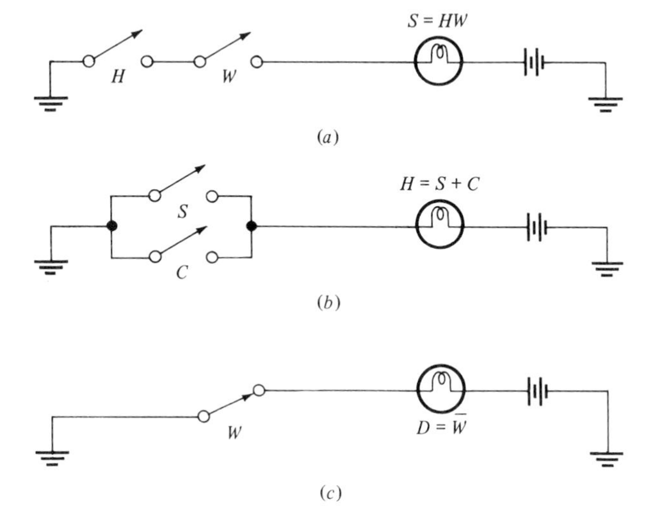

Fig. 12.1-2 Switching circuits for basic operations. (*a*) AND—series circuit; (*b*) OR—parallel circuit; (*c*) NOT—normally closed contact.

A little thought concerning an appropriate switching circuit for the OR operation leads to the parallel connection of the switches as shown in Fig. 12.1-2*b*.

NOT

The NOT operation is contained in the sentence "I shall go to the doctor if I am NOT well." The word NOT is represented by the overbar.* Thus the sentence given above is represented by

$$D = \bar{W} \tag{12.1-3}$$

where D = I shall go to the doctor, and W = I am well.

TABLE 12.1-2
Truth Table for
OR Operation

S	C	$H = S + C$
0	0	0
0	1	1
1	0	1
1	1	1

* Some texts use a prime instead of an overbar.

TABLE 12.1-3
Truth Table for
NOT Operation

W	$D = \bar{W}$
0	1
1	0

The word NOT implies a reversal of the truth or falsity of a proposition. Therefore, we have the truth table given in Table 12.1-3.

In looking for a circuit representation of the NOT operation we must differentiate between a switch's operation and its electrical state. The state of operation of a switch refers to its mechanical position; in our diagrams *up* is the unoperated state while *down* is the operated state. The electrical state of a switch refers to its electrical continuity; zero for an open circuit and one for a short circuit. Until now we have used only switches which are classified as normally open switches; their electrical state is open when they are unoperated and closed when they are operated. To obtain the NOT operation we must use a normally closed type of switch; its electrical state is opposite that of its mechanical state, i.e., the switch is a short circuit when it is unoperated and an open circuit when it is operated. The NOT circuit is shown in Fig. 12.1-2*c*.

COMBINATIONS OF LOGIC OPERATIONS

It is possible to combine a number of operations in a single expression. For example, the sentence

I will go swimming if it is hot and it is not raining, or if the pool is indoors.

is represented by

$$S = H\bar{R} + P \tag{12.1-4}$$

where S = I will go swimming.
H = it is hot.
R = it is raining.
P = the pool is indoors.

Combining the various circuit connections that we have discussed results in the circuit of Fig. 12.1-3. The reader should try various combinations of H, R, and P in order to convince himself that the logic is correct in Eq. (12.1-4) and that the switching circuit gives the expected values of S.

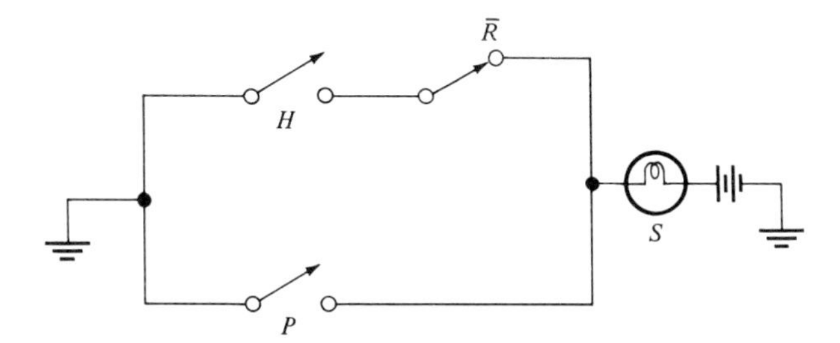

Fig. 12.1-3 Switching circuit for $S = H\bar{R} + P$.

12.2 SWITCHING CIRCUITS

We have discussed the fundamental concepts of the switching algebra and the relationship between the switching algebra and switching circuits. In the design of switching circuits two kinds of problems arise. Sometimes it is necessary to determine the algebraic description of a given circuit, which is an analysis problem, and sometimes it is necessary to find a switching circuit which is described by a given algebraic expression; this is a synthesis problem.

SWITCHING CIRCUIT ANALYSIS

The analysis of a switching circuit consists of the finding of an algebraic equation which describes the network. This is easily done using the following rules:

1 A normally open contact represents the specified variable. A normally closed contact represents the inverse (negation) of the variable which is indicated by an overbar.
2 A series connection corresponds to the operation of multiplication (AND).
3 A parallel connection corresponds to the operation of addition (OR).

These rules are simple extensions of the principles discussed in Sec. 12.1. The extension is the application of the operations of addition and multiplication to expressions, representing circuit segments, instead of simple variables, representing single switches. This extension is valid since the value of an expression must be 0 or 1 representing an open or closed circuit respectively. In fact, an entire expression may be equated to a single variable which may then be used in place of the circuit. Physically this corresponds to replacing a circuit segment by a single switch which is open whenever the circuit segment is an open circuit and shut whenever the circuit segment is a short circuit.

The following example illustrates the method for finding the algebraic description of a given switching circuit.

Example 12.2-1 Find an algebraic expression which describes each of the circuits of Fig. 12.2-1.

Solution Fig. 12.2-1*a* has two switches, X and Y, in series. Both switches are normally open resulting in the expression (check rules 1 and 2):

$$L_a = XY \tag{12.2-1}$$

Figure 12.2-1*b* is similar to Fig. 12.2-1*a*; the only difference is that the switch Y uses a normally closed contact rather than a normally open contact. Therefore, the algebraic expression is the same except that Y is replaced by $\bar{Y}$ (see rule 1). Therefore

$$L_b = X\bar{Y} \tag{12.2-2}$$

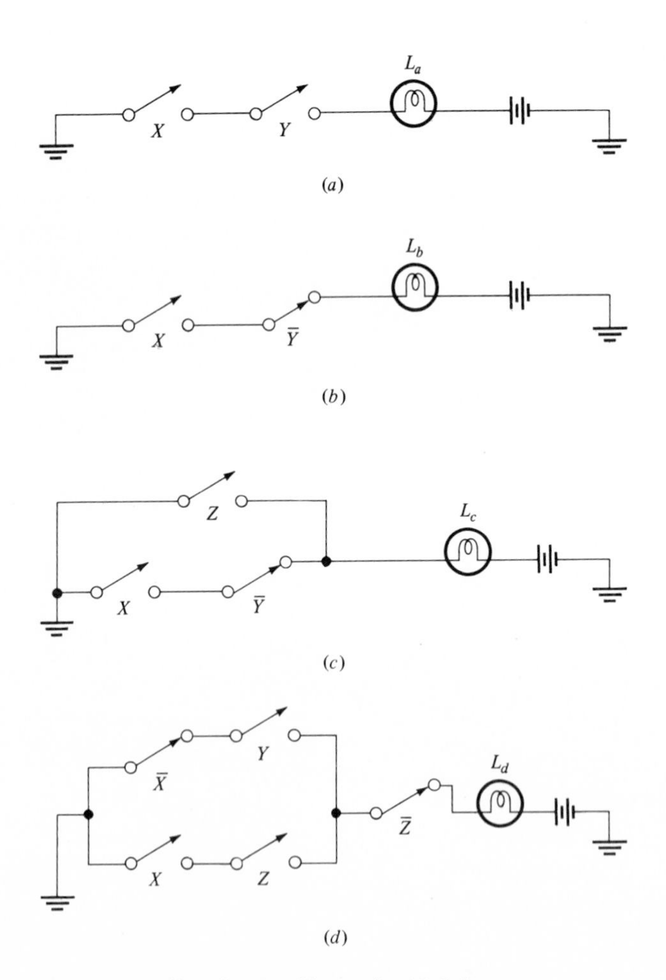

Fig. 12.2-1 Circuits for Example 12.2-1.

The circuit of Fig. 12.2-1*c* is like that of Fig. 12.2-1*b* but a normally open switch Z is in parallel with the X and Y switches. Therefore, using rule 3:

$$L_c = X\bar{Y} + Z \tag{12.2-3}$$

Figure 12.2-1*d* illustrates a more complex circuit. The analysis is, nevertheless, simple; the rules are applied to portions of the circuit and the expressions describing these circuit segments are combined according to the way in which the circuit segments are connected. This method results in

$$L_d = (\bar{X}Y + XZ)\bar{Z} \tag{12.2-4}$$

The first product $\bar{X}Y$ represents the upper left series connection; the product XZ represents the lower left series connection. These two products are added together since the two circuit segments are connected in parallel. Finally this sum is multiplied by $\bar{Z}$ because the left circuit segment is connected in series with a normally closed contact on the switch Z. ////

The preceding example led to expressions involving both addition and multiplication. In a complex expression the order in which operations are performed is indicated by parenthesis; expressions within the innermost set of parenthesis are evaluated first, the second innermost next, and so on to the outermost set of parenthesis. Within a set of parenthesis multiplications are done before additions, that is

$$AB + C = (AB) + C \tag{12.2-5}$$

The sequence of multiple additions or multiplications does not matter since the associative laws for addition and multiplication are both valid:

$$A + B + C = (A + B) + C = A + (B + C) \tag{12.2-6}$$

$$ABC = (AB)C = A(BC) \tag{12.2-7}$$

Thus, the rules concerning evaluation of expressions for the switching algebra are identical with those of ordinary algebra with the exception $1 + 1 = 1$ noted previously.

SWITCHING CIRCUIT SYNTHESIS

The synthesis of a switching circuit is the process of finding a circuit to satisfy a given specification. When the specification is given as an algebraic expression the problem is quite simple; it requires the application of the three rules given for analysis in the reverse direction. This is illustrated in the following example.

Example 12.2-2 Find switching circuits for each of the following algebraic expressions:

$$R_a = \bar{W} + X$$

$$R_b = W\bar{X} + YZ$$

$$R_c = VXY + W\bar{Y}(\bar{X} + \bar{Z})(X + \bar{W}Z)$$

Solution The circuit for R_a has a parallel connection of two switches by rule 3. By rule 1 the W switch is normally closed while the X switch is normally open. The circuit is shown in Fig. 12.2-2*a*.

The expression for R_b is more complicated. The complete circuit is shown in Fig. 12.2-2*b*; it is obtained by considering each product term separately and then the fact that they are added together. The circuit for the first product term, by rules 1 and 2 is shown in the dotted box 1; similarly the circuit segment for the second product term is shown in box 2. Since

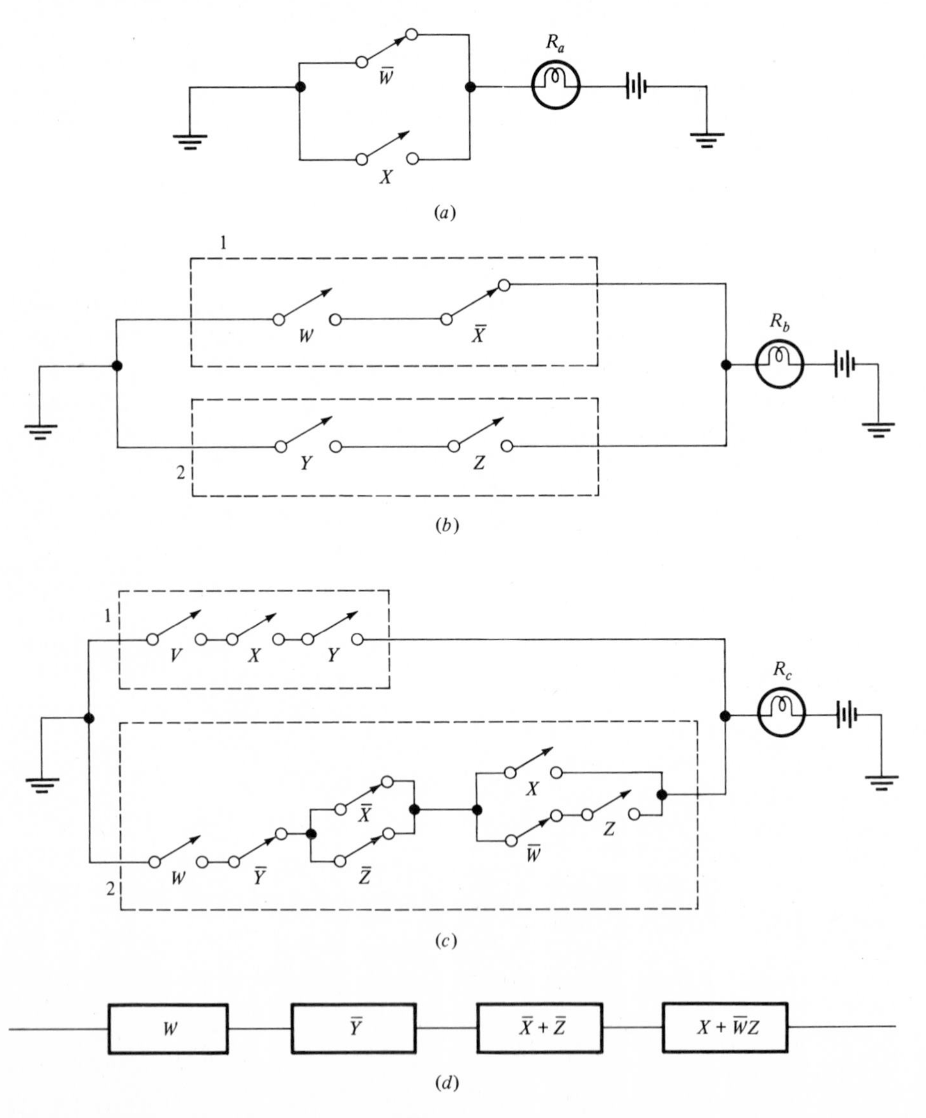

Fig. 12.2-2 Circuits for Example 12.2-2.

the two products are added together, the circuit segments in boxes 1 and 2 are connected in parallel by rule 3.

The circuit for R_c, shown in Fig. 12.2-2*c*, is obtained using the same method. Box 1 contains the circuit for the product VXY while box 2 contains the circuit for the expression $W\bar{Y}(\bar{X} + \bar{Z})(X + \bar{W}Z)$. Fig. 12.2-2*d* demonstrates the intermediate step in finding the circuit for this expression; it indicates that the circuit consists of four subcircuits connected in series. Each of the subcircuits implements one of the four factors in the expression. Each of these subexpressions is easily implemented as shown in Fig. 12.2-2*c*.

////

In a number of circuits in the preceding example we had multiple occurrences of switch contact pairs, some normally open and some normally closed, in a single circuit. The implication of such a situation is that all the contact pairs associated with a given variable are mechanically connected together so that all of the normally open contacts close and all of the normally closed contacts open simultaneously with the operation of the switch (to the "on" state) and vice versa when the switch is opened, i.e., placed in the "off" position. Such a switch is called a multipole switch. A switch which has only normally open or only normally closed contacts is called a single-throw switch; one with both normally open and normally closed contacts is called a double-throw switch. Some very common switch types, available in a large range of styles, sizes, and ratings, are the single-pole single-throw (SPST), single-pole double-throw (SPDT), double-pole single-throw (DPST), and double-pole double-throw (DPDT). There are

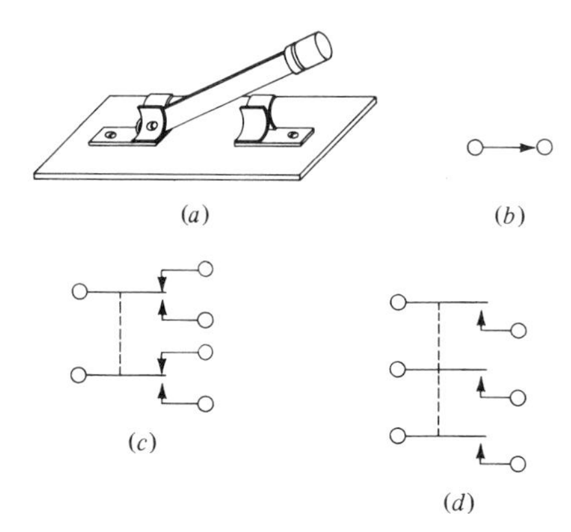

Fig. 12.2-3 Switch configurations. (*a*) SPST knife switch—pictorial representation; (*b*) SPST knife switch—schematic representation; (*c*) DPDT switch—schematic representation; (*d*) 3PST switch—schematic representation.

also a relatively large number of variations of the 3PST and 3PDT types. Switches with larger numbers of poles are also available but in a smaller variety of styles. Figure 12.2-3 shows a variety of switch types.

12.3 TRUTH TABLES

We have seen the correspondence between the switching circuit and the algebraic equation which describes that circuit. The algebraic equation enables us to determine the response of the circuit for given values of the input variables which corresponds to given states of the switches. The truth table is a tabular representation of the circuits behavior which can be derived from the algebraic equation. For a specific input combination it is necessary to substitute the values of the variables into the equation and to then evaluate it in order to determine the circuit response. On the other hand, the truth table lists the output value for every possible input combination. Therefore, it is only necessary to locate the given input combination and to then read off the corresponding output. We have already used such truth tables to define the AND ($\cdot$), OR ($+$), and NOT ($^-$) operations. The algebraic equation serves as a link between the truth table and the circuit. In the analysis problem we are given a circuit from which we find an algebraic equation which is then used to find the truth table. In the synthesis problem we are given a truth table from which we find an algebraic equation which is then used to find a circuit.

TRUTH TABLE STRUCTURE

The truth table has one column for each input variable and a column for the output variable (or one column for each output variable if there is more than one). There is a row for every possible combination of input values. If there are n input variables, then there are 2^n rows since each input variable can have one of the two possible values zero and one. For example, if there are three input variables, there are $2^3 = 8$ possible combinations: 000, 001, 010, 011, 100, 101, 110, 111. These values may be arranged in any order so long as each of them is included and there are no duplications. The order given above is a convenient one to use since it follows a logical pattern making it simple to check that there are no omissions or duplications. The numbers follow the sequence that would be indicated on a counter which used only the digits 0 and 1, that is, 000, 001, 010, 011, 100, 101, 110, 111.

ANALYSIS

In the analysis of a circuit we obtain the switching algebra equation for the circuit as described in Sec. 12.2. We then set up the truth table, filling in all the input and output variable column headings and all combinations of input variable values. We then substitute each combination of input values into the equation and evaluate it in order to determine the corresponding output value which is entered into the truth table. This process is illustrated in the following examples.

Example 12.3-1 A circuit is described by the logic equation

$$L = (A + \bar{B})C \tag{12.3-1}$$

construct a truth table which describes the circuit.

Solution The truth table has three input variables, A, B, and C, and a single output variable, L. Therefore it has four columns and $2^3 = 8$ rows. All combinations of input values are filled in for the variables A, B, and C. For each combination the value of L is computed using Eq. (12.3-1). For the first row, for example, $A = B = C = 0$, therefore

$$L = (0 + \bar{0})0 = (0 + 1)0 = (1)0 = 0$$

The resulting table is shown below.

A	B	C	L
0	0	0	0
0	0	1	1
0	1	0	0
0	1	1	0
1	0	0	0
1	0	1	1
1	1	0	0
1	1	1	1

////

Example 12.3-2 The circuit of Fig. 12.2-1d is described by Eq. (12.2-4) which is repeated below for convenience.

$$L_d = (\bar{X}Y + XZ)\bar{Z} \tag{12.2-4}$$

Construct the truth table which describes the circuit.

Solution There are three input variables, X, Y, and Z, and one output variable, L_d. Therefore there are four columns and $2^3 = 8$ rows in the table as illustrated in Table 12.3-1. We can save time and work by noting that $L_d = 0$ whenever $\bar{Z} = 0$, corresponding to $Z = 1$. (Since $\bar{X}Y + XZ$ must be either 0 or 1; in either case $L_d = 0$ if $\bar{Z} = 0$.) Therefore we need evaluate Eq. (12.2-4) separately only for the input combinations when $Z = 0$ in which case the equation reduces to:

$$L_d = (\bar{X}Y + X \cdot 0)1 = \bar{X}Y \tag{12.3-2}$$

since $X \cdot 0 = 0$ whether $X = 0$ or $X = 1$. Finally, we recognize that $\bar{X}Y = 1$

TABLE 12.3-1
Truth Table for
$L_d = (\bar{X}Y + XZ)\bar{Z}$

X	Y	Z	L_d
0	0	0	0
0	0	1	0
0	1	0	1
0	1	1	0
1	0	0	0
1	0	1	0
1	1	0	0
1	1	1	0

only when $\bar{X} = 1$ and $Y = 1$ which is the same as $X = 0$ and $Y = 1$. This permits us to complete the truth table.

This example illustrates the short cuts that are useful in evaluating the function. Rather than substitute each combination of input values into the equation and then evaluate it, we have noted certain simplifications which reduce the amount of work. There are no general procedures which can always be used toward this end but a few minutes spent in considering such possibilities will often save work in completing the truth table. ////

SYNTHESIS

The synthesis problem is one in which we are given a truth table and must find a corresponding circuit. This is done by obtaining an algebraic equation from the truth table from which we find a circuit using the methods of Sec. 12.2. The procedure for obtaining an algebraic equation from a truth table is best illustrated by an example.

Example 12.3-3 Find a logic equation corresponding to the truth table given in Table 12.3-2.

Solution Consider those rows of the table for which $L = 1$. The first row of the truth table implies that $L = 1$ if $X = 0$ AND $Y = 0$ AND $Z = 0$ which corresponds to $\bar{X} = 1$ AND $\bar{Y} = 1$ AND $\bar{Z} = 1$. This may be expressed by

$$L = \bar{X} \cdot \bar{Y} \cdot \bar{Z} \tag{12.3-3}$$

since multiplication represents AND and $1 \cdot 1 = 1$.

The fourth row indicates that $L = 1$ when $X = 0$ AND $Y = 1$ AND $Z = 1$ corresponding to $\bar{X} = 1$ AND $Y = 1$ AND $Z = 1$. This results in

$$L = \bar{X} \cdot Y \cdot Z \tag{12.3-4}$$

TABLE 12.3-2
Truth Table for
Example 12.3-3

X	Y	Z	L
0	0	0	1
0	0	1	0
0	1	0	0
0	1	1	1
1	0	0	1
1	0	1	0
1	1	0	0
1	1	1	1

In a similar manner the fifth and eighth rows result in

$$L = X \cdot \bar{Y} \cdot \bar{Z} \tag{12.3-5}$$

and

$$L = X \cdot Y \cdot Z \tag{12.3-6}$$

The complete truth table indicates that $L = 1$ if the inputs are those of the first row OR the fourth row OR the fifth row OR the eighth row. Thus, since addition represents OR, we have as the final expression for the truth table:

$$L = \bar{X}\bar{Y}\bar{Z} + \bar{X}YZ + X\bar{Y}\bar{Z} + XYZ \tag{12.3-7}$$

where the dots representing multiplication have been dropped although the meaning is the same. ////

From the preceding example we see that the *switching function*, which is the right-hand side of the algebraic equation describing a circuit or truth table, is obtained from the truth table as a *sum of products*, one product for each row for which the output variable is one. Each product is called a *complete product* because every input variable appears in it, either with a bar if its value is zero or without a bar if its value is one. This form of the switching function, the sum of complete products, is called a *canonical form* because it is a standard form into which any other form may be manipulated.

12.4 SWITCHING CIRCUIT SPECIFICATIONS

We have now considered the various steps in the analysis problem. Given a circuit we have determined its switching function, and from its switching function we have obtained its truth table. Since the purpose of analysis is to learn how something behaves we can accept the truth table as the final step in the analysis because it lists the output value for every possible combination of input values.

The synthesis problem is the inverse of the analysis problem. In the synthesis problem we are given a description of how a circuit should behave and we must design such a circuit. The description of the desired behavior is called the *specification.*

We have considered the solution to the synthesis problem if the specification is given as a switching function or a truth table. Generally however, the specification is given as a verbal description of how the circuit should function. It is then necessary for the designer to translate the verbal specification into a truth table from which he determines a switching function and then a circuit. He translates the verbal specification into a truth table by considering what the circuit output should be for each combination of input values. If he finds that the verbal specification does not provide sufficient information to permit him to do this, then he recognizes that the circuit is incompletely specified and he must get additional information from the person who provided the specification. Furthermore, he is aware of precisely what information is missing. The following example illustrates the entire synthesis procedure—from initial verbal specification to the circuit design.

Example 12.4-1 A company is owned by three partners A, B, and C. Partner A owns 45 percent of the business, partner B owns 30 percent, and partner C owns the remaining 25 percent. Decisions are made by voting with a total vote corresponding to 60 percent of the ownership required to pass a motion. Design a circuit with a switch for each partner and a lamp which will light if at least 60 percent of the ownership votes YES by putting their switches to the ON position.

Solution The three switches are denoted A, B, and C according to the partner who uses the switch. The lamp is denoted L and is 1 if the sum of the partners' percentages voting YES equals or exceeds 60 percent. A switch variable with a value of 1 indicates a YES vote, a 0 value indicates a NO vote. The truth table, annotated to show the percentage voting YES and the percentage owned by each partner is shown in Table 12.4-1.

From the last column we see that only three possible combinations will pass a measure. This could have been determined in advance from the specification without the truth table. However, the truth table organizes the procedure and allows us to handle much more complicated problems.

The algebraic equation obtained from the truth table is

$$L = A\bar{B}C + AB\bar{C} + ABC \tag{12.4-1}$$

The circuit is shown in Fig. 12.4-1. ////

TABLE 12.4-1 Truth Table for Example 12.4-1

45%	30%	25%	60%	
A	B	C	L	Percentage voting yes
0	0	0	0	0
0	0	1	0	25
0	1	0	0	30
0	1	1	0	55
1	0	0	0	45
1	0	1	1	70
1	1	0	1	75
1	1	1	1	100

12.5 THEOREMS

We are now able to convert a verbal specification for a switching circuit into a truth table, the truth table into a canonical switching function, and the switching function into a circuit. It is frequently possible however, to reduce the canonical switching function to a simpler form which results in a more economical circuit, i.e., one with fewer switch contacts, which is logically equivalent to the circuit corresponding to the canonical form. The circuit transformations corresponding to the reductions of the switching function could be made directly in the circuit diagram but with greater difficulty and increased likelihood of missing certain simplifications.

In order to reduce a switching function we must replace certain expressions by simpler equivalent expressions. In order to do this effectively we must develop a body of such substitutions, called theorems, based on the fundamental rules of the switching algebra and we must develop a degree of skill and facility in making these transformations through practice.

AXIOMS

The axioms are formalized statements of the fundamental rules already given in Sec. 12.1. The first axiom indicates the possible values of a switching

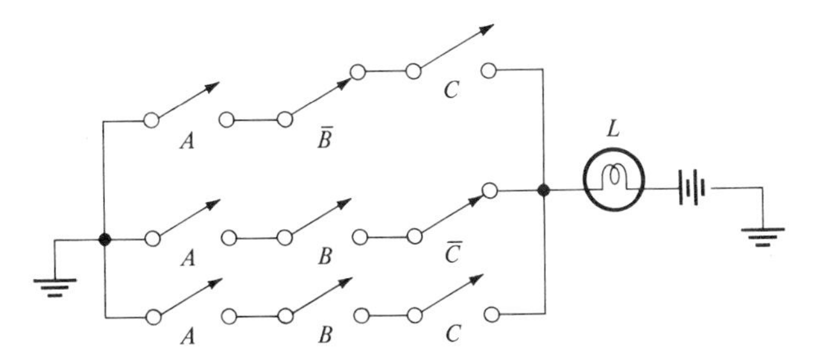

Fig. 12.4-1 Circuit for Example 12.4-1.

variable; the remaining ones, listed as dual pairs, define the operations. The dual of an equation is obtained by interchanging + and · and interchanging 0 and 1.

The concept of duality is useful because the dual of any valid theorem is also valid because each axiom used in proving the original theorem can be replaced by its dual thus proving the dual theorem. The axioms are

Either	$x = 0$	or	$x = 1$	(12.5-1)
$0 + 0 = 0$	(12.5-2*a*)		$1 \cdot 1 = 1$	(12.5-2*b*)
$1 + 1 = 1$	(12.5-3*a*)		$0 \cdot 0 = 0$	(12.5-3*b*)
$0 + 1 = 1 + 0 = 1$	(12.5-4*a*)		$1 \cdot 0 = 0 \cdot 1 = 0$	(12.5-4*b*)
$\bar{0} = 1$	(12.5-5*a*)		$\bar{1} = 0$	(12.5-5*b*)

THEOREMS

The theorems are all derived from the axioms, and therefore, do not contain any new information. Nevertheless, they are useful in simplifying switching functions.

There are two principle methods of proving the equivalence of two expressions in the switching algebra. The first is the *method of perfect induction* and the second is by the use of previously proved theorems or axioms. In general the second method is shorter for complex expressions while the method of perfect induction is useful in proving many of the elementary theorems. In this method each of the expressions is evaluated for every possible combination of values of the variables. If the values of the two expressions are equal to each other for every combination of variable values, they are equivalent.

The theorems are presented in dual pairs, as were the axioms. Since the dual of every axiom is true, it follows that the dual of every theorem is true. The following list includes a number, but by no means all, of the useful theorems for simplification of switching functions.

Some of the theorems are followed by proofs and in some cases the switching circuit equivalence is discussed. Some of the proofs are left as exercises for the reader.

The first theorem and its dual are

$$0 + x = x \quad (12.5\text{-}6a) \qquad 1 \cdot x = x \quad (12.5\text{-}6b)$$

The proof of Theorem (12.5-6*a*) is an application of the method of perfect induction. There are only two possible values of x by Theorem (12.5-1). Therefore it is only necessary to show that the two sides of Theorem (12.5-6*a*) are equal for both values of x. For $x = 0$ the left-side is

$$0 + x = 0 + 0 = 0 \qquad \text{by (12.5-2}a\text{)}$$

while the right-side is $x = 0$. For $x = 1$ the left-side is

$$0 + x = 0 + 1 = 1 \qquad \text{by (12.5-4}a\text{)}$$

while the right-side is $x = 1$. Therefore the two sides are equal for both values of x and the theorem is true. The dual theorem may be proved in a similar way. Alternatively, the dual theorem follows from this one and the fact that there is a dual axiom for each axiom used in the proof of this theorem. This alternative approach is better because it means that only one of each of the pairs of dual theorems to be presented need be proved.

The second theorem and its dual are:

$$1 + x = 1 \qquad (12.5\text{-}7a) \qquad 0 \cdot x = 0 \qquad (12.5\text{-}7b)$$

This theorem can be proved by perfect induction using the axioms as was Theorem (12.5-6a). For this theorem an alternative approach will be taken; the validity of the theorem will be demonstrated by showing the equivalence of the two circuits represented by the left and right side of Theorem (12.5-7a). The left side represents a parallel connection of a short circuit (represented by 1) and a contact x as shown in Fig. 12.5-1a. The right side represents a short circuit as shown in Fig. 12.5-1b. From the basic principles of electricity it is clear that the conduction between the two terminals of each circuit is the same whether the switch is open ($x = 0$) or closed ($x = 1$). Therefore the theorem is true. This type of proof is valid even though it is not mathematical in nature because of the one-to-one correspondence between the algebra and the circuit.

The third theorem represents the operation of taking an inverse of a variable. The theorem and its dual are

$$(\bar{x}) = \bar{x} \qquad (12.5\text{-}8a) \qquad (\bar{\bar{x}}) = x \qquad (12.5\text{-}8b)$$

Theorem (12.5-8b) is proved by perfect induction:

$$(\bar{\bar{0}}) = (\bar{1}) = \bar{1} = 0 \qquad \text{and} \qquad (\bar{\bar{1}}) = (\bar{0}) = \bar{0} = 1$$

Theorem (12.5-8a) may be proved in the same way or its validity may be based on the validity of its dual.

The next two theorems and their duals are:

$$x + \bar{x} = 1 \qquad (12.5\text{-}9a) \qquad x \cdot \bar{x} = 0 \qquad (12.5\text{-}9b)$$

$$x + x = x \qquad (12.5\text{-}10a) \qquad x \cdot x = x \qquad (12.5\text{-}10b)$$

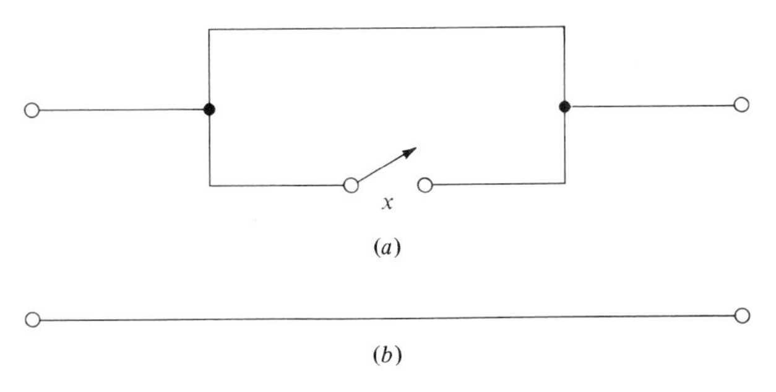

Fig. 12.5-1 Circuit representation of $1 + x = 1$. (a) Left side; (b) right side.

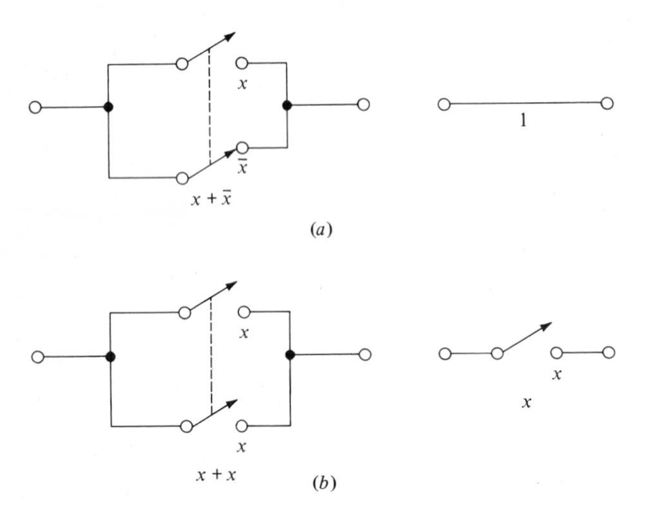

Fig. 12.5-2 Switching circuit representation of theorems. (*a*) $x + \bar{x} = 1$; (*b*) $x + x = x$.

The validity of these theorems is illustrated most meaningfully by their switching circuit representations shown in Fig. 12.5-2*a* and 12.5-2*b*.

The mechanical connection, indicated by the dotted line, between the normally open and normally closed contacts in the left hand circuit of Fig. 12.5-2*a* insures that when one contact is open the other is closed and vice-versa. Since the transient period while the contacts are being switched is not considered this means that the two terminals of the circuit are always connected. This circuit is, therefore, equivalent to the right-hand circuit. The two expressions represented by these circuits are the left and right expressions of Theorem (12.5-9*a*) thus proving the theorem.

The proof of Theorem (12.5-10*a*) is demonstrated by comparing the two circuits of Fig. 12.5-2*b* which are readily seen to be equivalent.

The next three theorems and their duals correspond to three of the fundamental theorems of ordinary algebra. As in ordinary algebra, multiplication takes precedence over addition, i.e., unless otherwise indicated by a parenthesis multiplications are performed before additions. The theorems are

Commutative laws:

$$x + y = y + x \qquad (12.5\text{-}11a) \qquad x \cdot y = y \cdot x \qquad (12.5\text{-}11b)$$

Associative laws:

$$(x + y) + z = x + (y + z) = x + y + z \qquad (12.5\text{-}12a)$$

$$(x \cdot y) \cdot z = x \cdot (y \cdot z) = x \cdot y \cdot z \qquad (12.5\text{-}12b)$$

Distributive laws:

$$x \cdot (y + z) = x \cdot y + x \cdot z \qquad (12.5\text{-}13a)$$

$$x + y \cdot z = (x + y) \cdot (x + z) \qquad (12.5\text{-}13b)$$

Of these six theorems, five are valid for ordinary algebra as well as the switching algebra. The only one which is not valid for ordinary algebra, but is valid for the switching algebra is Theorem (12.5-13*b*), the distributive law for addition over multiplication.

Theorems (12.5-11*a*), (12.5-12*a*), and (12.5-13*a*) can be proved using perfect induction. The second equality in Theorems (12.5-12*a*) and (12.5-12*b*) defines multiple additions and multiplications and therefore cannot be proved; the first equality in each case is, however, subject to proof.

The method of perfect induction is applied to Theorem (12.5-13*a*) as an example. A truth table having a column for both sides of the equality is set up in Table 12.5-1. It has $2^3 = 8$ rows since there are three variables. In addition to the columns for the three variables and the two expressions in the theorem there are also columns for subexpressions which facilitate finding the final expressions. The identity of the values in the last two columns representing the two expressions in the theorem proves its validity.

Another pair of theorems which are duals of each other and are sometimes useful are

$$xy + yz + \bar{x}z = xy + \bar{x}z \qquad (12.5\text{-}14a)$$

$$(x + y)(y + z)(\bar{x} + z) = (x + y)(\bar{x} + z) \qquad (12.5\text{-}14b)$$

The validity of Theorem (12.5-14*a*) is demonstrated using some of the preceding theorems and axioms. The proof is a bit unusual in that it starts out by increasing the complexity of the left-hand expression for reasons which only

TABLE 12.5-1 Truth Table for Proof of $x(y + z) = xy + xz$.

x	y	z	$y + z$	xy	xz	$x(y + z)$	$xy + xz$
0	0	0	0	0	0	0	0
0	0	1	1	0	0	0	0
0	1	0	1	0	0	0	0
0	1	1	1	0	0	0	0
1	0	0	0	0	0	0	0
1	0	1	1	0	1	1	1
1	1	0	1	1	0	1	1
1	1	1	1	1	1	1	1

become apparent a little later in the proof. Each theorem or axiom which is used is listed to the right of the step.

$$xy + yz + \bar{x}z = xy + 1yz + \bar{x}z \qquad (12.5\text{-}6b)$$

$$= xy + (x + \bar{x})yz + \bar{x}z \qquad (12.5\text{-}9a)$$

$$= xy + xyz + \bar{x}yz + \bar{x}z \qquad (12.5\text{-}11a),\ (12.5\text{-}13a)$$

$$= xy(1 + z) + \bar{x}z(1 + y) \qquad (12.5\text{-}11a),\ (12.5\text{-}13a)$$

$$= xy1 + \bar{x}z1 \qquad (12.5\text{-}7a)$$

$$= xy + \bar{x}z \qquad (12.5\text{-}6b),\ (12.5\text{-}11a)$$

This demonstrates the validity of Theorem (12.5-14a); since Theorem (12.5-14b) is its dual, it too is valid.

The final theorem is its own dual and is a very important theorem because it defines the inverse of any arbitrary switching function. The inverse of a function is another function whose value is opposite that of the original function for all values of the variables. The theorem is

$$\bar{f}(x_1, x_2, \ldots, x_n, +, \cdot, 0, 1) = f(\bar{x}_1 \bar{x}_2, \ldots, \bar{x}_n, \cdot, +, 1, 0) \qquad (12.5\text{-}15)$$

This theorem implies that the inverse of any function is found by replacing each variable by its inverse, interchanging all 0s and 1s and all additions and multiplications. In doing this it may be necessary to add parentheses to maintain the proper priority of operations, i.e., where a multiplication took precedence over addition it is now necessary to use parentheses to indicate that the addition should, in the inverse, be performed before the multiplication. This is illustrated in the following example.

Example 12.5-1 Find the inverse of the expression

$xy + z$

Solution The normal thing to do in applying Theorem (12.5-15) to the expression given above would be (inserting the multiplication sign for clarity):

$$\overline{x \cdot y + z} = \bar{x} + \bar{y} \cdot \bar{z}$$

which is an incorrect result. If we insert the parentheses which explicitly indicate the priority of operations we find the correct result.

$$\overline{(x \cdot y) + z} = (\bar{x} + \bar{y}) \cdot z$$

The student should check both of these by constructing their truth tables

////

From the preceding example we see that it is essential to keep the proper priority of operations in taking the inverse of an expression. This is most easily done by inserting parentheses to explicitly indicate the order of all operations. When the inverse expression has been found, parentheses which are in agreement with the implicit priorities may be eliminated. These will, in general, not be the same ones as were unnecessary in the original expression.

Two popular theorems in switching algebra, which are special cases of Theorem (12.5-15), are called De Morgan's laws:

$$\overline{(x + y)} = \bar{x} \cdot \bar{y} \qquad \overline{(x \cdot y)} = \bar{x} + \bar{y} \tag{12.5-16}$$

These theorems are easily proved by perfect induction or by the use of Theorem (12.5-15).

12.6 SIMPLIFICATION OF SWITCHING FUNCTIONS

The objective in simplifying a switching function is to transform it into an equivalent expression which includes as few literals (a variable with or without an overbar) and operations as possible. This will generally correspond to the most economical circuit for the given specification.

There are no set procedures for this phase of our work. Practice enables one to develop the ability to "see" what transformations and theorems will be useful in a given situation. The theorem

$$x\bar{y} + xy = x \tag{12.6-1}$$

is frequently useful, especially when x is treated as an expression rather than a single variable. Generally x will be the product of a number of literals. Theorem (12.6-1) is proved as follows:

$$x\bar{y} + xy = x(\bar{y} + y) = x1 = x$$

A second theorem which is frequently useful in conjunction with Theorem (12.6-1) is

$$x + x = x \tag{12.5-10a}$$

This theorem is often used in the reverse direction with x representing a product of terms to permit the same product to be combined with two different terms using Theorem (12.6-1) as illustrated in the following example.

Example 12.6-1 Simplify the expression

$$xyz + x\bar{y}z + xy\bar{z} \tag{12.6-2}$$

Solution We look for pairs of products which differ by a single literal; such a pair of products can then be combined using Eq. (12.6-1) into a single product with one less factor than either of the original ones. In our expression the first product can be combined with either the second or the third resulting in

$$xz(y + \bar{y}) + xy\bar{z}$$
$$= xz + xy\bar{z} \qquad (12.6\text{-}3)$$

or

$$xy(z + \bar{z}) + x\bar{y}z$$
$$= xy + x\bar{y}z \qquad (12.6\text{-}4)$$

Another way to proceed is to use Theorem (12.5-10*a*). With this we

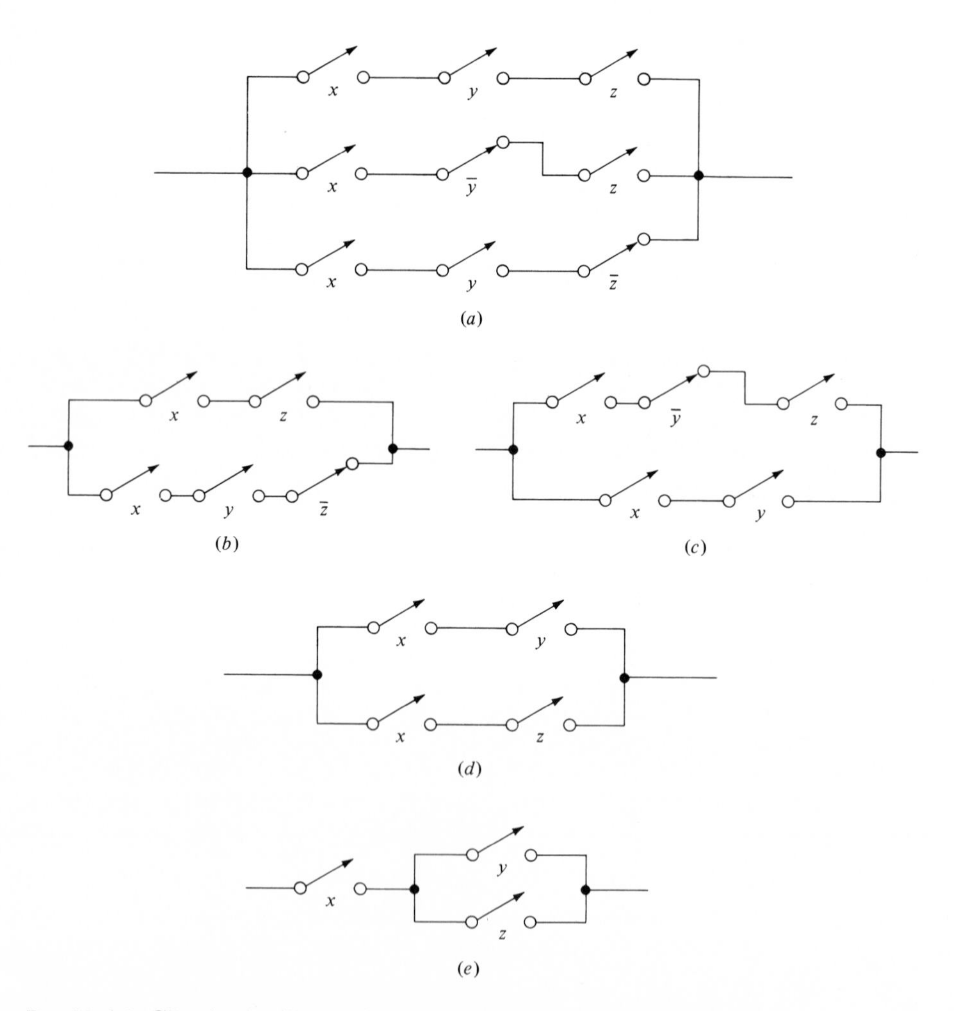

Fig. 12.6-1 Circuits for Example 12.6-1. (*a*) $xyz + x\bar{y}z + xy\bar{z}$, nine contact pairs; (*b*) $xz + xy\bar{z}$, five contact pairs; (*c*) $x\bar{y}z + xy$, five contact pairs; (*d*) $xy + xz$, four contact pairs; (*e*) $x(y + z)$, three contact pairs.

may rewrite the original expression as

$$xyz + xyz + x\bar{y}z + xy\bar{z}$$

We then combine the first and third terms and the second and fourth terms resulting in

$$xz(y + \bar{y}) + xy(z + \bar{z})$$

$$xz + xy \qquad (12.6\text{-}5)$$

which is simpler than Eq. (12.6-3) or (12.6-4). We now factor Eq. (12.6-5) to provide the final result

$$x(z + y) \qquad (12.6\text{-}6)$$

which, it is quite clear, cannot be further simplified. The circuits corresponding to Eqs. (12.6-2) through (12.6-6) are shown in Fig. 12.6-1 together with the number of contact pairs used in each circuit. We see that the final circuit uses one third the number of contact pairs used by the initial circuit; this would amount to a substantial reduction in cost. ////

We conclude this chapter with a final example illustrating the entire switching circuit design process.

Example 12.6-2 A buzzer in an automobile is to sound if someone is sitting in the driver's seat and doesn't have a seat belt on. There is a pressure sensor switch, p, in the seat which closes when someone sits in the seat and there is a seat belt switch, s, which closes when the seat belt is being worn. Draw the switching circuit which activates the buzzer, b.

Solution The first step is to construct the truth table from which the canonical switching function is determined. This switching function is simplified and the result is used to synthesize the circuit.

a Truth table. The truth table has two columns for the inputs, p and s, and one for the output, b. These are $2^2 = 4$ rows since there are 2 input variables.

p	s	b
0	0	0
0	1	0
1	0	1
1	1	0

b Canonical switching function. The canonical switching function consists of one complete product term since there is only one row for which the output is 1.

$$b = p\bar{s}$$

This corresponds to the situation where the pressure switch p is closed because someone is sitting in the seat AND the seat belt switch s is NOT closed because the seat belt is not being worn.

c Reduced switching function. The canonical switching function cannot be simplified. Therefore the reduced function is

$$b = p\bar{s}$$

d Circuit. The circuit consists of a series (for multiplication) connection of a normally open contact for p and a normally closed contact for $\bar{s}$ connected to the buzzer and battery as shown in Fig. 12.6-2. ////

At this point we have learned how to use the switching algebra for the design of switching circuits. Switching circuits have had wide application in the telephone industry and in some of the very early computers.

The need for higher speed and miniaturization coupled with the introduction of the transistor and integrated circuits has resulted in the replacement of switching circuits by electronic digital circuits. In the next chapter we will learn that the switching algebra and techniques developed in this chapter are readily adapted to the design of these electronic digital circuits.

SUMMARY

1 Logical and switching operations

AND	$x \cdot y$	series circuit
OR	$x + y$	parallel circuit
NOT	$\bar{x}$	normally closed contact

2 Switching axioms

Either $x = 0$	or	$x = 1$
$0 + 0 = 0$		$1 \cdot 1 = 1$
$1 + 1 = 1$		$0 \cdot 0 = 0$
$0 + 1 = 1 + 0 = 1$		$1 \cdot 0 = 0 \cdot 1 = 0$
$\bar{0} = 1$		$\bar{1} = 0$

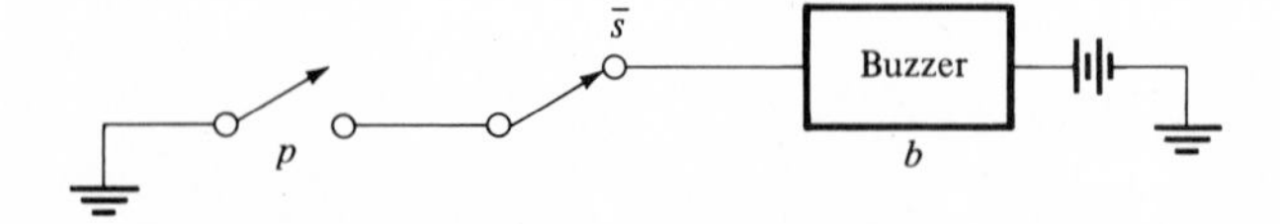

Fig. 12.6-2 Circuit for Example 12.6-2.

3 Theorems useful in simplification

$$x + x = x$$

$$x(y + z) = xy + xz$$

$$x\bar{y} + xy = x$$

4 A truth table for n input variables and one output variable has $n + 1$ columns and 2^n rows.

PROBLEMS

12.1-1 Assume

A = an animal
C = a cat
F = a feline
D = a domesticated animal
L = a lion
W = a wild animal

Write equations which represent the following statements:

a A cat is a feline and is a domesticated animal.
b An animal is a domesticated animal or a wild animal.
c A lion is a feline and is not a domesticated animal.

12.1-2 Draw a switching circuit for each of the statements of Prob. 12.1-1.

12.2-1 Write an algebraic equation which describes each of the circuits of Fig. P12.2-1.

12.2-2 Draw switching circuits for each of the following functions:

a $T = \bar{X}Y$
b $R = (W + \bar{S})\bar{T}P$
c $L = (M + P)\bar{R} + (\bar{P} + R)(K + M)$

12.3-1 Find the truth tables for each of the switching functions given in Prob. 12.2-2.

12.3-2 Find the switching function for each of the truth tables given below:

a

X	Y	Z	L
0	0	0	0
0	0	1	0
0	1	0	1
0	1	1	1
1	0	0	1
1	0	1	1
1	1	0	0
1	1	1	0

b

A	B	C	R
0	0	0	0
0	0	1	1
0	1	0	0
0	1	1	0
1	0	0	0
1	0	1	1
1	1	0	1
1	1	1	0

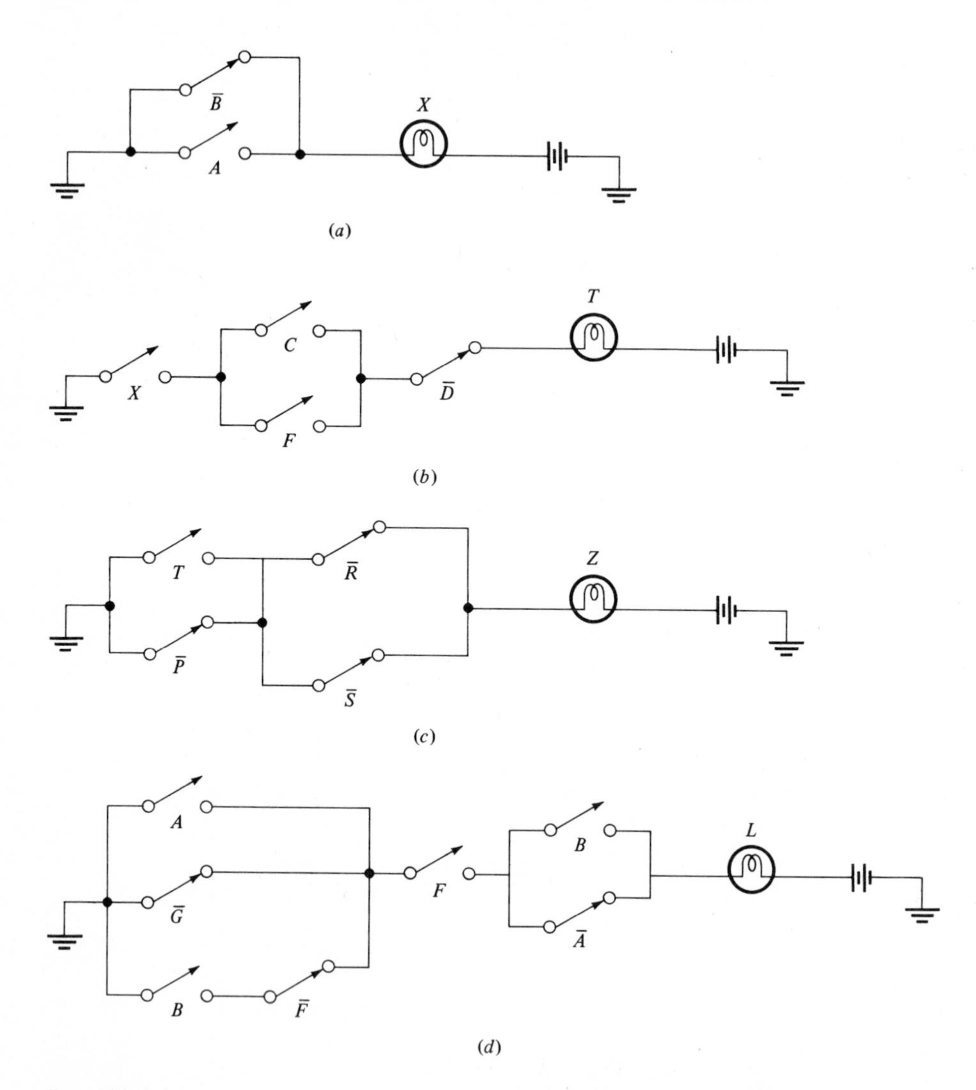

Fig. P12.2-1

12.4-1 A light on a stairway is to be controlled by two switches, one at the top of the stairs called T and one at the bottom of the stairs called B. Initially both switches and the light are off. If either switch is turned on the light is to go on. If the position of either switch is then changed the light is to go off. That is, either turning the "on" switch off or the "off" switch on should cause the light to go off. This permits the light to be controlled by either switch. Construct the truth table.

12.4-2 An automobile has a switch in each of the two front seats which goes on if someone sits in the seat; they are called S_1 and S_2. There is also a switch on each seat belt for these seats, called B_1 and B_2, which go on if the belt is buckled. A warning buzzer, W, is to go on if someone is sitting in a seat and the belt is not buckled. Construct the truth table.

12.4-3 Find a canonical switching function for the truth table found in Prob. 12.4-1 and draw the switching circuit.

12-4-4 Find a canonical switching function for the truth table found in Prob. 12.4-2.

12-5-1 Prove the following theorems using the method of perfect induction:

a Theorem (12.5-7a)
b Theorem (12.5-11*a*)

12.5-2 Prove the following theorems using the axioms and Theorems (12.5-6) through (12.5-15):

a $(x + y + z) \cdot (x + \bar{y} + z) = x + z$
b $x + \bar{x}z = x + z$ *Hint*: Prove the dual.

12.5-3 Prove the following theorems using their equivalent circuits:

a Theorem (12.5-11*a*)
b $\overline{x + y} = \bar{x} \cdot \bar{y}$

12.6-1 Simplify the following switching functions:

a $T = XY + \bar{X}YZ$
b $W = ABC̄ + A\bar{B}\bar{C} + B\bar{C}D$
c $L = PRT + \bar{P}ST + P\bar{S}T + P\bar{R}T$

12.6-2 Simplify the switching function of Prob. 12.4-4 and draw the corresponding circuit.

12.6-3 Design as economical as possible a switching circuit which will cause a lamp P to light when 65 percent of the ownership of a company votes yes on a motion if there are four owners A, B, C, and D who own 40, 25, 20, and 15 percent of the business respectively.

13
Combinational Digital Circuits

OBJECTIVES

Upon completion of this chapter, you should be able to

1 Distinguish between combinational and sequential digital circuits.
2 Draw the logic and circuit diagrams for the various diode and diode-transistor gates and state the switching functions they implement.
3 Demonstrate the advantage of diode-transistor gates over diode gates.
4 Convert between the binary and decimal number systems.
5 Design a digital circuit to satisfy a given specification.

INTRODUCTION

Digital logic circuits are becoming increasingly important in modern electronics. The digital computer, pocket calculator, and electronic telephone switching systems are just a few applications of digital electronics. The introduction of integrated circuits (ICs), and large scale integrated circuits (LSIs) more recently, has made it possible to place thousands of digital logic gates in a circuit the size of a postage stamp by fabricating the entire circuit as one semiconductor device. The devices already available are just the beginning of an avalanche of new devices and systems that couldn't have existed a few years ago.

The circuits we have studied up to this point, excluding the switching circuits introduced in Chap. 12, are characterized as *analog* circuits. In this chapter we introduce *digital* circuits which are the electronic counterpart of the switching circuits discussed in Chap. 12. The advantages of the digital circuit over the switching circuit are (1) much smaller size and lighter weight, (2) much faster operation, (3) lower cost, and (4) lower power requirements.

The signals (voltage and/or current) in an analog circuit vary continuously over a range of values and may assume any value in the range between the minimum and maximum levels as shown in Fig. 2-1*b*. The voltages in a digital circuit on the other hand can assume only two values, a high value and a low value, in theory. The signal changes very rapidly between these levels resulting in a graph which appears discontinuous as shown in Fig. 2-1*e*. In practice the digital signal can have values which fall into one of two ranges of values, any value in the higher range is treated as the high value and any value in the low range is treated as the low value. This is necessary because of the nonideal characteristics and the variability of the elements used in the construction of digital circuits.

The function of a digital circuit is the same as that of a switching circuit and it is designed using the switching algebra and truth tables in the same way. Since we are dealing with electronic circuits which have two voltage levels, rather than switches which may be open or closed, we must reinterpret the logical values 0 and 1 in terms of voltage values. We shall use the *positive logic convention* whereby a relatively low voltage represents logical 0 and a relatively high voltage represents logical 1.† Note that the logical values do not imply anything concerning the absolute voltage values. For example, logical 0 and 1 might be represented by −5 V and 0 V, respectively, or 3 V and 10 V, respectively; the only requirement is that the level representing logical 1 be greater than the level representing logical 0. Furthermore, the voltage level is a nominal value; we might have a voltage range from 0 V to 1 V representing logical 0 and a voltage range from 3.5 V to 5 V representing logical 1. These ideas are illustrated in Fig. 13-1.

Since the logical values are represented by voltage levels rather than open or

† In the negative logic convention a low voltage represents logical 1 while a high voltage represents logical 0.

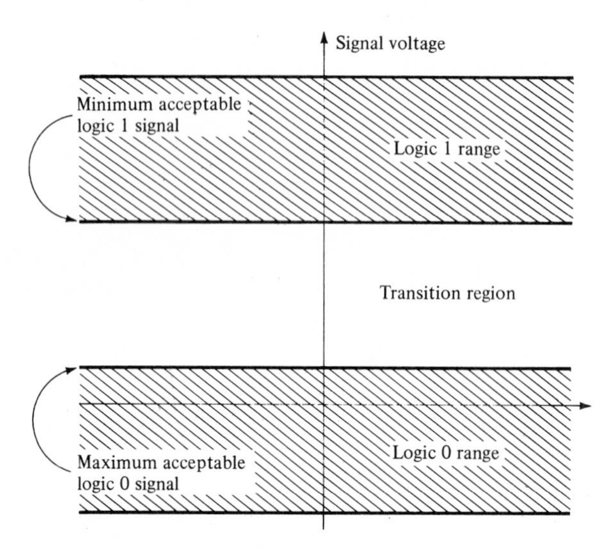

Fig. 13.1 Digital signal levels (positive logic).

closed switches, it is not possible to use series and parallel connections to implement the AND and OR functions nor is it possible to implement the NOT function using the same method used in switching circuits. In digital systems we use electronic circuits, called *logic gates,* which implement the various logic functions by providing an output voltage whose logical value is related to the logical values of the input voltages by the desired logic function. Thus, a logical variable in a digital circuit is associated with the voltage level of a given point rather than with the state of a given switch, as was the case in the switching circuit.

This chapter will describe circuits to implement the various gates and their logic symbols. After that the design of digital circuits will be described using some typical practical applications.

13.1 IDEAL DIODE GATES

In this section we will describe diode AND and OR gates. In order to emphasize the logical behavior of these circuits the diodes are treated as ideal diodes. In the next section we will investigate the effects of the actual diode characteristics.

Before we introduce the circuits it is important to agree upon a notation which will avoid confusing logical values and voltage levels, e.g., logical 1 as compared with 1 V. The notation which we will use is illustrated in Table 13.1-1.

TABLE 13.1-1 Notation

Quantity	Symbols
Logical variables	x, y, z, X, Y, Z
Corresponding voltages	$V_x, V_y, V_z, V_X, V_Y, V_Z$
Logical values	0, 1
Corresponding voltages	V_0, V_1
Voltage values	0 V, 1 V, 5 V

13.1-1 Ideal AND Gate

The ideal AND gate is shown in Fig. 13.1-1*a*. Logical 0 is represented by 0 V which is the ground potential. Logical 1 is represented by V_1 which is a positive voltage, normally between 5 V and 10 V.

There are four cases to consider corresponding to the four rows in the truth table for the AND function. As the first case, assume that $V_X = V_Y = V_1$ as

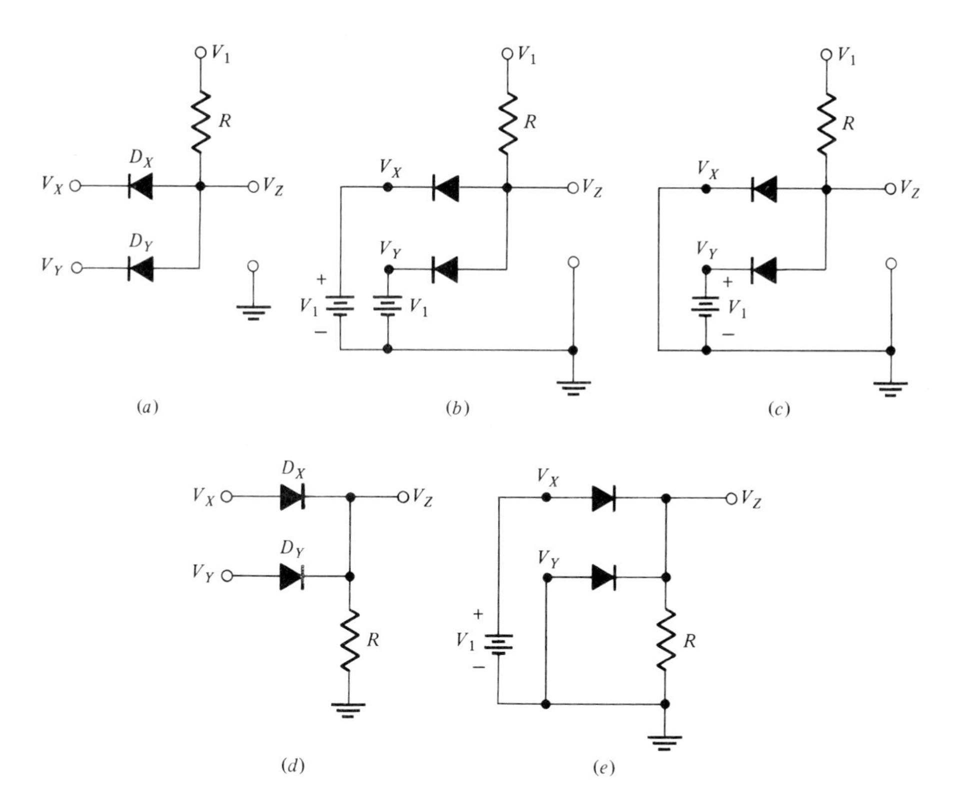

Fig. 13.1-1 Ideal gates. (*a*) AND gate, $Z = XY$; (*b*) AND gate with $X = Y = 1$ resulting in $Z = 1$; (*c*) AND gate with $X = 0$, $Y = 1$ resulting in $Z = 0$; (*d*) OR gate: $Z = X + Y$; (*e*) OR gate with $X = 1$, $Y = 0$ resulting in $Z = 1$.

shown in Fig. 13.1-1*b*. All voltages in the circuit are at the same level, no current flows through R, and $V_Z = V_1$. For the second case, if $V_X = 0$ V and $V_Y = V_1$ as shown in Fig. 13.1-1*c*, then D_X is forward-biased and acts like a short circuit resulting in $V_Z = 0$. Diode D_Y is reverse-biased and acts like an open circuit with the full voltage $V_Y = V_1$ across it.

If $V_X = V_1$ and $V_Y = 0$ V, the circuit behaves as described above except that the roles of D_X and D_Y are interchanged as are V_X and V_Y. In any case $V_Z = 0$ V.

If $V_X = V_Y = 0$ V, then both diodes behave like short circuits and $V_Z = 0$ V.

From this analysis we see that $V_Z = V_1$ if $V_X = V_Y = V_1$; otherwise $V_Z = 0$ V. This implies that $Z = 1$ if $X = Y = 1$; otherwise $Z = 0$. This is precisely the behavior of the AND function (see Chap. 12) and therefore the diode circuit of Fig. 13.1-1*a* is called an AND gate. It is possible to connect additional inputs, each through its own diode, to the circuit resulting in a gate with more than two inputs.

13.1-2 Ideal OR Gate

The circuit of Fig. 13.1-1*d* is an ideal OR gate. In this circuit, as in the ideal AND gate, logical 0 is represented by 0 V or ground potential and logical 1 is represented by the positive voltage V_1.

If $V_X = V_Y = 0$, then all voltages in the circuit are at the same level, no current flows through R, and $V_Z = 0$.

If either or both inputs are at the level V_1, then the diode in series with that input is forward-biased and acts like a short circuit resulting in $V_Z = V_1$. A current of value V_1/R flows through the diode, or diodes, and R so that a voltage of value V_1 appears across R. If only one of the inputs is at V_1 and the other is at 0 V as shown in Fig. 13.1-1*e*, the diode in series with the 0 V input will be reverse-biased since $V_Z = V_1$; this diode will act like an open circuit with the voltage $V_Z - 0 = V_1$ across it.

From this analysis we have found that $V_Z = V_1$ unless *both* V_X and V_Y are equal to 0 V. Thus $Z = 1$ if either $X = 1$ or $Y = 1$ or both, but $Z = 0$ if $X = Y = 0$. This is the definition of the OR function, as given in Chap. 12, and therefore the circuit of Fig. 13.1-1*d* is an OR gate. As with the AND gate the circuit can be extended with additional diodes to provide more than two inputs.

13.1-3 Logic Symbols

In designing digital systems the logical designer generally does not care how the gate is constructed internally; he is interested only in its logical description. A set of logic symbols are used to represent the various types of gates; those for the AND and OR gates are shown in Fig. 13.1-2. These symbols apply to gates no matter how they are built, not only to diode gates. Notice that inputs and outputs are represented by the logic variables rather than the voltages which repre-

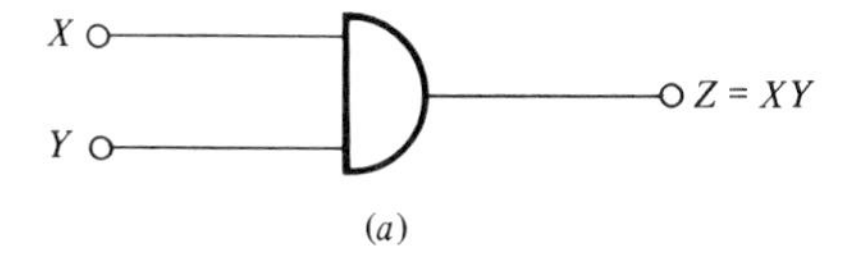

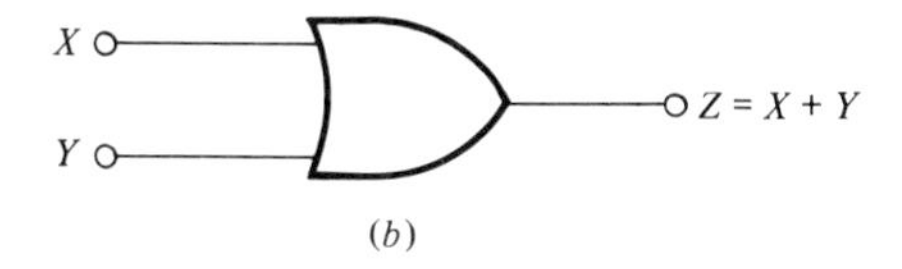

Fig. 13.1-2 Logic symbols. (*a*) AND gate; (*b*) OR gate.

sent them as was done in the circuit diagrams. Values in such a diagram would be represented by 0 and 1 for logical 0 and logical 1; there is no interest in the actual voltage values.

13.2 PRACTICAL DIODE GATES

To obtain a more realistic picture of the behavior of diode gate circuits we must take into account the parameters of the physical diodes used in building such a circuit. In addition to providing a more accurate analysis of the circuit we will obtain information useful in the design of such gates and we will find certain restrictions on their interconnection.

In order to simplify the analysis we will assume that the diodes reverse resistance, r_r, is much greater than its forward resistance, r_f, so that

$$r_r + r_f \approx r_r \qquad \text{and} \qquad r_r \,||\, r_f \approx r_f$$

Furthermore, it is assumed that the load resistance R used in the diode gates satisfies the pair of inequalities

$$r_f \ll R \ll r_r \tag{13.2-1}$$

which results in performance which best approximates that of ideal diode gate circuits. These inequalities are used to make the following simplifying approximations in the analysis of the practical gate circuits:

$$r_f \,||\, R \approx r_f$$

$$r_f + R \approx R$$

$$r_r \,||\, R \approx R$$

$$r_r + R \approx r_r$$

The optimum value of R to satisfy the inequalities in Eq. (13.2-1) provides the same ratio between r_f and R as the ratio between R and r_r:

$$\frac{r_f}{R} = \frac{R}{r_r}$$

The solution of this equation for R is

$$R = \sqrt{r_r r_f} \tag{13.2-2}$$

which states that the load resistance is the *geometric mean* of the forward and reverse resistances.

13.2-1 AND Gate

The practical diode AND gate is shown in Fig. 13.2-1*a* and its circuit model is shown in Fig. 13.2-1*b* where the forward conducting voltage drop is denoted V_f; its value is about 0.7 V for silicon diodes.

If $V_X = V_Y = V_1$ then both ideal diodes, D_X and D_Y, are reverse-biased by the sources. $V_Z = V_1$ in this situation and there is no need for any additional analysis.

If one of the inputs, say V_X, is equal to V_1 and the other, V_Y, is equal to 0 V, then from the analysis of the ideal diode AND gate we know that D_Y is forward-biased while D_X is reverse-biased. The model of Fig. 13.2-1*b* reduces to that of Fig. 13.2-1*c*. For this case V_X is connected to V_1 and V_Y is connected to 0 V. The parallel combination of r_r and R is replaced by the approximate equivalent resistance R. V_Z in the resulting circuit is easily found using superposition as indicated by the circuits of Figs. 13.2-1*d* and 13.2-1*e*, each of which is a simple voltage divider. This results in

$$V_Z = V_{Z1} + V_{Z2} = \frac{r_f}{R + r_f} V_1 + \frac{R}{R + r_f} V_f$$

$$= \frac{r_f V_1 + R V_f}{R + r_f} \approx V_f + \frac{r_f}{R} V_1 \tag{13.2-3}$$

The result is the same if the two inputs are interchanged; that is if $V_X = 0$ and $V_Y = V_1$ since the circuit is symmetrical with respect to these two inputs.

The only case left to consider is when both inputs are at 0 V. Then both diodes D_1 and D_2 will be forward-biased and the circuit of Fig. 13.2-1*b* reduces to the circuit model shown in Fig. 13.2-1*f*. In this figure the two points labeled *a* and *b* are both at the same voltage V_f and may therefore be connected together without changing the circuits behavior, i.e., inserting the connection results in no change because they are at the same voltage with or without it. This connection places the two sources V_f in parallel so they may be replaced by a single source of value V_f. The two resistances r_f are also in parallel and may be replaced by a single resistance of value $r_f \,||\, r_f = r_f/2$. The resulting circuit, shown in Fig. 13.2-1*g*, is identical to that of Fig. 13.2-1*c* except that $R \,||\, r_r = R$

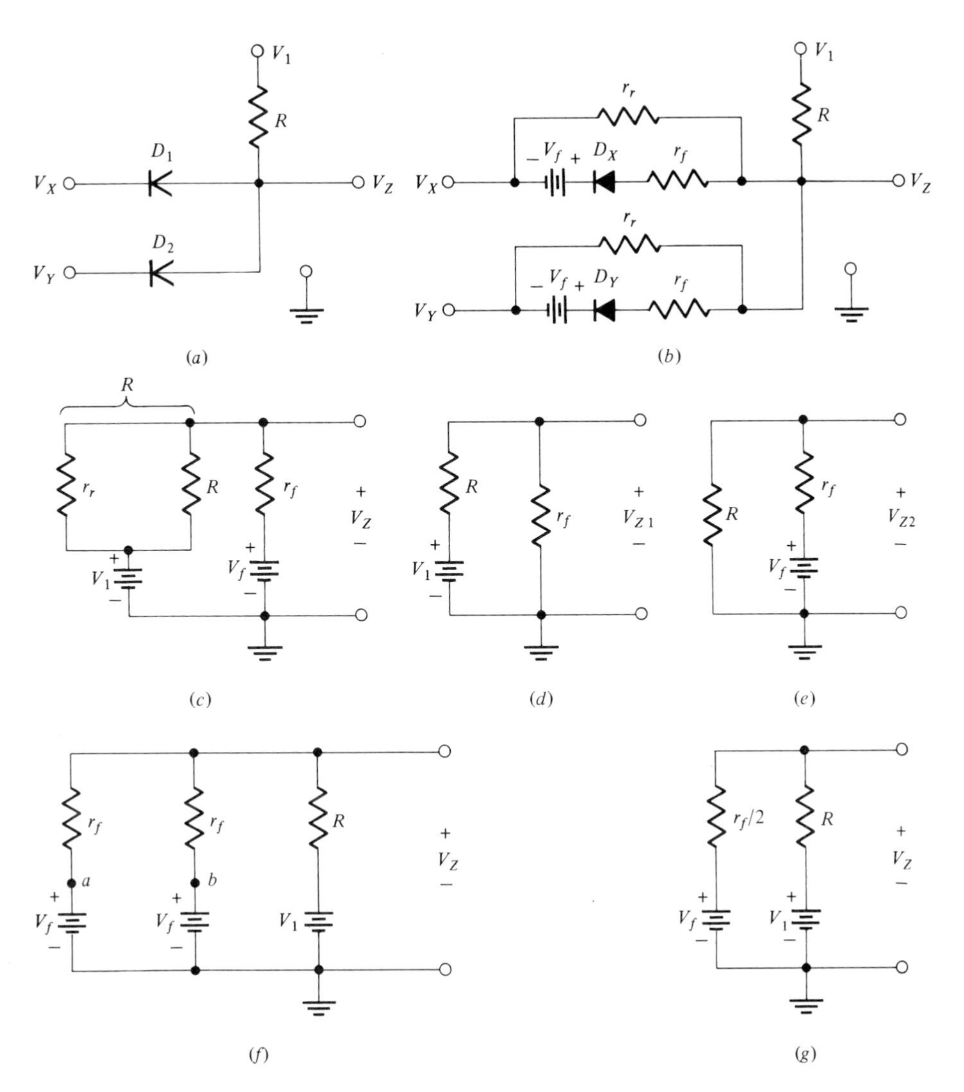

Fig. 13.2-1 AND gate analysis. (a) Practical AND gate; (b) circuit model; (c) equivalent circuit model for 0 V input and V_1 input; (d) equivalent circuit using superposition with $V_f = 0$; (e) equivalent circuit using superposition with $V_1 = 0$; (f) equivalent circuit model with both imputs at 0 V; (g) simplified equivalent circuit.

is replaced by R and r_f is replaced by $r_f/2$. Making these changes in Eq. (13.2-1), the solution for V_Z in the circuit of Fig. 13.2-1c, we obtain

$$V_Z = \frac{(r_f/2)V_1 + RV_f}{R + r_f/2} \approx \frac{r_f V_1 + 2RV_f}{2R} = V_f + \frac{r_f}{2R} V_1 \tag{13.2-4}$$

TABLE 13.2-1 Outputs for Ideal and Practical AND Gates.

V_X	V_Y	V_Z (ideal)	V_Z (practical)
0	0	0	$V_f + \frac{r_f}{2R} V_1$
0	V_1	0	$V_f + \frac{r_f}{R} V_1$
V_1	0	0	$V_f + \frac{r_f}{R} V_1$
V_1	V_1	V_1	V_1

A comparison of the outputs of the ideal and practical AND gates is shown in Table 13.2-1. Use of this table is illustrated in the example which follows.

Example 13.2-1 Design a two input AND gate using silicon diodes with $r_f = 20\ \Omega$ and $r_r = 1\ \text{M}\Omega$. Select the minimum supply voltage so that no output value differs from the ideal value by more than 10 percent of the supply voltage.

Solution From Eq. (13.2-2)

$$R = \sqrt{r_f r_r} = \sqrt{(20)(10^6)} = 4.47\ \text{k}\Omega$$

For silicon diodes, $V_f = 0.7$ V. From Table 13.2-1, the worst case occurs when one input is V_1 and the other is 0; the resulting output, which would be zero in the ideal case, is

$$\frac{r_f}{R} V_1 + V_f = \frac{V_1}{224} + 0.7$$

This must be less than or equal to 10 percent of the supply voltage, 0.1 V_1, therefore

$$\frac{V_1}{224} + 0.7 \leq 0.1\ V_1$$

$$0.00446\ V_1 + 0.7 \leq 0.1\ V_1$$

$$0.7 \leq 0.0955\ V_1$$

$$V_1 \geq 7.33\ \text{V}$$

The minimum value consistent with this requirement is thus $V_1 = 7.33$ V. We may also note that the V_1 term is much less than the V_f term and is frequently neglected. The error would be less than 5 percent in this example. ////

13.2-2 OR Gate

The OR gate is shown in Fig. 13.2*a* and its circuit model is shown in Fig. 13.2-2*b*. The diode parameters are the same as those in the AND gate.

If $V_X = V_Y = 0$ V, then ideal diodes D_X and D_Y are reverse-biased by the V_f sources and $V_Z = 0$ V.

If one of the inputs, say $V_X = V_1$, while the other, V_Y, is zero, then diode D_X is forward-biased while diode D_Y is reverse-biased and the model of Fig. 13.2-2*b* reduces to the circuit of Fig. 13.2-2*c* where $R \parallel r_r$ is replaced by its approximate equivalent, R. From this circuit we easily find, using the voltage divider relationship,

$$V_Z = \frac{R}{R + r_f}(V_1 - V_f) \approx V_1 - V_f \tag{13.2-5}$$

If the values of V_X and V_Y are interchanged, V_Z is unchanged; a consequence of the symmetry of the circuit.

The final case is when $V_X = V_Y = V_1$, then both diodes D_1 and D_2 are forward-biased and the model of Fig. 13.2-2*b* reduces to the circuit shown in Fig. 13.2-2*d*. Since points *a* and *b* in this circuit are at the same voltage, $V_1 - V_f$, they can be connected together. The four voltage sources may be replaced by a single source of value $V_1 - V_f$ and the two resistances of value r_f in parallel are replaced by the single-equivalent resistance of value $r_f/2$ as shown in Fig. 13.2-2*e* which is identical to the circuit of Fig. 13.2-2*c* except that r_f is replaced by $r_f/2$. Making these substitutions into Eq. (13.2-6) results in

$$V_Z = \frac{2R}{2R + r_f}(V_1 - V_f) \approx V_1 - V_f \tag{13.2-6}$$

A comparison of the outputs of the practical and ideal diode OR gates is shown in Table 13.2-2.

TABLE 13.2-2 Outputs for Practical and Ideal OR Gates.

V_X	V_Y	V_Z (practical)	V_Z (ideal)
0	0	0	0
0	V_1	$V_1 - V_f$	V_1
V_1	0	$V_1 - V_f$	V_1
V_1	V_1	$V_1 - V_f$	V_1

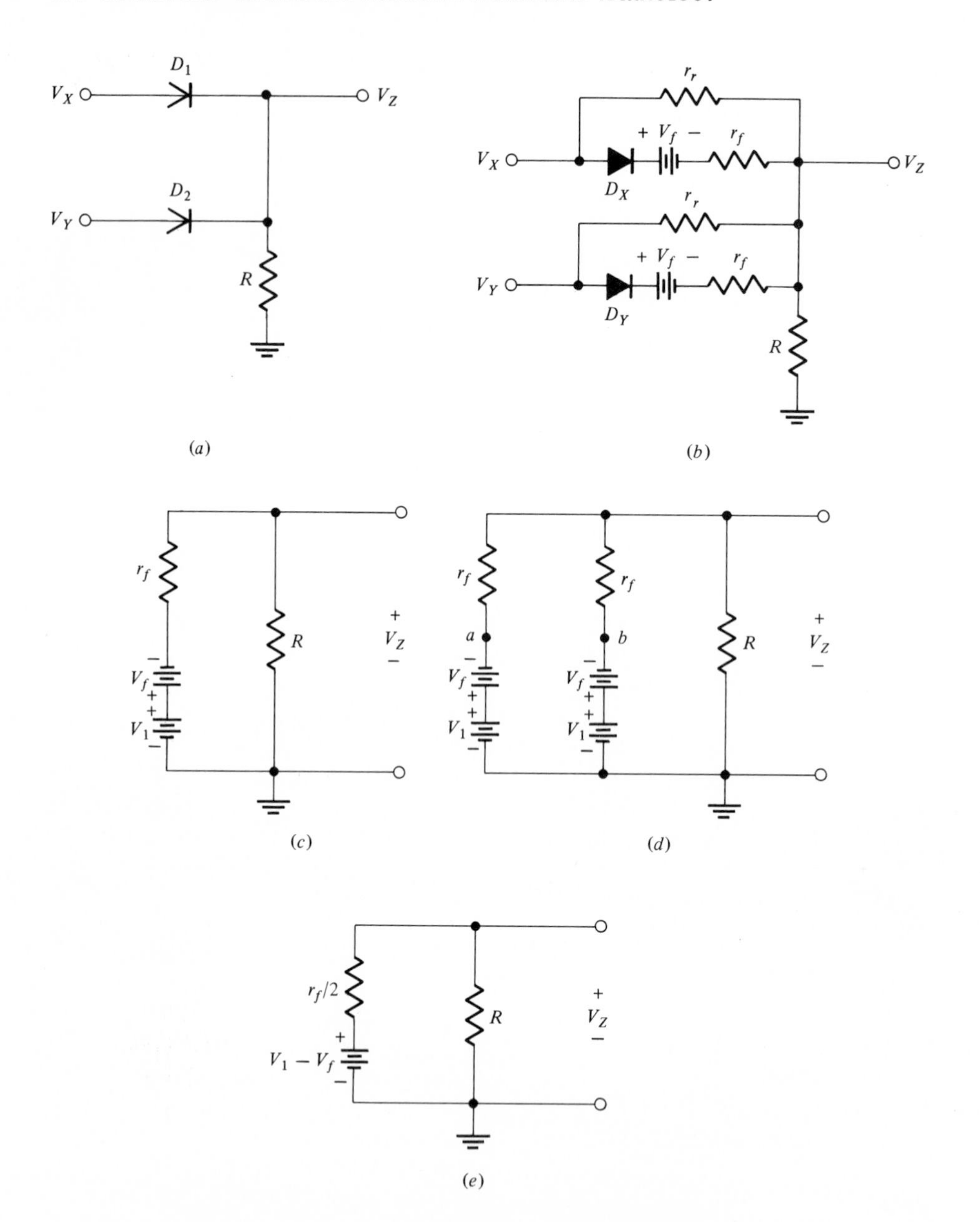

Fig. 13.2-2 OR gate analysis. (*a*) Practical OR gate; (*b*) circuit model; (*c*) equivalent model with 0 V input and V_1 input; (*d*) equivalent model with both inputs at V_1; (*e*) simplified equivalent circuit.

Example 13.2-2 Design a diode OR gate using silicon diodes with $r_f = 20\ \Omega$ and $r_r = 1\ \text{M}\Omega$. Use a 5-V power supply. Find the output values for all combinations of inputs.

Solution The only design value to be determined is that of R in Fig. 13.2-2*a*. It is

$$R = \sqrt{20 \times 10^6} = 4.47\ \text{k}\Omega$$

The following results are found from Table 13.2-2 using a value of $V_f = 0.7$ V for silicon diodes:

V_X, V	V_Y, V	V_Z, V
0	0	0
0	5	4.3
5	0	4.3
5	5	4.3

////

13.2-3 Multilevel Gate Circuits

From Tables 13.2-1 and 13.2-2 for typical values of r_r/r_f and V_1 we see that the output of the AND gate is approximately V_f instead of the ideal value of 0 V and for the OR gate the output is approximately $V_1 - V_f$ instead of the ideal value of V_1. Thus there is a signal degradation: the low voltage is higher than the ideal value and the high voltage is lower than the ideal value. This makes it more difficult to differentiate between the two signal levels. We might anticipate that this problem would be aggravated by connecting the output of one logic gate to the input of another one as shown in Fig. 13.2-3*a*. The circuit diagram using diode logic is shown in Fig. 13.2-3*b*.

Diodes D_1 and D_4 in this diagram will be reverse-biased and will therefore behave like resistances of value r_r. Each of these resistances, r_r, is in parallel with a much smaller resistance R and can be neglected. Diodes D_2 and D_3 will be forward-biased and will therefore each act like a resistance r_f in series with a voltage source V_f. The resulting equivalent circuit is shown in Fig. 13.2-3*c*.

The two parallel branches on the left are replaced by their Thevenin equivalent. The open circuit voltage, assuming $R \gg r_f$ and $V_1 \gg V_f$, was found to be approximately equal to V_f in the analysis of the AND gate. The Thevenin resistance is the parallel combination of R and r_f which is approximately equal to r_f since $R \gg r_f$. Combining this with the series branch consisting of r_f and V_f results in the circuit of Fig. 13.2-3*d*. This circuit is similar to the model of the AND circuit shown in Fig. 13.2-1*c*; the only differences are that V_f is replaced by $2V_f$ and r_f is replaced by $2r_f$. Making these substitutions in Eq. (13.2-1),

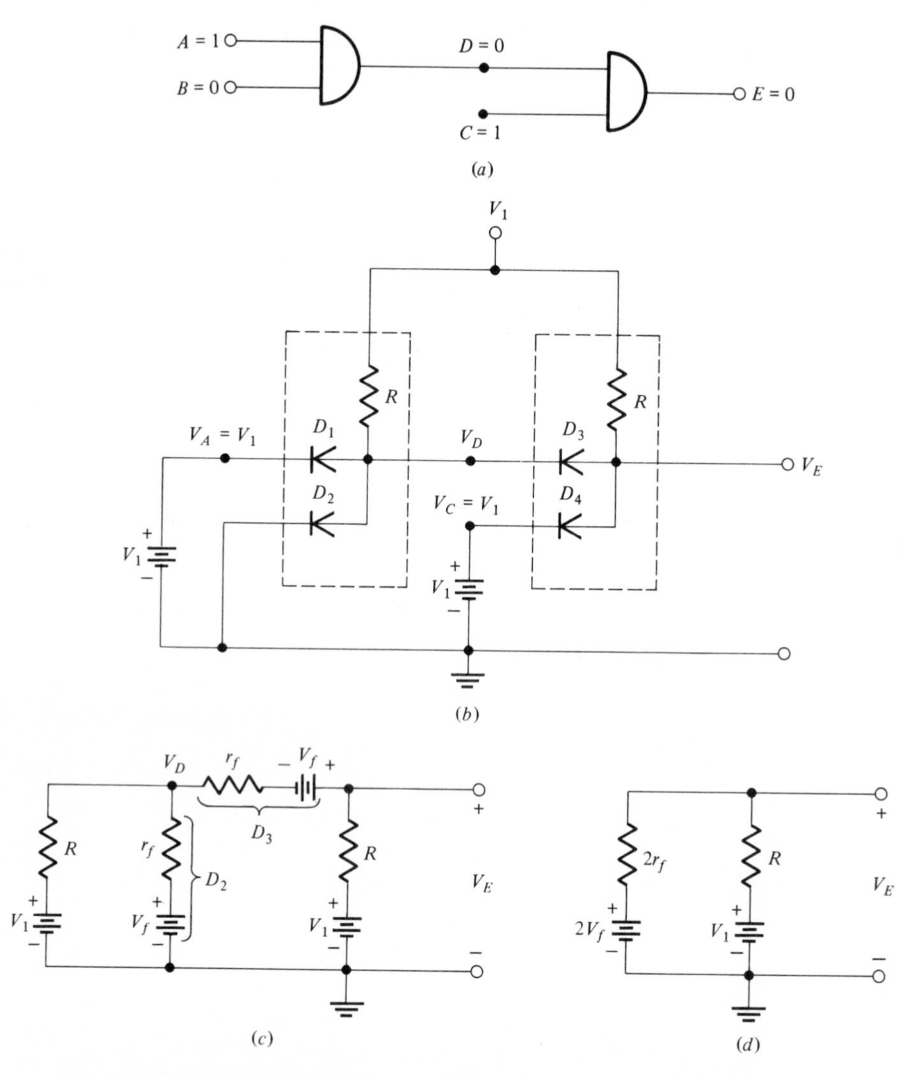

Fig. 13.2-3 Two level connection of AND gates. (*a*) Logic diagram; (*b*) circuit diagram; (*c*) equivalent circuit; (*d*) simplified equivalent circuit.

which is the result for Fig. 13.2-1*c*, we obtain

$$V_E = 2V_f + \frac{2r_f}{R} V_1 \tag{13.2-7}$$

For typical values of r_f, R, V_1, and V_f we find $V_E \approx 2V_f$ which is just double the output of a single AND gate and which would be 0 V in the ideal case.

From this analysis we see that the signal degradation is compounded when AND gates are connected in multilevel circuits. This can be demonstrated for

OR gates as well and it can also be shown that the problem becomes progressively worse as more levels are added until finally the voltage levels representing logical 0 and logical 1 are the same. As a consequence of this problem it is not possible to exceed a certain maximum number of gate levels in a diode gate circuit. The exact number depends on the relative values of V_1 and V_f; the larger V_1 is in comparison with V_f the larger the number. Typically, five levels are a reasonable maximum.

It is important to recognize that this problem is associated with diode gates and certain other types of gate circuits but it is not a universal problem. The root of the problem lies in the fact that the diode is a passive circuit element and the signal loses energy in each gate level until it is degraded beyond the point of recognition. We will see, later on in this chapter, that if active elements such as transistors are used this problem does not exist. The reason is that the transistor is able to increase the energy level of the signal by adding energy from its dc bias source to the time-varying signal; indeed, this is the primary function of a transistor amplifier.

13.3 TRANSISTOR INVERTER (NOT) GATE

The logic symbol for an inverter, or NOT, gate is shown in Fig. 13.3-1*a*. The triangle represents an amplifier while the circle at its vertex represents the inversion, i.e., high output for low input and low output for high input.

Unlike AND and OR gates, the inverter cannot be constructed from diodes because an active circuit element is required to provide the power gain required in the inversion process. The circuit diagram for a transistor inverter is shown in Fig. 13.3-1*b*. Its collector characteristics, load line, and operating points are shown in Fig. 13.3-1*c*. The circuit is often designed so that the transistor is either in the saturated state (point A in Fig. 13.3-1*c*) or the cutoff state (point B in Fig. 13.3-1*c*). This mode of operation results in more stable output values than does operation in the active region. The operation of the transistor inverter depends only on certain properties of the graphical characteristics. Most often, the designer has these parameters available instead of the characteristics. A model for the transistor, based on these parameters, is used in the design process. We shall develop such a model in this section and illustrate its use in the design process.

13.3-1 Base-to-Emitter Behavior

The base-to-emitter terminals of the transistor behave like a diode (p at the base, n at the emitter for an npn transistor). When the base current is positive this diode is conducting. The voltage across it changes very little with base current due to its forward resistance; the principal component of the base-to-emitter voltage is the forward voltage drop of the forward-biased diode. The

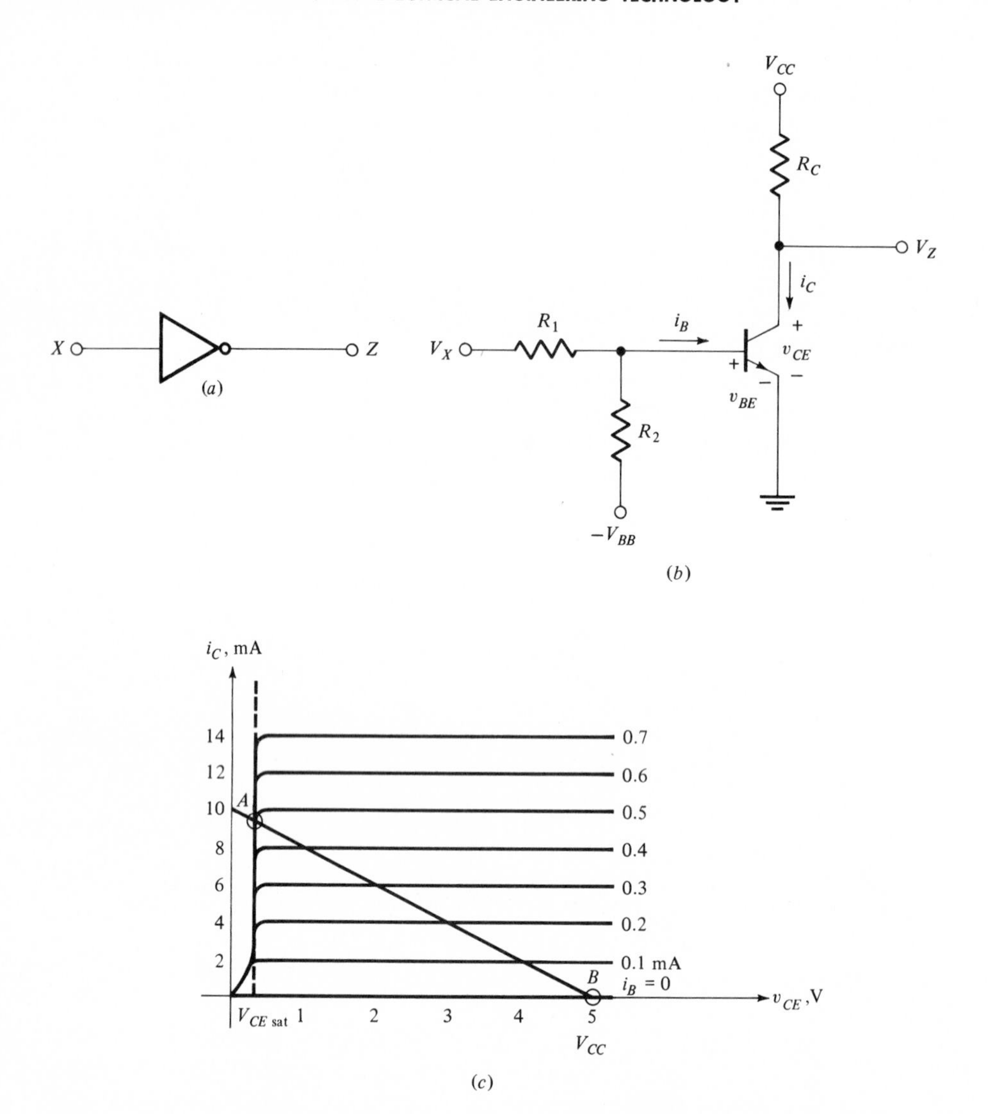

Fig. 13.3-1 Inverter (NOT) gate. (*a*) Logic symbol; (*b*) transistor circuit; (*c*) collector characteristic.

base-to-emitter voltage is quite constant throughout the saturation region and is a parameter which the manufacturer of a transistor generally specifies in his data sheets. Typical values for the base-to-emitter saturation voltage, $V_{BE,\text{sat}}$, are 0.7 V for silicon transistors and 0.3 V for germanium transistors.

Base-to-emitter voltages less than about 0 V for silicon and -0.1 V for germanium result in the transistor being at cutoff which corresponds to zero-emitter current. The base current remains very low until the base-to-emitter voltage reaches about 0.5 V for silicon and 0.1 V for germanium. The base-to-emitter voltage in the active region varies between this value and the saturation value.

From this discussion we conclude that the base-to-emitter terminal behavior can be represented by a voltage source of value $V_{BE,\text{sat}}$ in the saturated region and an open circuit in the cutoff region. The transistor becomes cut off when $V_{BE} < 0$ V for silicon and -0.1 V for germanium transistors.

13.3-2 Collector-to-Emitter Behavior

The behavior at the collector-to-emitter terminals is dependent on the base current; it is portrayed by the collector characteristics of Fig. 13.3-1*c*. From these characteristics we see that the collector-to-emitter voltage is relatively constant throughout the saturation region; its value is a parameter normally specified by the manufacturer as the collector-to-emitter saturation voltage, $V_{CE,\text{sat}}$. Typically values for $V_{CE,\text{sat}}$ are 0.3 V for silicon and 0.1 V for germanium transistors.

When $I_B = 0$ the collector current is very small; it is called the reverse collector-to-emitter leakage current, I_{CEO}. Manufacturers generally specify I_{CEO} or the reverse collector-to-base leakage current, I_{CBO}, which is related to I_{CEO} by the equation $I_{CEO} = (1 + \beta)I_{CBO}$. These currents are generally a few microamperes for germanium and fractions of a microampere for silicon. We shall neglect the effects of this current and treat it as zero. Therefore, the collector current is zero at cutoff.

In the active region the collector current is proportional to the base current with $i_C = \beta i_B$, β being the current amplification factor. For a given load line we find that the collector current ceases to increase once i_B exceeds a certain value. This occurs for $i_B = 0.5$ mA at point A in Fig. 13.3-1*d*. For i_B less than that amount we are in the active region for which $i_C = \beta i_B$; for $i_B > 0.5$ mA, i_C ceases to increase because all the characteristics for higher values of i_B have converged into the single saturation region curve. Thus the transistor enters saturation when $\beta i_B > i_C$. The collector-to-emitter terminal behavior can be represented by a voltage source of $V_{CE,\text{sat}}$ for the saturated transistor and by an open circuit for the cutoff region (since the collector current is essentially zero). In the active region the collector current is β times the base current. This discussion is summarized in Fig. 13.3-2 which tabulates models for the transistor in each state together with typical parameter values and the conditions required for cutoff and saturation for an *npn* transistor.

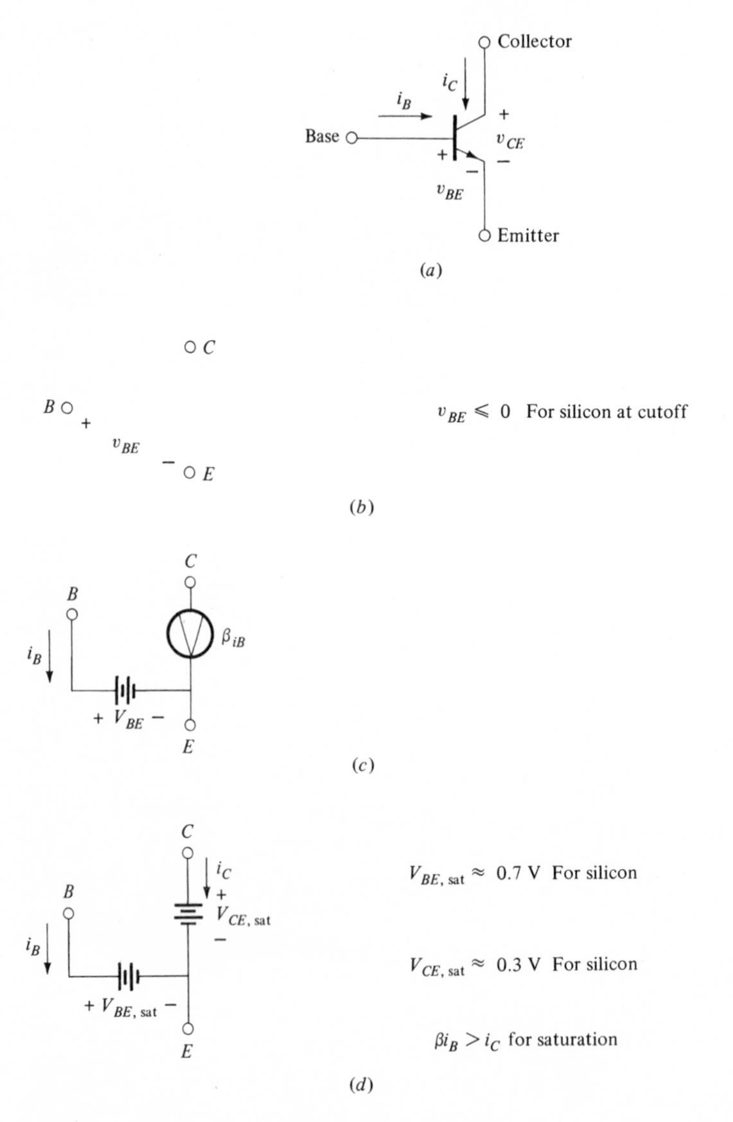

Fig. 13.3-2 Transistor models, typical parameters, and conditions for switching applications. (*a*) *npn* transistor diagram; (*b*) cutoff model; (*c*) active model; (*d*) saturated model.

13.3-3 Inverter Analysis

The analysis of an inverter circuit is relatively straightforward. The important steps are to determine whether the transistor enters the saturation and/or cutoff regions, the output voltage levels corresponding to the low and high input values, and the range of input values that will result in a high output as well as the input range that results in a low value. These steps are illustrated in the following example.

Example 13.3-1 The collector characteristics for the transistor in the inverter circuit of Fig. 13.3-3*a* are shown in Fig. 13.3-1*c*; $V_{BE,\text{sat}} = 0.7$ V for this transistor. Analyze the circuit.

Solution We first determine the output values corresponding to cutoff and saturation. The 5-V collector source voltage together with the 500-Ω collector resistance result in the load line shown in Fig. 13.3-1*c* from which we find that $V_Z \approx V_{CC} \approx 5$ V at cutoff and $V_Z = V_{CE,\text{sat}} = 0.3$ V in saturation. Alternatively, we could have arrived at the same result using the models of Fig. 13.3-2 by recognizing that there is zero-voltage drop across R_c at cutoff (since $i_C = 0$) and therefore $V_Z = V_{CC}$. The other value is obtained by noting that $V_Z = V_{CE,\text{sat}}$ at saturation.

The next step is to determine whether the transistor becomes saturated and/or cutoff. Cutoff, if it is going to occur at all, will occur for the low value of V_X. Since the input to a gate is frequently the output of a similar gate, we use $V_X = 0.3$ V. A simplified circuit for the inverter, in which the base circuitry is replaced by its Thevenin equivalent, is shown in Fig. 13.3-3*b*. Proof of this equivalence is left for the homework problems. (See Prob. 13.3-3.) Substituting $V_X = 0.3$ V results in $V_B = -0.182$ V. This negative base-to-emitter voltage indicates that the transistor is cut off for $V_X = 0.3$ V.

The higher input voltage will cause the transistor to become saturated if it becomes saturated at all. Using $V_X = 5$ V results in $V_B = 4.09$ V.

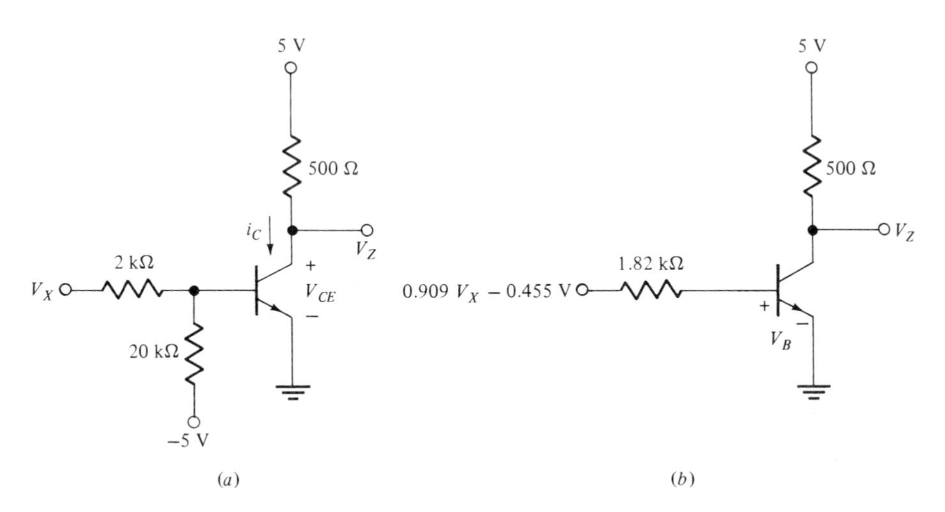

Fig. 13.3-3 Transistor inverter for Example 13.3-1. (*a*) Inverter circuit; (*b*) equivalent circuit with input circuit replaced by its Thevenin equivalent.

Assume the transistor is saturated for the moment; then $V_{BE} = V_{BE,\text{sat}} = 0.7$ V and $I_B = (4.09 - 0.7)/1.82 = 1.86$ mA. Since $V_{CE,\text{sat}} = 0.3$ V we have

$$I_{C,\text{sat}} = \frac{5 - 0.3}{0.5} = 9.4 \text{ mA}$$

From the collector characteristics

$$\beta = \left.\frac{\Delta i_C}{\Delta i_B}\right|_{V_{CE}=\text{constant}} = 20$$

Therefore $\beta I_B = 37.2$ mA. This is much greater than $I_{C,\text{sat}}$, hence the transistor is well into saturation.

The final part of the analysis is to find the range of input voltages that will result in proper inverter operation. Operating the transistor so it is at cutoff or saturated causes the output voltage to remain constant so long as the transistor remains at cutoff or saturated. Therefore any input voltage below the maximum voltage resulting in cutoff will lead to a high output and is an acceptable value for V_0; any value above the minimum required for saturation will result in a low output and is an acceptable value of V_1. For cutoff, assuming the transistor is silicon as implied by $V_{CE,\text{sat}}$ and $V_{BE,\text{sat}}$, the value of V_B must be less than zero. Therefore

$$0.909\ V_X - 0.455 < 0$$

which leads to $V_X < 0.501$ V for cutoff. For saturation we must have

$$I_B > \frac{I_C}{\beta} = \frac{V_{CC} - V_{CE,\text{sat}}}{\beta R_C} = \frac{5 - 0.3}{(20)\,(0.5)} = 0.47 \text{ mA}$$

From the base circuit we have

$$I_B = \frac{(0.909V_x - 0.455) - 0.7}{1.82} > 0.47 \text{ mA}$$

or $V_X > 2.21$ V for saturation. ////

13.3-4 Inverter Design

We conclude this section with a discussion of the design procedures for a transistor inverter. The designer must select values of V_{CC}, V_{BB}, R_C, R_1, R_2 and a transistor based on certain specifications which he is given. The values of V_{CC} and V_{BB} are frequently dictated by the supply voltages available in the system in which the inverter is to be used. The transistor selection is based on its being able to supply adequate power to drive the associated circuitry, its switching speed, its cost (frequently one of the most important factors), availability, and ability to withstand environmental conditions (an important requirement for military and aerospace applications). Once these decisions have been made the collector resistance is generally chosen to provide the largest collector current which does not exceed the maximum allowable collector power dissipation at saturation. This is done to provide a large input voltage difference between cutoff and saturation so that small noise voltages will be less likely to cause a false output value. Note that the collector power dissipation, i.e., $v_{CE}i_C$, is zero for cutoff. (Why?)

Finally the values of R_1 and R_2 are chosen so as to result in cutoff and saturation when the input voltages are the low and high output values, respectively. This is done based on the assumption that the input to the gate is the output of a similar gate. At this point the designer also ensures that there is a sufficient range to handle anticipated variations in the inputs. The following example illustrates this design procedure.

Example 13.3-2 Design a transistor inverter using a silicon transistor whose parameters are $V_{CE,\text{sat}} = 0.3$ V, $V_{BE,\text{sat}} = 0.7$ V, $\beta_{\min} = 40$ (we use the minimum value of β specified by the manufacturer so as to satisfy worst-case conditions) and whose maximum collector dissipation is 20 mW. Supply voltages of +5 V and −5 V are available. The circuit should not draw more than 20 mA from any of its voltage sources.

Solution The basic circuit configuration is shown in Fig. 13.3-4*a*. The output voltage at cutoff is 5 V since the collector current is 0. The output voltage at saturation is $V_{CE,\text{sat}} = 0.3$ V. The input voltages are assumed to be derived from a similar gate and are therefore 5 V and 0.3 V. Typical input and output waveforms are shown in Fig. 13.3-4*a*.

A simplified circuit, in which the input portion is replaced by its Thevenin equivalent, is shown in Fig. 13.3-4*b*. A model for the circuit at cutoff in which the input voltage is 0.3 V is shown in Fig. 13.3-4*c* while a model for the saturated circuit, with a 5 V input, is shown in Fig. 13.3-4*d*.

From the cutoff model it is seen that $i_C = 0$ and, as a result, $p_C = v_{CE}i_C = 0$. At saturation the collector dissipation is

$$p_C = v_{CE,\text{sat}}i_C = 0.3i_C$$

This value cannot exceed 20 mW, which corresponds to

$$i_C = \frac{20}{0.3} = 66.7 \text{ mA}$$

which is greater than the maximum allowable value of 20 mA. Therefore, the value $i_C = 20$ mA is used at saturation. The corresponding value of R_C, obtained from Fig. 13.3-4*d*, is

$$R_C = \frac{5 - 0.3}{20} = 0.235 \text{ k}\Omega$$

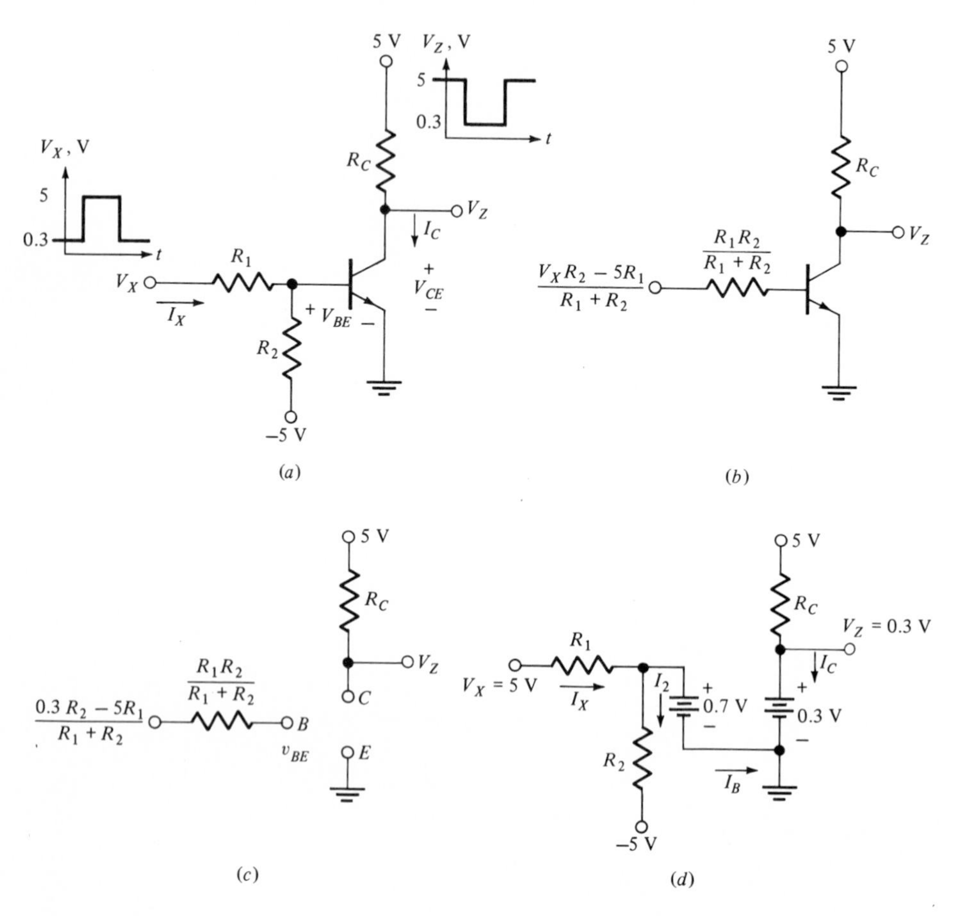

Fig. 13.3-4 Transistor inverter for Example 13.3-2. (*a*) Original circuit; (*b*) simplified circuit with input replaced by Thevenin equivalent; (*c*) equivalent circuit for cutoff region; (*d*) equivalent circuit for saturation region.

The values of R_1 and R_2 are chosen to result in cutoff and saturation for low and high inputs, respectively. From Fig. 13.3-4*c*, in order that $v_{BE} < 0$ (the condition for cutoff in silicon transistors), the following must be satisfied:

$$0.3R_2 < 5R_1 \tag{13.3-1}$$

The condition for saturation is $\beta I_B > I_C$. Substituting $\beta = 40$, $I_C = 20$ mA, and solving for I_B from Fig. 13.3-4*c* results in

$$40(I_X - I_2) = 40\left(\frac{5 - 0.7}{R_1} - \frac{0.7 + 5}{R_2}\right) > 20 \tag{13.3-2}$$

To satisfy Eq. (13.3-1), R_1 is selected so that $R_1 = 0.1R_2$, which is substituted into Eq. (13.3-2) and results in

$$40\left(\frac{4.3}{0.1R_2} - \frac{5.7}{R_2}\right) > 20$$

whose solution is $R_2 < 74.6$ kΩ. Choosing $R_2 = 68$ kΩ and $R_1 = 6.8$ kΩ satisfies Eqs. (13.3-1), and (13.3-2) and keeps the currents I_2 and I_X less than 20 mA. ////

13.4 DIODE-TRANSISTOR GATES

We have now provided electronic circuit implementations for the three switching algebra operations: AND, OR, and NOT. The AND and OR circuits are implemented using diodes; the NOT gate uses a transistor.

A disadvantage of the diode AND and OR circuits is the signal degradation inherent in them due to their passive nature. The inverter circuit is not subject to this problem because it is an active circuit. It would obviously be desirable to eliminate the signal degradation associated with the AND and OR gates. This can be done by using transistors instead of diodes as is indeed done in the case of integrated circuits in which a transistor is as cheap to produce and as small as a diode. In the case of discrete circuits, i.e., circuits in which each element is a separate device and not part of a single IC device, diodes are smaller and cheaper than transistors. For this reason it is advantageous to combine the diode AND and OR gates with the signal restoration properties of the transistor inverter. This is precisely the reason behind two additional kinds of electronic gate circuits called the NAND gate and NOR gate.

13.4-1 NAND Gate

The NAND gate is a combination of an AND gate and an inverter, the word NAND being a contraction for NOT and AND. It may be constructed by connecting the output of a diode AND gate to the input of a transistor inverter. This realizes the cost and size advantage of the diode circuit, and in addition restores the signal levels. The logic symbol for this gate is shown in Fig. 13.4-1*a*, and we see that it is a composite of the AND gate symbol followed by the circle which represents the inversion. The output of this gate is the inverse of the products of the inputs. If, as in Fig. 13.4-1, the inputs are X and Y and the output is Z, then

$$Z = \overline{X \cdot Y} \tag{13.4-1}$$

which, by DeMorgan's law, can also be written as

$$Z = \bar{X} + \bar{Y} \tag{13.4-2}$$

The circuit diagram for a diode-transistor NAND gate is shown in Fig.

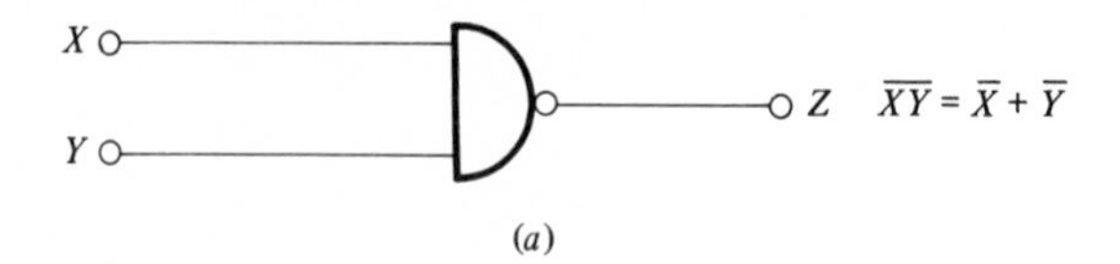

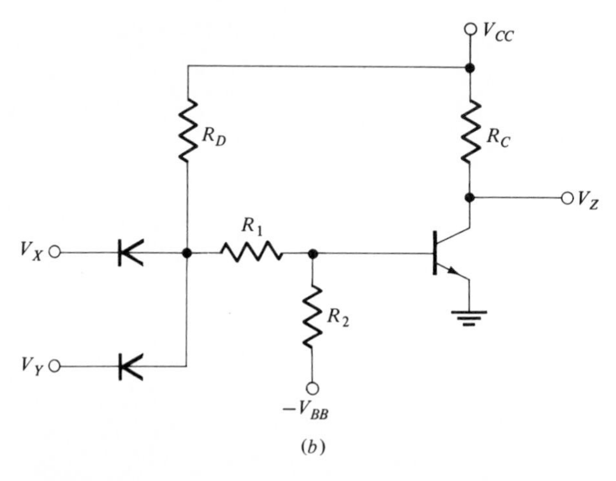

Fig. 13.4-1 NAND gate. (*a*) Logic symbol; (*b*) circuit diagram.

13.4-1*b*. The analysis performed for these circuits previously is valid with the one additional consideration of the loading effect on the AND gate by the inverter; this loading effect is frequently small enough to be neglected.

13.4-2 NOR Gate

The NOR gate is a combination of an OR gate and a NOT gate, the word NOR being a contraction of NOT and OR. Like the NAND gate, it provides the opportunity to realize the size and economy advantages of the diode OR circuit and the signal restoration advantages of the transistor inverter.

The logic symbol for the NOR gate is shown in Fig. 13.4-2*a*; it is an OR symbol followed by the NOT symbol. Its output, Z, is related to its inputs X and Y by:

$$Z = \overline{X + Y} \tag{13.4-3}$$

which, by DeMorgan's law, may also be written as

$$Z = \bar{X} \cdot \bar{Y} \tag{13.4-4}$$

The circuit diagram for a diode-transistor NOR gate is shown in Fig. 13.4-2*b*.

A few general comments are in order concerning electronic gates in general. The first is concerned with integrated circuits. The widespread use of ICs has

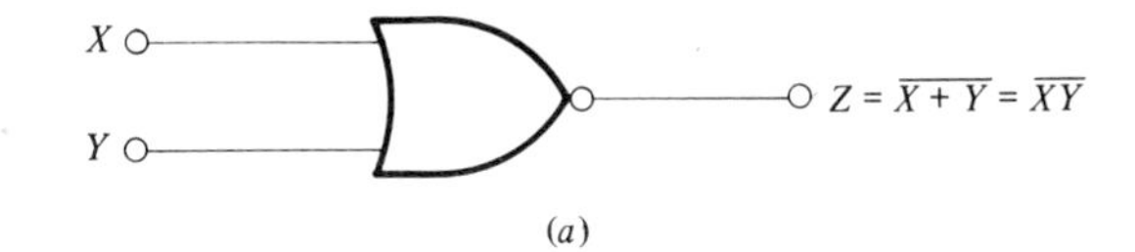

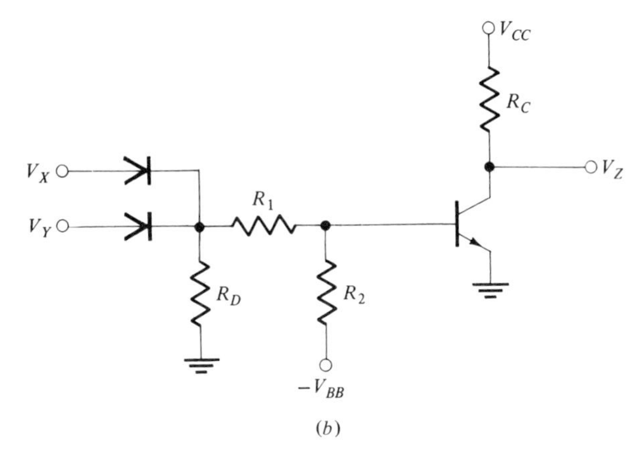

Fig. 13.4-2 NOR gate. (*a*) Logic symbol; (*b*) circuit diagram.

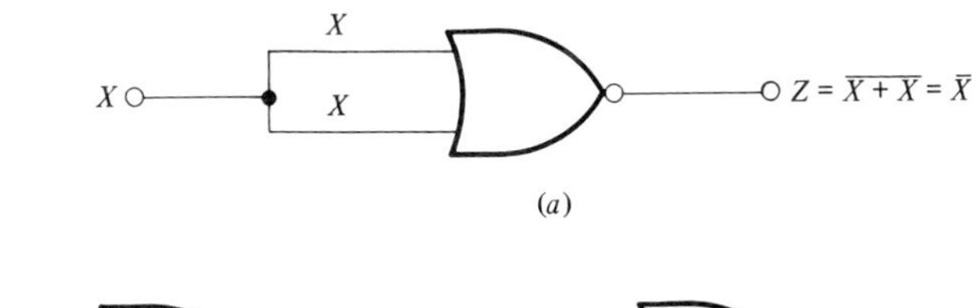

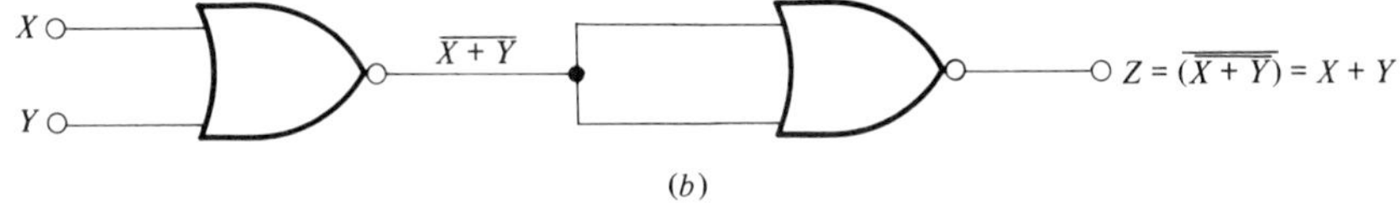

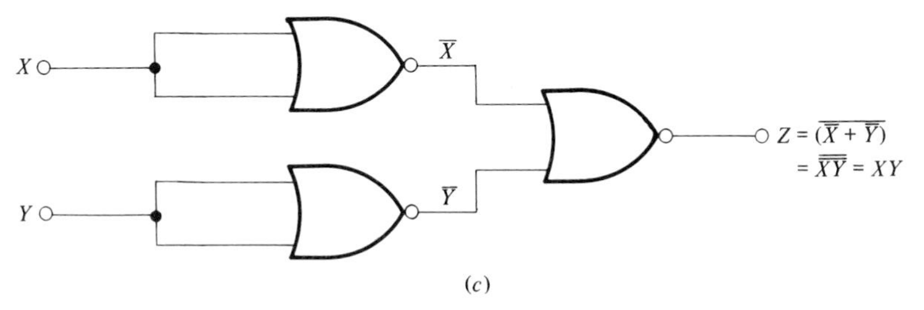

Fig. 13.4-3 Implementation of basic logic functions with NOR gates. (*a*) NOT; (*b*) OR; (*c*) AND.

pretty well displaced the use of diode-transistor logic gates by gates which are constructed entirely of transistors, or the newer field effect transistors (FET's). Thus, NAND, NOR, AND, and OR gates, as well as NOT gates, are constructed using transistors or FET's rather than diodes or diode-transistor combinations.

The second comment relates to NAND and NOR circuits. We have introduced them as a convenient means for combining the advantages of diode and transistor circuits. It would seem, then, that the NAND and NOR circuits would be less useful in IC's where we do not use diodes. However, there is another advantage of the NAND and the NOR circuit. Each of them is a universal element which can be used to implement any given logic function. Although it is true that a circuit realized using only one type of gate will not be as economical in terms of the number of transistors used, it provides the advantages of a single standardized building block, i.e., only one kind of element need be stocked, all gates are interchangeable, testing routines can be standardized, etc.

Figure 13.4-3 shows the implementation of the three basic logic functions using only NOR gates.

13.5 DESIGN OF ELECTRONIC DIGITAL CIRCUITS

The design of electronic digital circuits is a simple procedure once the design of switching circuits has been mastered. It is helpful to consider the analysis problem first.

In the analysis problem we are given a digital circuit and wish to find how it operates. We do this by writing an algebraic (switching algebra) description of the circuit from which we may, if we choose, construct a truth table. Since we have already considered the technique for obtaining a truth table from a switching function, we shall only consider the problem of finding the switching function for a given digital circuit.

The technique consists of labeling each intermediate signal in the circuit, writing an equation for each of these variables in terms of the inputs to the gate which this variable is the output of, and then substituting these equations into each other so as to find the final output variable in terms of the original inputs. Each of the switching functions derived along the way should be simplified so as to result in the simplest possible final result. This procedure is illustrated by the following example.

Example 13.5-1 The digital circuit of Fig. 13.5-1 has three inputs X_1, X_2, and X_3 and a single output Z. Find an expression for Z in terms of X_1, X_2, and X_3.

Solution We first label all the gate outputs $Y_1, Y_2, \ldots, Y_6$. We then

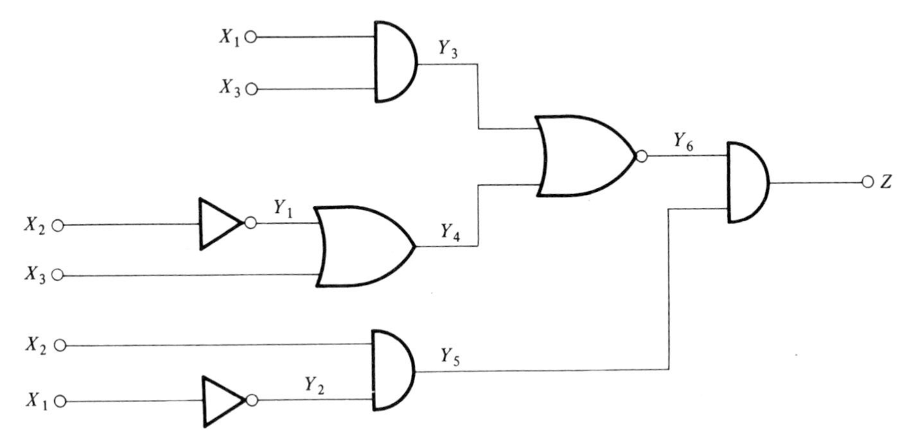

Fig. 13.5-1 Digital circuit for Example 13.5-1.

write the following equations:

$$Y_1 = \overline{X_2} \tag{13.5-1}$$

$$Y_2 = \overline{X_1} \tag{13.5-2}$$

$$Y_3 = X_1X_3 \tag{13.5-3}$$

$$Y_4 = Y_1 + X_3 \tag{13.5-4}$$

$$Y_5 = X_2Y_2 \tag{13.5-5}$$

$$Y_6 = \overline{Y_3 + Y_4} = \bar{Y}_3\bar{Y}_4 \tag{13.5-6}$$

$$Z = Y_6Y_5 \tag{13.5-7}$$

We then substitute Eq. (13.5-1) into Eq. (13.5-4), Eq. (13.5-2) into Eq. (13.5-5), Eqs. (13.5-3) and (13.5-4) into Eq. (13.5-6) and Eqs. (13.5-5) and (13.5-6) into Eq. (13.5-7) simplifying where possible:

$$Y_4 = \overline{X_2} + X_3 \tag{13.5-8}$$

$$Y_5 = X_2\overline{X_1} \tag{13.5-9}$$

$$Y_6 = \overline{X_1X_3}\,\overline{\overline{X_2} + X_3}$$

Using DeMorgan's law [Eq. (12.5-16)] we obtain

$$\begin{aligned} Y_6 &= (\overline{X_1} + \overline{X_3})(X_2\overline{X_3}) \\ &= \overline{X_1}X_2\overline{X_3} + X_2\overline{X_3} \\ &= (\overline{X_1} + 1)X_2\overline{X_3} = X_2\overline{X_3} \end{aligned} \tag{13.5-10}$$

Finally, using Eqs. (13.5-9) and (13.5-10),

$$Z = (X_2\overline{X_3})(X_2\overline{X_1}) = \overline{X_1}X_2\overline{X_3} \tag{13.5-11}$$

which is the final result. ////

In the design of an electronic digital circuit we are given some kind of specification for the circuit: a verbal description of its desired behavior, a truth table, or a switching function and we wish to obtain an electronic digital circuit which satisfies the given specification. The design procedure is similar to that for a switching circuit; we convert the specification from whatever form it is given into a minimal switching function. Then, instead of drawing the switching circuit, we draw the electronic digital circuit using electronic gates. This procedure is illustrated in the following example.

Example 13.5-2 Design an electronic digital circuit that is described by the truth table given below which defines the exclusive or (XOR) function, that is, $Z = 1$ if X or Y, but not both, are 1.

X	Y	Z
0	0	0
0	1	1
1	0	1
1	1	0

Solution We first find a switching function for Z (see Sec. 12.3):

$$Z = \bar{X}Y + X\bar{Y} \tag{13.5-12}$$

This switching function is minimal, i.e., cannot be simplified any further, and we may draw a circuit to implement it as shown in Fig. 13.5-2*a*.

Alternatively, we might perform the following algebraic manipulations. We begin by using the NOT operation twice on Eq. (13.5-12), which doesn't change it, since $\bar{\bar{Z}} = Z$:

$$Z = \overline{\overline{\bar{X}Y + X\bar{Y}}} \tag{13.5-13}$$

Using DeMorgan's theorems we obtain

$$Z = \overline{\overline{(\bar{X}Y)} \cdot \overline{(X\bar{Y})}} = \overline{(X + \bar{Y})(\bar{X} + Y)} \tag{13.5-14}$$

which when multiplied out produces

$$Z = \overline{X(\bar{X} + Y) + \bar{Y}(\bar{X} + Y)} \tag{13.5-15}$$

$$= \overline{X\bar{X} + XY + \bar{Y}\bar{X} + \bar{Y}Y} \tag{13.5-16}$$

Since $X\bar{X} = Y\bar{Y} = 0$

$$Z = \overline{XY + \bar{X}\bar{Y}} \tag{13.5-17}$$

Using DeMorgan's theorem again results in

$$Z = (\bar{X} + \bar{Y})(X + Y) \tag{13.5-18}$$

Finally, applying DeMorgan's theorem to the first factor provides the final

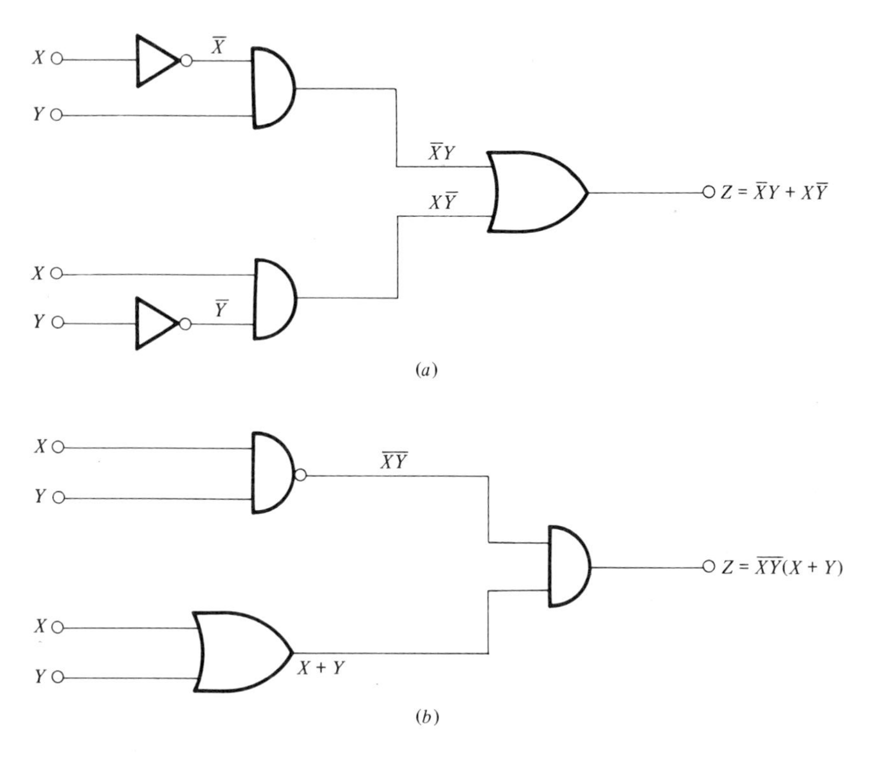

Fig. 13.5-2 Digital circuit for XOR function. (*a*) Circuit for $Z = \bar{X}Y + X\bar{Y}$; (*b*) circuit for $Z = \overline{XY}(X + Y)$.

result

$$Z = \overline{XY}(X + Y) \tag{13.5-19}$$

The idea to do this is certainly not self evident to the novice digital logic designer; one who is experienced in doing this sort of thing would recognize that such a manipulation might be helpful. The circuit for Eq. (13.5-19) is shown in Fig. 13.5-2*b*. ////

The preceding example illustrates an important aspect of electronic digital circuits related to the economy of these circuits.

In comparing the economy of two switching circuits it is possible to use the number of contacts as a measure of the cost of a circuit. It is found that this figure is minimized by reducing the switching function to its simplest form.

The situation with electronic circuits is not quite so simple since there are a variety of logic gates built with different types of elements and the cost is a function of the type of element, i.e., integrated circuit versus discrete circuit as well as the logic function. Despite these difficulties some measure of circuit cost must be adopted, even if it is somewhat arbitrary. A common measure is the number of gate inputs, since this is proportional to gate complexity for many types of circuits. Thus, the smaller the total number of gate inputs, the more economical the circuit. Comparing the circuits of Figs. 13.5-2*a* and 13.5-2*b* we find they have eight- and six-gate inputs, respectively. This makes the second circuit more economical despite the fact that both switching functions have the same number of literals and operations.

This lack of correlation between the simplicity of a switching function and economy of its corresponding circuit is a consequence of two factors: (1) the switching functions implemented by the gates are not necessarily the basic logical operations of AND, OR, and NOT, i.e., NAND and NOR functions are available, and (2) it is possible to synthesize subexpressions which appear in a function numerous times just once, and then connect the resulting output to many gate inputs. This technique is more limited in the case of switching circuits because of their topology, i.e., the physical interconnections of the switches. Furthermore, the optimum switching function is determined by the type of gates which may be used in a given circuit.

As a consequence of the discussion above, the minimization of an electronic digital circuit is not as straightforward as is that of a switching circuit. To a large extent it is based on the designer's experience, cleverness, and the amount of time he or she is willing to expend on looking into various possibilities. The remainder of this chapter is devoted to a few examples which, in addition to providing insight into this process, represent important types of digital systems.

13.6 CODE CONVERSION

It is very evident that numbers are of central importance in a digital computer. They are equally important in digital instrumentation and many other types of digital systems.

People are accustomed to using the decimal number system, a result of their being born with ten fingers. The decimal number system is not, in general, the most efficient system to use in digital systems, and in order to convert from one system to another code converters are used.

13.6-1 Binary Number System

The decimal number system is one example of a constant *radix,* or *base,* number system. Its radix or base is ten. This means that in counting, the units digit starts at 0 and is increased in steps of one to provide the numbers 1, 2, 3, 4, 5, 6, 7, 8, and 9. On the tenth count the units digit resets to 0, but the digit on its left is increased by one, resulting in the number 10. Once again the units digit is incremented through its range up to the number 9. On the next count it resets to 0, and the digit on its left is increased by one forming the number 20.

The binary number system has a base, or radix, of two and it uses only the digits 0 and 1. Counting in the binary number system follows the same pattern as in the decimal system subject to the different sets of digits. Starting with 0 the number is increased by one to form the number 1. The next count requires that this 1 reset to 0 since there are no larger digits. When this happens the digit to its left must increase to 1 thus forming the binary number 10 which corresponds to the decimal number 2. The binary numbers, starting from 0, are:

$$0, 1, 10, 11, 100, 101, 110, 111, 1000, 1001, \ldots$$

The value of a decimal number is found by multiplying each digit by its place value, which is a power of ten. For example:

$$\begin{aligned} 258 &= 2 \times 100 + 5 \times 10 + 8 \times 1 \\ &= 2 \times 10^2 + 5 \times 10^1 + 8 \times 10^0 \end{aligned}$$

The value of a binary number is found in the same way but the place values are powers of two. Thus, for example,

$$\begin{aligned} (1001011)_2 &= 1 \times 2^6 + 0 \times 2^5 + 0 \times 2^4 + 1 \times 2^3 \\ &\quad + 0 \times 2^2 + 1 \times 2^1 + 1 \times 2^0 \\ &= 1 \times 64 + 0 \times 32 + 0 \times 16 + 1 \times 8 \\ &\quad + 0 \times 4 + 1 \times 2 + 1 \times 1 \\ &= 64 + 0 + 0 + 8 + 0 + 2 + 1 = (75)_{10} \end{aligned}$$

where the parentheses around the number and the subscript 2 or 10 indicate binary or decimal numbers, respectively. If no base is indicated, a number is assumed to be decimal unless the context clearly indicates otherwise.

The fractional part of a decimal number is represented by digits to the right of the decimal point. Their place values are $10^{-1} = 1/10 = 0.1$, $10^{-2} = 1/100 = 0.01$, $10^{-3} = 1/1{,}000 = 0.001$, etc. In the binary system the fractional part of a number is represented by digits to the right of the *binary* point. Their place values are $2^{-1} = 1/2 = 0.5$, $2^{-2} = 1/4 = 0.25$, $2^{-3} = 1/8 = 0.125$, and so on. Thus

$$\begin{aligned} (0.101)_2 &= 0 \times 2^0 + 1 \times 2^{-1} + 0 \times 2^{-2} + 1 \times 2^{-3} \\ &= 0 \times 1 + 1 \times 0.5 + 0 \times 0.25 + 1 \times 0.125 \\ &= 0 + 0.5 + 0 + 0.125 = (0.625)_{10} \end{aligned}$$

The conversion of a binary number containing an integral and fractional part is illustrated in the following example.

Example 13.6-1 Find the decimal equivalent of $(101011.011)_2$.

Solution Proceeding as above, we obtain

$$(101011.011)_2 = 1 \times 2^5 + 0 \times 2^4 + 1 \times 2^3 + 0 \times 2^2 + 1 \times 2^1 + 1 \times 2^0$$
$$+ 0 \times 2^{-1} + 1 \times 2^{-2} + 1 \times 2^{-3}$$

We evaluate the right hand side, remembering that $2^{-n} = 1/2^n$,

$$\begin{aligned}(101011.011)_2 &= 1 \times 32 + 0 \times 16 + 1 \times 8 + 0 \times 4 + 1 \times 2 + 1 \times 1 \\ &\quad + 0/2^1 + 1/2^2 + 1/2^3 \\ &= 32 + 0 + 8 + 0 + 2 + 1 + 0/2 + 1/4 + 1/8 \\ &= 43 + 0 + 0.25 + 0.125 \\ &= (43.375)_{10}\end{aligned}$$

////

13.6-2 Binary Coded Decimal (BCD) Numbers

We have indicated that the binary number system is convenient for use in electronic digital circuits because of the large variety of bistable electronic devices. It is also possible to show that the binary number system permits a more economical representation of numbers than does the decimal system; that is, less physical equipment is required to represent any number in a given range using the binary system than is the case when the decimal system is used. The one big disadvantage of the binary system is its inconvenience for humans who have been trained to use the decimal system. Apart from their lack of familiarity with the binary system, people are prone to make errors using it because of the long strings of zeros and ones.

TABLE 13.6-1 Decimal—BCD Conversion

Decimal	BCD
0	0000
1	0001
2	0010
3	0011
4	0100
5	0101
6	0110
7	0111
8	1000
9	1001

A compromise solution to the problem is the use of the *binary coded decimal* (BCD) system in which each decimal digit is individually coded into a four-digit binary number. The ten-decimal digits and their four-digit binary representations are shown in Table 13.6-1. This permits the use of relatively simple code converters which convert between a single decimal digit and four bits instead of larger numbers of digits. The conversions are demonstrated in the following example.

Example 13.6-2 Perform the following conversions:

a $(25.4)_{10} = A_{\text{BCD}}$

b $(1001\ 0111.0011\ 0010)_{\text{BCD}} = B_{10}$

Solution

a Convert each decimal digit to a four-bit number of the form $a_3a_2a_1a_0$ where the decimal digit is equal to $8a_3 + 4a_2 + 2a_1 + a_0$ as shown in Table 13.6-1. Thus

$$(25.4)_{10} = (\overbrace{0010}^{2}\ \overbrace{0101}^{5}.\overbrace{0100}^{4})_{\text{BCD}}$$

b Group the BCD bits into groups of four working from the radix point to the left and right. Then convert each four-bit binary number into a single decimal digit.

$$(\overbrace{1001}^{9}\ \overbrace{0111}^{7}.\overbrace{0011}^{3}\ \overbrace{0010}^{2})_{\text{BCD}} = (97.32)_{10}$$

////

13.6-3 BCD to Decimal Code Converter

A converter as shown in Fig. 13.6-1*b* is frequently used at the output of a digital device to convert the internal BCD code to a decimal output. Such a converter has four inputs: X_3, X_2, X_1, and X_0, representing the binary values which are multiplied by the weights $2^3 = 8$, $2^2 = 4$, $2^1 = 2$, and $2^0 = 1$, respectively, to obtain the decimal output. The output of the code converter consists of ten different lines designated Z_0 through Z_9. For each decimal digit one of these has a logical 1 value while all of the others have logical 0 values.

The switching function for each output is a single complete product corresponding to the input combination for which it is equal to one. Thus, we have:

$$Z_0 = \overline{X_3}\,\overline{X_2}\,\overline{X_1}\,\overline{X_0}$$

if $X_0 = X_1 = X_2 = X_3 = 0$, then $Z_0 = 1$, which corresponds to decimal zero.

$$Z_1 = \overline{X_3}\,\overline{X_2}\,\overline{X_1}X_0$$

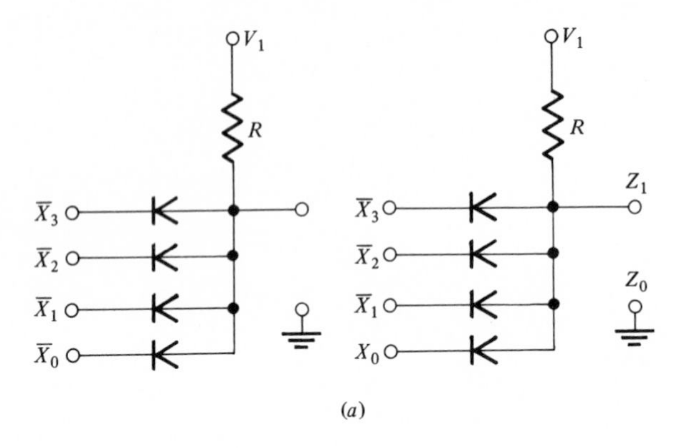

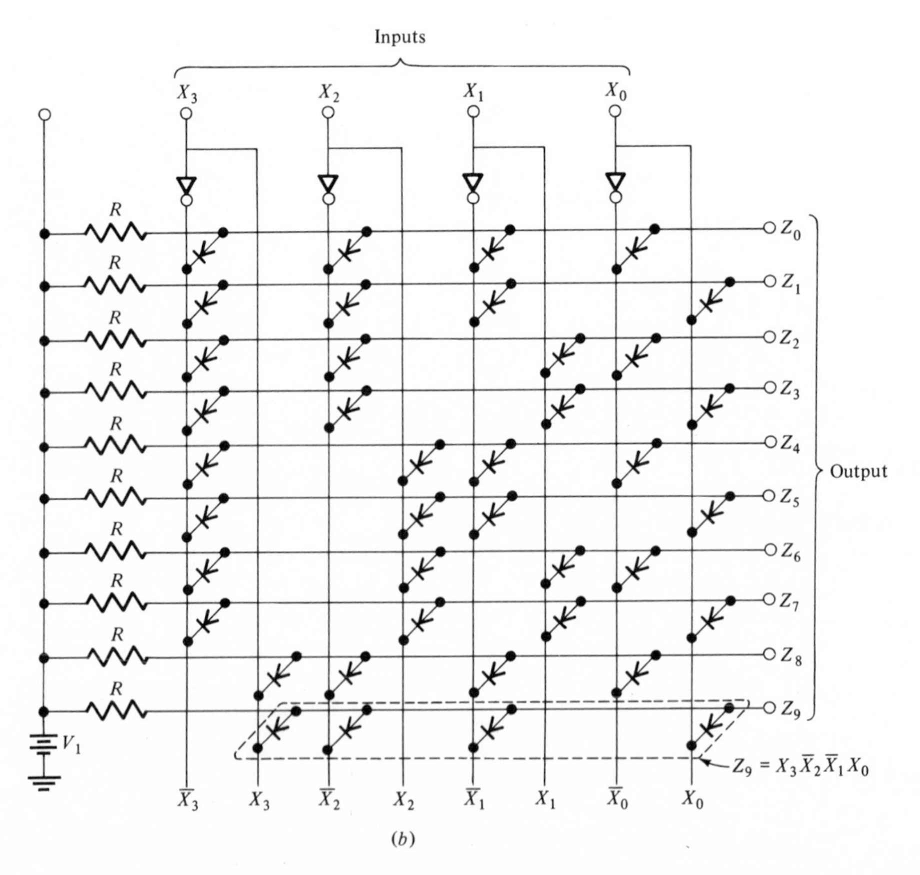

Fig. 13.6-1 BCD to decimal code converter. (*a*) AND circuit for decimal zero output; (*b*) inverters and diode matrix code converter.

if $X_0 = 1$, $X_1 = X_2 = X_3 = 0$, then $Z_1 = 1$, which corresponds to decimal 1.

$$Z_2 = \overline{X_3}\,\overline{X_2}X_1\overline{X_0}$$

$$Z_3 = \overline{X_3}\,\overline{X_2}X_1X_0$$

$$Z_4 = \overline{X_3}X_2\overline{X_1}\,\overline{X_0}$$

$$Z_5 = \overline{X_3}X_2\overline{X_1}X_0$$

$$Z_6 = \overline{X_3}X_2X_1\overline{X_0}$$

$$Z_7 = \overline{X_3}X_2X_1X_0$$

$$Z_8 = X_3\overline{X_2}\,\overline{X_1}\,\overline{X_0}$$

$$Z_9 = X_3\overline{X_2}\,\overline{X_1}X_0$$

Each of these outputs can be generated using a diode AND circuit if we have each of the input variables and its inverse available. The inverses can be obtained using four NOT gates.

The AND circuits for Z_0 and Z_1 are shown in Fig. 13.6-1*a*. The ten four-diode AND gates that are required are usually drawn as a diode matrix as shown in Fig. 13.6-1*b*. The student should take the time to carefully trace this circuit for several outputs in order to find the four inputs corresponding to each output and confirm the logic equation for that output. For example, the circuit for Z_9 is encircled with a dashed line and we see that $Z_9 = X_3\overline{X_2}\,\overline{X_1}X_0$.

13.7 BINARY ADDER

Addition is a basic arithmetic operation and one which must often be performed in calculators and instruments. In this section we investigate the design of binary adder circuits.

13.7-1 Binary Half-Adder

The first circuit we shall consider is one whose output is the sum of two single-digit binary inputs. Table 13.7-1 illustrates this addition. Note that when we

TABLE 13.7-1
Binary
Addition Table

A	*B*	Sum
0	0	0
0	1	1
1	0	1
1	1	10

TABLE 13.7-2
Truth Table for
Binary Half-Adder.

A	B	S	C
0	0	0	0
0	1	1	0
1	0	1	0
1	1	0	1

add 1 + 1 the sum is 0 and there is a *carry* of 1. For reasons which will become apparent later this circuit is called the binary *half-adder*.

We will designate the two inputs as A and B. The output consists of two bits: a sum bit designated S and a carry bit we will call C. From the binary addition table we obtain the truth table (Table 13.7-2) for the binary half-adder where the *logical* ones and zeros in the truth table correspond in this case to the *binary* ones and zeros in Table 13.7-1.

From Table 13.7-2 we obtain the two switching functions

$$S = \bar{A}B + A\bar{B} \tag{13.7-1}$$

and

$$C = AB \tag{13.7-2}$$

We recognize the function for S as the exclusive or (XOR) function which was synthesized in Example 13.5-2; the circuit is shown in Fig. 13.5-2*b*. If we use a separate AND gate and inverter, in place of the NAND gate, we obtain the carry output, C, as well as the sum output, S. This is shown in Fig. 13.7-1*a* while a block diagram symbol, which may be used to represent this circuit in diagrams of more complex circuits, is shown in Fig. 13.7-1*b*.

13.7-2 Binary Full Adder

When two binary numbers are to be added together a separate circuit is used for each digit position. A circuit which forms the sum for a single bit position is

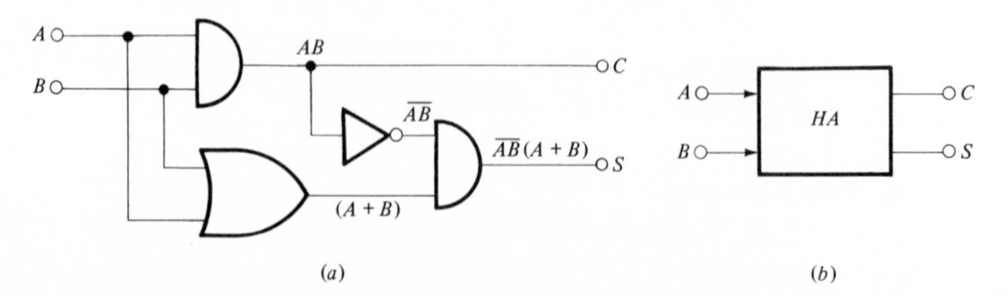

Fig. 13.7-1 Binary half-adder. (*a*) Logic circuit; (*b*) block diagram.

called a *binary full adder*. The binary full adder has three binary input digits: the two input bits to be added called the *addend* and the *augend*, and the carry from the previous bit position. For example, the binary digits 1 and 1 are added as follows

$$\begin{array}{r} 1 \leftarrow \text{augend} \\ +\ 1 \leftarrow \text{addend} \\ \hline \text{carry} \rightarrow 1\ \ 0 \leftarrow \text{sum} \end{array}$$

The full adder has two outputs; the sum bit and the carry bit to the next bit position.

The block diagram of a full adder is shown in Fig. 13.7-2*a*; A and B are the addend and augend input bits, D is the input carry from the previous bit position, S is the output sum, and C is the output carry.

The following binary additions can be verified by converting all numbers to decimal:

$$0 + 0 + 0 = 00$$

$$0 + 0 + 1 = 01$$

$$0 + 1 + 1 = 10$$

$$1 + 1 + 1 = 11$$

From these additions we see that the sum output of the full adder should be 1 when there are an odd number of 1 inputs and the carry output of a full adder should be 1 when there are two or more 1 inputs. The circuit of Fig. 13.7-2*b* satisfies these requirements. Recall that the carry output of a half-adder is 1 if both inputs are 1 and the sum output of a half-adder is 1 if a single one of its inputs is 1. Using this information we construct the truth table, Table 13.7-3, which verifies that the circuit of Fig. 13.7-2*b* does act like a full adder.

The student should check each row in this truth table using the properties of the half-adder and the circuit of Fig. 13.7-2*b*. It is now apparent that the half-

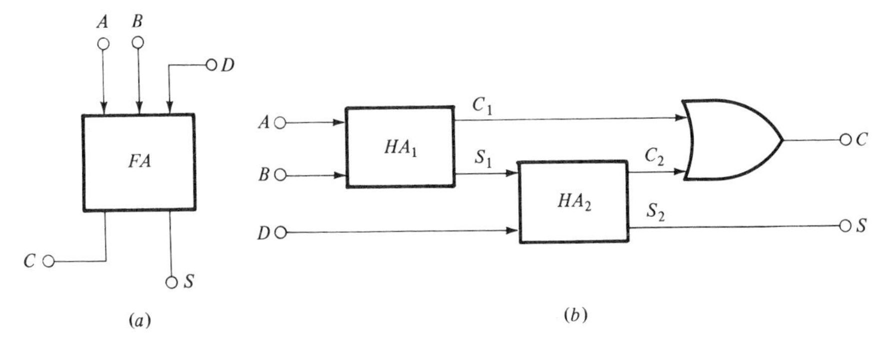

Fig. 13.7-2 Binary full adder. (*a*) Block diagram; (*b*) logic circuit.

TABLE 13.7-3 Truth Table for Full Adder

A	B	D	C_1	S_1	C_2	$S_2 = S$	$C = C_1 + C_2$
0	0	0	0	0	0	0	0
0	0	1	0	0	0	1	0
0	1	0	0	1	0	1	0
0	1	1	0	1	1	0	1
1	0	0	0	1	0	1	0
1	0	1	0	1	1	0	1
1	1	0	1	0	0	0	1
1	1	1	1	0	0	1	1

adder obtains its name from the fact that two of them are required to build a full adder.

13.7-3 Multibit Adder

A circuit which adds multibit binary numbers is constructed by connecting full adders together. Figure 13.7-3 shows the diagram of an adder which accepts two five-bit binary inputs, A and B, and produces a six-bit sum, S.

For example, if $A = 10011$ and $B = 11011$, then S is found as follows:

$$
\begin{aligned}
A &= 10011 \\
B &= \underline{11011} \\
S &= 101110
\end{aligned}
$$

The binary values throughout the circuit are shown in parentheses in Fig. 13.7-3.

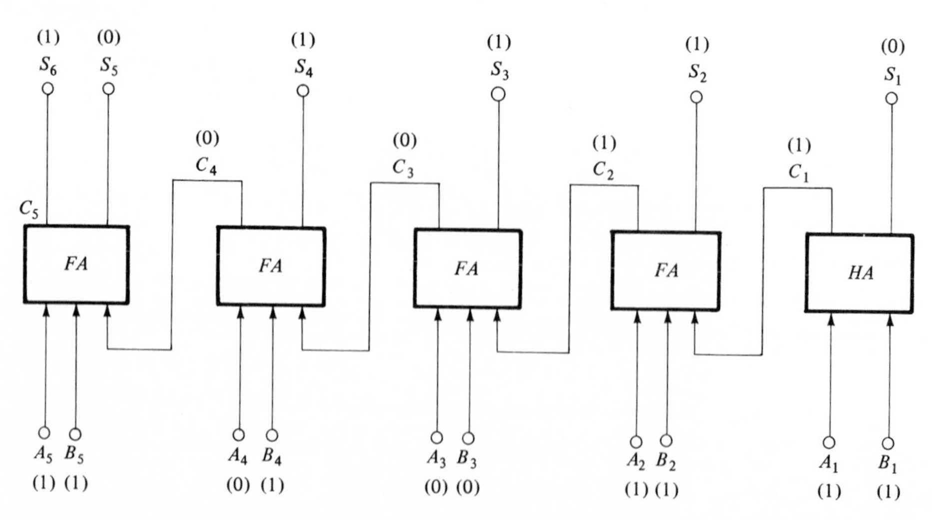

Fig. 13.7-3 Five-bit binary adder.

SUMMARY

1 Electronic digital circuits use voltage levels between a given point and ground to represent the value of a logic variable.

2 When the positive logic convention is used, 0 represents a relatively low voltage and 1 represents a relatively high voltage.

3 There are electronic circuits, called gates, which implement the logic operations AND, OR, NOT, NAND (NOT–AND), and NOR (NOT–OR).

4 AND and OR functions may be implemented using diode gates, but diode gates cause signal degradation because they are passive circuits. This limits the number of levels which may be used.

5 Inversion (the NOT function) requires an active circuit such as a transistor gate. Transistor gates, or other active gates, may be used to implement any function and restore output signal levels provided the input levels are not too far from their proper values.

6 The binary number system is a constant radix number system, like the decimal number system, but it uses a radix or base of 2 and is limited to digit values of 0 and 1.

7 The binary coded decimal (BCD) system represents each decimal digit by four bits (binary digits).

8 A multibit binary adder is built from full adders. Each full adder has three inputs: addend bit, augend bit, and carry bit from previous bit position, and two outputs: the sum and carry. A full adder is composed of two half-adders and an OR gate. Each half-adder adds two bits and provides a sum and carry output.

PROBLEMS

13.1-1 Draw the ideal diode circuits for the logic gates shown in Fig. P13.1-1. Assume logical 1 corresponds to 10 V and logical 0 to 0 V and 1 kΩ resistance values are used.

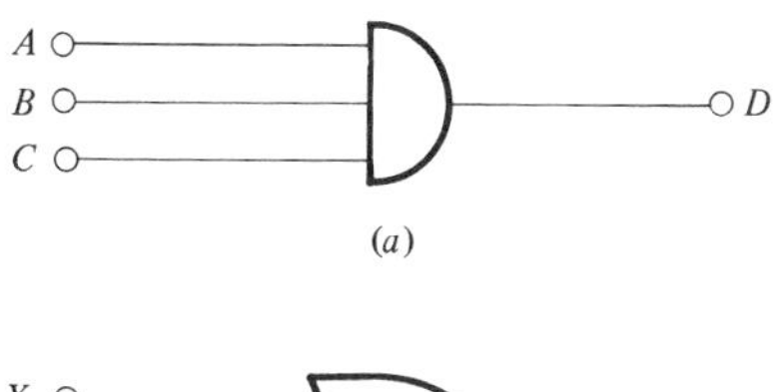

(a)

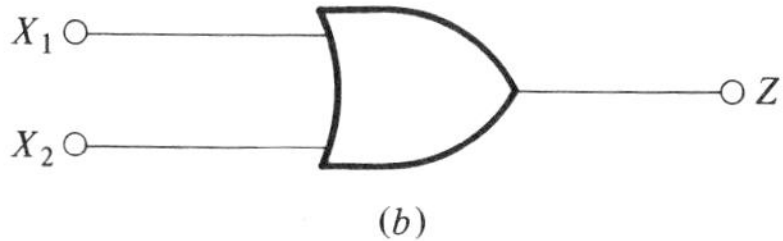

(b)

Fig. P13.1-1

13.1-2 Draw the logic gate circuits for the diode circuits shown in Fig. P13.1-2. What voltage levels correspond to the logical values 0 and 1?

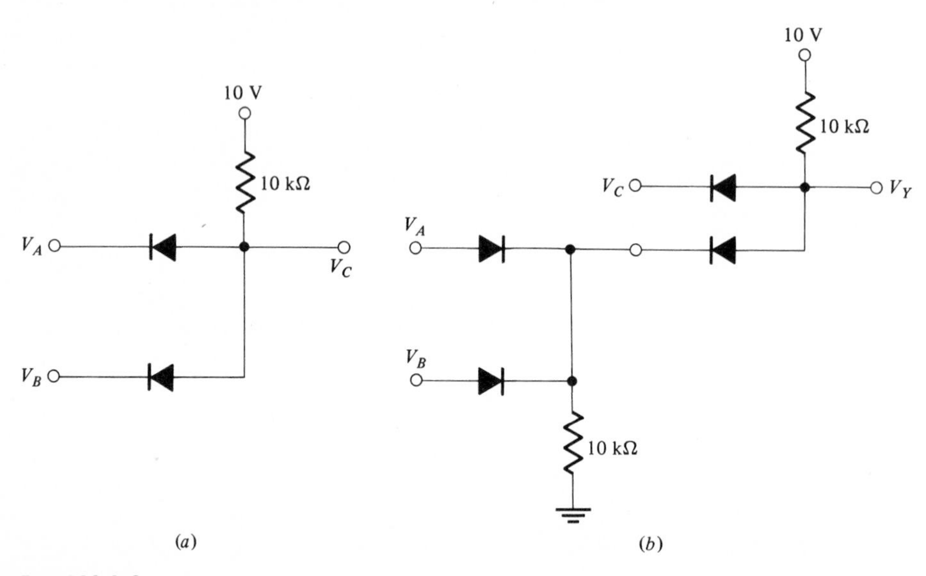

Fig. P13.1-2

13.2-1 The diode gates of Fig. P13.2-1 use a 10-V supply voltage and silicon diodes with $r_f = 50\ \Omega$ and $r_r = 2\ \text{M}\Omega$.

a Draw the complete diode gate circuit for each figure.

b Find the input voltage values corresponding to logical 0 and 1.

c Find the value of the load resistances in the gate circuits using the design formula given in the text.

d Find the values of the output voltage corresponding to Z for all combinations of input voltages representing X and Y values of 0 and 1.

e Compute the voltage values representing D and E for the indicated values of A, B, and C.

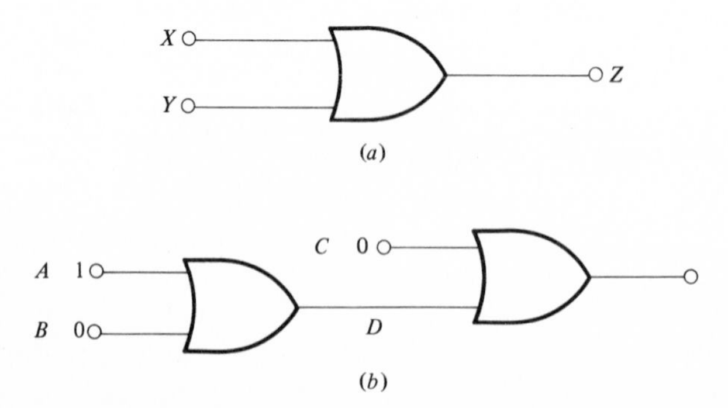

Fig. P13.2-1

13.2-2 Derive a general formula, making the usual approximations, for V_Z in the circuit of Fig. P13.2-2 if the gates are diode gates. What would the value of V_Z be if ideal diodes were used?

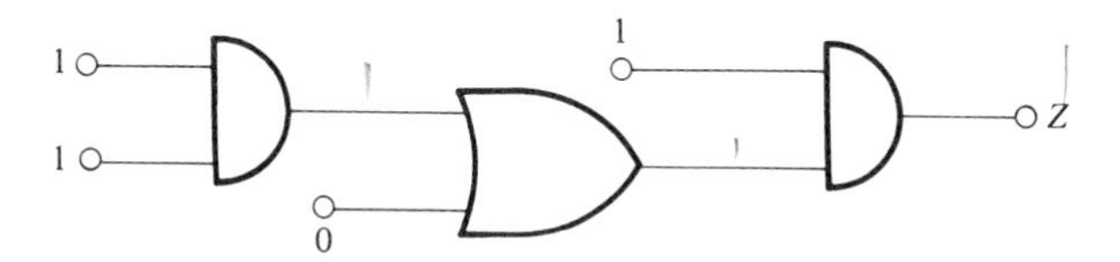

Fig. P13.2-2

13.2-3 The diode gates of Fig. P13.2-3 use a 10-V supply voltage and silicon diodes with $r_f = 100\ \Omega$ and $r_r = 1\ \text{M}\Omega$.

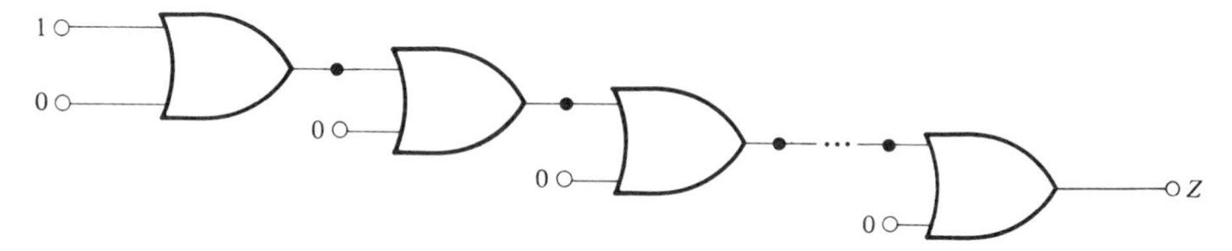

Fig. P13.2-3

a Determine the gate load resistance value using the design formula given in the text.

b What is the maximum number of gates that may be connected as shown while keeping the voltages representing logical 1 within 75 percent of their ideal value and keeping the voltages representing logical 0 less than 25 percent of the logical 1 value.

13.3-1 In the circuit of Fig. 13.3-1*b* $V_{CC} = 5$ V, $-V_{BB} = -5$ V, $R_C = 470\ \Omega$, $R_1 = 20\ \text{k}\Omega$, $R_2 = 100\ \text{k}\Omega$, $\beta = 100$, $V_{BE,\text{sat}} = 0.7$ V, and $V_{CE,\text{sat}} = 0.3$ V. Assuming the input V_X is the output of an identical stage, determine whether the circuit functions as an inverter.

13.3-2 The circuit of Fig. P13.3-2 is a transistor inverter which uses a diode in the base circuit instead of a negative supply voltage to ensure against false output signals due to small variations in the logical 0 input voltage. Assume the diode is a silicon diode with $r_f = 100\ \Omega$ and $r_r = 1\ \text{M}\Omega$. Assume the transistor is a silicon transistor with $\beta = 100$, $V_{CE,\text{sat}} = 0.3$ V and $V_{BE,\text{sat}} = 0.7$ V.

a If the transistor's maximum allowable collector power dissipation ($V_{CE}I_C$) is 50 mW, find the minimum value of R_C which may be used based on the saturation values. Set R_C equal to this value.

b Find V_0.

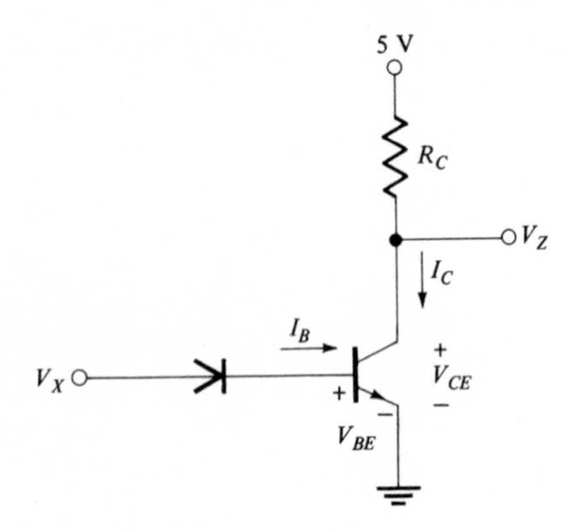

Fig. P13.3-2

c If $V_X = V_0$, the transistor is not cut off since there is no negative supply. Assuming $V_{BE} = 0$ for this condition, determine I_B.

d For the condition described in part *c* the transistor is in its active region. Determine I_C and V_Z. Is the transistor collector dissipation within the allowable range?

e Find V_Z if V_X is equal to the value of V_Z found in part *d*. Is the transistor in saturation?

f Find the minimum value of V_1 which, when applied at V_X, results in saturation of the transistor.

g Find the maximum value of V_0 which, when applied at V_X, results in the diode being reverse biased assuming $V_{BE} = 0$. What is the corresponding value of V_Z?

13.3-3 Design a transistor inverter using a +10-V collector supply and a −5-V base supply with a 1-kΩ collector resistance such that the nominal signal values are $V_0 = 0$ V and $V_1 = 10$ V, but the circuit will operate properly for $V_0 < 4$ V and $V_1 > 6$ V. The collector characteristics are shown in Fig. 13.3-1 and $V_{BE,\text{sat}} = 0.7$ V, $V_{CE,\text{sat}} = 0.3$ V.

13.3-4 Show the equivalence between the base circuits of Figs. 13.3-3*a* and 13.3-3*b*.

13.4-1 For the circuit of Fig. 13.4-1*b*, $R_D = R_1 = R_C = 1$ kΩ, $R_2 = 4$ kΩ, $V_{BB} = V_{CC} = 5$ V, the diodes are silicon and have $r_f = 50$ Ω and $r_r = 3$ MΩ. The transistor's characteristics are adequately approximated by those of Fig. 13.3-1 and $V_{BE,\text{sat}} = 0.7$ V.

a Find the output values V_0 and V_1 corresponding to logical 0 and 1, respectively.

b Find the output values corresponding to all combinations of input values using V_0 and V_1 as found in part *a*.

c Find $V_{1,\text{min}}$, the minimum input value of V_1, and $V_{0,\text{max}}$, the maximum input value of V_0, for which the circuit will provide the output values found in part *a*.

13.4-2 The circuit for a RTL (resistor-transistor logic) gate, which is sometimes used in place of DTL (diode-transistor logic) gates is shown in Fig. P13.4-2. This circuit can be adjusted to perform as either a NAND or a NOR gate. The transistor parameters are $V_{CE,\text{sat}} = 0.3$ V, $V_{BE,\text{sat}} = 0.7$ V, $\beta = 20$

a Find V_0 and V_1, the voltage levels corresponding to logical 0 and 1, for this circuit.

b Compute R so the transistor will be cut off with $V_{BE} = -0.2$ V when all the input voltages are equal to V_0.

c Compute R_c so the circuit is a NAND gate.

d Find a value of R_c which results in a NOR gate.

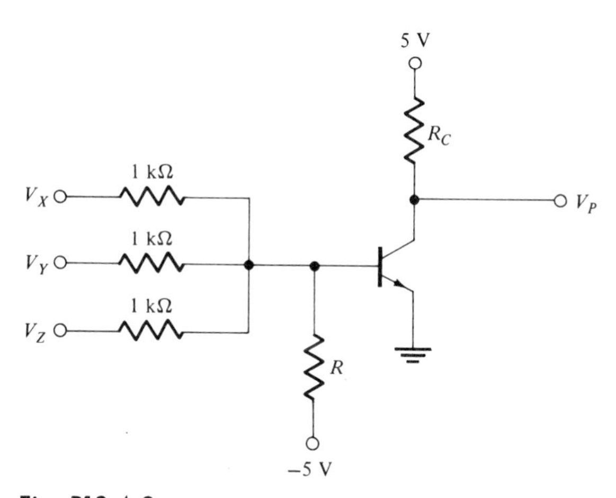

Fig. P13.4-2

13.4-3 Draw circuits to realize the AND, OR, and NOT functions using only NAND gates.

13.5-1 Design a voting machine which has four inputs, x_1, x_2, x_3, and x_4, and an output, z, which is 1 if a majority of the inputs are 1.

13.5-2 Three partners, A, B, and C, own 50, 30, and 20 percent, respectively, of a business. Build an electronic digital circuit which will provide a 1 output when 60 percent or more of the ownership votes yes on a motion. Each partner votes yes by pressing a button which results in a 1 input to the circuit.

13.5-3 An electronic digital circuit has four inputs and a single output, Z. Two of the inputs, V and W, form a binary number VW. The other two inputs, X and Y, form a second binary number XY. Each of these binary numbers has a decimal value between 0 and 3. The output is 1 if both numbers are even or both are odd; it is 0 if one number is even and the other is odd. Zero is considered an even number. Design the circuit described above.

13.6-1 Convert the following binary numbers to their decimal equivalents:

a 1011
b .011
c 1101001
d 101.101

13.6-2 Convert the binary numbers of Prob. 13.6-1 to their BCD equivalents.

13.6-3 Convert the following decimal numbers to their binary equivalents (do not go beyond four fractional binary places).

a 27
b 0.125
c 0.65
d 110.25

13.6-4 Convert the decimal numbers of Prob. 13.6-3 to BCD.

13.6-5 Convert the following BCD numbers to decimal.

a 1001 0110 0011
b 0111 0010 . 0100

13.6-6 Convert the BCD numbers of Prob. 13.6-5 to binary.

13.6-7 The octal number system is a constant radix number system with a

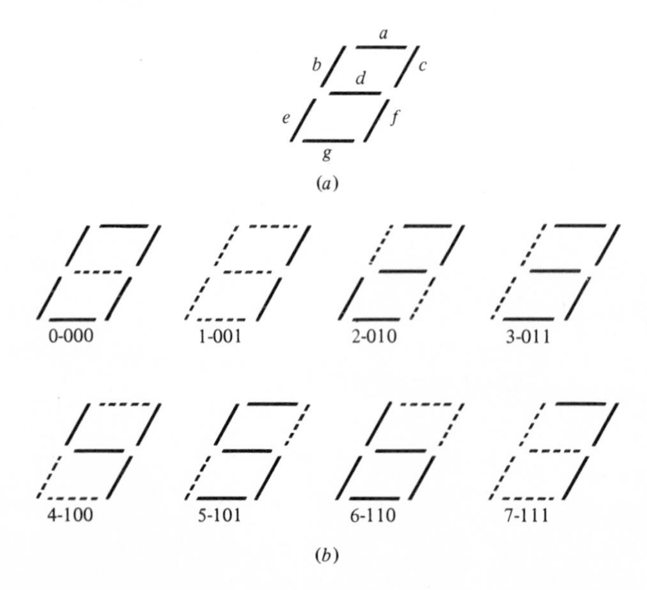

Fig. P13.6-7

base of 8. The digits in this system are 0, 1, 2, 3, 4, 5, 6, and 7. Since any octal digit can be represented by exactly three binary digits it is simple to convert between octal and binary; for this reason the octal system has been used extensively in the computer field, especially in minicomputers.

A display for octal numerals consists of seven LED (light emitting diode) segments which may be individually illuminated as shown in Fig. P13.6-7*a*. The segments which are illuminated to represent each octal digit are shown in Fig. P13.6-7*b* together with the octal number and its binary equivalent.

Design a circuit having the three inputs, X_1, X_2, and X_3, which represent the binary bits from left to right, and an output *e* which is 1 when the *e* segment of the LED display is to be illuminated.

13.7-1 The 2s complement of a binary number is obtained by changing all 0 bits to 1 bits, all 1 bits to 0 bits, and then adding 1 to the result. The 2s complement is frequently used in computers to implement the subtraction operation. Design a circuit using any of the logic gates we have discussed, as well as half- and full adders, whose output is the 2s complement of its four-bit input.

14
Calculus Essentials

OBJECTIVES

Upon completion of this chapter, you should be able to

1 Define average slope and find the average slope of a curve between given points.
2 State that instantaneous slope is the limit of the average slope between two points as the two points approach each other.
3 Find graphically the instantaneous slope of a curve at a given point.
4 Find average slope using the Δ method.
5 Find the derivatives of power, sinusoidal, and exponential functions.
6 Find the indefinite integral of power functions.
7 Use the definite integral to find area under a curve.
8 Find the running integral of a specified waveform.

INTRODUCTION

Up to this point all of the linear circuits considered have been resistive. This meant that all currents and voltages were related simply by a constant of proportionality. Thus, a step function input produced a step function response in all of the circuits we have thus far considered. A sinusoidal input produced sine-wave response with only the amplitude of the response differing from the input sinusoid; the input and output waveforms were in phase with each other.

At this point we are ready to begin consideration of the behavior of circuits containing inductors and capacitors. In order to understand the theory of such circuits in a meaningful way, it is necessary that the student first understand certain topics from the calculus. Specifically, these topics include the concepts of derivative and integral with respect to time. The treatment in this chapter is going to be on a physical basis and will be slanted towards those aspects of the topic which are most appropriate for beginning study of transient and sinusoidal response of simple RC and RL networks. The material is not meant to supplant that normally taught in a calculus course, but is presented here because many students using this text may not yet have reached the point in their math courses where this material is covered. In this "technological" treatment, we are going to emphasize physical interpretation of the derivative and integral along with graphical interpretations and methods.

14.1 RATE OF CHANGE AND SLOPE

In Chap. 1 we discussed linear equations of the form $y = mx + b$. The slope m is interpreted as the *rate of change* of y with respect to x ($\Delta y/\Delta x$) and is constant for linear equations, i.e., equations which plot as straight lines. In this chapter we are going to generalize this concept so that we can speak of the *instantaneous* slope of a curve. This leads us to the notion of the derivative, or instantaneous rate of change, which is used to describe the vi behavior of the inductor and capacitor. Before reading this section, the student should take a little time to review the material on linear equations in Sec. 1.2.

A useful way to think about rate of change (slope) is in terms of the familiar concept of speed as applied to an automobile. The speed, or velocity, of the auto in miles per hour, is actually the rate of change of the distance traveled. Thus it is the slope of the graph of distance traveled plotted against time. For example, consider how the distance traveled must vary in order for an auto to maintain a *constant* speed of 50 mi/h. A little thought will show that the graph of distance traveled *must* be a straight line which rises a distance of 50 mi in a time of 1 h, or 25 mi in 30 min, or any other combination that yields a *slope* of 50 mi/h. This kind of reasoning will cover a constant speed, but the question then arises as to what we call the speed when the graph of distance vs. time is a curved line. This leads to the need to define an *instantaneous* slope.

In order to arrive at the idea of instantaneous slope we consider first the

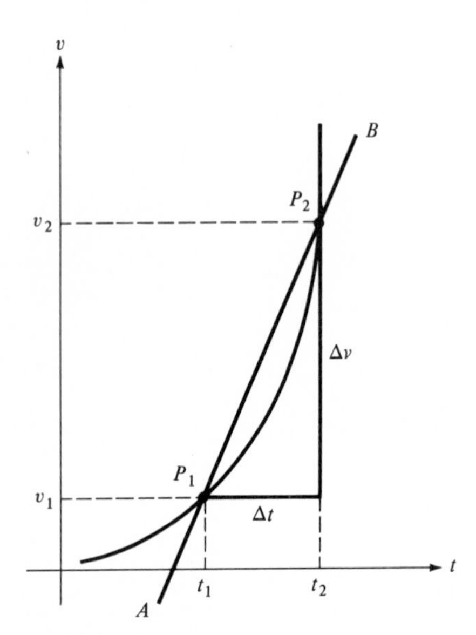

Fig. 14.1-1 Average slope of a curved line.

average slope of a curved line. In Fig. 14.1-1 we have shown a curved line which represents a graph of voltage v plotted against time t. Our aim is to find the slope of this curve between points P_1 and P_2. We note immediately that the slope is continuously changing between points P_1 and P_2. Thus we must begin by stating that we are going to find the *average* slope between these two points. This situation is similar to stating that, in a particular 5-h race, a driver averaged 121.32 mi/h. We know that he did not maintain this speed exactly for the full 5 h, but that his speed actually fluctuated between 0 mi/h and some value higher than the average speed of 121.32 mi/h.

The average slope of the curve between points P_1 and P_2 is defined as the slope of the straight line through these two points. In geometry, this line is called a *chord*. Thus, for the curve in Fig. 14.1-1, chord AB is passed through points P_1 and P_2 and we find its slope in terms of the increments of v and t;

$$\Delta v = v_2 - v_1$$

$$\Delta t = t_2 - t_1 \qquad (14.1\text{-}1)$$

and

$$\text{Average slope} = \frac{\Delta v}{\Delta t} = \frac{v_2 - v_1}{t_2 - t_1} \qquad (14.1\text{-}2)$$

Now let us allow point P_2 to move toward P_1. Several different positions are shown in Fig. 14.1-2. Each move from P_2 to P_2' and then to P_2'' produces a

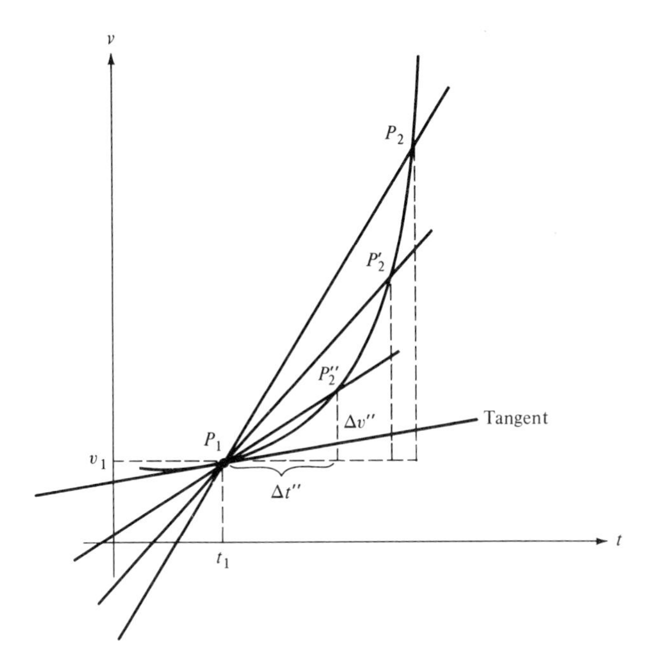

Fig. 14.1-2 Tangent as instantaneous slope.

smaller Δv and Δt than the last and we see that as Δt becomes smaller, the average slope line approaches the tangent line at P_1. Thus the tangent at P_1 is the *limiting position* of the line of average slope as P_2 approaches P_1. It is therefore logical that this will also be true if we approach from the other side of P_1 as long as the curve is continuous at P_1. The instantaneous slope of the curve at P_1 is then defined as the slope of the straight-line tangent to the curve at P_1. The mathematical definition involves the limiting process we have just discussed, that is,

$$\text{Instantaneous slope} = \lim_{\Delta t \to 0} \frac{\Delta v}{\Delta t} \tag{14.1-3}$$

The limit as $\Delta t \to 0$ is equivalent to our previous limiting process where we had $P_2 \to P_1$.

The following example illustrates the Δ method for finding the instantaneous slope using the theory we have developed up to this point.

Example 14.1-1 Find a general expression for the average slope of the function $v = t^2$ between any two points on the graph of the function. Then use Eq. (14.1-3) to find the instantaneous slope.

Solution Figure 14.1-3 shows a graph of the function $v = t^2$. The two general points are P_1 and P_2 and from the figure the coordinates of P_2

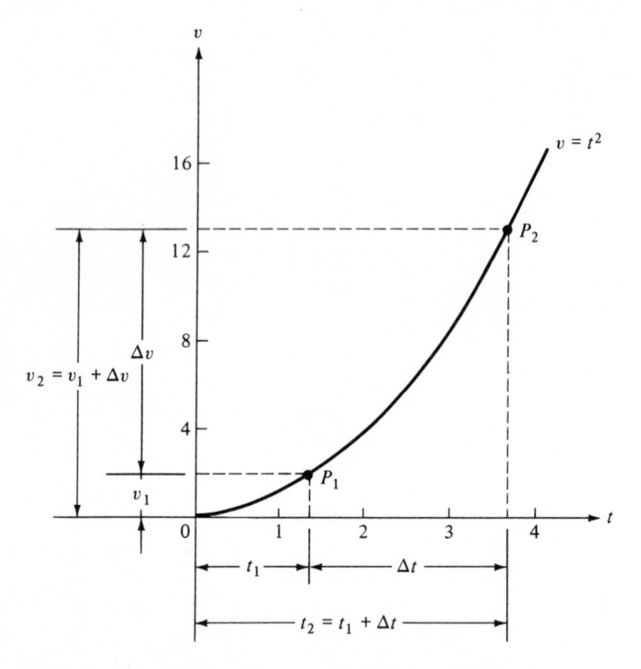

Fig. 14.1-3 Graph of $v = t^2$ for Example 14.1-1.

may be expressed as

$$v_2 = v_1 + \Delta v \tag{14.1-4}$$

$$t_2 = t_1 + \Delta t \tag{14.1-5}$$

In order to find the average slope $\Delta v/\Delta t$ we proceed as follows:

At point P_2:

$$v_2 = t_2^2 \tag{14.1-6}$$

Substituting Eqs. (14-1-4) and (14.1-5),

$$v_1 + \Delta v = (t_1 + \Delta t)^2$$

$$v_1 + \Delta v = t_1^2 + 2t_1\Delta t + (\Delta t)^2 \tag{14.1-7}$$

At point P_1:

$$v_1 = t_1^2 \tag{14.1-8}$$

Subtracting Eq. (14.1-8) from Eq. (14.1-7):

$$\Delta v = 2t_1\Delta t + (\Delta t)^2 \tag{14.1-9}$$

Dividing Eq. (14.1-9) by Δt:

$$\frac{\Delta v}{\Delta t} = 2t_1 + \Delta t \tag{14.1-10}$$

Equation (14.1-10) gives the average slope between any point P_1 on the curve and any other point P_2 on the curve which lies Δt seconds (s) from P_1.

Let us check this formula for the points $P_1(1,1)$ and $P_2(2,4)$ for which $\Delta v = 4 - 1 = 3$ and $\Delta t = 2 - 1 = 1$ so that $\Delta v/\Delta t = 3$. In order to use Eq. (14.1-10) we note that $t_1 = 1$ and $\Delta t = 2 - 1 = 1$ so that $\Delta v/\Delta t = 2(1) + 1 = 3$. This, of course, checks our previous answer.

To find the instantaneous slope we apply Eq. (14.1-3) to get

$$\begin{aligned}\text{Instantaneous slope at point } P_1 &= \lim_{\Delta t \to 0} \frac{\Delta v}{\Delta t} \\ &= \lim_{\Delta t \to 0} (2t_1 + \Delta t) \\ &= 2t_1 \qquad (14.1\text{-}11)\end{aligned}$$

In this example, finding the limiting value of the average slope is relatively easy, because of the nature of the function. ////

14.1-1 Instantaneous Slope—the Derivative

In calculus, the instantaneous slope is called the derivative and the usual symbol is dv/dt. Figure 14.1-4 shows graphically the difference between average slope and derivative, or instantaneous slope. The average slope shown in part *a* of the figure is $\Delta v/\Delta t$, the slope of the chord through the two points. Part *b* of the figure shows the instantaneous slope, or derivative at a point, to be the slope of

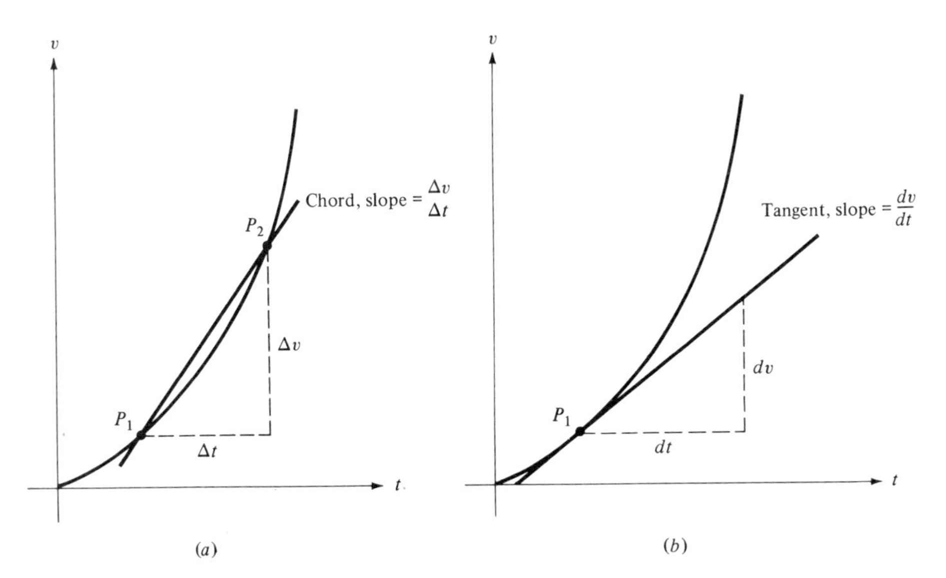

Fig. 14.1-4 (*a*) Average slope; (*b*) derivative.

the tangent to the curve at the point. This slope is dv/dt; dv and dt are called *differentials.*

The relation between the derivative and the average slope is then

$$\frac{dv}{dt} = \lim_{\Delta t \to 0} \frac{\Delta v}{\Delta t} \tag{14.1-12}$$

In Example 14.1 we found that the instantaneous slope for the function $v = t^2$ at point P_1 was $2t_1$. Since t_1 represents any arbitrary point, we may as well call it t, and we can then write

$$\frac{dv}{dt} = \frac{d}{dt} t^2 = 2t \tag{14.1-13}$$

A general formula can be stated for power functions of the form $y = t^n$. It is

$$\frac{d}{dt}(t^n) = nt^{n-1} \tag{14.1-14}$$

Let us find the derivative of $v = t^2$ at $t = \frac{1}{2}$ and at $t = \frac{3}{2}$.

At $t = \frac{1}{2}$:

$$\frac{dv}{dt} = 2\left(\frac{1}{2}\right) = 1$$

At $t = \frac{3}{2}$:

$$\frac{dv}{dt} = 2\left(\frac{3}{2}\right) = 3$$

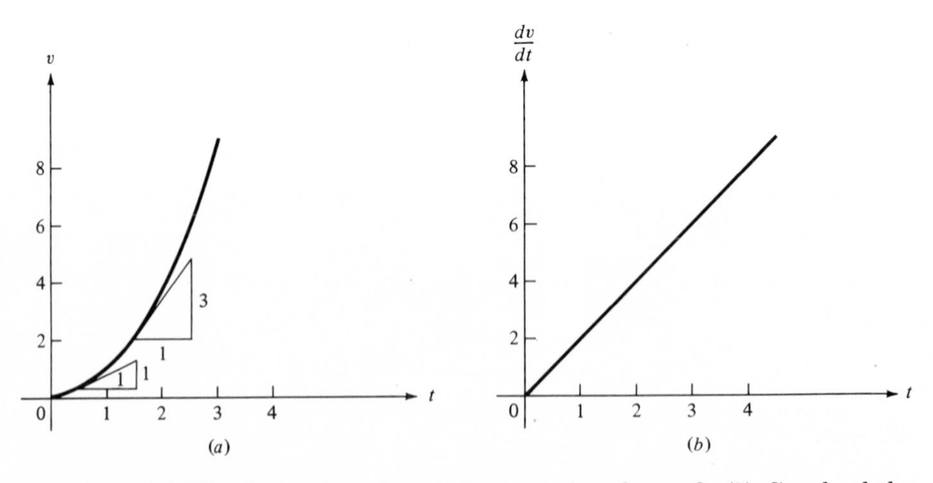

Fig. 14.1-5 (*a*) The derivative of $y = t^2$ at $t = \frac{1}{2}$ and $t = \frac{3}{2}$; (*b*) Graph of the derivative.

These points are shown in Fig. 14.1-5*a* along with the graphical interpretation in terms of the differentials dv and dt for the tangent line. In Fig. 14.1-5*b* we have plotted a graph of the derivative against time t. This illustrates a concept which is of primary importance in the study of circuits containing inductors and/or capacitors, i.e., the concept of the derivative as a function of time. In order to understand the response waveforms for such circuits we must be able to visualize the derivative plotted against time as well as the function.

As a further example, consider the following: In Sec. 3.4 we defined dc current as

$$I = \frac{\Delta q}{\Delta t}$$

In Sec. 3.5 the instantaneous current was defined by the same equation, but with the stipulation that Δt was a small interval at t:

$$i(t) = \frac{\Delta q}{\Delta t}$$

A more precise definition states that

$$i(t) = \frac{dq(t)}{dt} = \lim_{\Delta t \to 0} \frac{\Delta q}{\Delta t}$$

Now consider that an electric charge varies according to the equation $q = kt$. Let us use the Δ method to find the current corresponding to this charge. We proceed as before, noting that $q_2 = q_1 + \Delta q$ and $t_2 = t_1 + \Delta t$.

Then

$$q_2 = kt_2$$

from which

$$q_1 + \Delta q = k(t_1 + \Delta t)$$

also

$$q_1 = kt_1$$

Subtracting,

$$\Delta q = k\,\Delta t$$

Dividing by Δt,

$$\frac{\Delta q}{\Delta t} = k$$

Applying the definition

$$i = \frac{dq}{dt} = \lim_{\Delta t \to 0} \frac{\Delta q}{\Delta t} = k$$

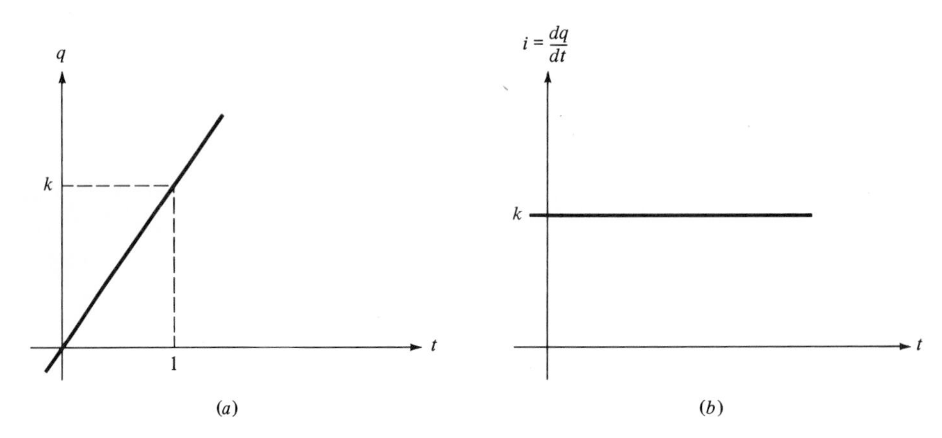

Fig. 14.1-6 (*a*) Charge function $q = kt$; (*b*) the current corresponding to charge $q = kt$.

Thus the current which is the derivative of $q = kt$ is a constant. This should not be surprising since $q = kt$ is the equation of a straight line and we know that the slope of a straight line is a constant. This particular charge function (often called a "ramp") and its derivative, the current, are plotted in Fig. 14.1-6.

DERIVATIVE OF EXPONENTIAL AND SINEWAVE

Two functions whose derivatives we will require frequently are the exponential and sinusoidal waveforms originally discussed in Chap. 2.

Consider first the decaying exponential shown in Fig. 14.1-7*a*. The expression for the waveform is

$$v(t) = V_0 e^{-t/\tau} \tag{14.1-15}$$

Finding the derivative by the delta method involves mathematics beyond the scope of this course so let us see what we can deduce directly from the graph. We observe that the slope of a mentally drawn tangent line near $t = 0$ is negative and its magnitude *decreases* as t increases, finally tending toward zero as t becomes very large. Thus the shape of the curve of the derivative should be similar to that of the exponential itself. It can be shown that the derivative is

$$\frac{dv}{dt} = V_0 \left(\frac{-1}{\tau} \right) e^{-t/\tau} \tag{14.1-16}$$

This is plotted in Fig. 14.1-7*b* and we see that the derivative is itself an exponential differing only in amplitude from the original exponential. It is interesting to note that the exponential, or functions expressible in terms of exponentials, is the only function for which this is true. It is for this reason that the exponential function arises in the solution to many problems.

The second waveform of interest is the sine function shown in Fig. 14.1-8*a*.

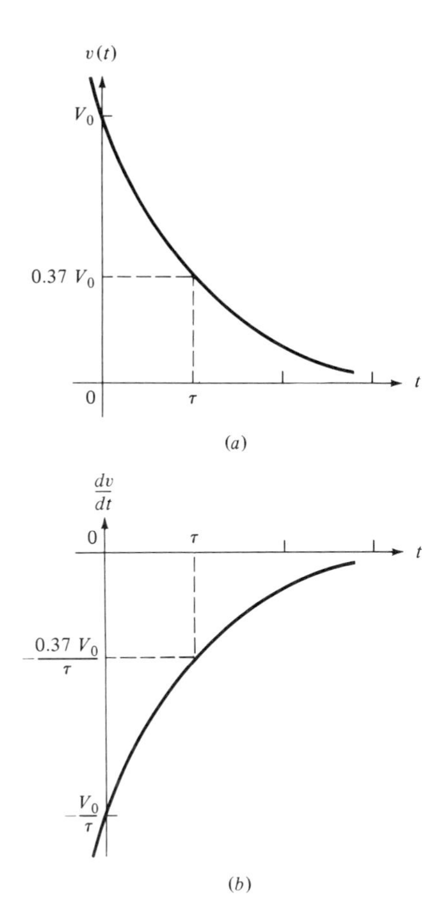

Fig. 14.1-7 (*a*) The decaying exponential; (*b*) the derivative.

This waveform was discussed in Sec. 2.4 and the expression for the curve is

$$v(t) = V_m \sin(\omega t + \theta) \tag{14.1-17}$$

Once again we do not have the mathematical tools available to find the derivative using the delta method but we can gain some insight as to its shape by considering the slope at various points in a qualitative way. We note that at $\omega t = 0$ the slope is positive and it *decreases* as $\omega t \rightarrow \pi/2$. Right at $\omega t = \pi/2$, where the sine curve is changing direction, the slope is zero since a tangent line drawn at this point is horizontal. Between $\omega t = \pi/2$ and $\omega t = \pi$ the slope increases negatively and reaches its maximum negative value at $\omega t = \pi$. From this point it rises toward zero at $\omega t = 3\pi/2$ and continues to increase in a positive direction up to $\omega t = 2\pi$ at which point the cycle repeats. The shape of the resulting curve ap-

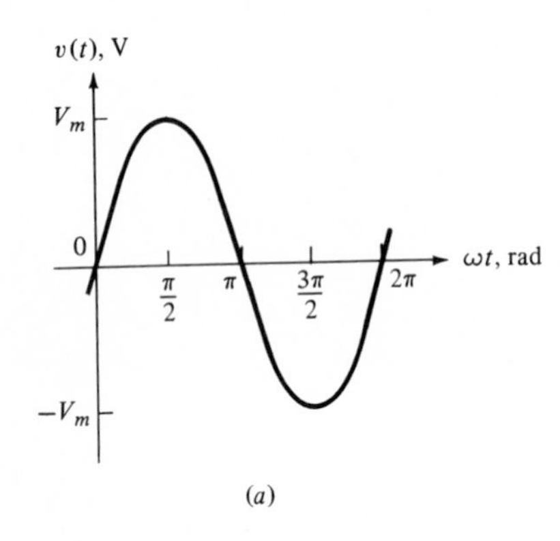

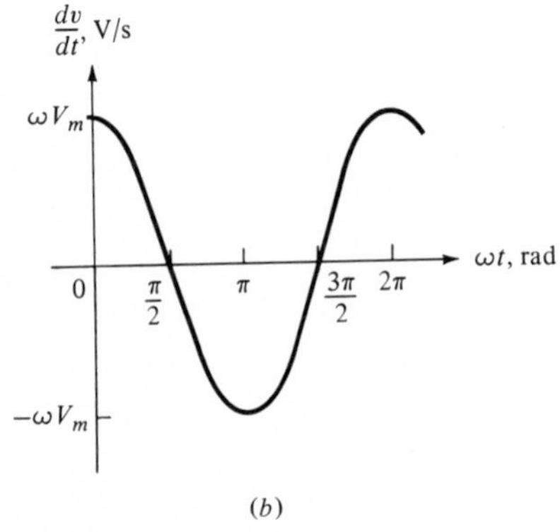

Fig. 14.1-8 (*a*) Sinewave, $\theta = 0$; (*b*) derivative.

pears to resemble the cosine function. The actual derivative is

$$\frac{dv}{dt} = \omega V_m \cos(\omega t + \theta)$$

$$= \omega V_m \sin\left(\omega t + \theta + \frac{\pi}{2}\right) \tag{14.1-18}$$

This is plotted in Fig. 14.1-8*b*. An important point to note is that the amplitude of the derivative is ω times the amplitude of the original sine function.

For the cosine function, the derivative is

$$\frac{d}{dt} V_m \cos(\omega t + \theta) = - V_m \sin(\omega t + \theta) \tag{14.1-19}$$

In the chapters to follow, we will be making use of these derivative expressions.

14.2 AREA AND THE INTEGRAL

In the previous sections we were concerned with the problem of finding the instantaneous slope, or derivative, dy/dx of a given function $y(x)$. Now that we have considered the derivative, we might ask the following question: Given a particular function, can we find another function whose derivative is the original function? For example, given a function t^2, does there exist a function $v(t)$ such that $dv/dt = t^2$? Reference to Eq. (14.1-14) will show that if $v(t) = t^3/3$ then $dv/dt = 3t^2/3 = t^2$. Further thought also shows that this is not the only possible function. If we add any constant to $v(t)$, that is,

$$v(t) = \frac{t^3}{3} + K$$

we get

$$\frac{dv}{dt} = \frac{d}{dt}\frac{t^3}{3} + \frac{d}{dt}K = \frac{3t^2}{3} + 0 = t^2$$

since the derivative of a constant is always zero.

Finding the function $v(t)$ is a process which is the inverse of differentiation; it is called *integration*. We will find in this section that just as the derivative is the instantaneous slope of a curve, the integral is related to the area under a curve.

14.2-1 The Indefinite Integral

In the opening paragraph, we pointed out that the process of integration (inverse of differentiation) when applied to the function $dv/dt = t^2$ yields $v(t) = t^3/3 + K$. The mathematical notation for this is as follows:

$$dv = t^2\,dt$$

$$v = \int dv = \int t^2\,dt$$

$$= \frac{t^3}{3} + K \qquad (14.2\text{-}1)$$

The symbol $\int$ is called the *integral* sign and the function inside the integral sign is the *integrand*. In this case the integral is indefinite because we are not sure of the constant. In practice, additional information will be required for the evaluation of K.

For the integral of the power function t^n we can generalize Eq. (14.2-1).

$$v = \int t^n\,dt$$

$$= \frac{t^{n+1}}{n+1} + K \qquad (14.2\text{-}2)$$

This formula does not apply when $n = -1$. We can check Eq. (14.2-2) by differentiating $v(t)$. This yields

$$\frac{dv}{dt} = \frac{(n+1)t^{n+1-1}}{n+1} + \frac{dK}{dt}$$

since $dK/dt = 0$, $dv/dt = t^n$, as expected.

Example 14.2-1 Find the indefinite integral of $7t^5 + t^3 + 3$ and check by differentiating the result.

Solution

$$v = \int (7t^5 + t^3 + 3)\, dt$$

The integral of a sum of factors is equal to the sum of the integrals, thus

$$v = \int 7t^5\, dt + \int t^3\, dt + \int 3dt$$

The multiplicative constants 7 in the first integral and 3 in the last can be moved outside the integral signs as follows

$$v = 7\int t^5\, dt + \int t^3\, dt + 3\int dt$$

Next we apply Eq. (14.2-2) noting that $n = 5$ in the first integral, 3 in the second, and 0 in the third. The result is

$$v = \frac{7t^6}{6} + K_1 + \frac{t^4}{4} + K_2 + 3t + K_3$$

Combining the constants so that $K = K_1 + K_2 + K_3$, the final result is

$$v = \frac{7t^6}{6} + \frac{t^4}{4} + 3t + K$$

Checking, we have

$$\frac{dv}{dt} = 7 \times \frac{6t^5}{6} + \frac{4t^3}{4} + 3 + 0 = 7t^5 + t^3 + 3$$

as expected. ////

Equation (14.2-2) is a rule for integrating one particular type of function. In calculus courses we will learn to integrate many other types of functions. For this course it is more important that we develop a feeling for the physical significance of the integral of a function in terms of area under a curve.

14.2-2 Area and the Definite Integral

The area under a curve is found by using the *definite integral.* This differs from the indefinite integral in that it applies only to the portion of the function lying between two points, called *limits.* The notation is

$$A = \int_{t_1}^{t_2} v(t)\, dt \qquad (14.2\text{-}3)$$

This can be explained in terms of Fig. 14.2-1. The point t_1 is called the *lower limit* and defines the beginning of the area calculated by the integral. The point t_2, called the *upper limit,* defines the end of the interval of interest. The area A is shown shaded in the figure and is seen to be that area bounded by the line $v = 0$, the upper and lower limits t_1 and t_2, and the curve $v(t)$.

When we evaluate a definite integral, we first find the indefinite integral. The indefinite integral evaluated at the lower limit is then subtracted from the value of the indefinite integral evaluated at the upper limit to give the area. In symbols, this appears as follows:

$$A = \int_{t_1}^{t_2} v(t)\, \mathrm{dt} = g(t)\Big|_{t_1}^{t_2} = g(t_2) - g(t_1) \qquad (14.2\text{-}4)$$

In this formula $g(t)$ is the indefinite integral of $v(t)$ and the vertical line with t_1 and t_2 at its ends signifies that we are to evaluate g at t_2 and subtract the value at t_1 as indicated. Note that the constant of integration does not appear at all in the definite integral since its appearance in $g(t_1)$ cancels its appearance in $g(t_2)$.

Let us apply this formula to some simple cases. First consider finding the area under the curve of voltage vs. time from $t = 0$ to $t = t_2$ shown in Fig. 14.2-2*a*. This is particularly simple because the voltage is constant and we see by inspection that the area in question is $V_0 t_2$. Applying Eq. (14.2-4) we note that

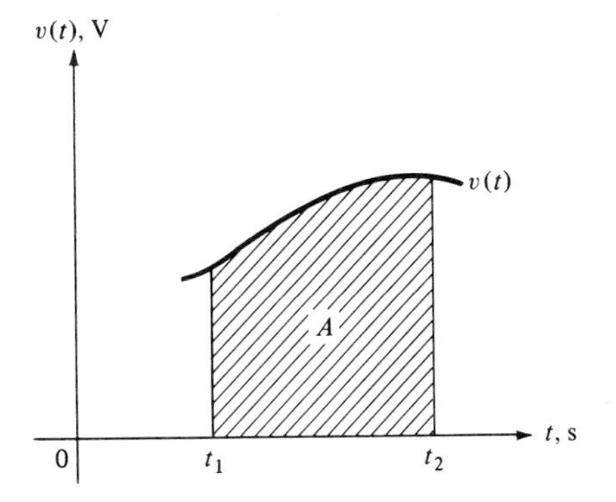

Fig. 14.2-1 Area found by the definite integral.

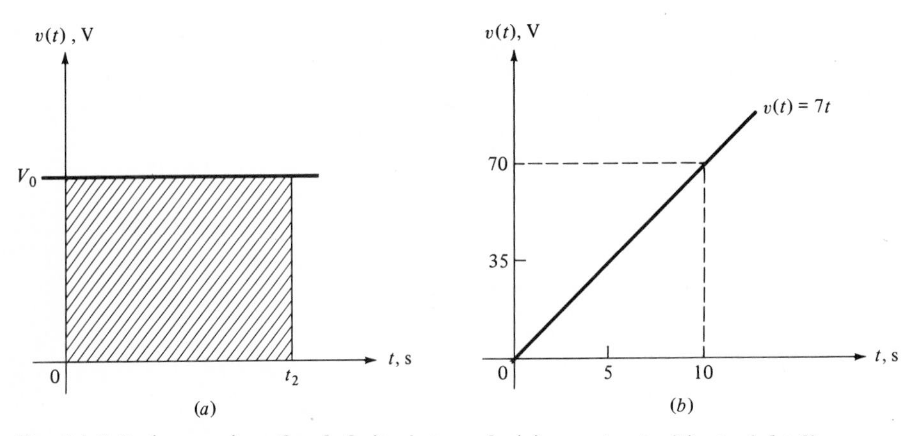

Fig. 14.2-2 Area using the definite integral. (*a*) constant; (*b*) straight line.

$v(t) = V_0$, and $t_1 = 0$. Then

$$A = \int_0^{t_2} V_0 \, dt = V_0 \int_0^{t_2} dt = V_0 t \Big|_0^{t_2} = V_0(t_2 - 0) = V_0 t_2$$

This agrees with the area calculated above by inspection.

For another example, consider the straight line voltage of Fig. 14.2-2*b* where $v(t) = 7t$. Let us find the area under this curve from $t = 0$ to $t = 10$ s using Eq. (14.2-4). We easily calculate in advance that the answer should be $(70)(10)/2 = 350$ volt-seconds (V·s). Now we apply Eq. (14.2-4) to get

$$A = \int_0^{10} 7t \, dt = 7 \int_0^{10} t \, dt = 7 \frac{t^2}{2} \Big|_0^{10} = 7\left(\frac{10^2}{2} - \frac{0^2}{2}\right) = 350 \text{ V·s}$$

As a variation of this problem, we calculate the area between $t = 5$ and $t = 10$ s.

$$A = \int_5^{10} 7t \, dt = 7 \frac{t^2}{2} \Big|_5^{10} = 7\left(\frac{10^2}{2} - \frac{5^2}{2}\right) = \frac{7}{2} \times 75 = \frac{525}{2} = 262.5 \text{ V·s}$$

The student should check this by using ordinary algebra to compute the given area.

14.2-3 Integrals of Standard Waveforms

In later chapters, we will require the integral of the exponential and the sinewave. These are:

$$\int V_0 e^{-at} \, dt = -V_0 \frac{e^{-at}}{a} + K \tag{14.2-5}$$

$$\int V_m \sin \omega t \, dt = \frac{-V_m}{\omega} \cos \omega t + K \tag{14.2-6}$$

$$\int V_m \cos \omega t \, dt = \frac{V_m}{\omega} \sin \omega t + K \tag{14.2-7}$$

As with the derivative of these functions, we note that the integral has the same form as the original function. In the case of the sine and cosine integrals, the amplitude is *divided* by ω. This is opposite to the derivative, where we found that the amplitude is *multiplied* by ω.

When analyzing circuits containing inductors and capacitors, we often encounter integrals in which the upper limit is a variable. Consider the following expression for the integral of a step-function voltage as shown in Fig. 14.2-3a:

$$g(t) = \int_{0^-}^{t} V_0 u(t)\, dt$$

The lower limit is written as 0^- to indicate that it occurs just before the step. When written in this form, the variable t in the integrand can be confused with the upper limit t so we replace the t in the integrand by x and write

$$g(t) = \int_{0^-}^{t} V_0 u(x)\, dx \tag{14.2-8}$$

Here x is called a dummy variable of integration because it will always disappear after the limits are substituted. Since $u(x) = 1$ for $x > 0$ we can replace $u(x)$ by 1 in the integrand and carry through the operation as follows

$$g(t) = \int_{0^+}^{t} V_0\, dx = V_0 x\Big|_{0^+}^{t} = V_0(t - 0^+) = V_0 t \tag{14.2-9}$$

This function, called a *ramp*, is plotted in Fig. 14.2-3b. Note that the function $g(t)$ actually measures the area under the step function between $t = 0$ and time t. As t increases from zero the area under the step increases linearly as shown by the ramp. We can see this without using the integral by dividing

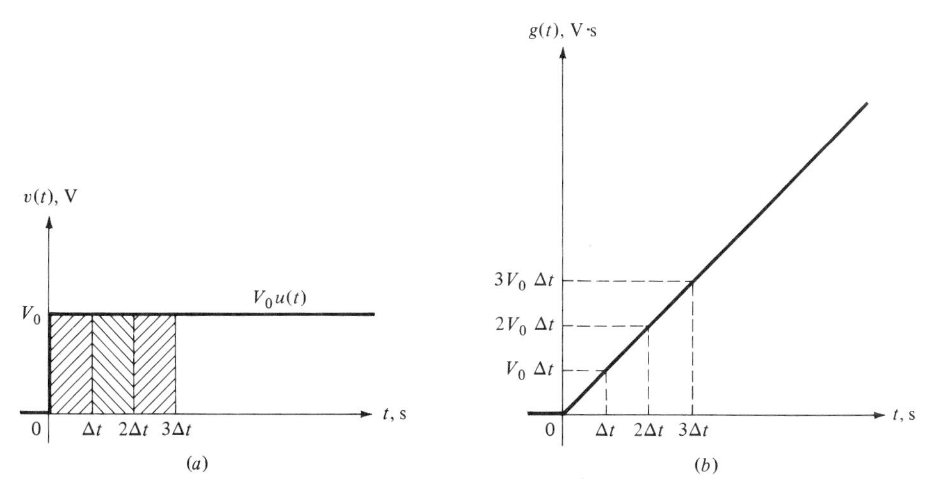

Fig. 14.2-3 Integral of a step. (a) Voltage step; (b) ramp.

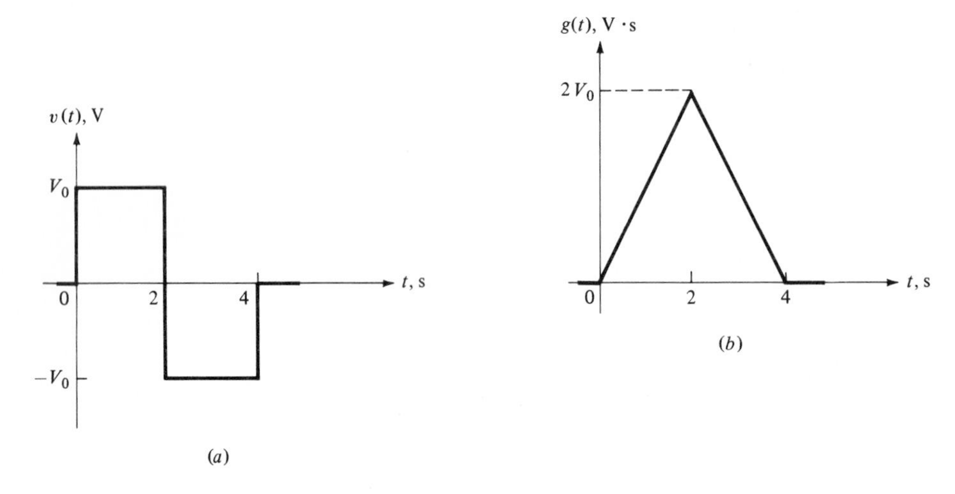

Fig. 14.2-4 Running integral of a pulse waveform. (*a*) Waveform; (*b*) integral.

the t axis into equal segments, each Δt-s wide. As we increase t from 0 to Δt, then Δt to $2\Delta t$, etc, we *add* an area $V_0\Delta t$ for each Δt step, so that the *area* continually increases as t increases.

To illustrate this idea further consider the pulse waveform shown in Fig. 14.2-4*a*. Let us find the running integral $g(t)$ of this waveform, that is,

$$g(t) = \int_{0^-}^{t} v(x)\, dx$$

This integral can be carried through formally, by expressing $v(x)$ in terms of a sum of step functions (see Sec. 2.1) or by reasoning in terms of the previous example. We will leave the former method for the homework problems and use reasoning for our solution here.

Consider the first pulse, which can be thought of as a step lasting from 0 to 2 s. The running integral over this interval will be a ramp which, at $t = 2$ s, will be at a value equal to the area under the pulse from 0 to 2 s, that is, $2V_0$ V·s. This is shown in Fig. 14.2-4*b*. Now, as our variable of integration passes through the point $t = 2$ s toward higher values of t, the area encountered is *below* the $v = 0$ line, and therefore is *negative*. As we move from $t = 2$ to $t = 4$ s we pick up more and more *negative* area which *subtracts* from the positive area picked up when we moved from $t = 0$ to $t = 2$ s. When we finally arrive at $t = 4$ s there is no additional area because $v(t) = 0$ for $t > 4$. From the $v(t)$ waveform we see that the positive area from 0 to 2 s is exactly balanced by the negative area from 2 to 4 s so that the *net* area is zero. Thus at $t = 4$ s the graph of $g(t)$ returns to zero.

This kind of reasoning, in which we interpret the integral in terms of area, will prove useful later on.

SUMMARY

1 The derivative of a function at a point measures the instantanteous slope of the graph of the function at the point. This is the same as the slope of the tangent to the curve at the point.
2 In mathematical terms the derivative is defined by the equation

$$\frac{dv}{dt} = \lim_{\Delta t \to 0} \frac{\Delta v}{\Delta t}$$

3 The mathematical operation of integration is the inverse of differentiation.
4 The definite integral measures the area under a curve.

PROBLEMS

14.1-1 Find the general equation for the average slope between any two points on the curve $v = t^3 + 2$. Use the Δ method.

14.1-2 Use the result of Prob. 14.1-1 to find the average slope between $t = 2$ and $t = 2.5$ for the curve $v = t^3 + 2$.

14.1-3 Repeat Prob. 14.1-1 for the function $v = t^3 - 2t$.

14.1-4 Find the average slope of the curve of the area of a circle as a function of the radius, $a = \pi r^2$. Calculate the change in area (Δa) if r increases from 5.0 to 5.3.

14.1-5 Find the derivative of $v = t^3 + 2$ and evaluate it at $t = 2$ and $t = 2.5$. Compare with the result of Prob. 14.1-2.

14.1-6 Plot the derivative of $v = t^3 + 2$ vs. t for values of t from 0 to 10 s.

14.1-7 Find the derivatives of the functions $v_1 = 5t^3$, $v_2 = 5t^{-3}$, and $v_3 = -5t^{-3}$.

14.1-8 Differentiate $v = 5t^{3/4}$

14.1-9 Differentiate $v = 4\sqrt{t}$. *Hint*: $\sqrt{t} = t^{1/2}$

14.1-10 Positive charge passes through a cross section of wire at the rate of 12 C/s beginning at $t = 0$ and ending at $t = 4$ s. Plot the graph of charge vs. time and find and plot the corresponding current.

14.1-11 Plot the exponential current $i(t) = 15e^{-t/10}$ mA. Find di/dt and plot it for $0 < t < 50$ s.

14.1-12 Plot two cycles of the voltage $v(t) = 3 \sin 5t$ V. Find dv/dt and plot two cycles on the same axes as the plot of $v(t)$.

14.2-1 Evaluate the following indefinite integrals:

a $\int 3t^4\, dt$

b $\int 6t^{-3}\, dt$

c $\int (5t^5 + 6t^{-2})\, dt$

d $\int (23t^7 + 15t^5 + 2t^2 + 7)\, dt$

Check all answers by differentiating the result.

14.2-2 Find the area under the curve $v = 2t + 3$ between $t = 2$ s and $t = 8$ s.

14.2-3 Find the area under the curve $i = 3t^2$ between $t = 1$ s and $t = 9$ s.

14.2-4 Find and plot

$$v(t) = \int_{0^-}^{t} (2x + 3)\, dx$$

in the range $0^- < t < 4$ s.

14.2-5 Express the pulse waveform of Fig. 14.2-4*a* as a sum of step functions. Find $\int_{0^-}^{t} v(x)\, dx$ by formally integrating the expression for $v(t)$. Check your result against Fig. 14.2-4*b*. *Hint*: Consider first the interval $0 < t < 2$ s. Any step functions which begin at times greater than 2 s will be zero in this interval. In the interval $2 < t < 4$ s any unit step functions which began at times less than 2 s will be 1.

14.2-6 For $v(t)$ as shown in Fig. P14.2-6 find

$$\int_{0^-}^{t} v(x)\, dx$$

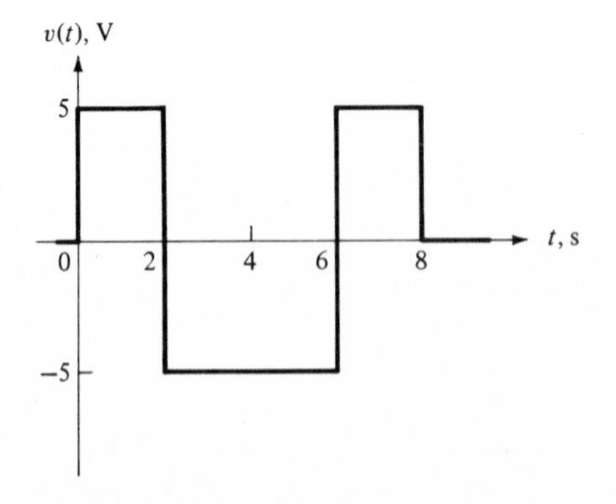

Fig. P14.2-6

15
Magnetic Circuits

OBJECTIVES

Upon completion of this chapter, you should be able to

1 List the basic magnetic quantities and the relationships that exist among them.
2 Find the analog electric circuit for a given magnetic circuit.
3 Analyze a single-loop magnetic circuit with or without an air gap.
4 Describe the electromagnetic interactions in a magnetic circuit.
5 Compute the force on a current-carrying conductor in a magnetic field.
6 Describe the operation of various devices which are based on electromagnetic interactions such as the transformer, galvanometer, generator, motor, and relay.

INTRODUCTION

Most of us are familiar with magnetism. We have experienced the forces that exist between permanent magnets and may be aware that there is a relationship between electricity and magnetism; we may have built an electromagnet or used one in a previous science class.

In this chapter we shall study the relationships between electricity and magnetism and magnetic forces. These ideas will then be used to show how these phenomena are put to use in a number of important electromagnetic devices including the galvanometer, motor, generator, relay, and transformer.

15.1 MAGNETIC QUANTITIES

In order to study magnetic phenomena we must identify the quantities used to measure magnetism. Magnetism results in two measurable quantities: forces and, under certain conditions, induced voltages. In order to explain these effects, a magnetic field is postulated. This field can be thought of as an invisible set of lines of force which interact with magnets and certain metals. These lines of force are measured by the force they exert on a small magnet which is used as a test probe; a compass needle is frequently used for this purpose. Iron filings sprinkled on a piece of cardboard under which there is a magnet align themselves along these lines of force, providing a picture of the field as shown in Fig. 15.1-1.

These magnetic fields are found to exist in the vicinity of current-carrying conductors as well as near permanent magnets. The theory which follows is based on numerous experiments and provides a unified explanation for all of the observed results.

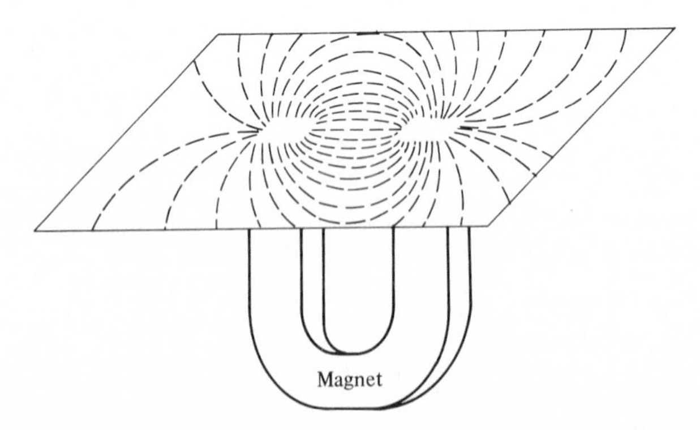

Fig. 15.1-1 Iron filings arranged along field lines of permanent magnet.

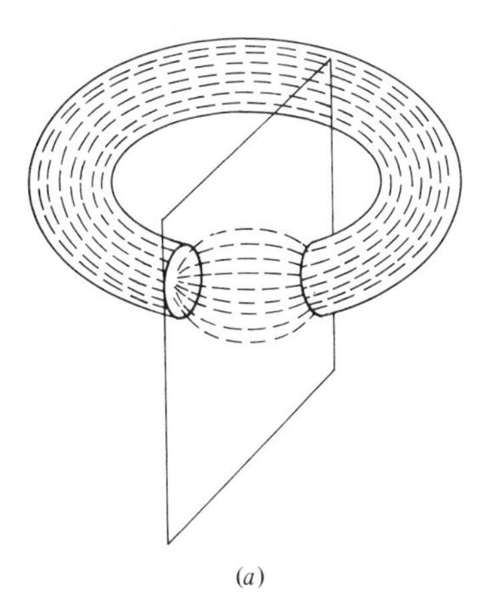

(a)

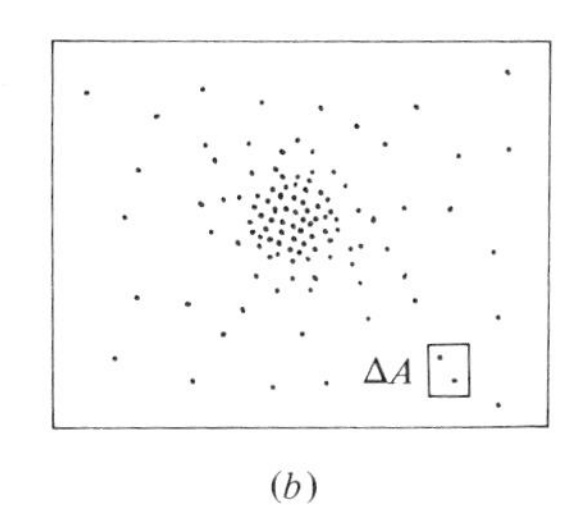

(b)

Fig. 15.1-2 Magnetic field produced by permanent magnet. (*a*) Three-dimensional perspective showing closed-loop flux lines; (*b*) illustration of points of intersection of flux lines with plane perpendicular to axis of magnet.

15.1-1 Magnetic Flux

The term *magnetic field* refers to a number of different but related quantities which may be measured. It is similar in this respect to the term *electricity* which refers to electric current and voltage, as well as electric power.

The *magnetic flux* is a measure of the total number of lines of force associated with the field; it is generally symbolized by the Greek letter phi (φ), and is measured in *webers* (Wb). Lines of magnetic flux always form closed loops. If we look at the flux lines associated with a magnet they seem to start on one pole and terminate on the other; the lines continue inside the magnet, however, and form a closed loop. These lines do not have an actual physical existence; we cannot isolate them and say there is a line passing through this particular point in space. They may be compared to light rays in this respect. We may show a beam of light as a group of rays emanating from a source even though the light really is a continuous beam. Figure 15.1-2*a* illustrates the flux associated with a magnet. The closed loops formed by the flux lines are shown along with the fringing, or spreading of the flux in the air gap of the magnet.

15.1-2 Magnetic Flux Density

A magnetic field exists in, and is distributed throughout, a region of space. The strength of the field is not necessarily uniform throughout this space. The *flux density*, generally symbolized by the letter B, is a measure of the field strength for a specific small region rather than the strength of the entire field which is given by the flux, φ. Figure 15.1-2*b* shows the intersection of the flux lines with a cardboard sheet placed between the poles of the magnet of Fig. 15.1-2*a*. From this figure we can see that the flux density is greatest near the axis of the magnet and decreases with distance away from the axis, especially when we pass beyond the cross-sectional boundary of the magnet. We can quantify this concept of flux density by considering an area ΔA, which is small enough so that the flux is relatively uniform over the area. Then, the flux density is defined by

$$B = \frac{\Delta\varphi}{\Delta A} \tag{15.1-1}$$

where $\Delta\varphi$ is the total amount of flux passing through the small area ΔA. For the case in which the flux is constant over the entire area A

$$B = \frac{\varphi}{A} \tag{15.1-2}$$

From Eqs. (15.1-1) or (15.1-2) we find the unit for flux density must be webers per square meter.

Example 15.1-1 The flux density inside the magnet of Fig. 15.1-1 is uniform with a value of 5,000 $\mathrm{Wb/m^2}$. The radius of the magnet's cross section is 0.5 cm. Find the total flux in the air gap.

Solution The total flux in the air gap is the same as the flux in the magnet since the flux lines are closed loops. (Note, in contrast, that the flux density is not the same inside and outside. Why?) Therefore, since the field is uniform,

$$\varphi = BA = 5{,}000\,\frac{\mathrm{Wb}}{\mathrm{m^2}} \times \pi(0.5\ \mathrm{cm})^2 \times \left(\frac{1\ \mathrm{m}}{100\ \mathrm{cm}}\right)^2$$

$$= 0.393\ \mathrm{Wb}$$ ////

15.1-3 Magnetic Field Intensity

Magnetic fields exist around conductors carrying electric currents, as well as around permanent magnets. We would like to relate the strength of the magnetic field surrounding a current-carrying conductor to the current which gives rise to it. The field depends not only on the current, however, but on the material in which the field exists as well. Furthermore, the relationship between the current and magnetic field strength is nonlinear for many materials.

The *magnetic field intensity*, generally symbolized by the letter H, is introduced to separate the effects of the current from the effects of the material on the magnetic field caused by the current. The magnetic field resulting from a current takes the form of a series of concentric circles surrounding the current,

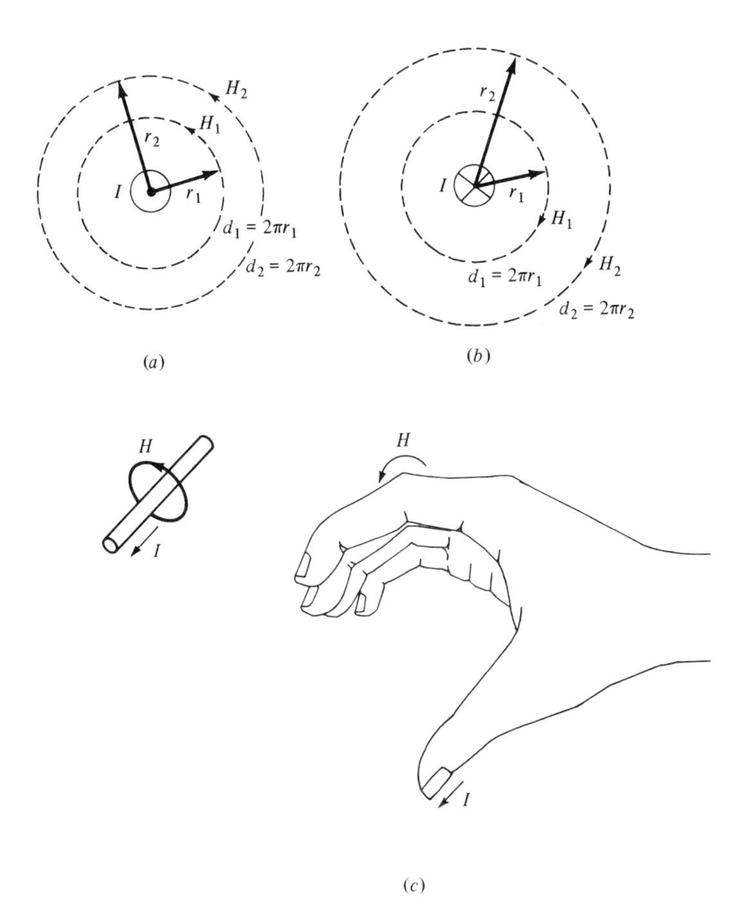

Fig. 15.1-3 Magnetic field intensity surrounding current carrying conductor. (*a*) View with current flowing out of page; (*b*) view with current flowing into page; (*c*) illustration showing use of right-hand rule.

as shown in Fig. 15.1-3*a* and *b*. Figure 15.1-3*a* illustrates the situation with the current flowing out of the page toward the observer as indicated by the dot in the circle representing the conductor. Figure 15.1-3*b* illustrates the same thing but this time the current is flowing into the page, away from the observer, as indicated by the cross in the circle representing the conductor.

The value of H is found using Ampere's circuital law:

$$Hd = I_{\text{enclosed}} \qquad (15.1\text{-}3)$$

where I_{enclosed} is the total current enclosed by the loop of length d, formed by H. Thus, for Fig. 15.1-3*a* or *b*

$$2\pi r_1 H_1 = I$$

or

$$H_1 = \frac{I}{2\pi r_1}$$

and similarly,

$$H_2 = \frac{I}{2\pi r_2}$$

Notice, from this result, that H decreases as we move farther away from the current. This is intuitively acceptable since we expect most effects to decrease as we get farther away from their source. In Eq. (15.1-3) the units for H are ampere turns per meter which is generally written At/m. The reason for this is that magnetic fields are frequently the result of current flowing in a coil and the total current enclosed by the loop H is the product of I and the number of turns in the coil; the dimensionless quantity turns (t) is a reminder of this.

The direction for H is found using the right-hand rule which states that if the thumb of the right hand is placed in the direction of the current, then the fingers will point in the direction of H as shown in Fig. 15.1-3*c*. This is sometimes called the right-hand screw rule because the thumb indicates the direction in which a right-hand screw will move if it is turned in the direction of the fingers.

15.1-4 Permeability

At this point in our description of the magnetic field it will be helpful to review what has been presented thus far. A magnetic field may be produced by a permanent magnet or by an electric current. The field may be detected by the force it exerts on a magnet and this effect is explained by assuming that there are invisible lines of force which are called magnetic flux, φ, which always form closed loops. The density of these lines at a particular point in space is called the flux density, B. A new quantity called the magnetic field intensity, H, is introduced in order to isolate the effect of the current from that of the material on the field. Ampere's circuital law is a linear relationship between H and I.

The relationship between B and H is presented next. It provides the final link in the relationship between the magnetic flux φ and the current I, which is its source. The relation between B and H is dependent on the material in which the magnetic field exists. The ratio of B to H is called the *permeability* of the material; it is generally symbolized by the Greek letter mu (μ). Therefore

$$B = \mu H \tag{15.1-4}$$

From this equation the unit of μ must be $(\text{Wb/m}^2)/(\text{A/m})$ or Wb/A·m. The unit of henry per meter (H/m) is often used for μ; both units are the same since 1 H = 1 Wb/A is the unit used to measure inductance (see Chap. 16).

The permeability is a property of the material in which the field exists. Most materials fall into one of two categories; they are either magnetic or nonmagnetic. The value of μ for nonmagnetic materials is very close to the value for free space (vacuum); this value is symbolized by μ_0 and its numeric value is

$$\mu_0 \approx 4\pi \times 10^{-7} \text{ Wb/A}\cdot\text{m} \tag{15.1-5}$$

The permeability of magnetic materials is much larger than that of free space. Sometimes the relative permeability, μ_r, is given where

$$\mu_r = \frac{\mu}{\mu_0} \tag{15.1-6}$$

μ_r is a dimensionless number.

The value of μ for most magnetic materials is not constant; it decreases with increasing values of H. For this reason the relationship between B and H is often presented by an experimentally determined BH curve for the material as shown in Fig. 15.1-4. Notice the saturation of B as H gets larger and larger, that is, B tends to approach a constant value, called the "saturation level," as H increases. The value of μ for a given material and value of H is found by dividing the corresponding value of B by the value of H; graphically, it is the slope of the line from the origin to the point on the curve corresponding to the given values of B and H.

Example 15.1-3 The magnetic field intensity in cast steel is 300 A/m. Find the flux density. Find the new flux densities if H is first doubled and then quadrupled.

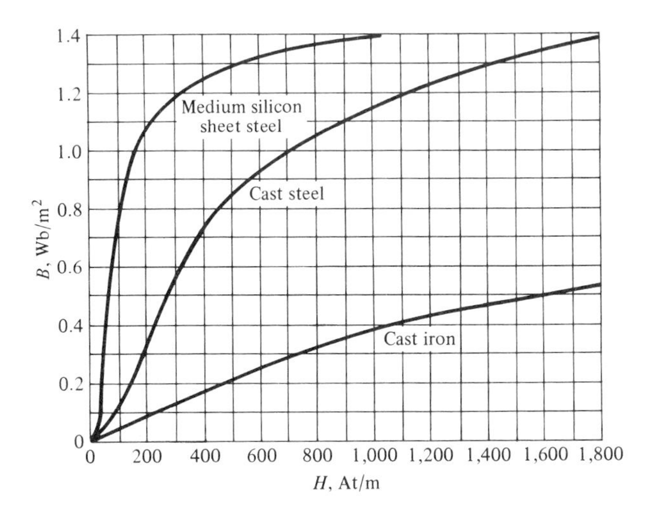

Fig. 15.1-4 BH curve for common ferromagnetic materials. (*Reprinted from Hammond and Gehmlich, "Electrical Engineering," 2d ed., McGraw-Hill Book Company, New York,* 1970.)

Solution From Fig. 15.1-4 we find the values tabulated below.

H, A/m	B, Wb/m².
300	0.55
600	0.93
1,200	1.22

Note the saturation effect. It is evidenced by the fact that a 300 A/m change in H (from 300 to 600 A/m) results in a change of 0.38 Wb/m² in B while a 600 A/m change in H (from 600 to 1,200 A/m), which is double the first change, results in a smaller change in B of only 0.29 Wb/m². ////

15.2 ELECTRIC CIRCUIT ANALOG

The solution of magnetic circuits is somewhat more complex than the solution of electric circuits because of two factors. One of these is the nonlinear BH curve—the additional complexity is in the evaluation of the solution, not in the theory required to analyze the circuit. We shall, in this section, demonstrate a direct analogy between the electric and magnetic circuits which will enable us to apply to magnetic circuits the concepts we have learned for electric circuits.

The second factor which complicates magnetic circuits is the fact that the magnetic fields are not contained within lumped circuit elements as are the electric currents and potential fields in electric circuits. In this book we avoid this problem by restricting our attention to magnetic systems in which the fields are contained within fixed boundaries, at least to a good approximation, so that we can use circuit-analysis techniques. Problems where this restriction is not satisfied must be solved using techniques which are beyond the scope of this text.

The simplest magnetic circuit consists of a closed loop of ferromagnetic material with a current-carrying coil of wire wrapped around part of it as shown in Fig. 15.2-1a. The coil has N turns and carries a current of value I. The mean length around the core is d and the cross-sectional area of the core is uniform and of value a while its permeability is μ.

The values of H will vary slightly from point to point within the cross-sectional area; by Eq. (15.1-3) those values which form smaller loops near the inner edge of the window will be slightly larger than those forming the larger loops near the outer edges. The average value is computed using the average length d in Eq. (15.1-3). It is necessary to replace I in Eq. (15.1-3) by NI in the present case because the flux loop encircles a total current of NI; N conductors each carrying I A. Thus, from Eq. (15.1-2)

$$Hd = NI \tag{15.2-1}$$

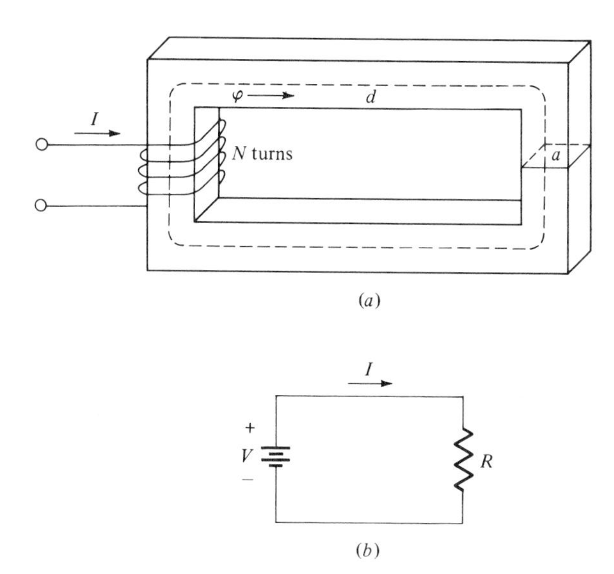

Fig. 15.2-1 Single-loop magnetic circuit and electric analog. (*a*) Magnetic circuit; (*b*) electric analog.

Furthermore, $B = \mu H$ and $\varphi = aB$; therefore $H = \varphi/\mu a$ and Eq. (15.2-1) becomes

$$\frac{d}{\mu a}\varphi = NI \tag{15.2-2}$$

This equation is of the same form as

$$RI = V \tag{15.2-3}$$

which corresponds to the electric circuit of Fig. 15.2-1*b*.

In the magnetic circuit, NI is called the magnetic force or magnetomotive force (MMF) since it is the source of the magnetic field; it is often symbolized by the script letter $\mathcal{F}$. In the analog circuit V is a voltage source which is the source of the current. Hd in the magnetic circuit is called the magnetic potential drop; it is analogous to the voltage, or electric potential, drop in the electric circuit. The flux φ flows around the magnetic circuit in closed loops just as its analog, I, does in the electric circuit. The reluctance, $\mathcal{R}$, of a magnetic circuit is defined as

$$\mathcal{R} = \frac{1}{\mu}\frac{d}{a} \tag{15.2-4}$$

which permits Eq. (15.2-2) to be written as

$$\mathcal{R}\varphi = \mathcal{F} \tag{15.2-5}$$

The analog of the reluctance is the electric resistance. The equation for resistance

$$R = \rho \frac{d}{a}$$

is the analog of Eq. (15.2-4) with the reciprocal of the resistivity, ρ, being the analog of permeability.

Thus, in theory at least, the solution of the magnetic circuit of Fig. 15.2-1*a* is no more difficult than the solution of its analog electric circuit shown in Fig. 15.2-1*b*. In practice, however, we are confronted with the nonlinear *BH* relationship. In order to solve a magnetic circuit, we first solve Eq. (15.2-1) for H, find B from the *BH* curve, and then multiply it by the cross-sectional area a to obtain the flux φ. Despite this difference in procedure, it is often helpful to consider the analog electric circuit in formulating the solution to a magnetic-circuit analysis problem.

The analysis of a single-loop magnetic circuit is demonstrated in the following example.

Example 15.2-1 A magnetic circuit has the form shown in Fig. 15.2-1*a* with $I = 2$ A, $N = 200$ turns, $d = 50$ cm, and $a = 2$ cm^2. The core is a steel casting. Find φ in the circuit.

Solution Using Eq. (15.2-1) we have, converting d to meters,

$$0.5H = 200 \times 2 = 400$$

$$H = 800 \text{ A/m}$$

From the *BH* curve, Fig. 15.1-4,

$$B \approx 1.05 \text{ Wb/m}^2$$

and

$$\varphi = BA = 2 \text{ cm}^2 \left(\frac{1 \text{ m}}{100 \text{ cm}}\right)^2 1.05 \text{ Wb/m}^2$$

$$= 1.10 \times 10^{-4} \text{ Wb}$$

////

The solution to Example 15.2-1 is predicated on an implicit assumption; namely, that all the flux remains within the cast steel core and none of it follows a path through the surrounding air. In actuality a small part of the flux does exist in the air, but it is a very small part of the total flux. The reason for this is that the air provides a parallel path for the flux. From Eq. (15.2-4) it is clear that the reluctance of the air will tend to be much larger than that of the iron core because its permeability is much smaller. Consideration of the analogous electric circuit indicates that most of the flux will travel through the smaller of the two parallel reluctances (just as most of the current will travel through the smaller of two parallel resistances in an electric circuit). This use of the electric analog to gain a qualitative understanding of a magnetic circuit is probably its most useful application.

15.3 DESIGN OF MAGNETIC CIRCUITS

It is frequently necessary to design a magnetic circuit which produces a given flux in a core which has a nonuniform cross section or an air gap in it. The procedure is described in terms of the following practical problem encountered in the design of a relay.

The relay is an electromechanical switch; basically a switch that is operated by an electromagnet whose input is an electrical signal. A diagram illustrating the operation of a relay is shown in Fig. 15.3-1. The construction of an actual relay may differ from that shown but the principles of operation are the same.

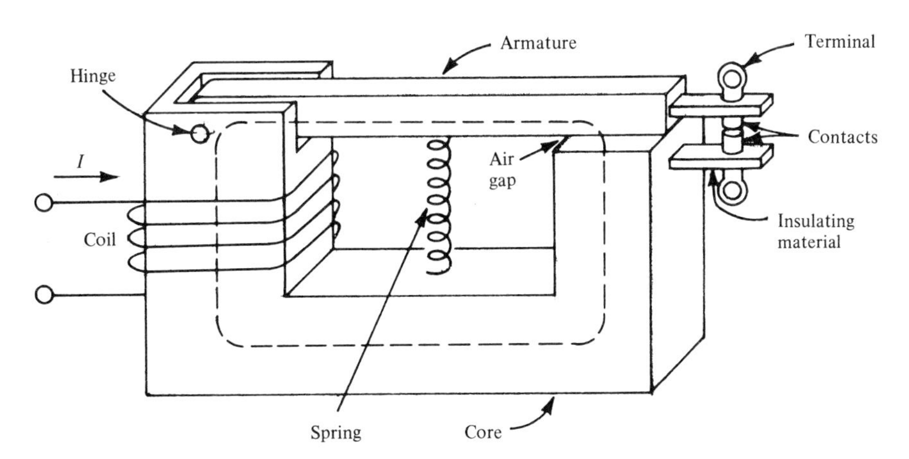

Fig. 15.3-1 Relay.

When current flows through the coil a magnetic flux flows through the core causing it to become magnetized. The magnetic force causes the movable part of the core, called the *armature*, to be attracted to the fixed part thereby closing the contacts. When the current is shut off, the magnetic flux becomes zero and there is no magnetic force. The spring pushes the armature of the core open, thereby opening the contacts.

The problem considered here is determination of the number of turns required in the coil to cause the relay to close for a given current. Suppose the specifications are as follows: mean core length = 8 cm, cross section of core = 1 cm × 1.5 cm, length of air gap = 2 mm, φ = 0.12 mWb, and the core is cast steel. Find the number of turns required if the pull-in current, i.e., the current required to close the relay, is to be 200 mA. Then determine the release current, i.e., the current at which the relay opens.

The steps in the determination of the number of turns required are

1 Compute the cross-sectional area of the core:

$$A_c = 1 \text{ cm} \times 1.5 \text{ cm} \times \left(\frac{1 \text{ m}}{100 \text{ cm}}\right)^2 = 1.5 \times 10^{-4} \text{ m}^2$$

2 Compute the flux density in the core:

$$B_c = \frac{\varphi}{A_c} = \frac{0.12 \times 10^{-3} \text{ Wb}}{1.5 \times 10^{-4} \text{ m}^2} = 0.8 \text{ Wb/m}^2$$

3 Compute the effective area of the air gap. This is greater than the area of the core due to fringing as shown in Fig. 15.3-2. If the length of the air gap is relatively small compared with the dimensions of its cross section, then a reasonable approximation to its effective cross-sectional area is found by adding its length to each cross-sectional dimension of the core. Thus

$$A_a = (1 + 0.2) \text{ cm} \times (1.5 + 0.2) \text{ cm} \times \left(\frac{1 \text{ m}}{100 \text{ cm}}\right)^2 = 2.04 \times 10^{-4} \text{ m}^2$$

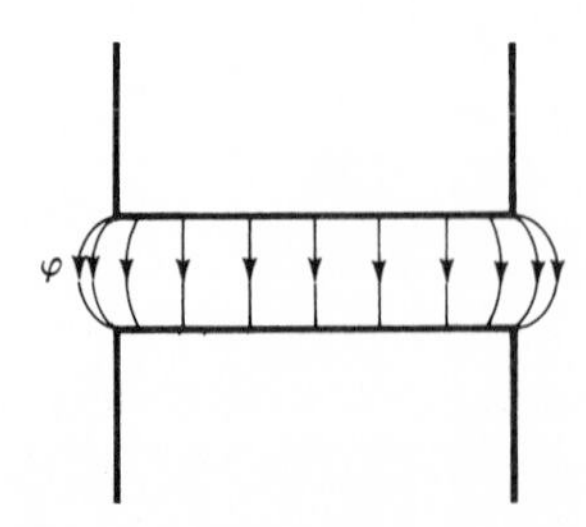

Fig. 15.3-2 Fringing effect at air gap.

4 Compute the flux density in the air gap:

$$B_a = \frac{\varphi}{A_a} = \frac{0.12 \times 10^{-3}\ \text{Wb}}{2.04 \times 10^{-4}\ \text{m}^2} = 0.588\ \text{Wb/m}^2$$

5 Find the magnetic field intensity in the core from the BH curve in Fig. 15.1-4:

$$H_c = 450\ \text{At/m}$$

6 Calculate the magnetic field intensity in the air gap using μ_0 from Eq. (15.1-5):

$$H_a = \frac{B_a}{\mu_0} = \frac{0.588\ \text{Wb/m}^2}{4\pi \times 10^{-7}\ \text{Wb/A}\cdot\text{m}} = 4.68 \times 10^{-5}\ \text{At/m}$$

7 Find the total magnetomotive force (MMF) required using Ampere's circuital law:

$$MMF = d_cH_c + d_aH_a = (0.08)(450) + (2 \times 10^{-3})(4.68 \times 10^{5})$$
$$= 972\ \text{At}$$

8 The number of turns required is found by equating the MMF to the product of the pull-in current and the number of turns, N.

$$N = \frac{MMF}{I_p} = \frac{972\ \text{At}}{0.200\ \text{A}} = 4{,}860\ \text{turns}$$

The release current is computed by following the same sequence of steps except that there is no air gap since the relay is closed and in the last step the MMF is divided by $N = 4{,}860$ turns to compute I_r. The steps are

9 The magnetic field intensity in the core is the same as calculated for the pull-in current in step 5: $H_c = 450$ At/m

10 The MMF is

$$MMF = d_cH_c = (0.08)(450) = 36\ \text{At}$$

11 The release current is

$$I_r = \frac{MMF}{N} = \frac{36\ \text{At}}{4{,}860\ \text{t}} = 7.41\ \text{mA}$$

Thus the current required to hold the relay closed once it has been closed is much less than that required to close it when it is open. This effect, i.e., the difference between the pull-in and release current values, is known as hysteresis. It is due to the presence and absence of the air gap in the two computations.

15.4 FARADAY'S LAW

One of the very important early discoveries in electricity was that changing the magnetic flux passing through a coil causes a voltage to be generated, or induced, between the two terminals of the coil. This voltage is found to be proportional to the rate of change of flux and the number of turns in the coil. The mathematical formula for this induced voltage is Faraday's law:

$$v(t) = -n\frac{d\varphi(t)}{dt} \qquad (15.4\text{-}1)$$

The negative sign is a consequence of the reference directions for v and φ which are those conventionally used in the literature; they are shown in the diagram of the coil in Fig. 15.4-1*a*. This coil is an air-core coil; it is wound on a tube of nonmagnetic material. Air-core coils normally produce smaller flux than coils wound on a core of magnetic material but have a linear *MMF*-flux relationship as a result of their constant permeability.

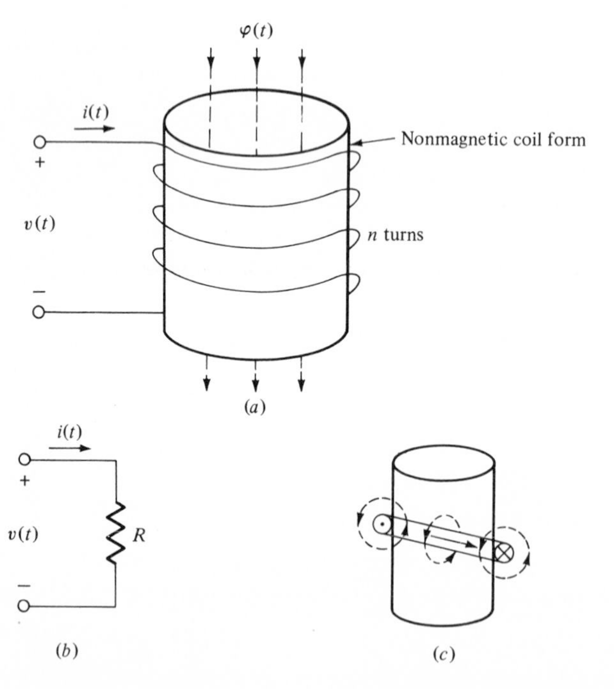

Fig. 15.4-1 Voltage induced by changing magnetic flux through coil. (*a*) Coil showing reference directions for $v(t)$ and $i(t)$; (*b*) imaginary resistance used to determine direction for $i(t)$; (*c*) typical coil turn used to find direction of positive flux for $i(t)$.

Lenz's law states that the polarity or direction of an induced quantity, such as the voltage in Faraday's law, is such that it will oppose the source that produces it, e.g., the flux in our case. This statement of the general principle is not sufficient to enable one to easily find the polarity of an induced voltage. The following discussion presents a method for accomplishing this. It is assumed that the direction of the flux is given and the polarity of the induced voltage is to be found. An arbitrary reference polarity for $v(t)$ is assumed as shown, for example, in Fig. 15.4-1*a*. A two-step procedure is used to determine the sign relating $v(t)$ to $n[d\varphi(t)/dt]$. The first step is to assign a direction for current flow such that the voltage and current are related by a positive sign; the second step is to find the sign of the current-flux relationship using the right-hand rule.

A simple way to assign the current direction is to mentally replace the coil by a resistance as shown in Fig. 15.4-1*b*. The current direction is then chosen such that $v(t) = Ri(t)$.

For the second step the direction of the flux produced by this current is determined using the right-hand rule. For the current direction of Fig. 15.4-1*b* positive flux would be directed upward inside the coil as shown by the dotted loops in Fig. 15.4-1*c*. Since this is opposite the reference direction for the flux in Fig. 15.4-1*a* there must be a minus sign in the relationship between v and φ in Eq. (15.4-1).

From Eq. (15.4-1) we find the following relationship among the units

$$1 \text{ V} = 1 \text{ Wb/s} \tag{15.4-2}$$

These concepts are applied in the following example.

Example 15.4-1 A flux varies as shown in Fig. 15.4-2a. It flows through a coil of 1,000 turns as shown in Fig. 15.4-1. Find the induced voltage $v(t)$.

Solution The voltage is given by

$$v(t) = -1{,}000 \frac{d\varphi(t)}{dt}$$

Since the derivative is the rate of change or slope of the function, the voltage is determined at each instant of time by the slope of $\varphi(t)$ at that instant. For example, between $t = 0$ ms and $t = 5$ s the flux increases linearly from 0 to 10 mWb. Thus, in this time interval,

$$\frac{d\varphi}{dt} = \frac{10 - 0}{5 - 0} = 2 \text{ mWb/s}$$

so that

$$v(t) = (-1{,}000)(2)(10^{-3}) = -2 \text{ V} \qquad 0 < t < 5 \text{ s}$$

Similar calculations lead to the complete curve of $v(t)$ shown in Fig. 15.4-2*b*. ////

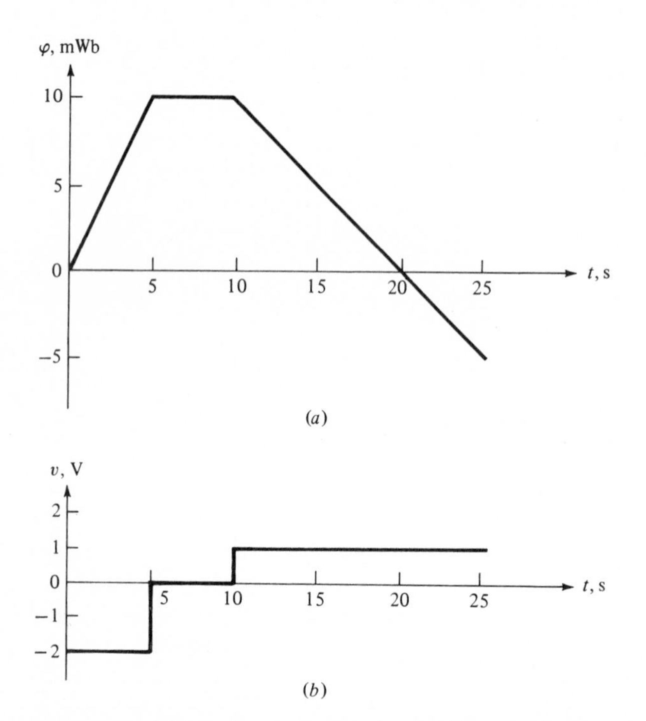

Fig. 15.4-2 Flux and voltage for Example 15.4-1. (*a*) Flux; (*b*) voltage.

15.5 FORCE ON CURRENT-CARRYING CONDUCTOR

If a conductor of length d carries a current i in a magnetic field with flux density B, as shown in Fig. 15.5-1*a*, then there will be a force f acting on the conductor whose magnitude is

$$f = Bid \tag{15.5-1}$$

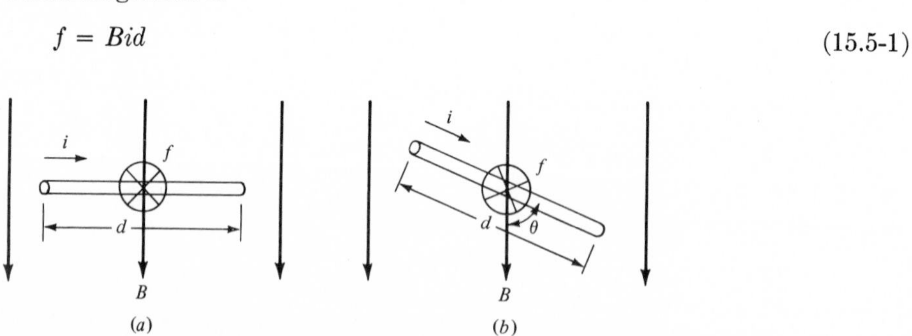

Fig. 15.5-1 Force on a current carrying conductor in a magnetic field. (*a*) i perpendicular to B; (*b*) angle θ between i and B.

If i and B are in the plane of the page, then f is directed into the page as shown in the figure. The rule for finding the direction of the force is often called the "right-hand screw rule." Imagine turning a right-hand screw so that the blade of the screwdriver rotates 90° from being parallel to i until it is parallel to B; then f will be in the direction that the screw moves. This will always be perpendicular to the plane formed by i and B as shown in Fig. 15.5-1a.

Equation (15.5-1) is valid if i and B are at right angles to each other, as is the case in Fig. 15.5-1a. The more general case is shown in Fig. 15.5-1b; in this diagram there is an angle between i and B. The formula for the force in this case is

$$f = Bid \sin \theta \tag{15.5-2}$$

while the direction is still given by the right-hand screw rule. In this case the screwdriver blade is rotated through the angle θ so as to turn a right-handed screw; the direction of the screw's movement is the direction of f.

The unit of force is the newton if B is in webers per square meter, i is in amperes, and d is in meters. Thus

$$1\ \text{N} = 1\ \text{A}\cdot\text{Wb/m} \tag{15.5-3}$$

Faraday's law and the law for the force on a current-carrying conductor in a magnetic field are used to explain the operation of many electrical devices. A number of examples are presented in the next section.

15.6 ELECTROMAGNETIC DEVICES

The devices to be described in this section perform a variety of functions; both electrical and electromechanical. The operation of each one of them is based on one or more of the principles of electromagnetism which have been discussed in previous sections. The description of their operation will not be detailed and no attempt will be made to provide a quantitative analysis. Our intent is to provide a qualitative understanding of the basic theory of operation of each device.

15.6-1 The Transformer

The transformer is a passive device which consists of two coils of wire wound on a single closed-loop magnetic core. The function of the transformer is to provide an output voltage which is some multiple of the input voltage without changing the power level. A diagram of an elementary transformer is shown in Fig. 15.6-1.

The transformer core is made of a high permeability metal so that its reluctance will be much lower than that of the surrounding air. As a result essentially all of the flux will remain within the core and will link both coils. A voltage applied to one coil will produce a current through it which results in a magnetic field intensity, H, and thus a magnetic flux, φ, flowing through the core.

By Faraday's law

$$v_1 = N_1 \frac{d\varphi}{dt} \qquad (15.6\text{-}1)$$

and since the same flux links both coils,

$$v_2 = N_2 \frac{d\varphi}{dt} \qquad (15.6\text{-}2)$$

The reader should check that the reference directions and polarities in Fig. 15.6-1 do not result in a minus sign in Eq. (15.6-1) or (15.6-2).

Comparing Eqs. (15.6-1) and (15.6-2),

$$\frac{v_1}{N_1} = \frac{d\varphi}{dt} = \frac{v_2}{N_2}$$

Therefore

$$\frac{v_2}{v_1} = \frac{N_2}{N_1} \qquad (15.6\text{-}3)$$

This is an extremely important relation governing transformer operation. By selecting the turns ratio N_2/N_1 properly, any desired voltage can be obtained. Frequently, one side is considered as the input side and the other side as the output side. The coil on the input side is called the primary coil and the one on the output side is called the secondary coil. A very important point to note in Eq. (15.6-3) is that both voltages are proportional to the *rate of change of flux with respect to time*. Therefore the transformer cannot be used with dc signals. One of the reasons for the widespread use of alternating current (ac) for power distribution is our ability to step its voltage up or down using transformers.

If a voltage v_1 is applied to a transformer primary coil and a load is connected to its secondary, then a current i_1 flows into the primary coil. As a result power is transferred into the transformer primary given by

$$p_1 = v_1 i_1 \qquad (15.6\text{-}4)$$

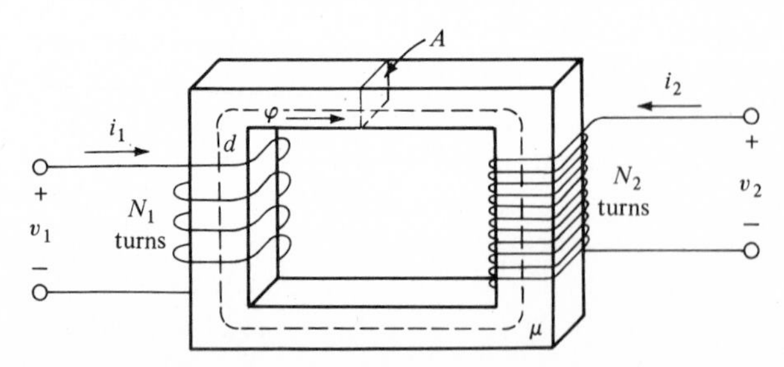

Fig. 15.6-1 Transformer.

This electric power is transformed into power in the magnetic field which is then transformed back into electric power at the secondary coil. Such a transformer is a lossless device; all the power into the primary coil comes out of the secondary coil and the net power absorbed by the transformer is zero.

The secondary, or load, power is

$$p_2 = v_2 i_2 \tag{15.6-5}$$

Since the net power into the transformer is zero,

$$v_2 i_2 + v_1 i_1 = 0 \tag{15.6-6}$$

or

$$i_2 = -\frac{v_1}{v_2} i_1 \tag{15.6-7}$$

Substituting for v_1/v_2 from Eq. (15.6-3a),

$$i_2 = -\frac{N_1}{N_2} i_1 \tag{15.6-8}$$

or

$$\frac{i_2}{i_1} = -\frac{N_1}{N_2} \tag{15.6-9}$$

The minus sign in Eq. (15.6-9) agrees with the direction of current flow if a resistive load is connected to the secondary of the transformer of Fig. 15.6-1.

The transformer described by Eqs. (15.6-3) and (15.6-9) is an *ideal transformer*; the power out of it is equal to the power in. We call the ratio of power out to power in the efficiency of the device and symbolize it by the Greek letter eta (η):

$$\eta = \frac{p_{\text{out}}}{p_{\text{in}}} \tag{15.6-10}$$

For the ideal transformer, $\eta = 1$.

Any real transformer has an efficiency less than 1 because there are power losses due to the fact that ideal transformers cannot be realized. These losses are due in part to the resistance of the coil windings which dissipate power in the form of heat and in part to other effects. Nevertheless, the efficiency of a well-designed transformer can be very high; values of 0.97 to 0.98 for η are not unusual.

Example 15.6-1 The primary coil of an ideal transformer has 100 turns and a voltage across it of 25 V. The secondary of the transformer has 500 turns and a 2-kΩ resistor is connected across it. Find the current in the primary coil.

Solution For the ideal transformer

$$v_2 = \frac{n_2}{n_1} V_1 = \frac{500}{100} 25 = 125 \text{ V}$$

The secondary current is

$$I_2 = \frac{V_2}{R} = \frac{125}{2} = 62.5 \text{ mA}$$

and the primary current is

$$I_1 = \frac{n_2}{n_1} I_2 = \frac{500}{100} 62.5 = 312.5 \text{ mA}$$ ////

15.6-2 Ac Generator

A rudimentary generator is shown in Fig. 15.6-2. It consists of a coil of wire which is rotated at a constant angular velocity ω rad/s in a uniform magnetic field provided by a permanent magnet. The coil is rotated by some mechanical means, such as a turbine. The leads of the coil are connected to slip rings which are attached to the armature on which the coil is wound so that the slip rings rotate with the coil. The armature is not shown in Fig. 15.6-2 in order to keep the diagram simple. Carbon rods, called brushes, are held against the slip rings by springs which are not shown in the diagram. These brushes are connected to the fixed terminals of the generator; the generated voltage, $v(t)$, is obtained across these terminals. The area enclosed by the coil is A. The flux lines follow a

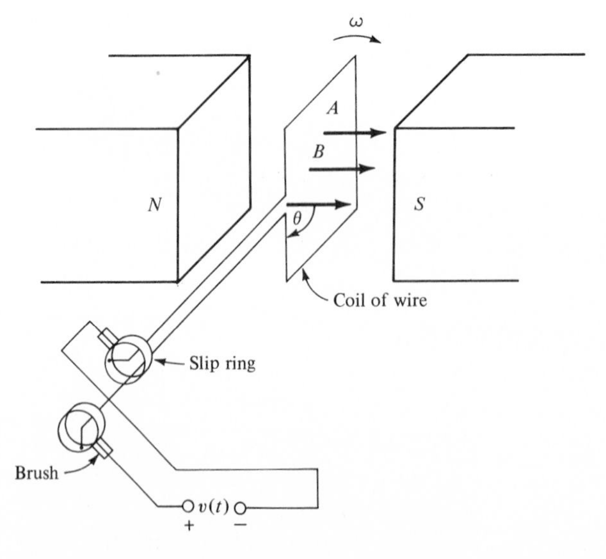

Fig. 15.6-2 Generator.

horizontal path from the north pole of the magnet to the south pole. The flux density is constant with a value of B in the area between the magnet's poles.

The angle between the horizontal flux lines and the radial edge of the coil is denoted by θ as shown in Fig. 15.6-2; as the coil rotates this angle varies between 0 and 360° for each rotation. For $\theta = 0$, the plane of the coil is parallel to the flux lines and the flux linking the coil is zero. For $\theta = 90°$, the plane of the coil is perpendicular to the flux lines and the flux linking the coil is maximum with value

$$\varphi_{\max} = BA \qquad (15.6\text{-}11)$$

As the coil rotates, $\varphi(t)$, the flux linking it, varies according to the equation

$$\varphi(t) = BA \sin \theta(t) \qquad (15.6\text{-}12)$$

in which the notation $\theta(t)$ explicitly indicates that θ varies with time. If $t = 0$ is chosen as an instant when $\theta(t) = 0$, then

$$\varphi(t) = BA \sin \omega t \qquad (15.6\text{-}13)$$

since the coil is rotating at constant angular velocity ω.

From Faraday's law we have (since $n = 1$)

$$v(t) = \frac{d\varphi(t)}{dt} = \omega BA \cos \omega t = V_m \cos \omega t \qquad (15.6\text{-}14)$$

Thus the magnetic device of Fig. 15.6-2 is a simple ac generator.

15.6-3 D'Arsonval Meter Movement

The diagram of a simple D'Arsonval meter movement is shown in Fig. 15.6-3*a*. This device was described in Sec. 7.1 in connection with dc measurements. It consists of an an armature pivoted between two bearings with a coil of wire wrapped parallel to the axis of the armature and a permanent magnet which provides a magnetic field. The armature is restrained from turning by a spring.

When current flows through the armature coil, a force F acts on each side of the coil as shown in Fig. 15.6-3*b*. Since B and I are perpendicular, this force is given by

$$F = BId \qquad (15.6\text{-}15)$$

where d is the length of the armature. The directions of the forces F are given by the right-hand rule and are as shown in Fig. 15.6-3*b*. Each of these forces has two components: one in the radial direction which is not shown and one in the tangential direction, shown as F_T in the diagram. The two tangential components provide a torque which rotates the armature clockwise; this rotation is resisted by the restraining force of the spring. The greater the current, the greater is the force tending to rotate the armature *clockwise*; the more the spring is stretched

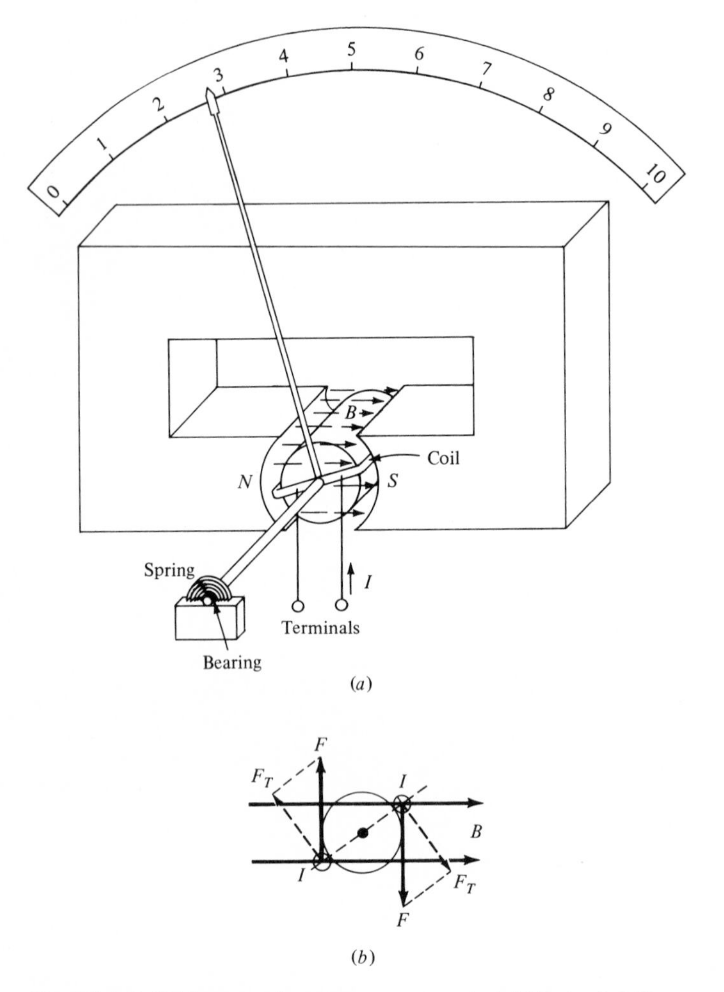

Fig. 15.6-3 D'Arsonval meter movement. (*a*) Pictorial diagram; (*b*) schematic showing forces.

the greater is its force tending to rotate the armature *counterclockwise*. The armature rotates clockwise until these forces are balanced, at which point it stops moving. The dial is calibrated to indicate the amount of current flowing through the armature winding.

15.6-4 Dc Motor

The motor is similar to a generator but operates in reverse fashion; its input is electrical energy and its output is mechanical rotation whereas for a generator the input is mechanical rotation and its output is electrical energy.

As would be expected the structure of a motor and generator are quite similar; some machines in fact, can be used as either a motor or generator. The motor we are going to describe will differ slightly from the generator of Sec. 15.6-2 because it is a dc motor whereas the generator was an ac machine.

The primary difference between the dc machine and the ac machine is the use of a commutator instead of slip rings to connect the stationary terminals to the rotating armature coil. The motor shown in Fig. 15.6-4*a* is highly simplified; an actual motor generally has multiple armature windings and uses electromagnets rather than permanent magnets to supply the magnetic field. Nevertheless, our simplified model demonstrates the basic concepts and has the advantage of simplicity.

The basic concept behind the operation of the dc motor is similar to that of the D'Arsonval meter movement; the forces on the current carrying conductors in the magnetic field cause the armature to rotate. In order to maintain the direction of force such that the direction of rotation remains the same it is necessary to reverse the current direction as the armature winding passes the vertical position. This is achieved by the use of a *commutator*; a split ring to which the armature winding is connected. The commutator is connected to the motor terminals by a pair of brushes held against the commutator by a pair of springs which are not shown in the diagram. As the armature winding passes the vertical position the splits in the commutator pass the brushes, as shown in Fig. 15.6-4*b*, thereby changing the direction of flow of armature current. The directions of the forces and the direction of rotation are also shown in Fig. 15.6-4*b*.

According to Faraday's law, a voltage is induced in the armature winding because of the change in flux linked as the motor armature rotates. The motor reaches a speed such that this induced voltage is just equal to the applied voltage.

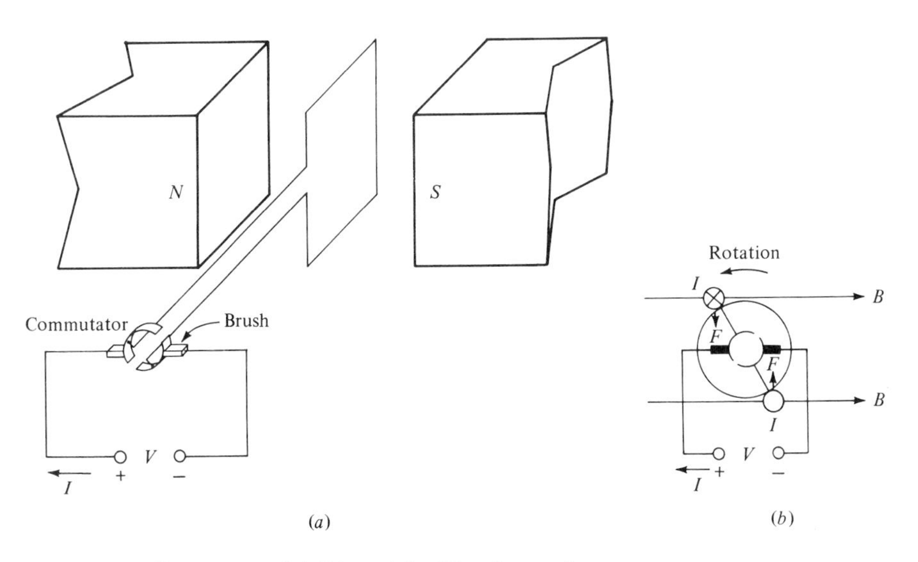

Fig. 15.6-4 Dc motor. (*a*) Pictorial; (*b*) schematic.

A quantitative analysis of this operation requires consideration of the voltage drop due to the armature winding resistance, the inertia of the armature, and frictional energy losses. Consideration of all these factors would take us well beyond the scope of this text; the qualitative description which we have presented is useful because of the insight it provides.

SUMMARY

1 Magnetic flux is measured in webers (Wb) and is generally symbolized by φ.

2 Magnetic flux density is the flux per unit area:

$$B = \frac{d\varphi}{da} \qquad \text{Wb/m}^2$$

3 Magnetomotive force is the source of magnetic flux. For a coil of n turns carrying I A, it is

$$MMF = nI \qquad \text{At (ampere turns)}$$

4 Magnetic field intensity is determined by the applied MMF and the length of the magnetic circuit, d, from the following relation

$$H = \frac{NI}{d} \qquad \text{At/m}$$

5 The permeability of a material relates B and H in the material:

$$\mu = \frac{B}{H} \qquad \text{H/m (henries/meter)}$$

6 The reluctance of a magnetic circuit is the analog of resistance in an electric circuit:

$$\mathcal{R} = \frac{Hd}{\varphi} = \frac{d}{\mu a} \qquad \text{A/Wb}$$

7 Faraday's law relates the voltages induced across a coil to the rate of change of flux linking the coil:

$$v(t) = -n\frac{d\varphi(t)}{dt} \qquad \text{V}$$

For appropriate reference directions see Fig. 15.4-1.

8 The force on a current-carrying conductor in a magnetic field is

$$F = BId \text{ N}$$

PROBLEMS

15.1-1 A magnetic core has a uniform flux density of 15 Wb/m^2 flowing through it. Its cross section is a 2×5 cm rectangle. Find the total flux φ.

15.1-2 A doughnut-shaped cast-steel magnetic core has a circular cross section whose radius is 1.5 cm. The diameter of the entire core from outer edge to outer edge is 10 cm. The total flux through it is 8×10^{-4} Wb. Find the flux density.

15.1-3 Find the magnetomotive force produced by 2.5 A flowing through a 200-turn coil.

15.1-4 A flux density of 1.0 Wb/m^2 flows through a cast steel core.

a Find the magnetic field intensity, H.
b Determine the permeability, μ.
c If the cross section of the core is 1×4 cm, find the flux, φ.

15.2-1 Find the reluctance of a 10 cm length of the core described in Prob. 15.1-4.

15.2-2 A cast-steel magnetic core has a circular cross section with a diameter of 1 cm. Its mean length is 20 cm. It is wrapped with a 100-turn coil carrying 250 mA. Find

a The applied MMF
b The magnetic field intensity
c The flux density
d The total flux

15.2-3 A 500-turn coil is wrapped around the core described in Prob. 15.1-2. Find the flux if a current of 1 A flows through it.

15.3-1 Find the current required in a 500-turn coil to produce a flux of 0.1 mWb in the core of Prob. 15.1-2 if a 0.1-cm air gap is cut in it.

15.3-2 The cast-steel magnetic core shown in Fig. P15.3-2 has a uniform thickness of 2 cm. Find the current I.

15.3-3 Find the current required in the magnetic circuit of Fig. P15.3-3 to produce a flux density of 0.5 Wb/m^2.

15.3-4 Repeat Prob. 15.2-3 if a 0.1-cm air gap is cut in the core.

15.4-1 The graph of the flux density passing through a 10-turn circular air-core coil of radius 2 cm shown in Fig. P15.4-1*a* is shown as a function of time in Fig. P15.4-1*b*. Draw a graph of the induced voltage $v(t)$.

15.6-1 An ideal transformer has 10 V across its primary coil and 25 V across its secondary. The secondary coil has 500 turns. How many turns does the primary coil have?

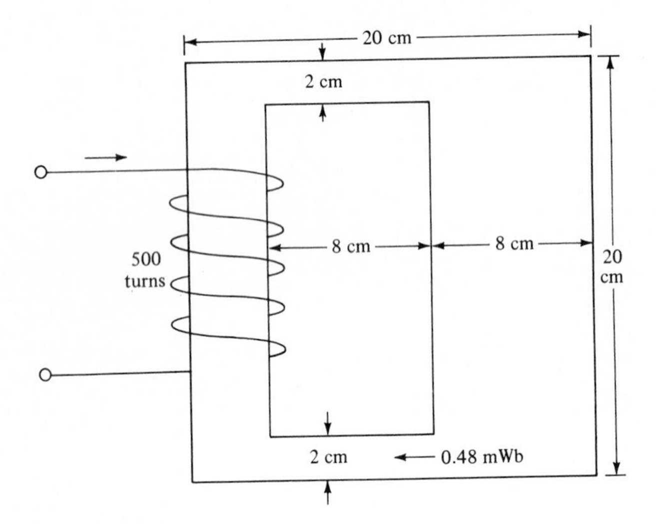

Fig. P15.3-2

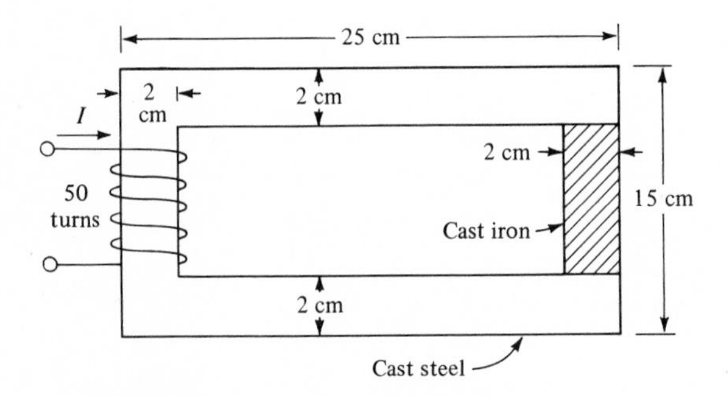

Fig. P15.3-3

15.6-2 An ideal transformer has a primary coil of 50 turns and a secondary coil of 1,000 turns. A resistance of 10 kΩ is connected across the secondary. Find the equivalent resistance seen at the primary coil. *Hint*: Assume a current of 1 A into the primary. Find the secondary current and then the secondary voltage. Finally find the primary voltage. The equivalent resistance is the ratio of primary voltage to primary current.

15.6-3 A certain transformer is ideal with a secondary to primary turns ratio of 5:1 except that it has a coil resistance of 1 Ω at the primary and 5 Ω at the secondary. These resistances may be modeled by resistances in series with the coils of an ideal transformer. Find the efficiency of the transformer if it is delivering 5 W to a 100 Ω load connected to the secondary.

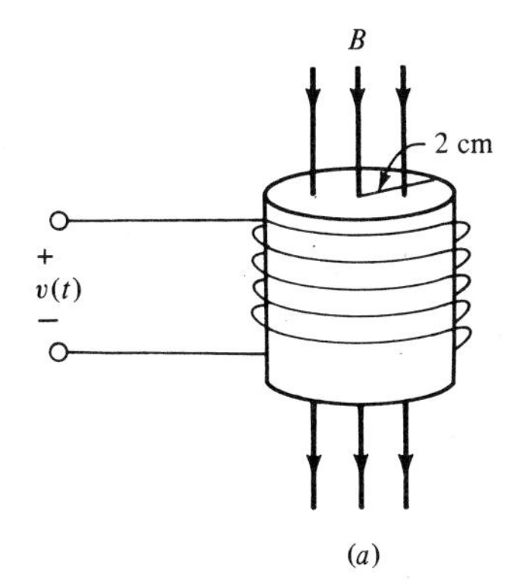

(a)

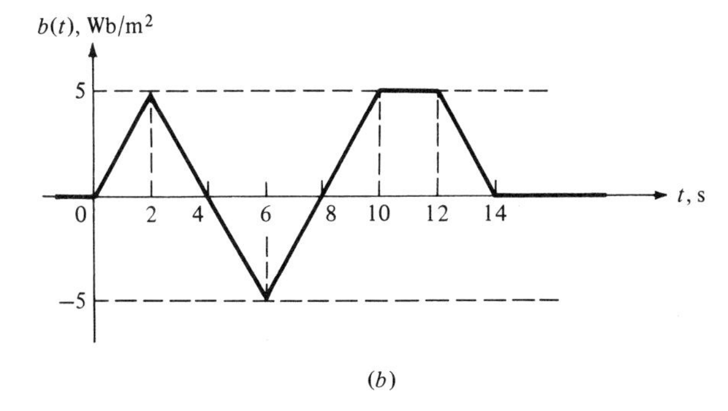

(b)

Fig. P15.4-1

16
Inductance and Capacitance

OBJECTIVES

Upon completion of this chapter you should be able to

1 State the derivative form of the *vi* relation for the inductor and use it to find the voltage response to various current waveforms.
2 State the integral form of the *vi* relation for the inductor and use it to find the current response to various voltage waveforms.
3 Find the equivalent inductance for series and parallel combinations of inductors.
4 Describe the phenomenon of mutual magnetic coupling.
5 Write KVL equations for circuits with magnetic coupling.
6 Explain the dot convention for coupled coils.
7 State the derivative form of the *vi* relation for the capacitor and use it to find the current response to various voltage waveforms.
8 State the integral form of the *vi* relation for the capacitor and use it to find the voltage response to various current waveforms.
9 Find the equivalent capacitance of series and parallel combinations of capacitors.
10 State the relations for energy stored in capacitors and inductors and use them to find the stored energy.

INTRODUCTION

Up to this point all of the circuits we have considered have consisted solely of resistors and sources for linear networks plus ideal diodes in the case of nonlinear networks. If the sources were other than dc, the responses throughout the network had exactly the same waveform as the source in the case of linear networks. The only effect that these resistive networks had on the waveforms was to change the relative level of currents and voltages; the wave shape remained the same. We are now ready to add to our catalog of linear elements two for which this behavior is no longer true. These are the inductor and capacitor.

To give you an idea of what is to come, we note that a capacitor is an element in which no current flows unless the voltage across the capacitance is changing with time, and an inductor is an element which has no voltage across its terminals unless the current through it is changing with time. Because of the fact that changes with time are involved, we are going to have to make considerable use of the concepts of derivative and integral which we studied in Chap. 14.

In this chapter we consider the fundamental behavior of these elements and their effect on basic waveforms.

16.1 INDUCTANCE

In this section we will describe the inductance element and derive its *vi* relation. The form usually taken by the inductance element is that of a long cylindrical coil of many turns of wire as shown pictorially in Fig. 16.1-1*a*. Often the coil is wound over a core of magnetic material, and the actual physical form which practical inductors take consists of many varieties, some of which are shown in Fig. 16.1-1*b*. The circuit symbol usually used is shown in Fig. 16.1-1*c*. Variations on this symbol sometimes indicate the presence of a magnetic core or a slug of magnetic material which can be moved in or out to vary the properties of the element. These variations are shown in Fig. 16.1-1*d*.

To derive the *vi* relation for the inductance we begin with Faraday's law (see Sec. 15.4) which states that a coil of N turns of wire placed in a magnetic flux field which is changing with time will have a voltage induced across its terminals according to the relation

$$v(t) = N\frac{d\varphi}{dt} \tag{16.1-1}$$

where $v(t)$ = voltage at coil terminals, V
N = number of turns
$d\varphi/dt$ = instantaneous rate of change of flux, Wb/s

Now consider the circuit shown in Fig. 16.1-2. When the switch is closed, the current begins to increase, and since flux φ is proportional to current i, the flux

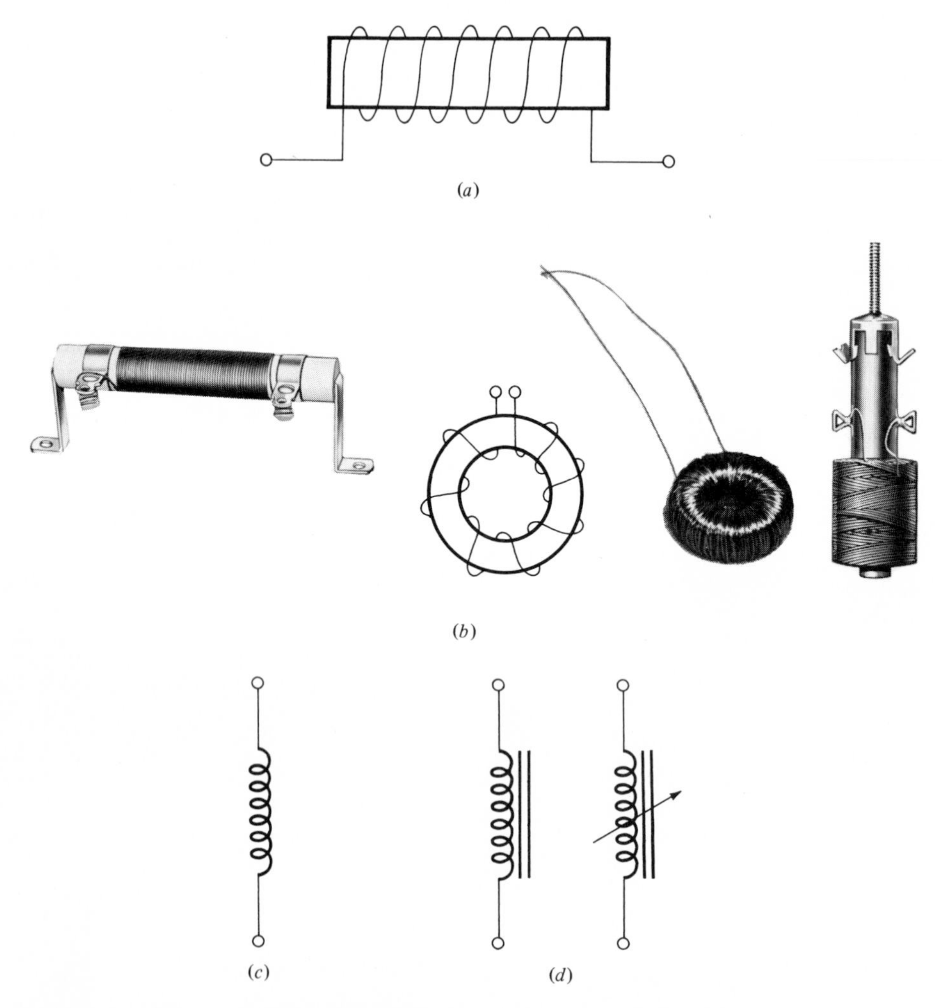

Fig. 16.1-1 Inductance. (*a*) Cylindrical coil; (*b*) types of inductor:—ceramic core inductor, toroidal inductor, variable inductor with moveable core (*J. W. Miller Co.*) (*c*) circuit symbol; (*d*) circuit symbol when magnetic core or moveable plug is present.

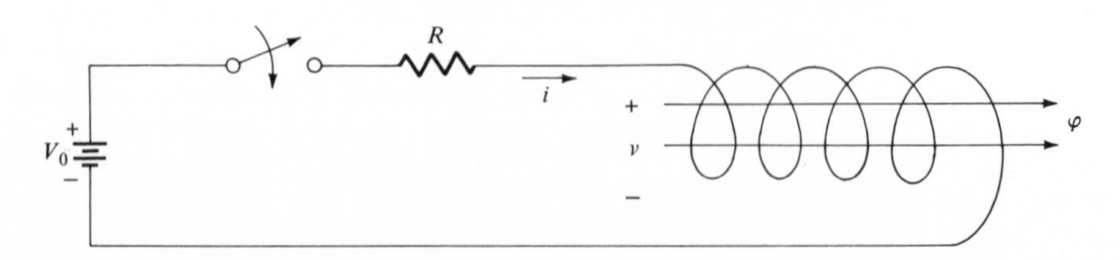

Fig. 16.1-2 Inductive circuit.

will also begin to increase. According to Faraday's law a voltage will be induced across the coil because the flux is changing. This voltage will be in such a direction as to *oppose* the change in current, an effect known as Lenz's law. We can relate the voltage across the coil to the current by noting that, from Eq. (15.2-5) the flux is

$$\varphi = \frac{Ni}{\mathcal{R}} \tag{16.1-2}$$

where φ = instantaneous flux, Wb
N = number of turns
i = instantaneous current, A
$\mathcal{R}$ = reluctance, A/Wb

N is constant for a given coil and if we make the reasonable assumption that $\mathcal{R}$ is constant we can differentiate both sides of Eq. (16.1-2) to get

$$\frac{d\varphi}{dt} = \frac{N}{\mathcal{R}} \frac{di}{dt} \tag{16.1-3}$$

Next we substitute $d\varphi/dt$ from Eq. (16.1-3) into Eq. (16.1-1). This yields

$$v(t) = \frac{N^2}{\mathcal{R}} \frac{di}{dt} \tag{16.1-4}$$

The constant $N^2/\mathcal{R}$ is called the inductance of the coil, symbolized by L, and the unit of inductance is the henry (H), in honor of Joseph Henry, an American physicist. Using this definition, Eq. (16.1-4) becomes

$$v(t) = L \frac{di}{dt} \tag{16.1-5}$$

This equation is the *vi* relation for the inductance. Note that if the current is constant, then $di/dt = 0$ and no voltage appears across the terminals. Thus the only time that voltage will appear across the terminals of an inductance is when the current is changing with time. This represents a major difference from the resistive circuits we have studied up to now. This difference can best be appreciated by considering several examples.

Example 16.1-1 The current waveform of Fig. 16.1-3*a* is applied to the *RL* circuit of Fig. 16.1-3*b*. Find the voltages across L and R, and the total voltage.

Solution Because of the nature of the current waveform, it is convenient to treat each segment separately. For $t < 0$, the current is zero, so $v(t) = 0$. For the first line segment we can represent the current by the straight-line equation

$$i(t) = t \qquad 0 < t < 2$$

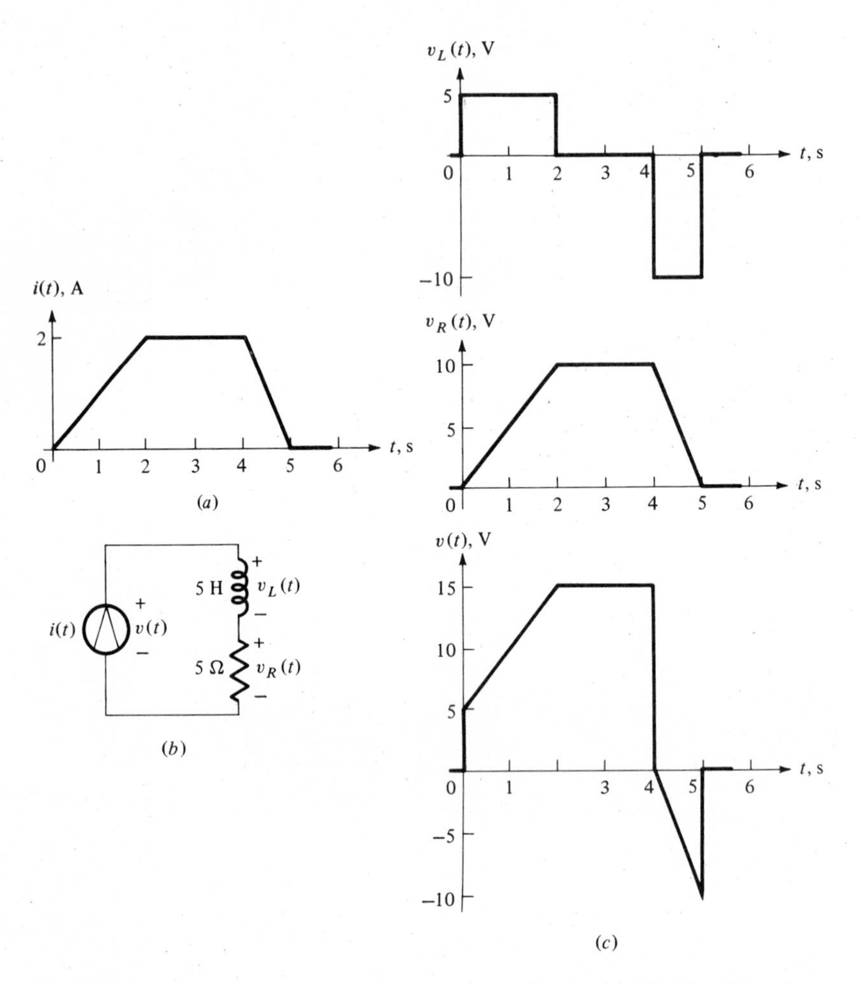

Fig. 16.1-3 Example 16.1-1 (*a*) Current waveform; (*b*) circuit; (*c*) voltage waveforms.

Then using Eq. (16.1-5),

$$v_L(t) = L\frac{di}{dt} = 5\frac{d}{dt}(t) = (5)(1) = 5 \text{ V} \qquad 0 < t < 2.$$

Thus, at $t = 0$, $v(t)$ jumps to 5 V where it remains until $t = 2$ s.

For the interval from $t = 2$ s, to $t = 4$ s, $i(t)$ is a constant so that $v_L(t) = 0$ for $2 < t < 4$. Thus the voltage drops from 5 V to 0 V at $t = 2$ s, where it remains until $t = 4$ s.

For the final interval, from $t = 4$ s to $t = 5$ s, the straight-line equation is (be sure to verify this)

$$i(t) = 10 - 2t \qquad 4 < t < 5$$

The voltage is

$$v_L(t) = L\frac{di}{dt} = 5\frac{d}{dt}(10 - 2t) = (5)(0 - 2) = -10\text{ V} \qquad \text{for} \quad 4 < t < 5$$

Thus at $t = 4$ s the voltage drops from 0 V to -10 V where it remains until $t = 5$ s when it returns to zero. The inductor voltage waveform is plotted in Fig. 16.1-3*c*.

The resistor voltage waveform is also plotted in Fig. 16.1-3*c*. It is obtained simply by noting that Ohm's law applies, that is,

$$v_R(t) = Ri(t) = 5i(t)$$

The total voltage is obtained using KVL,

$$v(t) = v_L(t) + v_R(t)$$

This is the final graph in Fig. 16.1-3*c*. The student should take careful note of the difference between the resistor voltage waveform and the inductor voltage waveform. ////

Example 16.1-2 An exponential current $i(t) = 3e^{-t/10}$ A is applied to a 5-mH inductor beginning at $t = 0$. Find and plot the resulting voltage across the inductor.

Solution From the *vi* relation for the inductance we have, referring to Eq. (14.1-16) for the derivative of the exponential,

$$v(t) = L\frac{di}{dt} = 5 \times 10^{-3}\frac{d}{dt}(3e^{-t/10}) = (5 \times 10^{-3})(3)\left(\frac{-1}{10}\right)e^{-t/10}$$

$$= -1.5 \times 10^{-3}\, e^{-t/10}\text{ V}$$

This is plotted in Fig. 16.1-4 along with the current waveform. ////

Example 16.1-3 A 20-mH inductor is to have a constant voltage across it of 3 V. At what rate must the current through it be changing in order for this voltage to be maintained across its terminals?

Solution Rearranging Eq. (16.1-5),

$$\frac{di}{dt} = \frac{v(t)}{L}$$

The derivative represents the rate at which the current is changing. Thus

$$\frac{di}{dt} = \frac{v}{L} = \frac{3}{0.02} = 150\text{ A/s}$$ ////

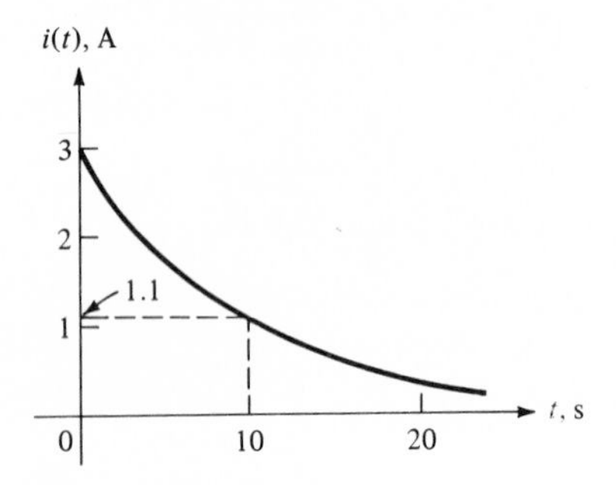

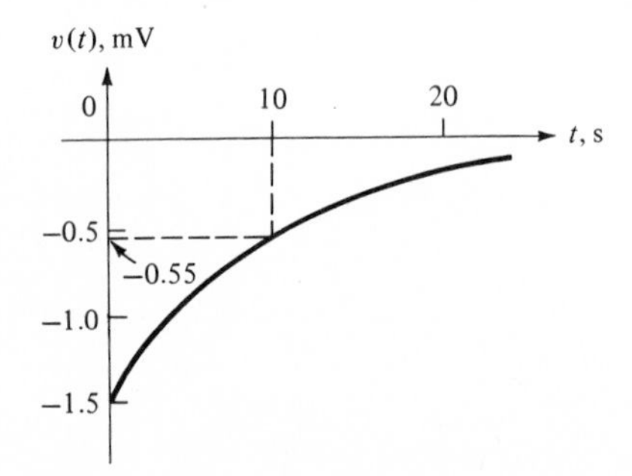

Fig. 16.1-4 Waveforms for Example 16.1-2.

SERIES AND PARALLEL INDUCTORS

If two inductors L_1 and L_2 are connected in series in such a way that none of the flux from L_1 links any of the turns of L_2, and vice versa, then the equivalent inductance is

$$L_{eq} = L_1 + L_2 \tag{16.1-6}$$

This can be proved using KVL and the *vi* relation for the inductance. The proof is left as an exercise.

If the same two inductors are connected in parallel, again with no common flux linkages, the equivalent inductance can be shown to be

$$L_{eq} = \frac{L_1 L_2}{L_1 + L_2} \tag{16.1-7}$$

It is evident from Eqs. (16.1-6) and (16.1-7) that the equivalent inductance of inductors in series or parallel is found in exactly the same way as the equivalent resistance of series or parallel combinations of resistors.

When the inductors are not magnetically isolated as above, the effect of the flux from each coil acting on the other must be taken into account. This is considered in the next section.

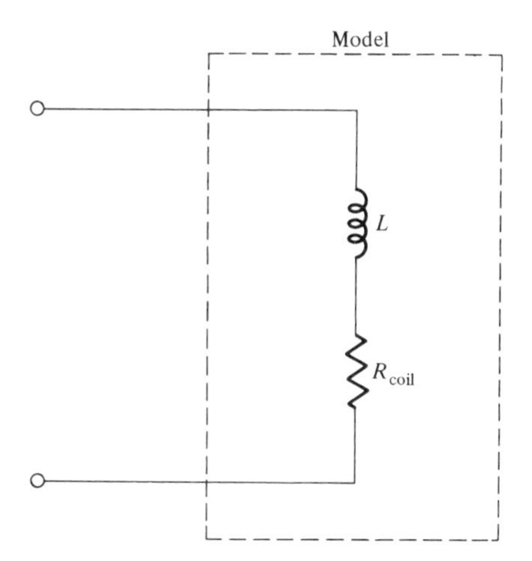

Fig. 16.1-5 Practical model for inductance coil.

COIL RESISTANCE

Practical inductors are made of numerous turns of wire which has finite resistance. This resistance cannot actually be separated from the ideal inductance of the coil and analysis taking account of this effect becomes very difficult. To simplify the analysis, we replace the actual coil by a *model,* in which the inductance and resistance are in series as shown in Fig. 16.1-5. This model will provide results that are sufficiently accurate for most purposes. When the resistance is negligible, the coil behaves like a pure inductance. We have made this assumption in the development of this section. In succeeding sections, the coil resistance will be assumed to be negligible unless otherwise stated.

16.2 MUTUAL INDUCTANCE

When two coils are placed so that some or all of the flux produced by the current in one coil passes through (links) the second coil, a voltage will be induced across the second coil due to changes in the flux produced by the first coil. This effect is called *mutual inductance.* In this section we will consider several aspects of mutual inductance coupling.

In the circuit of Fig. 16.2-1 we have two coils placed close together. The coils need not have an iron core as did the transformer of Sec. 15.6-1. The primary coil has N_1 turns and self-inductance L_1; the secondary coil has N_2 turns and self-inductance L_2. When voltage is applied to the primary, current will flow and primary flux φ_1 will be produced as shown. According to Faraday's law [see Eqs. (16.1-1) and (16.1-5)],

$$v_1 = N_1 \frac{d\varphi_1}{dt} = L_1 \frac{di}{dt} \tag{16.2-1}$$

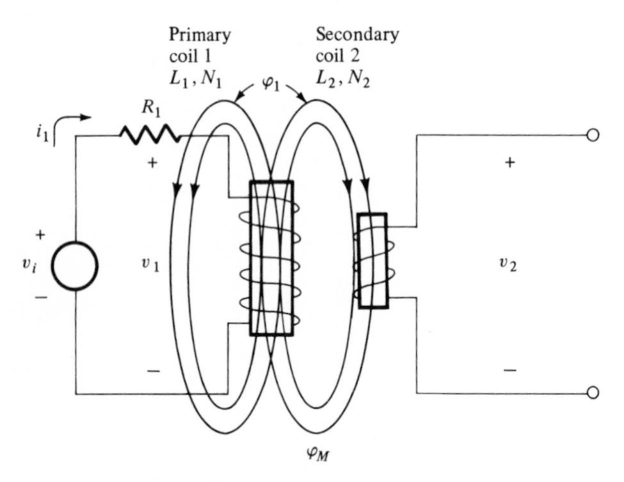

Fig. 16.2-1 Mutually coupled coils.

A portion of this flux will link the secondary and if it is changing with time a secondary voltage will be induced according to

$$v_2 = N_2 \frac{d\varphi_m}{dt} \tag{16.2-2}$$

where φ_m is the portion of φ_1 that links the secondary.

To measure the degree of coupling, a *coefficient of coupling* k is defined as follows:

$$k = \frac{\varphi_m}{\varphi_1} \tag{16.2-3}$$

If the coupling is perfect so that all of the primary flux links the secondary, then $\varphi_m = \varphi_1$ and $k = 1$. Values of k close to 1 are realized in iron-core transformers, and in configurations where the secondary is wound directly over the primary. Values of $k \ll 1$ (0.01 to 0.2) are typical of air-core coils.

Using the definition of Eq. (16.2-3), the secondary voltage of Eq. (16.2-2) can be written

$$v_2 = kN_2 \frac{d\varphi_1}{dt} \tag{16.2-4}$$

The *mutual inductance* M, measured in henries, between coils 1 and 2 is given by either of two formulas:

$$M_{12} = N_2 \frac{d\varphi_m}{di_1} = kN_2 \frac{d\varphi_1}{di_1}$$

or (16.2-5)

$$M_{21} = N_1 \frac{d\varphi_m}{di_2} = kN_1 \frac{d\varphi_2}{di_2}$$

In all cases $M_{12} = M_{21} = M$. Thus Eq. (16.2-5) becomes

$$M = kN_2 \frac{d\varphi_1}{di_1}$$
$$M = kN_1 \frac{d\varphi_2}{di_2} \tag{16.2-6}$$

Now we multiply these two equations together to get

$$M^2 = k^2 N_1 N_2 \frac{d\varphi_1}{di_1} \frac{d\varphi_2}{di_2}$$

Rearranging,

$$M^2 = k^2 \left(N_1 \frac{d\varphi_1}{di_1} \right) \left(N_2 \frac{d\varphi_2}{di_2} \right) \tag{16.2-7}$$

Using this equation, the mutual inductance can be related to the primary and secondary inductances as follows. From Eq. (16.2-1) the self-inductance of each coil can be written as

$$L_1 = N_1 \frac{d\varphi_1}{di_1} \qquad \text{and} \qquad L_2 = N_2 \frac{d\varphi_2}{di_2} \tag{16.2-8}$$

Using these relations Eq. (16.2-7) becomes

$$M^2 = k^2 L_1 L_2$$

Taking the square root of both sides,

$$M = k \sqrt{L_1 L_2} \tag{16.2-9}$$

This is an important relation. Typical calculations are shown in the following examples.

Example 16.2-1 In the coupled coils of Fig. 16.2-1 the following values apply: $k = 0.6$, $L_1 = 0.2$ H, $N_1 = 100$ turns, $L_2 = 0.4$ H, $N_2 = 200$ turns. Find

a The mutual inductance M

b The induced voltages across the primary and the secondary if the primary flux changes at the rate of 700 mWb/s

Solution

a From Eq. (16.2-6),

$$M = k \sqrt{L_1 L_2} = 0.6 \sqrt{0.2 \times 0.4} = 0.6 \sqrt{0.08} = 0.17 \text{ H}$$

b From Eq. (16.2-1),

$$v_1 = N_1 \frac{d\varphi_1}{dt} = (100)(0.7) = 70 \text{ V}$$

From Eq. (16.2-4),

$$v_2 = kN_2 \frac{d\varphi_1}{dt} = (0.6)(200)(0.7) = 84 \text{ V} \qquad ////$$

SERIES CONNECTION OF COUPLED COILS

The coupled coils of Fig. 16.2-1 may be connected in series in some applications. When this is the case, the total inductance is not simply $L_1 + L_2$ but is complicated by the presence of the mutual inductance. Because of the series connection, the same current flows in both coils as shown in Fig. 16.2-2*a* and *b* and the voltage induced in coil 1 will consist of a term due to the self-inductance L_1 and a term due to the current i flowing in coil 2 and mutual inductance M. The voltage is

$$v_1 = L_1 \frac{di}{dt} + M \frac{di}{dt}$$

$$= (L_1 + M) \frac{di}{dt} \qquad (16.2\text{-}10)$$

The sign of the mutual term is positive because of the winding directions shown

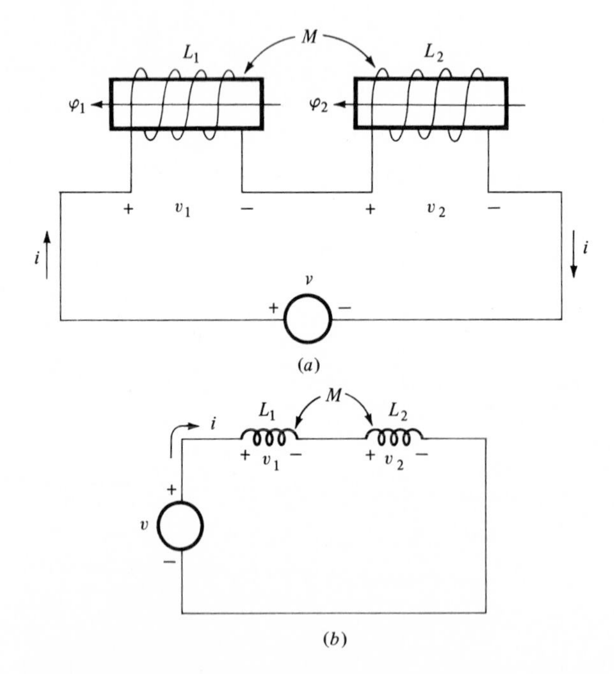

Fig. 16.2-2 Series-coupled coils. (*a*) Pictorial diagram; (*b*) circuit.

in Fig. 16.2-2a where it is seen that the fluxes due to the currents in the coils are additive. If the winding direction of one of the coils is changed, the components of flux due to the two coils would be in opposite directions, and the sign of the mutual term would be negative. This case will be considered shortly.

For the voltage across coil 2 we have

$$v_2 = (L_2 + M)\frac{di}{dt} \tag{16.2-11}$$

The total voltage v is, using KVL,

$$v = v_1 + v_2 = (L_1 + M)\frac{di}{dt} + (L_2 + M)\frac{di}{dt}$$

$$= (L_1 + L_2 + 2M)\frac{di}{dt}$$

This is the same as the vi relation of an equivalent inductance L_{eq}, where

$$L_{eq} = L_1 + L_2 + 2M \tag{16.2-12}$$

If one coil is reversed, this becomes

$$L_{eq} = L_1 + L_2 - 2M \tag{16.2-13}$$

It is interesting to note that Eqs. (16.2-12) and (16.2-13) can be used to measure M. We would proceed by first measuring L_{eq} both ways. M is then found by subtracting Eq. (16.2-13) from Eq. (16.2-12). This yields

$$L_{eq+} - L_{eq-} = L_1 + L_2 + 2M - (L_1 + L_2 - 2M)$$

$$= 4M$$

Thus

$$M = \frac{1}{4}(L_{eq+} - L_{eq-}) \tag{16.2-14}$$

where L_{eq+} is the larger of the measured equivalent inductances and L_{eq-} is the smaller.

Example 16.2-2 Two coils are wound so that mutual inductance is present between them. The self-inductances are 10 mH and 9 mH and the mutual inductance is 6 mH. Find the total inductance and coefficient of coupling if the coils are connected in the two possible arrangements of Fig. 16.2-2b.

Solution For the case where the fluxes add,

$$L_{eq} = L_1 + L_2 + 2M = 10 + 9 + 12 = 31 \text{ mH}$$

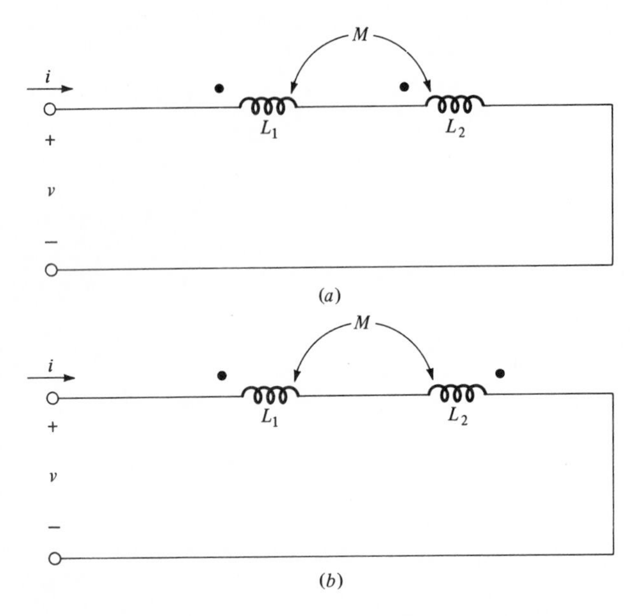

Fig. 16.2-3 Dot convention for coupled coils. (*a*) Series connection, M terms positive; (*b*) M terms negative.

The coefficient of coupling is

$$k = \frac{M}{\sqrt{L_1 L_2}} = \frac{6}{\sqrt{90}} = 0.632$$

For the other connection,

$$L_{eq} = L_1 + L_2 - 2M = 10 + 9 - 12 = 7 \text{ mH}$$

The coefficient of coupling is the same, $k = 0.632$. ////

In the derivation above we determined the sign of the mutual term by considering the actual coil-winding directions. It is seldom feasible to do this in practice because the normal circuit diagram does not show the winding direction pictorially so a system of dot markings is used to indicate the winding direction and, as a result, the sign of the mutual terms.

To illustrate the dot convention, consider Fig. 16.2-3*a* where we have shown a pair of coupled coils in series. The significance of the dots is as follows:

> Current flowing *in* at the dot on one winding induces a voltage across the other winding which is positive at the dotted end.

With the dots as shown in part *a* of the figure the current i flows *in* at the dot in coil 1 so the mutual voltage induced in coil 2 is more positive at the dotted end.

Also the current is flowing in at the dot in coil 2 so the mutual voltage induced in coil 1 is more positive at the dotted end. Since the self-inductance voltage across each coil is positive at the end at which the current enters, we see that the mutual voltages add, and the M terms are then positive. The KVL equation for the circuit with an explanation of each term is as follows:

$$v = L_1 \frac{di}{dt} + M \frac{di}{dt} + L_2 \frac{di}{dt} + M \frac{di}{dt} \tag{16.2-15}$$

$L_1(di/dt)$ is the voltage across coil 1 due to current i in coil 1

$M(di/dt)$ is the voltage across coil 1 due to current i in coil 2

$L_2(di/dt)$ is the voltage across coil 2 due to current i in coil 2

$M(di/dt)$ is the voltage across coil 2 due to current i in coil 1

For the other dot arrangement shown in Fig. 16.2-3*b*, the induced voltages in one coil due to the current flowing in the other are seen to have a polarity opposed to the self-inductance drops. Thus the KVL equation is identical to Eq. (16.2-15) except that the M turns are negative.

ESTABLISHING DOT MARKINGS

When the winding directions of the coils are not known and cannot be determined, the dot markings are easily established using the test circuit of Fig. 16.2-4. The battery is usually of relatively low voltage, and the momentary switch can be simply two pieces of bare wire. When the switch is closed the primary current will flow in the direction shown so a dot is placed at the end of the primary coil into which the current flows. If the voltmeter kicks in the positive direction, then the dot is placed where the positive terminal of the voltmeter is connected to the secondary. If the voltmeter kicks negative, the dot is connected to the other end of the secondary. Use of the dots in writing circuit equations when coupled coils are present is illustrated in the example which follows.

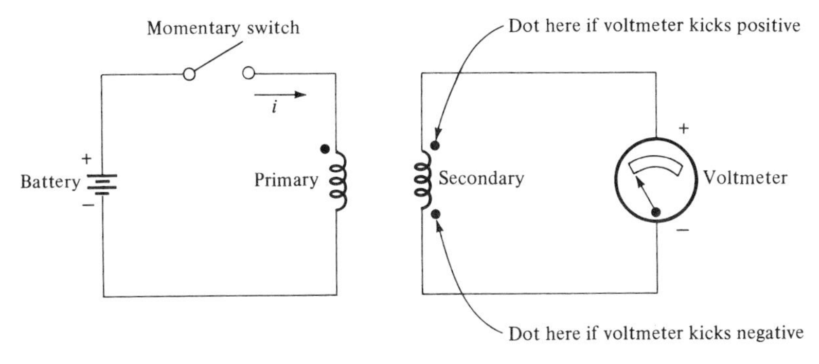

Fig. 16.2-4 Circuit for establishing dot positions.

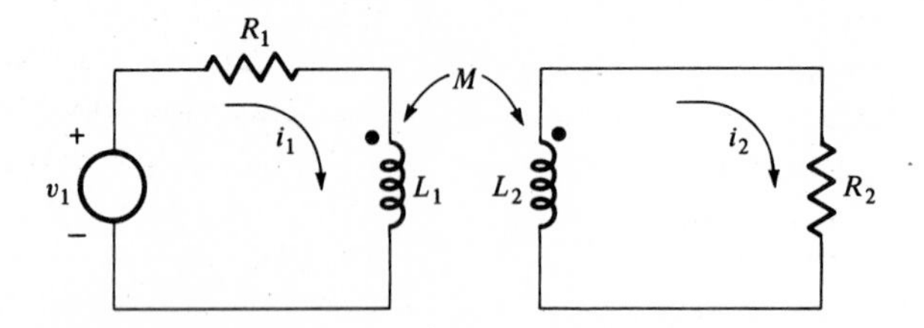

Fig. 16.2-5 Circuit for Example 16.2-3.

Example 16.2-3 Write the KVL equations for the circuit of Fig. 16.2-5.

Solution This is a two-loop network so we proceed as in Sec. 8.2. Loop currents i_1 and i_2 are assumed to flow clockwise as shown. The KVL equation for loop 1 is then

$$v_1 = R_1 i_1 + L_1 \frac{di_1}{dt} - M \frac{di_2}{dt}$$

The sign of the mutual term is found by noting that i_2 flows into coil 2 at the end *opposite* the dot. Thus the mutual voltage induced in coil 1 due to i_2 in coil 2 is more positive at the end of coil 1 opposite the dot. This is a voltage *rise* in the clockwise direction in which we are tracing loop 1 so the sign of the voltage is negative in accordance with our convention that voltage drops take a positive sign and voltage rises take a negative sign.

For loop 2 the KVL equation is

$$0 = -M \frac{di_1}{dt} + L_2 \frac{di_2}{dt} + R_2 i_2$$

The student should be sure that he understands the reasoning behind the negative sign of the mutual term in this equation. ////

16.3 CAPACITANCE

The inductance element considered in Sec. 16.1 differs from the previously studied resistance element in several ways. One difference is that while voltage and current are directly proportional in the resistance, voltage across an inductance is nonzero only when the current is changing with time. Another difference concerns energy; the resistance element dissipates energy in the form of heat while the inductance element cannot dissipate energy but can only store it in its magnetic field.

The capacitance element is closely related to the inductance. Current will flow through it only when the voltage across it is changing with time and it cannot dissipate energy but can only store it in its electric field. An actual capacitor usually consists of a pair of metal plates or foil separated by an insulting material, called a *dielectric*, as shown in Fig. 16.3-1*a* along with the circuit symbol for capacitance. To achieve reasonable values of capacitance, a large surface area is required so the assembly is often rolled into cylindrical form to conserve space. Various types of capacitors are shown in Fig. 16.3-1*b*. The

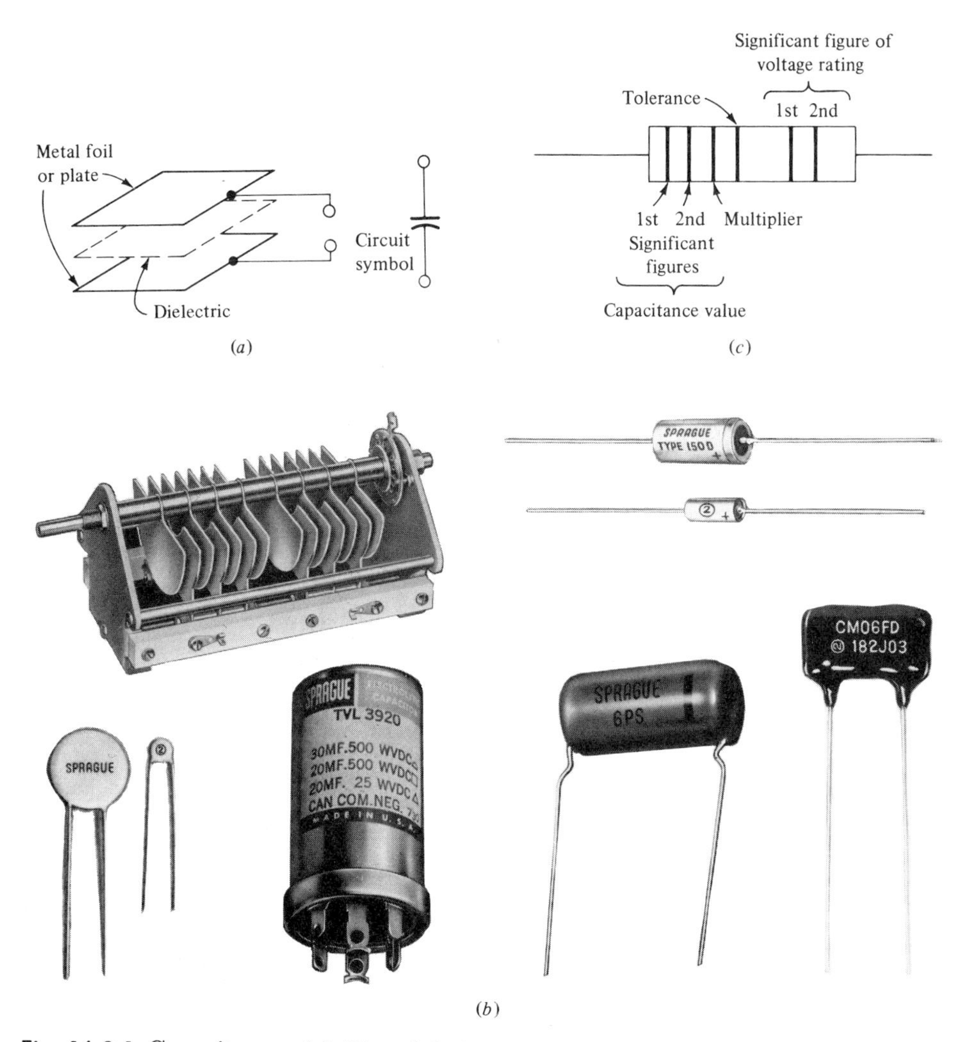

Fig. 16.3-1 Capacitance. (*a*) Pictorial sketch of a capacitor and circuit symbol. The curved line represents the plate which is connected to the lowest voltage point; (*b*) top row—variable air capacitor (*James Millen Mfg. Co. Inc.*), tantalum capacitors, bottom row—disc ceramic capacitors, electrolytic capacitor, paper capacitor, dipped mica capacitor (*Sprague Electric Co.*); (*c*) color-code bands for tubular capacitors.

ceramic capacitor consists of a ceramic disk of dielectric on each side of which a thin layer of silver is applied. Wires are attached to the silver and an outer insulation coating is applied. The *paper capacitor* consists of aluminum foil and wax-impregnated paper dielectric rolled into cylindrical form. Plastic-film dielectric capacitors using mylar, polyethylene, or polycarbonate films are also manufactured in cylindrical form. The *mica capacitor* consists of a stack of alternate layers of mica dielectric and conducting foil or silver fired onto the mica surface. *Variable capacitors* usually consist of sets of aluminum plates separated by air as the dielectric. Alternate plates are movable so that the capacitance can be varied by varying the area. *Electrolytic capacitors* utilize aluminum foil as one plate and an alkaline electrolyte as the other. Application of a dc voltage during manufacture causes a thin layer of aluminum oxide to be formed which acts as a dielectric. This type of capacitor exhibits very high values of capacitance but has the disadvantage that the voltage across it can only be of one polarity.

The color-coded band marking system for tubular paper capacitors is shown in Fig. 16.3-1*c*. For the capacitance value, given by the first three bands, the code is the same as for resistors given in Fig. 4.5-2. For the fourth (tolerance) band, green denotes 5 percent, white 10 percent, black 20 percent, orange 30 percent, and yellow 40 percent. The voltage rating is identified by a single digit number (fifth band) up to 900 V. The rating is obtained by multiplying the fifth band digit by 100. If the rating is above 900 V, a sixth band is added and the resulting two-digit number is multiplied by 100. Other types of capacitors may use dots rather than bands and the significance of the bands and dots varies somewhat for different types.

CAPACITOR CIRCUITS

Consider the circuit of Fig. 16.3-2. When the battery voltage is applied to the capacitor by closing the switch, it is found experimentally that after an initial transient period no current flows, and the plate connected to the positive battery terminal carries a charge $+q$, while the opposite plate carries a charge $-q$. The charge is found to be directly proportional to the voltage even if the voltage is varying with time and the current is not zero. Thus the defining

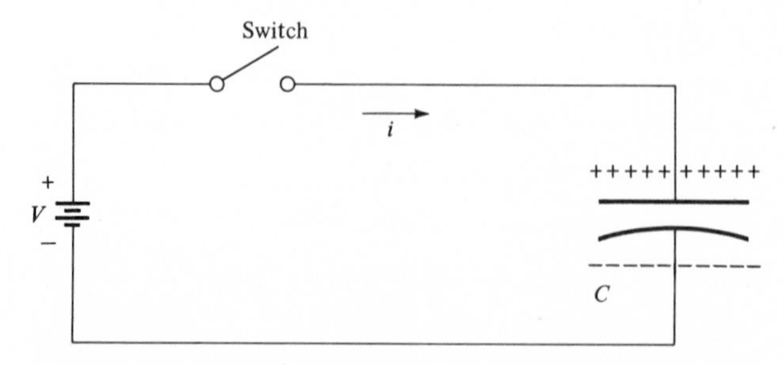

Fig. 16.3-2 Charge distribution on capacitor plates.

relation for the *ideal* capacitance is

$$q = Cv \tag{16.3-1}$$

where q = charge on plates, C
v = voltage across plates, V
C = capacitance, F (farads)

The farad is an extremely large unit and most actual capacitors range from picofarads (10^{-12} F) to microfarads (10^{-6} F). Typical values of mica capacitors range from 25 to 500 pF while paper capacitors range from about 0.0005 to 2 μF. Electrolytic capacitors range from about 1 to 2,500 μF with maximum allowable voltages from 3 to 450 V.

Referring to Eq. (16.3-1) we note that charge is usually an inconvenient variable to work with in circuit analysis. By returning to the definition of current we can convert the equation to the more convenient current and voltage variables. To accomplish this, we differentiate Eq. (16.3-1) with respect to time. This yields

$$\frac{dq}{dt} = C\frac{dv}{dt} \tag{16.3-2}$$

From Sec. 14.1 we have the definition of current: $i = dq/dt$. Thus

$$i = C\frac{dv}{dt} \tag{16.3-3}$$

This is the desired *vi* relation for the capacitor and we see that the current is zero unless the voltage is varying with time. It is instructive to compare this with the *vi* relation for the inductor:

$$v = L\frac{di}{dt} \tag{16.3-4}$$

If we interchange v and i, and replace L by C or C by L in either of these relations, we arrive at the other one. This is an example of *duality*. A useful consequence of duality is that all of the theory associated with Eq. (16.3-4) in Sec. 16.1 can be applied to Eq. (16.3-3) by merely interchanging v and i, and replacing L by C.

INTEGRAL FORM OF THE VI RELATION

It is often desirable to express the voltage explicity in terms of the current. To accomplish this, we return to the definition of current in the form

$$dq(t) = i(t)\,dt \tag{16.3-5}$$

To find the charge at the present time t, we integrate over all time, $-\infty$ to t. Thus (see Sec. 14.2-3)

$$q(t) = \int_{-\infty}^{t} i(x)\,dx \tag{16.3-6}$$

Where the dummy variable x has been used instead of t in the integrand to avoid confusion. For convenience the integral is separated into two integrals by dividing the time axis between the limits into two parts. Thus

$$q(t) = \int_{-\infty}^{0} i(x)\,dx + \int_{0}^{t} i(x)\,dx \tag{16.3-7}$$

The first integral is a constant which represents the net charge (designated as Q_0) deposited on the plates in the time interval $t = -\infty$ to $t = 0$. (It is assumed that the charge at $t = -\infty$ is zero.) With this, Eq. (16.3-7) becomes

$$q(t) = Q_0 + \int_{0}^{t} i(x)\,dx \tag{16.3-8}$$

Now, we return to Eq. (16.3-1) in the form

$$v(t) = \frac{q(t)}{C} \tag{16.3-9}$$

Substituting Eq. (16.3-8) into Eq. (16.3-9) we obtain the desired vi relation for the capacitor, with voltage expressed in terms of current:

$$v(t) = \frac{Q_0}{C} + \frac{1}{C}\int_{0}^{t} i(x)\,dx$$

$$= V_0 + \frac{1}{C}\int_{0}^{t} i(x)\,dx \tag{16.3-10}$$

where

$$V_0 = \frac{Q_0}{C} = \frac{1}{C}\int_{-\infty}^{0} i(x)\,dx \tag{16.3-11}$$

represents the voltage across the capacitor at time $t = 0$. The limits on the integral in Eq. (16.3-11) indicate that the capacitor's voltage depends on the complete past history of the current. Thus the capacitor has, in a sense, a "memory."

We can make use of *duality* to find the integral form of the vi relation for the inductor. In order to do this we replace v by i, i by v, and C by L in Eq. (16.3-10). This yields

$$i(t) = I_0 + \frac{1}{L}\int_{0}^{t} v(x)\,dx \tag{16.3-12}$$

where

$$I_0 = \frac{1}{L}\int_{-\infty}^{0} v(x)\,dx \tag{16.3-13}$$

represents the current through the coil at $t = 0$.

Example 16.3-1

a A 10-μF capacitor is to carry a constant current of 1 mA. At what rate must the voltage across it change in order that it carry this current?

b A constant current of 1 mA is applied to the positive terminal of a 10-μF capacitor which has an initial voltage of 10 V across it at $t = 0$. Find the capacitor voltage at $t = 100$ ms.

Solution

a From Eq. (16.3-3),

$$\frac{dv}{dt} = \frac{i}{C} = \frac{10^{-3}}{10 \times 10^{-6}} = 100 \text{ V/s}$$

b From Eq. (16.3-10),

$$v(t) = V_0 + \frac{1}{C}\int_0^t i(x)\,dx$$

$$V_0 = 10 \text{ V (given)}$$

$$t = 100 \text{ ms} = 0.1 \text{ s}$$

$$i(t) = 1 \text{ mA} = 10^{-3} \text{ A}$$

$$C = 10 \times 10^{-6} = 10^{-5} \text{ F}$$

Substituting these numbers,

$$v(0.1) = 10 + 10^5 \int_0^{0.1} 10^{-3}\,dx$$

$$= 10 + 10^5 \Big[(10^{-3}x) \Big]_0^{0.1}$$

$$= 10 + 10^5(10^{-4})$$

$$= 20 \text{ V}$$

////

STEP-FUNCTION RESPONSE

Consider the response of the ideal capacitor to a step function of current as shown in Fig. 16.3-3. The voltage across the capacitor, $v(t)$, is the response of this simple circuit to the current-source input. This voltage can be found using Eq. (16.3-10). For the case when $V_0 = 0$, we have $v(t) = 0$ for $t < 0$ and,

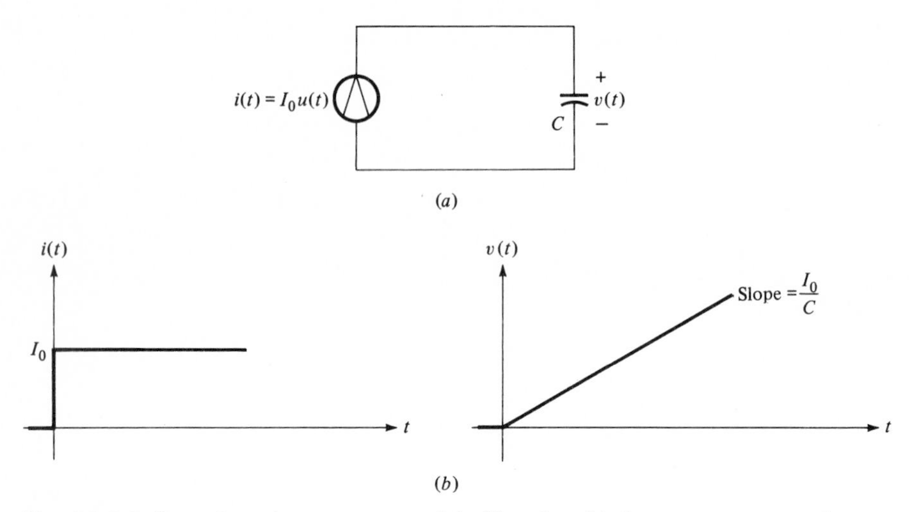

Fig. 16.3-3 Step-function response. (*a*) Circuit; (*b*) input current and output voltage.

for $t > 0$,

$$v(t) = \frac{1}{C}\int_0^t i(x)\,dx$$

$$= \frac{1}{C}\int_0^t I_0\,dx = \frac{I_0}{C}\int_0^t dx$$

$$= \frac{I_0}{C}\Big[x\Big]_0^t = \frac{I_0}{C}[t - 0]$$

$$= \frac{I_0}{C}t \qquad 0 < t < \infty \tag{16.3-14}$$

The current and voltage are sketched in Fig. 16.3-3*b*. The student may recall that this same problem was considered in Sec. 14.2-3 when we studied integration. As we pointed out at that time, the voltage waveform is called a *ramp* function. It is evident that the ramp is nonphysical because it theoretically increases without bound. However, ramp functions which last for a finite time are easily generated and are used to form "sawtooth" waveforms, which are often used in electronic circuits.

As in Sec. 14.2 the voltage waveform should be viewed as the running integral of the current waveform.

SERIES AND PARALLEL CAPACITORS

In this section we will find the equivalent capacitance for series and parallel capacitors. Consider first the parallel connection in Fig. 16.3-4*a*. The charge on

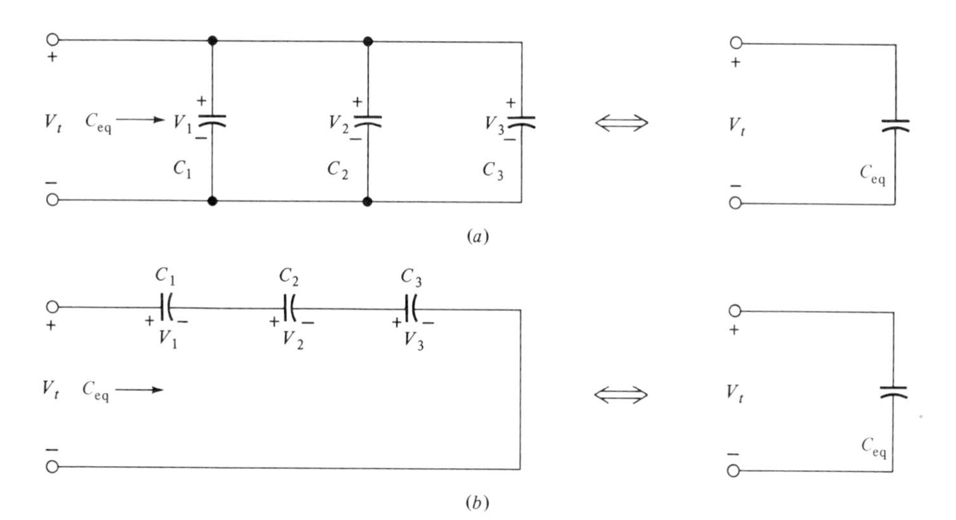

Fig. 16.3-4 Equivalent capacitance. (a) Parallel; (b) series.

the individual capacitors is

$$Q_1 = C_1V_1 \qquad Q_2 = C_2V_2 \qquad Q_3 = C_3V_3$$

However, because of the parallel connection

$$V_1 = V_2 = V_3 = V_t$$

The total charge is then

$$Q = C_{eq}V_t = Q_1 + Q_2 + Q_3 = C_1V_t + C_2V_t + C_3V_t$$

Dividing by V_t, we get

$$C_{eq} = C_1 + C_2 + C_3 \tag{16.3-15}$$

It is evident that the equivalent capacitance for a parallel combination of any number of capacitors is simply the sum of the individual capacitances.

Next we consider the series connection of Fig. 16.3-4b. Using KVL, we have

$$V_t = V_1 + V_2 + V_3 \tag{16.3-16}$$

where

$$V_1 = \frac{Q_1}{C_1} \qquad V_2 = \frac{Q_2}{C_2} \qquad V_3 = \frac{Q_3}{C_3} \tag{16.3-17}$$

In a series connection of capacitors, the charge on each must be the same (why?) so that $Q = Q_1 = Q_2 = Q_3$. Thus

$$V_t = \frac{Q}{C_{eq}} = \frac{Q}{C_1} + \frac{Q}{C_2} + \frac{Q}{C_3}$$

Dividing by Q,

$$\frac{1}{C_{eq}} = \frac{1}{C_1} + \frac{1}{C_2} + \frac{1}{C_3} \tag{6.3-18}$$

It is evident that the formula for capacitors in series is similar to that for resistors in parallel. For the important case of two capacitors in series

$$C_{eq} = \frac{C_1 C_2}{C_1 + C_2} \tag{6.3-19}$$

As with parallel resistors, the equivalent series capacitance is *always less than* any of the individual capacitances.

Example 16.3-2 For the circuit of Fig. 16.3-5 find

a The equivalent capacitance
b The charge on each capacitor
c The voltage across each capacitor

Solution

a Using Eq. (16.3-19),

$$C_2 \,||\, C_3 = 10 + 20 = 30\ \mu\text{F}$$

Using Eq. (16.3-15),

$$C_{eq} = \frac{C_1(C_2 \,||\, C_3)}{C_1 + (C_2 \,||\, C_3)}$$

$$= \frac{(5)\,(30)}{5 + 30} = 4.286\ \mu\text{F}$$

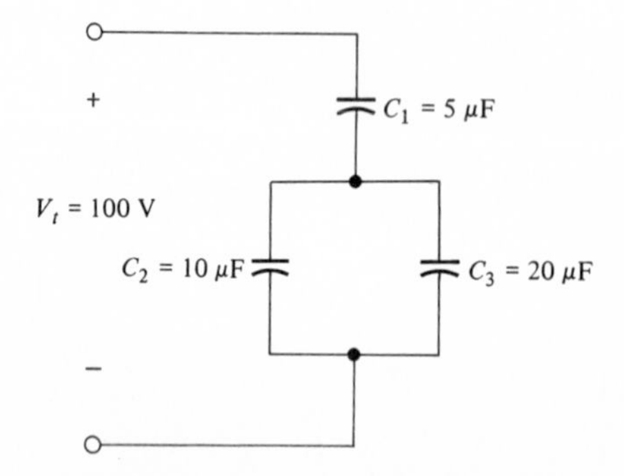

Fig. 16.3-5 Circuit for Example 16.3-2.

b To find the charge on each capacitor, we note that the total charge is

$$Q_t = C_{eq}V_t = (4.286 \times 10^{-6})(100)$$
$$= 428.6\ \mu C$$

Since C_1 is in series with $(C_2 + C_3)$, we have

$$Q_t = Q_1 = Q_2 + Q_3 = 428.6\ \mu C$$

To find Q_2 and Q_3 we need the capacitor voltages.

c $$V_1 = \frac{Q_1}{C_1} = \frac{428.6 \times 10^{-6}}{5 \times 10^{-6}} = 85.7\ V$$

Using KVL,

$$V_2 = V_3 = V_t - V_1 = 100 - 85.7 = 14.3\ V$$

Thus

$$Q_2 = C_2V_2 = (10 \times 10^{6})(14.3) = 143\ \mu C$$

$$Q_3 = C_3V_3 = (20 \times 10^{-6})(14.3) = 286\ \mu C$$

Note that the *larger* voltage appears across the *smaller* capacitance in the series connection. ////

16.4 ENERGY STORAGE

In earlier sections of this chapter it was mentioned that energy is stored in the magnetic field of an inductor in which current is flowing and also in the electric field of a capacitor which has a voltage across its terminals. If a capacitor is charged to a finite voltage by connecting it to a source of energy and then the source is suddenly removed, it is found that the voltage which existed across the capacitor at the instant when the source was removed is maintained for long periods of time as long as there is negligible leakage between the plates of the capacitor. During this time, the energy that was drawn from the source is stored in the electric field of the capacitor. If a resistance is connected across the plates of a previously charged capacitor, the stored energy will be dissipated as heat in the resistance. The energy stored in the electric field of a capacitor which has V_1 volts across it is

$$W_c = \tfrac{1}{2}CV_1^2 \tag{16.4-1}$$

where W_c is in joules (J), C is in farads (F), and V_1 is in volts (V).

The fact that this energy can be stored for long periods makes the capacitor useful where large currents are required for short periods of time as in electronic flash guns and electric spot welders. In these devices, storage capacitors are charged at low current through relatively high resistances. When the current is required for the flash tube or weld the charged capacitor is discharged through the low resistance of the flash tube or welding material. The current during the discharge is very high.

In the inductance, the energy stored in the magnetic field is

$$W_L = \tfrac{1}{2}LI_1^2 \qquad (16.4\text{-}2)$$

where W_L is in joules (J) when L is in henries (H) and I_1 is in amperes (A). This formula is clearly the *dual* of Eq. (16.4-1) for the capacitor.

The long-time energy storage feature of the capacitance element is not available in the inductance. This is because the current I_1 must be maintained in order to maintain the magnetic field. In order to maintain the current *without* an energy source, a zero resistance path for the current is required. Since every material has resistance (except at temperatures at or near absolute zero) this is not practical. If the current in an inductive circuit is interrupted by opening a switch, the magnetic field collapses very rapidly due to the sudden interruption. This rapidly collapsing field can generate very high voltages (see Sec.15.4) across the opening switch contacts, causing arcing and burning of the contacts. This problem can often be eliminated either by connecting a capacitor across the switch contacts to oppose the voltage change or by connecting a discharge resistor in parallel with the inductive circuit to absorb the energy in the collapsing field.

Example 16.4-1 An electronic flash gun has a 2,000-μF storage capacitor which is charged to 48 V for operation. The flash duration in use is 1/1,000 s.

a Find the charge on the capacitor and the energy stored.

b Assuming that all of the charge passes through the flashtube in the 1/1000-s flash duration, find the current through and the power delivered to the tube.

c After the flash the capacitor must be recharged. How long will this take if the charging power supply delivers a constant current of 100 mA?

Solution

a The charge is given by

$$Q = CV = (2 \times 10^{-3})(48) = 0.096 \text{ C}$$

The energy stored is, from Eq. (16.4-1)

$$W_c = \tfrac{1}{2}CV^2 = (\tfrac{1}{2})(2 \times 10^{-3})(48)^2 = 2.3 \text{ J}$$

b Using the definition of current,

$$I = \frac{\Delta Q}{\Delta t} = \frac{0.096\ C}{0.001 \text{ s}} = 96 \text{ A}$$

The power delivered is the time rate of change of energy (see Sec. 3.8):

$$P = \frac{\Delta W}{\Delta t} = \frac{2.3\ \text{J}}{0.001\ \text{s}} = 2{,}300\ \text{W}$$

Note that during the flash, the current and power are very high.

c The charging time is found from the definition of current

$$I_{\text{charge}} = \frac{\Delta Q}{\Delta t}$$

where ΔQ is the required charge of 0.096 C. Since the charging current is 100 mA, this gives

$$0.1 = \frac{0.096}{\Delta t}$$

from which

$$\Delta t = \frac{0.096}{0.1} = 0.96\ \text{s}$$

This should be compared to the 1 ms taken to discharge the flashtube.

////

SUMMARY

1 For the inductance, the *vi* relation is

$$v(t) = L\frac{di}{dt}$$

2 The integral form of the inductance *vi* relation is

$$i(t) = I_0 + \frac{1}{L}\int_0^t v(x)\,dx$$

where I_0 is the current through the coil at $t = 0$.

3 For *n* series inductors, the equivalent inductance is

$$L_{\text{eq}} = L_1 + L_2 + L_3 + \cdots + L_n$$

For two parallel inductors,

$$L_{\text{eq}} = \frac{L_1 L_2}{L_1 + L_2}$$

For *n* parallel inductors,

$$\frac{1}{L_{\text{eq}}} = \frac{1}{L_1} + \frac{1}{L_2} + \cdots + \frac{1}{L_n}$$

4 For magnetically coupled circuits, the coefficient of coupling is

$$k = \frac{M}{\sqrt{L_1 L_2}}$$

5 For magnetically coupled coils in series, the equivalent inductance is

$$L_{eq} = L_1 + L_2 \pm 2M$$

6 According to the dot convention for coupled coils, current flowing in at the dot in one coil causes the dotted end of the other coil to be at a positive voltage with respect to the other end.

7 The vi relation for the capacitance element is

$$i(t) = C\frac{dv}{dt}$$

8 The integral form of the capacitance vi relation is

$$v(t) = V_0 + \frac{1}{C}\int_0^t i(x)\,dx$$

Where V_0 is the voltage across the capacitor at $t = 0$.

9 For n parallel capacitors, the equivalent capacitance is

$$C_{eq} = C_1 + C_2 + \cdots + C_n$$

For two series capacitors,

$$C_{eq} = \frac{C_1 C_2}{C_1 + C_2}$$

For n series capacitors,

$$\frac{1}{C_{eq}} = \frac{1}{C_1} + \frac{1}{C_2} + \cdots + \frac{1}{C_n}$$

10 The energy stored in a capacitor is

$$W_c = \tfrac{1}{2}CV^2$$

11 The energy stored in an inductor is

$$W_L = \tfrac{1}{2}LI^2$$

PROBLEMS

16.1-1 A coil consists of 250 turns of wire wound on a closed steel core for which $\mu_r = 1{,}000$. The length of the core is 0.2 m and it has a cross-sectional area 0.04 m^2. Find the inductance of the coil.

16.1-2 Find the voltage induced in a 400 turn coil if the flux changes by 0.5 mWb in 0.05 s.

16.1-3 It is desired to induce a voltage of 6 V in a coil when the flux changes by 5 mWb in 0.02 s. How many turns should the coil have?

16.1-4 Current in a 10-mH coil changes uniformly from 10 to 100 mA in 5 ms. What is the voltage across the coil during this interval?

16.1-5 The current waveform of Fig. P16.1-5 is applied to a 10-mH inductance. Find and plot the resulting voltage.

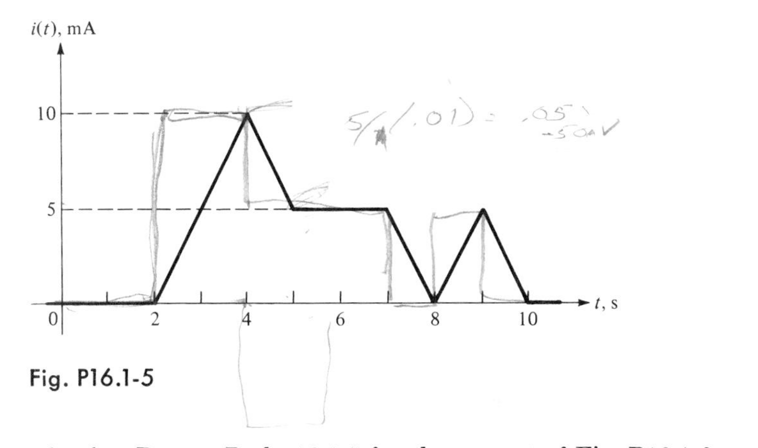

Fig. P16.1-5

16.1-6 Repeat Prob. 16.1-5 for the current of Fig. P16.1-6.

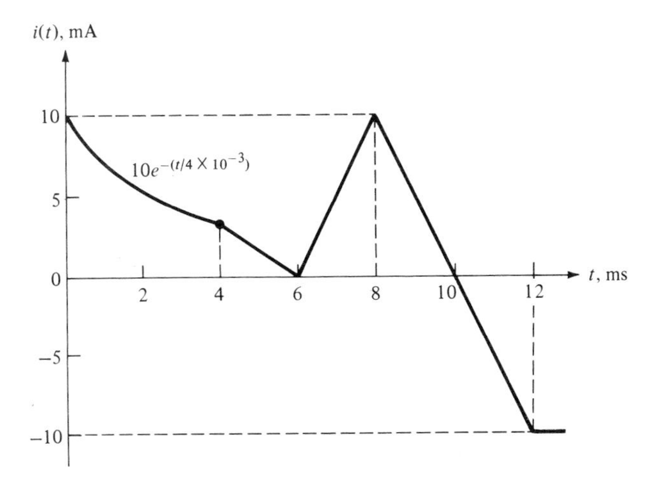

Fig. P16.1-6

16.1-7 Sketch two currents which will yield the voltage waveform of Fig. P16.1-7 when applied to a 5-H inductance.

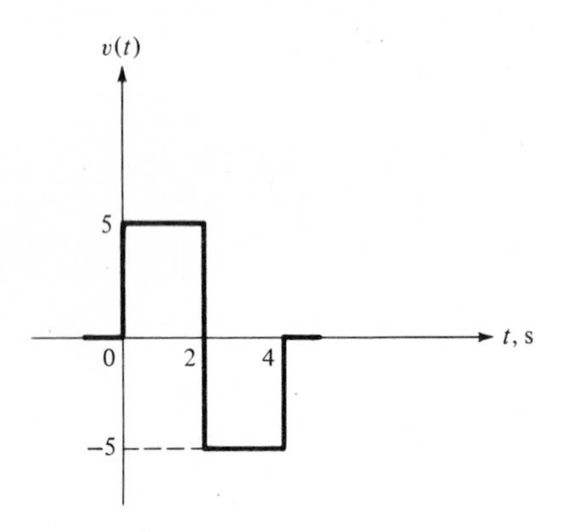

Fig. P16.1-7

16.1-8 The current waveform of Fig. P16.1-6 is applied to a series combination of resistance and inductance with $R = 2\ \text{k}\Omega$, $L = 4$ H. Find and sketch the voltage across L and R separately, and the total voltage.

16.1-9 Find the equivalent inductance of the two circuits of Fig. P16.1-9.

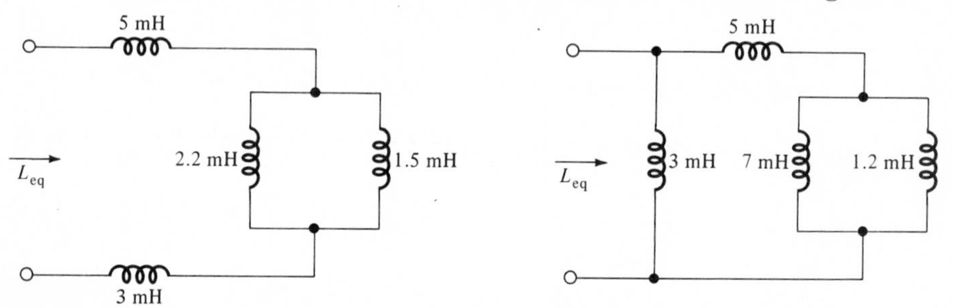

Fig. P16.1-9

16.2-1 In the coupled coils of Fig. 16.2-1 $k = 0.8$, the primary has 320 turns and 0.5 H inductance while the secondary has 80 turns and 0.2 H inductance. Find

a The mutual inductance M.

b The primary and secondary voltages if the primary current is changing at the rate of 20 A/s.

16.2-2 Repeat Prob. 16.2-1 if k is changed to 0.2. Compare.

16.2-3 For the circuit of Fig. P16.2-3 find the equivalent inductance and the coefficient of coupling.

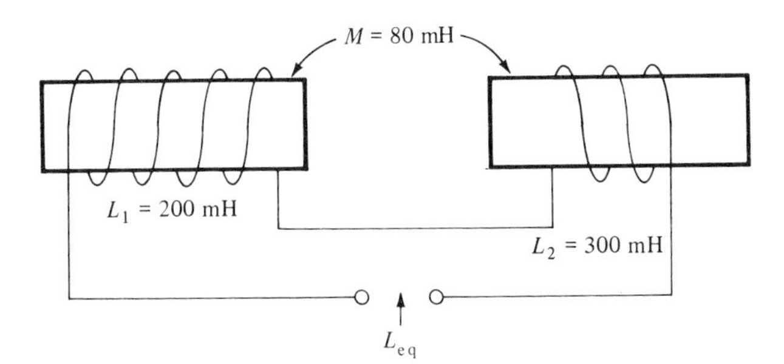

Fig. P16.2-3

16.2-4 For the series connection and dot placement of Fig. P16.2-4 find the equivalent inductance if the coefficient of coupling is (*a*) 1, (*b*) 0.5, (*c*) 0.

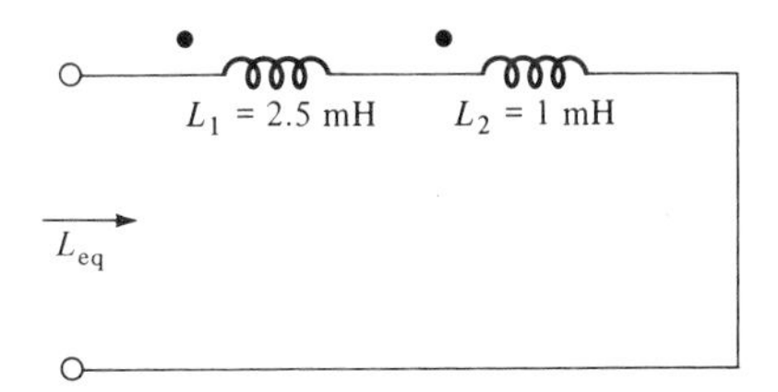

Fig. P16.2-4

16.2-5 Two coupled coils in series have a measured equivalent inductance of 1.6 H. One coil is reversed and the inductance is measured at 1.2 H. If one coil has a self-inductance of 0.5 H find the inductance of the other coil and the mutual inductance.

16.3-1 A 10-μF capacitor carries a charge of 450 μC. What is the voltage across the capacitor?

16.3-2 A capacitor carries a charge of 20 mC and the voltage across it is 60 V. Find the capacitance.

16.3-3 A 0.001-μF capacitor has a voltage of 220 V across it. How much charge is present?

16.3-4 A constant current of 2 mA is to be carried by a 0.05-μF capacitor. At what rate must the voltage across it be changing?

16.3-5 The voltage across a 5-μF capacitor decreases linearly from 10 to 2 V in 0.01 s. Find the current during this interval.

16.3-6 A constant current of 1 mA is to be maintained through a capacitor across which the voltage is changing at the rate of 500 V/s. Find the capacitance.

16.3-7 An initially uncharged 0.5-μF capacitor carries a constant current of 600 μA. Find the voltage across the capacitor at $t = 350$ ms.

16.3-8 A current $i(t) = 150u(t)$ μA is applied to an uncharged 0.05-μF capacitor. Find and sketch the voltage waveform in the interval $-10 < t < 100$ ms.

16.3-9 Repeat Prob 16.3-8 with an initial charge of -10 V on the capacitor.

16.3-10 The voltage waveform of Fig. P16.3-10 is applied to a 0.001-μF capacitor. Find and plot the current waveform.

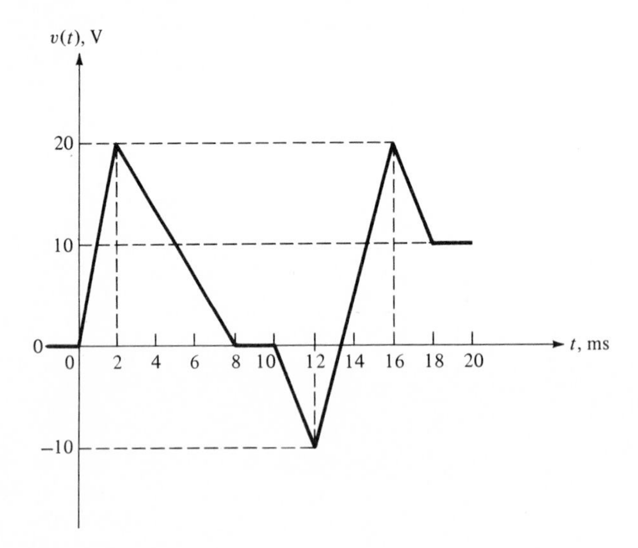

Fig. P16.3-10

16.3-11 The current waveform of Fig. P16.3-11 is flowing in a 2.5-μF capacitor. Find and sketch the resulting voltage waveform.

16.3-12 Find the equivalent capacitance of the circuits of Fig. P16.3-12.

16.3-13 Find the voltage across and the charge on each capacitor in the circuits of Fig. P16.3-13.

16.3-14 The voltage waveform shown in Fig. P16.3-14 is applied to a capacitor.

a Find and sketch the current waveform.
b Reduce the duration of the ramp to $\epsilon/2$ and repeat part *a*.
c Reduce the duration of the ramp to $\epsilon/4$ and repeat part *a*.

As the duration approaches zero, the current approaches an *impulse* function.

16.4-1 Find the energy stored in a 20-μF capacitor which has 120 V across its terminals.

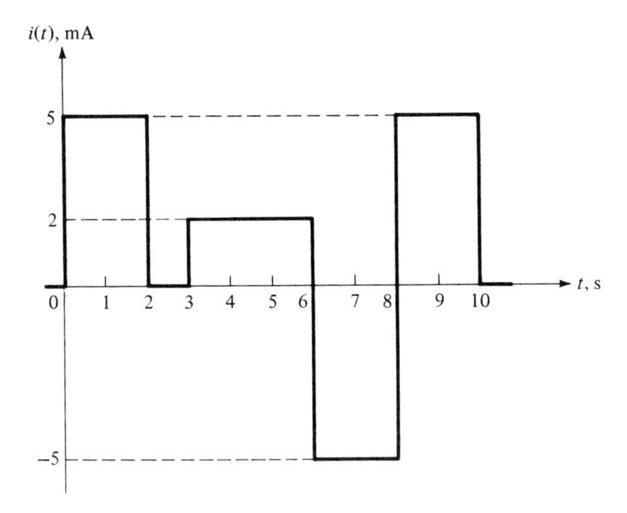

Fig. P16.3-11

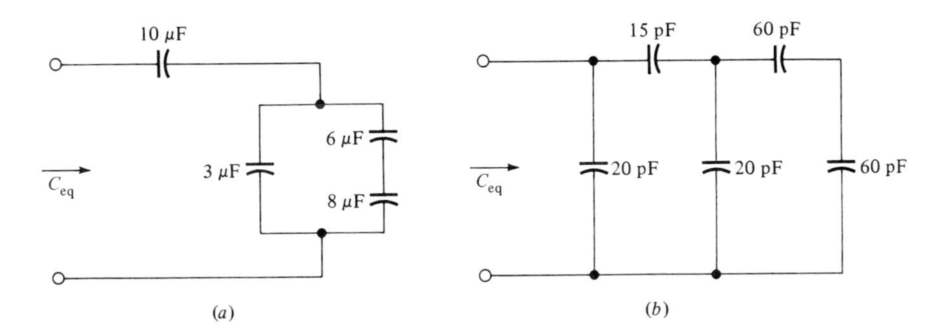

Fig. P16.3-12

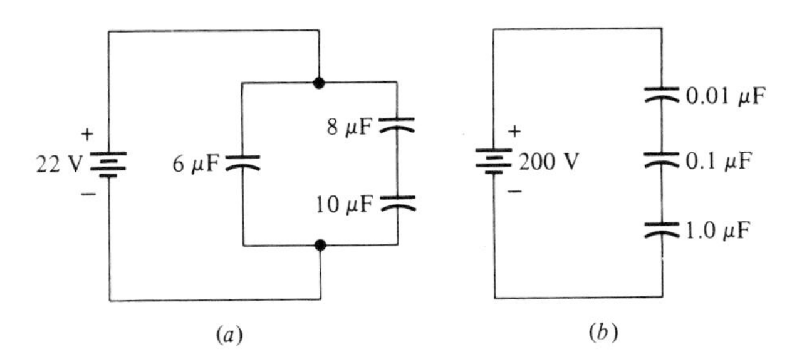

Fig. P16.3-13

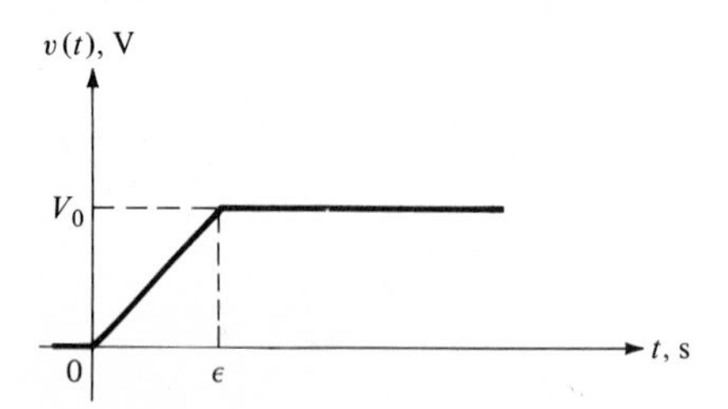

Fig. P16.3-14

16.4-2 The energy stored in a 0.1 μF capacitor is 4.3 J. Find the voltage across the capacitor and the charge on each plate.

16.4-3 A 10-μF and a 5-μF capacitor are connected in series across a 120-V source. Find the energy stored and the charge on each capacitor.

16.4-4 In order to spot-weld certain materials 5 J of energy are required in a time of 10 ms. A 100-μF storage capacitor is available.

a To what voltage must the capacitor be charged in order to supply the required energy?

b What is the current during the weld?

16.4-5 In a certain electromagnet, 60 A are required for the coil which has an inductance of 8.3 H. Find the energy stored in the magnetic field when the magnet is on.

17

Time Response of Circuits with Energy-Storage Elements

OBJECTIVES

Upon completion of this chapter you should be able to

1 Use Kirchhoff's laws and the *R*, *L*, and *C vi* relations to write differential equations for single energy-storage element circuits.
2 Solve a first-order differential equation using the step-by-step method.
3 State the general solution for a single energy-storage element circuit and define the three constants which completely define it.
4 State that the natural response takes the response from the initial condition to the final value.
5 Find the rise time of a typical *RC* circuit.
6 Use the general solution to find response of circuits when initial conditions are not zero.
7 Find the time constant of a network from an oscilloscope trace of its transient response.

INTRODUCTION

We have introduced the three linear passive elements: resistance, inductance, and capacitance, and their *vi* relations. In this chapter we will begin to study the response of circuits containing two or more different elements.

Since the *vi* relations of inductors and capacitors contain derivatives, our circuit equations will contain derivatives; they are called *differential equations.* The circuit responses which are solutions of these equations will vary with time. Because of the known nature of this time variation, such circuits are often used to generate precise time intervals which are, in turn, used for various applications. We will consider the important properties of the circuit response.

In the circuits of this chapter, the sources will be batteries or constant current sources applied to the circuit through switches. In this type of circuit, immediately after a switching action takes place, an unsettled state exists during which currents and voltages change with time. This is called the *transient state.* After sufficient time has elapsed, the transient state disappears, and all currents and voltages are constant (dc); these are called final, or *steady-state,* values. The transient states are usually associated with the opening and closing of switches but will also occur if the input voltage changes, as for example when a step function is applied. When considering the response to such signals, the procedure we will discuss in this chapter is often called *time-domain* analysis.

Most transient states are characterized by currents and voltages that vary *exponentially* (see Sec. 2.3) and our aim in this chapter is to characterize these exponentials in terms of the circuit elements and inputs.

17.1 FORMULATING THE CIRCUIT EQUATION

In this section we consider the formulation of the equations for circuits in which transients are occurring. Consider first the circuit of Fig. 17.1-1*a*. The battery is switched across the series RC combination at $t = 0$ as indicated by the arrow on the switch. The capacitor is uncharged before the switching. The circuit of Fig. 17.1-1*b* is completely equivalent to this because the step function $V_1 u(t)$ is zero for $t < 0$ and V_1 for $t > 0$. Thus the solution we are going to find applies to either circuit. We wish to find the voltage across the capacitor after the switch is closed, i.e., for $t > 0$. In order to do this we will need the value of the voltage across the capacitor immediately *after* the switch is thrown; we are given the fact that the capacitor is uncharged ($v_C = 0$) *before* the switch is thrown.

If the capacitor voltage were to change instantaneously, i.e., "jump," the graph of voltage vs. time would contain a vertical line at the point where the change occurs, as shown in Fig. 17.1–2. Since $i = C\,dv/dt$, the current at any time is proportional to the *slope* of the voltage-vs.-time curve. From the graph, we see that the slope is infinite at the point where the change occurs, so that an infinite current would be required. Since real currents cannot be infinite, it is

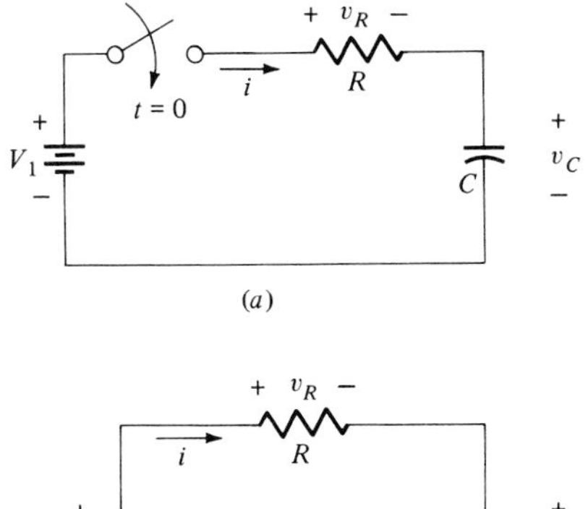

Fig. 17.1-1 Transient circuit. (*a*) Switched source; (*b*) step-function source.

not possible for the voltage across a capacitor to change instantaneously (as a step function), and the voltage across the capacitor in our circuit just *after* the switch is thrown remains at 0. The value of v_C just after the switch is thrown is called the *initial* condition and the fact that the capacitor voltage cannot change instantaneously is known as a continuity condition.*

To analyze the network, we apply KVL *after* the switch is thrown. This

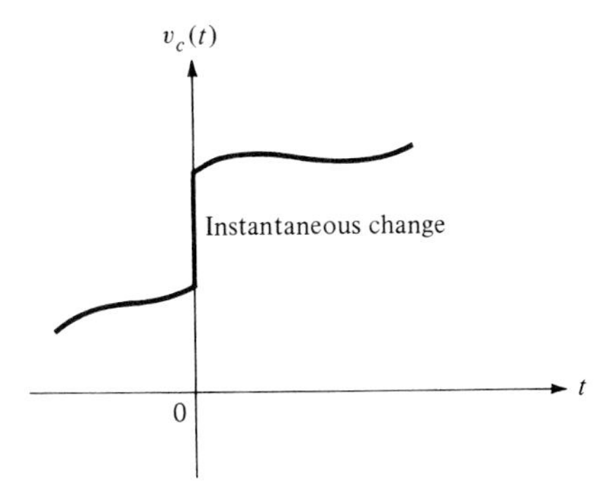

Fig. 17.1-2 Illustrating an instantaneous change in voltage.

* The continuity condition for capacitors applies only to voltage, *not* to current.

yields

$$V_1 = v_R + v_C$$
$$= Ri + v_C \tag{17.1-1}$$

where i, v_R, and v_C are unknown functions of t. For the capacitor we have the *vi* relation

$$i = C \frac{dv_C}{dt} \tag{17.1-2}$$

When this is substituted into Eq. (17.1-1) we obtain

$$RC \frac{dv_C}{dt} + v_C = V_1 \tag{17.1-3}$$

This linear *differential equation* holds *only* for $t > 0$. It relates the capacitor voltage v_C to the circuit elements R and C and the input voltage V_1. The equation alone is not sufficient to determine the entire solution for $v_C(t)$; additional information is required. In this case the additional information is most conveniently expressed as the *initial condition* $v_C(0) = 0$.

The solution to this equation will be considered in the next section. We can, however, find some of its properties in advance.

Consider Eq. (17.1-1) immediately after the switch is thrown. We know that $v_C = 0$ at this time from the continuity condition. Thus, for $t = 0$

$$V_1 = Ri(0) + v_C(0)$$
$$= Ri(0) + 0$$

and therefore

$$i(0) = \frac{V_1}{R} \tag{17.1-4}$$

Thus the current immediately after the switch is thrown is found using Ohm's law and considering the capacitor as a short circuit. Note that the current changes instantaneously from zero just prior to the switching to V_1/R just after the switching.

When the source is constant, as in this circuit, the capacitor voltage will eventually become constant. Then $dv_C/dt = 0$ and we have, from Eq. (17.1-3),

$$v_C(\infty) = V_1 \tag{17.1-5}$$

where $v_C(\infty)$ is used to indicate the value of v_C after a long time has elapsed.

From Eq. (17.1-2) we see that $i(\infty) = 0$ indicating that the capacitor "looks like" an *open circuit* in the steady state for dc sources. This fact can be used to simplify steady-state calculations in capacitive circuits with dc sources.

We have found the initial values $v_C(0) = 0$ and $i(0) = V_1/R$, and the

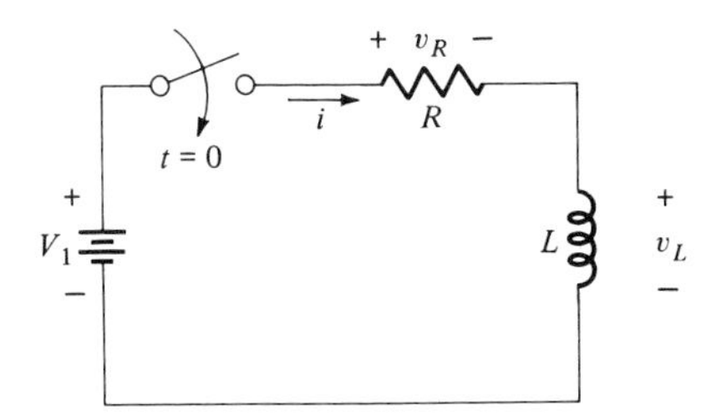

Fig. 17.1-3 Circuit for Example 17.1-1.

final values $v_C(\infty) = V_1$ and $i(\infty) = 0$ from physical considerations; the values in between, which determine the transient state, will be found when we solve the equation.

Application of these ideas to a circuit containing inductance will be found in the following example.

Example 17.1-1 Write the equation for the circuit of Fig. 17.1-3 and find the initial and final values of the current and voltages.

Solution Using KVL,

$$V_1 = v_L + v_R$$

$$= v_L + Ri \qquad (17.1\text{-}6)$$

Using the vi relation for the inductor, $v_L = L\, di/dt$,

$$L\frac{di}{dt} + Ri = V_1 \qquad (17.1\text{-}7)$$

This is the desired circuit equation for $t > 0$. In order to find the initial and final values we need the continuity condition for inductance. For the current in an inductance to change instantaneously, an infinite voltage is required. The explanation of this fact is the *dual* of that for the capacitor. Since real voltages cannot be infinite, the current through an inductor cannot change instantaneously. Thus, because the open switch forces the current to be 0 before $t = 0$, the current must also be 0 immediately after the switch is closed and we have $i(0) = 0$. Since $v_R = Ri$, we also have $v_R(0) = 0$ and consequently $v_L(0) = V_1$.

For the final values we note that after a long time the current must become constant so that for large t,

$$v_L(t \to \infty) = L\frac{di}{dt} \to 0$$

Thus $v_L(\infty) = 0$ and consequently $v_R(\infty) = V_1$. Finally,

$$i(\infty) = \frac{v_R(\infty)}{R} = \frac{V_1}{R}$$

The fact that $v_L(\infty) = 0$ indicates that in the steady state, for dc sources, the inductor looks like a *short circuit.* Note the duality to the capacitor, which looks like an *open circuit* under the same conditions. ////

17.2 STEP-BY-STEP SOLUTION OF THE DIFFERENTIAL EQUATION

In this section we will demonstrate a method of solution of the differential equation for the RC circuit which is very similar to the way in which the solution would be computed by a suitably programmed digital computer. The equation to be solved is Eq. (17.1-3) which describes the buildup of voltage across the capacitor in the circuit of Fig. 17.1-1. The equation is repeated for convenience:

$$RC\,\frac{dv_C(t)}{dt} + v_C(t) = V_1 \qquad (17.2\text{-}1)$$

This holds for $t > 0$ and is subject to the initial condition $v_C(0) = 0$.

Note that each term in this expression must have the units of volts (V). In order for the first term to be in volts the units of RC must be *time* (s). The product RC in this type of circuit is given the symbol τ (tau) and is called the time constant; it was discussed in some detail in Sec. 2.3. Using this definition, Eq. (17.2-1) can be written in the form

$$\frac{dv_C(t)}{dt} + \frac{v_C(t)}{\tau} = \frac{V_1}{\tau} \qquad (17.2\text{-}2)$$

In order to solve this equation step by step, we express the derivative in terms of increments (see Sec. 14.1-1):

$$\frac{\Delta v_C(t)}{\Delta t} + \frac{v_C(t)}{\tau} = \frac{V_1}{\tau} \qquad (17.2\text{-}3)$$

Noting that $\Delta v_C(t) = v_C(t + \Delta t) - v_C(t)$, this becomes

$$\frac{v_C(t + \Delta t) - v_C(t)}{\Delta t} + \frac{v_C(t)}{\tau} = \frac{V_1}{\tau}$$

Rearranging, we get

$$v_C(t + \Delta t) = v_C(t)\left(1 - \frac{\Delta t}{\tau}\right) + V_1\left(\frac{\Delta t}{\tau}\right) \qquad (17.2\text{-}4)$$

This is the desired relation which is called a *recursion formula.* It expresses v_C at a time $t + \Delta t$ in terms of v_C at time t (Δt seconds *earlier*) and in terms of circuit constants R, C, and V_1. If we can find v_C for *any time t*, we can use the equation to find its approximate value Δt seconds *later*. (The value is only approximate because $\Delta v/\Delta t$ is only an approximation to dv/dt.) This value can then be used to find the value another Δt seconds later, and so on in steps of Δt seconds. We illustrate by using this procedure for the values $R = 1\ \text{M}\Omega$, $C = 2\ \mu\text{F}$, and $V_1 = 10$ V. Now

$$\tau = RC = 10^6 \times 2 \times 10^{-6} = 2 \text{ s}$$

so that Eq. (17.2-4) becomes

$$v_C(t + \Delta t) = v_C(t)\left(1 - \frac{\Delta t}{2}\right) + 10\left(\frac{\Delta t}{2}\right) \tag{17.2-5}$$

The value chosen for the time step Δt should be one-tenth or less of the time constant; however, in order to simplify the illustration of the method we choose $\Delta t = 1$ s (this makes $\Delta t/\tau = \frac{1}{2}$). Then Eq. (17.2-5) becomes

$$v_C(t + \Delta t) = \frac{1}{2} v_C(t) + 5 \tag{17.2-6}$$

This holds for $t > 0$ and the initial condition is $v_C(0) = 0$. It is the initial condition that gives us the starting value for our step-by-step solution. We proceed as follows:

At $t = 0$

$$v_C(0) = 0 \text{ V} \qquad \text{(initial condition)}$$

Δt s later, Eq. (17.2-6) becomes

$$v_C(\Delta t) = \frac{1}{2} v_C(0) + 5 = 5 \text{ V}$$

2 Δt s later,

$$\begin{aligned} v_C(2\Delta t) &= \frac{1}{2} v_C(\Delta t) + 5 \\ &= \frac{1}{2}(5) + 5 \\ &= 7.5 \text{ V} \end{aligned}$$

$3\Delta t$ s later,

$$\begin{aligned} v_C(3\Delta t) &= \frac{1}{2} v_C(2\Delta t) + 5 \\ &= \frac{1}{2}(7.5) + 5 \\ &= 8.75 \text{ V} \end{aligned}$$

Continuing in this way, we get the results tabulated below:

t	$v_C(t)$
0	0
Δt	5
$2\Delta t$	7.5
$3\Delta t$	8.75
$4\Delta t$	9.375
$5\Delta t$	9.6875
$6\Delta t$	9.8438
$7\Delta t$	9.9219
$8\Delta t$	9.9609

These data are plotted in Fig. 17.2-1. The points are not joined together because our solution holds only at points Δt seconds apart. It must be emphasized that this solution is only an approximation to the correct solution of the original differential equation. The approximation becomes better as $\Delta t/\tau$ is taken smaller; however, even our choice of $\Delta t/\tau = 0.5$ leads to a solution remarkably similar to the actual solution. This can be seen by comparing the graph of Fig. 17.2-1 to the exponential curve of Fig. 2.3-4 which represents the actual solution.

The approximate solution does show how the capacitor voltage increases. As this increase is taking place, the resistor voltage is decreasing. This may be seen from the KVL equation for $t > 0$,

$$v_R(t) = V_1 - v_C(t)$$

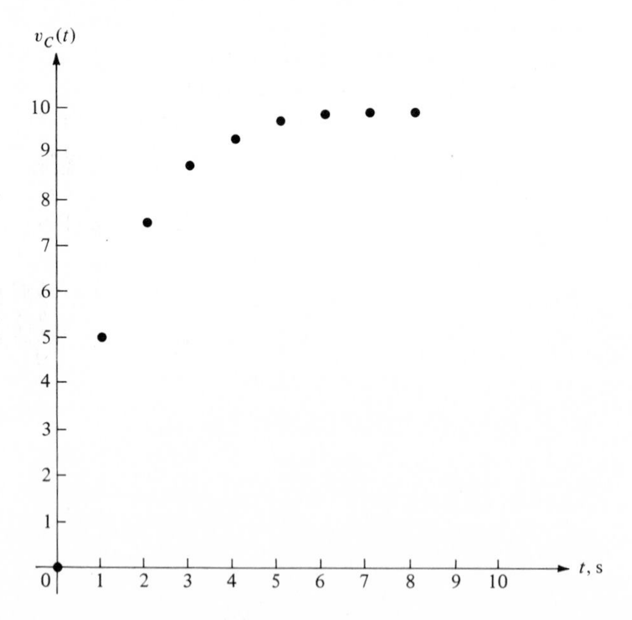

Fig. 17.2-1 Stepwise solution of differential equation.

Since the resistor voltage is decreasing, the current is also decreasing. This causes a decrease in the *rate* at which the capacitor voltage is increasing (recall that $i = C\ dv_C/dt$). As this rate of change approaches zero, all currents and voltages approach constants, i.e., steady-state values. In the next section we will set down the actual solution in general form and show how it can be applied to more complicated circuits. Additional examples of the stepwise method will be found in the homework problems.

17.3 NATURAL RESPONSE

In the last section we saw how the voltage builds up across the capacitor when a battery is switched across an RC circuit. This process is called *charging* and we say that the capacitor *charges up* to the final (steady-state) voltage. If a charged capacitor is disconnected from the charging circuit, it will ideally hold its charge indefinitely. If a resistor is connected across a charged capacitor, current will flow in the resistor and the energy stored in the electric field of the capacitor will be dissipated as heat in the resistor. This process is referred to as *discharging* the capacitor.

Consider the circuit of Fig. 17.3-1*a*. Assume that the switch has been in position 1 long enough so that the capacitor is charged to V_1 volts. At a time designated as $t = 0$ for convenience, the single-pole double-throw switch disconnects the battery and connects the resistor and capacitor together as shown in part *b* of the figure. After $t = 0$ the only energy in the circuit is that stored in the electric field of the capacitor and the response under these conditions is called the *natural* response since the response is determined by the network alone, i.e., with no external input. As noted in the previous section, the capacitor voltage cannot change instantaneously, and since it has been charged to V_1 volts before $t = 0$, it will remain at V_1 volts immediately after the switching and the initial condition for the circuit is $v_C(0) = V_1$.

To find the natural response, we first write the circuit differential equation. Using KCL at the single independent node,

$$i_C(t) + i_R(t) = 0 \qquad (17.3\text{-}1)$$

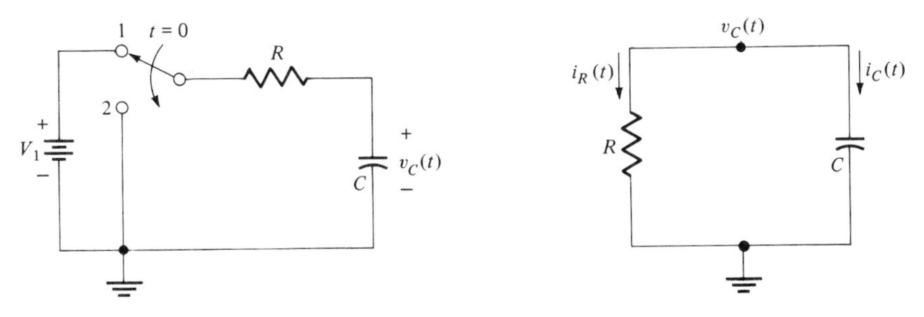

Fig. 17.3-1 Discharging a capacitor. (*a*) Circuit; (*b*) circuit for $t > 0$.

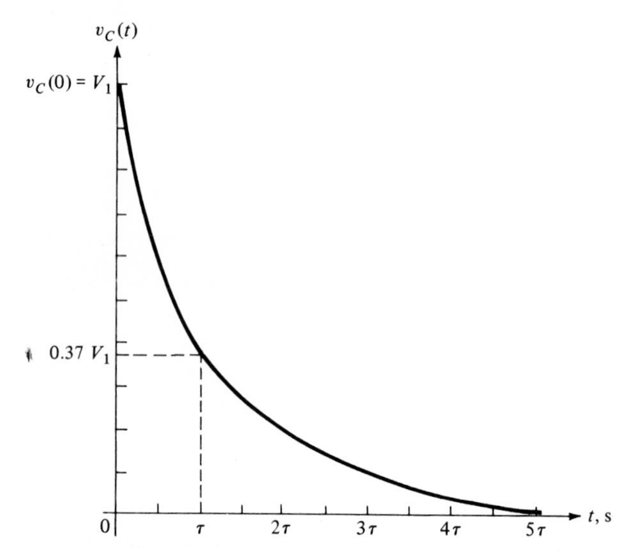

Fig. 17.3-2 Natural response of RC circuit.

Using the vi relations for C and R,

$$C\frac{dv_C}{dt} + \frac{v_C}{R} = 0$$

Dividing through by C and setting $\tau = RC$,

$$\frac{dv_C}{dt} + \frac{v_C}{\tau} = 0 \tag{17.3-2}$$

This equation holds for $t > 0$ and is subject to the initial condition $v_C(0) = V_1$. Its solution can be shown to be, for $t > 0$,

$$v_C(t) = v_C(0)e^{-t/\tau} \tag{17.3-3}$$

$$= V_1e^{-t/\tau} \tag{17.3-4}$$

A plot of this equation is shown in Fig. 17.3-2. It turns out that the natural response of *any* circuit containing only one capacitor or one inductor and any number of resistors has the same form, i.e., an exponential decay. In the next section we will present a general method for finding this response.

17.4 TOTAL RESPONSE—THE GENERAL SOLUTION

In this section we will present a general equation for a large class of transient responses. The responses are the currents and voltages in single energy-storage element circuits. For our purposes, such circuits consist of either one capacitor or one inductor, any number of resistors, and one source. The source may be a switched battery or constant current source, or a step function voltage or current source. All such circuits lead to a linear differential equation similar to Eq. (17.1-3).

The general solution, which holds for $t > 0$ is*

$$r(t) = r(\infty) + [r(0) - r(\infty)]e^{-t/\tau} \tag{17.4-1}$$

where $r(t)$ = response at time t, either current or voltage
$r(\infty)$ = final value of response, either a constant or zero
$r(0)$ = initial value of response, either a constant or zero
$e^{-t/\tau}$ = exponential factor (see Sec. 2.3)
τ = circuit time constant, to be explained shortly

The response is thus completely characterized by three numbers, $r(\infty)$, $r(0)$, and τ. These numbers are obtained from the circuit using basic laws as shown in the explanation and examples which follow.

1 $r(\infty)$ is the response a long time after the switching or step has taken place. This is found, as noted in Sec. 17.1, by considering capacitors as open circuits and inductors as short circuits.
2 $r(0)$ is the initial condition which has been discussed in connection with each example thus far. It is found by finding the response just prior to the switching and then applying the appropriate continuity condition to find the initial condition just after switching.
3 τ is the circuit time constant.

For RC circuits:

$$\tau = R_T C \tag{17.4-2}$$

For RL circuits:

$$\tau = \frac{L}{R_T} \tag{17.4-3}$$

L and C represent the inductance or capacitance. The Thevenin resistance R_T is found by setting the source to zero (short circuit for voltage sources, open circuit for current sources) and calculating the resistance looking into the circuit from the terminals of the inductor or capacitor. This is illustrated in the examples which follow.

Example 17.4-1 If the capacitor is initially uncharged in the circuit of Fig. 17.4-1a, find

a $v_C(t)$
b $i(t)$

* The student should check this for $t \to 0$ and $t \to \infty$ to see that it gives the expected solutions.

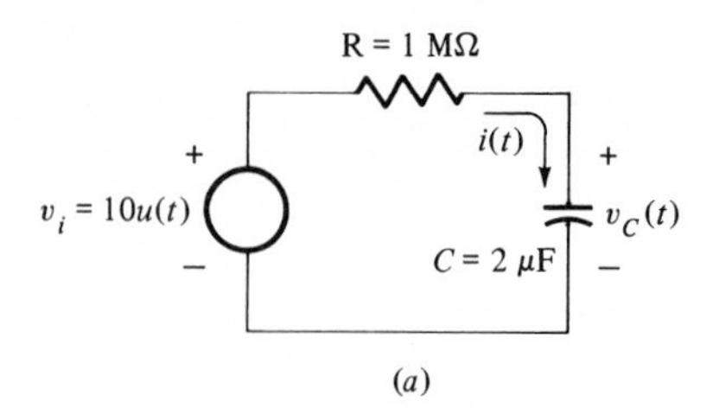

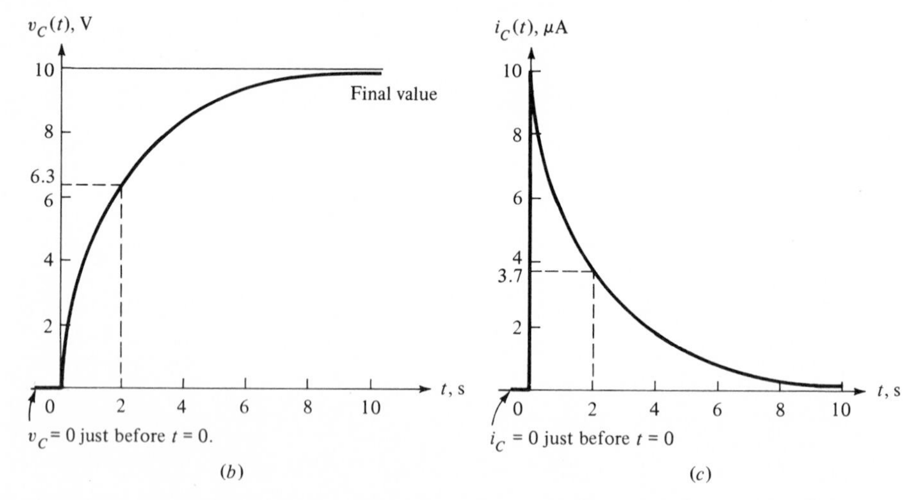

Fig. 17.4-1 Example 17.4-1 (*a*) Circuit; (*b*) capacitor voltage; (*c*) current.

Solution

a The general equation for $r(t) = v_C(t)$ becomes

$$v_C(t) = v_C(\infty) + [v_C(0) - v_C(\infty)]e^{-t/\tau} \tag{17.4-4}$$

The constants in this equation are found as follows:

1 After a long time the capacitor will charge to 10 V so that $v_C(\infty) = 10$.
2 Since the capacitor is initially uncharged $v_C = 0$ before the step begins and the continuity condition requires that $v_C(0) = 0$ immediately after the step begins.
3 For this simple circuit $\tau = RC = 2$ s by inspection.

Using these values in Eq. (17.4-4), we have

$$\begin{aligned} v_C(t) &= 10 + [0 - 10]e^{-t/2} \\ &= 10\,(1 - e^{-t/2}) \text{ V} \end{aligned} \tag{17.4-5}$$

This equation is plotted in Fig. 7.4-1*b*. Its important characteristics were discussed in Sec. 2.3. Briefly, they are that the voltage rises to 63 percent of the final value in a time equal to one time constant (2 s) and 99 percent of the final value in five time constants (10 s).

b The general equation for $r(t) = i(t)$ becomes

$$i(t) = i(\infty) + [i(0) - i(\infty)]e^{-t/\tau} \tag{17.4-6}$$

The constants are as follows:

1 After a time the capacitor appears as an open circuit so that $i(\infty) = 0$.
2 Immediately after the step begins we have $v_C(0) = 0$. Again using Ohm's law

$$i(0) = \frac{v_i(0) - v_C(0)}{R} = \frac{10 - 0}{10^6} = 10\ \mu\text{A}.$$

3 The time constant is, of course, the same as in part *a*, $\tau = 2$ s.

Substituting these values in Eq. (17.4-6) we obtain

$$i(t) = 0 + [10 - 0]e^{-t/2}\ \mu\text{A}$$
$$= 10\ e^{-t/2}\ \mu\text{A} \tag{17.4-7}$$

This is plotted in Fig. 17.4-1*c*. Note that the current is zero just before the step begins, and that the current *is* discontinuous (jumps) at $t = 0$. This does not violate the continuity condition which applies only to voltage for a capacitor. The exponential decay is 37 percent of its initial value after one time constant and is down to 1 percent after five time constants. ////

Example 17.4-2 Refer to the circuit of Fig. 17.1-1*a*. If $R = 39$ kΩ, $C = 0.05\ \mu$F, and $V_1 = 22$ V, find

a The voltage across the capacitor at $t = 4$ ms
b The time required for the voltage across the capacitor to reach 17 V

Solution

a The capacitor voltage is [from Eq. (17.4-4)]

$$v_C(t) = V_1(1 - e^{-t/\tau})$$

where $\tau = RC = 39 \times 10^3 \times 0.05 \times 10^{-6} = 1.95$ ms
Then, at $t = 4$ ms,

$$v_C(t) = 22(1 - e^{-4/1.95})$$
$$= 22(1 - e^{-2.05})$$

From Appendix C, $e^{-2.05} = 0.129$ so that

$$v_C(t) = 22(1 - 0.129)$$
$$= 19.2 \text{ V}$$

b $v_C(t) = 17 = 22(1 - e^{-t/\tau})$
Solving for t/τ, we find

$$1 - e^{-t/\tau} = \frac{17}{22}$$

$$e^{-t/\tau} = 1 - \frac{17}{22} = \frac{5}{22} = 0.227$$

From the exponential table in Appendix C we find that

$$e^{-1.48} = 0.228$$

thus

$$-\frac{t}{\tau} = -1.48$$

from which $t = 1.48\,\tau = (1.48)(1.95) = 2.89$ ms. ////

17.4-1 Natural and Forced Response

Let us return to the expression for the voltage rise across a capacitor given by Eq. 17.4-5. In general terms this equation can be written

$$v_C(t) = V_1(1 - e^{-t/\tau}) \tag{17.4-8}$$

In order to interpret this solution physically, we separate it into two parts. The first part consists of the constant term V_1. This is the *forced* part of the solution and we write

$$v_f(t) = V_1 \tag{17.4-9}$$

The second part of the solution, called the *natural response,* is

$$v_n(t) = -V_1 e^{-t/\tau} \tag{17.4-10}$$

These are plotted separately in Fig. 17.4-2 for $V_1 = 10$ V and $\tau = 2$ s.

The physical significance of the forced response is that it is the value which the output voltage assumes as $t \to \infty$. Thus, since the initial voltage is known to be zero and the final voltage, or forced response, is known to be 10, we see that the natural response *takes the solution from the initial condition to the final value.*

All circuit problems which lead to linear differential equations are characterized by this same type of solution, i.e., a forced response plus a natural response which provides the transition from the initial condition to the final value. In

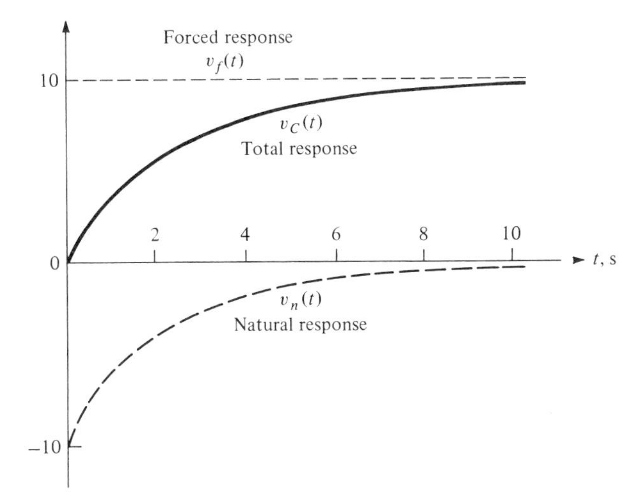

Fig. 17.4-2 Components of response of RC circuit to step input.

our example, the forced response is a constant, and has the same form as the input voltage. It is found that the forced response often has the same form as the input signal, whether it be a step function, sine wave, or some other wave form.

In single energy-storage element circuits, the natural response is a decaying exponential of the form $Ae^{-t/\tau}$. In more complicated circuits it has the form $Ae^{-t/\tau} \cos (\omega t + \theta)$ or even a sum of terms of this type, the numbers of terms depending on the complexity of the circuit.

17.5 PULSE RESPONSE AND RISE TIME

The pulse signals used in digital systems are often the input signals to circuits of the form shown in Fig. 17.5-1a. The capacitance in the circuit is usually *parasitic* in nature and is unintentionally introduced into the system because of the physical construction of the device. It is important that the pulse signals not be seriously distorted (i.e., made different from the input) in passing through these circuits. Thus it becomes important to characterize the distortion of the pulse in some measureable way as it passes through the circuit. A quantity called *rise time* is used as a measure of this distortion and is a very important digital system parameter.

In order to define rise time, we consider the circuit of Fig. 17.5-1a with a pulse input as shown. The output voltage is found by first finding the output voltage due to a step function of V_1 volts beginning at $t = 0$. The response equation is then used to find the output voltage at $t = T$ when the pulse switches off. After $t = T$, the input voltage is zero, so the input source appears as a short

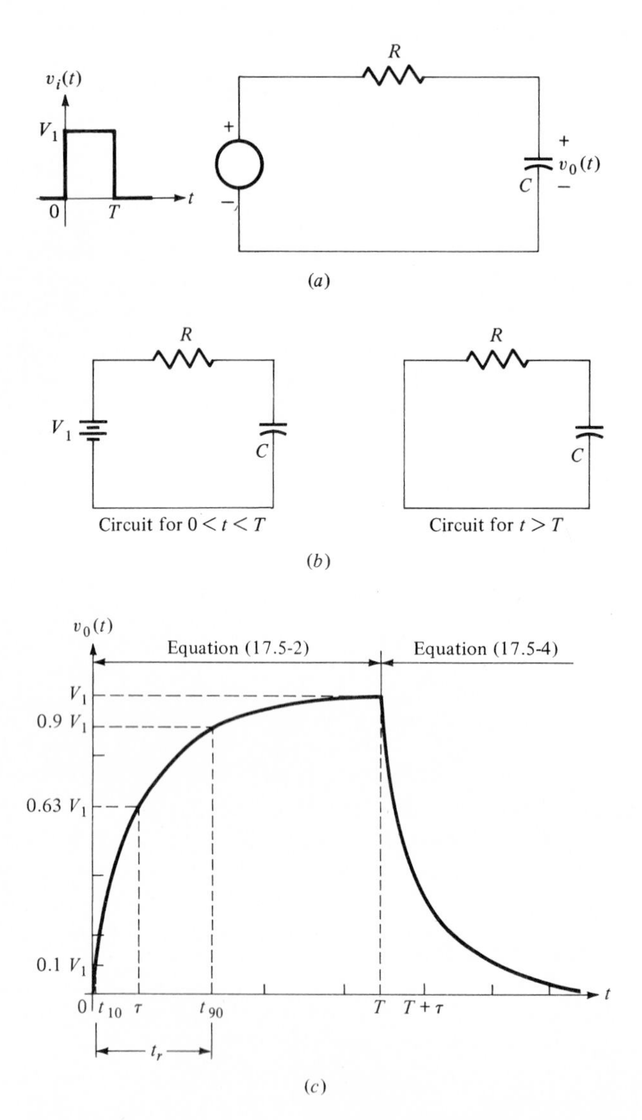

Fig. 17.5-1 Pulse response. (*a*) Circuit; (*b*) circuits during and after pulse; (*c*) response waveform.

circuit (Fig. 17.5-1*b*). Thus after $t = T$ the capacitor will discharge through the resistor. The voltage across the capacitor at $t = T$ is the initial condition for the discharge phase. For the charging phase, we have, from Eq. (17.4-8),

$$v_0(t) = V_1(1 - e^{-t/\tau}) \tag{17.5-1}$$

This expression is valid *only* while the pulse is on, i.e., for $0 < t < T$.

At time T the capacitor is charged to a voltage

$$v_0(T) = V_1(1 - e^{-T/\tau}) \tag{17.5-2}$$

If $T > 5\tau$ the exponential factor will be less than 0.01 and

$$v_0(T) \approx V_1 \tag{17.5-3}$$

Thus the voltage across the capacitor just *before* the pulse turns off is V_1 volts. After the pulse has turned off, the circuit shown in Fig. 17.5-1*b* for $t > T$ holds. The capacitor voltage *after* the pulse has turned off will be V_1 volts since it cannot change instantaneously, and this is the initial condition for the circuit for $t > T$. Since there is no source in this circuit the capacitor will discharge through the resistor and its voltage will decay according to the relation

$$v_C(t) = v_C(T)e^{-(t-T)/\tau} \tag{17.5-4}$$

This is valid *only* for $t > T$. Note that the exponent is in terms of a "shifted" time $(t - T)$ since the time origin for this part of the calculation is at $t = T$. When $t = T$

$$v_C(T) = v_C(T)e^{-(T-T)/\tau} = v_C(T)e^0 = v_C(T)$$

as we would expect.

Equations (17.5-2) and (17.5-4) are plotted in Fig. 17.5-1*c* and it is evident that the output waveshape differs from the input waveshape due to the curvature of both the leading and trailing edges. In a digital system we are usually concerned with the *speed* with which $v_0(t)$ rises toward its final value. We know that the rising exponential reaches 63 percent of its final value after one time constant has elapsed so that the time constant can be used as a measure of the speed of response of this circuit.

The criterion more often used for such circuits is the time it takes for $v_0(t)$ to rise from 10 to 90 percent of its final value. This is called the *rise time* t_r and we can find it as follows.

If we call t_{10} the time at which $v_0 = 0.1\ V_1$, then from Eq. (17.5-2)

$$v_0(t_{10}) = 0.1\ V_1 = V_1(1 - e^{-t_{10}/\tau})$$

Solving, this yields

$$e^{-t_{10}/\tau} = 0.9$$

From Appendix C, $e^{-0.11} = 0.896 \approx 0.9$. Thus

$$\frac{-t_{10}}{\tau} = -0.11 \qquad \text{and} \qquad t_{10} = 0.11\tau$$

If t_{90} is the time at which $v_0 = 0.9\ V_1$ the steps above yield

$$e^{-t_{90}/\tau} = 0.1$$

$$t_{90} = 2.3\ \tau$$

From the definition of rise time (see Fig. 17.5-1*c*),

$$t_r = t_{90} - t_{10} = 2.3\,\tau - 0.11\,\tau \approx 2.2\,\tau \tag{17.5-5}$$

Finally, for the *RC* circuit,

$$t_r = 2.2\,RC \tag{17.5-6}$$

This relates the 10 to 90 percent rise time to the element values of the circuit.

17.6 THE EFFECT OF NONZERO INITIAL CONDITIONS

When the capacitor in a transient circuit has an initial charge, care must be exercised in applying the general solution. This is illustrated in the example which follows.

Example 17.6-1 In the circuit of Fig. 17.5-1*a*, $R = 100$ kΩ, $C = 0.005$ μF, $V_1 = 10$ V, $T = 0.75$ ms and the capacitor is initially uncharged. Find and graph $v_0(t)$ and $i(t)$.

Solution We proceed as in Sec. 17.5. For the circuit

$$\tau = RC = 100 \times 10^3 \times 0.005 \times 10^{-6} = 0.5 \times 10^{-3}\text{ s} = 0.5\text{ ms}$$

While the pulse is on the capacitor charges according to Eq. (17.5-1),

$$v_0(t) = V_1(1 - e^{-t/\tau}) = 10(1 - e^{-t/0.5})$$

where t is in milliseconds.

When $t = T = 0.75$ ms the input pulse goes to zero and the capacitor has charged to [see Eq. (17.5-2)],

$$v_0(0.75) = 10(1 - e^{-0.75/0.5}) = 10(1 - e^{-1.5}) = 7.77\text{ V}$$

This is the final voltage for the charging phase which becomes the initial condition for the discharge phase

$$v_0(T) = 7.77\text{ V}$$

The equation which describes the discharge is Eq. (17.5-4)

$$v_0(t) = 7.77\,e^{-(t-0.75)/0.5}$$

The results are plotted in Fig. 17.6-1*a*. For the discharge phase, the voltage after one time constant is 37 percent of the initial value. Thus the voltage at $t = 0.75 + 0.5 = 1.25$ ms is $7.77 \times 0.37 = 2.87$ V.

The current can be found by several methods. We will use the general equation for exponential response. The student should check our results using the voltages found above to find the voltage across the resistor at $t = 0.25, 0.5, 0.75$, and 1.25 ms. Ohm's law is then used to find the current.

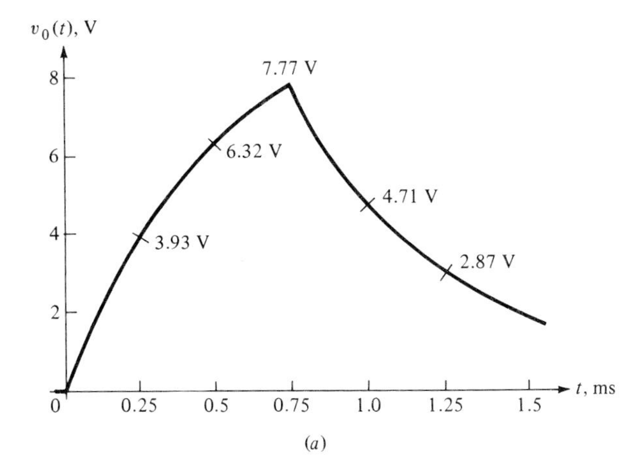

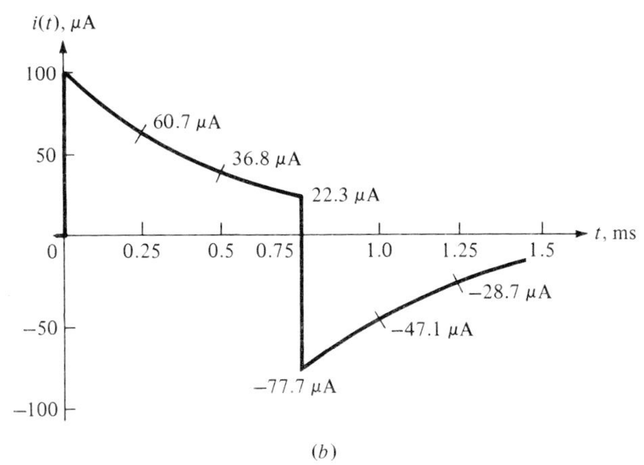

Fig. 17.6-1 Example 17.6-1 (*a*) Output voltage waveform; (*b*) Current waveform.

The general equation for the current during the time the pulse is on is

$$i(t) = i(0)e^{-t/\tau}$$

where

$$i(0) = \frac{V_1}{R} = \frac{10}{10^5} = 100\ \mu\text{A}$$

Thus

$$i(t) = 100\ e^{-t/0.5}\ \mu\text{A}$$

where again t is in milliseconds.

An instant before the pulse goes off ($t = 0.75^-$) we have

$$i(0.75^-) = 100\ e^{-0.75/0.5} = 22.3\ \mu\text{A}$$

When the pulse goes off ($t = 0.75^+$) the input voltage is zero and the capacitor voltage is 7.77 V. The voltage across the resistor is then -7.77 V so that the current is

$$i(0.75^+) = \frac{0 - 7.77}{R} = -77.7\ \mu\text{A}$$

From $t = 0.75^+$ on, the current will decay exponentially to zero according to

$$i(t) = -77.7\ e^{-(t-0.75)/0.5}\ \mu\text{A}$$

The current is plotted in Fig. 17.6-1*b*. ////

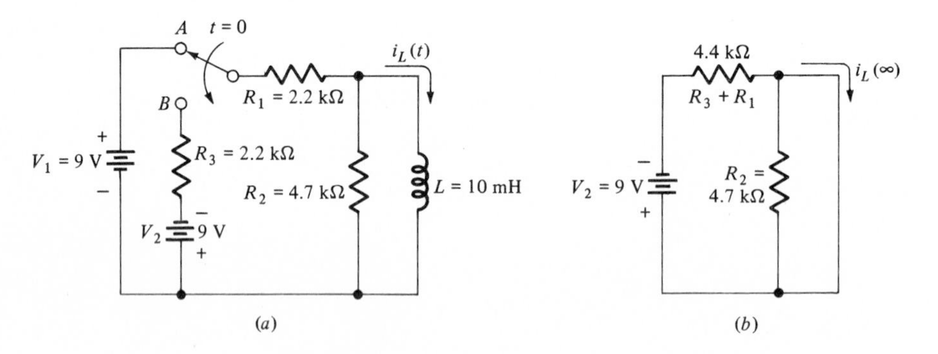

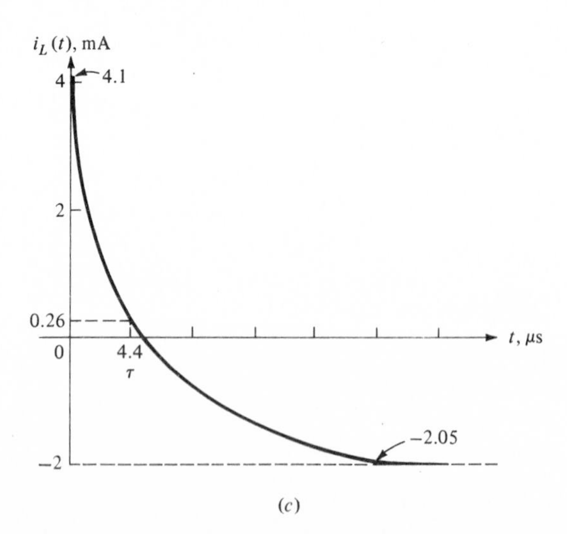

Fig. 17.6-2 Example 17.6-2 (*a*) Circuit; (*b*) circuit for $t \to \infty$; (*c*) response.

Example 17.6-2 In the circuit of Fig. 17.6-2*a* the switch is in position A for a long time. At $t = 0$ it is moved to position B. Find and plot $i_L(t)$ for $t > 0$.

Solution The general response equation for $i_L(t)$ is

$$i_L(t) = i_L(\infty) + [i_L(0) - i_L(\infty)]e^{-t/\tau}$$

For the circuit, $R_T = (R_1 + R_3) \parallel R_2$, and

$$\tau = \frac{L}{R_T} = \frac{10 \times 10^{-3}}{[(2.2 + 2.2) \parallel 4.7] \times 10^3} = \frac{10 \times 10^{-3}}{2.27 \times 10^3}$$

$$= 4.4 \times 10^{-6}\ \text{s} = 4.4\ \mu\text{s}$$

Just before the switch is thrown the current through the coil will be constant and the coil will appear as a short circuit because the switch has been in position A for a long time and all transients will have disappeared. This current is

$$i_L(0^-) = \frac{V_1}{R_1} = \frac{9}{2.2} = 4.1\ \text{mA}$$

The continuity condition for inductance requires that this current not change instantaneously. Thus, the current immediately after the switching is $i_L(0^+) = 4.1$ mA. This is our initial condition.

After the switch has been in position B for a long time the transient will have disappeared and the circuit for the calculation of $i_L(\infty)$ will be as shown in Fig. 17.6-2*b*. From this circuit

$$i_L(\infty) = \frac{-9}{4.4} = -2.05\ \text{mA}$$

The response is then

For $t < 0$,

$$i_L(t) = 4.1\ \text{mA}$$

For $t > 0$,

$$i_L(t) = -2.05 + [4.1 + 2.05]e^{-t/4.4}$$

$$= -2.05 + 6.15\ e^{-t/4.4}$$

with t in microseconds

These results are plotted in Fig. 17.6-2*c*. ////

17.7 TRANSIENT RESPONSE OF COMPLEX NETWORKS

When a network contains more than one capacitor or one inductor, the transient solution becomes much more complicated and the mathematics involved is beyond the scope of this course. However, as an illustration we will describe the response of the circuit of Fig. 17.7-1*a* qualitatively. This circuit contains one of each type of element: R, L, and C. The capacitor is charged to a voltage V_0 at time $t = 0$. The response of interest is the current for $t > 0$. The *form* of this response depends on the ratio $(R/2)\sqrt{(C/L)}$. If the resistance is chosen so that this ratio is greater than one, the response is said to be *overdamped* and will appear as shown in Fig. 17.7-1*b*. If the ratio is less than one, the circuit is said to be *underdamped*, or oscillatory, and the response is a damped sine wave, as shown in Fig. 17.7-1*c*. As the resistance is further reduced toward zero, the response approaches a pure sine wave, as shown in Fig. 17.7-1*d*, and the circuit is said to be *undamped.* The oscillations occur because energy is being interchanged between the electric field of the capacitor and the magnetic field of the inductor. These oscillations are eventually reduced to zero (unless $R = 0$) because some energy will be dissipated in the resistance during each cycle. In transistor oscillators, the transistors supply the lost energy, thus producing sustained oscillations. This type of circuit will be considered further in Chap. 19 in terms of frequency response.

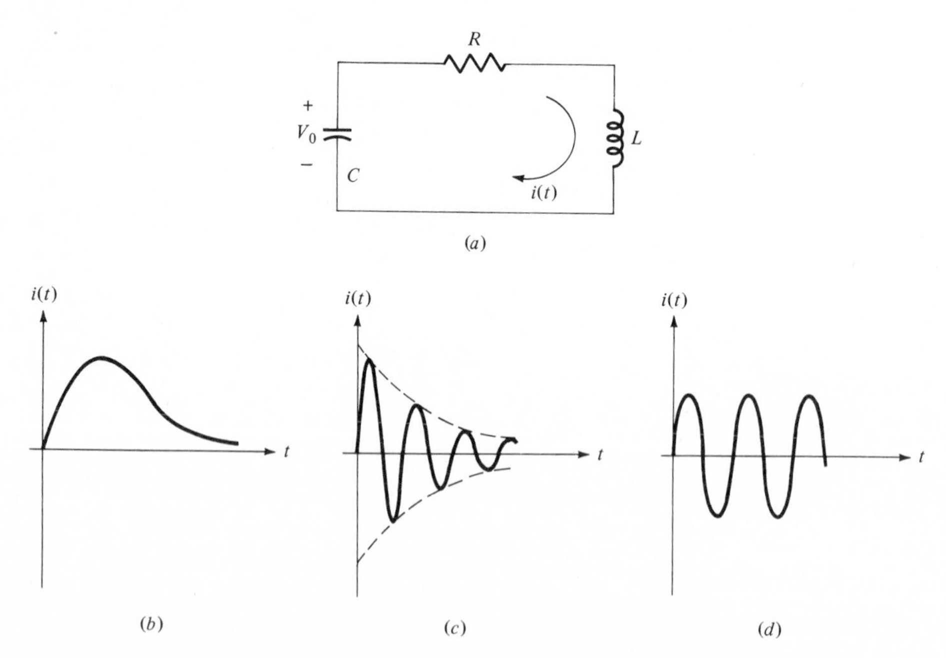

Fig. 17.7-1 Transient network. (*a*) Circuit; (*b*) overdamped response; (*c*) underdamped response; (*d*) undamped response.

17.8 SYSTEM TIME CONSTANT IDENTIFICATION FROM EXPERIMENTAL DATA

It is sometimes necessary to determine an unknown system time constant from experimental observation of the system transient response. This can be done by several different methods. Perhaps the easiest way to get a ballpark estimate of an unknown time constant is to apply pulses or a square wave to the system under study and to observe the response waveform on an oscilloscope. The pulse width or square wave frequency is varied until a pattern similar to that shown in Fig. 17.8-1*a* is obtained, where it is clear that the transient is completed within a half-cycle. The calibrated amplitude scale on the oscilloscope is then used to determine the point where the transient has risen (or decayed) 63 percent of the difference between its final and initial values. The time up to this point is one time constant, which can be read from the oscilloscope time axis calibration as indicated on the figure. The accuracy of this process is about 5 percent at best.

If more accuracy is desired, a plot on semilog paper may be used. This method has the advantage of averaging out errors made in individual readings.

Consider an exponential decay normalized to unity amplitude

$$r(t) = e^{-t/\tau} \qquad (17.8\text{-}1)$$

This is plotted on linear scales in Fig. 17.3-2. Taking the natural log of both sides,

$$\ln r(t) = \frac{-t}{\tau} \ln e = \frac{-t}{\tau} \qquad (17.8\text{-}2)$$

Since $\ln x = 2.3 \log x$ we have

$$\log r(t) = -\frac{t}{2.3\tau} \qquad (17.8\text{-}3)$$

If $r(t)$ is plotted versus t on semilog paper as shown in Fig. 17.8-1*b*, the result is a straight line which has a slope $-1/2.3\tau$.

If we have an experimental curve of $r(t)$ vs. t such as a tracing of an oscilloscope pattern and we replot the curve on semilog paper we should be able to fit a straight line to the points if the system being tested can be adequately described by a single time constant. Once we have fitted a straight line to the experimental points the easiest way to find τ is to note times t_2 and t_1 at which $r(t_2)$ and $r(t_1)$ are a decade apart, that is, $r(t_2) = 10r(t_1)$. Thus, using Eq. (17.8-3),

$$\log r(t_2) = -\frac{t_2}{2.3\tau} \qquad (17.8\text{-}4)$$

and

$$\log r(t_1) = -\frac{t_1}{2.3\tau} \qquad (17.8\text{-}5)$$

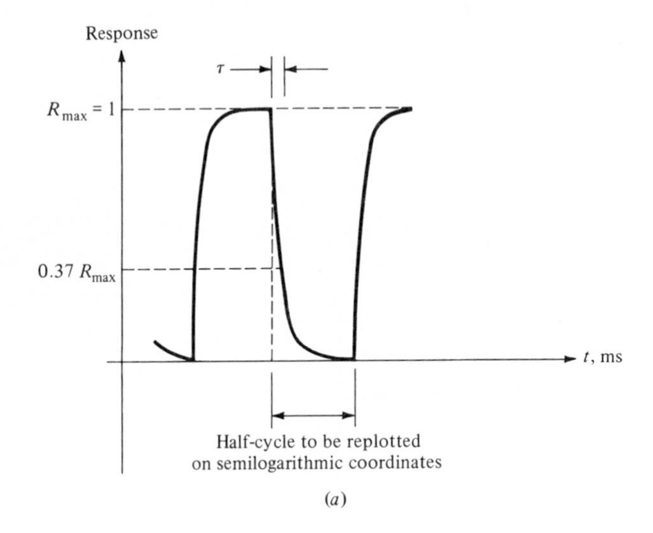

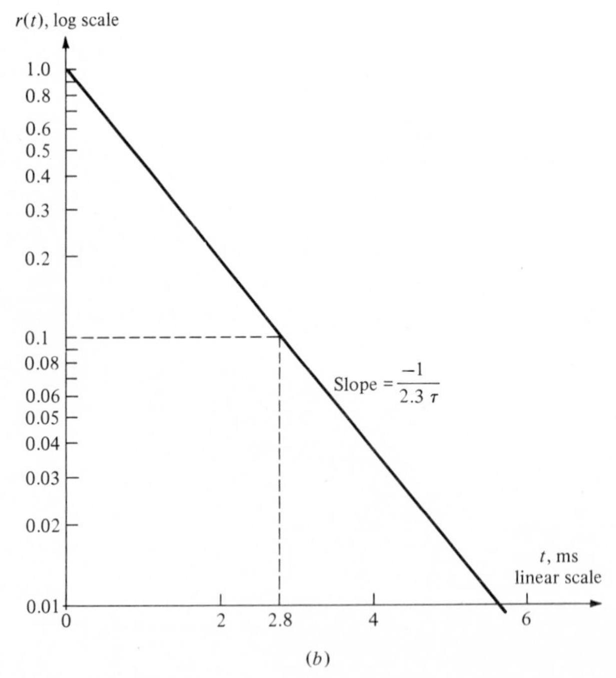

Fig. 17.8-1 Experimental determination of time constant. (*a*) Square-wave response; (*b*) semilog plot of one-half cycle.

Next we subtract Eq. (17.8-5) from Eq. (17.8-4) to get

$$\log r(t_2) - \log r(t_1) = \frac{1}{2.3\tau}(t_1 - t_2)$$

Since $\log x - \log y = \log (x/y)$, this becomes

$$\log \frac{r(t_2)}{r(t_1)} = \frac{1}{2.3\tau}(t_1 - t_2)$$

But

$$\frac{r(t_2)}{r(t_1)} = 10$$

Thus

$$\log 10 = 1 = \frac{1}{2.3\tau}(t_1 - t_2)$$

and

$$\tau = \frac{t_1 - t_2}{2.3} \tag{17.8-6}$$

It is usually convenient to take $t_2 = 0$ so that finally

$$\tau = \frac{t_1}{2.3} \tag{17.8-7}$$

where t_1 is the time at which the response is one-tenth of its value at $t = 0$. For the graph of Fig. 17.8-1*b*, $t_1 = 2.8$ ms so that $\tau = 2.8/2.3 = 1.22$ ms.

It is important to note that the amplitude of the exponential does not effect the *slope* of the semilog plot. If the amplitude is other than 1, the straight line will be shifted up or down but the slope will not be affected.

This technique can also be applied to rising exponentials. In this case it is convenient to subtract the final value before plotting on semilog paper. Examples will be left for the homework problems.

SUMMARY

1 Equations for circuits containing R and C or R and L elements are differential equations. Initial conditions are required in order to complete the solution.

2 The general solution of a first-order linear differential equation for the conditions of this chapter is

$$r(t) = r(\infty) + [r(0) - r(\infty)]e^{-t/\tau}$$

$r(0)$ is the initial value and $r(\infty)$ is the final value.

3 The total solution goes from the initial value to the final value. The natural response provides the transition.

4 The charge equation (exponential rise) is

$$r(t) = r(\infty)(1 - e^{-t/\tau})$$

The discharge equation (exponential decay) is

$$r(t) = r(0)e^{-t/\tau}$$

5 An unknown time constant is found from a straight line plot of the exponential decay on semilog paper using the equation $\tau = t_1/2.3$ where t_1 is the time at which the response is one-tenth of its value at $t = 0$.

PROBLEMS

17.1-1 Formulate the differential equation for $v_C(t)$ in the circuit of Fig. P17.1-1.

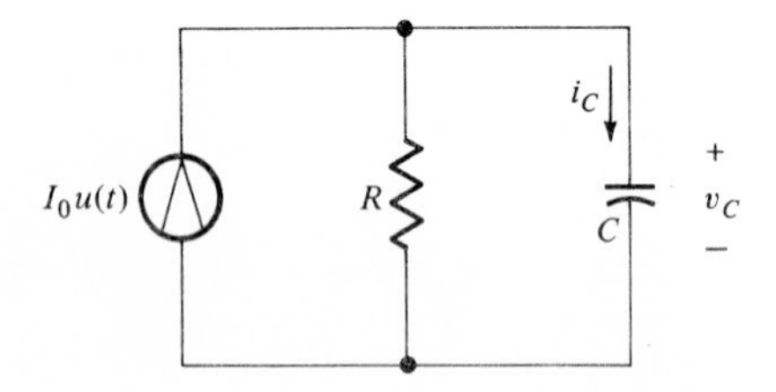

Fig. P17.1-1

17.1-2 Formulate the differential equation for $i_L(t)$ in the circuit of Fig. P17.1-2.

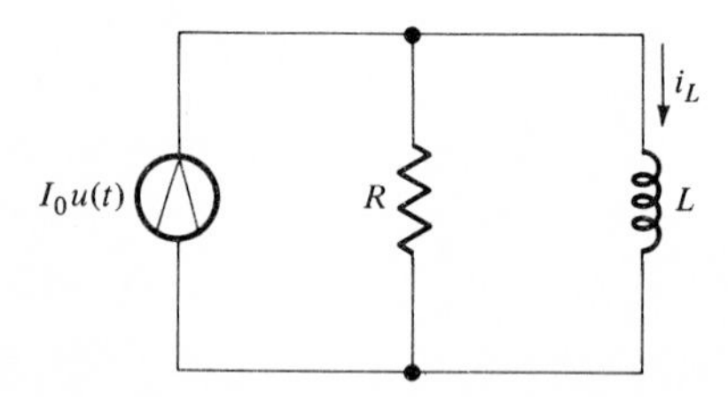

Fig. P17.1-2

17.1-3 In Fig. P17.1-1, $I_0 = 2$ mA, $R = 4.7$ kΩ, $C = 0.1$ μF, and $v_C(0) = 0$. Find all initial and final conditions.

17.2-1 For the values in Prob. 17.1-3, solve the differential equation using the step-by-step method. Choose $\Delta t = \tau/5$ and plot your results up to $t = 5\tau$.

17.4-1 Find and sketch $v_C(t)$ and $i_C(t)$ in the circuit of Fig. P17.4-1.

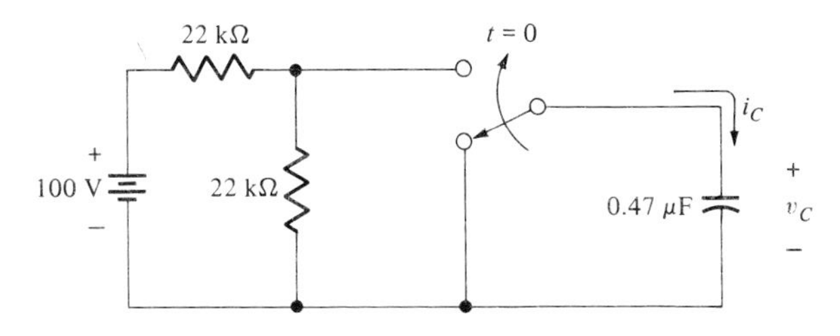

Fig. P17.4-1

17.4-2 In Prob. 17.4-1 find the time for the capacitor voltage to reach 25 V.

17.4-3 In the circuit of Fig. P17.4-3 the switch is initially in position A for a long time. At $t = 0$ it is moved to position B. Find and sketch $i_L(t)$ and $v_L(t)$ for $t > 0$.

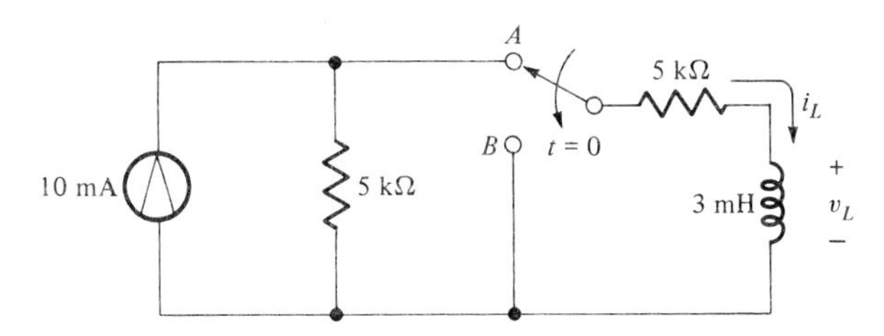

Fig. P17.4-3

17.4-4 In the circuit of Fig. P17.4-4 the switch is at A for a long time. At $t = 0$ it is moved to B. Find and plot $v_C(t)$ for $t > 0$.

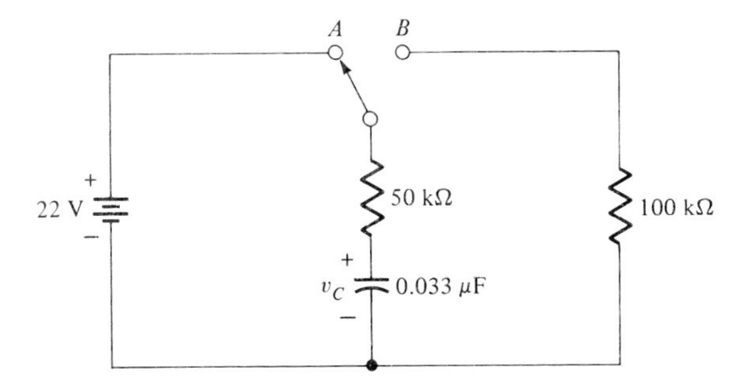

Fig. P17.4-4

17.5-1 In the circuit of Fig. 17.5-1 $T = 1$ ms, $R = 1.5$ kΩ, $C = 250$ pF. Find the rise time and sketch the output voltage waveform. Repeat if $R = 1.5$ MΩ.

17.6-1 In the circuit of Fig. P17.4-4 the switch is at A for a long time when it is moved to B at $t = 0$. At $t = 5$ ms it is moved back to A. Find and plot $v_C(t)$ for $0 < t < 30$ ms.

17.6-2 In the circuit of Fig. P17.1-1 the capacitor is charged to -10 V just prior to $t = 0$. If $I_0 = 2$ mA, $R = 5$ kΩ, and $C = 0.01$ μF, find and plot $v_C(t)$ and $i_C(t)$.

17.6-3 In the circuit of Fig. P17.4-3 the switch goes from A to B at $t = 0$. At $t = 0.3$ μs it goes back to A and at $t = 0.6$ μs it goes back to B where it remains. Find and plot $v_L(t)$ and $i_L(t)$ for $0 < t < 10$ μs.

17.8-1 Find the rise time of the circuits of Fig. P17.8-1.

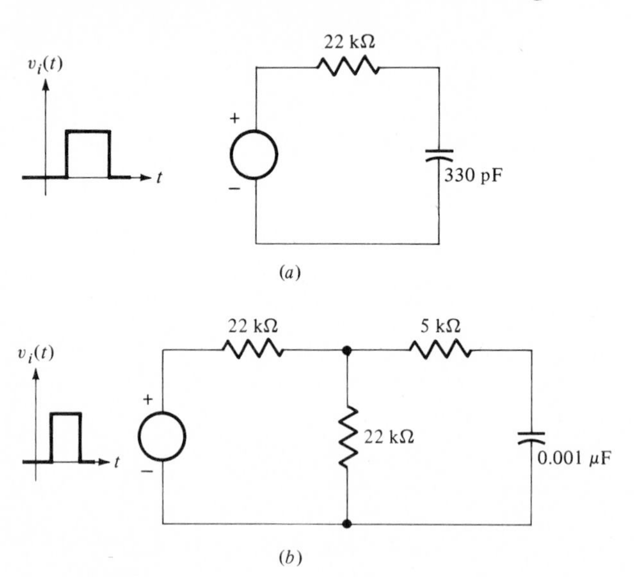

Fig. P17.8-1

17.8-2 Replot the oscilloscope trace of Fig. P17.8-2 on semilog paper and determine the time constant.

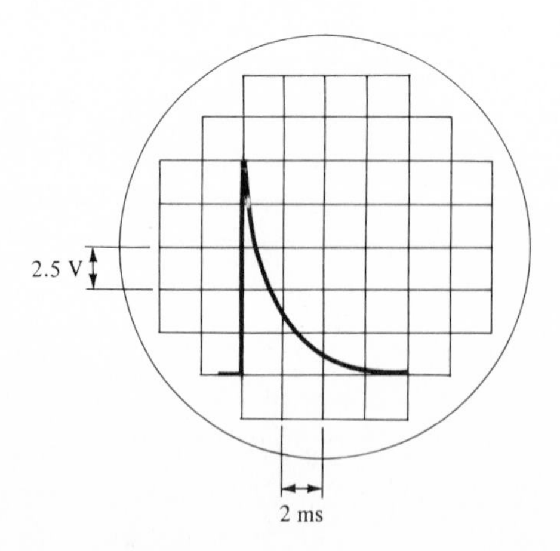

Fig. P17.8-2

18
Steady-State Sinusoidal Response

OBJECTIVES

Upon completion of this chapter, you should be able to

1 Convert between polar and rectangular representations of complex numbers and define the complex conjugate of a complex number.
2 Add, subtract, multiply, and divide complex numbers.
3 Find the phasor representation of a sinusoidal signal.
4 Calculate the sinusoidal impedances of R, L, and C circuits.
5 Calculate the steady-state sinusoidal response for a given circuit.
6 Compute the average power delivered to or by a circuit element.
7 Define effective value and calculate it for a sinusoidal signal.
8 Define power factor and explain why a high value is desirable and how to compensate for a low power-factor load.
9 Select a load value which provides maximum power transfer and use an ideal transformer to match the impedance level of a load to the circuit to which it is to be connected.

INTRODUCTION

In Chap. 17 we considered the response of *RC* and *RL* circuits to step-function inputs. The step is important because it is a standard signal for testing the *transient* response of a circuit, i.e., the time response to an abrupt change in the input signal. The steady-state response to a sinusoidal signal is another very important characteristic of a network. It is important because it determines a network's frequency response, which we will study in the next chapter, and because electric power is generated and transmitted in sinusoidal form. This means that all network computations in power distribution systems involve sinusoidal analysis.

We shall soon see that steady-state sinusoidal analysis requires extensive use of complex numbers. Therefore we devote the next section to the definition and manipulation of complex numbers.

18.1 COMPLEX NUMBERS

The real numbers may be represented graphically by the points along a straight line as shown in Fig. 18.1-1*a*. A point on the line is chosen as the origin and is

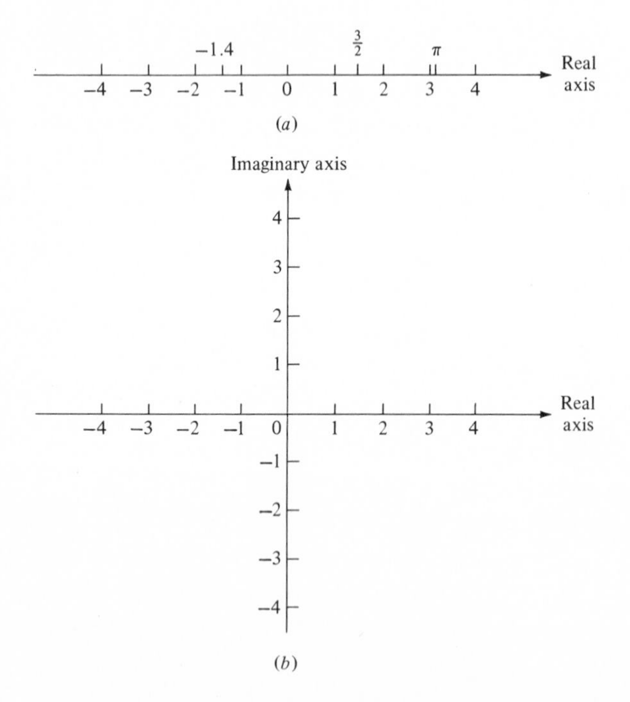

Fig. 18.1-1 Graphical representation of numbers. (*a*) Real-number axis; (*b*) real and imaginary axes.

associated with the number zero. Every other point on the line is associated with a numerical value proportional to its distance from the origin; the value is positive if it is to the right of the origin and negative if it is to the left.

The mathematical operations of addition, subtraction, multiplication, and division may all be defined in terms of the real numbers. When we consider the operation of taking the square root of a number, i.e., finding a value which when multiplied by itself produces a product equal to the given number, a problem arises; there is no real number which when multiplied by itself results in a negative product.

In order to circumvent this difficulty the unit imaginary number is introduced. Mathematicians generally use the letter i to represent the unit imaginary number while the letter j is used in the electrical and electronics field since i is customarily used to represent current; we shall use j. The unit imaginary number, j, has the property that its square is equal to -1:

$$j \times j = -1 \tag{18.1-1}$$

The square root of any negative number can now be found; if n is any positive real number, then $\sqrt{-n} = j\sqrt{n}$ or $\sqrt{-n} = -j\sqrt{n}$. These two possibilities are often combined in the shorthand expression

$$\sqrt{-n} = \pm j\sqrt{n} \tag{18.1-2}$$

as may be verified by squaring both sides of the two equations.

The imaginary numbers are graphically represented by points along the imaginary axis which is a line perpendicular to the real axis passing through its origin as shown in Fig. 18.1-1*b*. The real axis is customarily shown as a horizontal line; the imaginary axis is then a vertical line. Positive real values are to the right of the origin, negative real values to the left, positive imaginary values are above the origin, and negative imaginary values are below the origin.

The property of j expressed by Eq. (18.1-1) causes the complex plane representation of a number multiplied by j to be rotated 90° counterclockwise (CCW) from the original number. Thus since $+1$ is one unit to the right of the origin; $j \times 1 = j$ is one unit straight up, a rotation of 90° CCW; $j \times j = -1$ is one unit to the left, another rotation of 90° CCW; $j \times -1 = -j$ is straight down, another 90° CCW rotation; and finally $j \times -j = -j^2 = +1$ is again one unit to the right, the result of a fourth 90° CCW rotation. Some texts speak of the j operator as a number which rotates any number which it multiplies by 90° CCW. This concept of j is frequently useful. Note also that, from this point of view, -1 can be considered as an operator that rotates any number it multiplies by 180°.

18.1-1 Rectangular and Polar Forms

The real and imaginary axes lie in a plane. Every point in this plane corresponds to a *complex* number; a number which has a real part and an imaginary part. The point corresponding to the complex number $\mathbf{z}$ is shown in Fig. 18.1-2.†

† Boldface type is used for complex quantities.

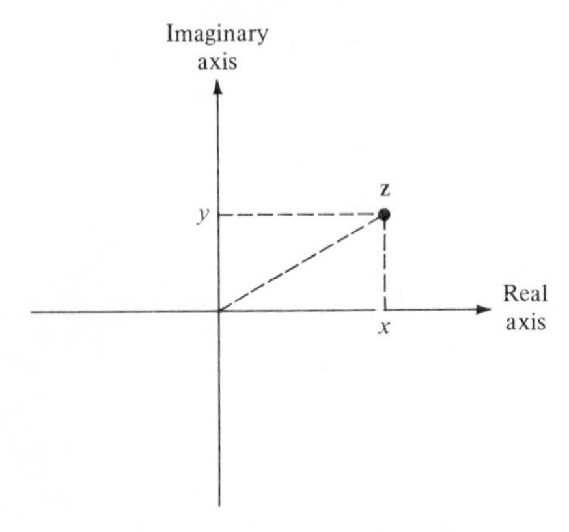

Fig. 18.1-2 Complex number showing its rectangular coordinates.

The coordinate of this point along the real axis, x, is called the *real part* of $\mathbf{z}$, symbolized

$$x = \operatorname{Re} \mathbf{z} \tag{18.1-3}$$

while the coordinate along the imaginary axis, y, is called the *imaginary part* of $\mathbf{z}$, symbolized

$$y = \operatorname{Im} \mathbf{z} \tag{18.1-4}$$

Note that the imaginary part of a complex number is a real number, it does not include the j by which it is multiplied in the complex number. The complex number is the sum of its real part and j times its imaginary part:

$$\mathbf{z} = x + jy \tag{18.1-5}$$

Graphically, $\mathbf{z}$ is obtained by drawing a rectangle one of whose sides is of length x along the real axis while a perpendicular side is of length y along the imaginary axis. The rectangle is completed by drawing the other two sides as shown in Fig. 18.1-2 and $\mathbf{z}$ is at the corner of the rectangle diagonally across from the origin.

Example 18.1-2

a Find the real and imaginary parts of the complex number $\mathbf{a}$ shown in Fig. 18.1-3*a*.

b Plot the point $\mathbf{b} = 2 - j3$ on the same set of axes. Find $\operatorname{Re} \mathbf{b}$ and $\operatorname{Im} \mathbf{b}$.

Solution

a From the graph we read the values

$$\operatorname{Re} \mathbf{a} = 5 \qquad \operatorname{Im} \mathbf{a} = 4$$

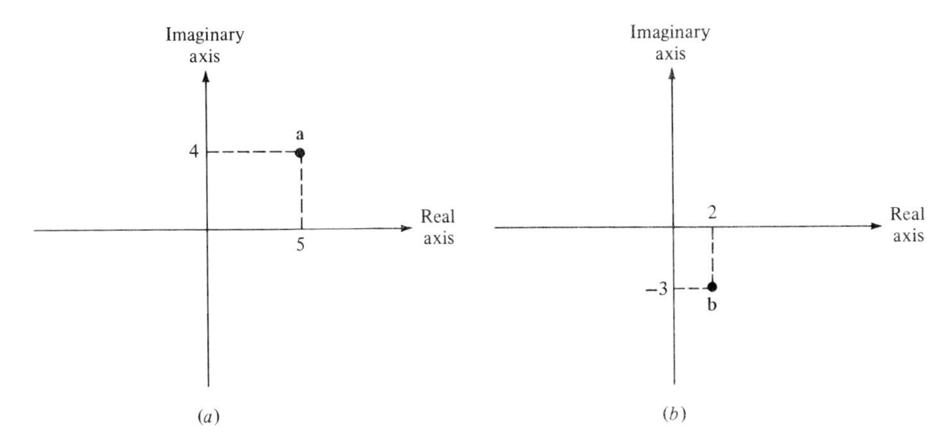

Fig. 18.1-3 Complex planes for Example 18.1-2. (*a*) Complex number a; (*b*) complex number $\mathbf{b} = 2 - j3$.

Thus $\mathbf{a} = 5 + j4$

b The complex number **b** is shown plotted in Fig. 18.1-3*b*. In order to plot the point we have used the values

$$\text{Re}\,\mathbf{b} = 2 \qquad \text{Im}\,\mathbf{b} = -3$$ ////

The coordinates of a point representing a complex number are its real part and imaginary part. A complex number written as the sum of the real and j times the imaginary coordinates, as in Eq. (18.1-5), is said to be expressed in *rectangular*, or *cartesian*, form.

The position of a point in a plane can be specified by its polar coordinates as well as its rectangular coordinates. The point representing the complex number **z**, is shown in Fig. 18.1-4 in terms of its *polar* coordinates. The distance r from the origin to the point **z** is called the *magnitude*, or *absolute value*, of **z** and is symbolized by a pair of vertical bars surrounding **z**:

$$r = |\mathbf{z}| \tag{18.1-6}$$

The absolute value is always a positive real number.

The angle, θ, measured counterclockwise from the positive real axis to the line connecting the point **z** and the origin is called the *argument* of **z**:

$$\theta = \arg \mathbf{z} \tag{18.1-7}$$

θ is a real number and is generally chosen between 0 and 360°, that is,

$$0° \leqslant \theta < 360° \qquad \text{or} \qquad 0 \leqslant \theta < 2\pi \text{ rad} \tag{18.1-8}$$

The magnitude r and argument θ of a complex number **z** are the polar coordinates of the point representing **z**. The *polar form* of **z** is written

$$\mathbf{z} = re^{j\theta} \tag{18.1-9}$$

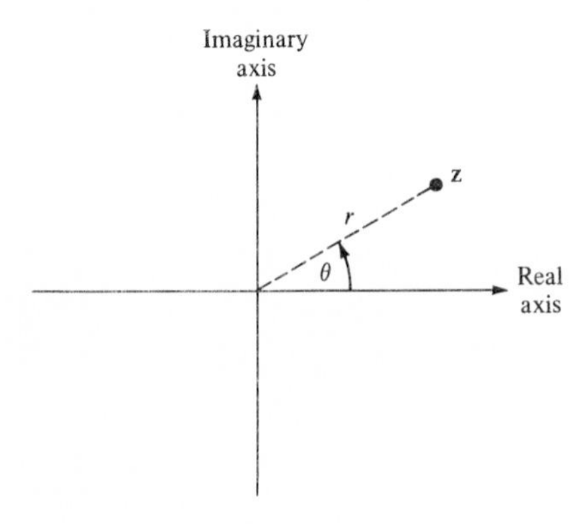

Fig. 18.1-4 Complex number showing its polar coordinates.

The validity of this expression will be established shortly. Sometimes the notation

$$\mathbf{z} = r \angle\theta \tag{18.1-10}$$

is used as an alternate to Eq. (18.1-9) but it does not have any mathematical meaning except as a shorthand for Eq. (18.1-9). It is to be read "**z** equals r at an angle θ."

Example 18.1-2 Plot the point corresponding to the complex number $2e^{j(\pi/4)}$.

Solution A line is drawn at an angle of $(\pi/4)$ rad $= 45°$ in the complex plane. The point two units from the origin along this line represents the given complex number; it is shown in Fig. 18.1-5. ////

POLAR-TO-RECTANGULAR CONVERSION

It is frequently necessary to convert between the polar and rectangular forms for a given complex number. The polar coordinates r and θ of a complex number **z** are shown in Fig. 18.1-6a. The line connecting the point **z** with the origin is the hypotenuse of a right triangle whose legs are the real and imaginary parts of the complex number. Using the trigonometric formulas for the right triangle we have

$$x = \operatorname{Re} \mathbf{z} = r \cos\theta \tag{18.1-11}$$

and

$$y = \operatorname{Im} \mathbf{z} = r \sin\theta \tag{18.1-12}$$

Multiplying Eq. (18.1-12) by j and adding Eq. (18.1-11) to Eq. (18.1-12), we get

$$x + jy = r(\cos\theta + j\sin\theta) = re^{j\theta}$$

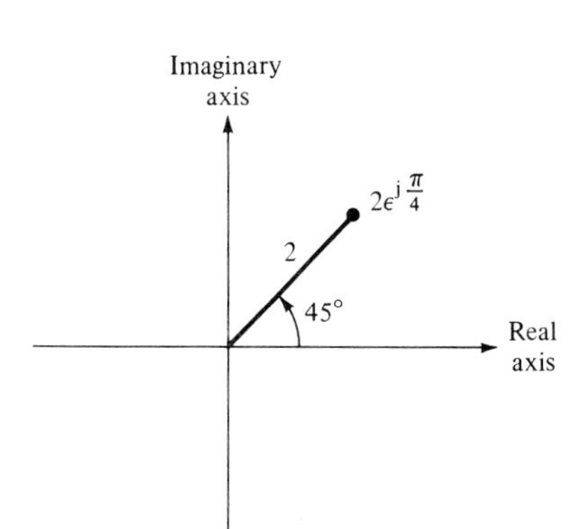

Fig. 18.1-5 Complex number $2e^{j(\pi/4)}$ for Example 18.1-3.

If we now set $r = 1$, this becomes

$$x + jy\,|_{r=1} = \cos\theta + j\sin\theta = e^{j\theta} \tag{18.1-13}$$

This is an important mathematical relation known as Euler's identity. From this relation we see that $e^{j\theta}$ is a complex number of unit magnitude at an angle of θ. The two formulas given in Eqs. (18.1-11) and (18.1-12) enable us to obtain the rectangular form of any complex number given in polar form as illustrated in the following example.

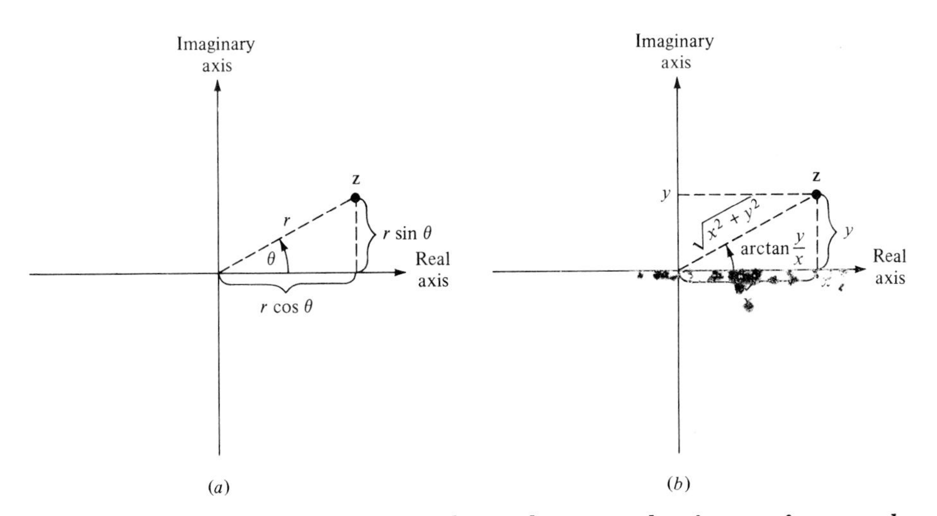

Fig. 18.1-6 Conversion between polar and rectangular forms of a complex number. (*a*) Conversion from polar to rectangular; (*b*) conversion from rectangular to polar.

Example 18.1-3 Find the rectangular form of the complex number $2e^{j(\pi/4)}$.

Solution For this complex number $r = 2$ and $\theta = (\pi/4)$ rad $= 45°$. By Eqs. (18.1-11) and (18.1-12) we have

$$x = 2 \cos 45° = 2 \times 0.707 = 1.414$$

and

$$y = 2 \sin 45° = 2 \times 0.707 = 1.414$$

Therefore $2e^{j(\pi/4)} = 1.414 + j1.414$. Compare this with Fig. 18.1-5. ////

RECTANGULAR-TO-POLAR CONVERSION

The rectangular coordinates of a complex number are shown in Fig. 18.1-6*b*. This time we use the formulas for a right triangle to obtain the polar coordinates. The pythagorean formula is used to find the hypotenuse of the triangle which is the magnitude of the complex number:

$$r = |\mathbf{z}| = \sqrt{x^2 + y^2} \tag{18.1-14}$$

The argument of $\mathbf{z}$ is given by

$$\theta = \arg \mathbf{z} = \arctan \frac{y}{x} \tag{18.1-15}$$

These two formulas enable us to convert from rectangular to polar form as illustrated by the following example.

Example 18.1-4 Find the polar representation of the complex number $1.414 + j1.414$.

Solution Using Eq. (18.1-14),

$$r = \sqrt{1.414^2 + 1.414^2} = \sqrt{2 + 2} = \sqrt{4} = 2$$

and from Eq. (18.1-15),

$$\theta = \arctan \left(\frac{1.414}{1.414}\right) = \arctan (1) = 45° = \frac{\pi}{4} \text{ rad}$$

Therefore

$$1.414 + j1.414 = 2e^{j(\pi/4)}$$

which is in agreement with the result of Example 18.1-3. ////

18.1-2 Addition and Subtraction of Complex Numbers

ADDITION

Two complex numbers, $\mathbf{z}_1 = x_1 + jy_1$ and $\mathbf{z}_2 = x_2 + jy_2$, are shown in Fig. 18.1-7*a*. The two are graphically added by moving one of them, $\mathbf{z}_2$ in our ex-

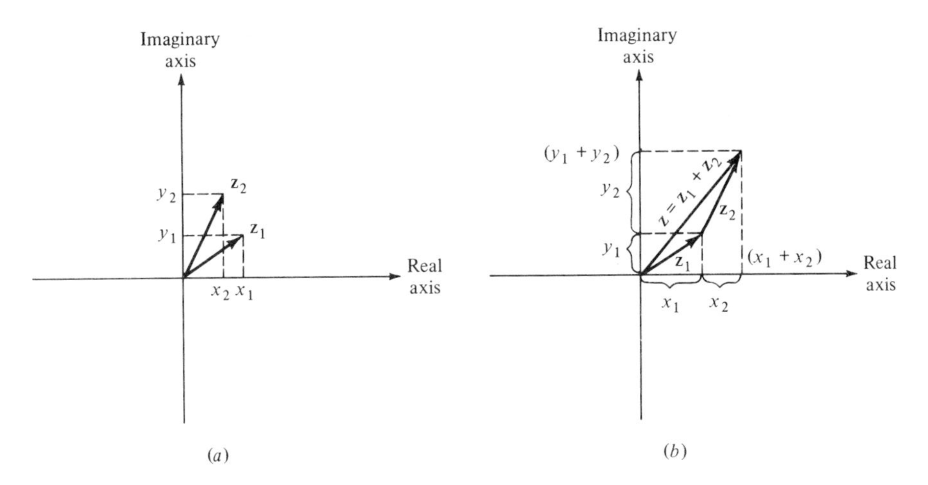

Fig. 18.1-7 Addition of complex numbers. (*a*) Complex plane representation of numbers; (*b*) graphical addition.

ample, so that it starts at the end of the other one rather than at the origin; its length and direction remain unchanged. This is shown in Fig. 18.1-7*b* where the sum $\mathbf{z} = \mathbf{z}_1 + \mathbf{z}_2$ is obtained by drawing a line from the origin to the end of the line representing $\mathbf{z}_2$. From the figure we see that

$$\begin{aligned} \mathbf{z} = \mathbf{z}_1 + \mathbf{z}_2 &= (x_1 + jy_1) + (x_2 + jy_2) \\ &= (x_1 + x_2) + j(y_1 + y_2) \end{aligned} \tag{18.1-16}$$

This shows that the real parts are added together and the imaginary parts are added together separately.

If $\mathbf{z}_1$ or $\mathbf{z}_2$ is in polar form it should be converted to rectangular form before performing the addition. If it is to be in polar form, the result should be converted after the addition. At this point, the reader should begin to notice, the frequent need for converting between the polar and rectangular forms; slide rules and electronic calculators intended for engineering applications generally have provisions to perform these conversions with relative ease.

SUBTRACTION

Subtraction is performed by separately subtracting the real and imaginary parts, just as they were separately added in addition. Thus

$$\begin{aligned} \mathbf{z} = \mathbf{z}_1 - \mathbf{z}_2 &= (x_1 + jy_1) - (x_2 + jy_2) \\ &= (x_1 - x_2) + j(y_1 - y_2) \end{aligned} \tag{18.1-17}$$

A graphical representation of this process is shown in Fig. 18.1-8, which indicates that the difference $\mathbf{z}_1 - \mathbf{z}_2$ is obtained by adding $-\mathbf{z}_2$ to $\mathbf{z}_1$. Note that $-\mathbf{z}_2$ has the same magnitude as $\mathbf{z}_2$ but the opposite direction. The addition and subtraction of complex numbers is illustrated in the following example.

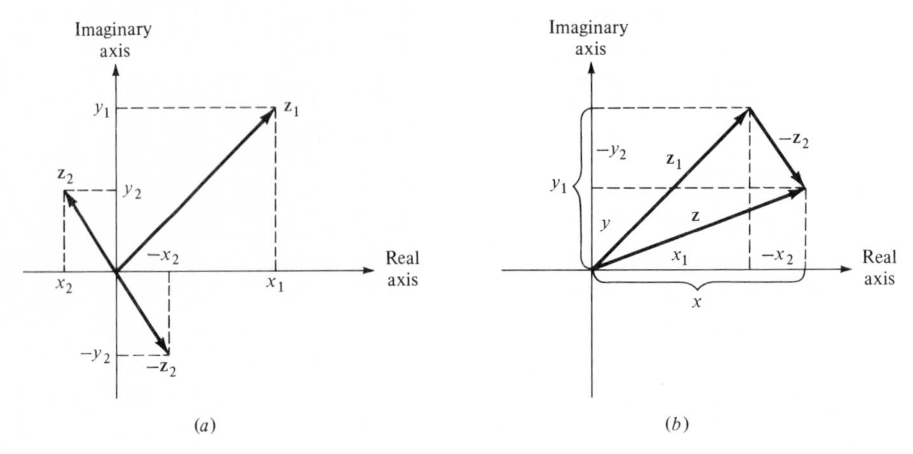

Fig. 18.1-8 Subtraction of complex numbers. (*a*) Complex plane representation of numbers; (*b*) graphical subtraction.

Example 18.1-5 Perform the indicated arithmetic operations on the following complex numbers expressing the result in the same form as the operands in each case.

a $(2) + (3 + j7) =$
b $(5 + j6) - (7 + j2) =$
c $2e^{j\pi/2} + 4e^{j\pi/6} =$
d $5e^{j60°} - 2e^{j300°} =$

The reader should try these on his own before looking at the solution.

Solution The first two are done using Eqs. (18.1-16) and (18.1-17):

a $(2) + (3 + j7) = (2 + 3) + (0 + j7) = 5 + j7$
b $(5 + j6) - (7 + j2) = (5 - 7) + j(6 - 2) = -2 + j4$
c In this one, the polar forms must first be converted to rectangular forms:

$$\mathbf{z} = 2e^{j\pi/2} + 4e^{j\pi/6} = \left(2\cos\frac{\pi}{2} + j2\sin\frac{\pi}{2}\right) + \left(4\cos\frac{\pi}{6} + j4\sin\frac{\pi}{6}\right)$$

$$= (0 + j2) + (3.46 + j2)$$

These expressions are added using Eq. (18.1-16) and the result is

$$\mathbf{z} = 3.46 + j4$$

Converting back to polar form,

$$\mathbf{z} = \sqrt{3.46^2 + 4^2}\, e^{j\,\arctan(4/3.46)} = 5.29\, e^{j\,\arctan 1.156} = 5.29\, e^{j\,49°}$$

d The steps in this part are the same as in the previous part except that the rectangular forms are subtracted using Eq. (18.1-17) rather than being added:

$$\begin{aligned} 5e^{j\,60°} - 2e^{j\,300°} &= (5 \cos 60° + j5 \sin 60°) \\ &\quad - (2 \cos 300° + j2 \sin 300°) \\ &= (2.5 + j4.33) - (1 - j1.73) \\ &= (2.5 - 1) + j(4.33 + 1.73) \\ &= 1.5 + j6.06 = \sqrt{1.5^2 + 6.06^2}\, e^{j \arctan(6.06/1.5)} \\ &= 6.23\, e^{j\,76°} \end{aligned}$$

////

18.1-3 Multiplication and Division of Complex Numbers

Complex numbers are most easily multiplied or divided if they are in polar form. The product of two complex numbers is

$$\begin{aligned} \mathbf{z} = \mathbf{z}_1\mathbf{z}_2 &= (r_1e^{j\theta_1})(r_2e^{j\theta_2}) \\ &= r_1r_2e^{j\theta_1}e^{j\theta_2} = r_1r_2e^{j(\theta_1+\theta_2)} \end{aligned} \qquad (18.1\text{-}18)$$

From Eq. (18.1-18) we see that the magnitude of the product of two complex numbers is the product of their magnitudes while the argument of the product of two complex numbers is the sum of their arguments.

The quotient of two complex numbers is

$$\begin{aligned} \mathbf{z} = \frac{\mathbf{z}_1}{\mathbf{z}_2} &= \frac{r_1e^{j\theta_1}}{r_2e^{j\theta_2}} = \frac{r_1}{r_2}e^{j\theta_1}\,e^{-j\theta_2} \\ &= \frac{r_1}{r_2}e^{j(\theta_1-\theta_2)} \end{aligned} \qquad (18.1\text{-}19)$$

From Eq. (18.1-19) we see that the magnitude of the quotient of two complex numbers is the quotient of their magnitudes while the argument of the quotient of two complex numbers is the difference between their arguments. Multiplication and division of complex numbers is illustrated in the following example.

Example 18.1-6 Perform the indicated operations, expressing the result in the same form as the operands.

a $25e^{j\pi} \times 3.5e^{-j\pi/4} =$
b $10e^{j\,20°} \div 4e^{j\,70°} =$
c $(3 + j4) \times (1 + j) =$
d $(5 - j2.5) \div (4 - j3) =$

Solution The first two parts are done using Eqs. (18.1-18) and (18.1-19):

a $25e^{j\pi} \times 3.5e^{-j(\pi/4)} = (25 \times 3.5)e^{j(\pi-\pi/4)} = 87.5e^{j(3\pi/4)}$

b $10e^{j\,20°} \div 4e^{j\,70°} = \dfrac{10}{4}\,e^{j(20°-70°)} = 2.5e^{-j\,50°}$

The third and fourth parts require conversions so the operations may be performed on the polar form:

c $(3 + j4) \times (1 + j) = 5e^{j\,53°} \times 1.414\,e^{j\,45°} = 7.07e^{j\,98°}$

$$= -0.98 + j7$$

d $(5 - j2.5) \div (4 - j3) = 5.59e^{-j\,26.6°} \div 5e^{-j\,37°}$

$$= \frac{5.59}{5}\,e^{j(-26.6°+37°)}$$

$$= 1.12e^{j\,10.4°} = 1.10 + j0.202$$ ////

Although it is easiest to multiply and divide numbers in polar form, these operations may also be done with numbers expressed in rectangular form. If both operands and the result are to be in rectangular form it may be easier to do the somewhat more complex arithmetic and not have to be bothered with the three conversions that would be required to use the polar form.

The formula for multiplication follows directly from the rules of algebra:

$$(x_1 + jy_1)(x_2 + jy_2) = x_1x_2 + jx_1y_2 + jx_2y_1 + j^2y_1y_2$$

since $j^2 = -1$, this results in

$$(x_1 + jy_1)(x_2 + jy_2) = (x_1x_2 - y_1y_2) + j(x_1y_2 + x_2y_1) \qquad (18.1\text{-}20)$$

The use of this formula is illustrated by the following example.

Example 18.1-7 Multiply $3 + j4$ by $1 + j$.

Solution

$$(3 + j4) \times (1 + j) = (3 \times 1 - 4 \times 1) + j(3 \times 1 + 4 \times 1)$$

$$= (3 - 4) + j(3 + 4) = -1 + j7$$

Compare this with Example 18.1-6, part *c*. The slight discrepancy is due to roundoff errors associated with the slide rule. ////

The division of complex numbers in rectangular form is presented in the following section because it requires the use of complex conjugate numbers.

18.1-4 Complex Conjugate Numbers

The complex conjugate of a complex number $\mathbf{z}$ is denoted $\mathbf{z}^*$. Graphically, it is the mirror image of $\mathbf{z}$ in the complex plane with the real axis being the axis of symmetry as shown in Fig. 18.1-9. From this figure we find that if $\mathbf{z} = re^{j\theta} = x + jy$, then

$$\mathbf{z}^* = re^{-j\theta} = x - jy \tag{18.1-21}$$

The product of a complex number with its complex conjugate is the square of the magnitude of the complex number as we prove in the following:

$$\mathbf{z}\mathbf{z}^* = (re^{j\theta})(re^{-j\theta}) = r^2e^{j(\theta-\theta)} = r^2 = |\,\mathbf{z}\,|^2 \tag{18.1-22}$$

The same result may be obtained using the rectangular form:

$$\mathbf{z}\mathbf{z}^* = (x + jy)(x - jy) = [x^2 - (-y^2)] + j(-xy + xy)$$

$$= x^2 + y^2 = |\,\mathbf{z}\,|^2$$

Using this result we may derive the formula for the quotient of two complex numbers expressed in rectangular form; the method is based on elimination of the complex denominator by multiplying it by its complex conjugate. In order to not change the value of the fraction, the numerator is multiplied by the same

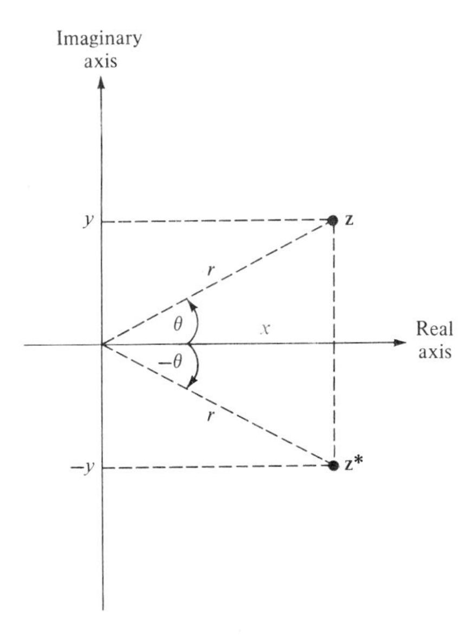

Fig. 18.1-9 Complex conjugate of a complex number.

value as follows:

$$\frac{\mathbf{z}_1}{\mathbf{z}_2} = \frac{\mathbf{z}_1}{\mathbf{z}_2} \times \frac{\mathbf{z}_2^*}{\mathbf{z}_2^*} = \frac{\mathbf{z}_1\mathbf{z}_2^*}{|\mathbf{z}_2|^2} = \frac{(x_1 + jy_1)(x_2 - jy_2)}{x_2^2 + y_2^2}$$

$$= \frac{x_1x_2 + y_1y_2}{x_2^2 + y_2^2} + j\frac{x_2y_1 - x_1y_2}{x_2^2 + y_2^2} \qquad (18.1\text{-}23)$$

Division of complex numbers in rectangular form is illustrated in the following example.

Example 18.1-8 Divide $5 - j2.5$ by $4 - j3$.

Solution

$$\frac{5 - j2.5}{4 - j3} = \frac{5 - j2.5}{4 - j3}\,\frac{4 + j3}{4 + j3}$$

$$= \frac{(5 \times 4 - j^2 2.5 \times 3) + j(5 \times 3 - 2.5 \times 4)}{4^2 + 3^2}$$

$$= \frac{(20 + 7.5) + j(15 - 10)}{16 + 9}$$

$$= \frac{27.5 + j5}{25} = 1.1 + j0.2$$

Compare this result with that of Example 18.1-6, part *d*. The limited precision of the slide rule accounts for the discrepancy. ////

18.2 PHASORS

Before continuing with the material in this chapter the reader should review Sec. 2.4 on the sinusoidal signal. The definitions of the terms amplitude, frequency (f), angular frequency (ω), and period (T) should be clearly understood as should the relationships

$$f = \frac{1}{T}$$

and

$$\omega = 2\pi f$$

In addition, the reader should be familiar with the following units: seconds, hertz, and radians per second.

EXPONENTIAL FORM OF SINUSOIDAL SIGNAL—THE PHASOR

Using Euler's identity, Eq. (18.1-13), we may relate the sinusoidal signals we are concerned with in this chapter to the exponential function. If we multiply both sides of Eq. (18.1-13) by V_m and let $\theta = \omega t + \varphi$, we obtain

$$V_m e^{j(\omega t+\varphi)} = V_m \cos(\omega t + \varphi) + jV_m \sin(\omega t + \varphi) \tag{18.2-1}$$

The expression $V_m e^{j(\omega t+\varphi)}$ is a complex function and therefore cannot be a physically obtainable signal. However, its real part,

$$Re(V_m e^{j(\omega t+\varphi)}) = V_m \cos(\omega t + \varphi) = v(t) \tag{18.2-2}$$

does represent a physical signal since it is a real sinusoidal function.

The complex function associated with the voltage signal $v(t) = V_m \cos(\omega t + \varphi)$ is $V_m e^{j(\omega t+\varphi)}$. This can be rewritten in the form

$$\begin{aligned} V_m e^{j(\omega t+\varphi)} &= V_m e^{j\varphi} e^{j\omega t} \\ &= \mathbf{V}_m e^{j\omega t} \end{aligned} \tag{18.2-3}$$

where

$$\mathbf{V}_m = V_m e^{j\varphi} = V_m \angle \varphi \tag{18.2-4}$$

Let us refer to Fig. 18.2-1 and consider the "complex" voltage $V_m \angle \varphi$. (Note that the complex plane has been rotated 90° from its usual orientation.) Suppose we let this vector rotate counterclockwise with constant angular velocity ω. Then the angle it makes with the real axis will be a function of time, i.e., if we denote the angle at time t by $\psi(t)$, then

$$\psi(t) = \omega t + \varphi$$

Consider the projection of this rotating vector on the positive real axis. This is

$$v(t) = V_m \cos \psi(t) = V_m \cos(\omega t + \varphi) \tag{18.2-5}$$

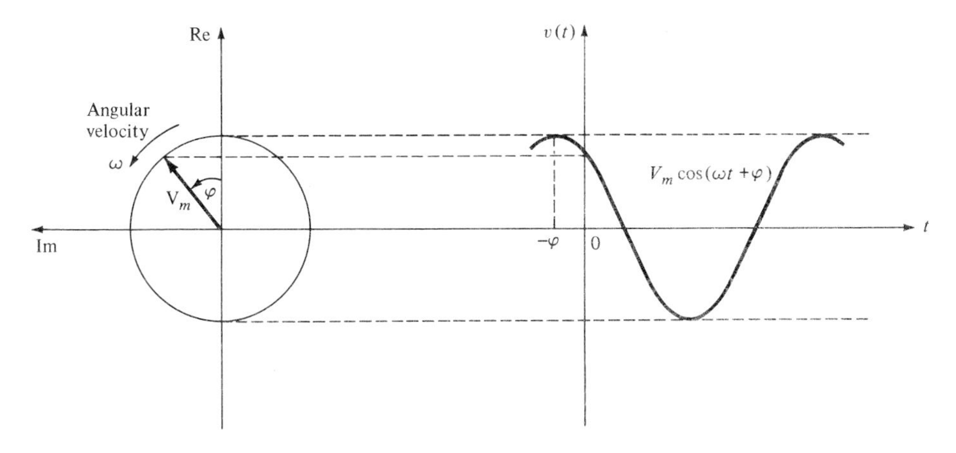

Fig. 18.2-1 Projection of rotating vector.

which is exactly the same as Eq. (2.4-2), the equation for the general sinusoid. The complex number $\mathbf{V}_m$ is called a phasor and should be thought of as the amplitude and phase of the rotating vector $\mathbf{V}_m e^{j\omega t}$ at $t = 0$. The key to the phasor method of ac circuit analysis is that the phasor of Eq. (18.2-4) contains all the information in Eq. (18.2-5) except for ω, in a shorthand form. Since, in a linear circuit driven by a sinusoidal source of angular frequency ω, all currents and voltages are of the same frequency, the information in the phasor voltages and currents is all that is required.

Example 18.2-1

a Find the phasor representation of

$$i(t) = 5 \sin (200\,\pi t + 30°) \text{ mA}$$

b Given a phasor voltage $\mathbf{V}_m = 15 \angle 30°$ V at a frequency of 60 Hz, write the expression for $v(t)$.

Solution

a Convert the sine function to a cosine function using the identity $\sin x = \cos (x - 90°)$.

$$i(t) = 5 \cos (200\pi t + 30° - 90°)$$

$$= 5 \cos (200\,\pi t - 60°) \text{ mA}$$

Then the phasor current is, using Eq. (18.2-2) and (18.2-4),

$$\mathbf{I}_m = 5 \angle -60° \text{ mA}$$

b Using Eq. (18.2-4) and then Eq. (18.2-2),

$$v(t) = 15 \cos (2\pi \times 60t + 30°)$$

$$= 15 \cos (120\pi t + 30°) \text{ V}$$

////

In ac circuit analysis we often have to add or subtract sinusoids of the same frequency but different phase angles. This sum (or difference) can be found by finding the sum (or difference) of their phasor representations, a considerably easier process than using the trigonometric form as shown in the following example.

Example 18.2-2 Given the two voltages,

$$v_1(t) = 10 \cos (\omega t + 30°)$$

$$v_2(t) = 15 \cos (\omega t + 45°)$$

Find $v(t) = v_1(t) + v_2(t)$ using (*a*) trigonometric identities and (*b*) phasors.

Solution

a Since $\cos(x + y) = \cos x \cos y - \sin x \sin y$,

$$\begin{aligned} v_1(t) &= 8.7 \cos \omega t - 5 \sin \omega t \\ v_2(t) &= 10.6 \cos \omega t - 10.6 \sin \omega t \\ \hline v(t) &= 19.3 \cos \omega t - 15.6 \sin \omega t \end{aligned}$$

To combine these terms we use the identity

$$a \cos \omega t + b \sin \omega t = \sqrt{a^2 + b^2} \cos\left(\omega t - \arctan \frac{b}{a}\right)$$

Thus

$$\begin{aligned} v(t) &= \sqrt{(19.3)^2 + (15.6)^2} \cos\left(\omega t - \arctan \frac{-15.6}{19.3}\right) \\ &= 24.8 \cos(\omega t + 39°) \end{aligned}$$

b Using phasors, we write

$$\begin{aligned} \mathbf{V} = \mathbf{V}_1 + \mathbf{V}_2 &= 10 \angle 30° + 15 \angle 45° \\ &= 8.7 + j5 + 10.6 + j10.6 \\ &= 19.3 + j15.6 \\ &= 24.8 \angle 39° \end{aligned}$$

Then converting back to the time domain,

$$v(t) = 24.8 \cos(\omega t + 39°)$$

Clearly, the phasor method is easier to apply. ////

18.3 SINUSOIDAL IMPEDANCE

In this section we find the sine-wave response of the capacitor. This will illustrate the difference between dc and ac circuit calculations.

Consider that the voltage across a capacitor is given by

$$v(t) = V_m \cos \omega t \tag{18.3-1}$$

The current is then

$$i(t) = C \frac{dv}{dt} = -\omega C V_m \sin \omega t \tag{18.3-2}$$

where we have used the formula for the derivative of cos ωt (Sec. 14.1):

$$\frac{d}{dt}\cos\omega t = -\omega\sin\omega t \qquad (18.3\text{-}3)$$

The waveforms of the current and voltage are shown in Fig. 18.3-1*a*.

The utility of phasor calculations lies in the fact that all of the laws and theorems we learned in dc circuit analysis carry over directly to ac circuit analysis. In particular Ohm's law and the two Kirchhoff laws are valid in phasor form as the following calculation carried through using phasors will show.

From Eq. (18.2-2) we have

$$v(t) = \text{Re}\,\mathbf{V}_m e^{j\omega t} \qquad (18.3\text{-}4)$$

where the phasor voltage is

$$\mathbf{V}_m = V_m \angle 0°$$

The current is

$$i(t) = C\frac{d}{dt}\text{Re}\,V_m e^{j\omega t}$$

$$= \text{Re}\,C\frac{d}{dt}V_m e^{j\omega t}$$

$$= \text{Re}\,j\omega C V_m e^{j\omega t} \qquad (18.3\text{-}5)$$

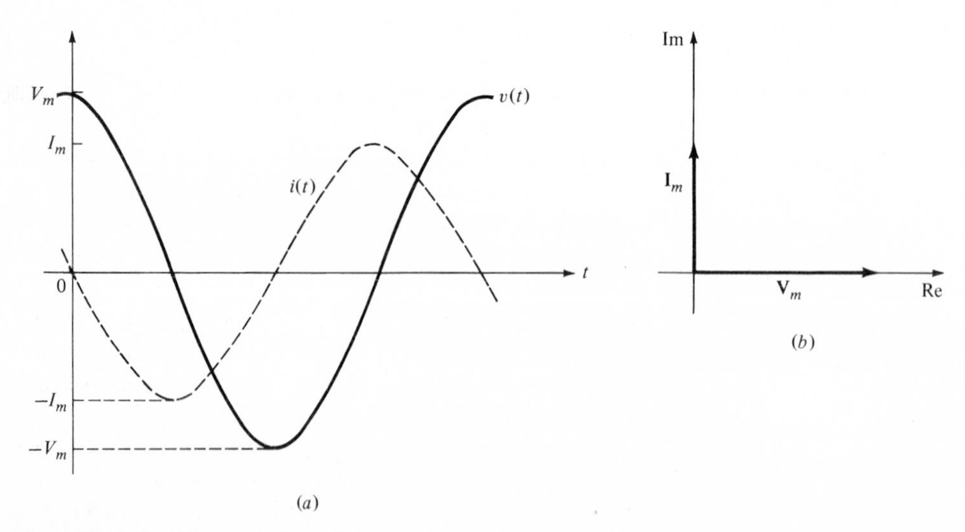

Fig. 18.3-1 Sinusoidal response of a capacitor. (*a*) Waveforms; (*b*) phasor diagram.

Using the fact that $j = e^{j\pi/2}$, Eq. (18.3-5) becomes

$$i(t) = \text{Re}\ \omega C V_m e^{j(\omega t+\pi/2)}$$

$$= \omega C V_m \cos\left(\omega t + \frac{\pi}{2}\right)$$

$$= -\omega C V_m \sin \omega t \qquad (18.3\text{-}6)$$

and we get exactly the same result as in (18.3-2). Most of the steps required can be omitted if we use Ohm's law in phasor form, that is,

$$\mathbf{V}_m = \mathbf{Z}\mathbf{I}_m \qquad (18.3\text{-}7)$$

where $\mathbf{Z}$, called the *impedance*, has the dimension of ohms and plays the same role as resistance in our original statement of the law, except that $\mathbf{Z}$ is complex. The reciprocal of $\mathbf{Z}$ is called the admittance $\mathbf{Y} = 1/\mathbf{Z}$. Admittance has the units of mhos and is analogous to conductance.

From Eq. (18.3-5) the phasor current is

$$\mathbf{I}_m = j\omega C \mathbf{V}_m \qquad (18.3\text{-}8)$$

Comparing Eqs. (18.3-7) with (18.3-8), we find that for the capacitor,

$$\mathbf{Z}_c = \frac{1}{j\omega C} = \frac{-j}{\omega C} = \frac{1}{\omega C} \angle -90^\circ \qquad (18.3\text{-}9)$$

The phasor voltage and current are usually plotted in the complex plane as shown in Fig. 18.3-1*b*. Except for the frequency, this plot contains all the information in the time-function plot of Fig. 18.3-1*a*.

Example 18.3-1 A 5-pF capacitor carries a current $i(t) = 2 \cos (\omega t + 30^\circ)$ mA with $\omega = 10^6$ rad/s. Find the voltage across the capacitor in phasor form and as a time function.

Solution

$$\mathbf{Z}_c = \frac{1}{j\omega C} = \frac{1}{\omega C} \angle -90^\circ = \frac{1 \angle -90^\circ}{(10^6)(5)(10^{-12})}$$

$$= 0.2 \times 10^6 \angle -90^\circ\ \Omega$$

From the problem statement,

$$\mathbf{I}_m = 2 \times 10^{-3} \angle 30^\circ \text{ A}$$

Then the phasor voltage is

$$\mathbf{V}_m = \mathbf{Z}_c\mathbf{I}_m = (0.2 \times 10^6 \angle -90^\circ)(2 \times 10^{-3} \angle 30^\circ)$$

$$= 0.4 \times 10^3 \angle -60^\circ \text{ V}$$

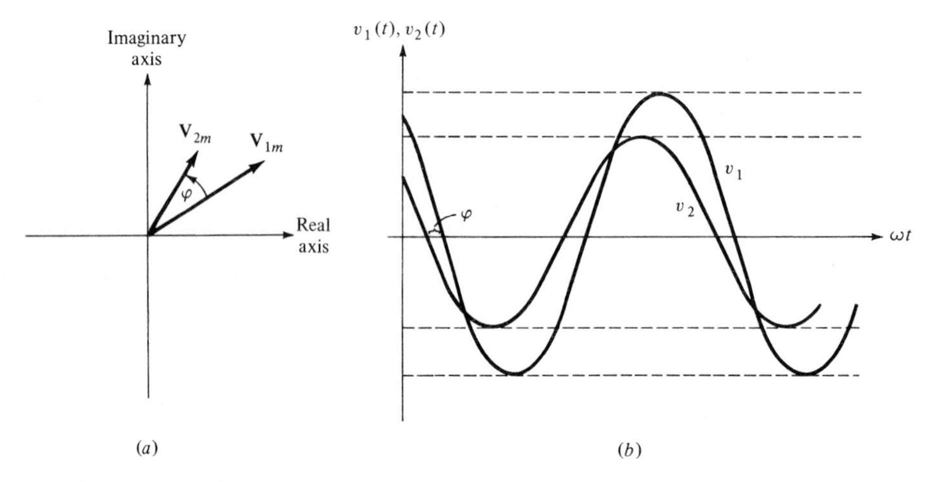

Fig. 18.3-2 Phase angle between two sinusoidal signals. (*a*) Phasor diagram; (*b*) graphs of time functions.

and the time function is

$$v(t) = 400 \cos (10^6 t - 60°) \text{ V} \qquad ////$$

The impedance of an inductance can be found in the same way as that of the capacitance (Prob. 18.3-3). The result is

$$\mathbf{Z}_L = j\omega L = \omega L \angle 90° \tag{18.3-10}$$

PHASE LEAD AND LAG

The phase angle between two sinusoidal signals of the same frequency is the distance between corresponding points on the two signals measured in degrees or radians. The corresponding points may be the positive peaks, the negative peaks, the zero crossing which precedes the positive portion, or any other convenient point.

The phasor diagram shows the phase angle even more clearly; it is the angle between the two phasors. Two phasors $\mathbf{V}_{1m}$ and $\mathbf{V}_{2m}$ are shown in Fig. 18.3-2*a*; the angle between them is φ. Recall from Sec. 18.2 that the phasor corresponds to the "rotating vector" representation of the sinusoidal signal at $t = 0$; as t increases the vectors rotate counterclockwise. With this in mind, we say that phasor $\mathbf{V}_{2m}$ is *leading* phasor $\mathbf{V}_{1m}$ by a phase angle of φ since it is further counterclockwise. An alternative statement to describe the situation is that phasor $\mathbf{V}_{1m}$ is lagging phasor $\mathbf{V}_{1m}$ by φ.

The situation is pictured for the time functions in Fig. 18.3-2*b*. In these graphs we see that a point on the $v_2(t)$ curve occurs $\omega t = \varphi$ before the corresponding point on $v_1(t)$; thus $v_2(t)$ leads $v_1(t)$ by φ or $v_1(t)$ lags $v_2(t)$ by φ.

IMPEDANCE COMBINATION

Series and parallel impedances may be combined as series and parallel resistances are. In general, the value of an impedance is a complex number. The real part of an impedance is called the *resistive* component and is generally symbolized by the letter R. The imaginary part of an impedance is called the *reactance* and is generally symbolized by the letter X. Thus

$$\mathbf{Z} = R + jX \tag{18.3-11}$$

which results in the equivalent circuit of Fig. 18.3-3. It is important to remember that this is an equivalent circuit—there is in general no physical element corresponding to either R or X.

The inductance and capacitance are purely reactive elements since the real parts of their impedances are zero. For the inductance element we find, from Eq. (18.3-10), the inductive reactance to be

$$X_L = \omega L = 2\pi f L \tag{18.3-12}$$

and the capacitive reactance is, from Eq. (18.3-9),

$$X_C = \frac{-1}{\omega C} = \frac{-1}{2\pi f C} \tag{18.3-13}$$

The following example illustrates computation of the reactance and impedance for energy storage elements.

Example 18.3-2 Compute the reactance and impedance for

a a 5-mH inductance

b a 0.01-μF capacitance

both for a 10-kHz signal.

Solution We compute the angular frequency

$$\omega = 2\pi f = 2\pi\ (10^4) = 62{,}800 \text{ rad/s}$$

a $X_L = \omega L = (62{,}800)(5 \times 10^{-3}) = 314\ \Omega$

$$\mathbf{Z}_L = jX_L = j\,314\ \Omega$$

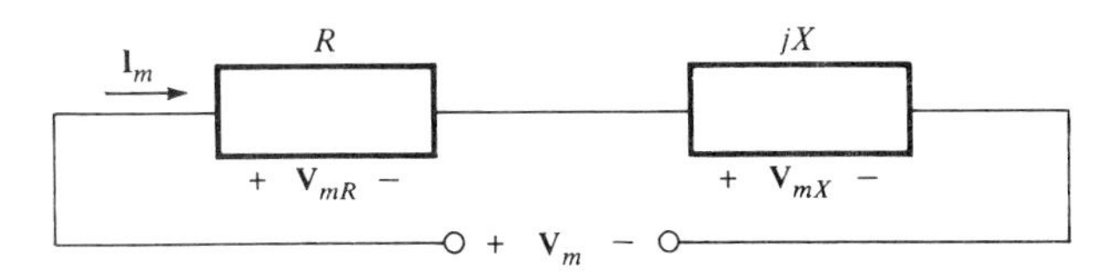

Fig. 18.3-3 Equivalent circuit of sinusoidal impedance.

b $X_C = \frac{-1}{\omega C} = \frac{-1}{(62{,}800)\,(0.01 \times 10^{-6})} = -1{,}590\ \Omega$

$\mathbf{Z}_C = jX_C = -j\,1{,}590\ \Omega$ ////

The current and voltage for the reactive elements have a phase angle of 90° between them since $\mathbf{Z} = jX$ and multiplication by j corresponds to a rotation of 90° counterclockwise. Specifically, since

$$\mathbf{V}_L = j\omega L \mathbf{I}_L$$

the voltage leads the current in an inductance. For a capacitance

$$\mathbf{V}_C = -j\left(\frac{1}{\omega C}\right)\mathbf{I}_C$$

and therefore the voltage lags the current for a capacitance. These relationships were previously illustrated in Fig. 18.3-1*b* for the capacitance.

IMPEDANCE CALCULATIONS

The impedance of circuits containing multiple elements is computed using the rules for series and parallel impedances. For two impedances $\mathbf{Z}_1$ and $\mathbf{Z}_2$ in series, the equivalent impedance is

$$\mathbf{Z} = \mathbf{Z}_1 + \mathbf{Z}_2 \tag{18.3-14}$$

and if the two impedances $\mathbf{Z}_1$ and $\mathbf{Z}_2$ are in parallel, the equivalent impedance is

$$\mathbf{Z} = \frac{\mathbf{Z}_1\mathbf{Z}_2}{\mathbf{Z}_1 + \mathbf{Z}_2} \tag{18.3-15}$$

The use of these formulas is illustrated in the following example.

Example 18.3-3 Find the impedance of the circuits of Fig. 18.3-4 if the voltage across each circuit is a 60-Hz sine wave.

Solution The angular frequency is

$$\omega = 2\pi f = 2\pi(60) = 377 \text{ rad/s}$$

a $\mathbf{Z}_a = j\omega L = j(377)(5) = j\,1.88\ \text{k}\Omega$

b $\mathbf{Z}_b = \frac{-j}{\omega C} = \frac{-j}{377 \times 10^{-6}} = -j\,2.65\ \text{k}\Omega$

c Using the results of *a* and the formula for impedances in series,

$$\mathbf{Z}_c = R + j\omega L = (2 + j\,1.88)\ \text{k}\Omega$$

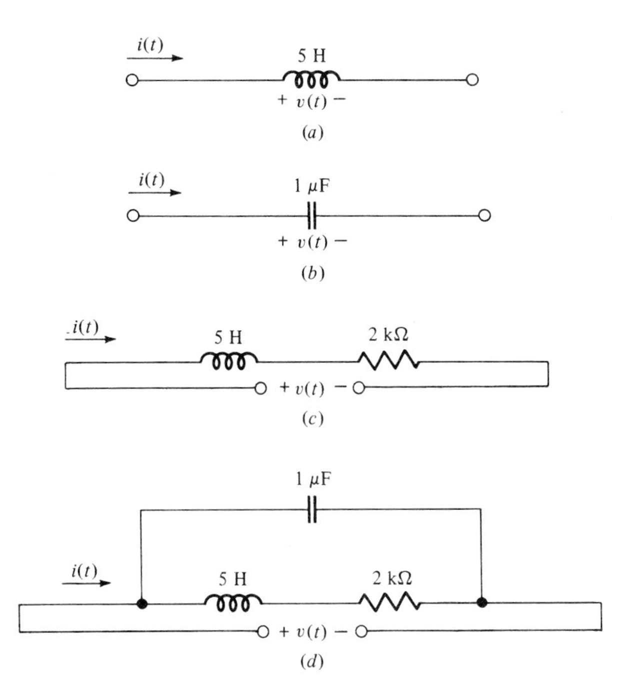

Fig. 18.3-4 Circuits for Example 18.3-3.

d Using the results of *b* and *c* and the formula for parallel impedances,

$$\mathbf{Z}_d = \frac{\mathbf{Z}_b\mathbf{Z}_c}{\mathbf{Z}_b + \mathbf{Z}_c} = \frac{(-j2.65)(2 + j1.88)}{2 + j1.88 - j2.65} = \frac{4.98 - j5.30}{2 - j0.77}$$

We rationalize the result, i.e., eliminate the complex denominator, by multiplying the numerator and denominator by the complex conjugate of the denominator:

$$\mathbf{Z}_d = \frac{4.98 - j5.30}{2 - j\,0.77}\,\frac{2 + j\,0.77}{2 + j\,0.77} = \frac{14.0 - j\,6.77}{4.59} = 3.05 - j\,1.47\text{ k}\Omega$$

////

It is frequently necessary to connect a resistive load R_L to a circuit which may be represented by a Thevenin voltage source V_T in series with a Thevenin resistance R_T. In Sec. 18.5-3, it is shown that maximum power is transferred from the source to the load if $R_L = R_T$. If R_L and R_T are both fixed, this *matching of impedance levels* may be accomplished by the use of an ideal transformer (Sec. 15.6-1), as illustrated in the following example.

Example 18.3-4 An impedance of 4 Ω is to be connected to a circuit whose Thevenin resistance is 100 Ω through an ideal transformer as shown

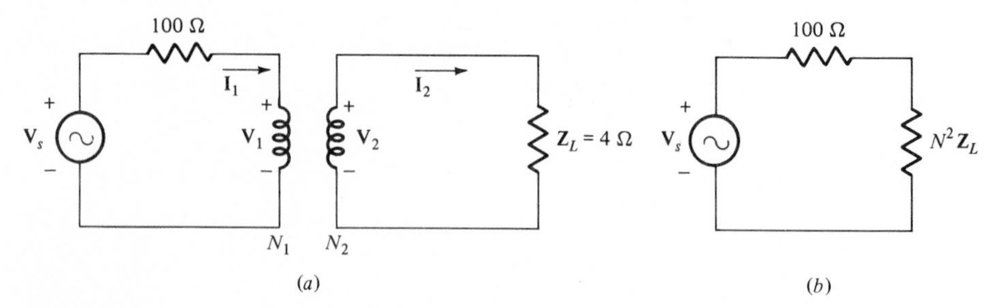

Fig. 18.3-5 Impedance matching. (*a*) Circuit with ideal transformer; (*b*) equivalent circuit with reflected impedance.

in Fig. 18.3-5*a*. Find the turns ratio of the transformer so that maximum power will be transferred from the source to the load.

Solution For the ideal transformer with primary to secondary turns ratio $N_1{:}N_2$, we have, from Sec. 15.6-1,

$$\mathbf{V}_1 = N\mathbf{V}_2$$

and

$$\mathbf{I}_1 = \frac{1}{N}\mathbf{I}_2$$

where $N = N_1/N_2$. If an impedance $\mathbf{Z}_L$ is connected to the secondary, then $\mathbf{V}_2 = \mathbf{Z}_L\mathbf{I}_2$ and the equivalent impedance seen looking into the primary side of the transformer, often called the reflected impedance, is

$$\mathbf{Z}_1 = \frac{\mathbf{V}_1}{\mathbf{I}_1} = \frac{N\mathbf{V}_2}{\frac{1}{N}\mathbf{I}_2} = N^2\frac{\mathbf{Z}_L\mathbf{I}_2}{\mathbf{I}_2} = N^2\mathbf{Z}_L \qquad (18.3\text{-}16)$$

as shown in Fig. 18.3-5*b*

In the problem, this impedance must equal the Thevenin resistance of 100 Ω. Since $\mathbf{Z}_L = 4\ \Omega$, we have, from Eq. (18.3-16),

$$4\,N^2 = 100$$

$$N^2 = 25$$

$$N = 5$$

Thus the turns ratio is $N_1/N_2 = 5{:}1$. Finally, since the ideal transformer is a lossless device, all the power into it is transmitted to the load, $\mathbf{Z}_L$. Note that this technique cannot be used with dc sources, since transformers operate only for time-varying signals. ////

18.4 SINUSOIDAL STEADY-STATE RESPONSE

The computation of the steady-state sinusoidal response of a network is similar to the computation of the response of a resistive network to a dc input. Conversions of the input signals to phasors and the phasor result back to time signals must be performed in addition to the usual steps. The sequence of steps for a network all of whose inputs are of the same frequency is as follows:

1 Convert all input signals to phasors.
2 Replace all elements by their equivalent impedances.
3 Compute the phasor outputs.
4 Convert the phasor outputs to time signals if required.

These steps are illustrated in the following example.

Example 18.4-1 Determine the steady-state current, $i(t)$, for the circuit of Fig. 18.4-1a.

Solution An equivalent circuit for that in Fig. 18.4-1a, in which the inductance is replaced by its impedance, $\mathbf{Z} = j\omega L$, and the sinusoidal source by its phasor equivalent is shown in Fig. 18.4-1b. Using Ohm's law for this circuit results in

$$\mathbf{I}_m = \frac{20 \angle 0^\circ}{j\,100} = 0.2 \angle -90^\circ$$

Therefore $i(t) = 0.2 \cos (10^3\, t - 90^\circ)$ A is the steady-state current. ////

The next example is slightly more complex; in it we demonstrate the use of Kirchhoff's voltage law for sinusoidal steady-state analysis.

Example 18.4-2 Find the steady-state current $i(t)$ and the voltage $v(t)$ in the circuit of Fig. 18.4-2a.

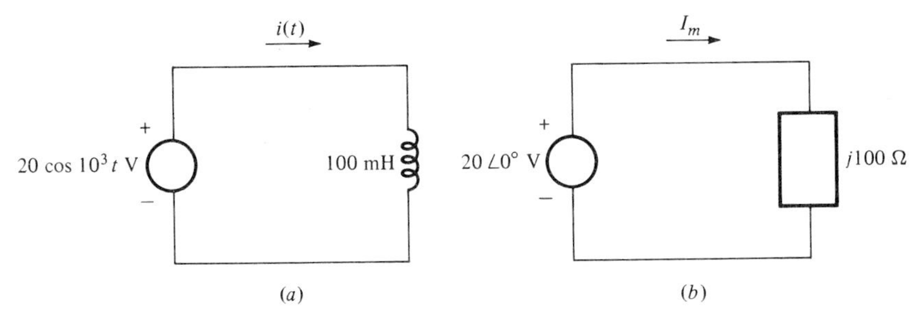

Fig. 18.4-1 Circuits for Example 18.4-1. (a) Original circuit; (b) equivalent impedance circuit.

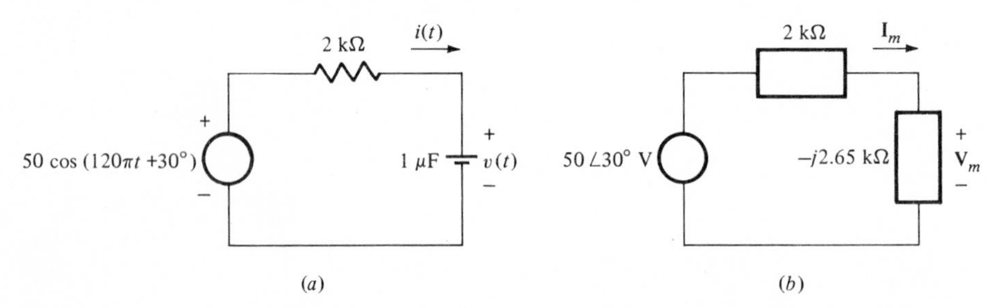

Fig. 18.4-2 Circuits for Example 18.4-2. (*a*) Original circuit; (*b*) equivalent impedance circuit.

Solution The capacitive impedance is

$$\mathbf{Z}_C = \frac{-j}{\omega C} = \frac{-j}{120\pi\ (10^{-6})} = -j\,2.65\ \text{k}\Omega$$

An equivalent circuit for the circuit of Fig. 18.4-2*a* in which phasors and impedances are used is shown in Fig. 18.4-2*b*. Using KVL,

$$\mathbf{V}_R + \mathbf{V}_C = \mathbf{V}_{\text{source}}$$

$$2\mathbf{I}_m - j\,2.65\,\mathbf{I}_m = 50\ \angle 30^\circ$$

$$\mathbf{I}_m = \frac{50\ \angle 30^\circ}{2 - j\,2.65} = \frac{50\ \angle 30^\circ}{3.32\ \angle -53^\circ} = 15.1\ \angle 83^\circ\ \text{mA}$$

and using Ohm's law,

$$\mathbf{V}_m = -j\,2.65\,\mathbf{I}_m = (2.65\ \angle -90^\circ)\,(15.1\ \angle 83^\circ) = 40.0\ \angle -7^\circ\ \text{V}$$

Therefore

$$i(t) = 15.1 \cos\,(120\pi\,t + 83^\circ)\ \text{mA}$$

and

$$v(t) = 40.0 \cos\,(120\pi\,t - 7^\circ)\ \text{V}$$

////

Using the other circuit-analysis laws previously derived with phasors and impedances it is possible to analyze even the most complex networks. The complex arithmetic does make the process rather tedious, however, and as a result computerized methods are often used.

18.4-1 Ac Analysis Using ECAP

As noted previously, steady-state ac calculations are carried through basically like dc calculations except that all the quantities dealt with are complex rather than real, so that both magnitudes and angles must be found. The arithmetic can become very tedious, and the chance of making an error becomes high when the slide rule is used.

The ECAP program can be used to perform steady-state ac analysis of circuits in essentially the same manner as it does for dc circuits. We illustrate this by means of the example which follows.

Example 18.4-3 Use ECAP to find the node voltage and branch currents in the circuit of Fig. 18.4-3*a*.

Solution Again an ECAP circuit with previously described branch and node numbering is prepared, as shown in Fig. 18.4-3*b*. The printout which includes the program is shown in Fig. 18.4-4*a*, and differences between it and previous examples are as follows:

Line 1 Tells the computer that an ECAP ac analysis is desired.
Line 2 The resistor and current source constitute branch 1. The resistance value is given in standard computer floating-point form. The number following the E indicates the power of 10 by which the number preceding the E is to be multiplied. Thus $R = 1\text{E}03$ means $R = 1 \times 10^3$ $(= 1\ \text{k}\Omega)$. The current source is specified as $1\text{E} - 03 = 1 \times 10^{-3}$ A in magnitude followed by a

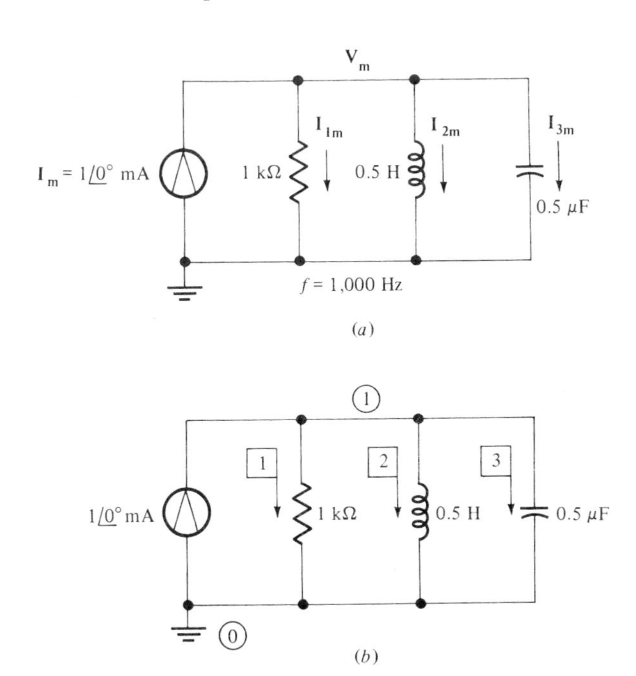

Fig. 18.4-3 ECAP ac analysis. (*a*) Circuit for Example 18.4-3; (*b*) ECAP circuit.

```
          1        AC ANALYSIS
          2   B1   N (1,0), R=1E 03, I=1E-03/0
              B2   N (1,0), L=0.5
          3   B3   N (1,0), C=0.5E-06
          4        FREQUENCY=1000
                   PRINT, NV, CA
                   EXECUTE

FREQ = 0.99999976E 03

 NODES                          NODE VOLTAGES

MAG 1-  1   0.33387309E 00
PHA        -0.70495956E 02


BRANCHES                      ELEMENT CURRENTS

MAG 1-  3   0.33387309E-03   0.10627507E-03   0.10488930E-02
PHA        -0.70495956E 02  -0.16049594E 03   0.19504013E 02
```

(a)

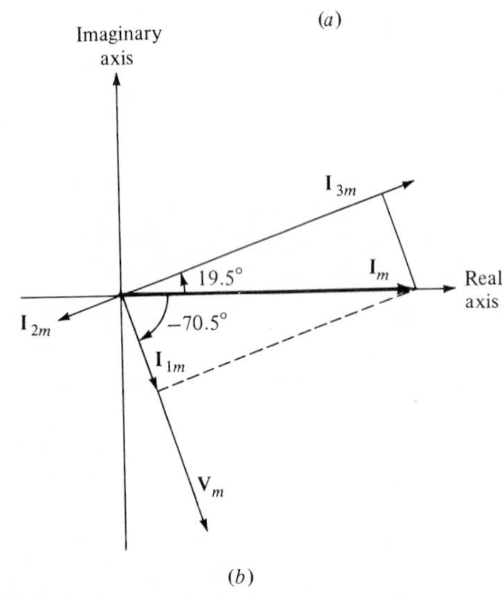

(b)

Fig. 18.4-4 ECAP ac analysis. (*a*) Computer printout for Example 18.4-3; (*b*) phasor diagram.

slash (/) and a phase angle of zero degrees. The phase angle of all sources must be specified (in degrees) even if they are zero.

Line 3 Floating-point notation is used to specify the capacitance

$$C = 0.5\text{E} - 06 = 0.5 \times 10^{-6} = 0.5\ \mu\text{F}$$

Line 4 A frequency card must be included for ac analysis. The frequency must be nonzero and is specified in units of hertz.

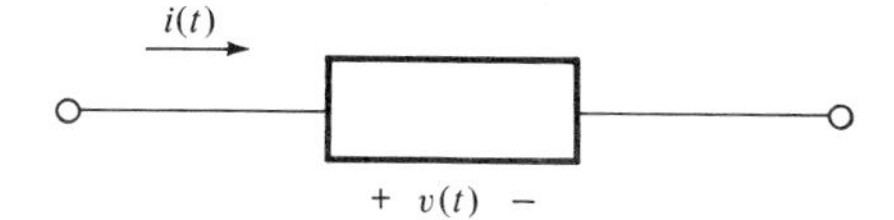

Fig. 18.5-1 Reference directions for computing power delivered to an element.

The results printed after the program include both magnitude and phase (in degrees). A phasor diagram of the results is shown in Fig. 18.4-4*b*.
////

18.5 POWER IN AC CIRCUITS

In this section we will discuss the computation of average power delivered to a circuit during sinusoidal steady-state operation. The instantaneous power delivered to the element shown in Fig. 18.5-1 is $p(t) = v(t)\ i(t)$. In general, it is the average power which is of primary concern. If the voltage and current are repetitive signals with period T, the average power is given by

$$P = \frac{1}{T}\int_0^T p(t)\ dt \tag{18.5-1}$$

The integral in this formula is the area under the curve of instantaneous power between 0 and T, shown by the cross hatched area in Fig. 18.5-2. Dividing this area by the time interval T results in a constant value, P, which forms a rectangle over the interval from 0 to T enclosing the same area. By definition, P is the average value of $p(t)$.

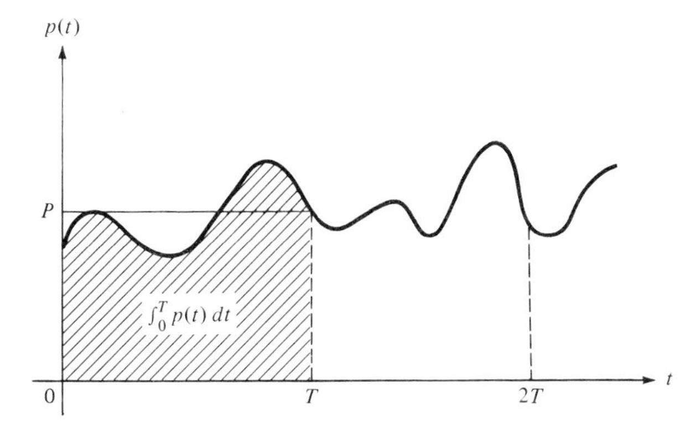

Fig. 18.5-2 Computation of average power.

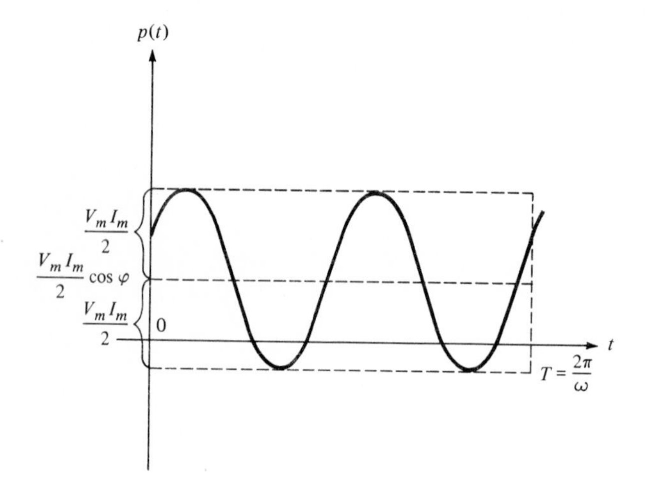

Fig. 18.5-3 Instantaneous power in ac circuit.

In this chapter we are specifically interested in sinusoidal signals. Therefore, we substitute $v(t) = V_m \cos(\omega t + \varphi)$ and $i(t) = I_m \cos \omega t$ into the formula for instantaneous power:

$$p(t) = V_m \cos(\omega t + \varphi) I_m \cos \omega t \tag{18.5-2}$$

The trigonometric identity

$$\cos x \cos y = \frac{1}{2}[\cos(x - y) + \cos(x + y)] \tag{18.5-3}$$

is applied to Eq. (18.5-2) resulting in

$$p(t) = \frac{V_m I_m}{2}[\cos \varphi + \cos(2\omega t + \varphi)] \tag{18.5-4}$$

which is plotted in Fig. 18.5-3. The instantaneous power consists of a constant component, $(V_m I_m/2) \cos \varphi$, plus a sinusoidal component, $(V_m I_m/2) \cos(2\omega t + \varphi)$, whose frequency is twice that of the current or voltage. When the instantaneous power is substituted into Eq. (18.5-1) to compute its average value, the sinusoidal component drops out since its average value is zero and we are left with

$$P = \frac{V_m I_m}{2} \cos \varphi \tag{18.5-5}$$

where φ is the phase angle between the voltage and current.

18.5-1 Effective Value

Equation (18.5-5) gives the average power delivered to an element by sinusoidal voltages and currents in terms of the peak values of these signals. In general, the amount of power delivered to an element is a function of the waveforms of the signals as well as the peak amplitudes. In order to avoid the use of different formulas for different waveforms, the *effective* value of a signal, rather than the peak value, is usually used. The effective value of a signal is, by definition, the value of a dc signal which when applied to a resistive load dissipates the same average power as the original signal. If the signal is a current, $i(t)$, and we are concerned with an interval from 0 to T, then according to the definition above,

$$P_{\text{ac}} = P_{\text{dc}}$$

$$\frac{1}{T}\int_0^T Ri^2(t)\,dt = RI_{\text{eff}}^2$$

$$\frac{1}{T}\int_0^T i^2(t)\,dt = I_{\text{eff}}^2$$

$$I_{\text{eff}} = \sqrt{\frac{1}{T}\int_0^T i^2(t)\,dt} \tag{18.5-6}$$

If the signal is a voltage, $v(t)$,

$$\frac{1}{T}\int_0^T \frac{v^2(t)}{R}\,dt = \frac{V_{\text{eff}}^2}{R}$$

which results in the same formula using voltages instead of currents:

$$V_{\text{eff}} = \sqrt{\frac{1}{T}\int_0^T v^2(t)\,dt}$$

From Eq. (18.5-6) we see that the effective value is computed as the square *root* of the *mean*, or average, value of the signal *squared*; for this reason it is frequently called the root-mean-square, or rms, value.

The effective value of a sinusoidal signal is the peak value divided by $\sqrt{2}$. This result may be obtained from Eq. (18.5-6). A simpler alternative is to apply Eq. (18.5-5) to a resistive load for which $V_m = RI_m$ and $\varphi = 0$ resulting in

$$P = \frac{RI_m{}^2}{2} = RI_{\text{eff}}{}^2 \qquad \text{or} \qquad I_{\text{eff}} = \frac{I_m}{\sqrt{2}}$$

Setting $I = V/R$, the result for voltages is

$$V_{\text{eff}} = \frac{V_m}{\sqrt{2}}$$

from which we conclude that the effective value of a sinusoidal signal is its peak value divided by $\sqrt{2}$. Representing the effective values by V and I the average power for a sinusoid is

$$P = VI\cos\varphi \tag{18.5-7}$$

Thus far we have used amplitude phasors; the amplitude phasor for $v(t) = V_m \cos(\omega t + \varphi)$ is $\mathbf{V}_m = V_m e^{j\varphi} = V_m \angle\varphi$. Most textbooks use *effective phasors*; the magnitude of an effective phasor is the effective value of the sinusoid rather than the peak value as is the case in the amplitude phasor. Thus the effective phasor for $v(t)$ is

$$\mathbf{V} = Ve^{j\varphi} = V \angle\varphi = \frac{V_m}{\sqrt{2}} \angle\varphi$$

We will generally use unsubscripted letters to represent effective values or phasors and we will generally use the subscript m to indicate peak values or amplitude phasors. The advantage of the effective phasor is that it eliminates the division by 2 in the power formulas.

Example 18.5-1 The standard line voltage in the United States is 110 V rms, 60 Hz. A 60-W light bulb is connected to this voltage. Find

a The effective current through it
b The peak current
c The peak voltage

Solution

a The light bulb is purely resistive so the voltage and current are in phase, that is, $\varphi = 0$. Therefore $\cos\varphi = 1$ and

$$I = \frac{P}{V} = \frac{60}{110} = 0.545 \text{ A}$$

b $I_m = \sqrt{2}\, I = (1.414)(0.545) = 0.771$ A
c $V_m = \sqrt{2}\, V = (1.414)(110) = 156$ V ////

18.5-2 Power Factor

The formula for the average power, Eq. (18.5-5), can be written in terms of the phasor values as

$$P = \text{Re}\,(\mathbf{VI}^*) = VI \cos\varphi \tag{18.5-8}$$

$\mathbf{I}^*$ must be used in this formula so that the phase angle of the current will be subtracted from that of the voltage rather than added to it as would be the case if $\mathbf{I}$ were used (see Prob. 18.5-6).

The product VI is called the *apparent power* and its units are volt-amperes (V·A). It is symbolized P_a.

The complex term $\mathbf{VI}^*$ is called the complex power and its units are also volt-amperes. Its real part is the power while its imaginary part, symbolized Q, is called the *reactive power* in units of reactive volt-amperes (VAR).

The reactive volt-amperes are a measure of the amount of energy transferred back and forth between the supply and energy storage elements of the circuit. The formula for the reactive volt-amperes is

$$Q = \mathrm{Im}(\mathbf{VI}^*) = VI \sin \varphi \tag{18.5-9}$$

This should be compared with Eq. (18.5-8).

The ratio of the average power to the apparent power delivered to a circuit is called the *power factor*, generally symbolized by pf. The power factor, from Eq. (18.5-6), is given by

$$\mathrm{pf} = \frac{P}{P_a} = \cos \varphi \tag{18.5-10}$$

where φ is the phase difference between the voltage and current. If $\mathbf{V}$ is the voltage across and $\mathbf{I}$ the current through an arbitrary impedance $\mathbf{Z}$, then the phase angle φ is the angle of the impedance $\mathbf{Z}$. For this case

$$\mathrm{pf} = \cos \varphi = \frac{\mathrm{Re}\,\mathbf{Z}}{|\mathbf{Z}|} = \frac{R}{\sqrt{R^2 + X^2}} \tag{18.5-11}$$

Typical calculations are illustrated in the example which follows.

Example 18.5-2 In the circuit of Fig. 18.5-4 find

a The power factor
b The current
c The real power
d The reactive power
e The apparent power

Solution

a $\mathbf{Z} = R + j(X_L - X_C) = 30 + j(60 - 20) = 30 + j40 = 50 \angle 53.2°$
From Eq. (18.5-11)

$$\mathrm{pf} = \cos \varphi = \cos 53.2° = 0.6$$

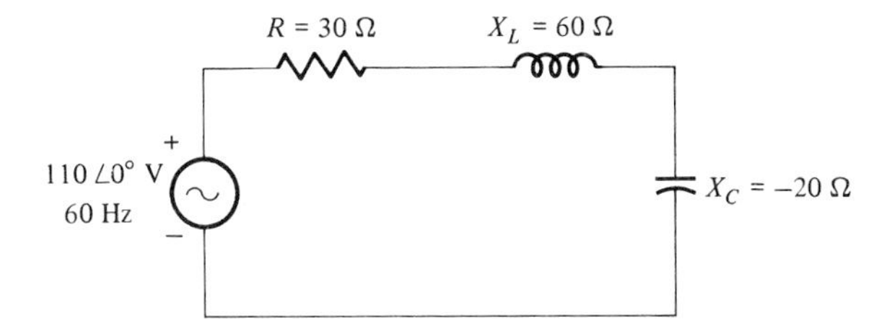

Fig. 18.5-4 Circuit for Example 18.5-2.

Note that **Z** is primarily inductive so that the phase angle is positive.

b Using Ohm's law in phasor form

$$\mathbf{I} = \frac{\mathbf{V}}{\mathbf{Z}} = \frac{110 \angle 0°}{50 \angle 53.2°} = 2.2 \angle -53.2° \text{ A.}$$

c The real power is found using Eq. (18.5-8):

$$P = VI \cos \varphi = (110)(2.2)(.6) = 145.2 \text{ W}$$

This can be checked by noting that only the resistance dissipates power so that

$$P = I^2R = (2.2)^2(30) = 145.2 \text{ W}$$

d The reactive power is found using Eq. (18.5-9):

$$Q = VI \sin \varphi = (110)(2.2) \sin 53.2° = 193.6 \text{ VAR}$$

This can be checked using the formula (see Prob. 18.5-5)

$$Q = I^2X = I^2(X_L - X_C) = (2.2)^2(40) = 193.6 \text{ VAR}$$

e The apparent power is

$$P_a = VI = (110)(2.2) = 242 \text{ V}\cdot\text{A}$$

////

POWER-FACTOR CORRECTION

Power companies generally provide an incentive in their rate structure to encourage customers to maximize the power factors of their loads because that reduces the power losses due to line resistance. The explanation is as follows: a higher current is required to provide the same amount of power when the power factor is low than when it is high. The losses due to line resistance are proportional to the square of the current and therefore are smaller when load power factors are higher. Power factors are corrected, i.e., made equal to unity, by inserting reactive elements in parallel with the load as illustrated in the following example.

Example 18.5-3 Most industrial loads are inductive so that they exhibit lagging power factors. Power-factor correction then involves connecting a capacitor in parallel with the load as shown in Fig. 18.5-5.

The load dissipates 2 kW at a power factor of 0.8 lagging. The supply voltage V is 440 V, 60 Hz. Find the capacitance required to correct the power factor to unity.

Solution This problem is most easily solved by noting that the power factor will be unity when the *net* reactive power is zero. Thus if we can find the positive inductive reactive power in the load, then the capacitor must absorb an equal negative amount so that the net reactive power is zero. To find the reactive power of the load we proceed as follows.

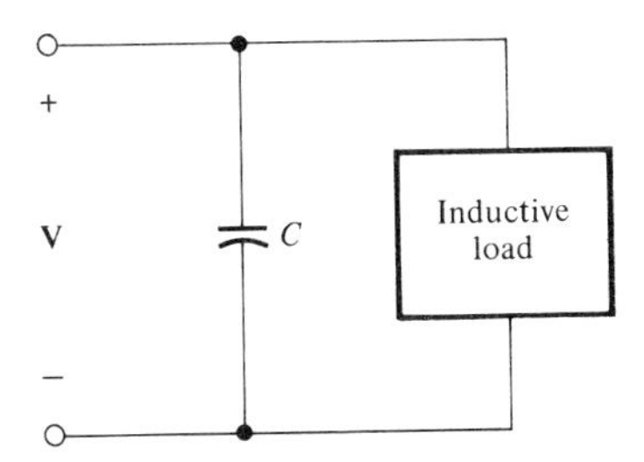

Fig. 18.5-5 Circuit for Example 18.5-3.

The power factor angle, which is the same as the angle of the load impedance, is

$$\varphi = \text{arccos } 0.8 = 36.9°$$

The power factor is

$$\text{pf} = \frac{P}{P_a}$$

from which the apparent power is

$$P_a = \frac{P}{\text{pf}} = \frac{2 \text{ kW}}{0.8} = 2.5 \text{ kVA}$$

Since $P_a = VI$, the reactive power of the load is

$$Q_L = VI \sin \varphi = P_a \sin \varphi = 2.5 \sin 36.9°$$
$$= 1.5 \text{ kVAR}$$

Thus the capacitor must absorb $Q_C = -1.5$ kVAR. It can be shown that for a capacitor (see Prob. 18.5-5)

$$Q_C = \frac{-V^2}{X_C}$$

from which

$$X_C = \frac{-V^2}{Q_C} = \frac{(440)^2}{1{,}500} = -129 \ \Omega$$

Finally, since $X_C = -1/\omega C$,

$$C = \frac{-1}{\omega X_C} = \frac{-1}{(2\pi)(60)(-129)} = 0.00002 = 20 \ \mu\text{F}$$

////

18.5-3 Maximum power transfer

The circuit designer is sometimes faced with the problem of selecting a load impedance to connect to a circuit so that maximum power is delivered to the load impedance. Assume a Thevenin voltage $\mathbf{V}_T$ and a Thevenin impedance $\mathbf{Z}_T = R_T + jX_T$. The problem is to determine the value of the load impedance $\mathbf{Z}_L = R_L + jX_L$ such that the power delivered to $\mathbf{Z}_L$ is maximum. The circuit, shown in Fig. 18.5-6a, leads to the following results:

$$\mathbf{I}_L = \frac{\mathbf{V}_T}{\mathbf{Z}_L + \mathbf{Z}_T} = \frac{\mathbf{V}_T}{R_L + R_T + j(X_L + X_T)} \tag{18.5-12}$$

From Eq (18.5-8),

$$P = \text{Re}\,(\mathbf{V}_L\mathbf{I}_L{}^*) = \text{Re}\,(Z_L\mathbf{I}_L\mathbf{I}_L{}^*) = \text{Re}\,(Z_L\,|\,\mathbf{I}_L\,|^2) = |\,\mathbf{I}_L\,|^2\text{Re}\,\mathbf{Z}_L$$

$$= \frac{V_T{}^2}{(R_L + R_T)^2 + (X_L + X_T)^2}\,R_L \tag{18.5-13}$$

It is clear that $X_L = -X_T$ is the value of X_L that maximizes P. Then

$$P = \frac{V_T{}^2 R_L}{(R_L + R_T)^2} = \frac{V_T{}^2}{R_T}\,\frac{R_L/R_T}{(R_L/R_T + 1)^2} \tag{18.5-14}$$

A plot of P vs. R_L/R_T is shown in Fig 18.5-6b. From this graph it is found that $R_L = R_T$ results in maximum power transfer to $\mathbf{Z}_L$.

It is interesting to note that when $\mathbf{Z}_L$ is chosen for maximum power transfer the efficiency, i.e., the ratio of power delivered to the load to that supplied by the Thevenin source, is 50 percent. This is easily determined by recognizing that all the power is dissipated by R_L and R_T. Since the two resistances are equal and carry the same current, they dissipate the same power and the source supplies twice the power delivered to R_L.

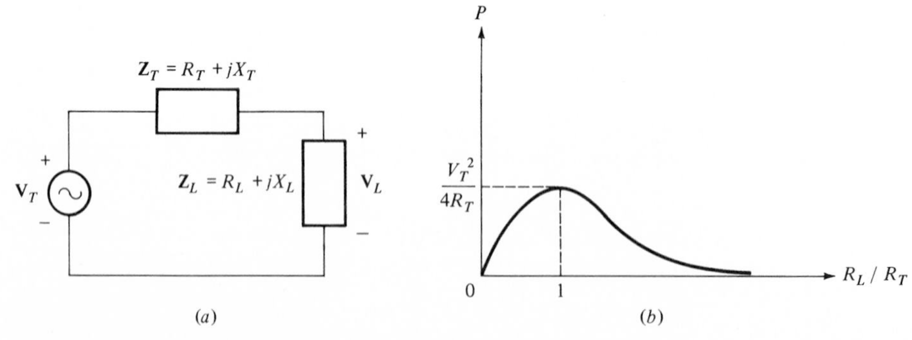

Fig. 18.5-6 Maximum power transfer. (a) Circuit; (b) power as a function of R_L/R_T for $X_L = -X_T$.

Many students fall into the trap of using this result to solve the problem of finding the source impedance required for maximum power transfer when the load is fixed at $\mathbf{Z}_L = R_L + jX_L$. The correct result for this problem is *not* $\mathbf{Z}_T = R_L - jX_L$ (see Prob. 18.5-7 for the correct solution).

SUMMARY

1 Complex numbers may be represented in rectangular or polar coordinates.

$$\mathbf{z} = x + jy = re^{j\varphi} = r \angle\varphi$$

a $x = \text{Re}\,\mathbf{z} = r\cos\varphi$
b $y = \text{Im}\,\mathbf{z} = r\sin\varphi$
c $r = |\mathbf{z}| = \sqrt{x^2 + y^2}$
d $\varphi = \arg z = \arctan(y/x)$
e $\mathbf{z}^* = x - jy = re^{-j\varphi} = r \angle -\varphi$

2 Complex arithmetic

a $(x_1 + jy_1) + (x_2 + jy_2) = (x_1 + x_2) + j(y_1 + y_2)$
b $(x_1 + jy_1) - (x_2 + jy_2) = (x_1 - x_2) + j(y_1 - y_2)$
c $r_1e^{j\theta_1} \times r_2e^{j\theta_2} = r_1r_2e^{j(\theta_1+\theta_2)}$

d $r_1e^{j\theta_1} \div r_2e^{j\theta_2} = \dfrac{r_1}{r_2}\, e^{j(\theta_1-\theta_2)}$

e $(x_1 + jy_1) \times (x_2 + jy_2) = (x_1x_2 - y_1y_2) + j(x_1y_2 + x_2y_1)$

f $\dfrac{x_1 + jy_1}{x_2 + jy_2}\dfrac{x_2 - jy_2}{x_2 - jy_2} = \dfrac{x_1x_2 + y_1y_2}{x_2^2 + y_2^2} + j\dfrac{x_2y_1 - x_1y_2}{x_2^2 + y_2^2}$

g $\mathbf{z}\mathbf{z}^* = |\mathbf{z}|^2$
h $\mathbf{z} + \mathbf{z}^* = 2\,\text{Re}(\mathbf{z})$
i $\mathbf{z} - \mathbf{z}^* = j2\,\text{Im}(\mathbf{z})$

3 A sinusoidal signal is a periodic signal characterized by an amplitude, period, and phase angle.

a Its amplitude is the peak value it attains.
b Its period T is the time interval for one complete cycle. Its frequency f, measured in hertz (Hz), is the reciprocal of the period measured in seconds (s). Its angular frequency is $\omega = 2\pi f$ and is measured in radians per second.
c Its phase angle is a measure of its position along the t axis.

4 The amplitude phasor representation of $V_m \cos(\omega t + \varphi)$ is $\mathbf{V}_m = V_m e^{j\varphi}$.

5 $V_m \cos(\omega t + \varphi) = V_m \sin\left(\omega t + \varphi + \frac{\pi}{2}\right)$.

6 Sinusoidal impedances

a $\mathbf{Z} = R + jX$; $\mathbf{Z}$ is the impedance, R is the resistive component, and X is the reactance.

b $Z_R = R$

$Z_L = j\omega L$

$$Z_C = \frac{1}{j\omega C} = \frac{-j}{\omega C}$$

c The voltage across an inductance leads the current through it by 90°. The voltage across a capacitance lags the current through it by 90°. The voltage across a resistance is in phase with the current through it.

7 The steady-state sinusoidal response is obtained by

a Replacing all elements by their impedances and all source values by their phasor representations.

b Computing the output phasors using the theorems and methods presented for resistive circuits.

c Converting the output phasors to the sinusoidal signals which they represent.

8 The effective phasor for a signal $V_m \sin(\omega t + \varphi)$ is $(V_m/\sqrt{2})\ \angle\varphi$; the effective value of the signal is $V_m/\sqrt{2}$.

9 The average power delivered to a load is $P = VI \cos\varphi$ where φ is the phase angle between $\mathbf{V}$ and $\mathbf{I}$.

PROBLEMS

18.1-1 Find both the rectangular and the polar forms for the complex numbers shown in Fig. P18.1-1.

18.1-2 Make a labeled sketch showing the complex plane representation of each of the following:

a $\mathbf{p} = 5 - j8$

b $\mathbf{i} = 10e^{j(\pi/2)} = 10\ \angle 90°$

c $\mathbf{z} = j10$

d $\mathbf{v} = 25e^{j(2\pi/3)} = 25\ \angle 120°$

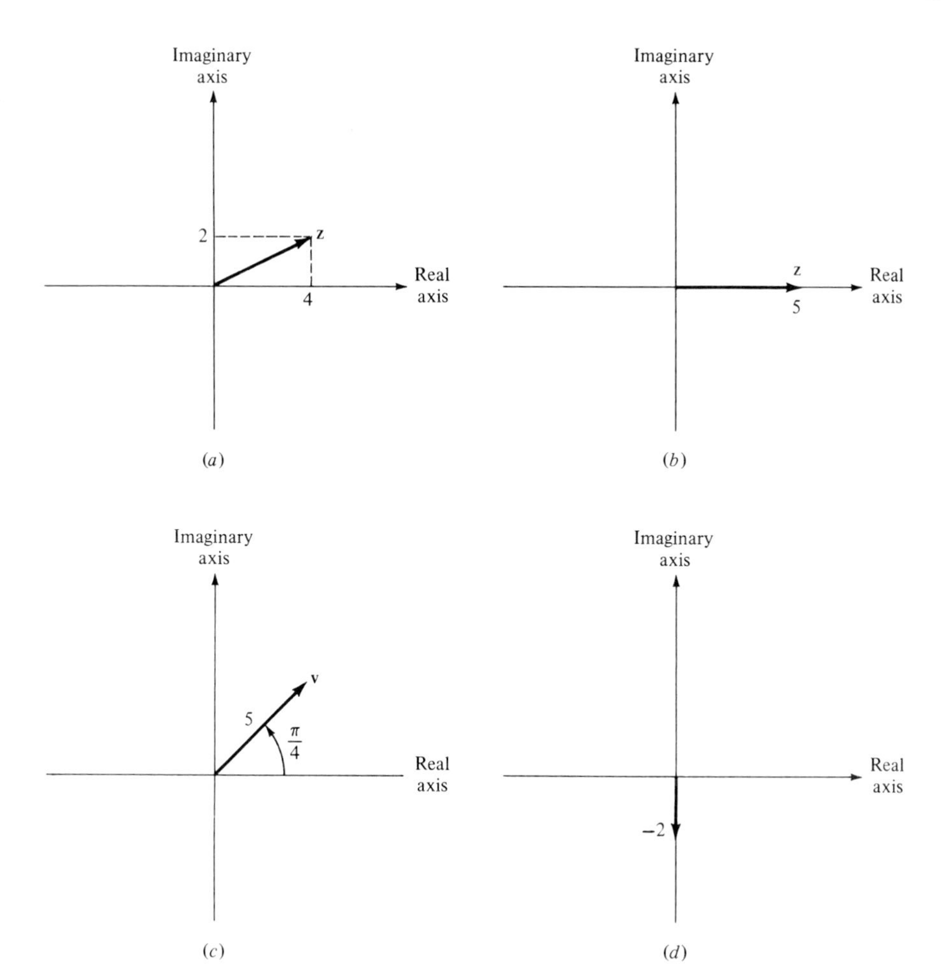

Fig. P18.1-1

18.1-3 Represent the following complex numbers in rectangular form:

a $10e^{j\pi} = 10 \angle \pi$
b $20e^{-j(\pi/6)} = 20 \angle -30°$
c $100e^{j225°} = 100 \angle 225°$

18.1-4 Represent each of the following numbers in polar form:

a $-3 + j4$
b $-10 - j10$
c $20 - j50$
d $j16$

18.1-5 Find the complex conjugates of the following numbers:

a $-4 - j2$
b $2 \angle -30°$

18.1-6 Perform the indicated addition and subtractions:

a $(2 + j3) + (5 - j4)$
b $25 \angle 90° + 100 \angle 143°$
c $(10 - j4) - (5 + j5)$
d $10 \angle 57.3° - 8 \angle -85.9°$

18-1-7 Perform the indicated multiplications and divisions:

a $(3 + j4)(2.5 - j1)$
b $18 \angle 20° \times 4.5 \angle 60°$
c $5 \angle 180° \div 40 \angle -45°$
d $(10 + j15) \div (6 + j2)$

18.1-8 Represent each of the following in terms of $|\mathbf{z}|$, arg $\mathbf{z}$, Re $\mathbf{z}$, and Im $\mathbf{z}$:

a $\mathbf{z}\mathbf{z}^*$
b $\mathbf{z} + \mathbf{z}^*$
c $\mathbf{z} - \mathbf{z}^*$

18.2-1 Express $\sin \theta$ in terms of the cosine.

18.2-2 Express $10 \sin (2 \times 10^5 \pi t + \pi/4)$ using the exponential function.

18.2-3 Find and draw the amplitude phasors for

a $v_1(t) = 200 \sin 2{,}000\pi t$ V
b $i(t) = 50 \cos 500\pi t$ mA
c $v_2(t) = 10 \sin (4{,}000\pi t + 25°)$ V
d $v_3(t) = 25 \cos (800t + \pi)$ V

18.2-4 List the following signals in order of leading phase, i.e., the one which leads all the others first, then the one which leads all the remaining ones except itself of course, and so on with the last one being the one which lags all the others. If any signals are in phase so indicate

$$v_1(t) = 10 \sin (100t + 30°)$$
$$v_2(t) = 50 \cos 100t$$
$$v_3(t) = 400 \cos (100t + 120°)$$
$$i(t) = -80 \sin (100t - \pi/2)$$

18.3-1 A current $i(t) = 10 \cos 2{,}000\pi t$ mA flows through each of the circuits of Fig. P18.3-1. For each circuit (*a*) find its impedance, (*b*) draw a phasor diagram showings its current and voltage phasors, and (*c*) find $v(t)$.

18.3-2 Repeat Prob. 18.3-1 if $i(t)$ is a 100-mA peak value, 60-Hz sinewave with a zero-phase angle.

0.1 μF

$i(t)$

10 kΩ

$+\ v(t)\ -$

(a)

$i(t)$ 10 mH 100 Ω

0.5 μF

$+\ v(t)\ -$

(b)

Fig. P18.3-1

18.3-3 Derive Eq. (18.3-10) using the same approach as was used for Eq. (18.3-9).

18.4-1 Find the sinusoidal steady-state outputs for each of the circuits shown in Fig. P18.4-1.

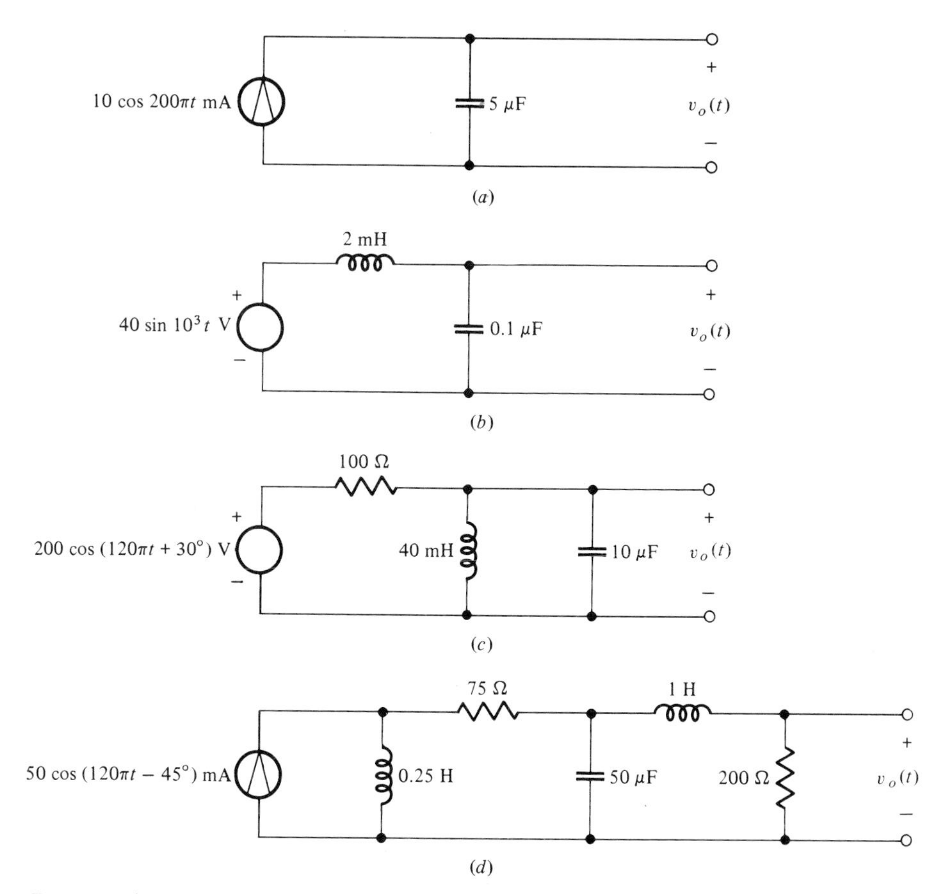

Fig. P18.4-1

18.4-2 Repeat Prob. 18.4-1 using a computer circuit analysis program.

18.5-1 Find the VAs, the VARs, and the power supplied by the sources in the circuits of Fig. P18.4-1.

18.5-2 Find and draw the effective phasors for each of the following signals.

a $v(t) = 100 \cos(120\pi t + 60°)$ V
b $i(t) = 25 \sin(120\pi t + 80°)$ mA

18.5-3 If $v(t)$ and $i(t)$ in Prob. 18.5-2 represent the voltage across and current through a circuit what is its power factor?

18.5-4 If the effective phasor voltage and current of Fig. P18.5-4 are associated with a given circuit element find

a The apparent power P_a delivered to the element
b The average power P delivered to the element
c The reactive volt-amperes Q delivered to the element
d Its power factor
e Its impedance. (inductive or capacitive?)

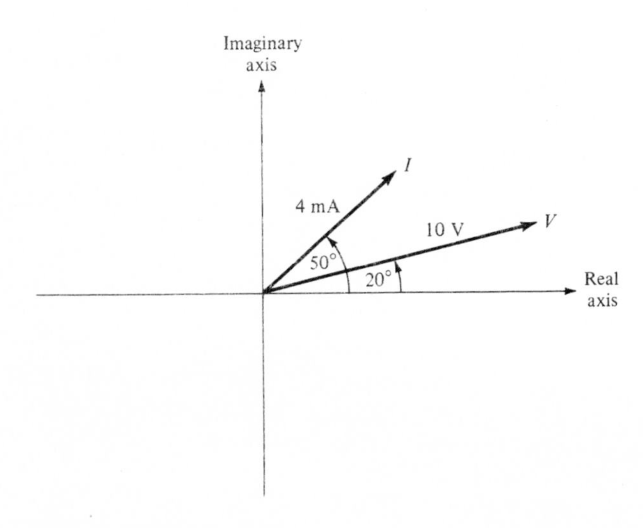

Fig. P18.5-4

18.5-5 Prove that the reactive power delivered to a circuit can be computed as

$$Q = I^2X = \frac{V^2}{X}$$

where X is the circuit reactance.

18.5-6 If $\mathbf{V} = V \angle \varphi_V$ and $\mathbf{I} = I \angle \varphi_I$ prove that $P = \text{Re}(\mathbf{VI}^*)$ agrees with $P = VI \cos \varphi$ where $\varphi = \varphi_V - \varphi_I$ is the phase angle between $\mathbf{V}$ and $\mathbf{I}$ while Re $(\mathbf{VI})$ is not equal to $VI \cos \varphi$.

18.5-7 Prove that the value of $\mathbf{Z}_T$ required to cause maximum power to be supplied to a given load impedance, $\mathbf{Z}_L = R_L + jX_L$, is $\mathbf{Z}_T = -jX_L$. What is the efficiency?

19
Frequency Response

OBJECTIVES

Upon completion of this chapter, the student should be able to

1 Define frequency spectrum of a signal.
2 Correlate the waveform of a signal and its frequency spectrum in a qualitative way.
3 Define the frequency response of a network.
4 Obtain and plot the frequency response of a network from its transfer function.
5 Describe resonance phenomena from a physical point of view.

INTRODUCTION

The analysis of linear electric networks has been separated into three categories according to the type of input: dc, steady-state ac (sinusoidal), and time response to single-occurrence waveforms, such as steps. The dc response of a network is important because of the prevalence of dc sources and their use as power supplies for all types of systems. The response to the step function is emphasized because the step represents a dc source being switched into a network and it leads to the pulse response which is important in digital systems. Some of the reasons for the importance of the sinusoidal waveform are

1 The sinusoidal signal occurs in many systems as part of the natural response. For example, the oscillatory motions of a mass bobbing on a spring or a swinging pendulum are both sinusoidal.
2 Electric energy is usually generated and transmitted in the form of sinusoidal signals because they are easily generated and voltage levels are readily changed by transformers. This permits efficient transmission at high voltage levels and step-down to relatively safe voltage levels prior to use.

The special study of the sinusoidal steady-state response was motivated by these considerations. This chapter is motivated by still another very important characteristic of the sinusoidal signal: *any signal can be represented as a sum of sinusoidal components of different frequencies.* It is the purpose of this chapter (1) to relate a signal to its frequency components and (2) to determine the way in which a network affects the various frequency components of its input signal. It is then possible, knowing the output frequency components, to determine the output waveform, at least in a qualitative way.

19.1 FREQUENCY SPECTRA OF SIGNALS

The purpose of this section is to establish the relation between a signal and the sum of the sinusoidal components which constitute the signal. This collection of sinusoidal components is called the *frequency spectrum* of the signal. Each frequency component has, in addition to its frequency, two other parameters: its amplitude and phase angle. The frequency spectrum is most easily understood in terms of a periodic signal.

19.1-1 Fourier Series

A periodic signal is one in which a waveform of some finite duration, called its *period,* recurs repeatedly. The square wave of Fig. 19.1-1a is an example of such a signal with period T.

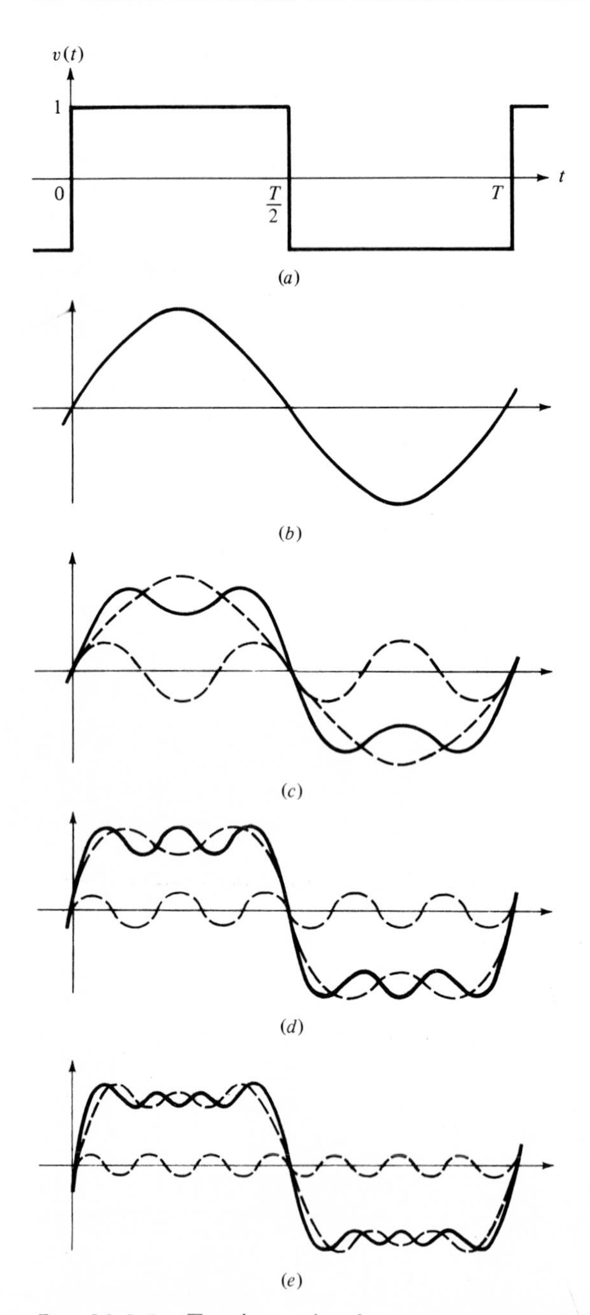

Fig. 19.1-1 Fourier series for square wave. (*a*) Square wave; (*b*) first harmonic; (*c*) sum of first and third harmonics; (*d*) sum of first, third, and fifth harmonics; (*e*) sum of first, third, fifth, and seventh harmonics.

Any such periodic signal may be represented by an infinite sum of sinusoidal components plus a constant term; this infinite sum is called the *Fourier series* of the signal. The formula for a periodic signal $f(t)$, with period T, is

$$f(t) = A_0 + \sum_{n=1}^{\infty} A_n \cos (n\omega_0 t) + \sum_{n=1}^{\infty} B_n \sin (n\omega_0 t) \tag{19.1-1}$$

where $\omega_0 = 2\pi f_0$ and $f_0 = 1/T$ is the *fundamental* frequency. The constants A_0, A_n, and B_n may all be computed using formulas[1] but the evaluation of various integrals makes the process rather difficult.

The two sums can be combined to produce the equivalent formula for the Fourier series:[2]

$$f(t) = C_0 + \sum_{n=1}^{\infty} C_n \cos (n\omega_0 t + \varphi_n) \tag{19.1-2}$$

C_0 is the dc term and is equal to the average value of $f(t)$. The term $C_n \cos (n\omega_0 t + \varphi_n)$ is called the nth *harmonic* of $f(t)$; its frequency is n times the fundamental. For most waveforms, the amplitudes of the harmonics, i.e., the values of C_n, become smaller and smaller as n increases beyond a certain point. As a result, it is possible to obtain a good approximation for the original function $f(t)$ using only the first few harmonics. This is illustrated for the square wave of Fig. 19.1-1*a* in Fig. 19.1-1*b* through *e*, which illustrates the sum of the harmonics through the first, third, fifth, and seventh. The even harmonics are all zero for the square wave due to its symmetry. The symmetry of the waveform also causes all the A_n coefficients, i.e., all the cosine function amplitudes, to be zero. The actual formula for the Fourier series of the square

[1] The formulas for the Fourier coefficients are

$$A_0 = \frac{1}{T}\int_0^T f(t)\,dt, \qquad A_n = \frac{2}{T}\int_0^T f(t)\cos (n\omega_0 t)\,dt$$

and

$$B_n = \frac{2}{T}\int_0^T f(t)\sin (n\omega_0 t)\,dt$$

where $n \neq 0$.

[2] Applying the appropriate trigonometric formula yields

$$C_n \cos (n\omega_0 t + \varphi_n) = C_n \cos \varphi_n \cos (n\omega_0 t) - C_n \sin \varphi_n \sin (n\omega_0 t)$$

which, when compared with Eqs. (19.1-1), results in

$$C_n = \sqrt{A_n^2 + B_n^2} \qquad \text{and} \qquad \varphi_n = \arctan \frac{-B_n}{A_n}$$

for $n \neq 0$ and $C_0 = A_0$.

wave is

$$f(t) = \frac{4}{\pi} \sin (\omega_0 t) + \frac{4}{3\pi} \sin (3\omega_0 t) + \frac{4}{5\pi} \sin (5\omega_0 t) + \frac{4}{7\pi} \sin (7\omega_0 t) + \cdots + \frac{4}{(2m+1)\pi} \sin [(2m+1)\omega_0 t] + \cdots \quad (19.1\text{-}3)$$

The frequency spectrum of a signal can be represented graphically by plotting the amplitude of each harmonic, i.e., the value of C_n, and the phase angle of the harmonic, i.e., the value of φ_n, as a function of the frequency of the harmonic. The frequency spectra of two square waves are shown, along with the square waves, in Fig. 19.1-2. From this figure the following may be seen:

1 The spectra are the same except for the frequencies of the harmonics.
2 The amplitudes of the harmonics decrease as the order of the harmonics increases.
3 The phase is constant at $-90°$ since $\sin \theta = \cos (\theta - 90°)$.
4 The frequency interval between harmonics decreases as the period of the square wave increases.

This last characteristic of the spectra is true no matter what the waveform is; the longer its period, the closer together the harmonics will be.

Greater insight into the relationship between a signal and its spectrum is obtained by comparing the spectra of the square wave and that of the triangular wave. These are shown in Fig. 19.1-3. The spectrum of the triangular wave is obtained from its Fourier series:

$$f(t) = \frac{8}{\pi^2} \sin \omega_0 t + \frac{8}{9\pi^2} \sin 3\omega_0 t + \frac{8}{25\pi^2} \sin 5\omega_0 t + \cdots + \frac{8}{(2n+1)^2\pi^2} \sin [(2n+1)\omega_0 t] + \cdots \quad (19.1\text{-}4)$$

The square wave and triangular wave in Fig. 19.1-3 are both of the same period, 1 s, and the same amplitude, 1. The scales of the amplitude spectra are different so that the maximum values will be the same height to facilitate comparison. From Fig. 19.1-3 the following may be seen:

1 The harmonics for the square wave and triangular wave occur at the same frequencies. This is due to their having the same period.
2 The phase spectra of the two waves are both constant at $-90°$ and the

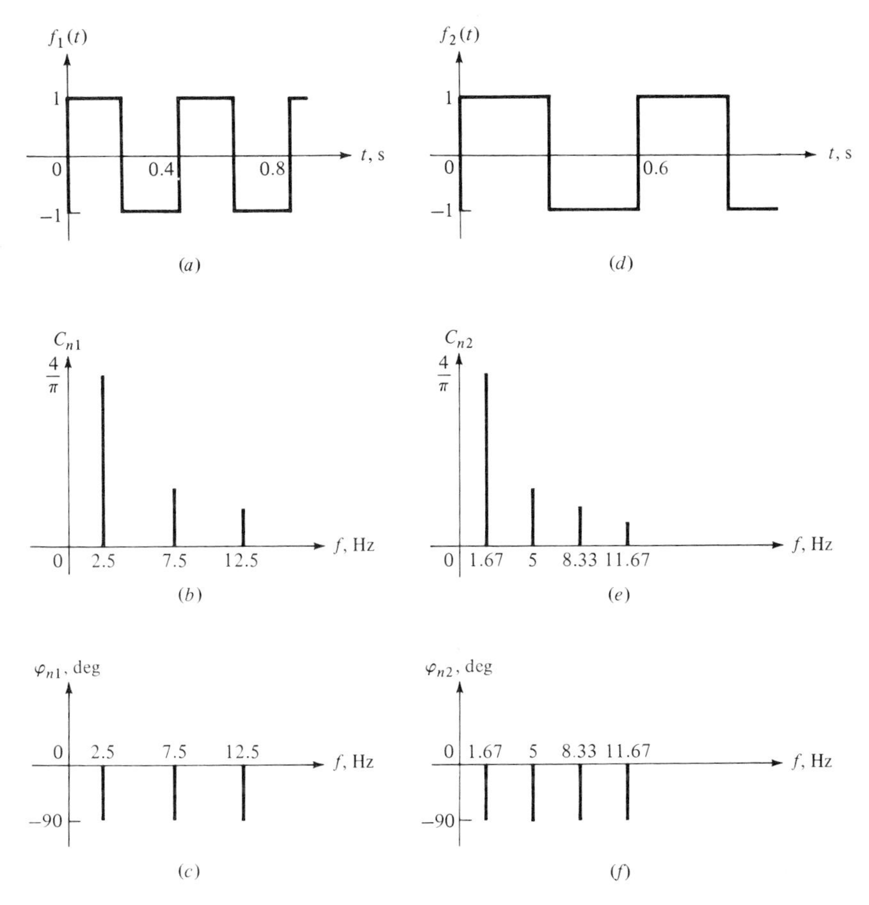

Fig. 19.1-2 Frequency spectra of square waves. (*a*) Square wave with 0.4-s period; (*b*) amplitude spectrum for 0.4-s square wave; (*c*) phase spectrum for 0.4-s square wave; (*d*) square wave with 0.6-s period; (*e*) amplitude spectrum for 0.6-s square wave; (*f*) phase spectrum for 0.6-s square wave.

even harmonics are all zero in both cases. This is due to both waves having the same type of symmetry.

3 The amplitudes of the harmonics decrease as n increases for both waves but the harmonic amplitudes of the triangular wave decrease more rapidly with increasing frequency than do the amplitudes of the square wave harmonics. This result is a specific case of a general principle concerning the waveform of a signal and its frequency spectrum. This principle can be stated as follows:

> The greater the rate of change of a function and the greater the discontinuities in a function or its derivatives, the greater the percentage of high-frequency content in its spectrum.

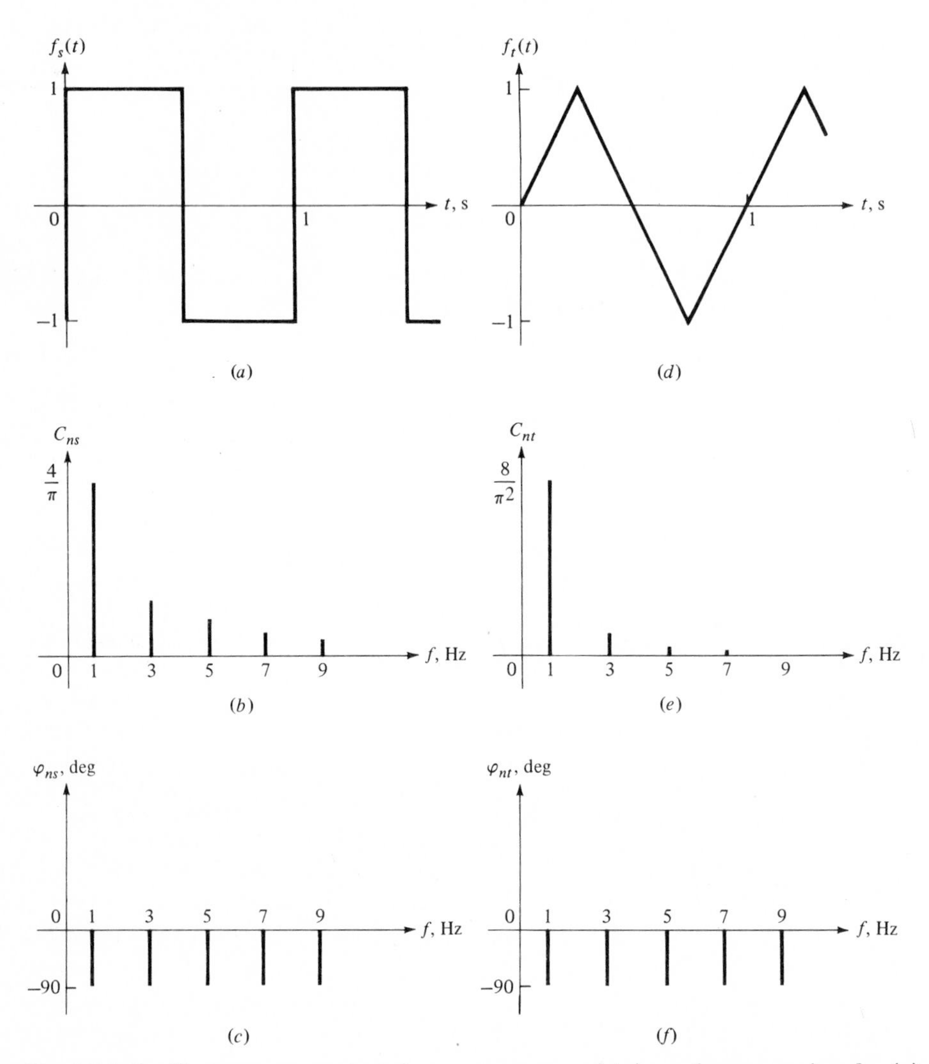

Fig. 19.1-3 Frequency spectra of square-wave and triangular-wave signals. (*a*) Square-wave signal; (*b*) amplitude spectrum of square-wave signal; (*c*) phase spectrum of square-wave signal; (*d*) triangular-wave signal; (*e*) amplitude spectrum of triangular-wave signal; (*f*) phase spectrum of triangular-wave signal.

As a consequence of this principle, it would be expected that the square wave would have a larger high-frequency content than the triangular wave, since the square wave has greater discontinuity and rates of change, specifically where it changes levels. From the various approximations of the square wave in Fig. 19.1-1, it is seen that the primary contribution of the high-frequency components is to the fast rise and drop of the signal between levels and to the achievement of the square corners of the original waveform.

19.2 TRANSFER FUNCTIONS

In this section we will extend the concept of the transfer function to include sinusoidal signals. This will enable us to find the steady-state ac response of networks containing R, L, and C elements in much the same way that we found the response of resistive networks to dc signals. In the case of ac signals, the solution requires the use of complex numbers so much more arithmetic is usually required to arrive at a numerical answer.

Consider the voltage divider network shown in Fig. 19.2-1*a*. In Sec. 6.3 we found that the output voltage V_2 was related to the input voltage V_1 by the voltage ratio transfer function

$$A_v = \frac{V_2}{V_1} = \frac{R_2}{R_1 + R_2} \tag{19.2-1}$$

Now we are led to consider the network shown in Fig. 19.2-1*b*, in which $\mathbf{V}_1$ is the phasor input voltage, and $\mathbf{Z}_1$ and $\mathbf{Z}_2$ may contain all three types of elements, R, L, and C. In order to find the output voltage phasor $\mathbf{V}_2$ we use the same voltage divider formula except that resistances are replaced by impedances. Thus

$$\mathbf{A}_v(\omega) = \frac{\mathbf{V}_2}{\mathbf{V}_1} = \frac{\mathbf{Z}_2}{\mathbf{Z}_1 + \mathbf{Z}_2} \tag{19.2-2}$$

If either impedance contains inductance or capacitance, the voltage ratio will usually be complex and will depend on the frequency of $\mathbf{V}_1$. We have indicated this by writing $\mathbf{A}_v(\omega)$. These ideas are illustrated in the example which follows.

> ***Example 19.2-1*** Use the transfer function concept to find expressions for the output voltage in the circuit of Fig. 19.2-2*a* and the output current in the circuit of Fig. 19.2-2*b*.

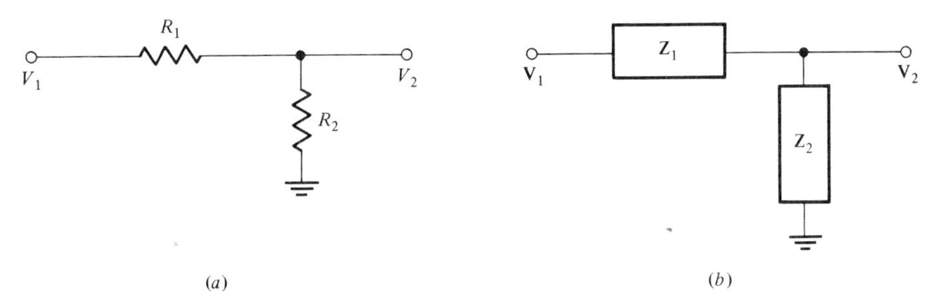

Fig. 19.2-1 Transfer functions. (*a*) dc; (*b*) ac.

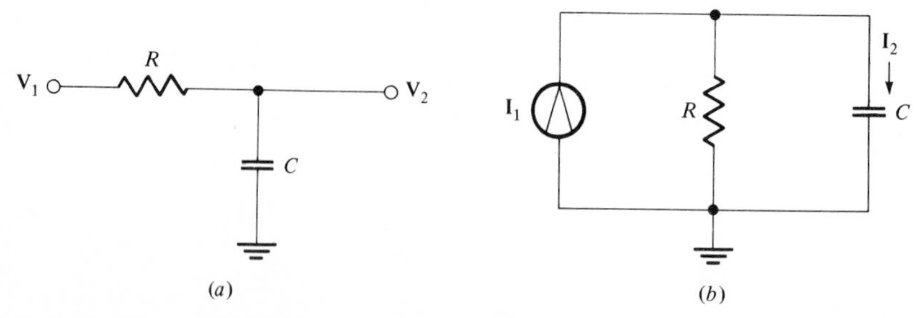

Fig. 19.2-2 Circuits for Example 19.2-1.

Solution Circuit a:

$$\mathbf{Z}_1 = R \qquad \mathbf{Z}_2 = \frac{1}{j\omega C}$$

$$\mathbf{A}_v(\omega) = \frac{1/j\omega C}{R + 1/j\omega C}$$

Multiplying numerator and denominator by $j\omega C$, this becomes

$$\mathbf{A}_v(\omega) = \frac{1}{1 + j\omega RC} \tag{19.2-3}$$

The output voltage can now be found:

$$\mathbf{V}_2 = \mathbf{A}_v(\omega)\mathbf{V}_1 \tag{19.2-4}$$

Circuit b: For this circuit the current divider is appropriate.

$$\mathbf{A}_i(\omega) = \frac{\mathbf{I}_2}{\mathbf{I}_1} = \frac{R}{R + 1/j\omega C} = \frac{j\omega RC}{1 + j\omega RC} \qquad (19.2\text{-}4)$$

The output current can be found from

$$\mathbf{I}_2 = \mathbf{A}_i(\omega)\mathbf{I}_1$$ ////

The reader will note that both transfer functions in the example are complex. In the next section we consider the dependence of these functions on frequency.

19.3 FREQUENCY RESPONSE

The term frequency response refers to information concerning the response of a system to sinusoids of all frequencies which lie in an appropriate range, often from dc (zero frequency) to very high frequencies. Frequency response can be

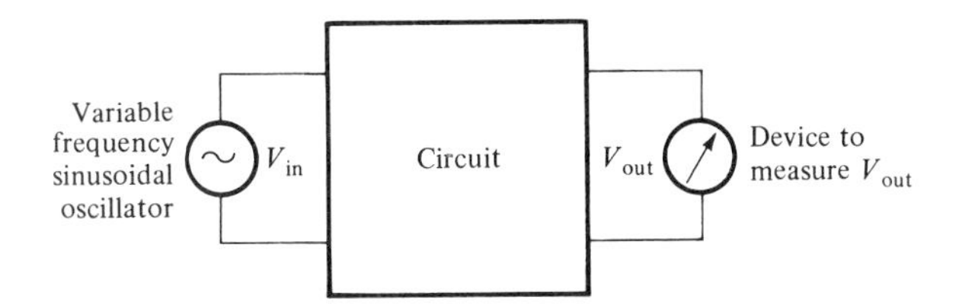

Fig. 19.3-1 Test circuit to determine frequency response.

determined experimentally using the circuit of Fig. 19.3-1. The frequency of the oscillator is varied over the range of interest and at each frequency setting the output amplitude and phase angle are measured. The input voltage is maintained at a convenient amplitude and phase, for example, $1 \angle 0°$ during the measurements. The frequency response is then displayed in the form of graphs of the output vs. frequency. The plot of output amplitude vs. frequency is called the amplitude response while the plot of phase angle vs. frequency is called the phase response. Sometimes the amplitude response conveys sufficient information and the phase response is omitted.

A typical amplitude response curve for a high-fidelity amplifier is shown in Fig. 19.3-2. The frequency scale is logarithmic and the transfer function which is plotted represents the voltage amplitude ratio

$$A_v = \frac{V_{\text{speaker}}}{V_{\text{input}}} = \frac{V_s}{V_i}$$

The graph should be interpreted in the following way:

> At $f = 5$ Hz the transfer function is $A_v = 71$. Thus if the input voltage amplitude is 0.1 V, then the output voltage amplitude will be $A_v V_i = (71)(0.1) = 7.1$ V.

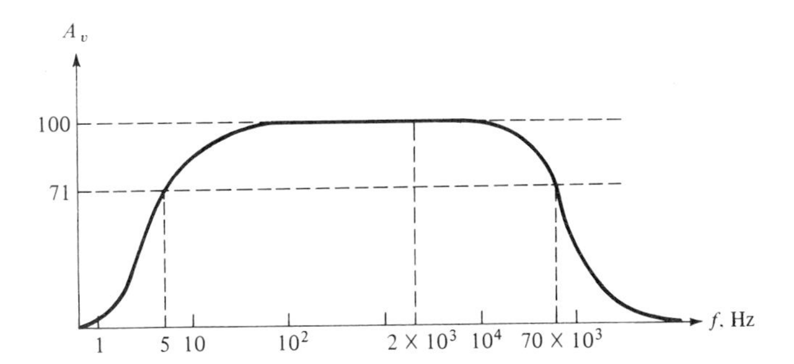

Fig. 19.3-2 High-fidelity amplifier response. $A_v = V_{\text{speaker}}/V_{\text{input}}$.

At $f = 2$ kHz the transfer function is $A_v = 100$. If the input is 0.1 V, the output will be $(100)(0.1) = 10$ V.

At $f = 70$ kHz the transfer function is again 71, so the output amplitude for 0.1 V input would be 7.1 V.

The reader will note that for frequencies below about 1 Hz and above about 100 kHz the response is very small compared with the response in the frequency band from 5 Hz to 70 kHz. Thus this amplifier is classified as a *bandpass* amplifier with a passband which extends from about 5 to 70 kHz. In a later section we will describe the specification of frequency passbands such as these in more detail.

19.3-1 Frequency Response from Transfer Functions

In the previous paragraph, an experimentally determined plot of frequency response was described. Frequency response information can also be obtained directly from a transfer function in the following way.

Consider the transfer function found in Example 19.2-1.

$$\mathbf{A}_v(\omega) = \frac{\mathbf{V}_2}{\mathbf{V}_1} = \frac{1}{1 + j\omega RC} \tag{19.2-3}$$

For a given frequency, $\mathbf{A}_v(\omega)$ is a complex value which relates the complex output phasor to the complex input phasor. Most often we are interested in the magnitude and phase of $\mathbf{A}_v(\omega)$. Since the magnitude of the quotient of two complex numbers is equal to the quotient of the individual magnitudes, we can write (recalling that $|x + jy| = \sqrt{x^2 + y^2}$)

$$A_v(\omega) = \frac{V_2}{V_1} = \frac{1}{\sqrt{1 + (\omega RC)^2}} \tag{19.3-1a}$$

The quantities $A_v(\omega)$, V_2, and V_1, when *not* printed with boldfaced type, represent the *magnitudes* of the original complex values, for example, $A_v(\omega) = |\mathbf{A}_v(\omega)|$.

The phase angle is [recall that $\arg(x + jy)^{-1} = -\arctan(y/x)$]

$$\varphi = \arg \mathbf{A}_v(\omega) = -\arctan(\omega RC) \tag{19.3-1b}$$

A graph of Eq. (19.3-1*a*) is the amplitude response and a graph of Eq. (19.3-1*b*) is the phase response. An example follows.

Example 19.3-1

a Plot the amplitude and phase response for the circuit of Fig. 19.3-3*a*.

b From the graphs find the output voltage when the input voltage is $2.5 \angle 30°$ V at a frequency of 250 Hz.

10 kΩ
$V_1 = 1\angle 0°$ V
0.1 μF
V_2

(a)

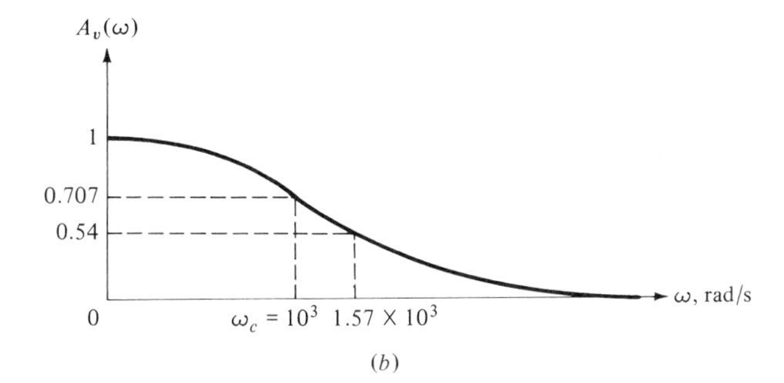

(b)

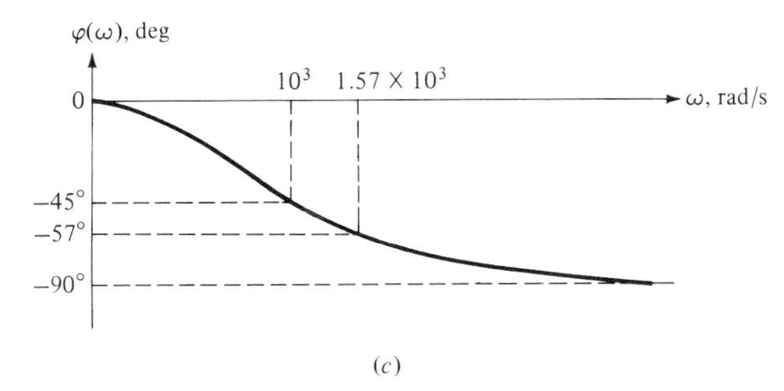

(c)

Fig. 19.3-3 *RC* circuit for Example 19.3-1. (*a*) Circuit; (*b*) amplitude-frequency response; (*c*) phase-frequency response.

Solution

a For the amplitude response we use Eq. (19.3-1*a*).

$$RC = (10 \times 10^3)(0.1 \times 10^{-6}) = 10^{-3}\ \text{s}$$

$$A_v(\omega) = \frac{1}{\sqrt{1 + (10^{-3}\,\omega)^2}} = \frac{1}{\sqrt{1 + (\omega/1{,}000)^2}}$$

For the phase response we use Eq. (19.3-1*b*):

$$\varphi = -\arctan\left(\frac{\omega}{1{,}000}\right)$$

The graphs are shown in Fig. 19.3-3*b* and *c*. Some key points are easily calculated:

At $\omega = 0$,

$$A_v = 1 \qquad \varphi = -\arctan 0 = 0°$$

At $\omega = 1{,}000$,

$$A_v = \frac{1}{\sqrt{1+1}} = 0.707 \qquad \varphi = -\arctan 1 = -45°$$

As $\omega \to \infty$,

$$A_v \to 0 \qquad \varphi \to -\arctan \infty = -90°$$

b At $f = 250$ Hz, $\omega = 2\pi f = 1{,}571$ rad/s. We read from the graphs the values $A_v(\omega) \approx 0.54$ and $\varphi \approx -57°$. Thus $\mathbf{A}_v(\omega) = 0.54 \angle -57°$. Now

$$\mathbf{A}_v(\omega) = \frac{\mathbf{V}_2}{\mathbf{V}_1}$$

so that

$$\mathbf{V}_2 = \mathbf{A}_v(\omega)\mathbf{V}_1 = (0.54 \angle -57°)(2.5 \angle 30°) = 1.35 \angle -27° \text{ V}$$

////

CUTOFF FREQUENCY

The amplitude response shown in Fig. 19.3-3*b* is typical of a large class of system responses which exhibit the same behavior. They all allow low-frequency signals to pass much more readily than high-frequency signals. The higher frequencies are said to be *attenuated.* Such systems are classified as *low-pass* and the circuit of Fig. 19.3-3*a* is called a low-pass filter. In order to distinguish between those frequencies that are passed and those that are stopped, a *cutoff* frequency f_c is defined as the frequency at which the amplitude response is $1/\sqrt{2} = 0.707$ times its maximum value. Thus the cutoff frequency of the low-pass filter of Fig. 19.3-3*a* is

$$f_c = \frac{1}{2\pi RC} = \frac{1{,}000}{2\pi} = 159 \text{ Hz} \tag{19.3-2}$$

Since power is proportional to the square of the voltage, the power at f_c is $(1/\sqrt{2})^2 = 1/2$ of the value at low frequencies. For this reason f_c is sometimes called the half-power frequency. An example of low-pass response to a square wave follows.

Example 19.3-2 The input voltage to the circuit of Fig. 19.3-4*a* is the square-wave signal shown in Fig. 19.3-4*b*. Find the output waveform.

Solution The first step is to represent the square wave by its Fourier series. Its fundamental angular frequency is $\omega_0 = 2\pi/T = 2\pi/0.02 = 314$

rad/s and its amplitude is 10 V. Using Eq. (19.1-3), its Fourier series is

$$v_1(t) \approx 12.7 \sin \omega_0 t + 4.24 \sin 3\omega_0 t + 2.55 \sin 5\omega_0 t + 1.82 \sin 7\omega_0 t \tag{19.3-3}$$

The second step is to compute the transfer function at each harmonic in the Fourier series. The transfer function magnitude and phase are given by Eqs. (19.3-1*a*) and (19.3-1*b*).

The general formulas for the amplitude and phase are

$$A_v(\omega) = \frac{1}{\sqrt{1 + (\omega/500)^2}} \quad \text{and} \quad \varphi = -\arctan \frac{\omega}{500}$$

A tabulation of the values at the odd harmonics of the square wave input is shown in the table below while sketches of the frequency response curves are identical to those in Fig. 19.3-3 with $\omega_c = 2\pi f_c$ changed to 500 rad/s.

ω	314	942	1,570	2,198
$A_v(\omega)$	0.847	0.469	0.303	0.222
$\varphi(\omega)$	−32.2°	−62.1°	−72.3°	−77.2°

The third step is to find the output waveform. This is done by multiplying the amplitude of each harmonic by the value of the transfer function at its frequency and adding the phase angle at the appropriate frequency. Using the values in the table above and the input Eq. (19.3-3), the output is

$$\begin{aligned} v_0(t) &= 10.8 \sin (\omega_0 t - 32.2^\circ) \\ &+ 1.99 \sin (3\omega_0 t - 62.1^\circ) \\ &+ 0.773 \sin (5\omega_0 t - 72.3^\circ) \\ &+ 0.404 \sin (7\omega_0 t - 77.2^\circ) \\ &+ \cdots \end{aligned} \tag{19.3-4}$$

A plot of this function is shown in Fig. 19.3-4*c*. ////

Example 19.3-3 . The circuit of Fig. 19.3-5*a* is a *high-pass* filter. Draw the frequency response plots for the circuit.

Solution The transfer function for the circuit is

$$\mathbf{A}_v(\omega) = \frac{\mathbf{V}_2}{\mathbf{V}_1} = \frac{j\omega L}{R + j\omega L} = \frac{j\omega(L/R)}{1 + j\omega(L/R)} \tag{19.3-5}$$

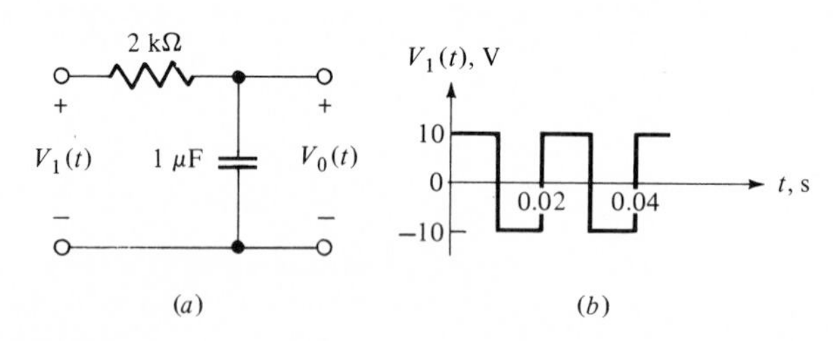

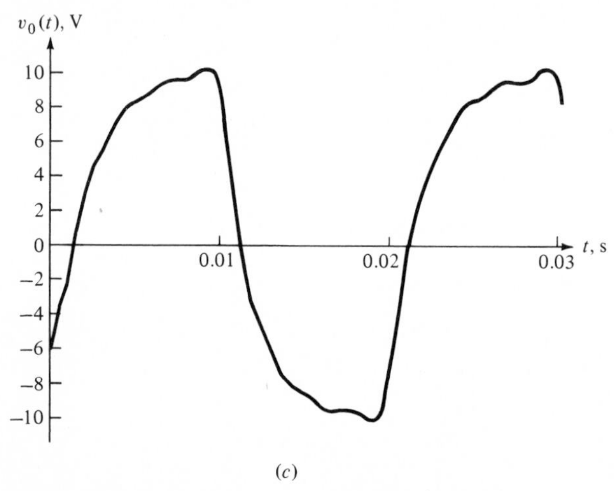

Fig. 19.3-4 Response of low-pass filter. (*a*) Circuit; (*b*) square-wave input; (*c*) output voltage.

The amplitude response is

$$A_v(\omega) = \frac{\omega(L/R)}{\sqrt{1 + (\omega(L/R))^2}} \tag{19.3-6}$$

and the phase response is

$$\varphi(\omega) = 90° - \arctan\left(\omega\frac{L}{R}\right) \tag{19.3-7}$$

Some key points are

At $\omega = 0$,

$$A_v(\omega) = 0 \qquad \varphi = 90°$$

At $\omega = R/L = \omega_c$,

$$A_v(\omega) = \frac{1}{\sqrt{2}} = 0.707 \qquad \varphi = 90° - \arctan 1 = 45°$$

As $\omega \to \infty$,

$$A_v(\omega) \to 1 \qquad \varphi \to 90° - 90° = 0°$$

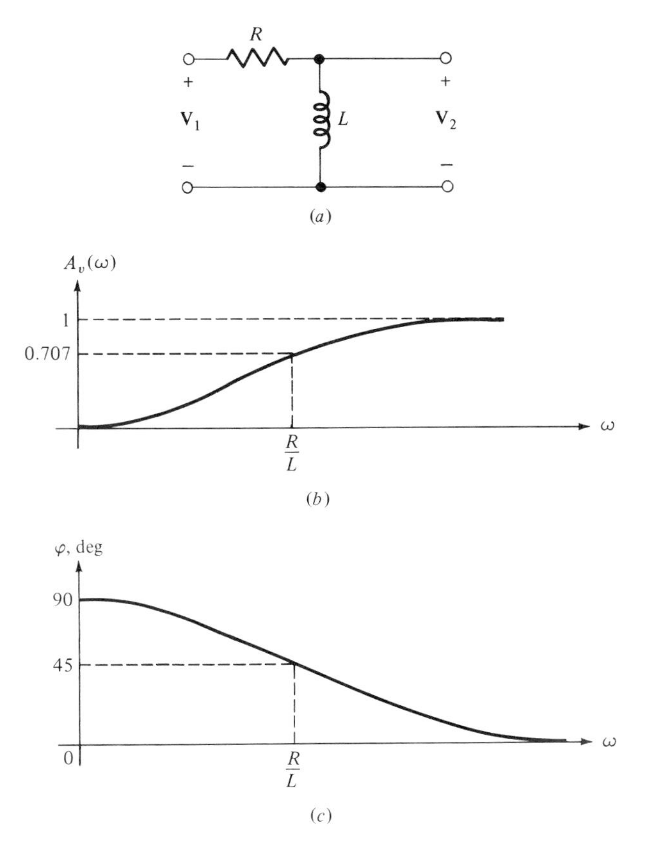

Fig. 19.3-5 High-pass filter frequency response. (*a*) Circuit; (*b*) amplitude response; (*c*) phase response.

The response curves are shown in Fig. 19.3-5*b* and *c*.

The circuit is called a high-pass filter because it attenuates low frequencies and passes high frequencies. Its cutoff frequency, which is the half-power frequency, is at $\omega = R/L$. ////

19.3-2 Frequency Response Using ECAP

Calculations of frequency response can become quite tedious. If ECAP (or a similar program) is available, they can be carried through with ease. The frequency is varied by using a MODIFY routine as illustrated in the following example.

Example 19.3-4 Use the ECAP program to determine the frequency response (amplitude and phase) of the circuit of Fig. 19.3-6*a* with $R = 1$ kΩ and $C = 1\ \mu$F. Cover the frequency range 40 to 300 Hz and plot the results on a logarithmic frequency scale.

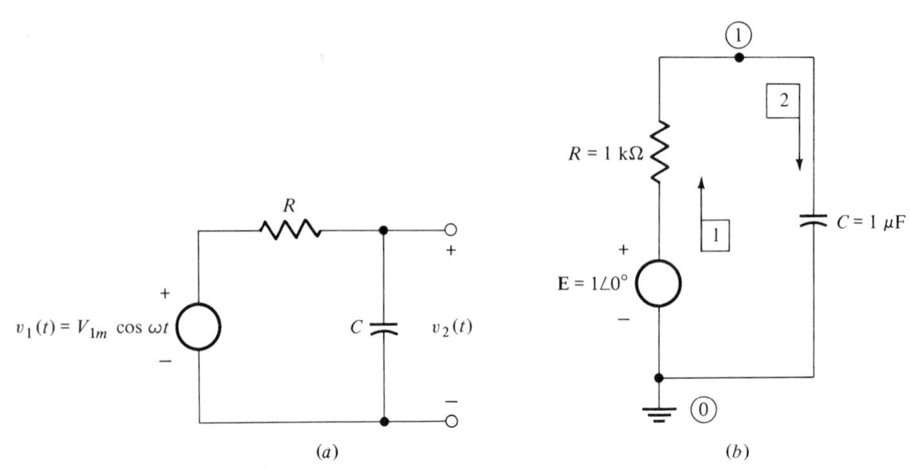

Fig. 19.3-6 ECAP frequency response. (*a*) Circuit for Example 19.3-4; (*b*) ECAP circuit.

Solution The ECAP circuit is shown in Fig. 19.3-6*b*, and the program in Fig. 19.3-7*a*. The only significant difference between this and previous ECAP examples is the MODIFY routine, which requests a frequency iteration, i.e., a calculation repeated at different frequencies. Two types of iteration are available:

Multiplicative This is used to obtain equally spaced points on a logarithmic scale. An example of the required frequency modification statement is

FREQ = 40(1.1)300

Here 40 Hz is the lower frequency limit, and 300 Hz the upper limit. The number 1.1 in parentheses is a *multiplication* factor. This statement requests calculation at the frequencies 40 Hz, (1.1)(40) Hz, $(1.1)^2(40)$ Hz, . . . , up to the nearest step greater than 300 Hz.

Additive This is used to obtain equally spaced points on a linear scale. The frequency statement is shown in Fig. 19.3-7*a* after the MODIFY statement. The lower frequency limit is 40 Hz, and the upper limit is 300 Hz. The +13 in parentheses requests calculations in 13 equal frequency steps between 40 and 300 Hz, i.e., at intervals of 20 Hz.

Note that the presence or absence of the plus sign in the parentheses tells the computer which type of iteration is desired.

A partial printout is shown in Fig. 19.3-7*b* and a plot of the results is shown in Fig. 19.3-8 using semilog coordinates. ////

19.4 RESONANT CIRCUITS, THE RLC BANDPASS FILTER

The parallel "tuned" circuit of Fig. 19.4-1*a* is often used as a bandpass filter when the signal occupies a narrow band of frequencies around a center frequency

```
C       FREQUENCY RESPONSE
C       NEW YORK INSTITUTE OF TECHNOLOGY
C
        AC ANALYSIS
    B1  N(0,1),R=1E 03,E=1E 00/0
    B2  N(1,0),C=1E-06
C
        FREQUENCY=40
C
        PRINT,NV
        EXECUTE


FREQ =   0.40000000E 02



        NODES                    NODE VOLTAGES


MAG       1-   1   0.96983886E 00
PHA               -0.14107789E 02
```

(*a*)

```
C
C
        MODIFY
        FREQUENCY=40(+13)300
        EXECUTE


FREQ =   0.40000000E 02



        NODES                    NODE VOLTAGES


MAG       1-   1   0.96983886E 00
PHA               -0.14107789E 02


FREQ =   0.59999985E 02



        NODES                    NODE VOLTAGES


MAG       1-   1   0.93571538E 00
PHA               -0.20655960E 02
```

(*b*)

Fig. 19.3-7 Example 19.3-4. (*a*) Program; (*b*) partial printout.

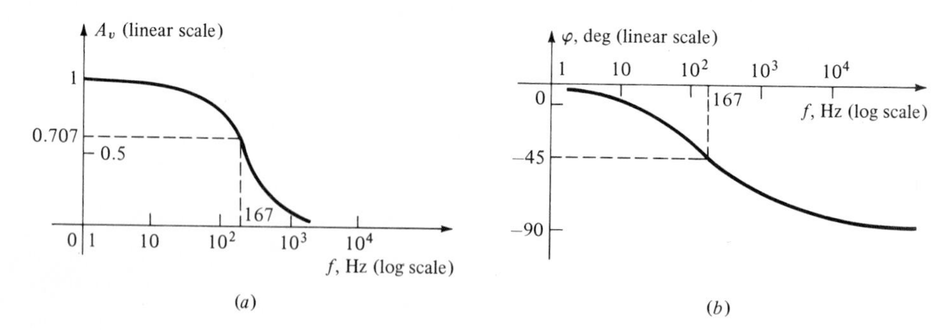

Fig. 19.3-8 Low-pass filter response. (*a*) Amplitude characteristic; (*b*) phase characteristic.

f_0. For this case there are two cutoff frequencies f_h and f_l (above and below the center frequency f_0) at which the response is $1/\sqrt{2}$ times the center-frequency value, as shown in Fig. 19.4-1*b*. The object of our analysis is to find these frequencies, and hence the bandwidth, in terms of R, L, and C.

We begin by writing the transfer function

$$\frac{\mathbf{V}_0}{\mathbf{I}_i} = \mathbf{Z}(\omega) = \frac{1}{1/R + j\omega C + 1/j\omega L} \tag{19.4-1a}$$

Note that the transfer function of interest is the impedance. To put this in a form suitable for interpretation, we first multiply numerator and denominator by R to get

$$\mathbf{Z}(\omega) = \frac{R}{1 + j\omega RC + R/j\omega L}$$

$$= \frac{R}{1 + jRC(\omega - 1/\omega LC)} \tag{19.4-1b}$$

Noting that LC must have the dimensions of seconds squared, we define

$$\omega_0^2 = \frac{1}{LC} \tag{19.4-2}$$

Substituting Eq. (19.4-2) into Eq. (19.4-1*b*),

$$\mathbf{Z}(\omega) = \frac{R}{1 + jRC(\omega - \omega_0^2/\omega)}$$

$$= \frac{R}{1 + j\omega_0 RC(\omega/\omega_0 - \omega_0/\omega)} \tag{19.4-3}$$

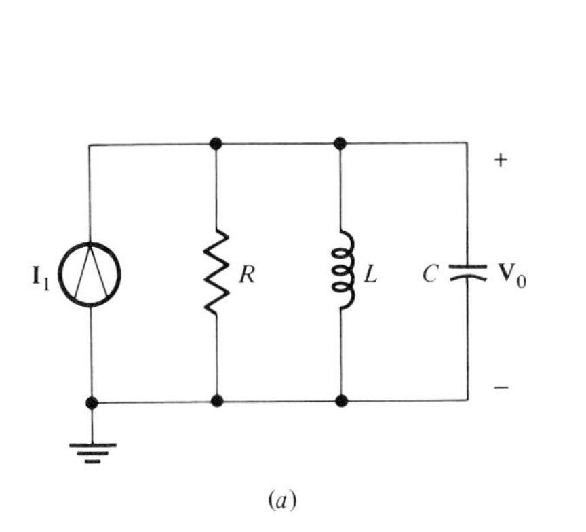

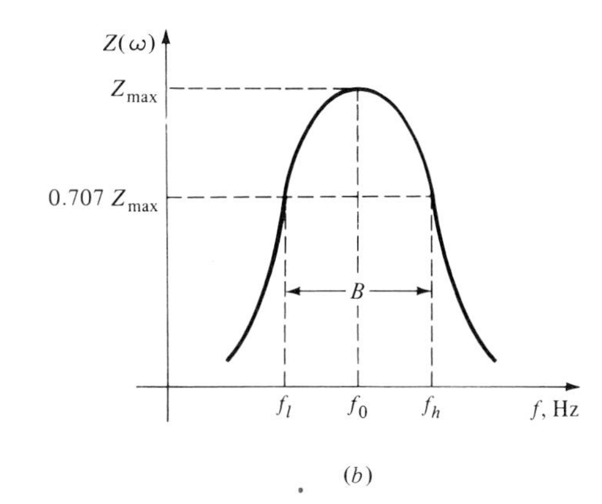

Fig. 19.4-1 Bandpass filter. (*a*) Circuit; (*b*) response.

It is convenient to define *normalized* frequency

$$x = \frac{\omega}{\omega_0} \tag{19.4-4}$$

and a *quality factor* (to be explained shortly)

$$Q = \omega_0 RC = \omega_0^2 \frac{RC}{\omega_0} = \frac{1}{LC} \frac{RC}{\omega_0} = \frac{R}{\omega_0 L} \tag{19.4-5}$$

With these definitions, the transfer function becomes

$$\mathbf{H}(x) = \frac{\mathbf{Z}(x)}{R} = \frac{1}{1 + jQ(x - 1/x)} \tag{19.4-6}$$

where we have divided through by R in order to *normalize* the expression.

Inspection of Eq. (19.4-6) indicates that the maximum amplitude of $\mathbf{H}(x)$ occurs at $x = 1$, that is, $\omega = \omega_0$. Thus $\omega_0/2\pi$ is the *center frequency*, and at this frequency the magnitude of $\mathbf{H}$ is unity and the impedance is simply R, a pure resistance. (Note that at $\omega = \omega_0$, the equivalent impedance of the inductor in parallel with the capacitor is infinite. Why?)

To find the cutoff frequencies, we look for those values of x for which the magnitude of $\mathbf{H}(jx) = 1/\sqrt{2}$, that is,

$$|\mathbf{H}(jx)| = \frac{1}{|1 + jQ(x - 1/x)|} = \frac{1}{\sqrt{2}} \tag{19.4-7}$$

Since $|1 \pm j1| = \sqrt{2}$, we see that the solution to this equation occurs where

$$\pm Q\left(x - \frac{1}{x}\right) = 1 \tag{19.4-8a}$$

Multiplying through by x and collecting terms yields

$$x^2 \mp \frac{x}{Q} - 1 = 0 \tag{19.4-8b}$$

Solving for x,

$$x = \pm \frac{1}{2Q} \pm \sqrt{\frac{1}{(2Q)^2} + 1} \tag{19.4-8c}$$

Thus there are four distinct values of x which satisfy Eq. (19.4-8c).

Since Q is always positive, we have

$$\sqrt{\frac{1}{(2Q)^2} + 1} > \frac{1}{2Q}$$

Also, $x = \omega/\omega_0$ must always be positive, so that we can discard the two negative values, leaving

$$x_h = \frac{\omega_h}{\omega_0} = \sqrt{1 + \frac{1}{(2Q)^2}} + \frac{1}{2Q} \tag{19.4-9a}$$

$$x_l = \frac{\omega_l}{\omega_0} = \sqrt{1 + \frac{1}{(2Q)^2}} - \frac{1}{2Q} \tag{19.4-9b}$$

These yield the required upper and lower cutoff frequencies.

The *bandwidth* B of the circuit is defined as the *difference* between the upper and lower cutoff frequencies. Thus, subtracting Eq. (19.4-9b) from (19.4-9a), we have

$$\omega_h - \omega_l = \frac{\omega_0}{Q} \qquad \text{rad/s} \tag{19.4-10a}$$

The bandwidth in hertz is then

$$B = \frac{\omega_0}{2\pi Q} = \frac{f_0}{Q} \qquad \text{Hz} \tag{19.4-10b}$$

The meaning of *quality factor* can now be explained. It is a measure of the *sharpness* of the filter. Higher values of Q lead to smaller relative bandwidths, B/f_0, or what is the same thing, sharper filters. Systems which exhibit the type of frequency response shown in Fig. 19.4-1b are often called *resonant* systems, and the response curve is called a *resonance* curve. The phenomenon of resonance is not significant unless the Q is at least 5 or more. For conventional *RLC* networks, values of Q in the range 10 to 100 are typical, while much higher Qs can be obtained by using transistors.

Specifications for the design of this type of network usually include f_0 and B.

From Eq. (19.4-2)

$$f_0 = \frac{1}{2\pi\sqrt{LC}} \qquad \text{Hz} \tag{19.4-11}$$

and using Eqs. (19.4-2) and (19.4-5),

$$B = \frac{1}{2\pi RC} = \frac{R}{2\pi L} \qquad \text{Hz} \tag{19.4-12}$$

These may be used as design equations, as shown in the example which follows.

Example 19.4-1 Design an *RLC* bandpass filter with a bandwidth of 3.5 kHz which is centered at 100 kHz. At the center frequency, the impedance is to be 10 kΩ.

Solution At the center frequency, Eq. (19.4-3) indicates that $\mathbf{Z}(\omega) = R$. Thus we choose $R = 10$ kΩ to meet the specification on impedance. Next, the bandwidth requirement of 3.5 kHz when substituted into Eq. (19.4-12) yields an *RC* product of

$$RC = \frac{1}{2\pi B} = \frac{1}{2\pi(3.5)(10^3)} \approx 0.045 \times 10^{-3}$$

Thus

$$C = \frac{0.045 \times 10^{-3}}{10^4} = 0.0045\ \mu\text{F}$$

Finally, from Eq. (19.4-2),

$$L = \frac{1}{\omega_0^2 C} = \frac{1}{(2\pi \times 10^5)^2(0.0045)(10^{-6})}$$

$$= 560\ \mu\text{H}$$

The circuit Q is, from Eq. (19.4-10*a*),

$$Q = \frac{f_0}{B} = \frac{10^5}{3.5 \times 10^3} = 29$$

The value is readily achieved with standard components. ////

Example 19.4-2 Use the ECAP program to determine the frequency response of the circuit designed in Example 19.4-1. Cover the frequency range 95 to 105 kHz on a linear scale.

Solution The ECAP circuit is shown in Fig. 19.4-2. The program is essentially the same as that used in Example 19.3-4. Since a *linear* frequency response calculation is required in this example, the additive iteration frequency statement follows the MODIFY statement in the program shown in Fig. 19.4-2*b* along with a partial printout. The lower frequency limit is 0.96 E 05(96 kHz)—95 kHz was done in the original analysis—and the upper limit is 0.105 E 06(105 kHz). The +9 in parentheses requests calculations in nine equal frequency steps between 96 and 105 kHz, that is, at intervals of 1 kHz.

A plot of the results is shown in Fig. 19.4-3. The center frequency is very close to 100 kHz, and the half-power frequencies occur at approximately 98.5 and 102 kHz. This agrees well with the original design values of Example 19.4-1, that is, $f_0 = 100$ kHz and $B = 3.5$ kHz. ////

19.5 CORRELATION OF TIME AND FREQUENCY RESPONSE

In Sec. 17.5, we studied the pulse response of the *RC* low-pass filter and found its 10 to 90 percent rise time to be

$$t_r = 2.2RC \tag{17.5-6}$$

In Sec. 19.3 we found the cutoff frequency of the same filter to be

$$f_c = \frac{1}{2\pi RC} \tag{19.3-2}$$

Solving for *RC* in Eq. (19.3-2) and substituting the result into Eq. (17.5-6) we obtain

$$t_r = \frac{2.2}{2\pi f_c} = \frac{0.35}{f_c} \tag{19.5-1}$$

Thus if we require a filter with a specific rise time, its cutoff frequency is then fixed by this relation.

In the design of digital systems it is often required to design for the shortest possible rise time. Analysis of different systems leads to relations similar to Eq. (19.5-1) indicating that rise time is inversely proportional to cutoff frequency. Thus in order to achieve short rise times the bandwidth must be as wide as possible.

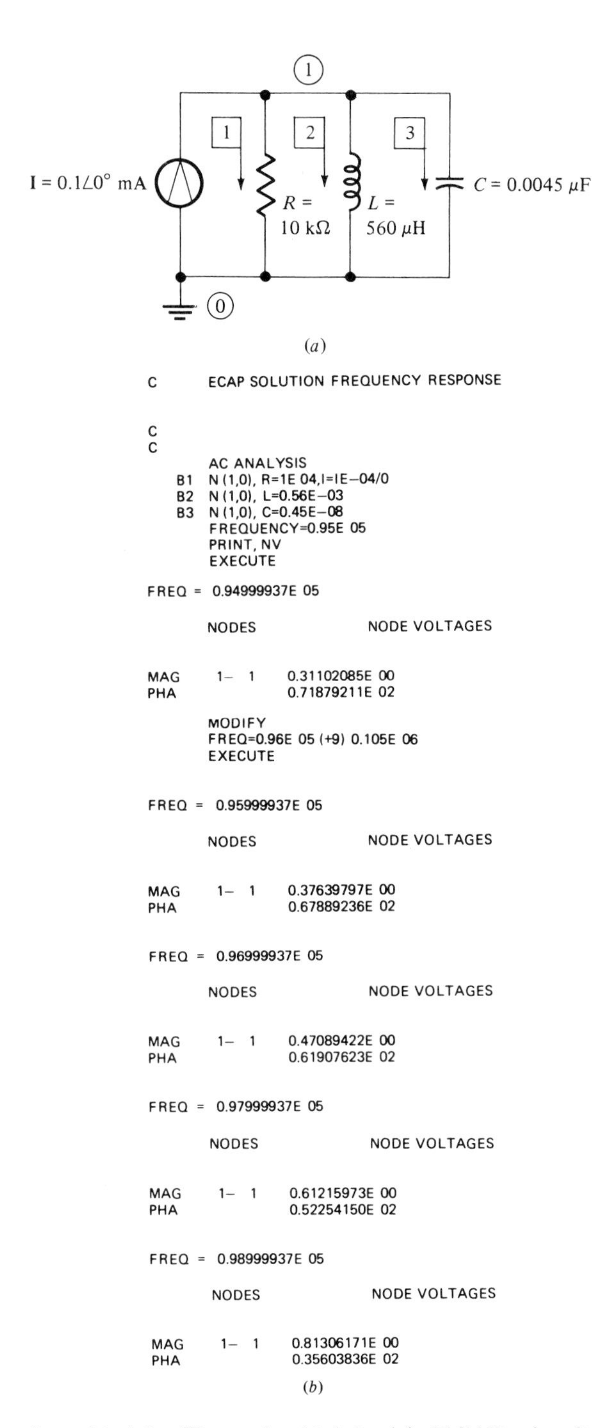


```
C          ECAP SOLUTION FREQUENCY RESPONSE


C
C
           AC ANALYSIS
     B1    N (1,0), R=1E 04,I=IE−04/0
     B2    N (1,0), L=0.56E−03
     B3    N (1,0), C=0.45E−08
           FREQUENCY=0.95E 05
           PRINT, NV
           EXECUTE

FREQ =  0.94999937E 05

           NODES                 NODE VOLTAGES


MAG       1−  1      0.31102085E 00
PHA                  0.71879211E 02

           MODIFY
           FREQ=0.96E 05 (+9) 0.105E 06
           EXECUTE

FREQ =  0.95999937E 05

           NODES                 NODE VOLTAGES


MAG       1−  1      0.37639797E 00
PHA                  0.67889236E 02

FREQ =  0.96999937E 05

           NODES                 NODE VOLTAGES


MAG       1−  1      0.47089422E 00
PHA                  0.61907623E 02

FREQ =  0.97999937E 05

           NODES                 NODE VOLTAGES


MAG       1−  1      0.61215973E 00
PHA                  0.52254150E 02

FREQ =  0.98999937E 05

           NODES                 NODE VOLTAGES


MAG       1−  1      0.81306171E 00
PHA                  0.35603836E 02
```

(b)

Fig. 19.4-2 Example 19.4-2. (*a*) ECAP circuit; (*b*) program and partial printout of results.

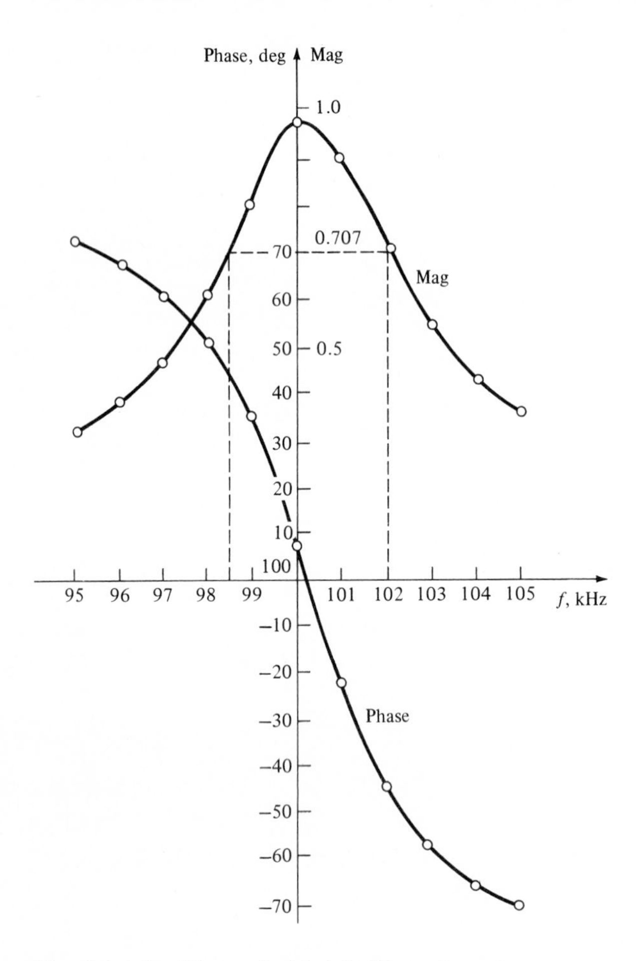

Fig. 19.4-3 Example 19.4-2. Plot of results.

SUMMARY

1 The *Fourier series* is an infinite sum of sinusoidal components which can be used to represent a periodic waveform.

a There is a dc component equal to the average value.
b The *fundamental,* or first, *harmonic* has a period equal to the period of the waveform.
c The frequencies of the other components are integral multiples of the fundamental frequency. The frequency of the nth harmonic is n times the fundamental frequency.

2 A signal with sharper and more rapid changes has relatively higher

amplitudes in its high-frequency components; smoother signals have relatively smaller high-frequency components.

3 The *frequency response* consists of two graphs:

a The *amplitude response* is a plot of the magnitude of the complex transfer function versus the angular frequency, ω.

b The *phase response* is a plot of the phase angle of the transfer function versus ω.

4 A *low-pass-filter* attenuates high-frequency components while passing low-frequency components.

A *high-pass filter* does the opposite. The *cutoff* frequency ω_c is that frequency at which the filter's amplitude response is equal to its maximum response divided by $\sqrt{2}$.

5 A *half-power frequency* is a frequency at which a circuit's amplitude response is $1/\sqrt{2}$ times its maximum value.

6 A *resonant* circuit has an amplitude response which peaks at a center frequency determined by its inductive and capacitive components. The *bandwidth* B is the frequency interval between the upper and lower cutoff frequencies. The *quality factor* Q measures the "sharpness" of the resonant peak, higher values of Q leading to sharper peaks.

PROBLEMS

19.2-1 Find the transfer functions, $\mathbf{A}_v(\omega) = \mathbf{V}_2/\mathbf{V}_1$, for each of the circuits in Fig. P19.2-1.

19.3-1 Find expressions for $A_v(\omega)$ and φ for the circuit of Fig. P19.2-1*a*. Determine $A_v(0)$, $A_v(\infty)$, $\varphi(0)$, and $\varphi(\infty)$. Find the half-power angular frequency and the phase angle at this frequency. Sketch A_v and φ as functions of ω.

19.3-2 The input for the circuit of Fig. P19.3-2*a* is shown in Fig. P19.3-2*b*.

a Find the circuit's transfer function.

b Sketch its frequency response plots. What kind of circuit is it?

c Select the value of R so that ω_c is equal to double the fundamental angular frequency of $v_1(t)$.

d Plot $v_2(t)$ for this case. *Hint*: Graphically subtract the voltage across the capacitor from $v_1(t)$.

19.4-1 For each of the circuits of Fig. P19.4-1 whose inputs are either $v_1(t)$ or $i_1(t)$ and whose outputs are either $v_0(t)$ or $i_0(t)$, determine

a The center frequency ω_0

b The bandwidth BW between the half-power points

c The circuit's Q

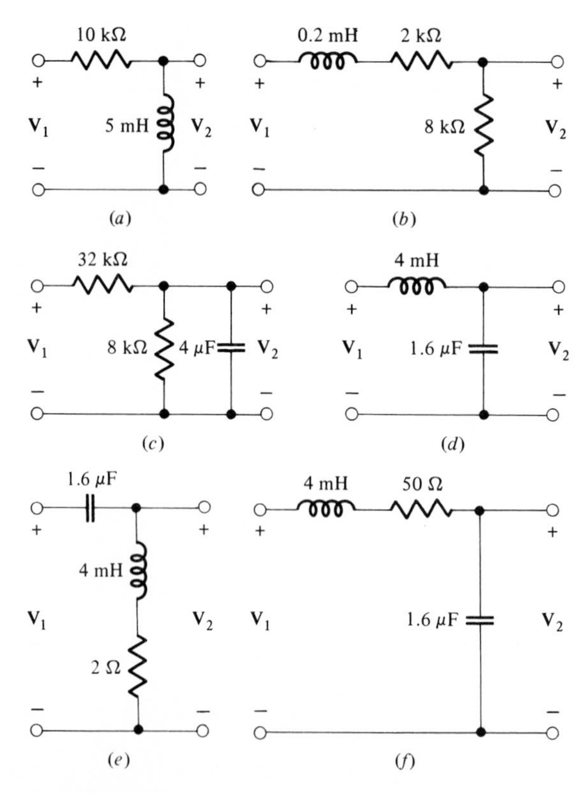

Fig. P19.2-1

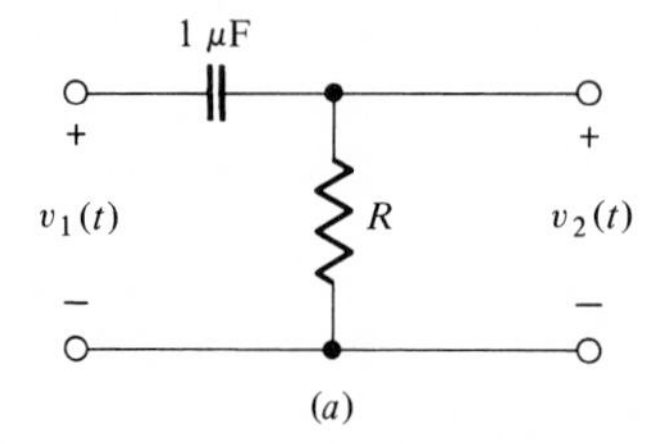

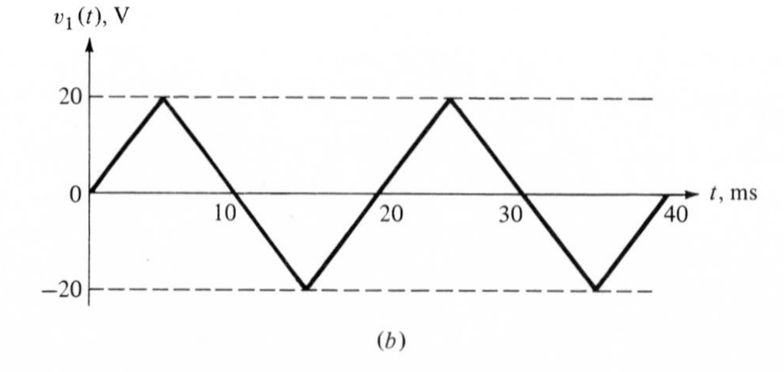

Fig. P19.3-2

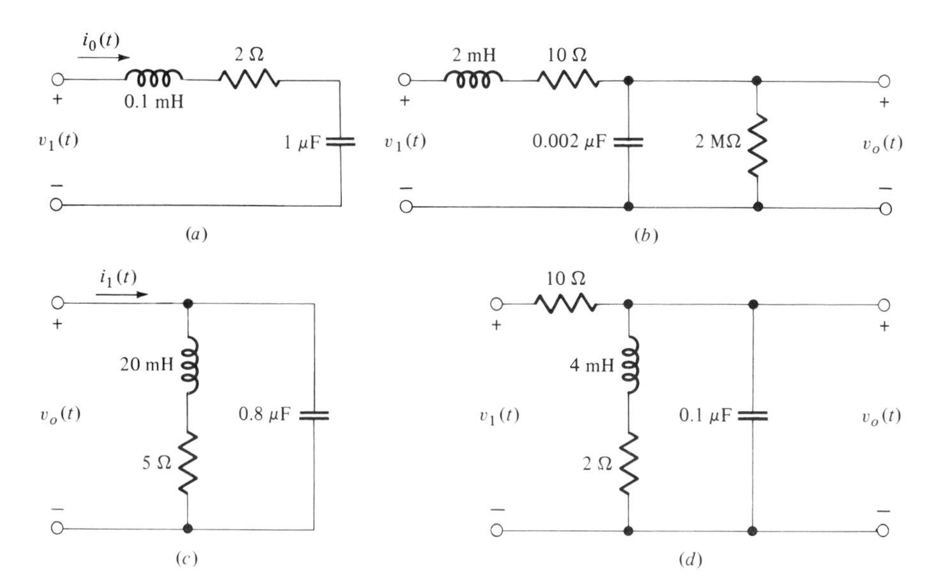

Fig. P19.4-1

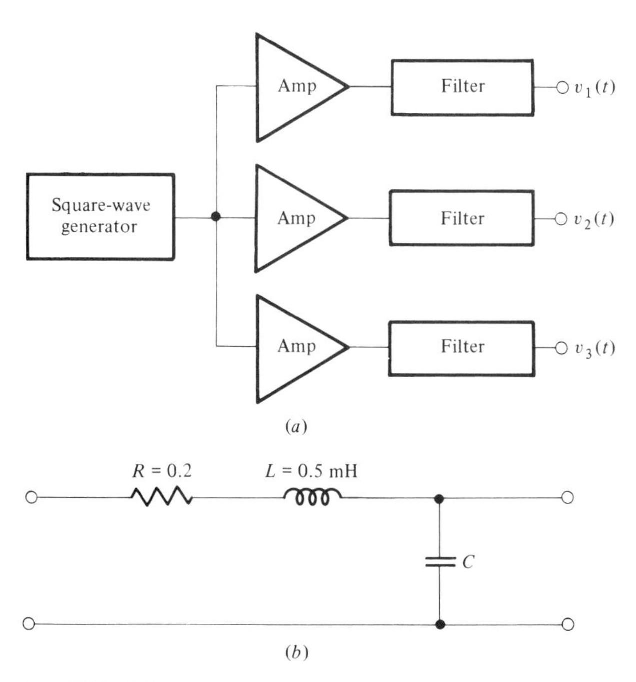

Fig. P19.4-3

For each high-Q circuit, sketch the amplitude and phase responses as a function of ω.

19.4-2 Design a series RLC circuit, i.e., find the values of R, L, and C, such that the resonant frequency is $f_0 = 1$ kHz (remember to convert to angular frequency), the bandwidth between half-power points is 100 Hz (which you must also convert to angular frequency), and the impedance at resonance should be 10 Ω.

19.4-3 Three sinusoidal signals are to be generated simultaneously using the system shown in Fig. P19.4-3*a*. Each filter has the form shown in Fig. P19.4-3*b*. They are resonant circuits being used as bandpass filters.

a If the square wave has a period of 1 s, what three frequencies would you try to obtain?

b What are the three values of C that you would use?

20
Ac Instruments

OBJECTIVES

Upon completion of this chapter you should be able to

1 Explain the operation of the half-wave and full-wave rectifier-type ac meter.
2 Find the conversion factor to be applied when using a rectifier-type meter for waveforms other than sinusoids.
3 Describe the short-shunt and long-shunt wattmeter connections and calculate true load power.
4 Describe the operation of a digital frequency meter.
5 Find the balance conditions for an ac bridge.
6 Describe the basic operation of the cathode-ray oscilloscope.

INTRODUCTION

In Chap. 7 we discussed the theory of dc instruments and the important concepts of instrument loading and measurement error. In this chapter we will consider instruments used for ac measurements. The theory of loading and error calculation given previously for dc circuits carries over directly to ac circuits.

As in dc measurements, two types of instruments are available: digital and analog. We will describe the most important of these. In addition, the oscilloscope, on which waveforms are displayed, will be described in some detail because of its importance. It may well be the most often used instrument in the typical electronics laboratory.

20.1 RECTIFIER-TYPE METERS

As noted in Chap. 7, the iron-vane and electrodynamometer meter movements will respond directly to ac waveforms, but their effective frequency range is limited to a few hundred hertz or less. Most ac voltmeters operate by converting some property of the ac waveform into a dc current which is passed through a D'Arsonval movement. The meter scale is calibrated in terms of the desired property of the ac waveform, even though the meter is actually responding to a dc current. A block diagram of the typical system is shown in Fig. 20.1-1.

Most often it is the rms value of the sine wave that is required, so most ac meters are calibrated to read rms. However, most types of ac meter do not actually respond to the rms value of the applied waveform so that incorrect values will be obtained unless the input is a sine wave.

Different types of meters respond to various waveform properties. These include the actual rms value, the average value of half-wave or full-wave rectified waveforms (see Chap. 9) and the peak or peak-to-peak value. These are shown for sine waves in Fig. 20.1-2. For each case, a different amount of dc current will flow through the D'Arsonval movement. In all but the true rms meter, the following problem arises. If, for example, a meter which responds to the full-wave rectified average is calibrated in terms of the rms value of a sine wave, it will not correctly indicate the rms value of any other waveform because the relationship between the rms value and the full-wave rectified average is different for sine waves than for other waveforms.

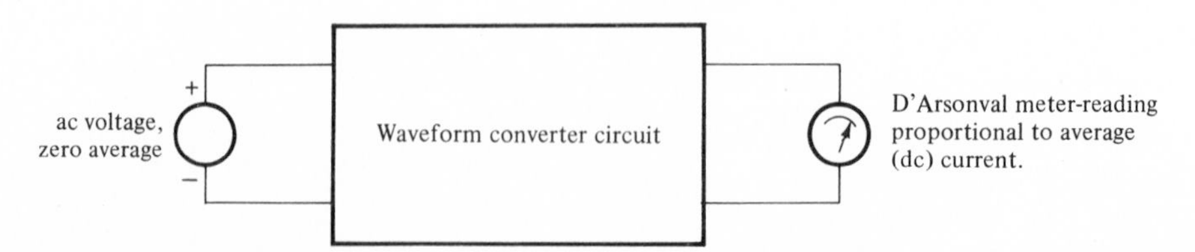

Fig. 20.1-1 Typical ac meter.

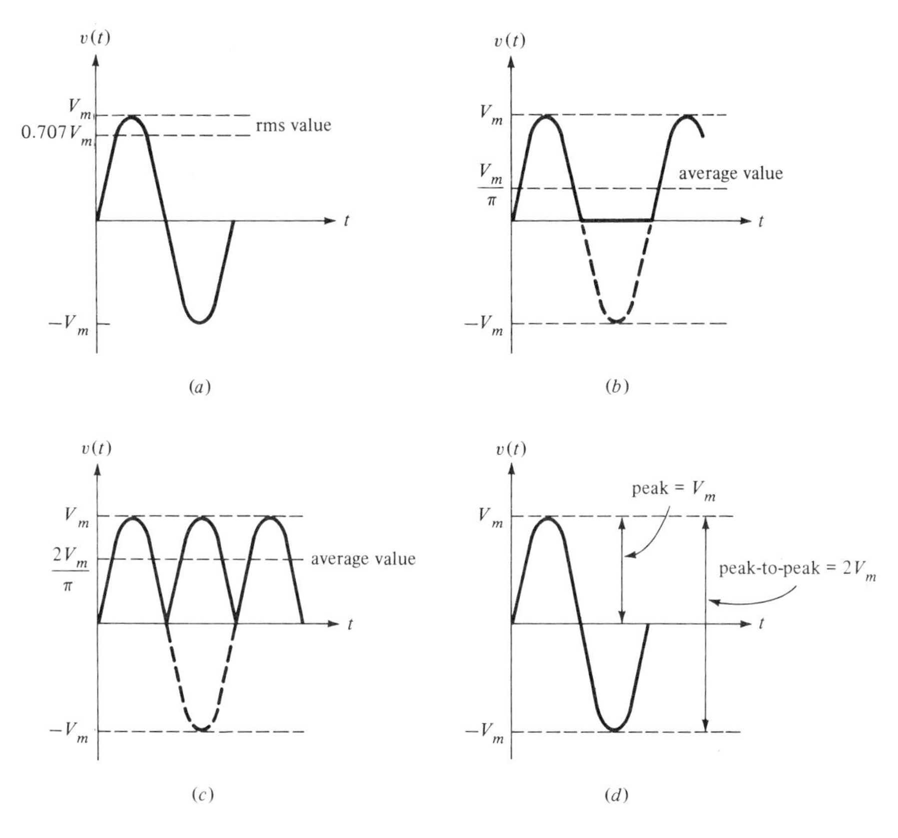

Fig. 20.1-2 Waveform properties used in converter circuits. (*a*) True rms value; (*b*) half-wave rectified waveform average value; (*c*) full-wave rectified waveform average value; (*d*) peak or peak-to-peak value.

In order to use such a meter for waveforms other than sine waves, the waveform to be measured must be known, and a correction factor must be applied. For example, consider that a meter with a full-wave rectifier is calibrated for sine waves but is to be used to measure the amplitude of a square wave. The rms value of the sine wave is $V_m/\sqrt{2}$ and the full-wave rectified average is $2V_m/\pi$. Thus the scale reading is $\pi/2\sqrt{2} = 1.11$ times the actual full-wave rectified average value of the input waveform. If the square wave to be measured has a peak value of V_0, then its full-wave rectified average is also V_0. The scale reading on the meter would be 1.11 times this or 1.11 V_0. Thus the scale reading must be multiplied by $1/1.11 = 0.9$ in order to obtain the amplitude of the square wave.

Sometimes the original ac waveform has a dc component. If the waveform converter has a capacitor in series to block the dc, the conversion factor may be found as in the preceding paragraph, otherwise the dc level must be taken into account.

The Simpson 260 VOM (schematic shown in Fig. 7.5-2) uses the full-wave rectified average and has a series capacitor present only when the OUTPUT terminal is used.

20.2 IRON-VANE AND ELECTRODYNAMOMETER METERS

The iron-vane and electrodynamometer meter movements were described in Sec. 7.6 and illustrated in Figs. 7.6-1 through 7.6-5. As we pointed out in that section, both of these meters can be used for either dc or ac. The dynamometer movement is inherently very accurate, and 0.1 percent accuracy is available. Because of the inductance of the coils the frequency response of this meter is limited to about 200 Hz for good accuracy. Another limitation is that 1 to 3 W of power is required to operate the movement. Available ranges are from 1 to 50 A and, using a multiplier resistor, from about 1 to 300 V.

For the iron-vane meter, the best accuracy available is about 0.5 percent, and about 1 W is usually required to operate the meter. Its frequency response is limited to about 125 Hz and available ranges are from about 10 mA to 50 A for current and as a voltmeter from about 1 to 750 V. This type is less expensive than the dynamometer movement and is used where extreme accuracy is not required.

20.3 THE WATTMETER

In order to measure power in a dc circuit, we can make use of the relation $P = VI$ and simply multiply voltmeter and ammeter readings. However, in an ac circuit, the problem is more difficult because the power factor angle (the phase angle between the voltage and the current) is involved. This would require a separate measurement of power factor or power factor angle. However, the electrodynamometer wattmeter, when connected as shown in Fig. 7.6-4 measures ac power directly, the mechanism automatically taking into account the power factor angle.

When using wattmeters, polarity must be carefully observed on both the current and potential coils as indicated in Fig. 7.6-4*b*. If either coil is reversed the meter deflection will be downscale.

There are two possible ways to connect the dynamometer wattmeter. The connection shown in Fig. 7.6-4*b* is called the *short-shunt* connection because the voltage coil is directly across the load. For this case the wattmeter reads the load power and the power lost in the voltage coil. In the *long-shunt* connection, the current coil is connected directly in series with the load, and the meter reads the load power plus the power lost in the current coil. If the resistance of the appropriate coil is known the meter reading can be corrected for this instrument power loss as shown in the example which follows.

Example 20.3-1 In the wattmeter circuit of Fig. 7.6-4*b* the meter reads 4 kW when the load voltage is 440 V. It is known that the resistance of the voltage coil is 5 kΩ. Find the true-load power.

Solution Since the voltage across the voltage coil is 440 V we have

$$P_{\text{lost}} = \frac{V^2}{R} = \frac{(440)^2}{5 \times 10^3} \approx 39 \text{ W}$$

Then

$$P_{\text{true}} = P_{\text{meter}} - P_{\text{lost}} = 4{,}000 - 39 \approx 3960 \text{ W} \qquad ////$$

The specifications for a wattmeter will include maximum ratings for both current and voltage coils. These ratings must not be exceeded: If they are, the meter may suffer physical damage. The problem of meter overload is not quite as simple for the wattmeter as it is for the ammeter or voltmeter because *either* the voltage or current rating may be exceeded even though the wattage reading is within range. For example, consider a wattmeter with a 1,000 W range, a maximum voltage rating of 100 V, and a maximum current rating of 10 A. If this wattmeter is connected to a circuit where the voltage is 60 V and the current is 15 A, the current reading will be exceeded by a large margin but the wattmeter reading will still be within range. This indicates that voltage and current levels should be approximately known before the wattmeter is connected. If it is known that the current rating will be exceeded, then a shunt can be used in parallel with the current coil. If the voltage rating will be exceeded, a multiplier resistor can be used in series with the voltage coil. In either case the wattmeter reading can be corrected by taking into account the shunt or multiplier resistor.

20.4 FREQUENCY METERS

The accurate measurement of frequency is often required in the laboratory. There are many methods for making this measurement, all of which fall into one of two categories: analog and digital.

Some common types of analog frequency meters include the vibrating-reed type, the resonant-circuit type, and the active filter type. The accuracy of these meters is about ±1 percent and they are usually limited to frequencies below a few MHz. The oscilloscope can be used for frequency measurement, but its accuracy is also limited to about ±1 percent.

In general, digital meters provide much higher accuracy and excellent stability along with their digital readout. They tend to be more expensive than their analog counterparts at present but their cost is dropping due to the use of integrated circuits.

The *digital frequency meter*, also called a *digital counter*, is a very accurate, easy to use instrument. Measurement ranges are from dc to over 40 GHz (1 GHz = 10^9 Hz), and typical accuracy is 1 part in 10^8 (= 10^{-6} percent).

Photographs of some commercial frequency meters are shown in Fig. 20.4-1 and a block diagram with waveforms is shown in Fig. 20.4-2. The input pulse generating circuit converts the input signal, waveform A, in Fig. 20.4-2b into a pulse train of one pulse per cycle, as in waveform B. The pulses are controlled by some property of the input waveform. In our case, each pulse occurs at the instant that the input waveform is going through zero with negative slope. This

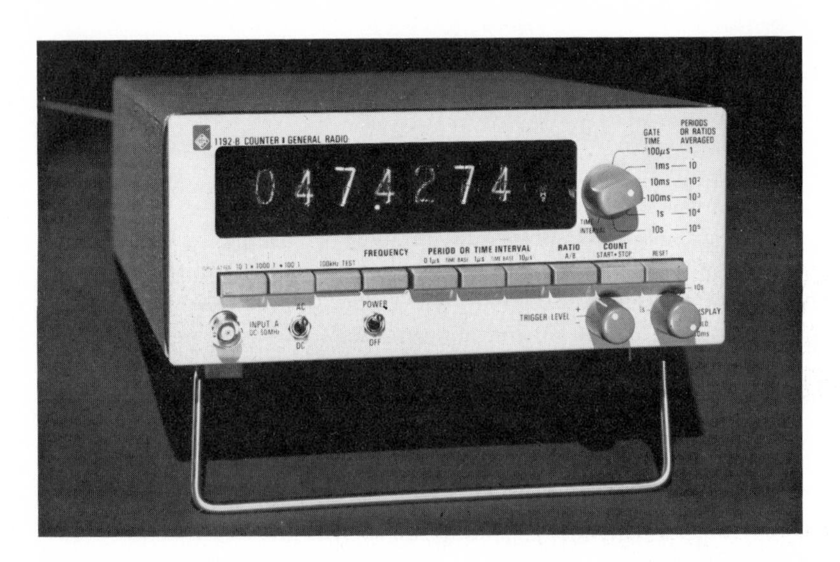

(a)

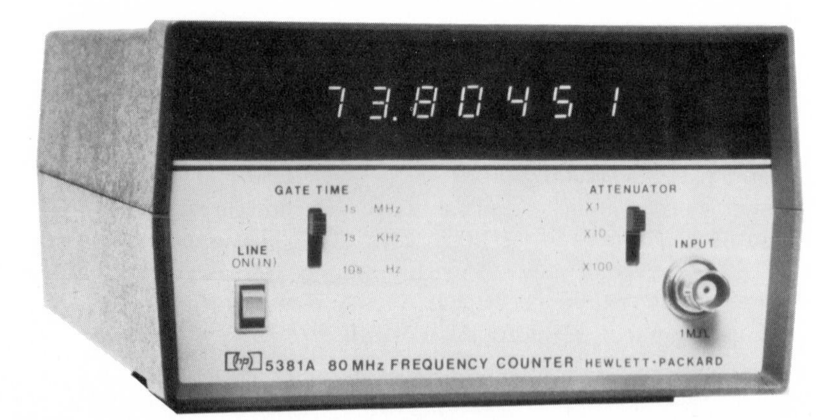

(b)

Fig. 20.4-1 Frequency meters. (*a, General Radio Co.*; *b, Hewlett-Packard.*)

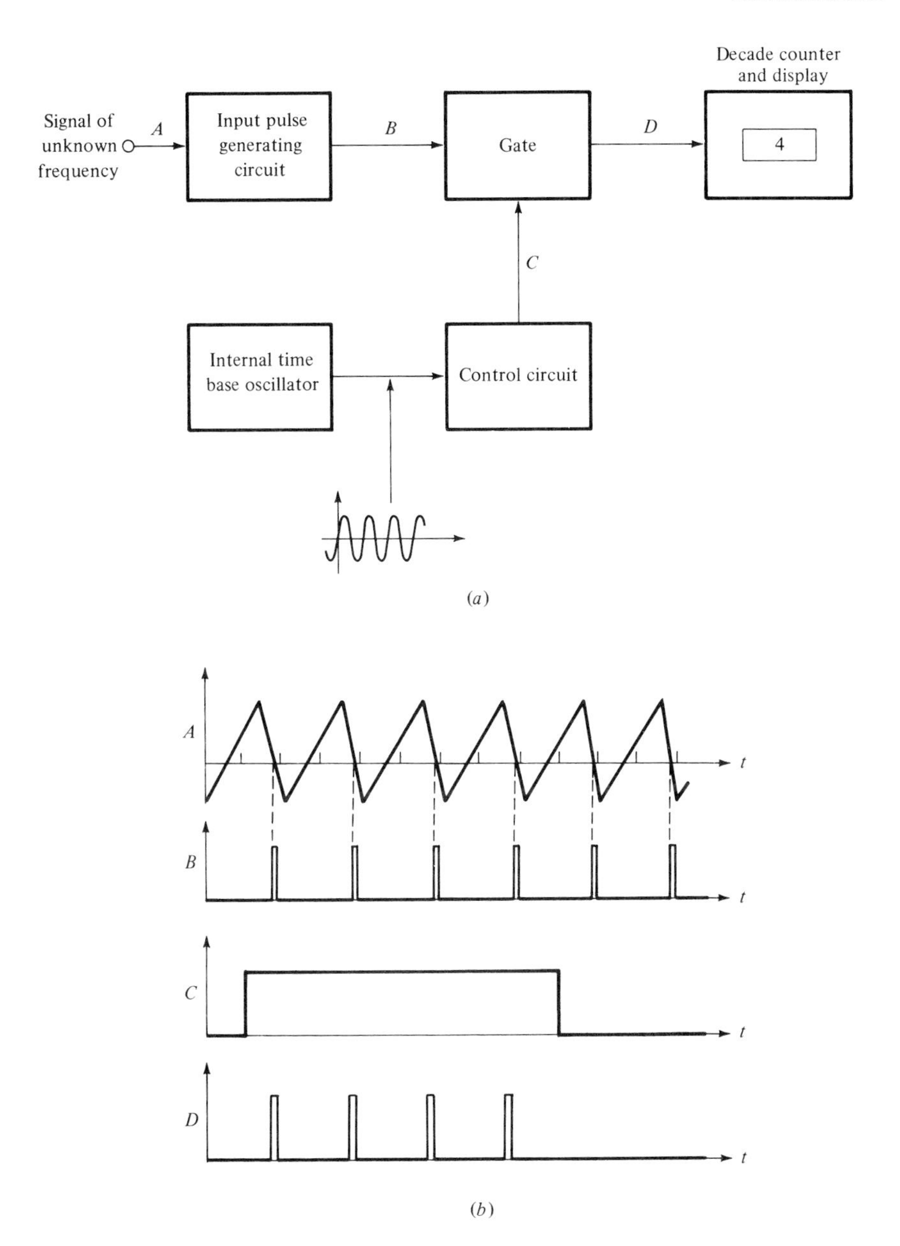

Fig. 20.4-2 Digital frequency meter. (*a*) Block diagram; (*b*) waveforms.

pulse train is applied to the gate circuit. The gating pulse (signal C) is derived from the internal time base oscillator by the control circuit and is 0.1 to 10 s in duration. This is called the counting time or sampling time. The action of the gate circuit is such that only those pulses from the input waveform which occur during the counting time are counted by the decade counter. Thus if the counting time is exactly 1 s the count displayed on the counter will be exactly equal to the number of cycles per second of the input waveform, i.e., the frequency, because we have exactly one pulse for each cycle. In Fig. 20.4-2*b*, four pulses will be counted during the counting time so the number 4 appears on the counter display, indicating that the frequency of the input waveform is 4 Hz.

The internal time-base oscillator is usually crystal-controlled and operates at a frequency of 100 kHz or 1 MHz. The control circuit is a frequency divider which counts a predetermined number of cycles of the time base oscillator output and from this generates gating signal C. If, for example, a gating signal of 1 s duration is desired and the time base frequency is 1 MHz, then the control circuit must count 10^6 cycles of the time-base output signal.

After each gating signal, there is a short time interval during which no measurements are made. The system is then reset, a gating pulse is generated, and a new measurement is made. While the new measurement is being made the previously measured value is displayed. The reset frequency is called the *sampling rate* or *display time* and is set by a control on the instrument.

The accuracy of the system is controlled by the accuracy of the counting time, which in turn depends on the stability of the time base oscillator. Typical accuracy is one part in 10^7 or 10^8. For higher accuracy, external time base signals may be fed into the instrument in place of the internal time base. The external signal would be obtained from a highly stable frequency standard.

Digital frequency meters may also be used in other ways. Some of these include direct display of the *period* of a waveform, display of the *ratio* of the frequencies of two different signals, or counting of events, either random or periodic.

20.5 IMPEDANCE BRIDGES

In Sec. 7.4 we showed how an unknown resistor could be measured by a *bridge* circuit. This idea can be extended to include the measurement of complex ac impedance as well. The circuit is shown in Fig. 20.5-1. It is similar to the Wheatstone bridge but the arms may be complex impedances, the source is usually a sinusoidal oscillator, and the null detector may be a sensitive ac voltmeter, or a set of earphones if the signal is in the audio range. As in the Wheatstone bridge, no current flows through the detector at balance, so the voltage across it is zero, and a development paralleling Eqs. (7.4-3) through (7.4-6) leads to the balance equation

$$\mathbf{Z}_x = \frac{\mathbf{Z}_3\mathbf{Z}_2}{\mathbf{Z}_1} \tag{20.5-1}$$

The unknown impedance $\mathbf{Z}_x$ will in general be complex, thus

$$\mathbf{Z}_x = R_x + jX_x = \frac{\mathbf{Z}_3\mathbf{Z}_2}{\mathbf{Z}_1} \tag{20.5-2}$$

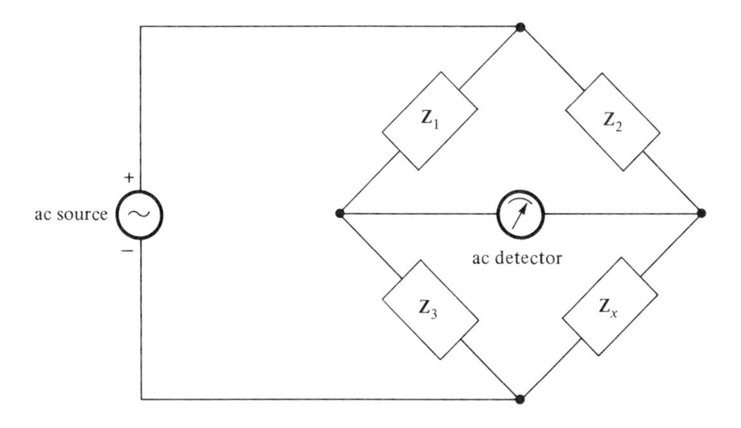

Fig. 20.5-1 Ac impedance bridge.

This is a complex equation, which will be satisfied only when the real parts of both sides are equal, and also the imaginary parts of both sides. It is sometimes convenient to choose $\mathbf{Z}_1$ and $\mathbf{Z}_2$ to be resistors such that their ratio is some appropriate power of 10. Then $\mathbf{Z}_3$ must be complex and both its resistive and reactive parts must be adjustable. The balance equation becomes

$$R_x + jX_x = (R_3 + jX_3)\frac{R_2}{R_1} \tag{20.5-3}$$

Equating real and imaginary parts separately, we have *two* balance equations

$$R_x = R_3\left(\frac{R_2}{R_1}\right) \tag{20.5-4}$$

$$X_x = X_3\left(\frac{R_2}{R_1}\right) \tag{20.5-5}$$

where R_2/R_1 is usually made equal to some convenient power of 10.

From the balance equation we see that two adjustments must be made in order to achieve balance, one to satisfy Eq. (20.5-4) and one to satisfy Eq. (20.5-5). Ideally, the two adjustments should be independent, i.e., when the real-part adjustment is being made to satisfy Eq. (20.5-4) there should be no effect on the balance in Eq. (20.5-5). If this is not so, then a number of successive adjustments of each component will have to be made in order to arrive at balance. Sometimes this process converges very slowly and special techniques must be used to speed up the procedure.

Example 20.5-1 The bridge circuit shown in Fig. 20.5-2 is called a Maxwell bridge and is used to measure inductance. Find the balance equations.

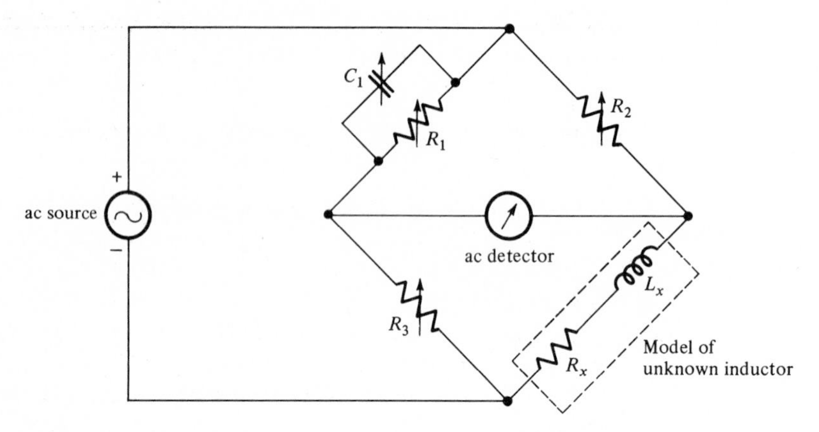

Fig. 20.5-2 Maxwell bridge.

Solution Because of the configuration of this bridge it is convenient to use Eq. (20.5-2) as the balance equation. The various impedances are

$$\mathbf{Z}_x = R_x + j\omega L_x$$

$$\mathbf{Z}_2 = R_2 \qquad \mathbf{Z}_3 = R_3$$

$$\mathbf{Z}_1 = R_1 \left\| \frac{1}{j\omega C_1} = \frac{R_1/j\omega C_1}{R_1 + 1/j\omega C_1} = \frac{R_1}{1 + j\omega R_1 C_1}\right.$$

Substituting into Eq. (20.5-2)

$$R_x + j\omega L_x = \frac{R_2 R_3}{R_1}(1 + j\omega R_1 C_1)$$

Equating real parts on both sides

$$R_x = \frac{R_2 R_3}{R_1} \tag{20.5-6}$$

Equating imaginary parts on both sides

$$\omega L_x = \frac{R_2 R_3}{R_1}(\omega R_1 C_1)$$

This reduces to

$$L_x = C_1 R_2 R_3 \tag{20.5-7}$$

When using this bridge R_2 and R_3 would first be set to give a convenient power of 10 multiplier. Next, R_1 would be adjusted for the best possible null. Then C_1 would be adjusted to improve this null. Then back to R_1 and then C_1 until the best possible null is obtained. ////

Additional bridge configurations will be considered in the homework problems.

20.6 THE OSCILLOSCOPE

In the study of physical devices and systems, we are usually concerned with input and output signals which vary, sometimes periodically, with time. It is very often desirable to view the actual time variations, and the cathode-ray oscilloscope ("scope" for short) is the instrument which provides the means for doing this if the signal can be transformed into a current or voltage. Modern-day oscilloscopes (see Fig. 20.6-1*a*) can operate over a voltage amplitude range from microvolts to kilovolts and a time range of minutes to nanoseconds.

In the oscilloscope the voltage is viewed on the face of the cathode-ray tube (CRT) which is the heart of the system (see Fig. 20.6-1*b*). Within the CRT there are three main components. An electron gun and focusing system provide a sharply focused beam of electrons which is directed at the fluorescent material coated on the inside of the CRT. This produces the spot which traces out the display. The deflecting system usually consists of two pairs of parallel plates, referred to as the vertical and horizontal plates, through which the electron beam passes. The system is usually arranged so that a positive voltage applied to the vertical input terminals deflects the beam upward in direct proportion to the applied voltage and a positive voltage on the horizontal input deflects the beam to the right in direct proportion to the applied voltage. Since the deflections are independent of each other, we can consider the tube face as an xy plane and the position of the spot can be expressed in terms of its x and y coordinates as follows:

$$x = K_x v_x \qquad y = K_y v_y \qquad \text{cm} \tag{20.6-1}$$

where v_x and v_y are the voltages applied to the input terminals of the scope and K_x and K_y are gain constants (units, centimeters per volt) relating spot position to input voltage. These gains include the effects of the amplifiers connected between the input terminals and the plates as shown in Fig. 20.6-1*c*.

We note at this point that the voltages v_x and v_y can be chosen so that the display is a TV picture, a graphical plot of a voltage at a particular point in a circuit, or even an artistic abstraction. In our work we shall most often be concerned with displaying voltage as a function of time. We next discuss means for accomplishing this.

DISPLAY OF VARIATIONS WITH RESPECT TO TIME

In order to display the variation of a voltage with time we start by applying a linearly increasing voltage to the x axis. This voltage can be expressed in the form

$$v_x = K_1 t \qquad (K_1 \text{ in volts per second}) \tag{20.6-2}$$

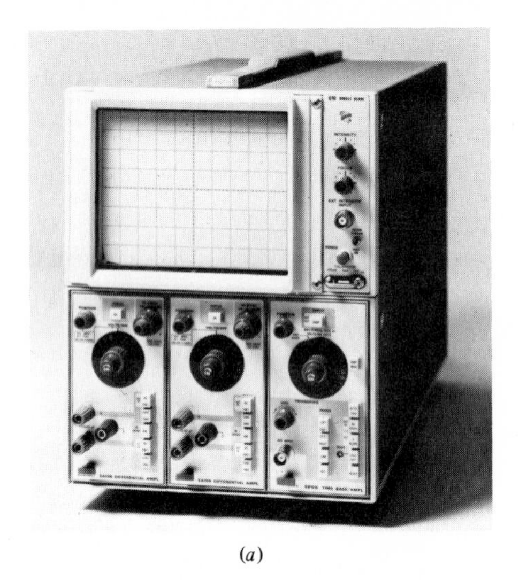

(a)

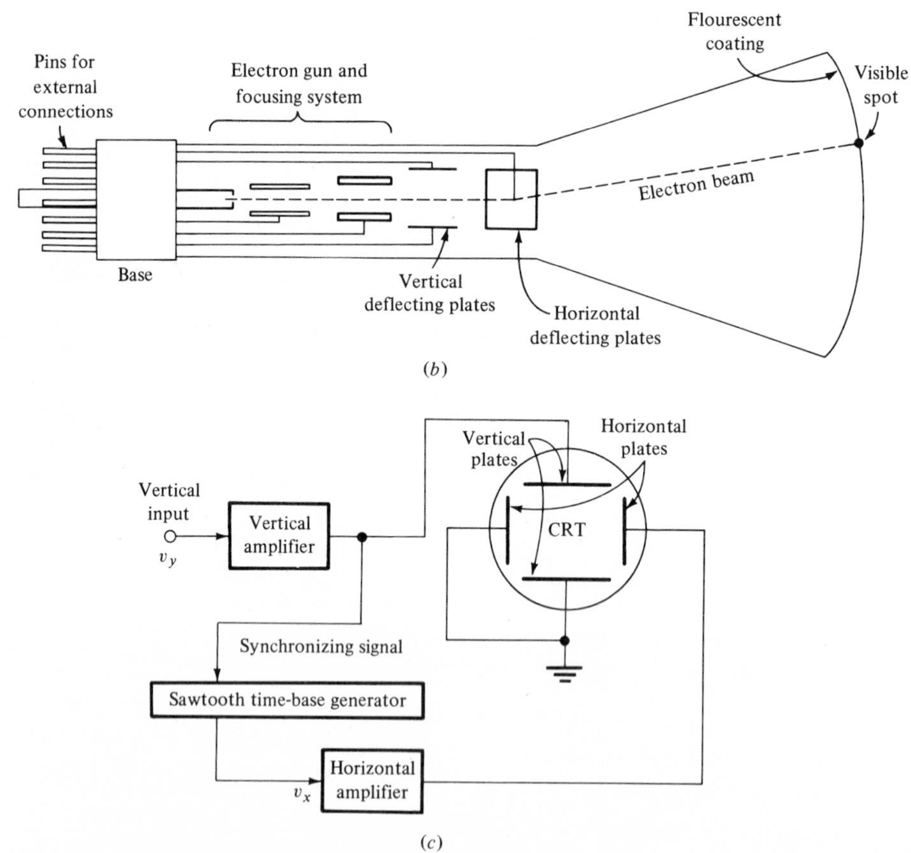

Fig. 20.6-1 The oscilloscope. (*a*) Tektronix type 5103N oscilloscope; (*b*) the cathode-ray tube; (*c*) simplified block diagram.

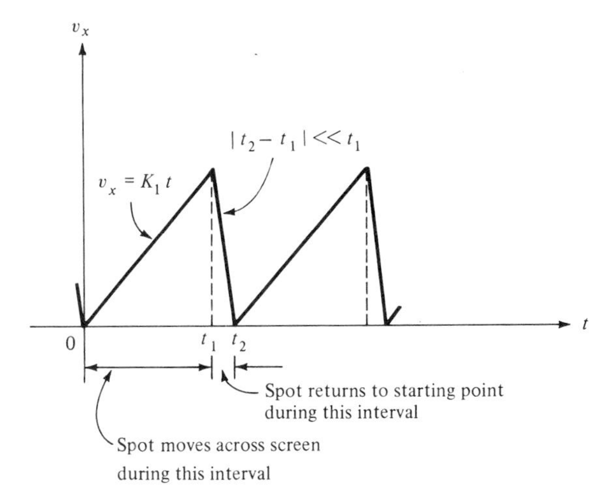

Fig. 20.6-2 Linear sweep waveform.

Thus the horizontal position of the spot is

$$x = K_x K_1 t$$

and if we combine constants so that $K_h = K_x K_1$ = spot velocity in centimeters per second, then

$$x = K_h t \tag{20.6-3}$$

and the beam sweeps across the face of the tube with constant velocity K_h. Clearly this is impractical because as soon as x becomes greater than the distance to the outer edge of the tube, our spot will disappear. This difficulty is overcome by using a *sawtooth* (called a linear sweep or time base), for v_x as shown in Fig. 20.6-2. With this waveform applied to the horizontal plates the beam *sweeps* back and forth across the tube face, with the return time interval very much less than the forward time interval. If v_y is a constant and the spot velocity is high enough, then persistence of vision will make the display appear to be a straight line.

Suppose now that v_y is a function of time $f(t)$. Let us determine the shape of the resulting display, first analytically and then graphically.

We have, assuming for convenience that $K_y = 1$ cm/V,

$$v_y = y = f(t) \tag{20.6-4}$$

Also, in the time interval from 0 to t_1 shown in Fig. 20.6-2, assuming $K_h = 1$ cm/s,

$$v_x = x = t \tag{20.6-5}$$

Substituting Eq. (20.6-5) into (20.6-4),

$$y = f(x) = f(t) \tag{20.6-6}$$

and the display is a normalized graph of the functional relation between v_y and t.

During the second and succeeding cycles of the sawtooth, the display may be completely different, depending on the function $f(t)$. This will usually produce an indecipherable jumble on the screen. If, on the other hand, $f(t)$ *repeats* itself one or any integral number of times during the 0 to t_1 interval, then the display will be the *same* during all sweep cycles. Such a function is called *periodic* and the display, due again to persistence of vision, will appear as a normalized graph of the number of *cycles* of $f(t)$ contained in the t_1 interval. Almost all oscilloscopes contain a sawtooth time-base generator and synchronizing circuits which can be set to automatically adjust K_h, the spot velocity, so that an integral number of cycles of the periodic waveform will appear on the screen.

GRAPHICAL DETERMINATION OF DISPLAY

The graphical method for determining the shape of the display is illustrated in Fig. 20.6-3 for the case where $f(t)$ is an exponential function.

The interval t_1 is divided into equal subintervals Δt. At $t = 0$, the spot will be at $x = y = 0$. At $t = \Delta t$ the spot will move to the point P_1, at the intersection of the projections from $v_y(\Delta t)$ and $v_x(\Delta t)$. At $t = 2\Delta t$ the spot will be at P_2, the intersection of the projections from $v_y(2\Delta t)$, and $v_x(2\Delta t)$, and so on. Clearly,

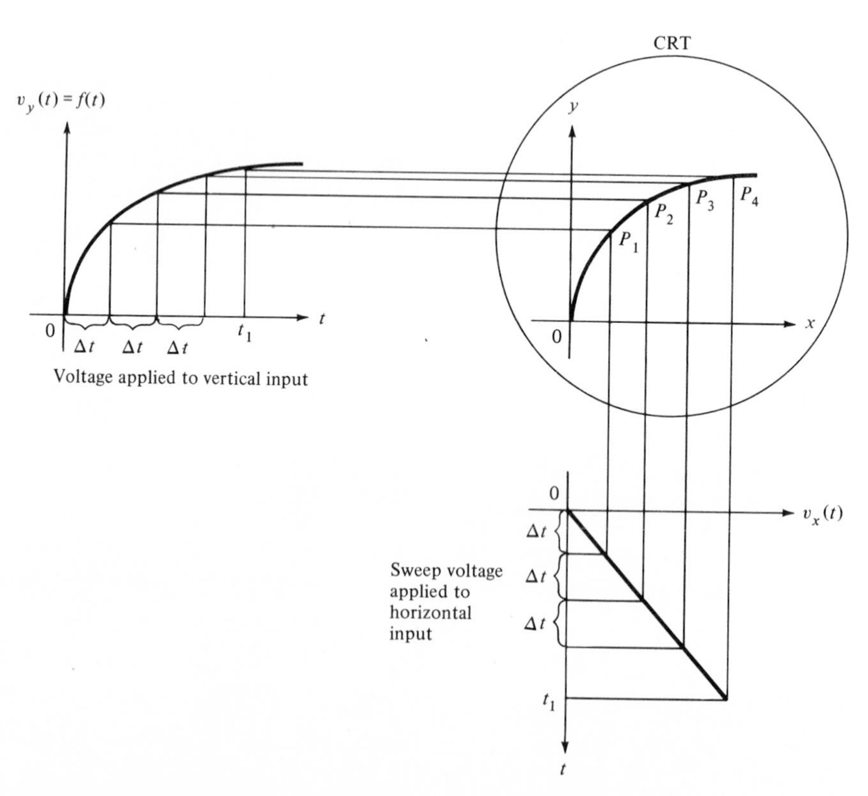

Fig. 20.6-3 Graphical construction for CRT display.

the curve generated by the sequence of points 0, P_2, . . . is the same as $v_y(t)$, because of the linear relation between v_x and t. At $t = t_1$ the spot rapidly returns to $x = y = 0$. Note carefully that this graphical procedure was applied only for the time interval 0 to t_1. Unless $v_y(t)$ repeats itself from t_1 to $2t_1$, the display will change as succeeding sweep cycles come into view. Many oscilloscopes have very slow sweep speeds available (1 s to traverse 2 cm and slower) so that slowly varying nonperiodic waveforms can be observed. In addition, long-persistence cathode-ray tubes are available so that a single sweep may be retained on the face of the tube for long periods of time.

OSCILLOSCOPE MEASUREMENT OF SINUSOIDS

A typical display of voltage vs. time is shown in Fig. 20.6-4, for a sinusoidal voltage applied to the vertical input terminals and a sawtooth sweep to the horizontal.

Most modern scopes have calibrated voltage and time axes. Thus it is a relatively simple matter to read voltage amplitude and time information directly from the display. The time information can be used to calculate frequency and phase.

The actual voltage is obtained by applying the calibration factor indicated on the *vertical sensitivity* dial. For example, if the vertical sensitivity is set at 100 mV/cm, then the waveform of Fig. 20.6-4 has a peak-to-peak voltage of

$$V_{p-p} = 4 \text{ cm} \times 100 \text{ mV/cm} = 400 \text{ mV}$$

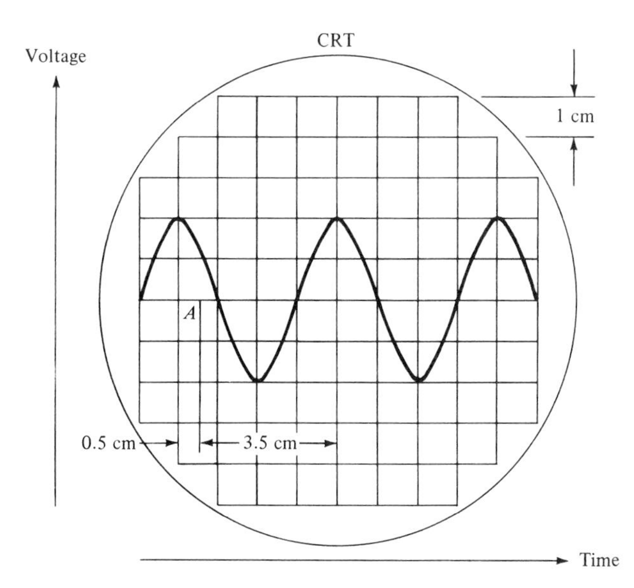

Fig. 20.6-4 Typical display.

Thus

$$V_m = \frac{1}{2} V_{p-p} = 200 \text{ mV}$$

The calibration of the time scale is given by the sweep speed on the *horizontal display* dial. If in Fig. 20.6-4, the sweep speed is 5 ms/cm, then one complete cycle occurs in

$$T = 4 \text{ cm} \times 5 \text{ ms/cm} = 20 \text{ ms} = 0.02 \text{ s}$$

The frequency is then

$$f = \frac{1}{t} = \frac{1}{0.02} = 50 \text{ Hz}$$

and the angular frequency is $2\pi f \approx 314$ rad/s.

In order to determine the phase we must know the location of the origin of the time scale. For present purposes, let us assume that the origin of the time scale is at point A in Fig. 20.6-4. Then, noting that 2 cm on the time scale is equivalent to 180° of phase shift (i.e., one-half of a complete cycle) we have

$$\theta = \frac{3.5 \text{ cm}}{2 \text{ cm}} \times 180° = 315°$$

We now have sufficient information to write the equation for the voltage being displayed. It is

$$v(t) = 200 \cos (2\pi \times 50t - 315°) \text{ mV} \qquad t \text{ in seconds}$$

In the example above, the accuracy of the result depends on the accuracy with which we can read the dimensions of the display and on the accuracy of the calibration factors as indicated on the dials. Typically, the overall accuracy of this type of measurement ranges from ± 3 to ± 10 percent.

Modern day oscilloscopes are extremely versatile and can be used to measure many properties of signals other than those described above. The basic principles, however, remain the same, the signal processing taking place in the amplifier circuits. Because of the differences in control designations, etc. among scopes of different manufacturers, additional details will be left for laboratory courses.

SUMMARY

1 In the usual rectifier-type ac meter, the rectifier converts the ac to pulsating dc and the D'Arsonval movement responds to the average value of the rectified waveform. The scale is usually calibrated in terms of the rms value of a sine wave so corrections must be applied for other waveforms.

2 Iron-vane and electrodynamometer meters are more expensive than

rectifier types but somewhat more accurate. They are used at low frequencies in some applications.

3 In the dynamometer wattmeter, the current I flows through one coil and the voltage V is applied to the other coil. The reading of the meter is then $VI \cos \theta$, where θ is the phase angle.

4 In a digital frequency meter, the number of cycles in a precisely known interval of time, e.g., 1 s, is counted and displayed. Such meters are characterized by excellent accuracy and negligible loading.

5 Impedance bridges for ac require two separate balancing operations to account for the real and imaginary parts of the unknown impedance.

6 An oscilloscope is used to view the actual time variation of the signal applied to its terminals. Calibrated controls allow for easy amplitude and time measurements.

REFERENCES

1 Wedlock, Bruce D. and J. K. Roberge: "Electronic Components and Measurements," Prentice-Hall, Inc., Englewood Cliffs, N.J., 1969.

2 Tiedemann, Arthur T.: "Elements of Electrical Measurements," Allyn and Bacon, Inc., Boston, Mass., 1967.

PROBLEMS

20.1-1 A D'Arsonval meter with a half-wave rectifier, calibrated to read rms for sine waves, is to be used to measure the peak value of a square wave. There is no dc component. Find the factor by which the scale reading must be multiplied to obtain the correct peak value of the square wave.

20.1-2 The full-wave rectifier meter described in Sec. 20.1 is connected to a 30-V peak-to-peak square wave which has no dc component. What will the meter reading be? The half-wave rectifier meter of Prob. 20.1-1 is connected to the same square wave. What will it read?

20.3-1 In the short-shunt wattmeter circuit of Fig. 7.6-4*b* the meter reads 10 kW when the load voltage is 660 V. The resistance of the voltage coil is 10 kΩ. Find the true load power. The meter is now connected long shunt and the resistance of the current coil is 0.1 Ω. Find the new meter reading.

20.5-1 Derive the balance equation for the inductance bridge shown in Fig. P20.5-1.

20.5-2 The bridge circuit shown in Fig. P20.5-2 is called a Hay bridge. Show

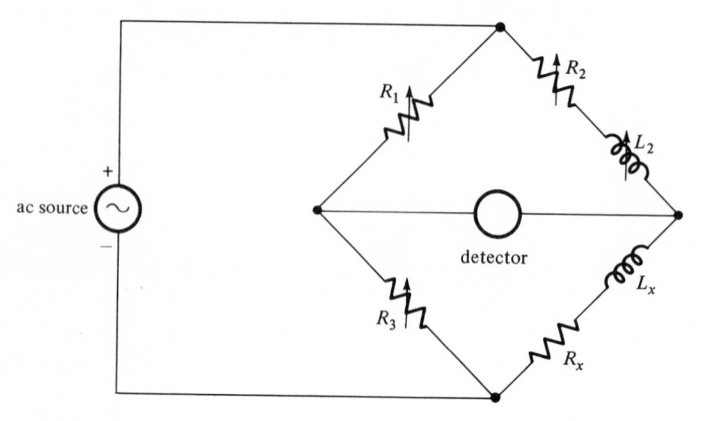

Fig. P20.5-1

that the balance equations are

$$L_x = \frac{C_1R_2R_3}{1 + \omega^2C_1^2R_1^2}$$

and

$$R_x = \frac{\omega^2C_1^2R_1R_2R_3}{1 + \omega^2C_1^2R_1^2}$$

Find L_x and R_x for the values indicated.

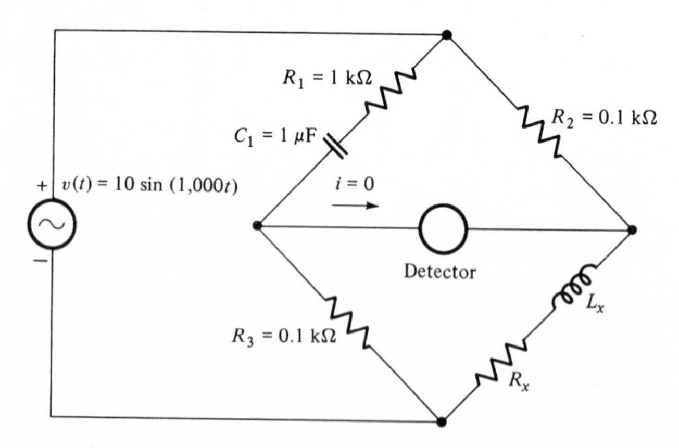

Fig. P20.5-2

20.5-3 In the Maxwell bridge of Fig. 20.5-2 $R_1 = 2$ kΩ, $C_1 = 3$ μF, $R_2 = 0.5$ kΩ, and $R_3 = 4$ kΩ. Find L_x and R_x.

20.6-1 For the display shown in Fig. P20.6-1, the vertical amplifier control is set at 0.05 V/cm and the sweep speed is 0.02 ms/cm. Find the amplitude and frequency of the sine wave.

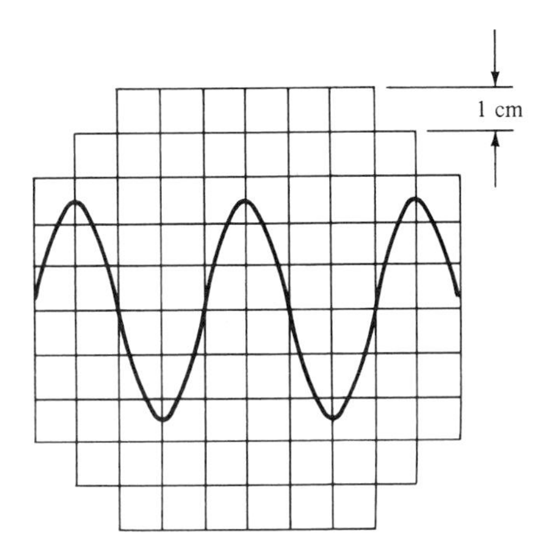

Fig. P20.6-1

20.6-2 In Fig. P20.6-2 we have shown one of many pairs of waveforms which generate the letter L on the scope face. Find a pair of waveforms which will generate a capital T. Assume that no discernible indication will appear during any required retracing.

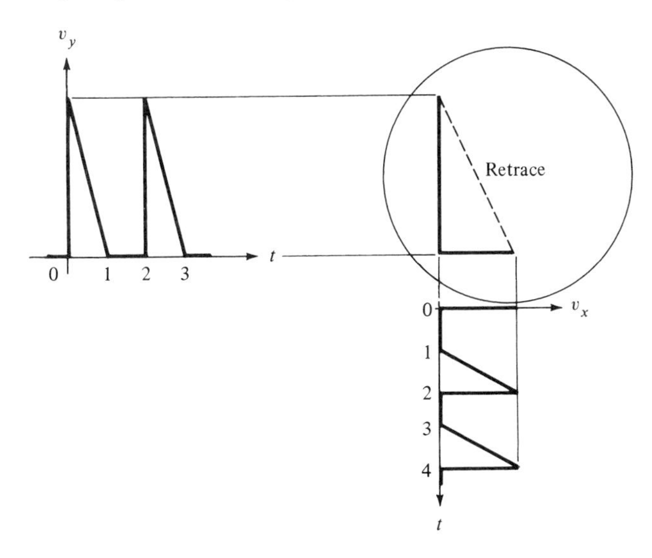

Fig. P20.6-2

20.6-3 Repeat Prob. 20.6-2 for a capital V.

21
Feedback and Operational Amplifiers

OBJECTIVES

Upon completion of this chapter, you should be able to

1 List the main characteristics of an operational amplifier.
2 Draw the operational amplifier circuits for a scaler, adder, and integrator.
3 Discuss the basic concepts of feedback.
4 Describe the effects of feedback on amplification, input resistance, output resistance, bandwidth, and stability.

INTRODUCTION

A *feedback* system is one in which a portion of the output signal is fed back and combined with the input. This arrangement is used to improve the system's performance, for example, to compensate for ambient temperature changes in a home heating system. In this system a thermostat measures the room temperature, which is the output signal, and turns the furnace on or off depending on whether this temperature is below or above the desired temperature.

In this chapter the effects of feedback on various system characteristics are investigated. The basic concepts of feedback are presented in terms of *operational amplifier* circuits.

21.1 FEEDBACK

Feedback is an essential ingredient in many engineering systems. It is used in guidance systems for missiles and space exploration rockets, in transistor amplifiers for stabilization, in home heating systems, and by the human body itself. Feedback makes it possible to compensate for external disturbances over which the designer has no control (the outside temperature in a home heating system for example) or to achieve precise operation of a system built with low precision components. We can illustrate this point and present the basic concepts of feedback with an example using one of the feedback systems in the human being. If a person is asked to make a dot in the circle of Fig. 21.1-1*a* with a pencil they can

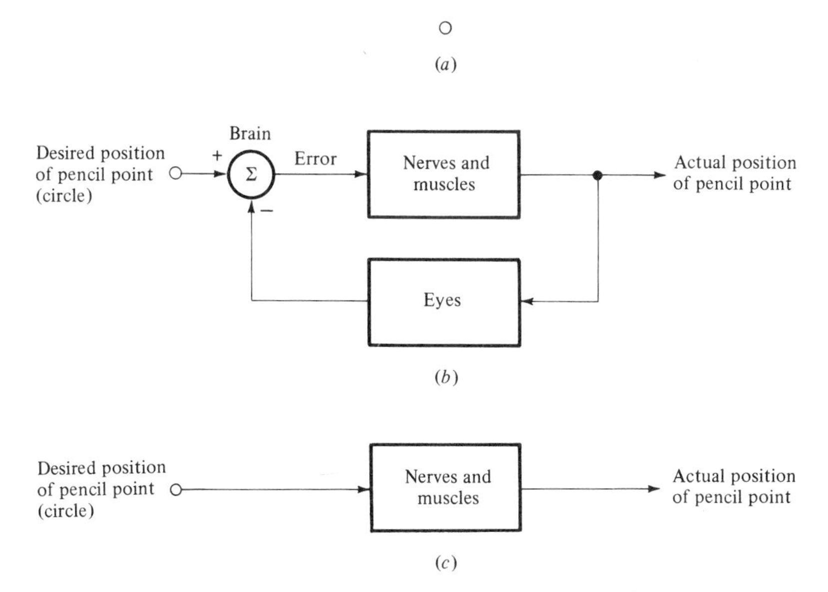

Fig. 21.1-1 Feedback. (*a*) Circle for experiment; (*b*) human feedback system; (*c*) human system without feedback.

generally do it rather easily. If they are asked to do it again, but to close their eyes after aiming the pencil from about one foot above the circle and to then make the dot, they generally miss the circle. The reason for this is that the system with the eyes open is a feedback system; when the eyes are closed the feedback path has been removed and the components of the system are simply not of sufficient precision to be able to place the pencil point in the circle.

A diagram of the system used to place the pencil point in the circle is shown in Fig. 21.1-1*b*. The feedback is provided by the eyes which indicate the actual position of the pencil point. The brain determines the error, i.e., the difference between the desired and actual positions of the pencil point, and activates the nerves which control the muscles accordingly. As demonstrated in the experiment, the human is capable of very precise operation using feedback.

In the second trial the nerves and muscles are the same, but the feedback path is no longer present, as shown in Fig. 21.1-1*c*. The inability to achieve the same precision of performance indicates that the basic components are not especially precise; it is only by providing feedback that the high precision performance can be obtained.

The system input in both cases is the same, the desired position of the pencil point which is the circle. The system output is also the same in both cases—the actual position of the pencil point. The elements in the forward path, the nerves and muscles, are the same in both cases. The only difference is that there is a *closed-loop* feedback system in the first case; the eyes feed the output signal back to the input. The second case is an *open-loop* system; there is no feedback.

Now that the basic concept of feedback has been presented the general model of a feedback system will be discussed. This will be done in terms of block diagrams for simplicity.

21.1-1 General Model

The general model for a feedback system is shown in block diagram form in Fig. 21.1-2. The arrows in the diagram indicate the direction of signal flow. The forward path contains an element with transfer function A while the feedback path has an element with transfer function β. Usually, in amplifier applications, the β element is a passive network, often simply one or two resistances.

The output (voltage v_0 in this case) is *fed back* through the β network which

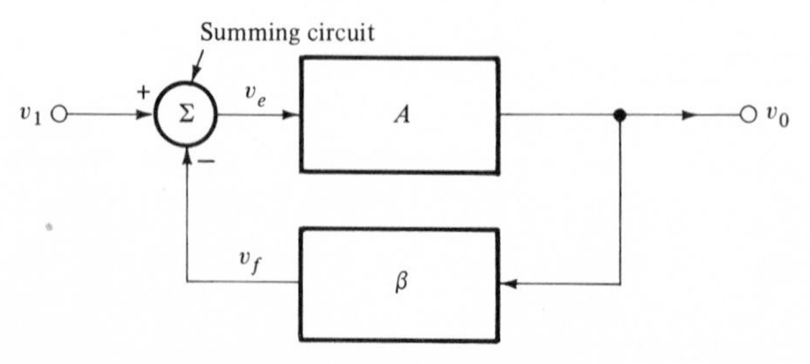

Fig. 21.1-2 Feedback system.

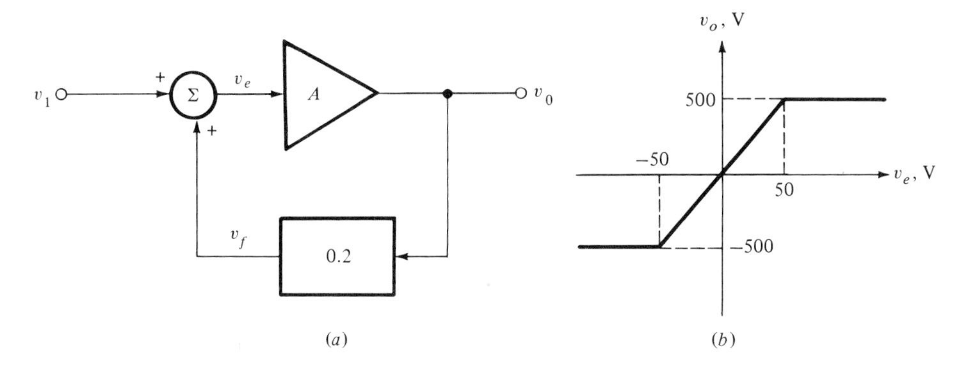

Fig. 21.1-3 Positive feedback system for Example 21.1-1. (*a*) Block diagram; (*b*) amplifier characteristic.

has as its output $v_f = \beta v_0$. This fed-back voltage, v_f, is compared with the input, v_1, in the summing circuit. The comparison yields an "error" voltage $v_e = v_1 - v_f$ which is the input to the element with transfer function A (usually an amplifier).

The behavior of this system can be described by assuming a step function input at v_1. This causes an initial step increase in v_e which results in a step increase of A times the input step in v_0. This, in turn, results in an increase of β times the increase in v_0 in v_f which is subtracted from the original increase in v_1 thereby cancelling out some of the increase in v_e which, in turn, decreases the change in v_0.

The system of Fig. 21.1-2 is called a *negative feedback system* for $A\beta > 0$ because the feedback tends to reduce the effect of input changes on the output. If v_f were added to v_1, instead of being subtracted from it, the system would be a *positive feedback system* because the feedback would cause large changes in the output for small input changes. In practice, the output of a positive feedback system will frequently increase until the system reaches saturation at which point it is no longer in its linear region of operation; such a system is said to be *unstable*. The behavior of such a system is illustrated in the following example.

Example 21.1-1 The positive feedback system of Fig. 21.1-3*a* has $\beta = 0.2$ and an amplifier whose input/output characteristic is shown in Fig. 21.1-3*b*. Find the output for a unit step input.

Solution The amplifier has an amplification of $500/50 = 10$ and it saturates at $v_0 = \pm 500$ V. Prior to the onset of the step input all signals are zero. Table 21.1-1 below illustrates the buildup of the output voltage. Note that the output signal remains at the saturation value of 500 V because the amplifier is saturated. This behavior is generally not desirable. ////

The use of positive feedback systems is restricted to special kinds of applications because of their inherent instability. One application in which this

TABLE 21.1-1 Behavior of Positive Feedback System.

Notes	v_1	$v_f = 0.2v_0$	$v_e = v_1 + v_f$	$v_o = 10v_e$ or 500
Prior to step	0	0	0	0
Onset of step	1	0	1	10
	1	2	3	30
	1	6	7	70
	1	14	15	150
	1	30	31	310
Saturation	1	62	63	500
	1	100	101	500
Final output	1	100	101	500

characteristic is used is in the construction of *oscillators,* i.e., circuits which generate cyclical time-varying signals.

Negative feedback is more widely used than positive feedback because it tends to result in stable performance. It is used in feedback control systems and in electronic circuits. The emphasis in this chapter is on negative feedback.

21.1-2 Closed-Loop Transfer Function

The overall transfer function of a feedback system is called the *closed-loop transfer function.* It is calculated from Fig. 21.1-4*a* as follows:

$$v_e = v_1 - v_f \tag{21.1-1}$$

$$v_f = \beta v_0 \tag{21.1-2}$$

$$v_0 = A v_e \tag{21.1-3}$$

Substitution of Eq. (21.1-2) into (21.1-1) yields

$$v_e = v_1 - \beta v_0 \tag{21.1-4}$$

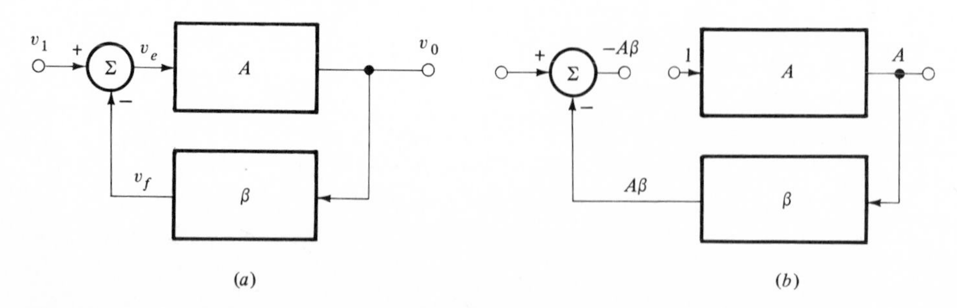

Fig. 21.1-4 Closed-loop transfer-function computation. (*a*) Closed-loop system; (*b*) closed-loop system with break in loop.

which when substituted into Eq. (21.1-3) results in

$$v_0 = Av_1 - A\beta v_0 \tag{21.1-5}$$

The solution of Eq. (21.1-5) is the closed-loop transfer function:

$$A_f = \frac{v_0}{v_1} = \frac{A}{1 + A\beta} \tag{21.1-6}$$

The closed-loop transfer function is frequently expressed in terms of a quantity called the *loop transmittance*, T_L. The loop transmittance is the amplification around the closed loop. It may be computed by breaking the feedback loop, "injecting" a unit signal at the break and computing the signal returned to the break. This is illustrated in Fig. 21.1-4*b* which shows a unit signal injected into the element with transfer function A and a signal $-A\beta$ returned to the break. The loop transmittance, also called the *loop gain*, is $T_L = -A\beta$. A negative value for T_L, as in this case (assuming $A\beta > 0$), implies negative feedback while a positive value for T_L corresponds to positive feedback. The closed-loop transfer function is

$$A_f = \frac{A}{1 - T_L} \tag{21.1-7}$$

where A is, in general, the forward transmittance from v_1 to v_0. These concepts are illustrated in the following example.

Example 21.1-2 The amplifier in the circuit of Fig. 21.1-5*a* is an ideal

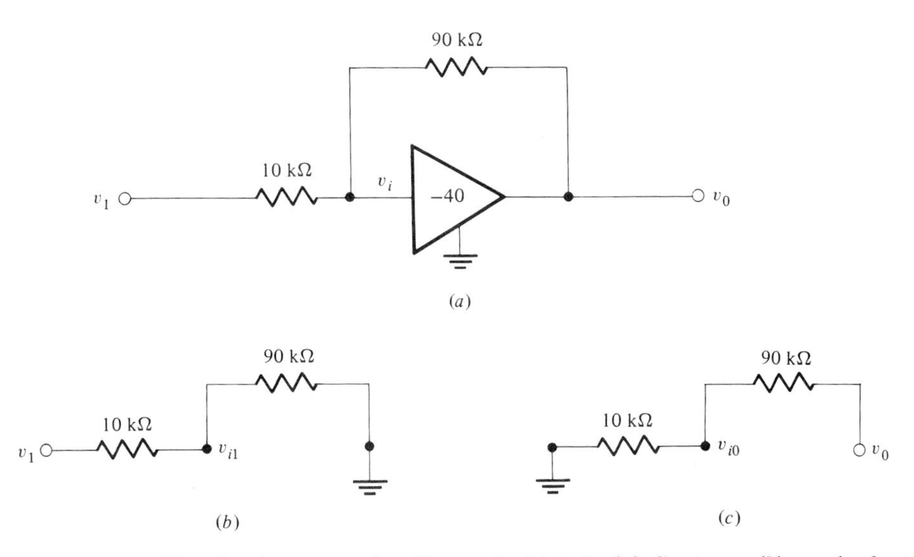

Fig. 21.1-5 Feedback system for Example 21.1-2. (*a*) System; (*b*) equivalent circuit for determining contribution to v_i by v_1; (*c*) equivalent circuit for determining contribution to v_i by v_0.

voltage amplifier (infinite input impedance and zero output impedance) with a voltage amplification of −40. Determine the closed-loop transfer function. Is the feedback positive or negative?

Solution Although v_1 is the only input to the circuit of Fig. 21.1-5*a*, it is convenient, for purposes of computing v_i only, to treat v_1 and v_0 as if they were two independent inputs. Using superposition, the contribution of each of these voltages to v_i is computed separately and the two components are added to find v_i.

The contribution of v_1 to v_i, denoted v_{i1}, is computed using the equivalent circuit of Fig. 21.1-5*b* in which v_0 is replaced by a short to ground. From this circuit v_{i1} is found using the voltage divider relationship:

$$v_{i1} = \frac{90}{10 + 90} v_1 = 0.9\, v_1 \tag{21.1-8}$$

The contribution of v_0 to v_i, denoted v_{i0}, is computed from the equivalent circuit of Fig. 21.1-5*c* in which v_1 is replaced by a short to ground. Using the voltage divider relationship,

$$v_{i0} = \frac{10}{10 + 90} v_0 = 0.1\, v_0 \tag{21.1-9}$$

Adding the results, Eqs. (21.1-8) and (21.1-9) results in

$$v_i = v_{i1} + v_{i0} = 0.9\, v_1 + 0.1\, v_0 \tag{21.1-10}$$

From Fig. 21.1-5*a*,

$$v_0 = -40\, v_i = -36\, v_1 - 4\, v_0 \tag{21.1-11}$$

whose solution is the closed-loop transfer function:

$$\frac{v_0}{v_1} = A_f = \frac{-36}{1 + 4} = -7.2 \tag{21.1-12}$$

Comparison of Eq. (21.1-12) with Eq. (21.1-7) implies that $T_L = -4$ which means that the system is a *negative feedback system.* ////

21.2 OPERATIONAL AMPLIFIERS

The operational amplifier was originally developed for use in analog computers. It was subsequently found to be a very useful analog component for many instrumentation and other systems applications. These new applications, together with the developments of integrated circuit technology, have resulted in present day models of operational amplifiers which feature many improvements. The use of integrated circuits instead of vacuum tubes has resulted in tremendous miniaturization. It is now possible to obtain two operational amplifiers in a standard 14-pin dual in-line package (DIP) which measures approximately

0.75 in × 0.25 in × 0.20 in. Another major change has been in cost; it is possible to obtain an operational amplifier for as little as one dollar although high quality operational amplifiers, such as those required for precision instrumentation or computational purposes are considerably more expensive.

The typical integrated circuit operational amplifier is a high gain, differential input voltage amplifier which is designed to approximate an ideal voltage amplifier. This means that its input resistance is very high, typically 50 MΩ, and its output resistance is very low, typically 50 Ω to 100 Ω.

One of the most important characteristics of the operational amplifier is its high gain; a voltage amplification of 10^6 is common.

In many systems the input voltage is connected between an input terminal and ground. The ground terminal is the reference for all of the voltages, both ac and dc, in the amplifier or system of which the amplifier is a part. This configuration is called a single-ended input, and it places constraints on the input to the amplifier because one side is grounded. An alternate configuration is called a differential input and consists of two input terminals, neither of which is grounded, to which the input voltage is applied. The amplifier output is the amplification factor multiplied by the voltage difference between these two terminals.

$$v_0 = Av_i = A(v_a - v_b) \tag{21.2-1}$$

This input configuration is also called a double-ended input.

An amplifier's output can be single-ended or double-ended depending on whether or not one of the two output terminals is grounded. Most operational amplifiers have a single-ended output. The symbol and an equivalent circuit for the operational amplifier are shown in Fig. 21.2-1. As indicated already, A is very large, R_{in} is very large, and R_{out} is rather small.

There are other features of the operational amplifier which have not been discussed and which are not indicated in the equivalent circuit. One important characteristic in this category is the amplifier's frequency response. The operational amplifier is basically a low frequency device; its amplification is relatively constant from dc to a rather low cutoff frequency, typically 1 Hz. Above the cutoff frequency, the amplification decreases with increasing frequency. The frequency response does not have a sharp break; rather it is a gradual decrease in amplitude that becomes more rapid after the cutoff frequency. Generally, the

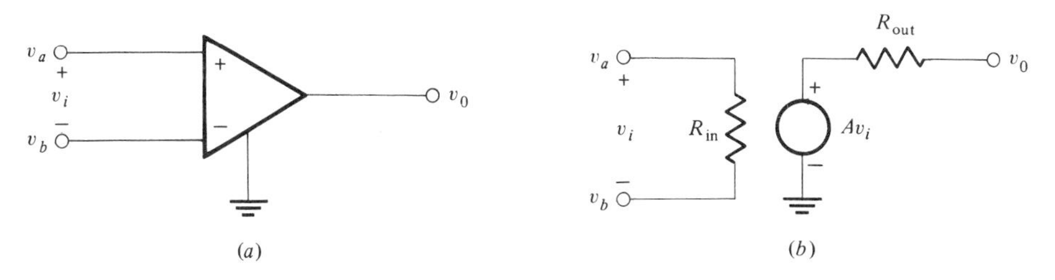

Fig. 21.2-1 Operational amplifier. (a) Symbol; (b) equivalent circuit.

cutoff frequency is chosen as that frequency at which the amplification is $1/\sqrt{2}$ times its maximum value. The bandwidth is equal to the cutoff frequency since the amplification does not fall off at the low frequency end.

Another general characteristic of operational amplifiers which is important is their *stability*. This refers to the constancy of their amplification and their bias voltages. Early operational amplifiers for analog computers used a very expensive special technique, called *chopper stabilization*, to achieve good dc stability. Modern solid state techniques have made it possible to achieve relatively good stability without the need for chopper stabilization and this is one of the reasons for the great reduction in cost of these amplifiers. Chopper stabilization is still used where very high precision is needed.

Other amplifier characteristics which are of importance include drift, common-mode rejection ratio, and noise, but a discussion of all of these is beyond the scope of this text and is not necessary for the discussion which follows.

21.3 OPERATIONAL AMPLIFIER APPLICATIONS

The original applications of the operational amplifier in analog computers were in circuits intended to provide the basic linear mathematical functions: multiplication by a constant or scaling, addition, and integration. The availability of operational amplifiers at lower prices and in smaller packages has spurred their usage as "universal analog systems components;" in this context their primary usage is in signal processing, interfacing, instrumentation amplifiers, and waveshaping.

The applications described in this section are circuits which perform the three most useful linear mathematical operations. The basic concepts are presented for the simplest circuit, the scaler.

21.3-1 Scaler

The scaler circuit functions as a linear amplifier, thereby scaling the input voltage or multiplying it by a constant. The circuit is shown in Fig 21.3-1*a*. The operational amplifier is assumed to have a very large input impedance so that the input current i_i is essentially zero. The amplification A is assumed to be very high. The input voltage v_1 is connected, through resistance R_1, to the negative input of the amplifier; and the output voltage v_0 is connected, through resistance R_f, to the same terminal. The voltage at this terminal is denoted v_i. Because it is the negative input terminal, the loop gain is negative and the scaler is therefore a *negative feedback amplifier*. All the operational amplifier circuits discussed in this chapter use negative feedback.

VIRTUAL GROUND

A key concept in the analysis of the scaler, as well as the other operational amplifier circuits, is that of the *virtual ground*. The essence of this concept is

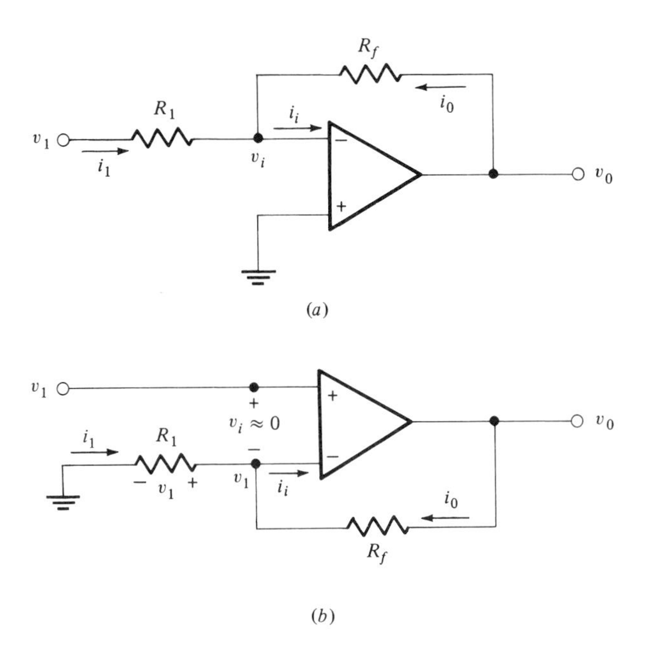

Fig. 21.3-1 Scaler circuits. (*a*) Negative scaling; (*b*) positive scaling.

that the amplifier input voltage v_i is forced to such a small value that for all practical purposes it may be assumed to be zero. The negative input terminal is therefore essentially at ground voltage and is referred to as a virtual ground.

The action of the circuit which results in this behavior may be explained as follows. Assume the input voltage v_1 is initially zero resulting in $v_i = 0$ and $v_0 = 0$. Now v_1 is increased to a positive value; this causes v_i to start to increase. The large negative amplification $v_0/v_i = -A$ causes v_0 to become negative. The negative value of v_0 is fed back to v_i and balances the original positive input v_1. A very small value of v_i is sufficient to result in a substantial value for v_0 because A is so large (in the order of one million). Thus v_i is not exactly equal to zero (if it were, v_0 would be equal to zero also) but is sufficiently small compared with v_1 and v_0 to be neglected in the computation of v_0. Once v_0 is calculated, the value of v_i can be computed as $-v_0/A$ and it can be checked that it is indeed negligible compared with v_1 and v_0.

ANALYSIS OF THE SCALER

Consider the scaler circuit shown in Fig. 21.3-1*a*. As noted earlier, i_i is essentially zero owing to the very high input impedance of the amplifier. Furthermore, v_i is essentially zero since it is a virtual ground. Therefore

$$i_1 = \frac{v_1 - v_i}{R_1} \approx \frac{v_1}{R_1} \tag{21.3-1}$$

and

$$i_0 = \frac{v_0 - v_i}{R_f} \approx \frac{v_0}{R_f} \tag{21.3-2}$$

also

$$i_1 + i_0 = i_i \approx 0 \tag{21.3-3}$$

Substituting Eqs. (21.3-1) and (21.3-2) into Eq. (21.3-3) results in

$$\frac{v_1}{R_1} + \frac{v_0}{R_f} \approx 0$$

which is rearranged to result in

$$\frac{v_0}{v_1} \approx -\frac{R_f}{R_1} \tag{21.3-4}$$

This is the desired result for the scaler. It indicates that the output is the input multiplied by minus the ratio of the two resistance values; it is independent of the amplification of the amplifier so long as the amplification is very large.

How large does the amplification have to be for the result given in Eq. (21.3-4) to be valid? The answer to this question is found in Eqs. (21.3-1) and (21.3-2); the amplification A must be large enough to validate the approximations. Substituting $v_i = -v_0/A$ into Eq. (21.3-2) results in

$$i_0 = \frac{v_0 + v_0/A}{R_f} = \frac{v_0(1 + 1/A)}{R_f} \approx \frac{v_0}{R_f}$$

This is valid if $A \gg 1$, which holds for most operational amplifiers for frequencies up to about 10 kHz. Substituting

$$v_i = -\frac{v_0}{A} = \frac{R_f}{AR_1} v_1$$

into Eq. (21.3-1) results in

$$i_1 = \frac{v_1 - (R_f/AR_1)(v_1)}{R_1} = \frac{(1 - R_f/AR_1)v_1}{R_1} \approx \frac{v_1}{R_1}$$

which is valid if $R_f/AR_1 \ll 1$ or $R_f/R_1 \ll A$. Thus, the condition on A is that it must be much larger than the magnitude of the scaler transfer function, R_f/R_1, also called the closed-loop amplification. This analysis points up the fact that the designer of a scaler must limit the multiplication factor of the circuit to a value which is much less than the amplifier's amplification.

POSITIVE SCALER

The scaler of Fig. 21.3-1a provides an output equal to the input multiplied by a *negative* constant. It is often necessary to multiply by a *positive* constant.

The first impulse is to interchange the amplifier terminals so that the negative input terminal is at ground and the positive one is used for the junction of R_1 and R_f. This is a mistake, however, because it introduces positive, rather than negative, feedback which tends to result in instability.

A scaler which uses negative feedback and provides an output which is equal to the input multiplied by a positive constant is shown in Fig. 21.3-1*b*. The junction of R_1 and R_f is still at the amplifier's negative input terminal (the other end of R_1 is grounded) but the *circuit's* input signal is applied to the positive input terminal.

The amplifier input v_i is driven to an essentially zero value by the high amplification and negative feedback as in the circuit of Fig. 21.3-1*a*. In this case it is not a virtual ground, since the other terminal is not grounded; instead the voltage at the negative input terminal is approximately equal to the voltage at the positive input terminal which is v_1. As a result the voltages across R_1 and R_f are v_1 and $v_0 - v_1$, as shown in Fig. 21.3-1*d*, respectively. The current i_i is essentially zero due to the very high input resistance and small value of v_i. Using Kirchhoff's current law at the junction of R_1 and R_f results in

$$i_1 + i_0 = -\frac{v_1}{R_1} + \frac{v_0 - v_1}{R_f} = i_i \approx 0 \tag{21.3-5}$$

which yields

$$\frac{v_0}{v_1} = 1 + \frac{R_f}{R_1} \tag{21.3-6}$$

Note the difference between this expression and the one for the scaler with a negative multiplier; not only is there a change in sign but the value 1 is added to the constant.

21.3-2 Adder

The adder circuit is similar in principle to the scaler but multiple input resistances are used as shown in Fig. 21.3-2.

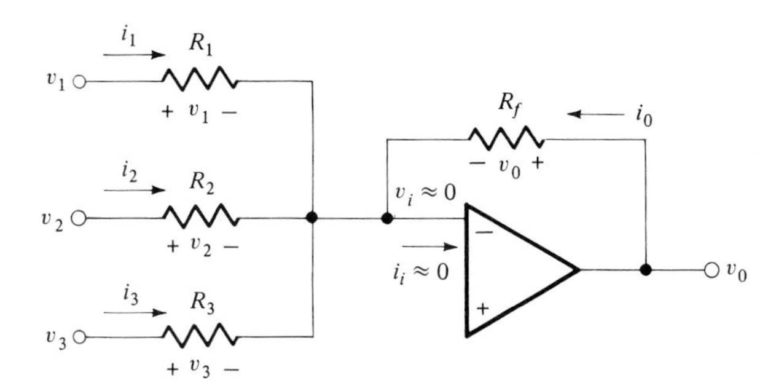

Fig. 21.3-2 Adder circuit.

The amplifier input voltage is driven to "virtual zero" so that the negative input terminal can be considered to be at ground. The input current to the amplifier, i_i, is essentially zero as a result of the low input voltage and high input resistance. This results in the voltages v_1, v_2, v_3, and v_0 across R_1, R_2, R_3, and R_f respectively as shown in the figure. Kirchhoff's current law applied to the amplifier input results in

$$i_1 + i_2 + i_3 + i_0 = \frac{v_1}{R_1} + \frac{v_2}{R_2} + \frac{v_3}{R_3} + \frac{v_0}{R_f} \approx 0 \tag{21.3-7}$$

This is solved for v_0:

$$v_0 = -\left(\frac{R_f}{R_1} v_1 + \frac{R_f}{R_2} v_2 + \frac{R_f}{R_3} v_3\right) \tag{21.3-8}$$

This result indicates that each term in the sum can be multiplied by a different constant. The minus sign which multiplies the entire expression is unavoidable because the negative input terminal of the amplifier must be used to provide negative feedback. The following example illustrates the steps in the design of an adder.

Example 21.3-1 Design a circuit whose output is

$$v_0 = 2v_1 + 5v_2 \tag{21.3-9}$$

Solution Two operational amplifiers must be used; one to add the two signals, v_1 and v_2, and the other to change the sign back to a positive one. Thus the overall input/output relationship is the product of two terms; one relating v_3 to v_1 and v_2 for the first portion of the circuit and the other relating v_0 to v_3 for the second part of the circuit as shown in Fig. 21.3-3. The loading effect of the second circuit on the first can be neglected because the output impedance of the operational amplifier is very small.

All the resistances are chosen, arbitrarily, to be 10 kΩ except for the two input resistances R_1 and R_2 as shown in Fig. 21.3-3. Using Eqs. (21.3-4)

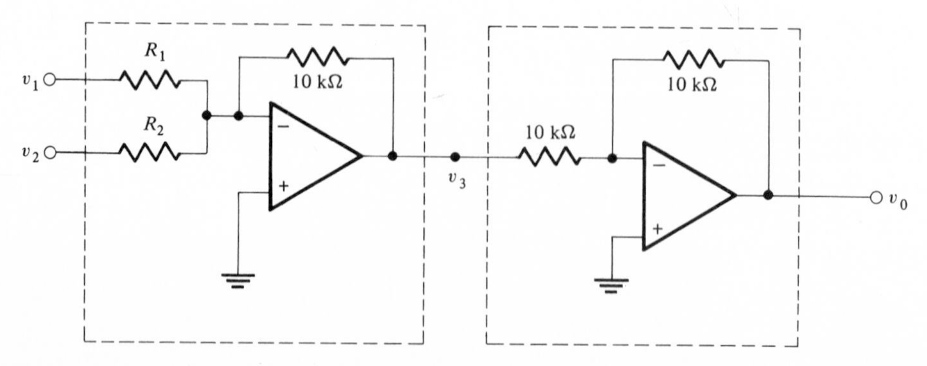

Fig. 21.3-3 Adder circuit for Example 21.3-1.

and (21.3-8) the output, v_0, is

$$v_0 = -\left(\frac{10}{R_1} v_1 + \frac{10}{R_2} v_2\right)\left(-\frac{10}{10}\right)$$

$$v_0 = \frac{10}{R_1} v_1 + \frac{10}{R_2} v_2 \qquad (21.3\text{-}10)$$

Equating Eq. (21.3-9) with Eq. (21.3-10) we see that in order to meet the specifications we must have $R_1 = 5\ \mathrm{k}\Omega$ and $R_2 = 2\ \mathrm{k}\Omega$. ////

21.3-3 Integrator

The output of an integrator circuit is proportional to the integral of its input. An integrator circuit is shown in Fig. 21.3-4. The analysis is similar to that of the scaler circuit with one exception; the terminal characteristic equation for a capacitance must be used in place of Ohm's law to relate v_0 and i_0. This equation is

$$v_0 = \frac{1}{C} \int_{-\infty}^{t} i_0(x)\, dx \qquad (21.3\text{-}11)$$

The other equation, based on zero input current and the virtual ground at the junction of R and C, is

$$i_1 + i_0 = \frac{v_1}{R} + i_0 \approx 0 \qquad (21.3\text{-}12)$$

Solving Eq. (21.3-12) for i_0 and substituting into Eq. (21.3-11) yields

$$v_0 \approx -\frac{1}{RC} \int_{-\infty}^{t} v_1(x)\, dx \qquad (21.3\text{-}13)$$

which shows the output to be proportional to the integral of the input.

There are many more applications of the operational amplifier. The basic ones that we have described are frequently used and should lead to an under-

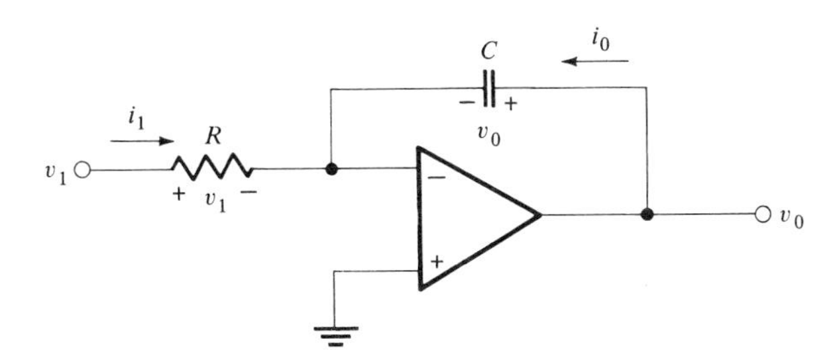

Fig. 21.3-4 Integrator circuit.

standing of the basic concepts which can be used to design other circuits. The following examples illustrate typical applications of the integrator circuit.

Example 21.3-2 An operational amplifier circuit is to be used to obtain the triangular wave shown in Fig. 21.3-5*b* from the square wave shown in Fig. 21.3-5*a*. The amplitude, V_t, of the triangular wave should be variable between 1 and 10 V. Design such a circuit.

Solution The triangular wave is the integral of the square wave, so an integrator circuit is used. The output of the integrator is

$$v_t(t) = -\frac{1}{RC}\int_0^t v_s(x)\,dx$$

since it is assumed that $v_s(t) = 0$ for $t < 0$. Substituting $t = 0.01$ s in the

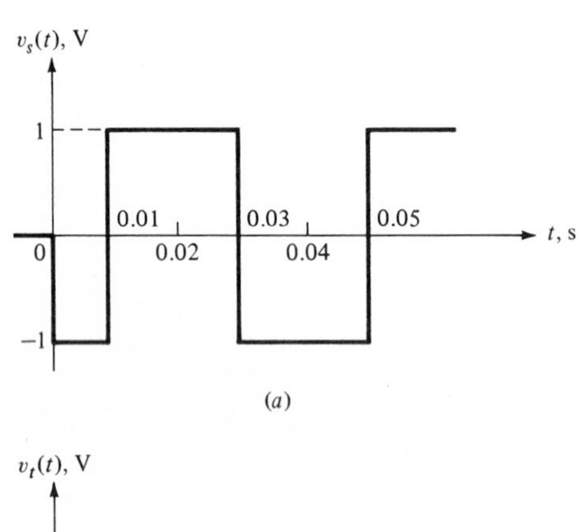

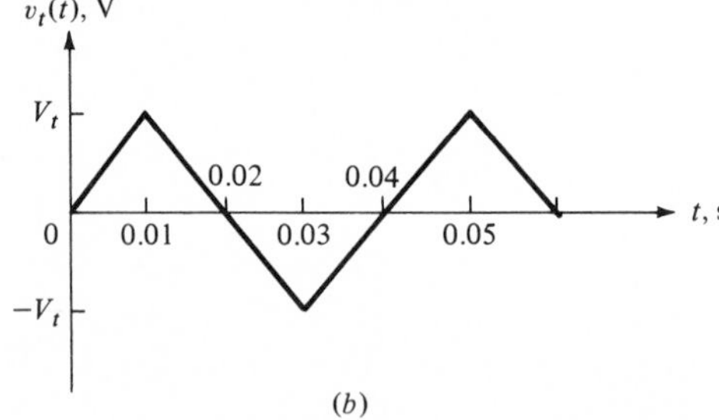

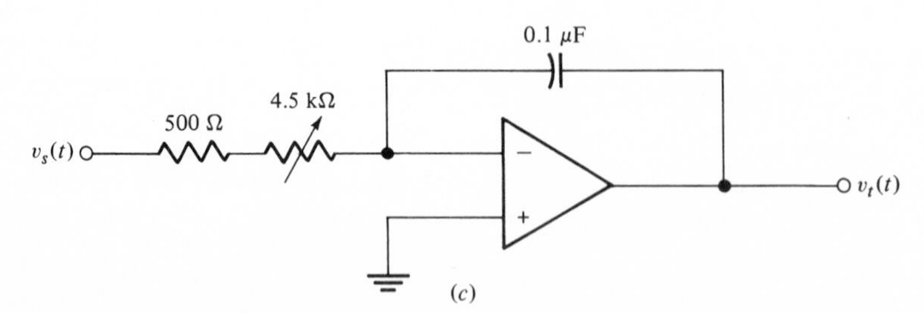

Fig. 21.3-5 Waveforms and circuit for Example 21.3-2. (*a*) Square wave input; (*b*) triangular wave output; (*c*) integrator circuit.

equation above (see Prob. 21.3-5) results in

$$v_t(0.01) = V_t = \frac{0.005}{RC} \tag{21.3-12}$$

In order to vary the amplitude a variable resistor is used since it is cheaper than a variable capacitor. A capacitance value of 1 μF is chosen as a convenient value; substituting this value into the previous equation and solving for R results in

$$R = \frac{5{,}000}{V_t}$$

For $V_t = 1$ V, $R = 5{,}000\ \Omega$ and for $V_t = 10$ V, $R = 500\ \Omega$. Therefore a 500 Ω fixed resistor in series with a 4.5 kΩ variable resistor is used. The circuit is shown in Fig. 21.3-5*c*. ////

Example 21.3-3 The quantity of water that is delivered to a 500 gallon vat through a pipe with a 3 in inside diameter is to be measured remotely using a 10-V full scale voltmeter connected to circuitry which is connected to a sensor which generates a voltage proportional to the speed of the water in the pipe. As shown in Fig. 21.3-6, the sensor consists of a small pinwheel connected to a permanent magnet which rotates in the space inside a coil of wire within a sealed housing. This assembly is placed in the pipe so the angular velocity of the pinwheel is proportional to the speed of the water.

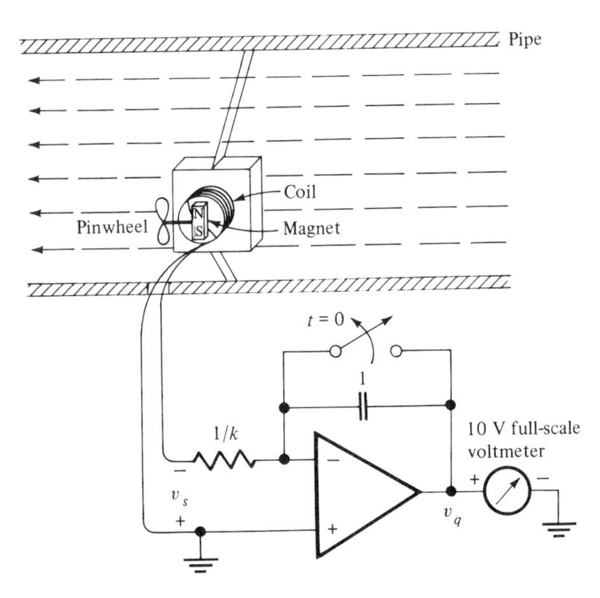

Fig. 21.3-6 System for measuring quantity of fluid flowing through pipe.

A voltage is generated which is proportional to the angular velocity of the magnet. The parameters of this pinwheel-generator are such that a flow velocity of 1 ft/s produces a voltage of 0.5 V.

Design the circuitry which must be inserted between the pinwheel-generator and voltmeter in order to measure a quantity between 0 and 500 gallons.

Solution The pinwheel-generator voltage is proportional to the *water flow*; the meter is to measure the *quantity* of water delivered from 0 to time t which is the integral of the flow over this interval. Therefore an integrator is required. The output of the pinwheel-generator is connected to the input of the integrator so that its positive side is grounded as shown in Fig. 21.3-6. Thus, the input to the integrator is $-v_s$ where v_s is proportional to the flow. A switch is placed across the integrator's capacitance which is opened at $t = 0$ thus starting the integration at $t = 0$. The output of the integrator is

$$v_q = -\frac{1}{RC}\int_0^t - v_s(x)dx = k\int_0^t v_s(x)\,dx$$

if $R = 1/k$ and $C = 1$ as shown in Fig. 21.3-6. By choosing k properly a full scale voltage reading corresponds to a full vat. The remainder of this example is devoted to determining k.

The volume of water transported per minute at a flow of 1 ft/s through the pipe is

$$\pi\left(\frac{3}{2}\,\text{in}\right)^2 \times 1\,\frac{\text{ft}}{\text{s}} \times 12\,\frac{\text{in}}{\text{ft}} \times \frac{60\ \text{s}}{\text{min}} = 5089.38\,\frac{\text{in}^3}{\text{min}}$$

There are 231 in^3 in 1 gal; therefore the flow is

$$5089.38\,\frac{\text{in}^3}{\text{min}} \times \frac{1\ \text{gal}}{231\ \text{in}^3} = 22.03\ \text{gal/min}$$

A flow velocity of 1 ft/s causes the pinwheel-generator to produce 0.5 V. Therefore a flow of 44.06 gal/min generates 1 V which means that a flow of f gal/min generates $f/44.06 = 0.0227f$ V. The output of the integrator is

$$v_q = k\int_0^t 0.0227f\,dx = 0.0227kft\ \text{V}$$

if f is constant. The total quantity in t min is ft. Therefore

$$v_q = 0.0227kq\ \text{V}$$

where q is the quantity in gallons. A quantity of 500 gal should correspond to a full-scale voltage reading of 10 V. Therefore

$$0.0227k(500) = 10$$

which results in

$$k = \frac{10}{(500)(0.0227)} = 0.881$$

Hence C = 1 μF and R = $1/k$ = 1.13 MΩ will provide proper readings of the quantity of water. ////

21.4 EFFECTS OF FEEDBACK

The introduction of feedback into a system affects many of the system's characteristics in predictable ways. In many cases the feedback is introduced specifically in order to affect one or more system characteristics in a beneficial way; sometimes the effects are undesirable but unavoidable. This section is concerned with some of these effects. For convenience, the discussion will be developed in terms of feedback as applied to an operational amplifier. The resulting principles are quite general and equally applicable to other systems.

The effects of feedback will be limited to those resulting from the more commonly used type, negative feedback. The effects of positive feedback are, in general, just opposite those of negative feedback.

21.4-1 Amplification

The effect of feedback on amplification has already been determined when the scaler circuit was covered in Sec. 21.3-1. The open-loop amplification of the operational amplifier is $-A$, a very large negative value. The amplification with negative feedback is approximately $-(R_f/R_1)$ for very large values of A. This is usually a much smaller value; therefore, negative feedback reduces the amplification.

In order to quantitatively relate the decrease in amplification to the amount of feedback, it is helpful to represent the negative feedback amplifier shown in Fig. 21.4-1*a*, in the block diagram form of Fig. 21.1-2.
For the circuit of Fig. 21.4-1*a*,

$$i_i = i_1 + i_0 = \frac{v_1 - v_i}{R_1} + \frac{v_0 - v_i}{R_f} \qquad (21\text{-}4\text{-}1)$$

As discussed earlier, $i_i \approx 0$. Using this fact in Eq. (21.4-1), v_i becomes

$$v_i = \frac{R_f}{R_1 + R_f} v_1 + \frac{R_1}{R_1 + R_f} v_0 \qquad (21.4\text{-}2)$$

This, together with the circuit diagram of Fig. 21.4-1*a* leads to the block diagram of Fig. 21.4-1*b*. If the feedback loop is opened the system reduces to that of Fig. 21.4-1*c*; it is called the *open-loop* system.

The presence of a plus sign on both inputs to the summer circuit in the block

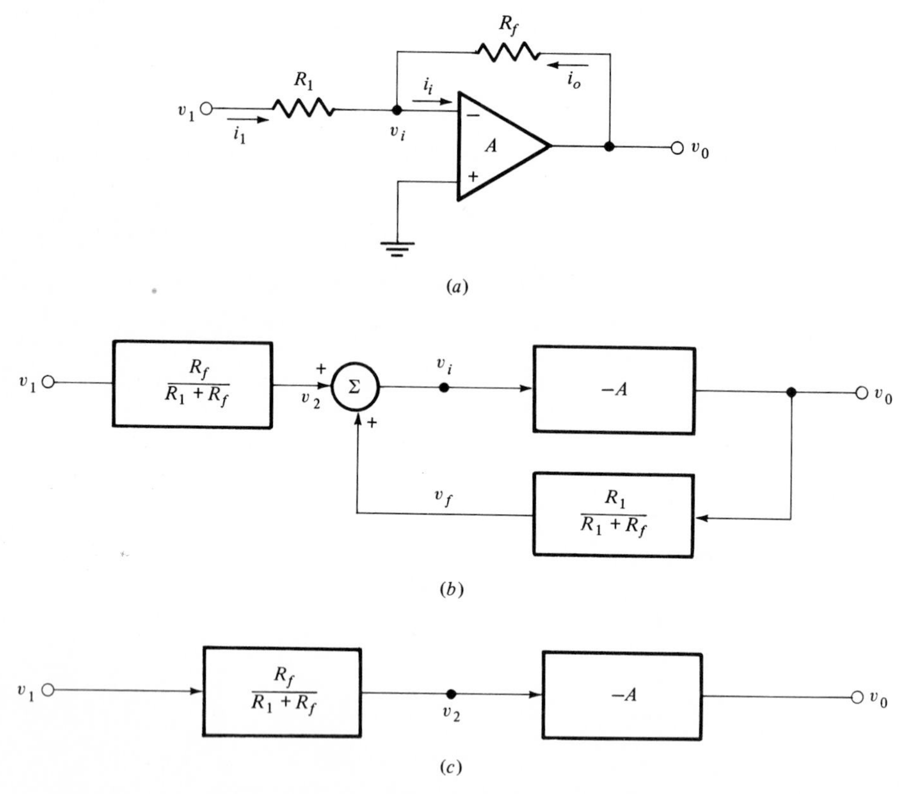

Fig. 21.4-1 Amplifier with negative feedback. (*a*) Circuit diagram; (*b*) block diagram; (*c*) open-loop system.

diagram of Fig. 21.4-1*b* might lead to the conclusion that it is a positive feedback system. Computation of the loop gain

$$T_L = \frac{-AR_1}{R_1 + R_f} \tag{21.4-3}$$

reveals that the system is actually a negative feedback system, since the loop gain is negative.

The open-loop amplification, from Fig. 21.4-1*c*, is the product of two transfer functions:

$$A_0 = \frac{v_2}{v_1}\frac{v_0}{v_2} = \frac{R_f}{R_1 + R_f}(-A) = \frac{-AR_f}{R_1 + R_f} \tag{21.4-4}$$

From Fig. 21.4-1*b* and Eq. (21.1-6) the closed-loop amplification is

$$A_f = \frac{v_2}{v_1}\frac{v_0}{v_2} = \frac{R_f}{R_1 + R_f}\frac{-A}{1 + AR_1/(R_1 + R_f)} = \frac{-AR_f}{AR_1 + R_1 + R_f} \tag{21.4-5}$$

Substituting Eqs. (21.4-3) and (21.4-4) into Eq. (21.4-5) (use the next-to-last expression) results in

$$A_f = \frac{1}{1 - T_L} A_0 \qquad (12.4\text{-}6)$$

which indicates the reduction in amplification resulting from negative feedback, i.e., negative values of T_L. The greater the magnitude of T_L, the greater the reduction in amplification. These concepts are used in the following example.

Example 21.4-1 An amplifier has a voltage amplification of -20 and essentially infinite input resistance and zero output resistance. Negative feedback is applied by connecting resistances R_1 and R_f as shown in Fig. 21.4-2. Find R_f if $R_1 = 5$ kΩ and the closed-loop amplification is to be -10.

Solution The voltage amplification is not large enough to use the approximate formula for operational amplifiers. Instead, Eq. (21.4-5) is solved for R_f with $A_f = -10$ as follows:

$$\frac{-20R_f}{(+20)(5) + 5 + R_f} = -10$$

$$-20R_f = -1{,}000 - 50 - 10R_f$$

$$R_f = \frac{-1{,}050}{-10} = 105\ \text{k}\Omega \qquad ////$$

21.4-2 Input resistance

The effects of feedback on an amplifier's input resistance are demonstrated using the diagram of Fig. 21.4-3*a*. This shows the model of an operational amplifier with input resistance R_{in} connected as a scaler with external resistances R_f and R_1.

The open-loop equivalent of the circuit, which retains the effects of R_1 and R_f on the input, is shown in Fig. 21.4-3*b*. It is obtained by changing the circuit so as to replace the v_0 contribution to v_i by zero volts without changing anything

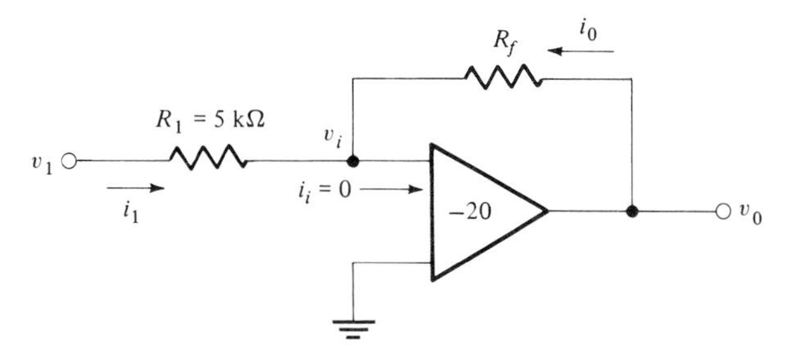

Fig. 21.4-2 Feedback system for Example 21.4-1.

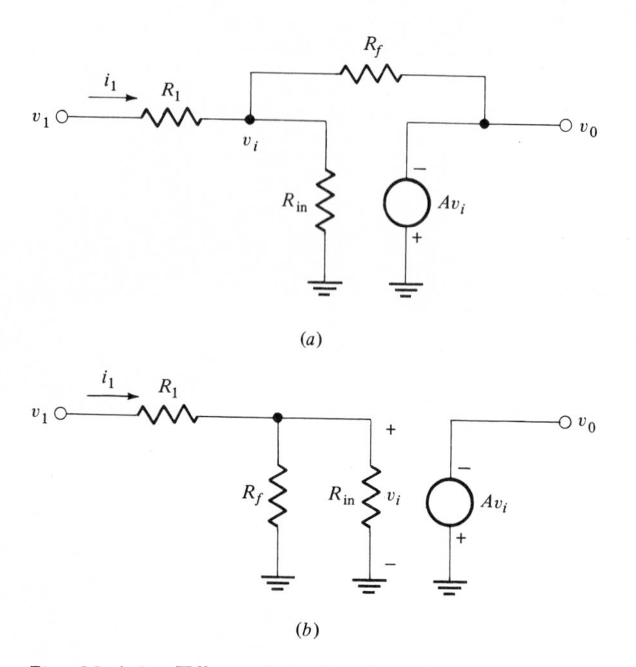

Fig. 21.4-3 Effect of feedback on input resistance. (*a*) Closed-loop system; (*b*) open-loop system, i.e., without the feedback.

else. The input resistance for this circuit is computed first as a basis for comparison. By inspection, this resistance is R_1 in series with the parallel connection of R_f and R_{in}. Therefore, the open-loop input resistance is

$$R_{in,0} = R_1 + \frac{R_f R_{in}}{R_f + R_{in}} \tag{21.4-7}$$

The closed-loop input resistance is the ratio of input voltage v_1 to the input current i_1. If A is very large, the amplifier input voltage v_i is at "virtual ground." Therefore

$$R_{in,f} \approx \frac{v_1}{i_1} = R_1 \tag{21.4-8}$$

which is less than the open-loop input resistance. Therefore, negative feedback reduces the input resistance of an amplifier. This derivation is based on a very large amplification but the result is qualitatively valid in all cases although the numerical formula is not. This is demonstrated in Prob. 21.4-1.

21.4-3 Output resistance

Negative feedback usually causes a reduction in the output impedance. Figure 21.4-4*a* shows the model of an amplifier with output resistance R_0 connected to

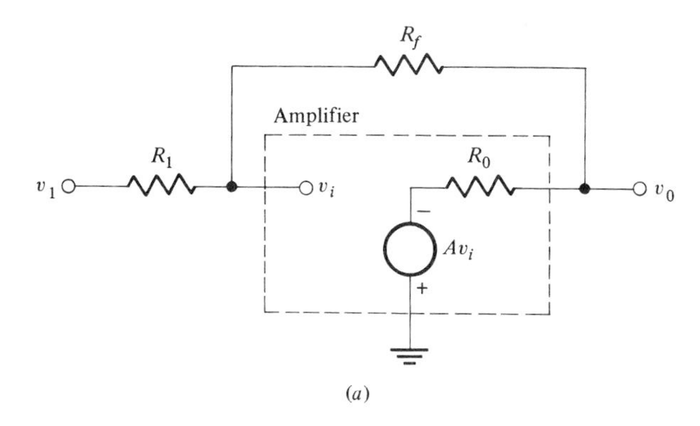

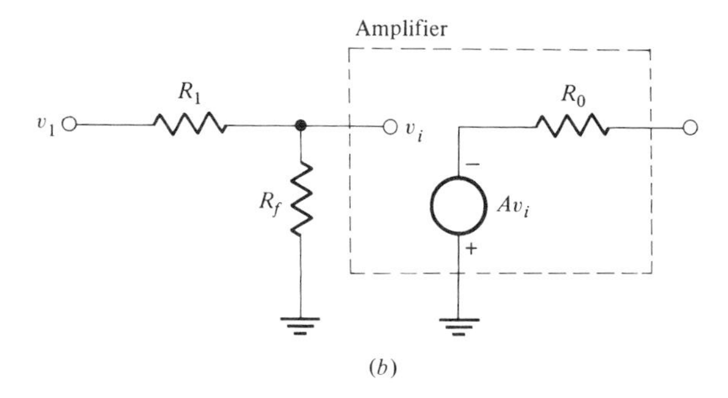

Fig. 21.4-4 Effects of feedback on output resistance of amplifier. (*a*) Amplifier with negative feedback; (*b*) equivalent open-loop amplifier circuit (for $R_f + R_1 \gg R_0$).

external resistances R_1 and R_f which provide negative feedback. Figure 21.4-4*b* shows an approximate open-loop equivalent, i.e., the circuit with no feedback, which is valid for $R_1 + R_f \gg R_0$.

The output resistance of the open-loop circuit is, from Fig. 21.4-4*b*, seen to be R_0. An analysis of the closed-loop circuit reveals that the output resistance with feedback is

$$R_{\text{out},f} = \frac{R_0}{1 - T_L} \tag{21.4-9}$$

where T_L is the loop gain already computed as

$$T_L = \frac{-AR_1}{R_1 + R_f} \tag{21.4-3}$$

The result, Eq. (21.4-9), is valid only if $R_1 + R_f \gg R_0/(1 - T_L)$ which is normally the case.

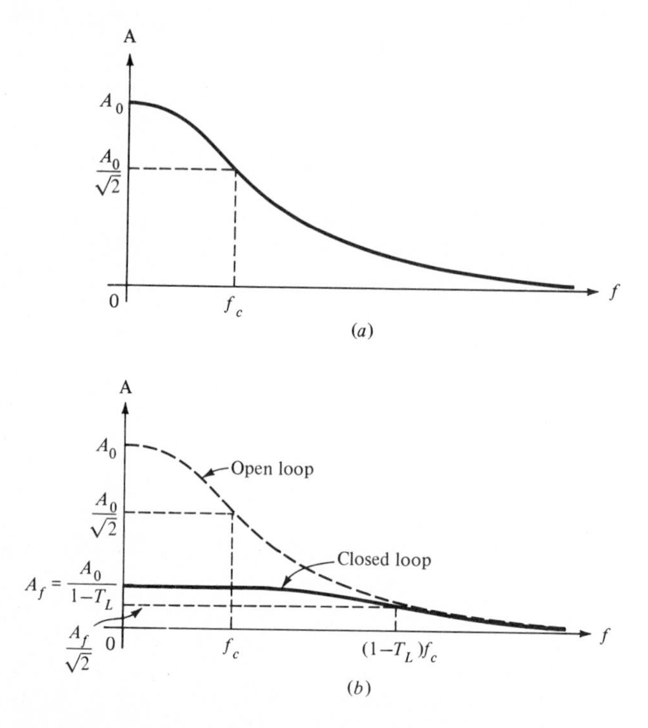

Fig. 21.4-5 Amplitude frequency response of amplifier circuits. (*a*) Open-loop; (*b*) closed-loop.

21.4-4 Bandwidth

The amplitude frequency response of a typical operational amplifier is shown in Fig. 21.4-5*a*. The amplitude decreases from a maximum value of A_0 at dc as the frequency increases. The cutoff frequency, f_c, is the frequency at which the amplitude is $A_0/\sqrt{2}$, that is, the half-power point. The open-loop bandwidth is the range from dc to f_c and is, therefore, equal to the half-power frequency:

$$BW_0 = f_c \tag{21.4-10}$$

A typical value of f_c for an operational amplifier is 1 Hz.

When negative feedback is applied to the amplifier its amplification is decreased by $1 - T_L$ where T_L is the loop gain, that is,

$$A_f = \frac{A_0}{1 - T_L} \tag{21.4-6}$$

The amplitude frequency response of the closed-loop amplifier circuit is shown in Fig. 21.4-5*b*. It shows that the closed-loop cutoff frequency and bandwidth are

$$BW_f = f_{cf} = (1 - T_L)f_c \tag{21.4-11}$$

Thus, the bandwidth is increased by the same factor that the amplification is decreased. The product of the amplification and bandwidth is called the *gain-bandwidth product.* From Eqs. (21.4-6) and (21.4-1) it is seen that it is a constant:

$$\mathrm{GBW} = A_f BW_f = A_0 f_c \tag{21.4-12}$$

The very high amplification, typically about 10^6, at dc makes it possible to design an amplifier with moderately large amplification and bandwidth despite the low value of f_c. This is illustrated in the following examples.

Example 21.4-2 The amplifier in the circuit of Fig. 21.4-1*a* has a cutoff frequency of 5 Hz, a dc amplification of 10^4, input resistance of 100 MΩ, and output resistance of 100 Ω. $R_1 = 500\ \Omega$ and $R_f = 5\ \mathrm{k}\Omega$.

a Find the dc amplification of the closed-loop circuit.
b Find its bandwidth.
c Find its input resistance.
d Find its output resistance.

Solution The loop gain is required for many of the computations in this problem. It is

$$T_L = \frac{-A_0 R_1}{R_1 + R_f} = \frac{-(10^4)(500)}{500 + 5{,}000} = \frac{-10^4}{11}$$

The required factor is $1 - T_L = 910$.

a The closed-loop dc amplification is

$$A_f = \frac{A_0}{1 - T_L} = \frac{10^4}{910} = 11$$

b The closed-loop bandwidth is

$$\mathrm{BW}_f = (1 - T_L)f_c = (910)(5) = 4.55\ \mathrm{kHz}$$

c The closed-loop input resistance is

$$R_{\mathrm{in},f} \approx R_1 = 500\ \Omega$$

d The closed-loop output resistance is

$$R_{\mathrm{out},f} \approx \frac{R_0}{1 - T_L} \approx \frac{100}{910} = 0.110\ \Omega$$

provided that $R_1 + R_f = 5{,}500 \gg \dfrac{R_0}{1 - T_L} = 0.110$

The inequality is obviously satisfied. ////

Example 21.4-3 An operational amplifier has an amplification at dc of 10^6, a gain-bandwidth product of 10^6, an input resistance of 20 MΩ and an output resistance of 70 Ω.

a Use this operational amplifier in a feedback configuration to design an amplifier with a bandwidth of 10 kHz and an input resistance of 50 kΩ.
b What is the dc amplification of this circuit?
c What is its output resistance?

Solution

a Negative feedback must be used to increase the bandwidth. The circuit configuration is that of Fig. 21.4-1*a*. The open-loop bandwidth is 1 Hz since $A_0 = 10^6$ and GBW $= 10^6 = A_0 f_c$. The closed-loop bandwidth is $(1 - T_L)$ 1 Hz $=$ 10,000 Hz. Therefore

$$T_L \approx -10^4 = \frac{-A_0 R_1}{R_1 + R_f} = \frac{-10^6 R_1}{R_1 + R_f}$$

from which we find $R_f = 99R_1$.

The input resistance is, from Sec. 21.4-2,

$$R_{\text{in},f} = R_1 \tag{21.4-8}$$

Therefore $R_1 = 50$ kΩ and $R_f = 4.95$ MΩ ≈ 5 MΩ

b The dc amplification is reduced by the same factor that the bandwidth is increased, in this case, 10^4. Therefore

$$A_f = \frac{A_0}{10^4} = \frac{10^6}{10^4} = 100$$

c The output resistance is reduced by $1 - T_L$ if $R_1 + R_f \gg R_0$, which it is in this case. Therefore

$$R_{0f} = \frac{R_0}{1 - T_L} = \frac{70}{10^4} = 7 \text{ m}\Omega$$ ////

21.4-5 Stabilization

One benefit of negative feedback is that the system automatically compensates for internal changes. For example, the amplification of an amplifier depends on the characteristics of the transistors used in its manufacture. These characteristics are subject to change with aging and with temperature variations. In addition to these problems there are additional problems if an amplifier is to be mass-produced due to very large variations from unit to unit of the same transistor, often as large as 100 percent or more. To calibrate each amplifier individually is a very expensive solution to this problem and still leaves the problem of variations due to aging and temperature changes.

The application of negative feedback reduces the variation in overall performance due to relatively large changes in open-loop performance. The closed-loop amplification of the amplifier of Fig. 21.4-1*a*, for example, is

$$A_f \approx -\frac{R_f}{R_1}$$

as long as the operational amplifier amplification A is much larger than R_f/R_1. Hence, the closed-loop performance is essentially independent of the open-loop amplification. This is a remarkable conclusion. It states that the closed-loop gain of the system depends *only* on two passive elements, R_1 and R_f, regardless of how A fluctuates, as long as $|A\beta|$ remains large.

It is true that the closed-loop performance does depend critically on the values of R_f and R_1, but it is easier to control these characteristics than the characteristics of a transistor. In fact, it is possible to get resistors which are precise to 0.1 percent and whose temperature coefficients are matched so that as the temperature changes the ratio of R_f to R_1 remains fixed within a few parts per million, assuming they are kept at identical temperatures. The following example illustrates the self-compensation of an amplifier with negative feedback.

Example 21.4-4 In the feedback amplifier of Fig. 21.4-1*a*, $i_i = 0$, $R_1 = 10$ kΩ and $R_f = 5$ MΩ. Assume that the op-amp's amplification A may vary between 10^5 and 10^7.

a Find the total percentage variation in the op-amp amplification using 10^6 as the nominal value.
b Find the total variation in closed-loop amplification.
c Find the percentage variation in closed-loop amplification.

Solution

a The total percentage variation about the nominal value is:

$$100\,\frac{10^7 - 10^5}{10^6} = 990 \text{ percent}$$

b From Eq. (21.4-5) the closed-loop amplification for $R_1 = 10$ kΩ and $R_f = 5$ MΩ is

$$A_f = \frac{-5 \times 10^3 A}{10 + 5 \times 10^3 + 10A} = \frac{-500}{1 + 501/A}$$

For $A = 10^5$,

$$A_f = \frac{-500}{1 + 0.00501} \approx -497.51$$

for $A = 10^6$,

$$A_f = \frac{-500}{1 + 0.000501} \approx -499.75$$

For $A = 10^7$,

$$A_f = \frac{-500}{1 + 0.0000501}$$

$$\approx -499.97$$

Therefore the magnitude of the total variation is

$$499.97 - 497.51 = 2.46$$

c The magnitude of the total percentage variation about the nominal value -499.75 is

$$100 \frac{499.97 - 497.51}{499.75} = 0.492 \text{ percent}$$

The effect of the negative feedback has been to reduce an open-loop variation of 990 percent to less than 1/2 percent. ////

SUMMARY

1 Feedback is a process whereby part of a system's output is fed back to be combined with its input by means of a closed loop.
2 The polarity of the loop gain determines whether the feedback is positive or negative.

a Positive feedback is primarily useful in circuits for the generation of signals (oscillators).
b Negative feedback is generally used for self-compensation and control applications.

3 An operational amplifier (op-amp) is a differential input voltage amplifier with

a Very high voltage amplification
b High input resistance
c Low output resistance
d Low-pass amplitude frequency response essentially flat to about 1 Hz
e Stable performance

4 Operational amplifiers are used as scalers, adders, and integrators in analog computers. They are also used in these and other ways for various instrumentation and systems applications.
5 Negative feedback used in an amplifier circuit

a Decreases its amplification
b Decreases its input resistance[1]
c Decreases its output resistance[1]
d Increases its bandwidth
e Stabilizes its performance relative to variations in the amplifier characteristics

Positive feedback tends to have the opposite effect on a system.

Problems

21.1-1 Draw a block diagram illustrating the feedback system used in a home heating system. Is it a positive or negative feedback system?

21.2-1 Draw the circuit model for a differential input operational amplifier whose voltage amplification is 5×10^5, input resistance is 20 MΩ, and output resistance is 100 Ω.

21.2-2 Modify the circuit model of Prob. 21.2-1 to incorporate a bandwidth of 1 Hz. *Hint*: See Example 19.3-1 on the low-pass filter.

21.3-1 Using as many 1 kΩ, 10 kΩ, 100 kΩ, and 1 MΩ resistances as required, but no more than required, and no more than one 0.1 μF capacitance per circuit, draw circuits to implement the following functions using one operational amplifier per circuit.

a $v_0 = -10\, v_1$
b $v_0 = 21\, v_1$

c $v_0(t) = -100 \int_{-\infty}^{t} v_1(x)\, dx$

21.3-2 Is it possible to design a noninverting integrator using a single operational amplifier? If it is, draw the circuit.

21.3-3 Draw a circuit using a single operational amplifier with two inputs v_1

[1] These statements are true for the circuits discussed in this chapter but are not universally true for all negative feedback configurations.

and v_2 whose output is

$$v_0(t) = -\int_{-\infty}^{t} [v_1(x) + 2\, v_2(x)]\, dx$$

21.3-4 Draw a circuit using an operational amplifier and resistances and capacitances whose output is proportional to the derivative of its input. Give the formula for its output, v_0, in terms of its input, v_1.

21.3-5 Prove that the value given in Eq. (21.3-12) is correct.

21.4-1 An amplifier has an input resistance of 1 MΩ and a voltage amplification of −10. It is connected as shown in Fig. P21.4-1. Find the circuit input resistance, $R_1 = v_1/i_1$, and the closed-loop amplification, $A = v_0/v_1$.

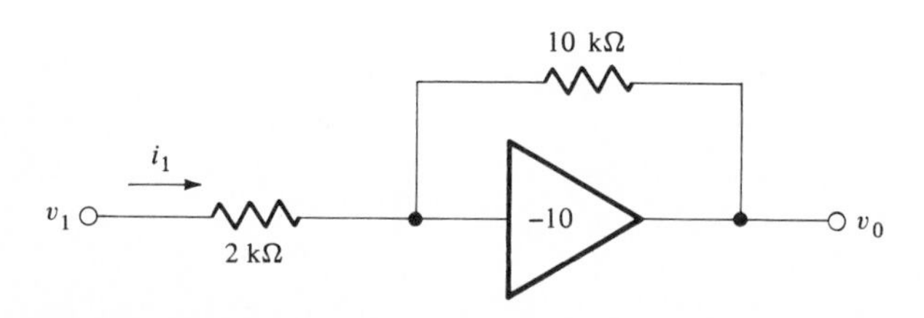

Fig. P21.4-1

21.4-2 An operational amplifier has a voltage amplification of -10^6, an input resistance of 50 MΩ, and an output resistance of 100 Ω. It is connected into a negative feedback configuration as shown in Fig. 21.4-1*a*. Find R_1 and R_f so that the overall input resistance is 1 kΩ and the output resistance is 0.1 Ω. Find the closed-loop amplification with these values.

21.4-3 An amplifier has a nominal amplification of -10^4 which may vary by ±10 percent. Its input resistance is very large and its output resistance is very small. Negative feedback is to be applied to the amplifier so the closed-loop amplification will not vary by more than ±0.1 percent from its nominal value. What is the greatest possible value of the nominal closed-loop amplification?

22
Sequential Digital Circuits and Hybrid Systems

OBJECTIVES

Upon completion of this chapter, you should be able to

1 Differentiate a digital sequential circuit from a digital combinational circuit.
2 Compare synchronous and asynchronous pulse sequential circuits and level sequential circuits.
3 Define level mode, pulse mode, and clock mode.
4 Describe the operation of bistable, astable, and monostable multivibrators and their functions in digital sequential circuits.
5 Draw logic diagrams for storage and shift registers.
6 Draw logic diagrams for binary counters.
7 Describe the operation of parallel and sequential binary adders.
8 Define hybrid systems.
9 Describe the operation of A/D and D/A converters.

INTRODUCTION

In Chap. 13 we introduced digital electronic circuits. All of the circuits considered in that chapter were *combinational digital circuits* because the outputs of the circuits at any instant of time depended only on the inputs to the circuits at the *same* instant of time. There are many applications in which it is desirable for a circuit output to depend on *previous* input values as well as the present values; a digital circuit for which this is true is called a *sequential digital circuit.*

In this chapter we shall introduce the basic concepts of sequential circuits, the elements used in their construction, and some of the most widely used circuits. Designers have available today a wide variety of digital circuits in integrated circuit form. Because of the microscopic size of the basic elements and the use of medium-scale integration (MSI) and large-scale integration (LSI), it is possible to obtain a complex digital system in a package no larger than a matchbook. As a result, designers must take a systems point of view, rather than the narrower circuit design point of view, in approaching the problems given to them. Today's technologist must think of a multilevel shift register as just one more system element just as yesterday's technologist thought of the transistor as a circuit element. This chapter should serve as a starting point in the development of this attitude.

The chapter concludes with a discussion of hybrid systems: systems which incorporate both analog and digital components. A major topic in this discussion is analog-to-digital and digital-to-analog conversion. These are the techniques the modern systems oriented technologist must use to interface the digital systems he designs to other systems which are analog in nature.

22.1 SEQUENTIAL CIRCUIT CONCEPTS

The output of a sequential circuit depends, in general, on previous inputs, as well as the present input. This indicates that a sequential circuit must have *memory*, or the ability to *store* information. A general block diagram for a sequential circuit is shown in Fig. 22.1-1; the blocks in this diagram indicate functional parts of the system rather than physical components. We see from the diagram that the sequential circuit includes a combinational circuit as one of its two elements, the other element being a memory.

The signal paths shown in Fig. 22.1-1 are shown as single lines for simplicity but represent multiple parallel lines; the signal labeled y, for example, may actually represent n lines: $y_1, y_2, \ldots, y_n$. The system input is x and z is the system output.

An additional input to the combinational circuit is the output of the memory. These signals are called the *state variables* and are collectively designated y. They represent the effect of the sequential circuit's past history on its present output. A sequential circuit is sometimes called a *sequential machine* and the state variables are said to determine the *state* of the sequential machine. The

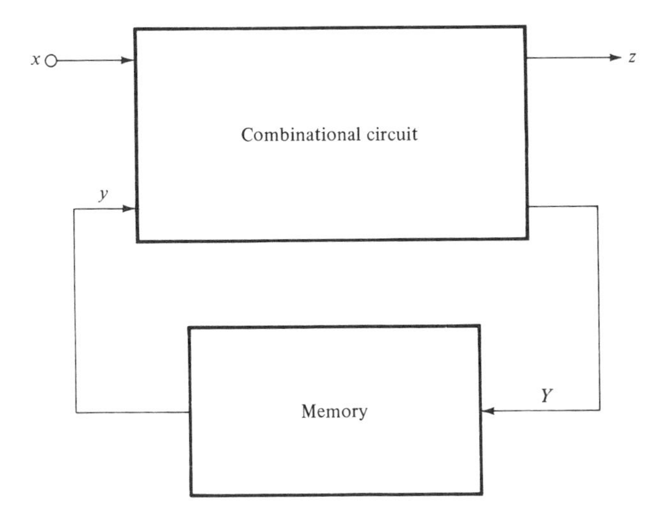

Fig. 22.1-1 Functional block diagram of a sequential digital circuit.

state of the machine is its internal configuration; its output at any time is determined by its inputs and its state, i.e., the values of y. Thus the output of the sequential circuit, z, depends only on the present inputs to the combinational circuit, the system input x and the state variables y.

The combinational circuit has a second set of outputs, Y, called the excitation variables. These variables constitute the input to the memory and determine the future values of the state variables, y.

The purpose of the next sections is to describe in greater detail, and exemplify, the way in which the sequential mode of operation is obtained by combining a combinational circuit with the memory function.

Sequential circuits can be subdivided into three types depending on the nature of the signals in the system. The three types are

1 *Level mode,* in which a signal may change at any time maintaining its new level until the next change
2 *Pulse mode,* a special case of level mode in which signals tend to be at the logical zero level except for brief periods during which they may be at the logical one level
3 *Clock mode,* a special case of pulse mode in which a master synchronizing signal called the *clock* is used

22.1-1 Level mode

The signals in a level mode circuit may change at any time and in general an input change is followed by an output change after a time delay during which the various intermediate signals change. Figure 22.1-2a shows typical level mode

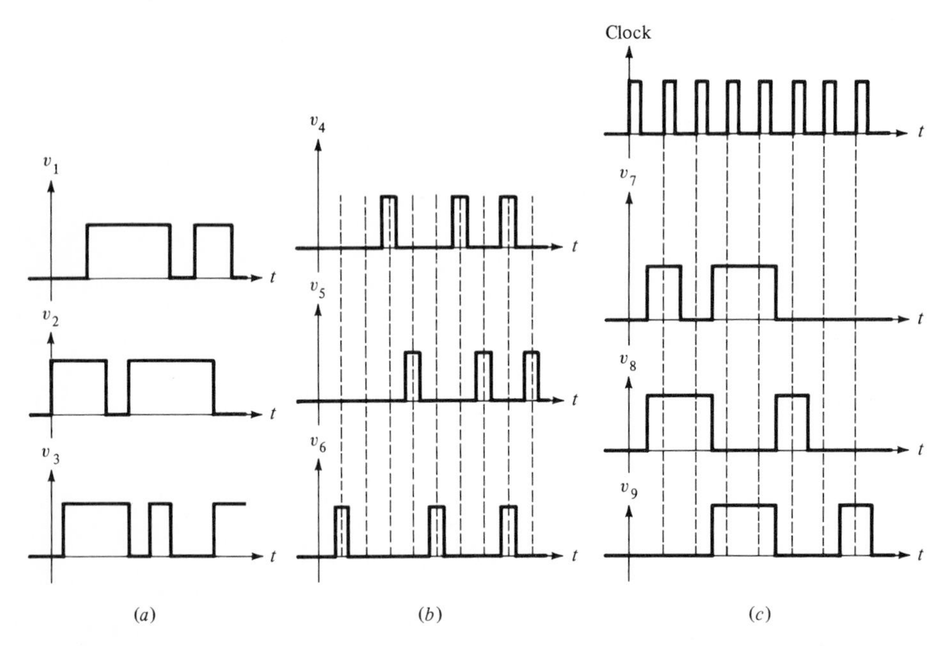

Fig. 22.1-2 Sequential circuit signals for various modes. (*a*) Level mode; (*b*) pulse mode; (*c*) clock mode.

signals in a system. Each level of circuitry between the input and output introduces a delay; therefore, the number of levels of circuitry should be kept to a minimum if high speed operation is important. Level mode circuitry is *asynchronous*, i.e., the changes in the signals do not occur simultaneously, because signals can change at arbitrary times.

22.1-2 Pulse mode

A pulse mode circuit is a special type of level mode circuit in which the inputs are pulses, that is, an input is at a high level (for positive logic) for a short period of time and it then returns to its normal low level. Signals of this type are shown in Fig. 22.1-2*b*. The pulse duration must be long enough for the circuit to respond, but short enough so that the pulse doesn't continue to keep causing changes of state. The period between pulses must be long enough for the circuit to reach a steady state condition, i.e., a condition in which no signal changes are occurring. The inputs to a pulse mode circuit are restricted so that only one input has a pulse at any moment as shown in Fig. 22.1-2*b*. This is done to avoid the problems that would arise in trying to obtain simultaneous pulse inputs because imperfect components could result in one pulse starting or ending slightly before the other. This could result in improper circuit operation. The output of a pulse mode circuit may be pulses or level signals. A pulse mode circuit is asynchronous because the input pulses can occur at any time (so long as there is only one at a time).

22.1-3 Clock mode

The clock mode sequential circuit is a special case of the pulse mode circuit in which a special signal, called the clock signal or simply clock, is used to synchronize the circuit operation. Figure 22.1-2c illustrates signals of this type. The clock is a train of pulses which enable the logic gates to respond to their inputs. Clock mode circuits are designed so that all transient changes occur between pulses when the circuit does not respond to the signals; the clock pulses are made short enough so that gate inputs do not change during a clock pulse. In some circuits the gates are enabled only by a rapid change in the clock signal thus causing the circuit to respond to input signals only at the moment of the pulse rise or the pulse drop, depending on how the circuit is designed. Furthermore, the outputs of the circuit are sensed only during the clock pulses or their rise or fall. This mode of operation is termed *synchronous* because the circuit can respond to inputs only at discrete instants. The synchronous mode of operation eliminates many of the timing problems inherent in asynchronous operation but it simply is not applicable to all applications. A clock mode circuit is sometimes called a synchronous-pulsed sequential circuit while a pulse mode circuit is called an asynchronous-pulsed sequential circuit.

22.2 MULTIVIBRATORS

We have seen that a sequential circuit must contain a memory, but we have not as yet discussed the physical implementations of memory in digital circuits. In this section we shall introduce the bistable multivibrator or, as it is better known, the *flip-flop*. This circuit is commonly used to provide the memory function. We will also discuss two closely related circuits, the astable and monostable multivibrators, which are widely used in digital systems to generate pulse trains and for providing time delays and pulse shaping.

22.2-1 Flip-flops

The basic flip-flop, or bistable multivibrator circuit, is shown in Fig. 22.2-1a. Comparison of this circuit with that of a transistor inverter circuit, shown in Fig. 22.2-1b, shows that the bistable multivibrator consists of two separate inverter circuits with the output of one connected to the input of the other as shown in Fig. 22.2-1c. No external inputs are shown in this diagram because it is used only to explain the essential bistable operation of the flip-flop. A bistable circuit has two stable states; these states correspond to $Q = 0$ and $Q = 1$ in the flip-flop. Figure 22.2-1c illustrates that if $Q = 0$, the input to the left inverter is 0;

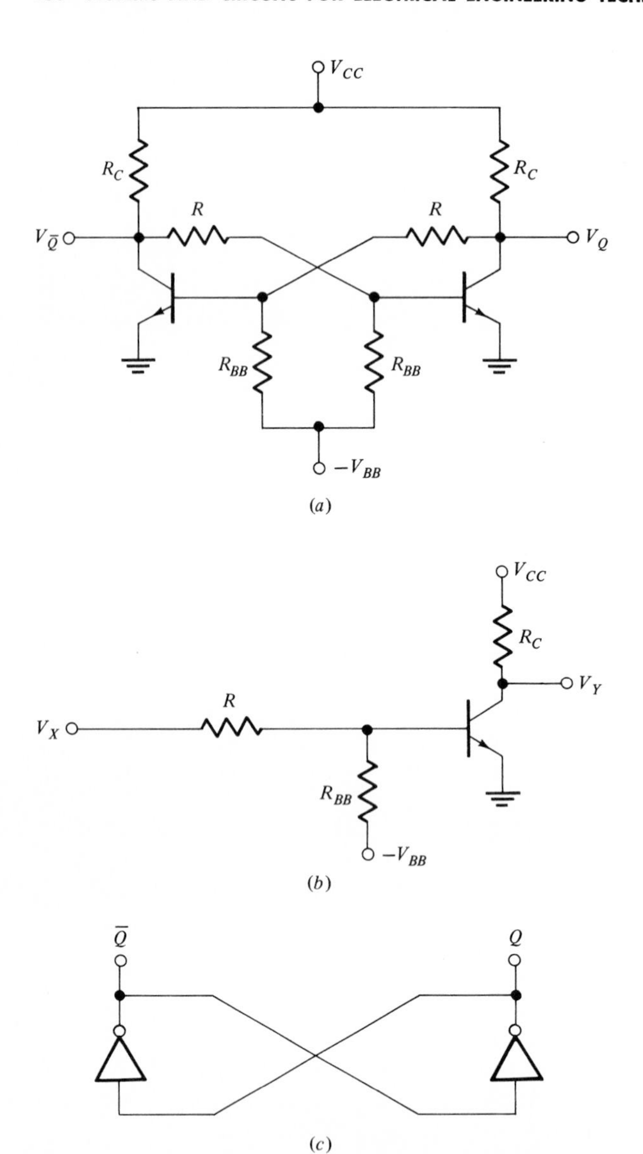

Fig. 22.2-1 Flip-flop. (*a*) Basic bistable multivibrator circuit; (*b*) inverter circuit; (*c*) logic diagram of basic bistable multivibrator.

its output is $\bar{Q} = 1$; the input to the right inverter is 1; and finally the ouput of the right inverter is $Q = 0$. Thus, the circuit can remain with $Q = 0$ indefinitely and that is a stable state. On the other hand, if $Q = 1$, then the input to the left flip-flop is 1; its output is $\bar{Q} = 0$; the input to the right flip-flop is 0; and the output of the right flip-flop is $Q = 1$. Therefore the circuit can remain indefinitely with $Q = 1$ and that is a stable state. The output of the circuit of Fig. 22.2-1*c* will remain at 0 or 1 according to what it initially was without ever changing. The other output is, in each case, the inverse of Q and is therefore denoted $\bar{Q}$.

Unlike the basic flip-flop of Fig. 22.2-1, an actual flip-flop has inputs which enable us to change the output values. The inputs to a flip-flop are called *triggers*. There are various ways of adding trigger inputs to the basic flip-flop. A few of the more popular types of flip-flops and their logical equivalents are shown in Figs. 22.2-2 through 22.2-5. The logic equivalents are only equivalent in terms of logic function; they are not intended to imply anything about the electronic construction of the flip-flop.

SR FLIP-FLOP

The set-reset (SR) flip-flop is shown in Fig. 22.2-2; an equivalent logic diagram is shown in Fig. 22.2-2*a* and the block diagram is shown in Fig. 22.2-2*b*. The operation of the flip-flop is summarized in the truth table of Fig. 22.2-2*c*. In this table Q^{ν} represents the initial output prior to the input and $Q^{\nu+1}$ represents the output right after the flip-flop has responded to the input. The following discussion contains an analysis of the circuit of Fig. 22.2-2*a*, which provides the output values for the truth table.

1 For $S = R = 0$. The NOR gates behave like inverters in this case so the circuit is equivalent to that of Fig. 22.2-1*c*. The output remains at whatever value it has initially.

2 For $S = 1$, $R = 0$. The output $\bar{Q}$ of the left NOR gate is

$$\bar{Q} = \overline{S + Q} = \overline{1 + Q}$$

Whatever the value of Q, $1 + Q = 1$ since $1 + 0 = 1$ and $1 + 1 = 1$. Therefore

$$\bar{Q} = \overline{1 + Q} = \bar{1} = 0$$

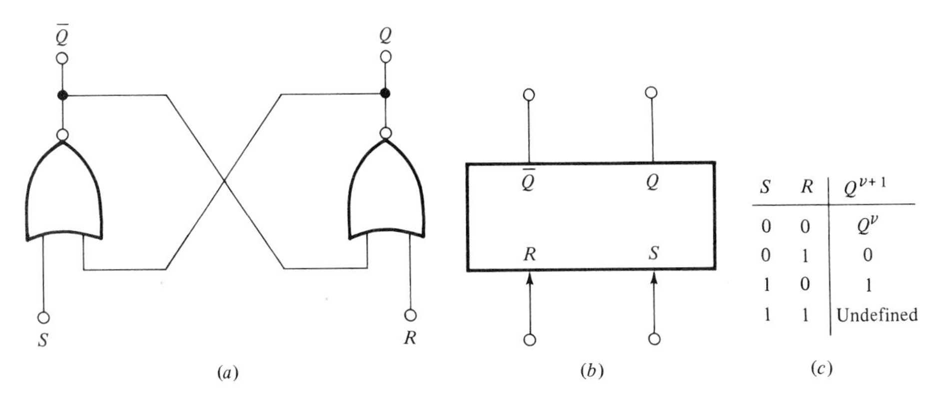

S	R	$Q^{\nu+1}$
0	0	Q^{ν}
0	1	0
1	0	1
1	1	Undefined

Fig. 22.2-2 Set-reset (SR) flip-flop. (*a*) Logic equivalent; (*b*) block diagram; (*c*) truth table.

Using this value of $\bar{Q}$, the output Q of the right NOR gate is

$$Q = \overline{\bar{Q} + R} = \overline{0 + 0} = \bar{0} = 1$$

Thus $S = 1$, $R = 0$ sets Q to 1 and $\bar{Q}$ to 0.

3 For $S = 0$, $R = 1$. The output Q of the right NOR gate is

$$Q = \overline{\bar{Q} + R} = \overline{\bar{Q} + 1}$$

Whatever the value of $\bar{Q}$, $\bar{Q} + 1 = 1$ since $0 + 1 = 1$ and $1 + 1 = 1$. Therefore

$$Q = \overline{\bar{Q} + 1} = \bar{1} = 0$$

Using this value of Q, the output $\bar{Q}$ of the left NOR gate is

$$\bar{Q} = \overline{S + Q} = \overline{0 + 0} = \bar{0} = 1$$

Thus $S = 0$, $R = 1$ resets Q to 0 and $\bar{Q}$ to 1.

4 For $S = R = 1$

The output $\bar{Q}$ of the left NOR gate is

$$\bar{Q} = \overline{S + Q} = \overline{1 + Q} = \bar{1} = 0$$

The output of the right NOR gate is

$$Q = \overline{\bar{Q} + R} = \overline{\bar{Q} + 1} = \bar{1} = 0$$

Thus $S = R = 1$ sets both outputs to zero. This result violates the logical requirement that the value of $\bar{Q}$ be the complement of the value of Q. Furthermore, the output value for a subsequent input of $S = R = 0$ will depend on which of the inputs changes first and the response times of the circuit components; it will not be predictable on the basis of the circuit logic. For these reasons the output for $S = R = 1$ is said to be undefined. The designer of logic systems should ensure that this input does not arise when using the SR flip-flop.

CLOCKED SR FLIP-FLOP

The clocked set-reset flip-flop is useful in synchronous sequential circuits. It performs in much the same way as the SR but the output changes only in response to the SR inputs when the clock input C is 1. A logically equivalent circuit is shown in Fig. 22.2-3*a*. The two AND gates transmit the R and S inputs to the unclocked SR flip-flop only when the C input has a logical 1 value.

A more common type of clocked flip-flop is represented in Fig. 22.2-3*b* where additional circuitry is included which transforms each clock pulse C into a spike, i.e., a very narrow pulse, during the falling (trailing) edge of C. This is illustrated in Fig. 22.2-3*c*. (It is possible to design the circuit to result in a spike on the rising, or leading, edge of C, but this is not as common. We will assume the operation indicated in Fig. 22.2-3*c* unless it is explicitly noted otherwise. The flip-flop illustrated in Fig. 22.2-3*b* is most accurately described as an edge-

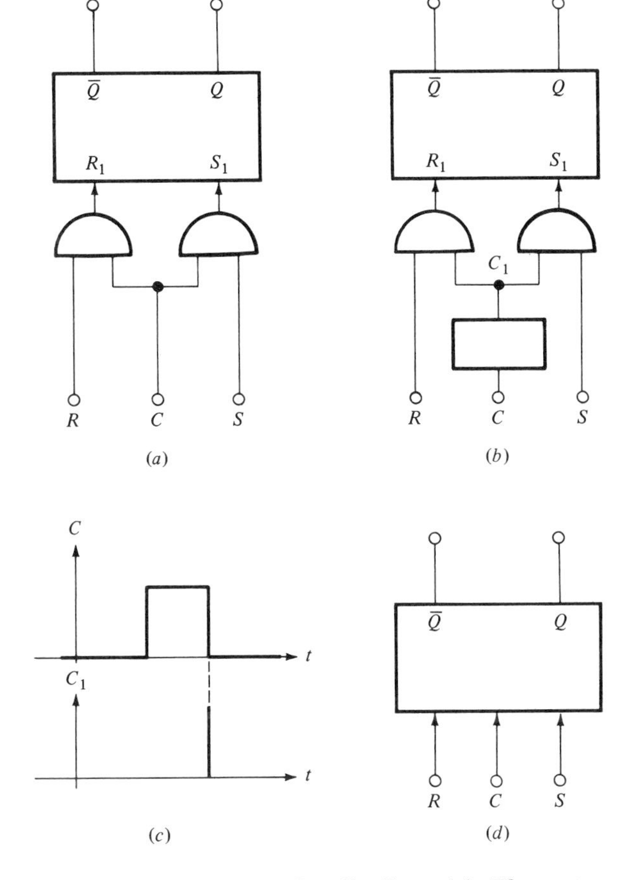

Fig. 22.2-3 Clocked *SR* flip-flop. (*a*) Elementary clocked *SR* flip-flop; (*b*) edge-triggered clocked *SR* flip-flop; (*c*) clock waveform in edge-triggered clocked *SR* flip-flop; (*d*) block diagram of clocked *SR* flip-flop.

triggered clocked *SR* flip-flop, but since this type is so common it is generally referred to simply as a clocked *SR* flip-flop; its block diagram is shown in Fig. 22.2-3*d*.

JK FLIP-FLOP

The *JK* flip-flop is similar to the clocked *SR* flip-flop with one exception; it is designed to respond in a predictable way when $J = K = 1$. For other combinations of inputs the *JK* flip-flop responds the same as the clocked *SR* flip-flop with *S* replaced by *J* and *R* by *K*. For $J = K = 1$, the output flips to the opposite of its present value on the falling edge of the clock pulse.

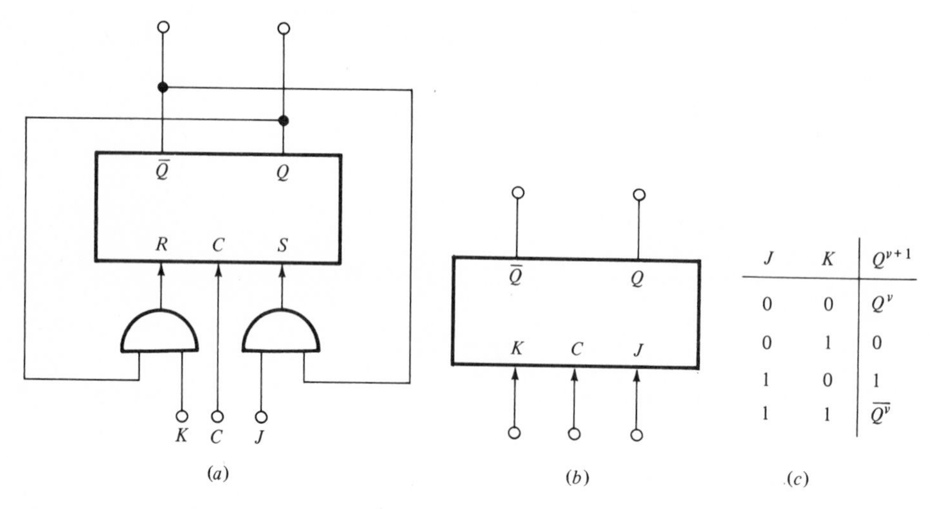

J	K	$Q^{\nu+1}$
0	0	Q^ν
0	1	0
1	0	1
1	1	$\overline{Q^\nu}$

Fig. 22.2-4 *JK* flip-flop. (*a*) Logical equivalent; (*b*) block diagram; (*c*) truth table.

The logical equivalent of the *JK* flip-flop is shown in Fig. 22.2-4*a*, its block diagram is shown in Fig. 22.2-4*b*, and its truth table is shown in Fig. 22.2-4*c*. The truth table applies only during the falling edge of the clock pulse, however; at other times the output remains unchanged. The truth table for the *JK* flip-flop is obtained from its logical equivalent circuit as illustrated in the following example.

Example 22.2-1 Derive the truth table shown in Fig. 22.2-4*c* using the logic circuit of Fig. 22.2-4*a*.

Solution From Fig. 22.2-4*a*,

$$R = KQ \qquad \text{and} \qquad S = J\bar{Q}$$

a For $J = K = 0$, $R = S = 0$; and from Fig. 22.2-2*c*, $Q^{\nu+1} = Q^\nu$.

b For $J = 0$ and $K = 1$, $R = Q^\nu$ and $S = 0$. If $Q^\nu = 0$, $R = 0$ and $S = 0$, so that $Q^{\nu+1} = Q^\nu = 0$. If $Q^\nu = 1$, $R = 1$ and $S = 0$ so that $Q^{\nu+1} = 0$ (see Fig. 22.2-2*c*).

c For $J = 1$ and $K = 0$, $R = 0$ and $S = \overline{Q^\nu}$. If $Q^\nu = 0$, $R = 0$ and $S = 1$, so that $Q^{\nu+1} = 1$ (see Fig. 22.2-2*c*). If $Q^\nu = 1$, $R = 0$ and $S = 0$, so that $Q^{\nu+1} = Q^\nu = 1$.

d For $J = K = 1$, $R = Q^\nu$ and $S = \overline{Q^\nu}$. If $Q^\nu = 0$, $R = 0$ and $S = 1$ so that $Q^{\nu+1} = 1$ (see Fig. 22.2-2*c*). If $Q^\nu = 1$, $R = 1$ and $S = 0$ so that $Q^{\nu+1} = 0$ (see Fig. 22.2-2*c*).

The truth table of Fig. 22.2-4*c* agrees with these results which include all possible situations. ////

TOGGLE

The toggle, or T flip-flop, is a single-input flip-flop whose output reverses each time the input is pulsed. The logical equivalent of the toggle is shown in Fig. 22.2-5*a* and its block diagram is shown in Fig. 22.2-5*b*. Verification of its operation is left as an exercise.

DELAY FLIP-FLOP

The delay (D) flip-flop is frequently used in clocked sequential systems to provide a delay of one clock pulse, i.e., its output at each clock pulse is always equal to its input during the previous clock pulse. A logical equivalent of the delay flip-flop is shown in Fig. 22.2-5*c*. while its block diagram is shown in Fig. 22.2-5*d*. Typical input and output waveforms, together with the clock waveform, for the delay flip-flop are shown in Fig. 22.2-5*e*.

This completes the description of the most popular flip-flops. There are other variations which are combinations of these types. The particular type to be

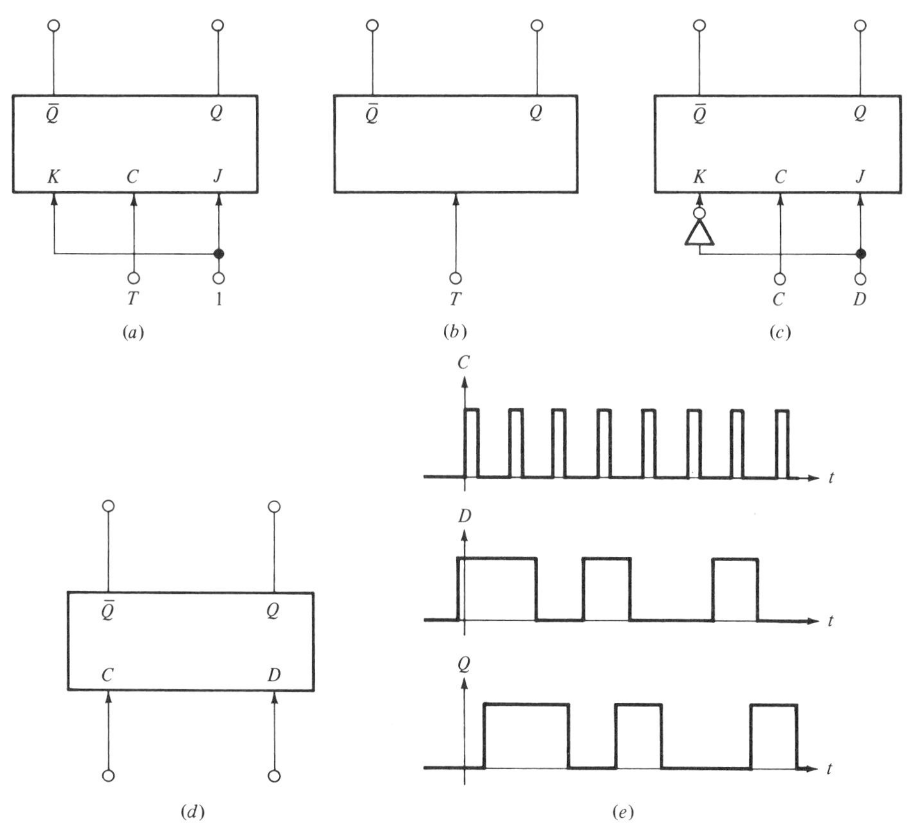

Fig. 22.2-5 Flip-flops. (*a*) Toggle logical equivalent; (*b*) toggle block diagram; (*c*) delay logical equivalent; (*d*) delay block diagram; (*e*) waveforms for delay flip-flop.

used is determined by many factors most of which are of a practical rather than theoretical nature, e.g., cost, availability.

The important characteristic common to all of these flip-flops is that the state of the flip-flop at any time is determined by the last input, that is the input during the last clock pulse or the last nonzero input for a nonclocked type of flip-flop. Therefore, the flip-flop can remember, or store, a single binary digit (bit). By using a number of flip-flops we can store multidigit binary numbers. These may represent simple numbers or coded information of any type; for example, alphabetic information or control signals for an industrial processing plant.

22.2-2 Free-running multivibrator

The free-running, or *astable*, multivibrator provides a pulse train output which is often used as a clock pulse. The circuit of the astable multi is shown in Fig. 22.2-6*a*; like the bistable multivibrator it consists of a pair of transistors each of whose input is derived from the other's output. The coupling is through capacitors rather than resistors and the biasing of the inputs is different. The analysis of this circuit is somewhat beyond the scope of this text but the output waveforms are shown in Fig. 22.2-6*b*, where the logical 1 level corresponds to V_{CC} V. Each output consists of a train of pulses; $\bar{Q} = 0$ when $Q = 1$ and vice versa. The duration of each of the two portions of the pulse train, i.e., the high and low amplitude segments, can be adjusted individually by varying R_1C_1 and R_2C_2. The pulse width can be shown to be $0.694R_1C_1$ or $0.694R_2C_2$ ($0.694 = \ln 2$), as shown in Fig. 22.2-6*b*.

Many relatively simple, small scale systems or circuits require a string of pulses to initiate sequential steps in a process. If the timing requirements are not critical, the astable multivibrator provides an inexpensive solution to the problem. Where precise timing is needed, more expensive crystal-controlled oscillators are used.

22.2-3 One-shot multivibrator

The one-shot, or monostable, multivibrator is used to provide time delays and to reshape pulses. The one-shot has a single stable state and an unstable state; the circuit is a combination of the bistable and astable circuits with a single input and two outputs, one of which is the inverse of the other. When a pulse is applied to the input, the output which is normally low goes high for a period of time after which it returns to the low voltage. The duration of the pulse is determined by an RC combination in the multivibrator circuit. A block diagram and the waveforms for a one shot are shown in Fig. 22.2-7.

By inserting a one-shot in a circuit, a time delay equal to the pulse width of the one shot can be obtained. Another application is the reshaping of pulses.

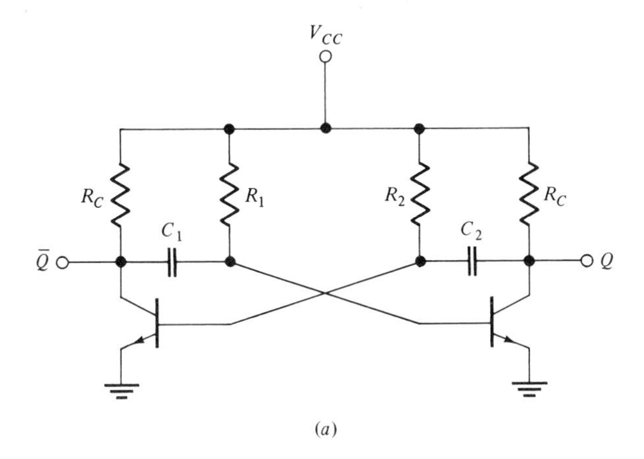

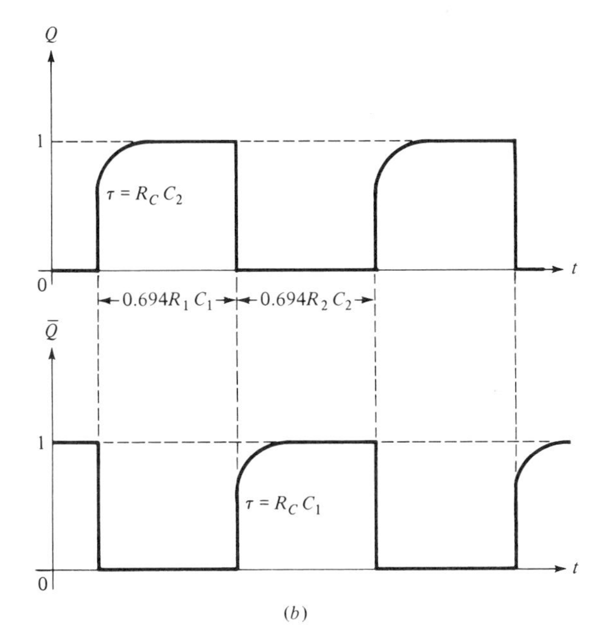

Fig. 22.2-6 Astable multivibrator. (*a*) Circuit diagram; (*b*) output voltages.

This is required because the transmission of a pulse through multiple logic gates tends to round its corners, diminish its amplitude, and delay its edges. This can result in delayed and unreliable responses of gates and multivibrators. To prevent these problems the one-shot is inserted into the circuit to restore the pulse before it is badly distorted.

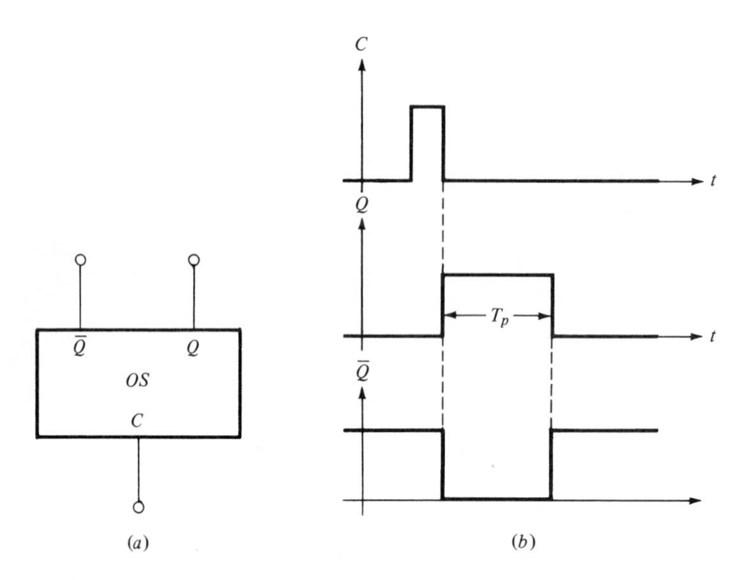

Fig. 22.2-7 One-shot multivibrator. (*a*) Block diagram; (*b*) waveforms.

This concludes the discussion of the three types of multivibrators. It is important to be familiar with them because they are widely used in digital systems. The following example illustrates the use of some of these components.

Example 22.2-2 Design an electronic door lock which consists of three buttons A, B, and C such that the lock opens only if A is pushed momentarily and then C and B, in that order and only if they are both operated so that C is released and B pushed at least 2 s and no more than 3 s after A is released.

Solution The timing is performed by one-shot multivibrators. Two of them are required; one for the 2-s interval following A and the second for the subsequent 1-s interval. The first one-shot is triggered by the drop in voltage when A is released. Its output is a positive pulse of 2-s duration whose trailing edge triggers a second one shot whose output is a positive pulse of 1-s duration.

In order to open the lock it is necessary to push C and then B while the output of the second one-shot is high. AND gates are used to ensure that the output of the second one-shot is high. In order to ensure that C is pressed before B, a flip-flop is used which is reset by release of A or B and set by the combination of release of C and the high output from the second one-shot. Then, in order to open the lock, it is necessary that B be closed and the flip-flop and second one-shot both be high. The logic diagram is shown in Fig. 22.2-8*a* while the signal waveforms are shown in Fig. 22.2-8*b*. ////

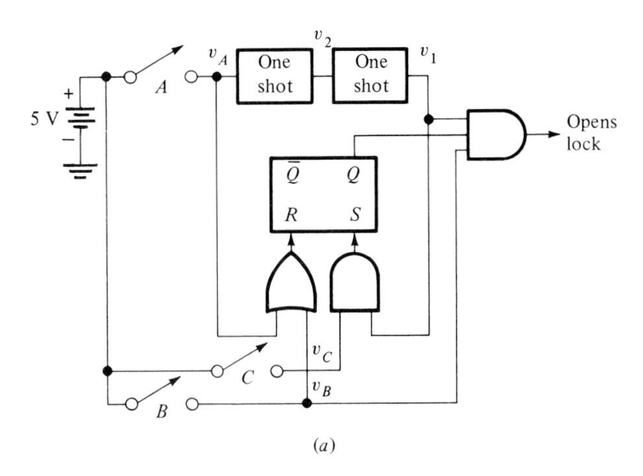

(*a*)

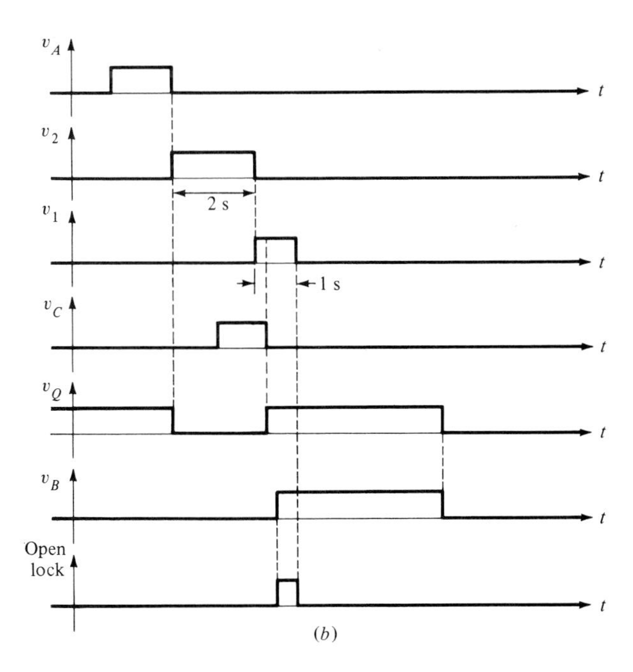

(*b*)

Fig. 22.2-8 Electronic lock for Example 22.2-2. (*a*) Logic diagram; (*b*) signal waveforms.

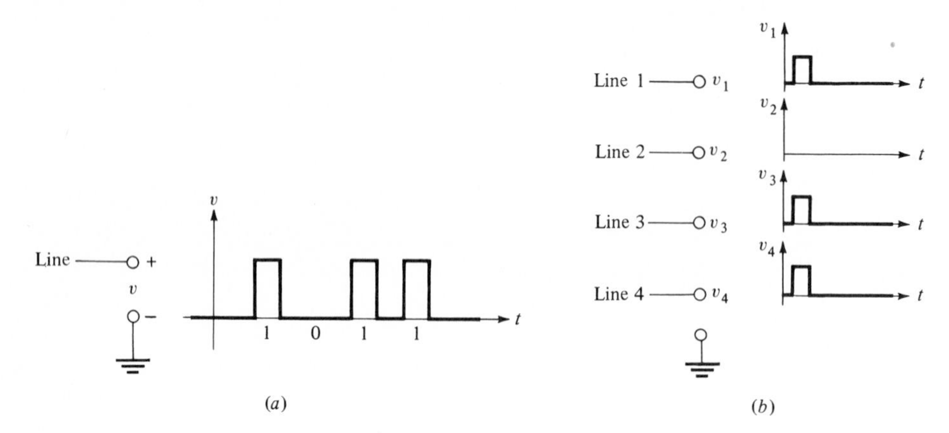

Fig. 22.3-1 Comparison of serial- and parallel-data formats. (*a*) Serial; (*b*) parallel.

22.3 REGISTERS

A register is a group of flip-flops which can store a multibit collection of data. There are two main types: the first type is called a *storage register* and is used simply for storage while the other, called a *shift register,* is useful for parallel-to-serial and serial-to-parallel conversions. Data is said to be in *parallel* form when it is represented by simultaneous pulses on multiple lines; it is in *serial* form when it occurs as a sequence of pulses on a single line. A comparison of serial and parallel data is shown in Fig. 22.3-1.

22.3-1 Storage register

The storage register consists of a number of parallel flip-flops connected with additional circuitry for clearing and loading. There are various types of register which differ in the mode of clearing and loading.

One type requires that the register be cleared before information is stored in it. The diagram for a four-bit register of this type is shown in Fig. 22.3-2*a*. Its operation is as follows: a 1 level is applied to the CLEAR line; next the clock pulse resets the flip-flops. This sets P_1, P_2, P_3, and P_4 to the 0 level. The CLEAR line is then reset to 0. When the desired signals are on the input lines, S_1, S_2, S_3, and S_4, a 1 level is applied to the LOAD line. This causes those flip-flops with 1 inputs to be set during the next clock pulse so that their outputs are at the 1 level. After this the LOAD level is reset to 0. This causes the value of each P_i at the end of this sequence of operation to equal the value of the corresponding S_i when the LOAD line and the clock pulse were both at the 1 level. Notice that the 1 level applied to the LOAD line *gates* the inputs to the flip-flops which respond only during the clock pulse.

The storage register discussed above has a disadvantage if it is intended for use in a system in which high speed is desired because it takes at least two clock pulses to change the information stored: one to clear or reset it and the second to load it. This disadvantage is overcome in the four-bit register shown in Fig. 22.3-2*b* which requires only a single clock cycle to load information. When the data is available on the input lines S_1, S_2, S_3, and S_4 a 1 level is applied to the LOAD line. During the next clock pulse each flip-flop is either set (if the input is

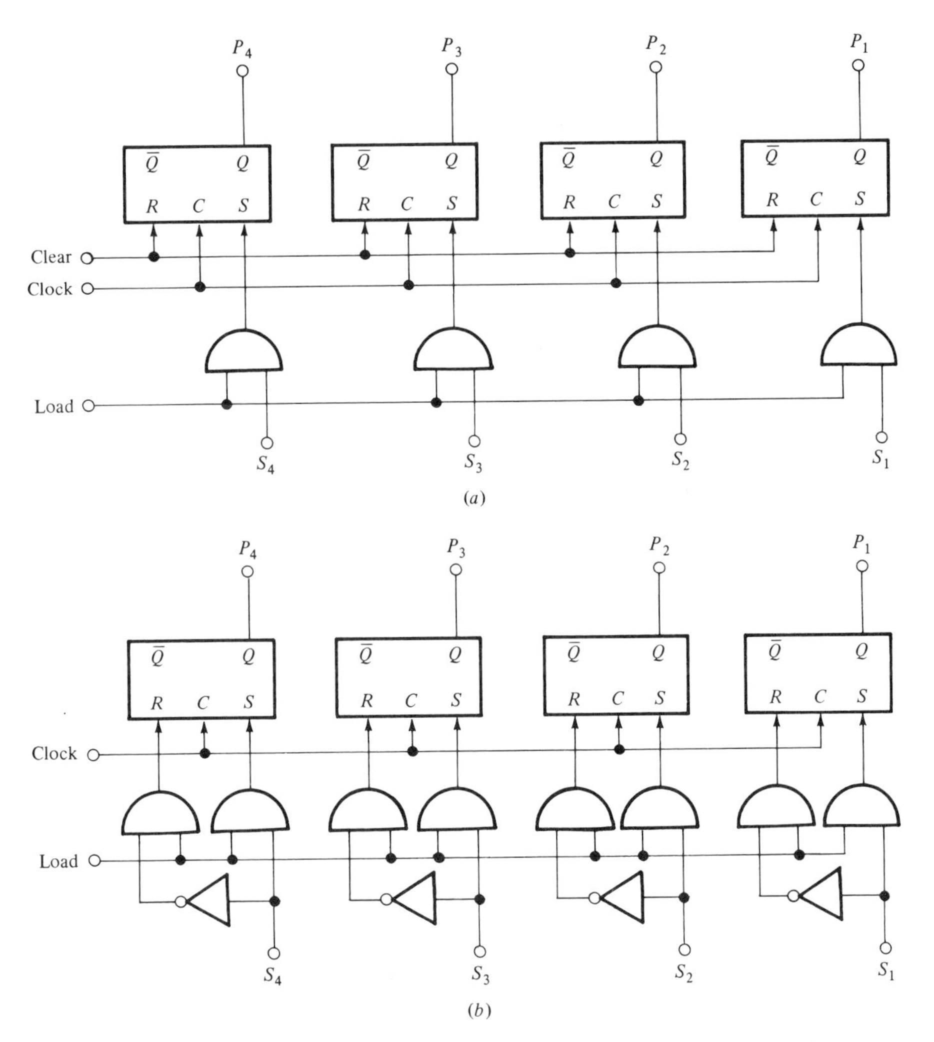

Fig. 22.3-2 Storage registers. (*a*) Two-cycle load; (*b*) single-cycle load.

a 1) or reset (if the input is a 0). Notice that this circuit requires twice as many AND gates as the one in Fig. 22.3-2*a* in addition to an inverter for each bit. Therefore it is more complex and more expensive; this is the price that must be paid to achieve the higher speed of operation. If often happens that the designer can make a trade-off between economy and speed. One of the big design problems is to strike a reasonable balance between these two criteria.

22.3-2 Shift registers

One way in which the designer can control the speed-economy balance is by his choice of the serial versus parallel modes of operation. Consider, for example, the problem of adding two eight bit binary numbers. In the *parallel mode*, eight full adders are used which simultaneously add the eight pairs of digits. In the *serial mode*, a single full adder is used; the first pair of digits is added, then the second pair, then the third, and so on, until the eighth pair of digits is added. This takes approximately eight times as long as the parallel adder. It is obvious that the cost of the single full adder used in the serial mode is about one-eighth the cost of the eight full adders required in the parallel mode but the time required in the serial mode is eight times that needed in the parallel mode. In general, a compromise is made in which a combination of serial and parallel modes is used. For example, four parallel adders might be used with two sequential operations required to add the eight bits. Furthermore, certain operations may be done in parallel and others in serial fashion. As a result, it is necessary to convert data between serial format, in which successive bits occur on the same line at successive instants of time, and parallel format, in which successive bits occur on different lines all at the same instant of time (see Fig. 22.3-1).

The shift register can be used to convert from serial to parallel format or vice versa. The diagram of a basic four bit shift register to be used for serial to parallel conversion is shown in Fig. 22.3-3*a*. It is seen to be a cascade connection of four delay (D) flip-flops. Also shown in the diagram is an input signal and the SHIFT signal which is the clock pulse train. Figure 22.2-3*b* also shows the parallel outputs as functions of time. The outputs have values before the first clock pulse which have resulted from previous serial inputs not shown in the diagram. The first SHIFT pulse coincides with no pulse on the SIGNAL line and causes P_1 to become 0. The initial value $P_1 = 1$ when applied with this first SHIFT pulse to the second flip-flop results in the new value $P_2 = 1$. The initial value $P_2 = 0$ is shifted into the third flip-flop by the first SHIFT pulse resulting

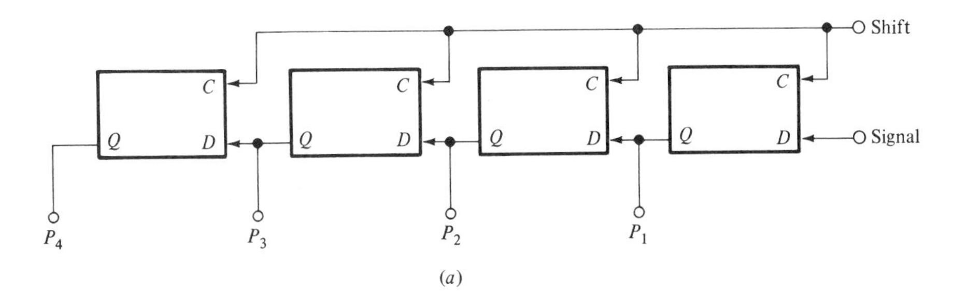

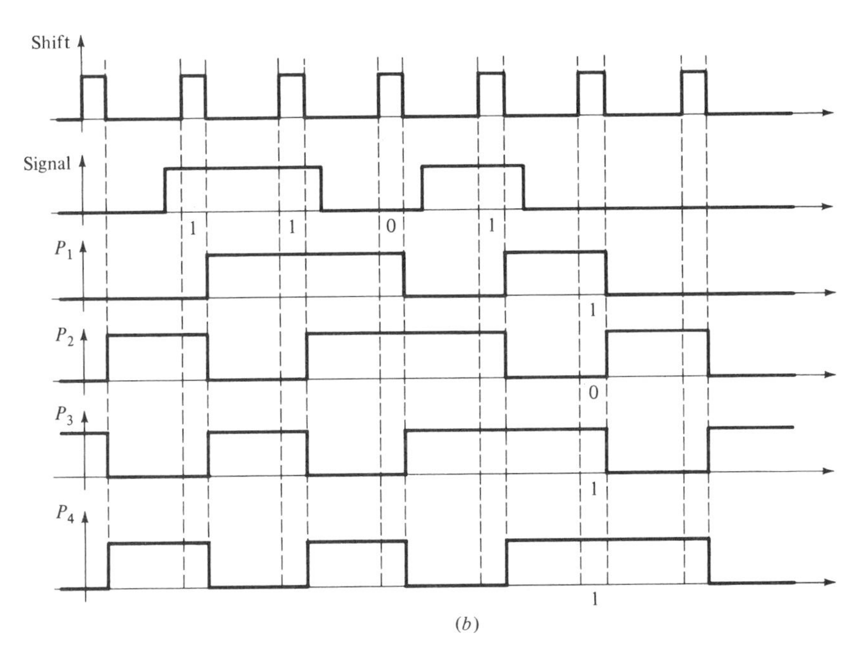

Fig. 22.3-3 Serial to parallel conversion. (*a*) Basic four-bit shift register; (*b*) outputs at successive clock pulses.

in the new value $P_3 = 0$ while the initial value $P_3 = 1$ is simultaneously shifted into the final flip-flop resulting in the new value $P_4 = 1$. The initial value $P_4 = 0$ is shifted out of the shift register and lost. Each subsequent shift pulse causes the content of each flip-flop to be transferred to the next flip-flop to the left. The content of the leftmost flip-flop is lost while the rightmost flip-flop is loaded with the value on the signal input line. With this in mind the reader can check the sequence of parallel ouputs P_1, P_2, P_3, P_4 shown in Fig. 22.3-3*b*. Notice that the values following the final shift pulse of Fig. 22.3-3*a* are the parallel equivalent of the last four serial inputs to the SIGNAL input. Thus, the shift register provides a serial-to-parallel conversion.

In order to obtain a parallel-to-serial conversion, it is necessary to use a parallel load shift register such as the one shown in Fig. 22.3-4*a*. There are four parallel inputs, P_1, P_2, P_3, and P_4, in this circuit. These inputs are gated into the D flip-flops by the combination of a 1 level on the LOAD input and a pulse on the SHIFT line; while this is happening the outputs, Q, of the flip-flops are blocked from the inputs, D, of the next flip-flop by the inverted LOAD signal applied to the AND gates between the flip-flops. When the flip-flops have been loaded, the LOAD signal is reset to 0 which blocks the parallel inputs from the flip-flops while the inverted LOAD signal, which will now be a 1, gates the output, Q, of each flip-flop to the input, D, of the next flip-flop on the left. While the LOAD signal is 0 the input, D_1, to the rightmost flip-flop is 0 due to the AND gate feeding it. This causes the contents of the flip-flops to be shifted left one position for each SHIFT pulse with zeros entering at the right end and the content of the leftmost flip-flop being gated to the serial output S. Waveforms illustrating the circuit operation are shown in Fig. 22.3-4*b* where all flip-flops are assumed to initially be in the 0 state, that is, $Q = 0$. The first shift pulse results in $S = 0$ since $Q_4 = 0$. The second shift pulse results in loading the flip-flops with $Q_4Q_3Q_2Q_1 = P_4P_3P_2P_1 = 1101$ since the LOAD signal is high. Thus $D_i = P_i$ for $i = 1, 2, 3$, and 4. For the third and every subsequent shift pulse, the LOAD signal is low resulting in $D_1 = 0$, and the inverse of the LOAD signal is high resulting in $D_{i+1} = Q_i$ for $i = 1, 2$, and 3. This results in one shift to the left for each SHIFT pulse. During each shift pulse Q_4 is gated to S resulting in the serial output shown in Fig. 22.3-4*b*. This serial output during SHIFT pulses 3, 4, 5, and 6 is equal to the parallel input loaded during SHIFT pulse 2.

22.4 BINARY COUNTER

The binary counter is another widely used digital circuit. Its output is the binary representation of the number of input pulses since it was last reset. Like the storage and shift registers, counters are available in various sizes (numbers of bits) and design variations as medium-scale integrated circuits (MSI). A four-bit counter MSI circuit is available in a package which is about 0.25 in $\times$ 0.75 in $\times$ 0.2 in thick.

The basic component of the counter is the toggle, or T-type, flip-flop, shown in Fig. 22.4-1*a*. This circuit may be designed to trigger on either the positive or the negative-going edge of a pulse. Which of these two edges it triggers on is determined by the particular design, but the negative-going edge is probably the more common choice. This means that for positive logic, the flip-flop will trigger on the trailing edge as in the typical input and output signals shown in Fig.

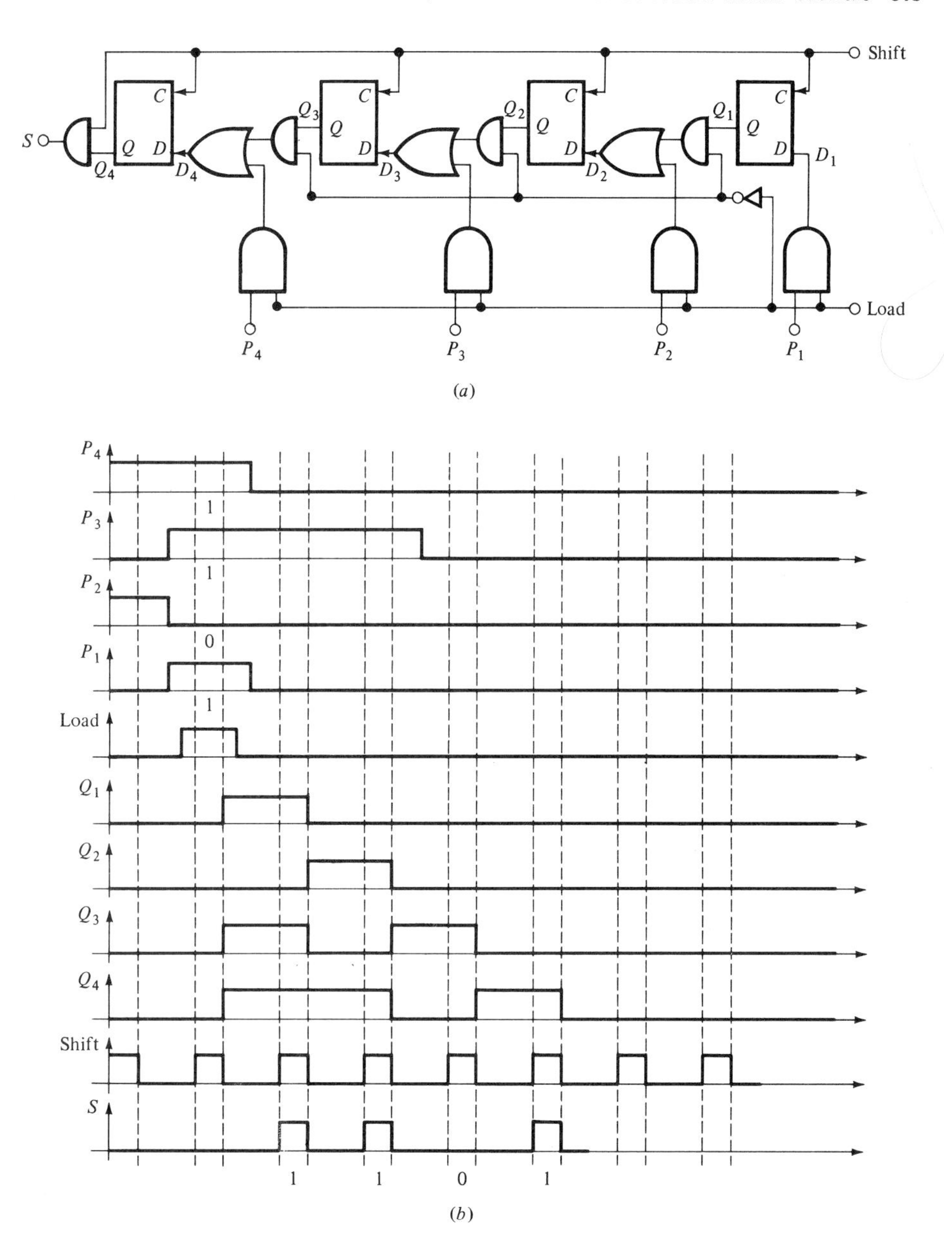

Fig. 22.3-4 Parallel-to-serial conversion. (*a*) Parallel-load shift register; (*b*) waveforms.

22.4-1*b*. An additional input R is provided to reset the toggle so that $Q = 0$. This input is a level input and it overrides the toggle input T so that $Q = 0$ whenever $R = 1$. The waveform shows that the output changes shortly after each negative going, or trailing, edge of the input. The diagram of a three-bit binary counter is shown in Fig. 22.4-1*c*; it is constructed of three negative-going

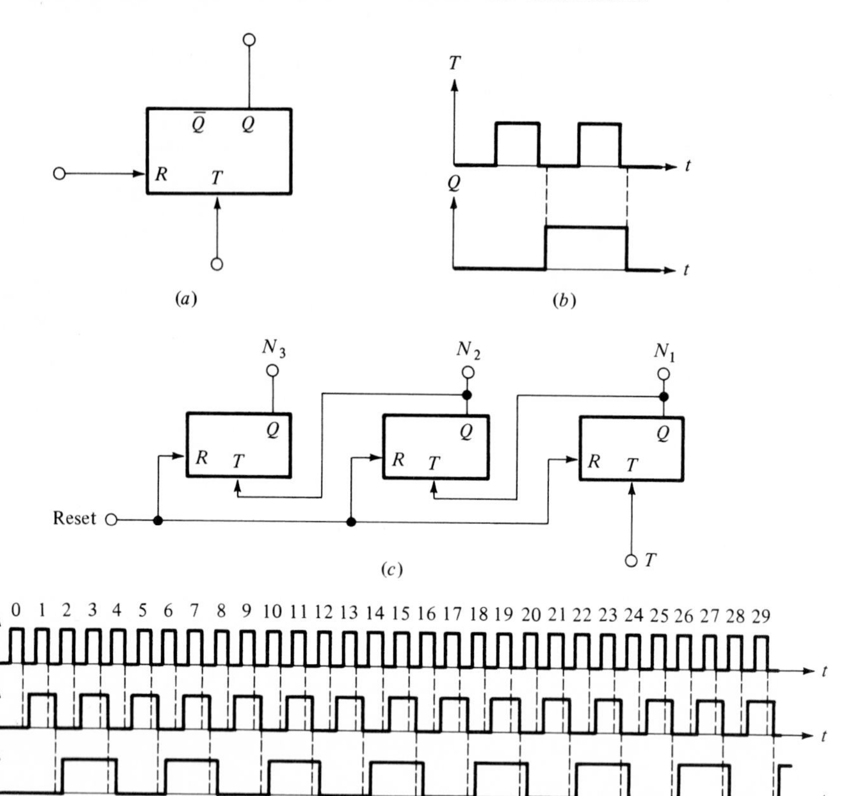

Fig. 22.4-1 Binary counter. (*a*) Edge-triggered toggle (T) flip-flop; (*b*) signals illustrating switching operation of toggle; (*c*) three-bit binary counter; (*d*) signals illustrating operation of binary counter.

edge-triggered toggles. The input signal (T) and output signals (N_1, N_2, and N_3) of the counter are shown in Fig, 22.4-1*d*. It is assumed that $N_1 = N_2 = N_3 = 0$ initially. Each flip-flop changes state when its input drops in value (on the trailing edge of the pulse). The binary counter is sometimes called a frequency divider because the output of each stage has half as many pulses as the output of the stage to the right; thus N_2 has half the number of pulses that N_1 has and N_3 has half the number of pulses that N_2 has.

The binary count is the binary number $N_3N_2N_1$; therefore N_1 has a weight of $2^0 = 1$, N_2 a weight of $2^1 = 2$, and N_3 a weight of $2^2 = 4$ as indicated in Fig. 22.4-1*d*. The decimal equivalent is found by multiplying each bit value (1 for a high voltage and 0 for a low voltage) by the bit's weight and adding the results.

These values are shown below the N_3 signal while the consecutively numbered input pulses are shown at the top of this figure. Note that the count agrees with the number of input pulses counted from 0 to 7 but the count then starts over; this is a consequence of using only three bits which provide a maximum value of 7.

22.5 BINARY ADDERS

Addition is the most basic arithmetic operation and is therefore present in almost any digital system which does arithmetic processing of information. For this reason, and in order to illustrate the use of the digital elements described in this chapter in the design of a digital system, we shall take a detailed look at both parallel and serial adders.

22.5-1 Parallel adder

The adder we are going to describe adds two four-bit numbers together. The adder performs the operation

$$\begin{array}{r} A_4\,A_3\,A_2\,A_1 \\ +\,B_4\,B_3\,B_2\,B_1 \\ \hline S_5\,S_4\,S_3\,S_2\,S_1 \end{array}$$

where each of the symbols A_i, B_i, and S_i represents a single bit. The input to the adder includes, in addition to the eight numerical bits, an additional single bit signal called a *strobe* signal. The strobe signal is required because the inputs to the adder circuit are not valid data on every clock pulse. This is because the circuit supplying this data takes time to perform its operations and may be getting data at a variable rate. The strobe signal is set to the 1 level when valid data is coming into the adder and is otherwise zero. The adder circuit accepts and adds the input data when the strobe is high but not when it is low; in other words the strobe is a control signal. It is analogous to the LOAD signals used in the storage registers of Fig. 22.3-2. It may be supplied by the same circuit that supplies the input data to the adder or it may be generated by a central control which controls both the adder and the circuit supplying its input data.

The adder circuit shown in Fig. 22.5-1 consists of two four-bit storage registers which are loaded in a single pulse period under control of the strobe input. The remainder of the circuit consists of three full adders and a half-adder (the rightmost bit position has no carry input so a full adder is not needed). The full-adder and half-adder circuits were discussed in Sec. 13.7.

The circuits shown within the dotted boxes are available as single chip MSI integrated circuits which makes it very simple to put together such a system.

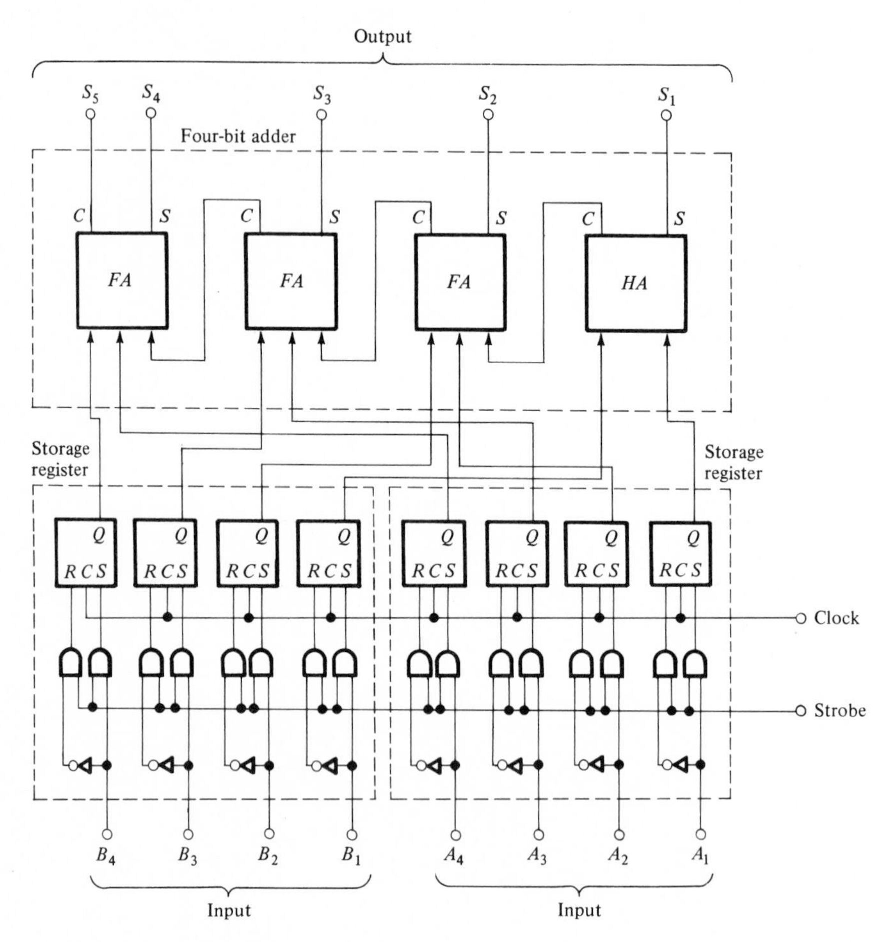

Fig. 22.5-1 Four-bit parallel binary adder.

22.5-2 Serial adder

As discussed in Sec. 22.3-2 serial adders can be used where speed is not important and economy is a factor. Furthermore, if the data arrives in serial fashion there is no advantage to using the parallel adder. The circuit of a serial adder is shown in Fig. 22.5-2*a*. Successive data bits arrive in synchronization with successive clock pulses and are stored in the two lower delay (D) flip-flops. The outputs of these flip-flops do not change until shortly after the clock pulse due to the response time of the flip-flop. During the clock pulse the sum output, S, of the full adder is gated by the clock pulse to the output SUM. The carry output, NEW CARRY, of the full adder is input to the upper delay flip-flop which causes its output level, LAST CARRY, to change to this input value shortly after the clock pulse is complete. This output level serves as the third input to the full

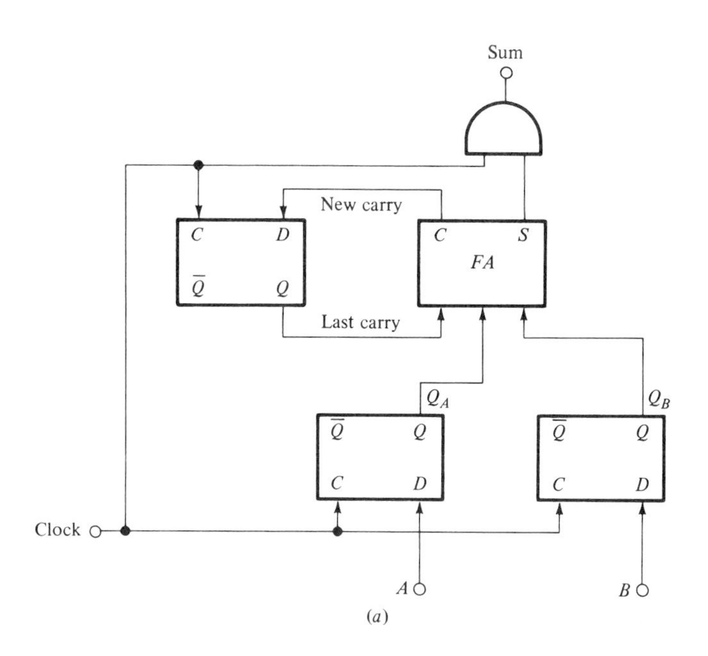

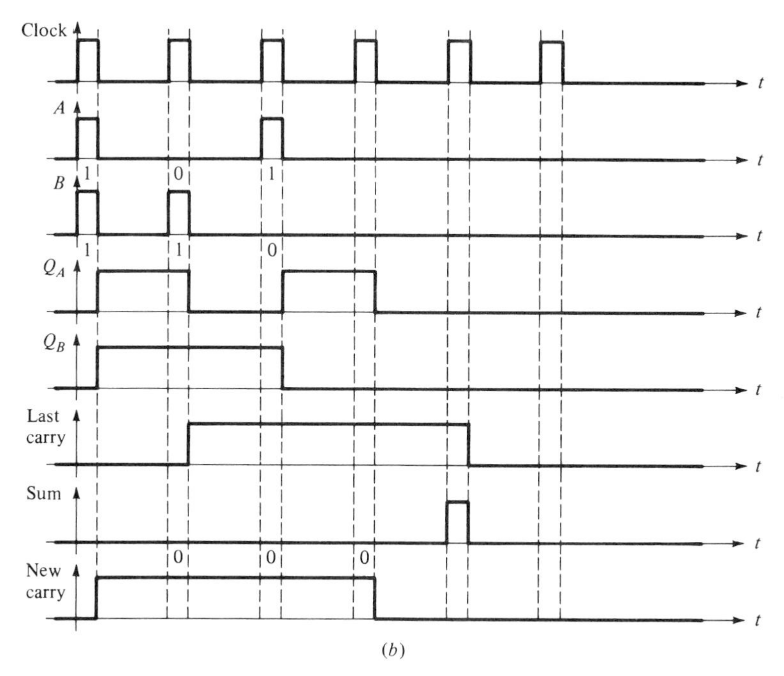

Fig. 22.5-2 Serial binary adder. (*a*) Circuit; (*b*) waveforms illustrating 101 + 011 = 1000.

adder. Note that the output signal must be considered for *two* clock pulses beyond the last input pulses; during the first clock pulse the sum of the last two inputs appears at SUM while the possible carry bit appears at SUM during the second clock pulse. A series of waveforms representing the addition $101 + 011 = 1000$ is shown in Fig. 22.5-2*b*. Note that the least significant bits must occur first in time so that their carry will be added to the more significant bits. This results in the numbers appearing to be in reverse order in the figure because earlier times are to the left.

This concludes the discussion of sequential digital systems. The reader will find it both interesting and informative to look through catalogs of digital integrated circuits. They contain a large variety of circuits performing many different functions using different types of semiconductor technology each with its characteristic speed, power requirement, noise immunity, and cost.

22.6 HYBRID SYSTEMS

A *hybrid* electronic system is one which contains both analog and digital elements. Despite the great emphasis on digital electronics in today's technology there are still many functions that are better, more easily, or more economically performed using analog circuits and systems. Furthermore, many systems are inherently analog in nature; the speed of an automobile, for example, is an analog quantity (it is continuously variable) even though it may be measured with a digital meter. As a consequence of these considerations many systems are hybrid in nature.

A very important consideration in the design of hybrid systems is the interfacing of the analog and digital subsystems. This is done by means of analog-to-digital (A/D) and digital-to-analog (D/A) converters.

22.6-1 Digital-to-analog conversion

A digital-to-analog (D/A) converter is a device whose input is a digital signal and whose output is an analog signal with a value proportional to the numerical value of the input signal. The input signal may be in parallel or serial mode. There are many different ways in which to design a D/A converter; they differ in terms of speed, economy, precision, size, power requirements, and other characteristics.

The D/A converter shown in Fig. 22.6-1 is a rather simple six-bit parallel input converter which uses an operational amplifier adder circuit (refer to Sec. 21.3-2). The nominal voltage levels for the digital signals are +5 V for logical 1 and 0 V for logical 0. The range of linear operation for the operational amplifier is −10 V to +10 V.

The converter consists of three subsystems: a six-bit storage register constructed of six *D* flip-flops, an analog adder, and an analog inverter. The operation of this D/A converter is as follows: The input to the converter is the binary number $d_6d_5d_4d_3d_2d_1$ where each d is logical 1 or 0. These six bits are entered into

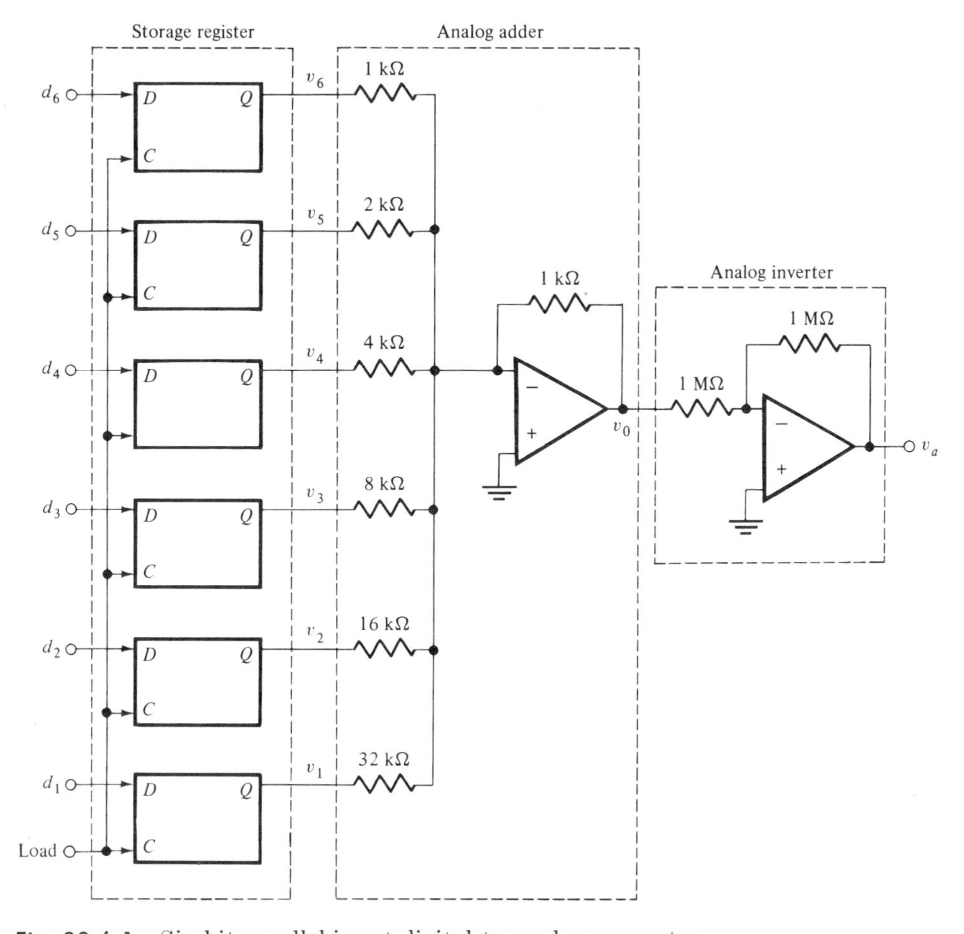

Fig. 22.6-1 Six-bit parallel input digital-to-analog converter.

the storage register by providing a LOAD pulse. The function of the storage register is to provide uniform amplitude output pulses. This is necessary because there is a range of voltages around the nominal values which constitute acceptable values of the logical signals but would result in incorrect analog outputs. The storage register is designed so that the output voltage levels are very precise; +5 V for a logical 1 input and 0 V for a logical 0 input.

The output voltages, v_1 through v_6, of the storage register are the inputs to the analog adder. This adder is constructed of a group of resistors and a single operational amplifier. From the theory covered in Sec. 21.3-2 the output of the analog adder is

$$v_0 = -1\left(\frac{v_6}{1} + \frac{v_5}{2} + \frac{v_4}{4} + \frac{v_3}{8} + \frac{v_2}{16} + \frac{v_1}{32}\right) \tag{22.6-1}$$

If $d_6d_5d_4d_3d_2d_1 = 000000$, then $v_0 = 0$ which is its minimum magnitude. If

$d_6d_5d_4d_3d_2d_1 = 111111$, then all the v's are 5 V, which results in

$$v_0 = -1\left(\frac{5}{1} + \frac{5}{2} + \frac{5}{4} + \frac{5}{8} + \frac{5}{16} + \frac{5}{32}\right)$$

$$= \frac{-5}{32}(32 + 16 + 8 + 4 + 2 + 1) = \frac{-5(63)}{32} = -9.84375 \text{ V}$$

which is its maximum magnitude. Note that the choice of values utilizes essentially the entire linear range of operation thereby maximizing the precision of the converter. The minus sign is due to the inversion in the operational amplifier which is required to provide the negative feedback.

The final stage in the D/A converter is the analog inverter required to eliminate the minus sign introduced by the inverter. This gives $v_a = -v_0$ resulting in a range of values from 0 V to 9.84375 V for v_a corresponding to digital input values ranging from 0 to 63 (binary 111 111).

22.6-2 Analog-to-digital conversion

An *analog-to-digital (A/D) converter* is a device which converts an analog signal to a digital signal whose numerical value is proportional to the value of the analog voltage.

There are many different ways to design an A/D converter, each with its own particular characteristics. The basic block diagram of one popular type is shown in Fig. 22.6-2*a*. For simplicity, it is assumed that the analog input voltage, v_a, is always positive or zero. In practice this can be achieved by adding a positive dc voltage to the input so that the resulting voltage is always positive; this technique is often used.

The A/D converter consists of an AND gate and four subassemblies: a comparator, a binary counter, an astable multi, and a D/A converter. The AND gate was described in Sec. 13.2-1 while the binary counter, astable multi, and D/A converter were discussed in earlier sections of this chapter.

Operation of the comparator is as follows: It consists of an operational amplifier with an amplification (for $R_f \gg R_1$)

$$A \approx \frac{R_f}{R_1} = \frac{10^6}{100} = 10^4$$

The differential input is used so that the amplifier output is (see Sec. 21.3-1)

$$v_e \approx 10^4(v_a - v_d) \tag{22.6-2}$$

where v_a is the analog input voltage to the converter and v_d is an analog voltage which is proportional to the digital ouput. The output, v_e, of the amplifier is applied to a voltage divider consisting of a 1-kΩ resistor and a 5-V zener diode. If $v_e < 0$, the zener diode conducts resulting in the GATE voltage across it being

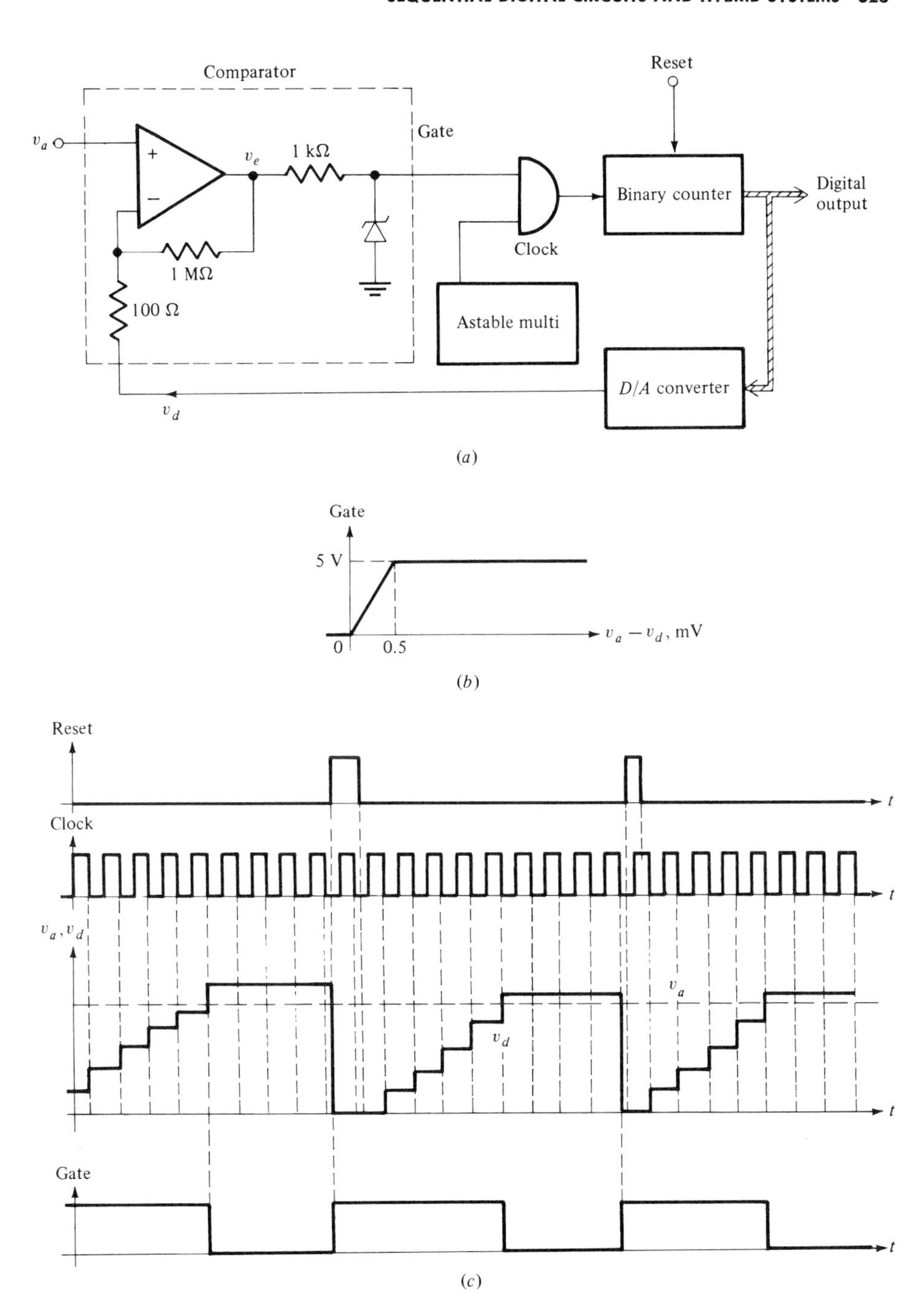

Fig. 22.6-2 Parallel output analog-to-digital converter. (*a*) Converter diagram; (*b*) comparator characteristics; (*c*) waveforms.

approximately zero. For $v_e > 5$ V, the zener diode is in the zener region resulting in a constant GATE voltage equal to about 5 V. For $0 < v_e < 5$ V, the zener diode is not conducting and the GATE voltage is approximately equal to v_e. The behavior in the various ranges of v_a to v_d results in the complete comparator characteristic shown in Fig. 22.6-2*b*.

It is assumed that the nominal voltage levels are 0 and 5 V for logical 0 and 1, respectively. This means that the GATE signal will be indeterminate for $v_a - v_d$ somewhere between 0 and 0.5 mV.

When $v_a > v_d$ the GATE signal is high thereby gating the CLOCK signal, i.e., transmitting the CLOCK signal through the AND gate, to the binary counter. The DIGITAL OUTPUT, which is the counter output and consists of a number of parallel lines representing the binary count, is incremented by one for each clock pulse gated through to the counter. This digital output is the input to the D/A converter which generates an analog output voltage v_d which is proportional to the digital output. The voltage v_d is incremented for each count resulting in a staircase waveform as shown in Fig. 22.6-2*c*, which illustrates the operation of the A/D converter. When v_d reaches a level such that $v_a \leq v_d$, the GATE signal goes low blocking the CLOCK pulses from the binary counter and maintaining the DIGITAL OUTPUT at a fixed level until a RESET pulse is applied to the counter resetting it to digital zero and causing $v_d = 0$ (analog 0 from the D/A converter) which results in $v_a > v_d$. This condition causes the GATE signal to go high causing the DIGITAL OUTPUT and v_d to increase and begin its staircase form again. All of this is illustrated in Fig. 22.6-2*c*. The digital output represents the analog voltage correctly only when the GATE voltage is low; when the gate voltage is high the digital output is being incremented up so as to reach the analog voltage and the two are not equal. The digital level, v_d, is incremented in finite size steps so that the digital level, v_d, may exceed the analog level, v_a, by a fraction of a step. This is an unavoidable error in A/D conversion; it may be reduced by the use of smaller steps which requires a larger number of digits in the counter to cover the same range of analog voltages.

22.6-3 Digital voltmeter

This section concludes the chapter with an example of a hybrid system: the digital voltmeter.

A simple form of the digital voltmeter is shown in Fig. 22.6-3. It is little more, in principle, than the A/D converter just discussed. The analog voltage is converted to a 12-bit binary number using an A/D converter similar to the one shown in Fig. 22.6-2. The 12-bit binary number can represent decimal values from 0 to $2^{12} - 1 = 4095$. Therefore, the binary number is converted to 4 four-bit BCD numbers (see Sec. 13.6-2) each of which is further converted to a seven-bit signal used to drive a seven-segment light-emitting diode (LED) display. This display consists of seven segments which can be individually lighted so as to form any of the ten digits. The layout of the segments is shown in Fig. 22.6-3.

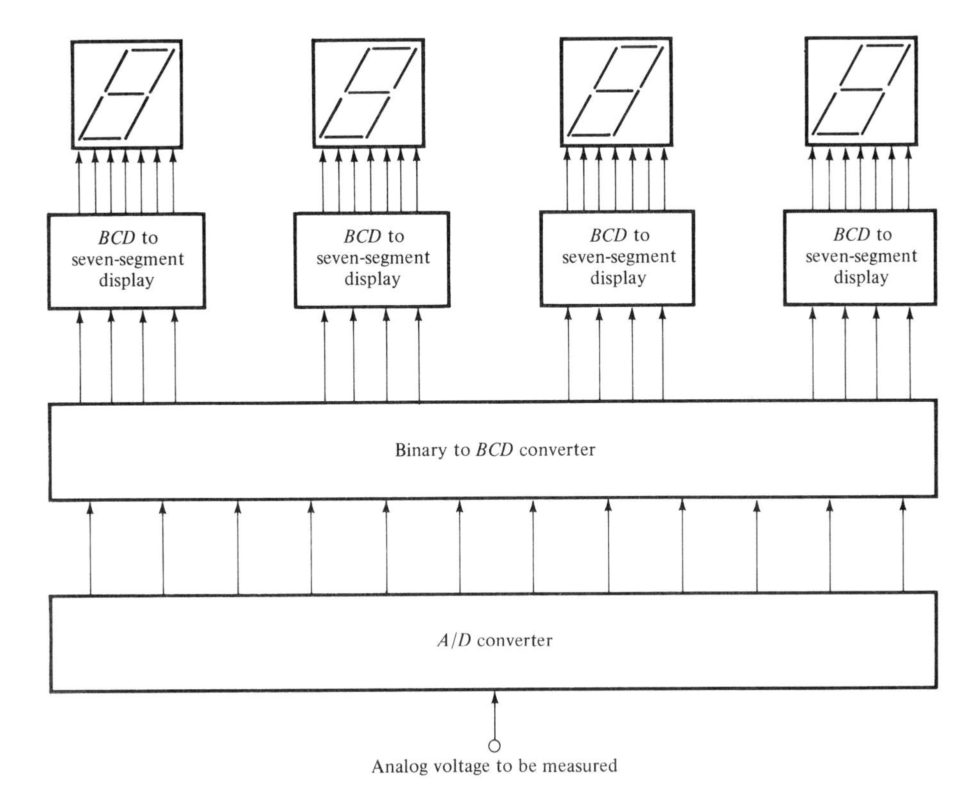

Fig. 22.6-3 Digital voltmeter.

This meter is described as a $3\frac{1}{2}$ digit meter because its output is four digits but the value of the most significant digit cannot exceed 4. The precision of the meter is limited to one part in 4095 which is slightly less than 0.025%. The meter can reach this level of accuracy only if its other components are of sufficient precision.

SUMMARY

1 The output of a combinational circuit depends only on its present inputs while the output of a sequential circuit depends on its previous as well as its present inputs.
2 There are three types of sequential circuits: level mode is an asynchronous type, pulse mode is an asynchronous type, and clocked mode is a synchronous type.
3 A flip-flop, or bistable multivibrator, stores one bit of data. Different triggering variations result in *SR*, clocked *SR*, *JK*, *T*, and *D* flip-flops.
4 A one-shot, or monostable multivibrator, is used as a delay element or pulse shaper.

5 A clock, also called a free-running or astable multivibrator, provides a train of pulses.
6 In the serial mode, digital data is represented by a sequential string of pulses in time on a single line; in the parallel mode, digital data is represented by a set of simultaneous pulses on parallel lines.
7 A storage register is a group of flip-flops used to store data.
8 A shift register is used for serial-to-parallel or parallel-to-serial conversions.
9 Binary counters can be designed using a series of edge-triggered toggles. (T-type flip-flops).
10 A hybrid system has both analog and digital components.
11 An analog-to-digital (A/D) converter transforms an analog signal to an equivalent digital signal. A digital-to-analog (D/A) converter transforms a digital signal to an equivalent analog signal.

PROBLEMS

22.1-1 What differentiates a sequential digital circuit from a combinational digital circuit?

22.1-2 What essential function is present in a sequential circuit but not in a combinational circuit?

22.1-3 List the three modes of sequential circuits. For each mode, indicate whether it is synchronous or asynchronous.

22.1-4 For each of the following digital systems indicate whether it is most likely to be combinational or sequential, and if it is sequential what its most probable mode of operation is.

a Railroad signal system.
b Large-scale digital computer.

22.2-1 List the three types of multivibrators and their main functions in digital systems.

22.2-2 List four types of clocked flip-flops and describe the triggering conditions for each.

22.2-3 Draw a logic diagram for a clocked *SR* flip-flop using combinational gates.

22.2-4 Derive a truth table to describe the operation of a *T* (toggle) flip-flop from the equivalent circuit of Fig. 22.2-5*a*. Does it agree with the description given in this chapter?

22.2-5 Repeat Prob. 22.2-4 for the *D* (delay) flip-flop using Fig. 22.2-5*c*.

22.3-1 Draw a three-bit storage register that can be loaded in a single-clock pulse period.

22.3-2 What important applications are shift registers used for? Describe one of these applications in detail.

22.4-1 Draw the circuit for a positive logic binary counter that counts up to 15 and uses edge-triggered flip-flops that trigger on the *positive-going* edge.

22.5-1 Draw the diagram of a four-bit parallel binary adder as shown in Fig. 22.5-1. Show the logic values at every point when $A = 1001$ and $B = 1101$ are being added.

22.5-2 Draw the waveforms, as in Fig. 22.5-2*b*, when the values given in Prob. 22.5-1 are being added in the serial adder of Fig. 22.5-2*a*.

appendix A The Slide Rule

Through the ages, man has developed many ingenious methods for reducing the drudgery of numerical calculation. These efforts have culminated today in the electronic digital computer, which can perform millions of calculations in a fraction of a second. At the time of this writing, miniature portable electronic calculators are available at reasonable prices, and it is assumed that many students using this text will have such calculators. However, many engineers and technologists use the slide rule for routine day-to-day calculations.

In this section, we present the basic operations of multiplication and division as performed on the slide rule (the slide rule cannot be used for addition and subtraction). Most modern slide rules can do other operations, such as finding common or natural logarithms, trigonometric functions, square roots, and exponential functions. Since these operations are somewhat different from one slide rule to another, while multiplication and division are identical on all slide rules, we shall concentrate on these two operations. Additional operations can then best be learned from the instruction manual which comes with the slide rule.

CONSTRUCTION OF THE SLIDE RULE

The principle of operation of the slide rule is based on the fact that logarithms convert the operation of multiplication into addition, that is, $\log(xy) = \log x + \log y$. Thus, the slide rule is constructed so that the distances between numbers (which are proportional to the differences between their logarithms) are *added* by mechanical manipulation, and the *sum* of the distances is then proportional to the *product* of the numbers. This product is read directly from the scales.

We shall be concerned only with the C and D scales on a typical 5- or 10-in slide rule. These scales are identical and are illustrated in Fig. A-1a. Note that the scale is divided into 10 major divisions which are *not* equal in length. As noted previously, the distances are proportional to the logarithm of the number.

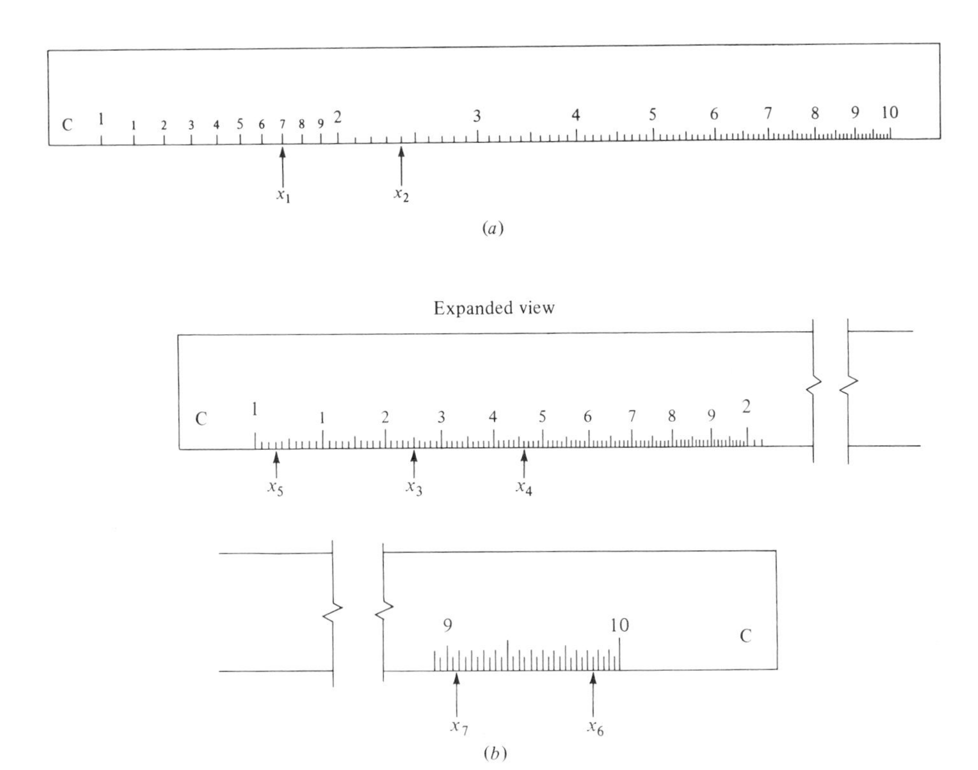

Fig. A-1 The slide rule. (*a*) Major divisions; (*b*) examples at extremes of C scale.

Thus the first marker (number 1) is placed at the left end of the scale (since $\log_{10} 1 = 0$, this is considered to be the beginning of the scale). The second marker (2) is placed at a distance proportional to $\log_{10} 2 = 0.301$ from the 1 marker. Thus on a 10-in scale, the number 2 marker is 3.01-in from the number 1 marker. The remainder of the major markers are placed accordingly.

Now observe that the interval between major markers 1 and 2 is further subdivided in the same manner, as is the interval between major markers 2 and 3. Each of these subdivisions of the numbers is thus one-tenth or 0.1 unit, the distance between subdivisions being proportional to the logarithm. Thus the point marked x_1 in the figure corresponds to 1.7, while x_2 corresponds to 2.4. Note that x_1 may be 0.0017, 0.017, 0.17, 1.7, 17, 170, or 1.7 times any power of 10. The slide rule *does not* locate the decimal point. We must do this separately as discussed in Sec. 1.3-3. On a 10-in slide rule, each of the subdivisions between 1 and 2 is further subdivided but not marked. Usually the line marking each fifth unmarked division is longer than the other lines. This makes it easier to read the scale. Clearly, the major divisions at the right end of the scale cannot be subdivided as far as those on the left end because of the small distances involved so that readings are less precise at the right end.

Before we begin to discuss multiplication and division we need to practice reading the scale as in the following example.

Example A-1 Without looking at the answers below, read the numbers marked x_3 through x_7 in Fig. A-1*b*.

Solution

x_3 The arrow is at the fifth subdivision between 1.2 and 1.3. Thus the answer is $x_3 = 1.25$.

x_4 The arrow is at the sixth subdivision between 1.4 and 1.5. Thus the answer is $x_4 = 1.46$.

x_5 This one is very often read wrongly by beginners. The arrow is at the third subdivision between 1.0 and 1.1. Thus the answer is $x_5 = 1.03$. The common mistake is to read this as 1.3.

Note that all three of the above are easily read to three significant figures. The next two cannot be read to this accuracy

x_6 The arrow is approximately midway between 9.7 and 9.8. Thus $x_6 \approx 9.75$ if we assume that the log scale is linear over this short distance.

x_7 The arrow is approximately 80 percent of the distance between 9.0 and 9.1. Thus $x_7 \approx 9.08$.

In these last two examples the third significant figure is only an estimate, because of the small distances involved. ////

The student should practice reading the scale until some proficiency is attained before proceeding to the next topic.

MULTIPLYING WITH THE SLIDE RULE

As noted previously, two scales are required for multiplication. These are usually called the C and D scales and are identical in layout. They are mounted so that one can be moved relative to the other, the usual arrangement being shown in Fig. A-2. To multiply two numbers on the slide rule we move the slide so as to *add* the distances corresponding to the logarithms of the numbers. This is

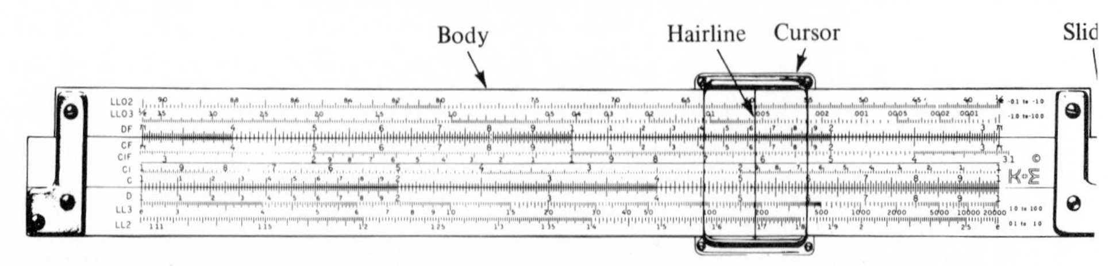

Fig. A-2 Slide rule (*Keuffel and Esser Co.*)

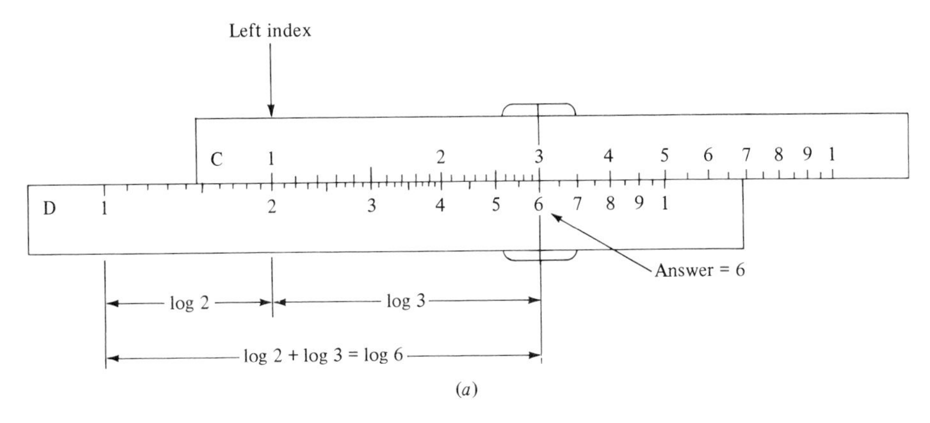

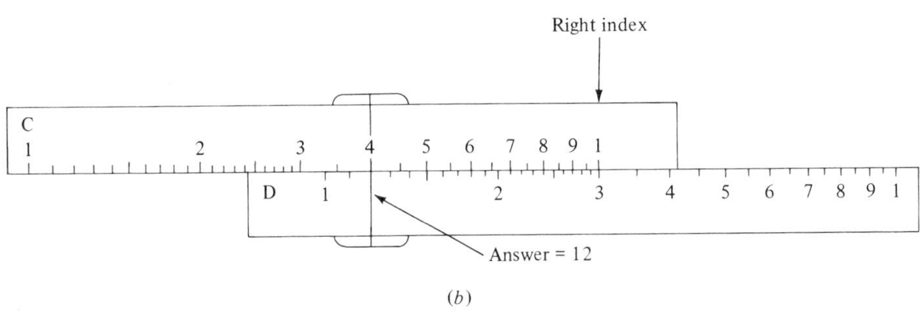

Fig. A-3 Multiplying with the slide rule. (*a*) 2×3; (*b*) 3×4.

illustrated in Fig. A-3*a* where the slide rule settings for carrying out the multiplication 2×3 are shown. The sequence of steps to be followed is

1 Set the left-hand *C* scale index (1) opposite 2 on the *D* scale. This establishes the distance proportional to log 2.
2 Move the cursor until the hairline is directly over 3 on the *C* scale. This adds the distance proportional to log 3 to that proportional to log 2.
3 Read the answer 6 under the hairline on the *D* scale.

Note that 0.2×0.03 or 20×300 would be handled the same way, with the proper decimal point placement determined separately.

The procedure above works fine, unless the answer happens to appear to the left of the *C* index, in which case the right hand index should be used. This is illustrated in Fig. A-3*b* where the settings for finding 3×4 are shown.

The only way for a beginner to become proficient at this is for him to *practice*, first with simple integer multiplications, that is, 5×6, 7×3, etc., and then with more complex numbers.

DIVISION WITH THE SLIDE RULE

Since division and multiplication are inverse operations, their implementations on the slide rule are very similar. In order to divide two numbers the slide is moved so as to *subtract* the distance corresponding to the logarithm of the divisor from the distance corresponding to the logarithm of the dividend. This is illustrated in Fig. A-4 where the settings required for dividing 9 by 5 are shown. The steps required for division are as follows:

1 Set the hairline over the dividend (9 in our example above) on the *D* scale.
2 Move the slide until the divisor (5 in the example) is directly under the hairline.
3 Read the answer under either the left or right *C* scale index, whichever appears within the D scale.

COMBINED MULTIPLICATION AND DIVISION

After the student has gained facility in multiplying and dividing two numbers at a time, he will find it very easy to carry out a continuous string of multiplications and divisions on the slide rule. The method for doing this is illustrated in the example which follows.

Example A-2 Use the slide rule to find

$$x = \frac{(2.1)(36)}{(150)(84)}$$

Solution

a We first find 2.1/150 as shown in Fig. A-5*a*. The hairline is placed over 2.1 on the *D* scale and the slide is moved so that 150 on the *C* scale is under the hairline as shown in position 1. The result which appears under the left *C* index is not required so we don't bother to read it.
b We note that this result must next be multiplied by 36. This is done by

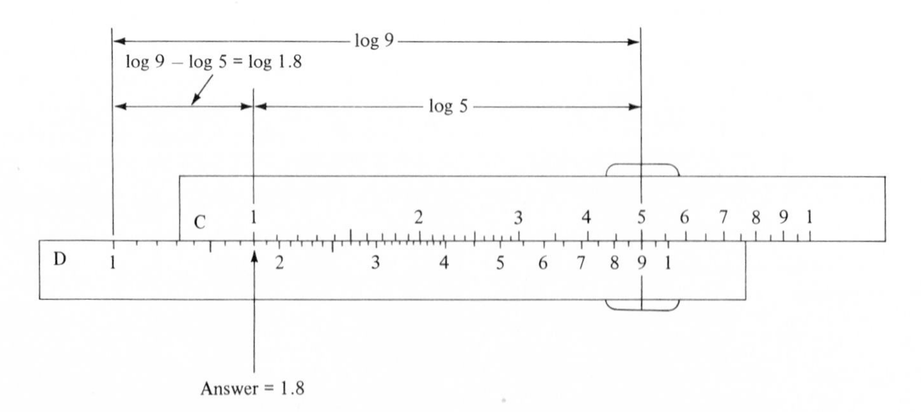

Fig. A-4 Dividing with the slide rule.

(a)

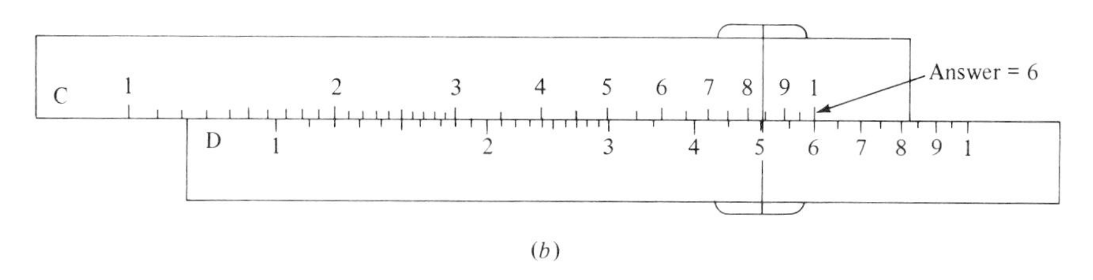

(b)

Fig. A-5 Slide rule positions for Example A-2.

moving the hairline to position 2 over 36 on the C scale. Under the hairline on the D scale we now have $(2.1)(36)/150$. This, when divided by 84 will give the desired result.

c To accomplish this last division we leave the hairline in position 2 and move the slide so that 84 on the C scale appears under the hairline as shown in Fig. A-5*b*. The significant digits of the final answer, 60, then appears under the right C index.

d In order to place the decimal point we proceed as in Sec. 1.3-3 using scientific notation.

$$x = \frac{(2.1)(3.6)(10^1)}{(1.5)(8.4)(10^2)(10^1)} = 0.6 \times 10^{-2}$$

Finally $x = 6 \times 10^{-3}$.
This procedure can be carried out as many times as required. ////

Occasionally, the C scale index may have to be changed from right to left but this should pose no problem. Estimating the answer before carrying through a long sequence of operations such as these should become a habit since it is relatively easy to make mistakes along the way.

appendix B

INTRODUCTION

In Sec. 8.2 we considered two node-pair networks analyzed by the node-voltage method and two-loop networks analyzed by the loop-current method. In this appendix we will extend these ideas to more complicated networks, i.e., those which lead to more than two simultaneous equations. In addition, we will describe the determinant method of solution for simultaneous linear equations.

B-1 NODAL ANALYSIS

In order to illustrate the node-voltage method we consider the three node-pair network shown in Fig. B-1. We will set up the three simultaneous equations for this network by following the sequence of steps set down in Sec. 8.2.

Step 1 Choose a reference node and identify and label the voltage at each independent node. This is done on the circuit diagram. The voltages are V_1, V_2, and V_3.
Step 2 The *assumed* branch currents are shown in the figure.
Step 3 Expressing branch currents in terms of node voltages;

$$I_1 = \frac{V_1}{2.2}$$

$$I_2 = \frac{V_1 - V_2}{6}$$

$$I_3 = \frac{V_1 - V_3}{3}$$

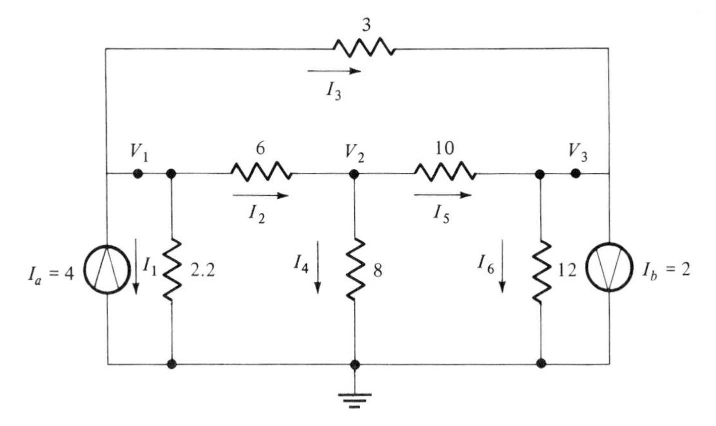

Fig. B-1 Node-voltage network. Units are kilohms and milliamperes.

$$I_4 = \frac{V_2}{8}$$

$$I_5 = \frac{V_2 - V_3}{10}$$

$$I_6 = \frac{V_3}{12}$$

Step 4 KCL equations at each node:

Node 1

$$I_1 + I_2 + I_3 = I_a$$

Node 2

$$-I_2 + I_4 + I_5 = 0$$

Node 3

$$-I_5 + I_6 - I_3 = -I_b$$

The student should carefully check the sign of each term in these equations in order to be sure that he understands the sign convention.
Step 5 Substituting step 3 into step 4.

$$\frac{V_1}{2.2} + \frac{V_1 - V_2}{6} + \frac{V_1 - V_3}{3} = 4$$

$$\frac{-(V_1 - V_2)}{6} + \frac{V_2}{8} + \frac{V_2 - V_3}{10} = 0$$

$$\frac{-(V_2 - V_3)}{10} + \frac{V_3}{12} - \frac{(V_1 - V_3)}{3} = -2$$

Combining terms

$$V_1\left(\frac{1}{2.2} + \frac{1}{6} + \frac{1}{3}\right) - V_2\left(\frac{1}{6}\right) - V_3\left(\frac{1}{3}\right) = 4 \qquad \text{(B-1)}$$

$$-V_1\left(\frac{1}{6}\right) + V_2\left(\frac{1}{6} + \frac{1}{8} + \frac{1}{10}\right) - V_3\left(\frac{1}{10}\right) = 0 \qquad \text{(B-2)}$$

$$-V_1\left(\frac{1}{3}\right) - V_2\left(\frac{1}{10}\right) + V_3\left(\frac{1}{10} + \frac{1}{12} + \frac{1}{3}\right) = -2 \qquad \text{(B-3)}$$

The student at this point should check the physical interpretation of each of the coefficients as outlined in Sec. 8.2. For example, the coefficient of V_3 in Eq. (B-3) should be the total conductance connected to node 3. A glance at the circuit indicates that this is correct.

Step 6 Solve for V_1, V_2, and V_3. Simplifying the coefficients, we get the following set of equations:

$$0.955V_1 - 0.167V_2 - 0.333V_3 = 4$$

$$-0.167V_1 + 0.392V_2 - 0.1V_3 = 0$$

$$-0.333V_1 - 0.1V_2 + 0.517V_3 = -2$$

They are best solved by the determinant method if a computer is not available. The process will be explained in Sec. B-3.

The example above was done in terms of dc. If the problem involved an ac circuit, the procedure is the same, but in place of the resistances, we would have complex impedances, and the currents and voltages would be in phasor form. The *arithmetic* involved can become extremely tedious and computer solutions would generally be much more practical. For networks with more than three independent node pairs, the procedure is similar to that given above.

B-2 LOOP ANALYSIS

The loop-current method will be applied to the four-loop network shown in Fig. B-2. As before, we will follow the sequence of steps recommended in Sec. 8.2.

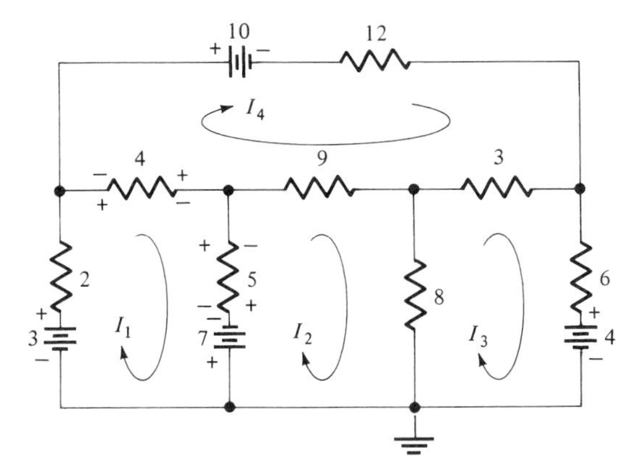

Fig. B-2 Network for loop current analysis. Units are volts and kilohms.

Step 1 Assign clockwise currents to each "window." This is done on the circuit diagram for the four "windows".
Step 2 Assign polarities to the voltages resulting from the loop currents flowing in each element. This is done on the circuit diagram for those resistances on the perimeter of loop 1. For the remainder of the resistances, the student should visualize the polarities since marking them should no longer be necessary.
Step 3 Write a KVL equation for each loop

Loop 1

$$-3 + 2I_1 + 4I_1 - 4I_4 + 5I_1 - 5I_2 - 7 = 0$$

Loop 2

$$7 + 5I_2 - 5I_1 + 9I_2 - 9I_4 + 8I_2 - 8I_3 = 0$$

Loop 3

$$8I_3 - 8I_2 + 3I_3 - 3I_4 + 6I_3 + 4 = 0$$

Loop 4

$$10 + 12I_4 + 3I_4 - 3I_3 + 9I_4 - 9I_2 + 4I_4 - 4I_1 = 0$$

Rearranging these equations in tabular form, we get

Loop	I_1	I_2	I_3	I_4	
1	+11	−5	0	−4	= 10
2	−5	+22	−8	−9	= −7
3	0	−8	+17	−3	= −4
4	−4	−9	−3	+28	= −10

In this form, the coefficients are easily checked and verified according to their physical interpretation. For example, the coefficient of I_3 in the equation for loop 3 is in the third-column, third-row position in the table. This coefficient represents the total resistance around the perimeter of loop 3, that is, $5 + 9 + 8 = 22\ \Omega$. As another example, the second-row, fourth-column entry is a mutual term, representing the negative of the resistance common to loops 2 and 4. From the circuit, this resistance is 9 Ω, and the coefficient is thus verified. The fifth column of the table represents the right-hand sides of the equations; they are the net source voltage rises in the tracing direction for each loop.

Step 4 The table formulated in step 3 is in a form which makes it easy to set up the determinant solution for the equation. This will be considered in the next section.

B-3 DETERMINANT SOLUTION OF SIMULTANEOUS EQUATIONS

The method of determinants is an organized "mechanical" method for solving sets of simultaneous equations. In this section we will present the method in sufficient detail so that the reader will be able to apply it to sets of two or three equations. Extension to sets of four or more equations require additional theory which will be covered in mathematics courses.

SECOND-ORDER DETERMINANTS

Consider the loop equations

$$3I_1 - 2I_2 = 10$$
$$-2I_1 + 5I_2 = 20$$

The determinant of the coefficients of these equations is an array consisting of two rows and two columns containing the coefficients in the following form:

$$\begin{vmatrix} 3 & -2 \\ -2 & 5 \end{vmatrix}$$

This is called a *second*-order determinant because it has two rows and two columns. Determinants have a numerical value which is determined by a process of *diagonal multiplication* shown diagrammatically below:

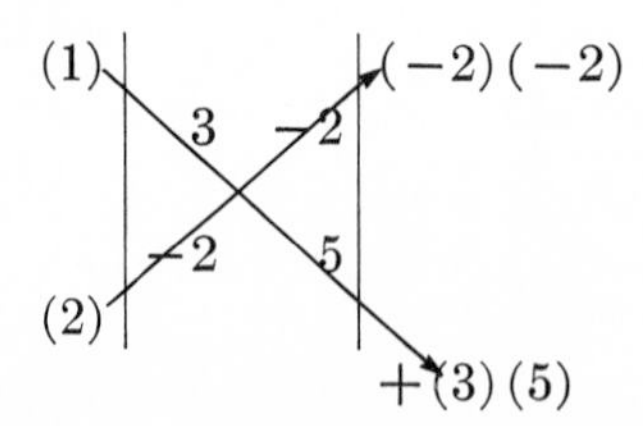

The product of the terms along each diagonal is computed. The value of the determinant is the difference of the two products, the product for diagonal (1) minus the product for diagonal (2):

$$+(3)(5) - (-2)(-2) = 15 - 4 = 11$$

The solutions for I_1 and I_2 are expressed as ratios of second-order determinants, the denominator being the determinant of the coefficients. The rule for formulating the numerator determinant is quite simple: It is the same as the denominator except for the column corresponding to the variable for which we are solving (I_1 in the example below). The coefficients in this column are replaced by the column of constants, i.e., the right-hand sides of the equations. For example, I_1 is given by

$$I_1 = \frac{\begin{vmatrix} 10 & -2 \\ 20 & 5 \end{vmatrix}}{\begin{vmatrix} 3 & -2 \\ -2 & 5 \end{vmatrix}} = \frac{(10)(5) - (20)(-2)}{(3)(5) - (-2)(-2)} = \frac{50 + 40}{15 - 4} = \frac{90}{11} \text{ A}$$

The current I_2 is given by

$$I_2 = \frac{\begin{vmatrix} 3 & 10 \\ -2 & 20 \end{vmatrix}}{\begin{vmatrix} 3 & -2 \\ -2 & 5 \end{vmatrix}} = \frac{(3)(20) - (-2)(10)}{11} = \frac{60 + 20}{11} = \frac{80}{11} \text{ A}$$

As a further example, we solve the set of equations

$$3x + 2y = 8$$

$$5x + y = 11$$

$$x = \frac{\begin{vmatrix} 8 & 2 \\ 11 & 1 \end{vmatrix}}{\begin{vmatrix} 3 & 2 \\ 5 & 1 \end{vmatrix}} = \frac{(8)(1) - (11)(2)}{(3)(1) - (5)(2)} = \frac{8 - 22}{3 - 10} \quad \frac{-14}{-7} = 2$$

$$y = \frac{\begin{vmatrix} 3 & 8 \\ 5 & 11 \end{vmatrix}}{-7} = \frac{(3)(11) - (5)(8)}{-7} = \frac{33 - 40}{-7} = 1$$

THIRD-ORDER DETERMINANTS

When we have three simultaneous equations to solve, we arrive at third-order determinants each of which has three rows and three columns. For example, consider the set of equations

$$x + y + z = 15$$

$$2x - y - z = 10$$

$$3x + 2y - 5z = 14$$

The determinant of the coefficients is

$$\begin{vmatrix} 1 & 1 & 1 \\ 2 & -1 & -1 \\ 3 & 2 & -5 \end{vmatrix}$$

In order to evaluate this determinant using diagonal multiplication we rewrite it with the first two columns repeated to the right as follows:

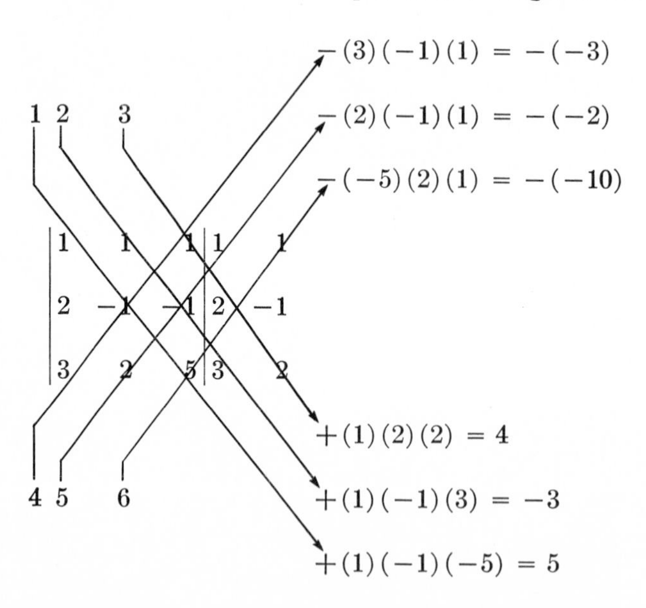

The diagonal multiplication is then performed as shown. For this case diagonals 1, 2, and 3 take a positive sign while diagonals 4, 5, and 6 take a negative sign. The result is

$$4 - 3 + 5 - (-3) - (-2) - (-10) = 21$$

The next step in the solution is to formulate the numerator determinants according to the rule previously stated. The complete expression for x is then

$$x = \frac{\begin{vmatrix} 15 & 1 & 1 \\ 0 & -1 & -1 \\ 14 & 2 & -5 \end{vmatrix} \begin{matrix} 15 & 1 \\ 0 & -1 \\ 14 & 2 \end{matrix}}{\begin{vmatrix} 1 & 1 & 1 \\ 2 & -1 & -1 \\ 3 & 2 & 5 \end{vmatrix}} = \frac{+75 - 14 + 0 - (-14) - (-30) - (0)}{21}$$

$$= \frac{+105}{21} = 5$$

The proof that $y = 7$ and $z = 3$ is left for the reader.

As a further example we evaluate the node voltages in the set of equations found in Sec. B-1. The first step is to evaluate the determinant of the coefficients.

$$\begin{vmatrix} 0.955 & -.0167 & -0.333 \\ -0.167 & +0.392 & -0.1 \\ -0.333 & -0.1 & +0.517 \end{vmatrix} \begin{matrix} 0.955 & -0.167 \\ -0.167 & +0.392 \\ -0.333 & -0.1 \end{matrix}$$

$$= +(0.196) + (-0.006) + (-0.006) - (0.043) - (0.009) - (0.014)$$

$$= 0.118$$

The numerator determinant for V_1 is

$$\begin{vmatrix} 4 & -0.167 & -0.333 \\ 0 & +0.392 & -0.1 \\ -2 & -0.1 & +0.517 \end{vmatrix} \begin{matrix} 4 & -0.167 \\ 0 & +0.392 \\ -2 & -0.1 \end{matrix}$$

$$= +(0.811) + (-0.033) + (0) - (0.261) - (0.040) - (0)$$

$$= 0.477$$

Thus

$$V_1 = \frac{0.477}{0.118} = 4.04 \text{ V}$$

V_2 and V_3 are found in a similar fashion

$$V_2 = \frac{0.176}{0.118} = 1.49 \text{ V}$$

$$V_3 = \frac{-0.104}{0.118} = -0.881 \text{ V}$$

The student should check these answers by finding all of the currents in the circuit of Fig. B-1 and verifying KCL at each node.

appendix C
Tables

TABLE C-1 Natural Trigonometric Functions

deg	*function*	0.0°	0.1°	0.2°	0.3°	0.4°	0.5°	0.6°	0.7°	0.8°	0.9°
0	sin	0.0000	0.0017	0.0035	0.0052	0.0070	0.0087	0.0105	0.0122	0.0140	0.0157
	cos	1.0000	1.0000	1.0000	1.0000	1.0000	1.0000	0.9999	0.9999	0.9999	0.9999
	tan	0.0000	0.0017	0.0035	0.0052	0.0070	0.0087	0.0105	0.0122	0.0140	0.0157
1	sin	0.0175	0.0192	0.0209	0.0227	0.0244	0.0262	0.0279	0.0297	0.0314	0.0332
	cos	0.9998	0.9998	0.9998	0.9997	0.9997	0.9997	0.9996	0.9996	0.9995	0.9995
	tan	0.0175	0.0192	0.0209	0.0227	0.0244	0.0262	0.0279	0.0297	0.0314	0.0332
2	sin	0.0349	0.0366	0.0384	0.0401	0.0419	0.0436	0.0454	0.0471	0.0488	0.0506
	cos	0.9994	0.9993	0.9993	0.9992	0.9991	0.9990	0.9990	0.9989	0.9988	0.9987
	tan	0.0349	0.0367	0.0384	0.0402	0.0419	0.0437	0.0454	0.0472	0.0489	0.0507
3	sin	0.0523	0.0541	0.0558	0.0576	0.0593	0.0610	0.0628	0.0645	0.0663	0.0680
	cos	0.9986	0.9985	0.9984	0.9983	0.9982	0.9981	0.9980	0.9979	0.9978	0.9977
	tan	0.0524	0.0542	0.0559	0.0577	0.0594	0.0612	0.0629	0.0647	0.0664	0.0682
4	sin	0.0698	0.0715	0.0732	0.0750	0.0767	0.0785	0.0802	0.0819	0.0837	0.0854
	cos	0.9976	0.9974	0.9973	0.9972	0.9971	0.9969	0.9968	0.9966	0.9965	0.9963
	tan	0.0699	0.0717	0.0734	0.0752	0.0769	0.0787	0.0805	0.0822	0.0840	0.0857
5	sin	0.0872	0.0889	0.0906	0.0924	0.0941	0.0958	0.0976	0.0993	0.1011	0.1028
	cos	0.9962	0.9960	0.9959	0.9957	0.9956	0.9954	0.9952	0.9951	0.9949	0.9947
	tan	0.0875	0.0892	0.0910	0.0928	0.0945	0.0963	0.0981	0.0998	0.1016	0.1033
6	sin	0.1045	0.1063	0.1080	0.1097	0.1115	0.1132	0.1149	0.1167	0.1184	0.1201
	cos	0.9945	0.9943	0.9942	0.9940	0.9938	0.9936	0.9934	0.9932	0.9930	0.9928
	tan	0.1051	0.1069	0.1086	0.1104	0.1122	0.1139	0.1157	0.1175	0.1192	0.1210
7	sin	0.1219	0.1236	0.1253	0.1271	0.1288	0.1305	0.1323	0.1340	0.1357	0.1374
	cos	0.9925	0.9923	0.9921	0.9919	0.9917	0.9914	0.9912	0.9910	0.9907	0.9905
	tan	0.1228	0.1246	0.1263	0.1281	0.1299	0.1317	0.1334	0.1352	0.1370	0.1388
8	sin	0.1392	0.1409	0.1426	0.1444	0.1461	0.1478	0.1495	0.1513	0.1530	0.1547
	cos	0.9903	0.9900	0.9898	0.9895	0.9893	0.9890	0.9888	0.9885	0.9882	0.9880
	tan	0.1405	0.1423	0.1441	0.1459	0.1477	0.1495	0.1512	0.1530	0.1548	0.1566
9	sin	0.1564	0.1582	0.1599	0.1616	0.1633	0.1650	0.1668	0.1685	0.1702	0.1719
	cos	0.9877	0.9874	0.9871	0.9869	0.9866	0.9863	0.9860	0.9857	0.9854	0.9851
	tan	0.1584	0.1602	0.1620	0.1638	0.1655	0.1673	0.1691	0.1709	0.1727	0.1745
10	sin	0.1736	0.1754	0.1771	0.1778	0.1805	0.1822	0.1840	0.1857	0.1874	0.1891
	cos	0.9848	0.9845	0.9842	0.9839	0.9836	0.9833	0.9829	0.9826	0.9823	0.9820
	tan	0.1763	0.1781	0.1799	0.1817	0.1835	0.1853	0.1871	0.1890	0.1908	0.1926
11	sin	0.1908	0.1925	0.1942	0.1959	0.1977	0.1994	0.2011	0.2028	0.2045	0.2062
	cos	0.9816	0.9813	0.9810	0.9806	0.9803	0.9799	0.9796	0.9792	0.9789	0.9785
	tan	0.1944	0.1962	0.1980	0.1998	0.2016	0.2035	0.2053	0.2071	0.2089	0.2107
12	sin	0.2079	0.2096	0.2113	0.2130	0.2147	0.2164	0.2181	0.2198	0.2215	0.2232
	cos	0.9781	0.9778	0.9774	0.9770	0.9767	0.9763	0.9759	0.9755	0.9751	0.9748
	tan	0.2126	0.2144	0.2162	0.2180	0.2199	0.2217	0.2235	0.2254	0.2272	0.2290
13	sin	0.2250	0.2267	0.2284	0.2300	0.2318	0.2334	0.2351	0.2368	0.2385	0.2402
	cos	0.9744	0.9740	0.9736	0.9732	0.9728	0.9724	0.9720	0.9715	0.9711	0.9707
	tan	0.2309	0.2327	0.2345	0.2364	0.2382	0.2401	0.2419	0.2438	0.2456	0.2475
14	sin	0.2419	0.2436	0.2453	0.2470	0.2487	0.2504	0.2521	0.2538	0.2554	0.2571
	cos	0.9703	0.9699	0.9694	0.9690	0.9686	0.9681	0.9677	0.9673	0.9668	0.9664
	tan	0.2493	0.2512	0.2530	0.2549	0.2568	0.2586	0.2605	0.2623	0.2642	0.2661
15	sin	0.2588	0.2605	0.2622	0.2639	0.2656	0.2672	0.2689	0.2706	0.2723	0.2740
	cos	0.9659	0.9655	0.9650	0.9646	0.9641	0.9636	0.9632	0.9627	0.9622	0.9617
	tan	0.2679	0.2698	0.2717	0.2736	0.2754	0.2773	0.2792	0.2811	0.2830	0.2849
16	sin	0.2756	0.2773	0.2790	0.2807	0.2823	0.2840	0.2857	0.2874	0.2890	0.2907
	cos	0.9613	0.9608	0.9603	0.9598	0.9593	0.9588	0.9583	0.9578	0.9573	0.9568
	tan	0.2867	0.2886	0.2905	0.2924	0.2943	0.2962	0.2981	0.3000	0.3019	0.3038
17	sin	0.2924	0.2940	0.2957	0.2974	0.2990	0.3007	0.3024	0.3040	0.3057	0.3074
	cos	0.9563	0.9558	0.9553	0.9548	0.9542	0.9537	0.9532	0.9527	0.9521	0.9516
	tan	0.3057	0.3076	0.3096	0.3115	0.3134	0.3153	0.3172	0.3191	0.3211	0.3230
18	sin	0.3090	0.3107	0.3123	0.3140	0.3156	0.3173	0.3190	0.3206	0.3223	0.3239
	cos	0.9511	0.9505	0.9500	0.9494	0.9489	0.9483	0.9478	0.9472	0.9466	0.9461
	tan	0.3249	0.3269	0.3288	0.3307	0.3327	0.3346	0.3365	0.3385	0.3404	0.3424
19	sin	0.3256	0.3272	0.3289	0.3305	0.3322	0.3338	0.3355	0.3371	0.3387	0.3404
	cos	0.9455	0.9449	0.9444	0.9438	0.9432	0.9426	0.9421	0.9415	0.9409	0.9403
	tan	0.3443	0.3463	0.3482	0.3502	0.3522	0.3541	0.3561	0.3581	0.3600	0.3620
20	sin	0.3420	0.3437	0.3453	0.3469	0.3486	0.3502	0.3518	0.3535	0.3551	0.3567
	cos	0.9397	0.9391	0.9385	0.9379	0.9373	0.9367	0.9361	0.9354	0.9348	0.9342
	tan	0.3640	0.3659	0.3679	0.3699	0.3719	0.3739	0.3759	0.3779	0.3799	0.3819
deg	*function*	0′	6′	12′	18′	24′	30′	36′	42′	48′	54′

deg	function	0.0°	0.1°	0.2°	0.3°	0.4°	0.5°	0.6°	0.7°	0.8°	0.9°
21	sin	0.3584	0.3600	0.3616	0.3633	0.3649	0.3665	0.3681	0.3697	0.3714	0.3730
	cos	0.9336	0.9330	0.9323	0.9317	0.9311	0.9304	0.9298	0.9291	0.9285	0.9278
	tan	0.3839	0.3859	0.3879	0.3899	0.3919	0.3939	0.3959	0.3979	0.4000	0.4020
22	sin	0.3746	0.3762	0.3778	0.3795	0.3811	0.3827	0.3843	0.3859	0.3875	0.3891
	cos	0.9272	0.9265	0.9259	0.9252	0.9245	0.9239	0.9232	0.9225	0.9219	0.9212
	tan	0.4040	0.4061	0.4081	0.4101	0.4122	0.4142	0.4163	0.4183	0.4204	0.4224
23	sin	0.3907	0.3923	0.3939	0.3955	0.3971	0.3987	0.4003	0.4019	0.4035	0.4051
	cos	0.9205	0.9198	0.9191	0.9184	0.9178	0.9171	0.9164	0.9157	0.9150	0.9143
	tan	0.4245	0.4265	0.4286	0.4307	0.4327	0.4348	0.4369	0.4390	0.4411	0.4431
24	sin	0.4067	0.4083	0.4099	0.4115	0.4131	0.4147	0.4163	0.4179	0.4195	0.4210
	cos	0.9135	0.9128	0.9121	0.9114	0.9107	0.9100	0.9092	0.9085	0.9078	0.9070
	tan	0.4452	0.4473	0.4494	0.4515	0.4536	0.4557	0.4578	0.4599	0.4621	0.4642
25	sin	0.4226	0.4242	0.4258	0.4274	0.4289	0.4305	0.4321	0.4337	0.4352	0.4368
	cos	0.9063	0.9056	0.9048	0.9041	0.9033	0.9026	0.9018	0.9011	0.9003	0.8996
	tan	0.4663	0.4684	0.4706	0.4727	0.4748	0.4770	0.4791	0.4813	0.4834	0.4856
26	sin	0.4384	0.4399	0.4415	0.4431	0.4446	0.4462	0.4478	0.4493	0.4509	0.4524
	cos	0.8988	0.8980	0.8973	0.8965	0.8957	0.8949	0.8942	0.8934	0.8926	0.8918
	tan	0.4877	0.4899	0.4921	0.4942	0.4964	0.4986	0.5008	0.5029	0.5051	0.5073
27	sin	0.4540	0.4555	0.4571	0.4586	0.4602	0.4617	0.4633	0.4648	0.4664	0.4679
	cos	0.8910	0.8902	0.8894	0.8886	0.8878	0.8870	0.8862	0.8854	0.8846	0.8838
	tan	0.5095	0.5117	0.5139	0.5161	0.5184	0.5206	0.5228	0.5250	0.5272	0.5295
28	sin	0.4695	0.4710	0.4726	0.4741	0.4756	0.4772	0.4787	0.4802	0.4818	0.4833
	cos	0.8829	0.8821	0.8813	0.8805	0.8796	0.8788	0.8780	0.8771	0.8763	0.8755
	tan	0.5317	0.5340	0.5362	0.5384	0.5407	0.5430	0.5452	0.5475	0.5498	0.5520
29	sin	0.4848	0.4863	0.4879	0.4894	0.4909	0.4924	0.4939	0.4955	0.4970	0.4985
	cos	0.8746	0.8738	0.8729	0.8721	0.8712	0.8704	0.8695	0.8686	0.8678	0.8669
	tan	0.5543	0.5566	0.5589	0.5612	0.5635	0.5658	0.5681	0.5704	0.5727	0.5750
30	sin	0.5000	0.5015	0.5030	0.5045	0.5060	0.5075	0.5090	0.5105	0.5120	0.5135
	cos	0.8660	0.8652	0.8643	0.8634	0.8625	0.8616	0.8607	0.8599	0.8590	0.8581
	tan	0.5774	0.5797	0.5820	0.5844	0.5967	0.5890	0.5914	0.5938	0.5961	0.5985
31	sin	0.5150	0.5165	0.5180	0.5195	0.5210	0.5225	0.5240	0.5255	0.5270	0.5284
	cos	0.8572	0.8563	0.8554	0.8545	0.8536	0.8526	0.8517	0.8508	0.8499	0.8490
	tan	0.6009	0.6032	0.6056	0.6080	0.6104	0.6128	0.6152	0.6176	0.6200	0.6224
32	sin	0.5299	0.5314	0.5329	0.5344	0.5358	0.5373	0.5388	0.5402	0.5417	0.5432
	cos	0.8480	0.8471	0.8462	0.8453	0.8443	0.8434	0.8425	0.8415	0.8406	0.8396
	tan	0.6249	0.6273	0.6297	0.6322	0.6346	0.6371	0.6395	0.6420	0.6445	0.6469
33	sin	0.5446	0.5461	0.5476	0.5490	0.5505	0.5519	0.5534	0.5548	0.5563	0.5577
	cos	0.8387	0.8377	0.8368	0.8358	0.8348	0.8339	0.8329	0.8320	0.8310	0.8300
	tan	0.6494	0.6519	0.6544	0.6569	0.6594	0.6619	0.6644	0.6669	0.6694	0.6720
34	sin	0.5592	0.5606	0.5621	0.5635	0.5650	0.5664	0.5678	0.5693	0.5707	0.5721
	cos	0.8290	0.8281	0.8271	0.8261	0.8251	0.8241	0.8231	0.8221	0.8211	0.8202
	tan	0.6745	0.6771	0.6796	0.6822	0.6847	0.6873	0.6899	0.6924	0.6950	0.6976
35	sin	0.5736	0.5750	0.5764	0.5779	0.5793	0.5807	0.5821	0.5835	0.5850	0.5864
	cos	0.8192	0.8181	0.8171	0.8161	0.8151	0.8141	0.8131	0.8121	0.8111	0.8100
	tan	0.7002	0.7028	0.7054	0.7080	0.7107	0.7133	0.7159	0.7186	0.7212	0.7239
36	sin	0.5878	0.5892	0.5906	0.5920	0.5934	0.5948	0.5962	0.5976	0.5990	0.6004
	cos	0.8090	0.8080	0.8070	0.8059	0.8049	0.8039	0.8028	0.8018	0.8007	0.7997
	tan	0.7265	0.7292	0.7319	0.7346	0.7373	0.7400	0.7427	0.7454	0.7481	0.7508
37	sin	0.6018	0.6032	0.6046	0.6060	0.6074	0.6088	0.6101	0.6115	0.6129	0.6143
	cos	0.7986	0.7976	0.7965	0.7955	0.7944	0.7934	0.7923	0.7912	0.7902	0.7891
	tan	0.7536	0.7563	0.7590	0.7618	0.7646	0.7673	0.7701	0.7729	0.7757	0.7785
38	sin	0.6157	0.6170	0.6184	0.6198	0.6211	0.6225	0.6239	0.6252	0.6266	0.6280
	cos	0.7880	0.7869	0.7859	0.7848	0.7837	0.7826	0.7815	0.7804	0.7793	0.7782
	tan	0.7813	0.7841	0.7869	0.7898	0.7926	0.7954	0.7983	0.8012	0.8040	0.8069
39	sin	0.6293	0.6307	0.6320	0.6334	0.6347	0.6361	0.6374	0.6388	0.6401	0.6414
	cos	0.7771	0.7760	0.7749	0.7738	0.7727	0.7716	0.7705	0.7694	0.7683	0.7672
	tan	0.8098	0.8127	0.8156	0.8185	0.8214	0.8243	0.8273	0.8302	0.8332	0.8361
40	sin	0.6428	9.6441	0.6455	0.6468	0.6481	0.6494	0.6508	0.6521	0.6534	0.6547
	cos	0.7660	0.7649	0.7638	0.7627	0.7615	0.7604	0.7593	0.7581	0.7570	0.7559
	tan	0.8391	0.8421	0.8451	0.8481	0.8511	0.8541	0.8571	0.8601	0.8632	0.8662
41	sin	0.6561	0.6574	0.6587	0.6600	0.6613	0.6626	0.6639	0.6652	0.6665	0.6678
	cos	0.7547	0.7536	0.7524	0.7513	0.7501	0.7490	0.7478	0.7466	0.7455	0.7443
	tan	0.8693	0.8724	0.8754	0.8785	0.8816	0.8847	0.8878	0.8910	0.8941	0.8972
deg	*function*	**0′**	**6′**	**12′**	**18′**	**24′**	**30′**	**36′**	**42′**	**48′**	**54′**

deg	function	0.0°	0.1°	0.2°	0.3°	0.4°	0.5°	0.6°	0.7°	0.8°	0.9°
42	sin	0.6691	0.6704	0.6717	0.6730	0.6743	0.6756	0.6769	0.6782	0.6794	0.6807
	cos	0.7431	0.7420	0.7408	0.7396	0.7385	0.7373	0.7361	0.7349	0.7337	0.7325
	tan	0.9004	0.9036	0.9067	0.9099	0.9131	0.9163	0.9195	0.9228	0.9260	0.9293
43	sin	0.6820	0.6833	0.6845	0.6858	0.6871	0.6884	0.6896	0.6909	0.6921	0.6934
	cos	0.7314	0.7302	0.7290	0.7278	0.7266	0.7254	0.7242	0.7230	0.7218	0.7206
	tan	0.9325	0.9358	0.9391	0.9424	0.9457	0.9490	0.9523	0.9556	0.9590	0.9623
44	sin	0.6947	0.6959	0.6972	0.6984	0.6997	0.7009	0.7022	0.7034	0.7046	0.7059
	cos	0.7193	0.7181	0.7169	0.7157	0.7145	0.7133	0.7120	0.7108	0.7096	0.7083
	tan	0.9657	0.9691	0.9725	0.9759	0.9793	0.9827	0.9861	0.9896	0.9930	0.9965
45	sin	0.7071	0.7083	0.7096	0.7108	0.7120	0.7133	0.7145	0.7157	0.7169	0.7181
	cos	0.7071	0.7059	0.7046	0.7034	0.7022	0.7009	0.6997	0.6984	0.6972	0.6959
	tan	1.0000	1.0035	1.0070	1.0105	1.0141	1.0176	1.0212	1.0247	1.0283	1.0319
46	sin	0.7193	0.7206	0.7218	0.7230	0.7242	0.7254	0.7266	0.7278	0.7290	0.7302
	cos	0.6947	0.6934	0.6921	0.6909	0.6896	0.6884	0.6871	0.6858	0.6845	0.6833
	tan	1.0355	1.0392	1.0428	1.0464	1.0501	1.0538	1.0575	1.0612	1.0649	1.0686
47	sin	0.7314	0.7325	0.7337	0.7349	0.7361	0.7373	0.7385	0.7396	0.7408	0.7420
	cos	0.6820	0.6807	0.6794	0.6782	0.6769	0.6756	0.6743	0.6730	0.6717	0.6704
	tan	1.0724	1.0761	1.0799	1.0837	1.0875	1.0913	1.0951	1.0990	1.1028	1.1067
48	sin	0.7431	0.7443	0.7455	0.7466	0.7478	0.7490	0.7501	0.7513	0.7524	0.7536
	cos	0.6691	0.6678	0.6665	0.6652	0.6639	0.6626	0.6613	0.6600	0.6587	0.6574
	tan	1.1106	1.1145	1.1184	1.1224	1.1263	1.1303	1.1343	1.1383	1.1423	1.1463
49	sin	0.7547	0.7559	0.7570	0.7581	0.7593	0.7604	0.7615	0.7627	0.7638	0.7649
	cos	0.6561	0.6547	0.6534	0.6521	0.6508	0.6494	0.6481	0.6468	0.6455	0.6441
	tan	1.1504	1.1544	1.1585	1.1626	1.1667	1.1708	1.1750	1.1792	1.1833	1.1875
50	sin	0.7660	0.7672	0.7683	0.7694	0.7705	0.7716	0.7727	0.7738	0.7749	0.7760
	cos	0.6428	0.6414	0.6401	0.6388	0.6374	0.6361	0.6347	0.6334	0.6320	0.6307
	tan	1.1918	1.1960	1.2002	1.2045	1.2088	1.2131	1.2174	1.2218	1.2261	1.2305
51	sin	0.7771	0.7782	0.7793	0.7804	0.7815	0.7826	0.7837	0.7848	0.7859	0.7869
	cos	0.6293	0.6280	0.6266	0.6252	0.6239	0.6225	0.6211	0.6198	0.6184	0.6170
	tan	1.2349	1.2393	1.2437	1.2482	1.2527	1.2572	1.2617	1.2662	1.2708	1.2753
52	sin	0.7880	0.7891	0.7902	0.7912	0.7923	0.7934	0.7944	0.7955	0.7965	0.7976
	cos	0.6157	0.6143	0.6129	0.6115	0.6101	0.6088	0.6074	0.6060	0.6046	0.6032
	tan	1.2799	1.2846	1.2892	1.2938	1.2985	1.3032	1.3079	1.3127	1.3175	1.3222
53	sin	0.7986	0.7997	0.8007	0.8018	0.8028	0.8039	0.8049	0.8059	0.8070	0.8080
	cos	0.6018	0.6004	0.5990	0.5976	0.5962	0.5948	0.5934	0.5920	0.5906	0.5892
	tan	1.3270	1.3319	1.3367	1.3416	1.3465	1.3514	1.3564	1.3613	1.3663	1.3713
54	sin	0.8090	0.8100	0.8111	0.8121	0.8131	0.8141	0.8151	0.8161	0.8171	0.8181
	cos	0.5878	0.5864	0.5850	0.5835	0.5821	0.5807	0.5793	0.5779	0.5764	0.5750
	tan	1.3764	1.3814	1.3865	1.3916	1.3968	1.4019	1.4071	1.4124	1.4176	1.4229
55	sin	0.8192	0.8202	0.8211	0.8221	0.8231	0.8241	0.8251	0.8261	0.8271	0.8281
	cos	0.5736	0.5721	0.5707	0.5693	0.5678	0.5664	0.5650	0.5635	0.5621	0.5606
	tan	1.4281	1.4335	1.4388	1.4442	1.4496	1.4550	1.4605	1.4659	1.4715	1.4770
56	sin	0.8290	0.8300	0.8310	0.8320	0.8329	0.8339	0.8348	0.8358	0.8368	0.8377
	cos	0.5592	0.5577	0.5563	0.5548	0.5534	0.5519	0.5505	0.5490	0.5476	0.5461
	tan	1.4826	1.4882	1.4938	1.4994	1.5051	1.5108	1.5166	1.5224	1.5282	1.5340
57	sin	0.8387	0.8396	0.8406	0.8415	0.8425	0.8434	0.8443	0.8453	0.8462	0.8471
	cos	0.5446	0.5432	0.5417	0.5402	0.5388	0.5373	0.5358	0.5344	0.5329	0.5314
	tan	1.5399	1.5458	1.5517	1.5577	1.5637	1.5697	1.5757	1.5818	1.5880	1.5941
58	sin	0.8480	0.8490	0.8499	0.8508	0.8517	0.8526	0.8536	0.8545	0.8554	0.8563
	cos	0.5299	0.5284	0.5270	0.5255	0.5240	0.5225	0.5210	0.5195	0.5180	0.5165
	tan	1.6003	1.6066	1.6128	1.6191	1.6255	1.6319	1.6383	1.6447	1.6512	1.6577
59	sin	0.8572	0.8581	0.8590	0.8599	0.8607	0.8616	0.8625	0.8634	0.8643	0.8652
	cos	0.5150	0.5135	0.5120	0.5105	0.5090	0.5075	0.5060	0.5045	0.5030	0.5015
	tan	1.6643	1.6709	1.6775	1.6842	1.6909	1.6977	1.7045	1.7113	1.7182	1.7251
60	sin	0.8660	0.8669	0.8678	0.8686	0.8695	0.8704	0.8712	0.8721	0.8729	0.8738
	cos	0.5000	0.4985	0.4970	0.4955	0.4939	0.4924	0.4909	0.4894	0.4879	0.4863
	tan	1.7321	1.7391	1.7461	1.7532	1.7603	1.7675	1.7747	1.7820	1.7893	1.7966
61	sin	0.8746	0.8755	0.8763	0.8771	0.8780	0.8788	0.8796	0.8805	0.8813	0.8821
	cos	0.4848	0.4833	0.4818	0.4802	0.4787	0.4772	0.4756	0.4741	0.4726	0.4710
	tan	1.8040	1.8115	1.8190	1.8265	1.8341	1.8418	1.8495	1.8572	1.8650	1.8728
62	sin	0.8829	0.8838	0.8846	0.8854	0.8862	0.8870	0.8878	0.8886	0.8894	0.8902
	cos	0.4695	0.4679	0.4664	0.4648	0.4633	0.4617	0.4602	0.4586	0.4571	0.4555
	tan	1.8807	1.8887	1.8967	1.9047	1.9128	1.9210	1.9292	1.9375	1.9458	1.9542
deg	*function*	**0′**	**6′**	**12′**	**18′**	**24′**	**30′**	**36′**	**42′**	**48′**	**54′**

deg	*function*	**0.0°**	**0.1°**	**0.2°**	**0.3°**	**0.4°**	**0.5°**	**0.6°**	**0.7°**	**0.8°**	**0.9°**
63	sin	0.8910	0.8918	0.8926	0.8934	0.8942	0.8949	0.8957	0.8965	0.8973	0.8980
	cos	0.4540	0.4524	0.4509	0.4493	0.4478	0.4462	0.4446	0.4431	0.4415	0.4399
	tan	1.9626	1.9711	1.9797	1.9883	1.9970	2.0057	2.0145	2.0233	2.0323	2.0413
64	sin	0.8988	0.8996	0.9003	0.9011	0.9018	0.9026	0.9033	0.9041	0.9048	0.9056
	cos	0.4384	0.4368	0.4352	0.4337	0.4321	0.4305	0.4289	0.4274	0.4258	0.4242
	tan	2.0503	2.0594	2.0686	2.0778	2.0872	2.0965	2.1060	2.1155	2.1251	2.1348
65	sin	0.9063	0.9070	0.9078	0.9085	0.9092	0.9100	0.9107	0.9114	0.9121	0.9128
	cos	0.4226	0.4210	0.4195	0.4179	0.4163	0.4147	0.4131	0.4115	0.4099	0.4083
	tan	2.1445	2.1543	2.1642	2.1742	2.1842	2.1943	2.2045	2.2148	2.2251	2.2355
66	sin	0.9135	0.9143	0.9150	0.9157	0.9164	0.9171	0.9178	0.9184	0.9191	0.9198
	cos	0.4067	0.4051	0.4035	0.4019	0.4003	0.3987	0.3971	0.3955	0.3939	0.3923
	tan	2.2460	2.2566	2.2673	2.2781	2.2889	2.2998	2.3109	2.3220	2.3332	2.3445
67	sin	0.9205	0.9212	0.9219	0.9225	0.9232	0.9239	0.9245	0.9252	0.9259	0.9265
	cos	0.3907	0.3891	0.3875	0.3859	0.3843	0.3827	0.3811	0.3795	0.3778	0.3762
	tan	2.3559	2.3673	2.3789	2.3906	2.4023	2.4142	2.4262	2.4383	2.4504	2.4627
68	sin	0.9272	0.9278	0.9285	0.9291	0.9298	0.9304	0.9311	0.9317	0.9323	0.9330
	cos	0.3746	0.3730	0.3714	0.3697	0.3681	0.3665	0.3649	0.3633	0.3616	0.3600
	tan	2.4751	2.4876	2.5002	2.5129	2.5257	2.5386	2.5517	2.5649	2.5782	2.5916
69	sin	0.9336	0.9342	0.9348	0.9354	0.9361	0.9367	0.9373	0.9379	0.9385	0.9391
	cos	0.3584	0.3567	0.3551	0.3535	0.3518	0.3502	0.3486	0.3469	0.3453	0.3437
	tan	2.6051	2.6187	2.6325	2.6464	2.6605	2.6746	2.6889	2.7034	2.7179	2.7326
70	sin	0.9397	0.9403	0.9409	0.9415	0.9421	0.9426	0.9432	0.9438	0.9444	0.9449
	cos	0.3420	0.3404	0.3387	0.3371	0.3355	0.3338	0.3322	0.3305	0.3289	0.3272
	tan	2.7475	2.7625	2.7776	2.7929	2.8083	2.8239	2.8397	2.8556	2.8716	2.8878
71	sin	0.9455	0.9461	0.9466	0.9472	0.9478	0.9483	0.9489	0.9494	0.9500	0.9505
	cos	0.3256	0.3239	0.3223	0.3206	0.3190	0.3173	0.3156	0.3140	0.3123	0.3107
	tan	2.9042	2.9208	2.9375	2.9544	2.9714	2.9887	3.0061	3.0237	3.0415	3.0595
72	sin	0.9511	0.9516	0.9521	0.9527	0.9532	0.9537	0.9542	0.9548	0.9553	0.9558
	cos	0.3090	0.3074	0.3057	0.3040	0.3024	0.3007	0.2990	0.2974	0.2957	0.2940
	tan	3.0777	3.0961	3.1146	3.1334	3.1524	3.1716	3.1910	3.2106	3.2305	3.2506
73	sin	0.9563	0.9568	0.9573	0.9578	0.9583	0.9588	0.9593	0.9598	0.9603	0.9608
	cos	0.2924	0.2907	0.2890	0.2874	0.2857	0.2840	0.2823	0.2807	0.2790	0.2773
	tan	3.2709	3.2914	3.3122	3.3332	3.3544	3.3759	3.3977	3.4197	3.4420	3.4646
74	sin	0.9613	0.9617	0.9622	0.9627	0.9632	0.9636	0.9641	0.9646	0.9650	0.9655
	cos	0.2756	0.2740	0.2723	0.2706	0.2689	0.2672	0.2656	0.2639	0.2622	0.2605
	tan	3.4874	3.5105	3.5339	3.5576	3.5816	3.6059	3.6305	3.6554	3.6806	3.7062
75	sin	0.9659	0.9664	0.9668	0.9673	0.9677	0.9681	0.9686	0.9690	0.9694	0.9699
	cos	0.2588	0.2571	0.2554	0.2538	0.2521	0.2504	0.2487	0.2470	0.2453	0.2436
	tan	3.7321	3.7583	3.7848	3.8118	3.8391	3.8667	3.8947	3.9232	3.9520	3.9812
76	sin	0.9703	0.9707	0.9711	0.9715	0.9720	0.9724	0.9728	0.9732	0.9736	0.9740
	cos	0.2419	0.2402	0.2385	0.2368	0.2351	0.2334	0.2317	0.2300	0.2284	0.2267
	tan	4.0108	4.0408	4.0713	4.1022	4.1335	4.1653	4.1976	4.2303	4.2635	4.2972
77	sin	0.9744	0.9748	0.9751	0.9755	0.9759	0.9763	0.9767	0.9770	0.9774	0.9778
	cos	0.2250	0.2232	0.2215	0.2198	0.2181	0.2164	0.2147	0.2130	0.2113	0.2096
	tan	4.3315	4.3662	4.4015	4.4374	4.4737	4.5107	4.5483	4.5864	4.6252	4.6646
78	sin	0.9781	0.9785	0.9789	0.9792	0.9796	0.9799	0.9803	0.9806	0.9810	0.9813
	cos	0.2079	0.2062	0.2045	0.2028	0.2011	0.1994	0.1977	0.1959	0.1942	0.1925
	tan	4.7046	4.7453	4.7867	4.8288	4.8716	4.9152	4.9594	5.0045	5.0504	5.0970
79	sin	0.9816	0.9820	0.9823	0.9826	0.9829	0.9833	0.9836	0.9839	0.9842	0.9845
	cos	0.1908	0.1891	0.1874	0.1857	0.1840	0.1822	0.1805	0.1788	0.1771	0.1754
	tan	5.1446	5.1929	5.2422	5.2924	5.3435	5.3955	5.4486	5.5026	5.5578	5.6140
80	sin	0.9848	0.9851	0.9854	0.9857	0.9860	0.9863	0.9866	0.9869	0.9871	0.9874
	cos	0.1736	0.1719	0.1702	0.1685	0.1668	0.1650	0.1633	0.1616	0.1599	0.1582
	tan	5.6713	5.7297	5.7894	5.8502	5.9124	5.9758	6.0405	6.1066	6.1742	6.2432
81	sin	0.9877	0.9880	0.9882	0.9885	0.9888	0.9890	0.9893	0.9895	0.9898	0.9900
	cos	0.1564	0.1547	0.1530	0.1513	0.1495	0.1478	0.1461	0.1444	0.1426	0.1409
	tan	6.3138	6.3859	6.4596	6.5350	6.6122	6.6912	6.7720	6.8548	6.9395	7.0264
82	sin	0.9903	0.9905	0.9907	0.9910	0.9912	0.9914	0.9917	0.9919	0.9921	0.9923
	cos	0.1392	0.1374	0.1357	0.1340	0.1323	0.1305	0.1288	0.1271	0.1253	0.1236
	tan	7.1154	7.2066	7.3002	7.3962	7.4947	7.5958	7.6996	7.8062	7.9158	8.0285
83	sin	0.9925	0.9928	0.9930	0.9932	0.9934	0.9936	0.9938	0.9940	0.9942	0.9943
	cos	0.1219	0.1201	0.1184	0.1167	0.1149	0.1132	0.1115	0.1097	0.1080	0.1063
	tan	8.1443	8.2636	8.3863	8.5126	8.6427	8.7769	8.9152	9.0579	9.2052	9.3572
deg	*function*	**0′**	**6′**	**12′**	**18′**	**24′**	**30′**	**36′**	**42′**	**48′**	**54′**

deg	function	0.0°	0.1°	0.2°	0.3°	0.4°	0.5°	0.6°	0.7°	0.8°	0.9°
84	sin	0.9945	0.9947	0.9949	0.9951	0.9952	0.9954	0.9956	0.9957	0.9959	0.9960
	cos	0.1045	0.1028	0.1011	0.0993	0.0976	0.0958	0.0941	0.0924	0.0906	0.0889
	tan	9.5144	9.6768	9.8448	10.02	10.20	10.39	10.58	10.78	10.99	11.20
85	sin	0.9962	0.9963	0.9965	0.9966	0.9968	0.9969	0.9971	0.9972	0.9973	0.9974
	cos	0.0872	0.0854	0.0837	0.0819	0.0802	0.0785	0.0767	0.0750	0.0732	0.0715
	tan	11.43	11.66	11.91	12.16	12.43	12.71	13.00	13.30	13.62	13.95
86	sin	0.9976	0.9977	0.9978	0.9979	0.9980	0.9981	0.9982	0.9983	0.9984	0.9985
	cos	0.0698	0.0680	0.0663	0.0645	0.0628	0.0610	0.0593	0.0576	0.0558	0.0541
	tan	14.30	14.67	15.06	15.46	15.89	16.35	16.83	17.34	17.89	18.46
87	sin	0.9986	0.9987	0.9988	0.9989	0.9990	0.9990	0.9991	0.9992	0.9993	0.9993
	cos	0.0523	0.0506	0.0488	0.0471	0.0454	0.0436	0.0419	0.0401	0.0384	0.0366
	tan	19.08	19.74	20.45	21.20	22.02	22.90	23.86	24.90	26.03	27.27
88	sin	0.9994	0.9995	0.9995	0.9996	0.9996	0.9997	0.9997	0.9997	0.9998	0.9998
	cos	0.0349	0.0332	0.0314	0.0297	0.0279	0.0262	0.0244	0.0227	0.0209	0.0192
	tan	28.64	30.14	31.82	33.69	35.80	38.19	40.92	44.07	47.74	52.08
89	sin	0.9998	0.9999	0.9999	0.9999	0.9999	1.000	1.000	1.000	1.000	1.000
	cos	0.0175	0.0157	0.0140	0.0122	0.0105	0.0087	0.0070	0.0052	0.0035	0.0017
	tan	57.29	63.66	71.62	81.85	95.49	114.6	143.2	191.0	286.5	573.0
deg	function	0′	6′	12′	18′	24′	30′	36′	42′	48′	54′

SOURCE: N. M. Cooke and H. F. R. Adams, "Basic Mathematics for Electronics," 3d ed., McGraw-Hill Book Company, 1970.

TABLE C-2 The Exponential Function exp (-x)

X	EXP(-X)
0.00	1.000
0.01	0.990
0.02	0.980
0.03	0.970
0.04	0.961
0.05	0.951
0.06	0.942
0.07	0.932
0.08	0.923
0.09	0.914
0.10	0.905
0.11	0.896
0.12	0.887
0.13	0.878
0.14	0.869
0.15	0.861
0.16	0.852
0.17	0.844
0.18	0.835
0.19	0.827
0.20	0.819
0.21	0.811
0.22	0.803
0.23	0.795
0.24	0.787
0.25	0.779
0.26	0.771
0.27	0.763
0.28	0.756
0.29	0.748
0.30	0.741
0.31	0.733
0.32	0.726
0.33	0.719
0.34	0.712
0.35	0.705
0.36	0.698
0.37	0.691
0.38	0.684
0.39	0.677
0.40	0.670
0.41	0.664
0.42	0.657
0.43	0.651
0.44	0.644
0.45	0.638
0.46	0.631
0.47	0.625
0.48	0.619
0.49	0.613
0.50	0.607
0.51	0.600
0.52	0.595
0.53	0.589
0.54	0.583
0.55	0.577
0.56	0.571
0.57	0.566
0.58	0.560
0.59	0.554
0.60	0.549
0.61	0.543
0.62	0.538
0.63	0.533
0.64	0.527
0.65	0.522
0.66	0.517
0.67	0.512
0.68	0.507
0.69	0.502
0.70	0.497
0.71	0.492
0.72	0.487
0.73	0.482
0.74	0.477
0.75	0.472
0.76	0.468
0.77	0.463
0.78	0.458
0.79	0.454
0.80	0.449
0.81	0.445
0.82	0.440
0.83	0.436
0.84	0.432
0.85	0.427
0.86	0.423
0.87	0.419
0.88	0.415
0.89	0.411
0.90	0.407
0.91	0.403
0.92	0.399
0.93	0.395
0.94	0.391
0.95	0.387
0.96	0.383
0.97	0.379
0.98	0.375
0.99	0.372
1.00	0.368
1.01	0.364
1.02	0.361
1.03	0.357
1.04	0.353
1.05	0.350
1.06	0.346
1.07	0.343
1.08	0.340
1.09	0.336
1.10	0.333
1.11	0.330
1.12	0.326
1.13	0.323
1.14	0.320
1.15	0.317
1.16	0.313
1.17	0.310
1.18	0.307
1.19	0.304
1.20	0.301
1.21	0.298
1.22	0.295
1.23	0.292
1.24	0.289
1.25	0.287
1.26	0.284
1.27	0.281
1.28	0.278
1.29	0.275
1.30	0.273
1.31	0.270
1.32	0.267
1.33	0.264
1.34	0.262
1.35	0.259
1.36	0.257
1.37	0.254
1.38	0.252
1.39	0.249
1.40	0.247
1.41	0.244
1.42	0.242
1.43	0.239
1.44	0.237
1.45	0.235
1.46	0.232
1.47	0.230
1.48	0.228
1.49	0.225

X	EXP(-X)	X	EXP(-X)	X	EXP(-X)
1.50---	0.223	2.00---	0.135	2.5---	0.082
1.51---	0.221	2.01---	0.134	2.6---	0.074
1.52---	0.219	2.02---	0.133	2.7---	0.067
1.53---	0.217	2.03---	0.131	2.8---	0.061
1.54---	0.214	2.04---	0.130	2.9---	0.055
1.55---	0.212	2.05---	0.129	3.0---	0.050
1.56---	0.210	2.06---	0.127	3.1---	0.045
1.57---	0.208	2.07---	0.126	3.2---	0.041
1.58---	0.206	2.08---	0.125	3.3---	0.037
1.59---	0.204	2.09---	0.124	3.4---	0.033
1.60---	0.202	2.10---	0.122	3.5---	0.030
1.61---	0.200	2.11---	0.121	3.6---	0.027
1.62---	0.198	2.12---	0.120	3.7---	0.025
1.63---	0.196	2.13---	0.119	3.8---	0.022
1.64---	0.194	2.14---	0.118	3.9---	0.020
1.65---	0.192	2.15---	0.116	4.0---	0.018
1.66---	0.190	2.16---	0.115	4.1---	0.017
1.67---	0.188	2.17---	0.114	4.2---	0.015
1.68---	0.186	2.18---	0.113	4.3---	0.014
1.69---	0.185	2.19---	0.112	4.4---	0.012
1.70---	0.183	2.20---	0.111	4.5---	0.011
1.71---	0.181	2.21---	0.110	4.6---	0.010
1.72---	0.179	2.22---	0.109	4.7---	0.009
1.73---	0.177	2.23---	0.108	4.8---	0.008
1.74---	0.176	2.24---	0.106	4.9---	0.007
1.75---	0.174	2.25---	0.105	5.0---	0.007
1.76---	0.172	2.26---	0.104		
1.77---	0.170	2.27---	0.103		
1.78---	0.169	2.28---	0.102		
1.79---	0.167	2.29---	0.101		
1.80---	0.165	2.30---	0.100		
1.81---	0.164	2.31---	0.099		
1.82---	0.162	2.32---	0.098		
1.83---	0.160	2.33---	0.097		
1.84---	0.159	2.34---	0.096		
1.85---	0.157	2.35---	0.095		
1.86---	0.156	2.36---	0.094		
1.87---	0.154	2.37---	0.093		
1.88---	0.153	2.38---	0.093		
1.89---	0.151	2.39---	0.092		
1.90---	0.150	2.40---	0.091		
1.91---	0.148	2.41---	0.090		
1.92---	0.147	2.42---	0.089		
1.93---	0.145	2.43---	0.088		
1.94---	0.144	2.44---	0.087		
1.95---	0.142	2.45---	0.086		
1.96---	0.141	2.46---	0.085		
1.97---	0.139	2.47---	0.085		
1.98---	0.138	2.48---	0.084		
1.99---	0.137	2.49---	0.083		

TABLE C-3 Standard Annealed Copper Wire Solid. American Wire Gage

gage number	diameter, mils	area, cir mils	resistance, Ω/1000 ft 25°C (77°F)	weight, lb/1000 ft	allowable current capacity, A: rubber insulation	varnished cambric insulation	other insulations
0000	460.0	211,600.0	0.0500	641.0	225	270	325
000	410.0	167,800.0	0.0630	508.0	175	210	275
00	365.0	133,100.0	0.0795	403.0	150	180	225
0	325.0	105,500.0	0.100	319.0	125	150	200
1	289.0	83,690.0	0.126	253.0	100	120	150
2	258.0	66,370.0	0.159	201.0	90	110	125
3	229.0	52,640.0	0.201	159.0	80	95	100
4	204.0	41,740.0	0.253	126.0	70	85	90
5	182.0	33,100.0	0.319	100.0	55	65	80
6	162.0	26,250.0	0.403	79.5	50	60	70
7	144.0	20,820.0	0.508	63.0			
8	128.0	16,510.0	0.641	50.0	35	40	50
9	114.0	13,090.0	0.808	39.6			
10	102.0	10,380.0	1.02	31.4	25	30	30
11	91.0	8,234.0	1.28	24.9			
12	81.0	6,530.0	1.62	19.8	20	25	25
13	72.0	5,178.0	2.04	15.7			
14	64.0	4,107.0	2.58	12.4	15	18	20
15	57.0	3,257.0	3.25	9.86			
16	51.0	2,583.0	4.09	7.82	6		
17	45.0	2,048.0	5.16	6.20			
18	40.0	1,624.0	6.51	4.92	3		
19	36.0	1,288.0	8.21	3.90			
20	32.0	1,022.0	10.4	3.09			

21	28.5	810.0	13.1	2.45
22	25.3	642.0	16.5	1.95
23	22.6	509.0	20.8	1.54
24	20.1	404.0	26.2	1.22
25	17.9	320.0	33.0	0.970
26	15.9	254.0	41.6	0.769
27	14.2	202.0	52.5	0.610
28	12.6	160.0	66.2	0.484
29	11.3	127.0	83.4	0.384
30	10.0	100.0	105.0	0.304
31	8.9	79.7	133.0	0.241
32	8.0	63.2	167.0	0.191
33	7.1	50.1	211.0	0.152
34	6.3	39.8	266.0	0.120
35	5.6	31.5	335.0	0.0954
36	5.0	25.0	423.0	0.0757
37	4.5	19.8	533.0	0.0600
38	4.0	15.7	673.0	0.0476
39	3.5	12.5	848.0	0.0377
40	3.1	9.9	1070.0	0.0299

TABLE C-4 Resistivity of Various Materials

Material	Resistivity ρ, at 20°C units of $\Omega \cdot m$
Silver	1.6×10^{-8}
Copper	1.72×10^{-8}
Gold	2.44×10^{-8}
Aluminum	2.83×10^{-8}
Tungsten	5.51×10^{-8}
Nickel	7.8×10^{-8}
Iron	12×10^{-8}
Constantan	49×10^{-8}
Nichrome	100×10^{-8}
Carbon	$21{,}000 \times 10^{-8}$

TABLE C-5 Temperature Coefficients of Resistance at 20°C.

Material	Temperature coefficient α
Silver	0.0038
Copper	0.0039
Gold	0.0034
Aluminum	0.0039
Tungsten	0.0051
Nickel	0.006
Iron	0.0055
Constantan	0.000008
Nichrome	0.00044
Carbon	−0.0005

Index

Index